Applied Calculus

THIRD EDITION

Geoffrey C. Berresford

Long Island University

Andrew M. Rockett

Long Island University

Houghton Mifflin Company
Boston   New York

Publisher: Jack Shira
Sponsoring Editor: Lauren Schultz
Development Manager: Maureen Ross
Assistant Editor: Jennifer King
Editorial Assistant: Kasey McGarrigle
Senior Project Editor: Tracy Patruno
Senior Manufacturing Coordinator: Marie Barnes
Senior Marketing Manager: Danielle Potvin
Marketing Associate: Nicole Mollica

Cover image: Dunes in Desert © Jake Wyman/Photonica

Photo credits:
Chapter 1: © AP Photo/Amy Sancetta; Chapter 2: © Firefly Productions/
Corbis; Chapter 3: AP/Wide World Photos; Chapter 4: © Index Stock Imagery,
Inc.; page 270: Shroud of Turin, Bettmann; Chapter 5: © Jeremy Walker/
Getty Images, Inc.; Chapter 6: © Alan Schein Photography/Corbis;
Chapter 7: © AP Photo/Insurance Institute for Highway Safety; Chapter 8:
© LWA-Dann Tardiff/Corbis; Chapter 9: Photo courtesy of Danielle Potvin;
Chapter 10: Lee White/Corbis; Chapter 11: AFP/Scott Olson/Corbis.

Printed in the U.S.A.

Library of Congress Control Number: 2002116635

ISBN: 0-618-29342-6

23456789 DOC 07 06 05 04 03

Contents

Preface

A scientific study of yawning found that more yawns occurred in calculus class than anywhere else.* This book hopes to remedy that situation. Rather than being another dry recitation of standard results, our presentation exhibits some of the many fascinating and useful applications of mathematics in business, the sciences, and everyday life. Even beyond its utility, however, there is a beauty to calculus, and we hope to convey some of its elegance and simplicity.

This book is an introduction to calculus and its applications to the management, social, behavioral, and biomedical sciences, and other fields. The seven-chapter *Brief Applied Calculus* contains more than enough material for a one-semester course, and the eleven-chapter *Applied Calculus* contains additional chapters on trignometry, differential equations, sequences and series, and probability for a two-semester course. The only prerequisites are some knowledge of algebra, functions, and graphing, which are reviewed in Chapter 1.

CHANGES IN THE THIRD EDITION

First, what has *not* changed is the basic character of the book: simple, clear, and mathematically correct explanations of calculus, alternating with relevant and engaging examples. While we have not omitted any topics, we have tightened the exposition and repositioned many diagrams to significantly shorten the text.

We have updated almost every exercise and example that uses real-world data, while also adding many new exercises in every area, especially Biomedical Science.

Section 2.2 (Rates of Change, Slopes, and Derivatives) has been rewritten to develop rates of change before slopes, as seems appropriate to a book featuring so many applications.

We have moved the **Projects and Essays** to the Internet at **math.college.hmco.com** along with most of the "Application Previews" of the second edition. For courses using MicroSoft Excel or similar software, we have created several (optional) **Spreadsheet Explorations.**

*Ronald Baenninger, "Some Comparative Aspects of Yawning in Betta splendens, Homo sapiens, Panthera leo, and Papoi spinx," *Journal of Comparative Psychology,* 4, vol. 101, 349–354, 1987.

We have removed the scientific calculator symbol (but kept the graphing calculator symbol) since calculator use is now universal among today's college students. We have added spaces around every equation to make the text more readable.

FEATURES

Realistic Applications. The basic nature of courses using this book is very "applied," therefore, this book contains an unusually large number of applications, many appearing in no other textbook. These applications show that calculus is not just the manipulation of abstract symbols but is deeply connected to everyday life.

Graphing Calculators (Optional). Reading this book does *not* require a graphing calculator, but having one will simplify the calculations in many problems, and at the same time deepen understanding of mathematics by permitting students to concentrate on *concepts*. Throughout the book are **Graphing Calculator Explorations** and **Exercises,** which explore new topics, carry out "messy" calculations, or show the limitations of technology. A discussion of the essentials of graphing calculators follows this preface. For those not using a graphing calculator, the Graphing Calculator Explorations are boxed so that they can be read for enrichment. Students, however, *will* need calculators with keys like y^x and $\ln x$ for powers and natural logarithms.

Spreadsheets (Optional). Each of the first seven chapters now includes an Excel **Spreadsheet Exploration.**

Application Previews. Each chapter begins with an Application Preview that presents an interesting application of the mathematics being developed. They are self-contained (although some exercises are based on them), and serve to motivate the coming material. Topics include world records in the mile run, Stevens' Law of Psychophysics, cigarette smoking, and predicting personal wealth.

Practice Problems. Learning requiries active participation—"mathematics is not a spectator sport." Throughout the reading are short pencil-and-paper **Practice Problems** designed to consolidate understanding of one topic before another is introduced. Complete solutions to all practice problems are given at the end of each section.

Annotations. Notes to the right of many mathematical formulas and manipulations state the results in words, emphasizing the important skill of "reading mathematics." They also provide explanation and justification for the steps in calculations, and interpretation of the results.

Extensive Exercises. Anyone who has learned mathematics did so by solving many problems, and the exercises are the most essential part of the learning process. Exercises are graded from routine drill to significant applications. Most **Applied Exercises** have both general and specific titles, such as "Environmental Science: Pollution Control." At the end of the book are answers to the odd-numbered exercises, and answers to *all* Chapter Review Exercises and Cumulative Review Exercises.

Constant Reinforcement. Summaries and reviews occur frequently. **Section Summaries** briefly state essential formulas and key concepts. **Chapter Summaries** review the major developments (keyed to particular review exercises) and offer **Hints and Suggestions** that unify the chapter, give helpful reminders, and list a selection of **Review Exercises for a Practice Test** for the chapter. **Cumulative Reviews** contain exercises from groups of related chapters.

Accuracy and Proofs. All of the answers and other mathematics have been checked carefully by several mathematicians. The statements of definitions and theorems are mathematically accurate. Because the treatment is applied rather than theoretical, intuitive and geometric justifications have often been preferred to formal proofs. Such a justification or proof accompanies every important mathematical idea; we never resort to phrases like "it can be shown that . . .". When proofs are given, they are correct and "honest."

Philosophy. We wrote this book with several principles in mind. One is that to learn something, it is best to begin doing it as soon as possible. Therefore the preliminary material is brief, so that students begin calculus without delay. An early start allows more time during the course for interesting applications and necessary review. Another principle is that the mathematics should be done together with the applications. Consequently every section contains applications (there are no "pure math" sections).

HOW TO OBTAIN GRAPHING CALCULATOR PROGRAMS

The optional graphing calculator programs used in the text have been written for a variety of Texas Instruments Graphing Calculators (including the *TI-82, TI-83, TI-85, TI-86, TI-89,* and *TI-92*), and may be obtained for free, in any of the following ways:

- If you know someone who already has the programs on a Texas Instruments graphing calculator like yours, you can easily transfer the programs from their calculator to yours using the black cable that came with the calculator and the LINK button.

- You may download the programs and instructions from the Houghton Mifflin website at **math.college.hmco.com** onto a computer. Then use TI-GRAPH LINK™ (available for purchase in stores) to transfer the programs to your calculator.

- You may send a formatted 3-1/2 "floppy" disk to the authors at the following address, specifying the title of your textbook, the type of your computer (IBM-compatible or Macintosh), and your type of Texas Instruments calculator. We will return a disk containing the appropriate programs and descriptive information. You then use TI-GRAPH LINK™ to transfer the programs from your computer to your calculator.

- You may send your calculator (*TI-82, TI-83, TI-85, TI-86, TI-89,* or *TI-92*) to the authors at the following address, and we will return it loaded with the appropriate programs and a packet of descriptive information.

Authors' Address: Dr. G. C. Berresford and Dr. A. M. Rockett
Department of Mathematics
C. W. Post Campus of Long Island University
720 Northern Boulevard
Brookville, New York 11548-1300

HOW TO OBTAIN EXCEL SPREADSHEET EXPLORATIONS

The MicroSoft Excel spreadsheets used in the Spreadsheet Explorations may be obtained for free in either of the following ways:

- You may download the spreadsheet files from the Houghton Mifflin website at **math.college.hmco.com** onto a computer.

- You may send a formatted 3-1/2 "floppy" disk to the authors, specifying both the title of your textbook and the type of your computer (IBM-compatible or Macintosh), and we will return it to you with the appropriate Excel files.

SUPPLEMENTS FOR THE INSTRUCTOR

Instructor's Resource Manual (with Solutions and Printed Test Bank) Contains complete solutions to all problems in the text and a printed Test Bank available in computerized form on the CD. Two Chapter Tests for each chapter are also included.

HM ClassPrep with HM Testing CD ROM combines all the resources of ClassPrep plus the algorithmically generated tests and gradebook features of HM Testing. This CD ROM includes a variety of supplements: an electronic Instructor's Solutions Manual, Digital Figures and Tables from the book, chapter tests, the resources available on the student CD, and end of chapter Projects and Essays.

- Create and print quizzes and tests quickly and easily
- Deliver tests and record results via a LAN or the Internet
- Record and tabulate student test grades in an electronic gradebook

eduSpace® is a text-specific online learning environment that combines algorithmic tutorials with homework capabilities and classroom management functions.

- Electronically grade and record student results
- Manage a lecture-based or distance learning course online

Companion Website for Instructors offers additional teaching resources such as Digital Lesson Slides. Visit **math.college.hmco.com/instructors** and choose this textbook from the list provided.

SUPPLEMENTS FOR THE STUDENT

HM mathSpace™ Tutorial CD ROM is a new tutorial CD that allows students to practice calculus skills and review concepts. The CD contains algorithmically generated exercises and step-by-step solutions for student practice, along with a brief Introduction to Excel, Graphing Calculator Guide, and end of chapter Projects and Essays. Also included is an algebra tutorial review that provides a skill refresher.

SMARTHINKING™ live online tutoring allows students to communicate in real-time with instructors who will help them understand difficult concepts and guide them through the problem solving process.

eduSpace® is a text-specific online learning environment that combines algorithmic tutorials with homework capabilities.

- Practice skills with additional exercises and quizzes
- Get tutorial help outside class
- Complete homework assignments online if assigned by the instructor

Companion Website for Students offers additional learning resources such as the Graphing Calculator Programs that accompany this text. Visit the text-specific website at **math.college.hmco.com/students** and choose this textbook from the list provided.

Houghton Mifflin Mathematics Instructional Videos and DVDs—a lecture series featuring Dana Mosely in which he provides careful explanations of key concepts, examples, exercises, and applications in a lecture-based format.

Student Solutions Manual contains complete solutions to all odd-numbered problems in the text and solutions to all Chapter Test questions.

ACKNOWLEDGMENTS

We are indebted to many people for their useful suggestions, conversations, and correspondence during the writing and revising of this book. We thank Chris and Lee Berresford, Anne Burns, Richard Cavaliere, Ruth Enoch, Theodore Faticoni, Jeff Goodman, Susan Halter, Brita and Ed Immergut, Ethel Matin, Gary Patric, Shelly Rothman, Charlene Russert, Stuart Saal, Bob Sickles, Michael Simon, John Stevenson, and all of our "Math 6" students at C. W. Post over past years for serving as proofreaders and critics.

We had the good fortune to have had supportive and expert editors at Houghton Mifflin: Lauren Schultz (Sponsoring Editor), Maureen Ross (Development Manager), Jennifer M. King (Assistant Editor), and Tracy Patruno (Senior Project Editor). They made the difficult tasks seem easy, and helped beyond words. We also express our gratitude to the many others at Houghton Mifflin who made important contributions too numerous to mention.

The following reviewers have contributed greatly to the development of this text:

John A. Blake, *Oakwood College,* Alabama; Dave Bregenzer, *Utah State University;* Donald O. Clayton, *Madisonville Community College,* Kentucky; Charles C. Clever, *South Dakota State University;* Dale L. Craft, *South Florida Community College;* Kent Craghead, *Colby Community College,* Kansas; John Haverhals, *Bradley University,* Illinois; Randall Helmstutler, *University of Virginia;* David Hutchison, *Indiana State University;* Alan S. Jian, *Solano Community College,* California; Dr. Hilbert Johs, *Wayne State College,* Nebraska; Michael Longfritz, *Rensselear Polytechnic Institute,* New York; Dr. Hank Martel, *Broward Community College,* Florida; Donna Mills, *Frederick Community College,* Maryland; Pat Moreland, *Cowley College,* Kansas; Sue Neal, *Wichita State University,* Kansas; Catherine A. Roberts, *University of Rhode Island;* George W. Schultz, *St. Petersburg College,* Florida; Jaak Vilms, *Colorado State University;* Elizabeth White, *Trident Technical College,* Charleston, South Carolina; Kenneth J. Word, *Central Texas College*

Finally, and most importantly, we thank our wives, Barbara and Kathryn, for their encouragement and support.

COMMENTS WELCOMED

With the knowledge that any book can always be improved, we welcome corrections, constructive criticisms, and suggestions from every reader.

A User's Guide to Features

▼ Applications Preview

Found on every chapter opener page, Application Previews serve to motivate the chapter. They offer a unique "mathematics in your world" application or an interesting historical note. A page with further information on the topic, and often a related exercise number, is referenced.

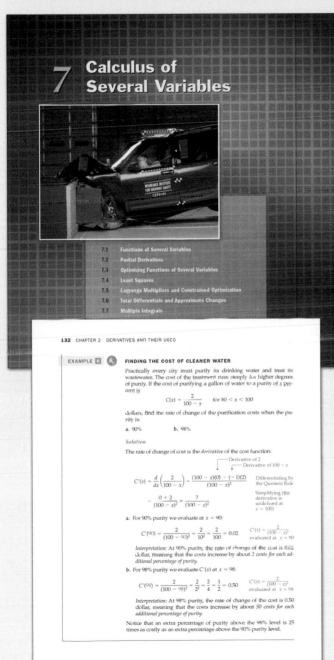

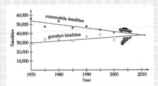

◄ Real World Icon

This globe icon marks examples in which calculus is connected to every-day life.

▼ Graphing Calculator Explorations

To allow for optional use of the graphing calculator, the Explorations are boxed. Exercises and examples that are designed to be done with a graphing calculator are marked with an icon.

▼ Spreadsheet Explorations

Boxed for optional use, these spreadsheets will enhance students' understanding of the material using Excel, an alternative for those who prefer spreadsheet technology.

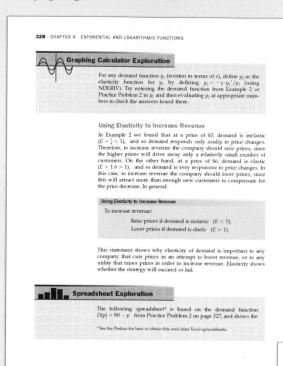

Practice Problems ➤

Students can check their understanding of a topic as they read the text or do homework by working out a Practice Problem. *Complete solutions are found at the end of each section, just after the Section Summary.*

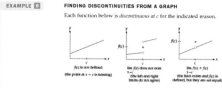

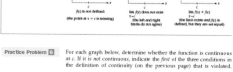

▼ Section Summary

Found at the end of every section, these summaries briefly state the main ideas of the section, providing a study tool or reminder for students. They are followed by complete solutions for the Practice Problems within that section.

Calculator-drawn graphs of the functions in Example 9 are shown below. The polynomial (on the left) and the exponential function (on the right) are continuous, although you can't really tell from such graphs. The rational function (in the middle) exhibits the discontinuity at $x = -1$.

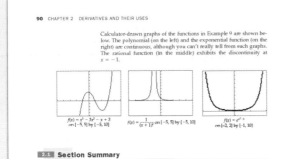

$f(x) = x^3 - 3x^2 - x + 3$ on $[-5, 5]$ by $[-5, 10]$

$f(x) = \dfrac{1}{(x+1)^2}$ on $[-5, 5]$ by $[-5, 10]$

$f(x) = e^{x^2 - 4}$ on $[-2, 2]$ by $[-1, 10]$

2.1 Section Summary

The limit $\lim_{x \to c} f(x)$ is the number (if it exists) that $f(x)$ approaches as x approaches c ($x \neq c$). Left and right limits are found by letting x approach c from one side only. If the left or right limit (or both) do not exist, or do not agree, then the limit does not exist. Limits can often be found from tables of function values for x near c, or by direct substitution at $x = c$ (provided that the function consists of additions, subtractions, multiplications, divisions, roots, and powers, and also provided that the resulting expression is defined). Sometimes it helps to simplify before direct substitution.

Continuity can be understood *geometrically* and *analytically*. Geometrically, a function is continuous at c if its graph has neither a hole nor a jump at $x = c$ (which is equivalent to the curve's passing through the point at c). Analytically, a function f is continuous at c if both sides of the following equation are defined and equal:

$$\lim_{x \to c} f(x) = f(c)$$

This equation may be interpreted as saying that continuity is equivalent to being able to *move the limit operation through the function*:

$$\lim_{x \to c} f(x) = f(\lim_{x \to c} x) = f(c)$$

A polynomial is continuous everywhere, and a rational function is continuous everywhere except where the denominator is zero.

Solutions to Practice Problems

1. $\lim_{x \to 0}(1 + x)^{1/x} \approx 2.71828$

2. $\lim_{x \to 3}(2x^2 - 4x + 1) = 2 \cdot 3^2 - 4 \cdot 3 + 1$
   $= 18 - 12 + 1 = 7$

3. $\lim_{x \to 5}\dfrac{2x^2 - 10x}{x - 5} = \lim_{x \to 5}\dfrac{2x(x - 5)}{x - 5}$
   $= \lim_{x \to 5}\dfrac{2x(x - 5)}{x - 5} = \lim_{x \to 5} 2x = 10$

4. $x \to 3^+$ means: x approaches (positive) 3 *from the left.*
   $x \to -3$ means x approaches -3 (the ordinary two-sided limit).
   $x \to 3^-$ means: x approaches -3 *from the left.*

5. $\lim_{h \to 0}(3x^2 + 5xh + 1) = 3x^2 + 5x \cdot 0 + 1$ Using direct substitution
   $= 3x^2 + 1$

6. a. Discontinuous, $f(c)$ not defined
   b. Discontinuous, $\lim_{x \to c} f(x) \neq f(c)$
   c. Continuous

2.1 Exercises

Complete the tables and use them to find the given limit. Round calculations to three decimal places. A graphing calculator with a TABLE feature will be very helpful.

x	$5x - 7$	x	$5x - 7$

1. $\lim_{x \to 2}(5x - 7)$

x			
1.9		2.1	
1.99		2.01	
1.999		2.001	

x	$2x + 1$	x	$2x + 1$

2. $\lim_{x \to 2}(2x + 1)$

x			
3.9		4.1	
3.99		4.01	
3.999		4.001	

3. $\lim_{x \to 1}\dfrac{x^5 - 1}{x - 1}$

x	$\dfrac{x^5 - 1}{x - 1}$	x	$\dfrac{x^5 - 1}{x - 1}$
0.9		1.1	
0.99		1.01	
0.999		1.001	

4. $\lim_{x \to 1}\dfrac{x^4 - 1}{x - 1}$

x	$\dfrac{x^4 - 1}{x - 1}$	x	$\dfrac{x^4 - 1}{x - 1}$
0.9		1.1	
0.99		1.01	
0.999		1.001	

per year (for $0 \le x \le 16$). Find the total weight gain from age 1 to age 9.

81–82. GENERAL: Repetitive Tasks After t hours of work, a bank clerk can process checks at the rate of $r(t)$ checks per hour for the function $r(t)$ given below. How many checks will the clerk process during the first three hours (time 0 to time 3)?

81. $r(t) = t^2 + 9t + 5$

82. $r(t) = -t^2 + 60t + 9$

83–84. BUSINESS: Cost A company's marginal cost function is $MC(x)$ (given below), where x is the number of units. Find the total cost of the first hundred units ($x = 0$ to $x = 100$).

83. $MC(x) = 6e^{-0.02x}$ 84. $MC(x) = 8e^{-0.01x}$

85. GENERAL: Price Increase The price of a double-dip ice cream cone is increasing at the rate of $18e^{0.08t}$ cents per year, where t is measured in years and $t = 0$ corresponds to 2000. Find the total change in price between the years 2000 and 2010.

86. BUSINESS: Sales An automobile dealer estimates that the newest model car will sell at the rate of $30/t$ cars per month, where t is measured in months and $t = 1$ corresponds to the beginning of January. Find the number of cars that will be sold from the beginning of January to the beginning of May.

87. BUSINESS: Tin Consumption World consumption of tin is running at the rate of $0.24e^{0.01t}$ million tons per year, where t is measured in years and $t = 0$ corresponds to the beginning of 2000. Find the total consumption of tin from the beginning of 2000 to the beginning of 2010.

88. SOCIOLOGY: Marriages There are approximately $2.2e^{0.01t}$ million marriages per year in the United States, where t is the number of years since 2000. Assuming that this rate continues, find the number of marriages from the year 2000 to the year 2010.

89. BEHAVIORAL SCIENCE: Learning A student can memorize words at the rate of $6t^{-1/5}$ words per minute after t minutes. Find the total number of words that the student can memorize in the first 10 minutes.

90. BIOMEDICAL: Epidemics An epidemic is spreading at the rate of $12t^{1.2}$ new cases per day, where t is the number of days since the epidemic began. Find the total number of new cases in the first 10 days of the epidemic.

91. ECONOMICS: Pareto's Law The economist Vilfredo Pareto estimated that the number of people who have an income between A and B dollars ($A < B$) is given by a definite integral of the form

$$N = \int_A^B ax^{-b}\, dx \qquad (b \neq 1)$$

where a and b are constants. Solve this integral.

92. BIOMEDICAL: Poiseuille's Law According to Poiseuille's law, the speed of blood in a blood vessel is given by $V = \dfrac{p}{4Lv}(R^2 - r^2)$, where R is the radius of the blood vessel, r is the distance of the blood from the center of the blood vessel, and p, L, and v are constants determined by the pressure and viscosity of the blood and the length of the vessel. The total blood flow is then given by

$$\left(\begin{array}{c}\text{Total} \\ \text{blood flow}\end{array}\right) = \int_0^R 2\pi \frac{p}{4Lv}(R^2 - r^2)r\, dr$$

Find the total blood flow by finding this integral (p, L, v, and R are constants).

93–94. BUSINESS: Capital Value of an Asset The *capital value* of an asset (such as an oil well) that produces a continuous stream of income is the sum of the present value of all future earnings from the asset. Therefore, the capital value of an asset that produces income at the rate of $r(t)$ dollars per year (at a continuous interest rate i) is

$$\left(\begin{array}{c}\text{Capital} \\ \text{value}\end{array}\right) = \int_0^T r(t)e^{-it}\, dt$$

where T is the expected life (in years) of the asset.

Applied Exercises ➤

The applications in the exercise sets are labeled with a general title as well as a more specific title. The instructor can easily assign homework that is relevant to the students in that class.

End of Chapter Material

To help students study, each chapter ends with a **Chapter Summary with Hints and Suggestions** and **Review Exercises**. The last bullet of the Hints and Suggestions lists the Review Exercises that a student could use to self-test. Both even and odd answers are supplied in the back of the book.

Cumulative Review

There is a Cumulative Review after every 3-4 chapters. Even and odd answers are supplied in the back of the book.

Graphing Calculator Terminology

The graphing calculator applications have been kept as generic as possible for use with any of the popular graphing calculators. Certain standard calculator terms are capitalized in this book and are described below. Your calculator may use slightly different terminology. The viewing or graphing **WINDOW** is the part of the Cartesian plane shown in the display screen of your graphing calculator. **XMIN** and **XMAX** are the smallest and largest *x*-values shown, and **YMIN** and **YMAX** are the smallest and largest *y*-values shown. These values can be set by using the **WINDOW** or **RANGE** command and are changed automatically by using any of the **ZOOM** operations. **XSCALE** and **YSCALE** define the distance between tick marks on the *x*- and *y*-axes.

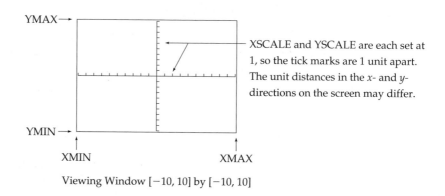

XSCALE and YSCALE are each set at 1, so the tick marks are 1 unit apart. The unit distances in the *x*- and *y*-directions on the screen may differ.

Viewing Window $[-10, 10]$ by $[-10, 10]$

The viewing window is always [XMIN, XMAX] by [YMIN, YMAX]. We will set XSCALE and YSCALE so that there are a reasonable number of tick marks (generally 2 to 20) on each axis. The *x*- and *y*-axes will not be visible if the viewing window does not include the origin.

Pixel, an abbreviation for *pic*ture *el*ement, refers to a tiny rectangle on the screen that can be darkened to represent a dot on a graph. Pixels are arranged in a rectangular array on the screen. In the above window, the axes and tick marks are formed by darkened pixels. The size of the screen and number of pixels varies with different calculators.

TRACE allows you to move a flashing pixel, or *cursor,* along a curve in the viewing window with the *x*- and *y*-coordinates shown at the bottom of the screen.

Useful Hint: To make the *x*-values in TRACE take simple values like .1, .2, and .3, choose XMIN and XMAX to be multiples of one less than the number of pixels across the screen. For example, on the *TI-82* and *TI-83,* which have 95 pixels across the screen, using an *x*-window like [–9.4, 9.4] or [–4.7, 4.7] or [940, –940] will TRACE with simpler *x*-values that the standard windows stated in this book.

ZOOM IN allows you to magnify any part of the viewing window to see finer detail around a chosen point. **ZOOM OUT** does the opposite, like stepping back to see a larger portion of the plane but with less detail. These and other **ZOOM** commands change the viewing window.

VALUE or **EVALUATE** finds the value of a previously entered expression at a specified *x*-value.

SOLVE or **ROOT** finds the *x*-value that solves $f(x) = 0$, equivalently, the *x*-intercepts of a curve. When applied to a difference $f(x) - g(x)$, it finds the *x*-value where the two curves meet (also done by the **INTERSECT** command).

MAX and **MIN** find the maximum and minimum values of a previously entered curve between specified *x*-values.

NDERIV or **DERIV** or dy/dx approximates the *derivative* of a function at a point. **FnInt** or $\int f(x)dx$ approximates the definite integral of a function on an interval.

In **CONNECTED MODE** your calculator will darken pixels to connect calculated points on a graph to show it as a continuous or "unbroken" curve. However, this may lead to "false lines" in a graph that should have breaks or "jumps." False lines can be eliminated by using **DOT MODE.**

The **TABLE** command lists in table form the values of a function, just as you have probably done when graphing a curve. The *x*-values may be chosen by you or by the calculator.

The **Order of Operations** used by most calculators evaluates operations in the following order: first powers and roots, then operations like **LN** and **LOG,** then multiplication and division, then addition and subtraction—left to right within each level. For example, $5 \wedge 2x$ means $(5 \wedge 2)x$, *not* $5 \wedge (2x)$. Also, $1/x + 1$ means $(1/x) + 1$, *not* $1/(x + 1)$. See your calculator's instruction manual for further information. *Be careful:* Some calculators evaluate $1/2x$ as $(1/2)x$ and some as $1/(2x)$. When in doubt, use parentheses to clarify the expression.

Much more information can be found in the manual for your graphing calculator. Other features will be discussed later as needed.

1 Functions

Moroccan runner Hicham El Guerrouj, current world record holder for the mile run, bested the record set 6 years earlier by 1.26 seconds.

Application Preview

World Record Mile Runs

The dots on the graph below show the world record times for the mile run from 1865 to the 1999 world record of 3 minutes 43.13 seconds, set by the Moroccan runner Hicham El Guerrouj. These points fall roughly along a line, called the **regression line.** The regression line is easily found using a graphing calculator, based on a method called **least squares,** which is explained in Chapter 7. Notice that the times do not level off as you might expect, but continue to decrease.

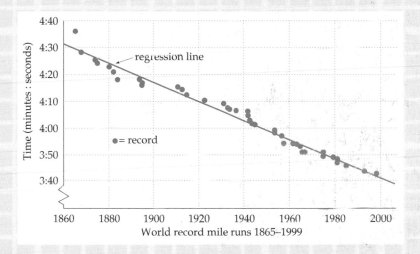

World record mile runs 1865–1999

History of the Record for the Mile Run

Time	Year	Athlete	Time	Year	Athlete
4:36.5	1865	Richard Webster	4:09.2	1931	Jules Ladoumegue
4:29.0	1868	William Chinnery	4:07.6	1933	Jack Lovelock
4.28.8	1868	Walter Gibbs	4:06.8	1934	Glenn Cunningham
4:26.0	1874	Walter Slade	4:06.4	1937	Sydney Wooderson
4:24.5	1875	Walter Slade	4:06.2	1942	Gunder Hägg
4:23.2	1880	Walter George	4:06.2	1942	Arne Andersson
4:21.4	1882	Walter George	4:04.6	1942	Gunder Hägg
4:18.4	1884	Walter George	4:02.6	1943	Arne Andersson
4:18.2	1894	Fred Bacon	4:01.6	1944	Arne Andersson
4:17.0	1895	Fred Bacon	4:01.4	1945	Gunder Hägg
4:15.6	1895	Thomas Conneff	3:59.4	1954	Roger Bannister
4:15.4	1911	John Paul Jones	3:58.0	1954	John Landy
4:14.4	1913	John Paul Jones	3:57.2	1957	Derek Ibbotson
4:12.6	1915	Norman Taber	3:54.5	1958	Herb Elliott
4:10.4	1923	Paavo Nurmi	3:54.4	1962	Peter Snell

World Record Mile Runs	Time	Year	Athlete	Time	Year	Athlete
(continued)	3:54.1	1964	Peter Snell	3:48.8	1980	Steve Ovett
	3:53.6	1965	Michel Jazy	3:48.53	1981	Sebastian Coe
	3:51.3	1966	Jim Ryun	3:48.40	1981	Steve Ovett
	3:51.1	1967	Jim Ryun	3:47.33	1981	Sebastian Coe
	3:51.0	1975	Filbert Bayi	3:46.31	1985	Steve Cram
	3:49.4	1975	John Walker	3:44.39	1993	Noureddine Morceli
	3:49.0	1979	Sebastian Coe	3:43.13	1999	Hicham El Guerrouj

Source: USA Track & Field

The equation of the regression line shown in the graph is $y = -0.356x + 257.44$, where x represents years after 1900 and y is in seconds. The regression line can be used to predict the world mile record in future years. Notice that the most recent world record would have been predicted quite accurately by this line, since the rightmost dot falls almost exactly on the line. Linear trends, however, must not be extended too far. The downward slope of this line means that it will eventually "predict" mile runs in a fraction of a second, or even in *negative* time (see page 16). *Moral:* In the real world, linear trends do not continue indefinitely. This and other topics in "linear" mathematics will be developed in Section 1.1.

1.1 REAL NUMBERS, INEQUALITIES, AND LINES

Introduction

Quite simply, calculus is the study of rates of change. We will use calculus to analyze rates of inflation, rates of learning, rates of population growth, and rates of natural resource consumption.

In this first section we will study *linear* relationships between two variable quantities—that is, relationships that can be represented by *lines.* In later sections we will study *nonlinear* relationships, which can be represented by *curves.*

When reading this book, it will be helpful (but not necessary) to have a graphing calculator. The **Graphing Calculator Explorations** show how to use a graphing calculator to explore a concept more deeply or to analyze an application in more detail. The parts of the book that require graphing calculators are marked by the symbol ▦ . If you do not have a *graphing* calculator, you will need a *scientific* or *business* calculator with keys like $\boxed{y^x}$ and $\boxed{\ln x}$ for powers and logarithms. For those with access to a computer, the **Spreadsheet Explorations** show how spreadsheets can be used for some of the operations of a graphing calculator.

Real Numbers and Inequalities

In this book the word "number" means *real* number, a number that can be represented by a point on the number line (also called the *real line*).

The *order* of the real numbers is expressed by inequalities, with $a < b$ meaning "a is to the *left* of b," and $a > b$ meaning "a is to the *right* of b."

Inequalities	
Inequality	*In Words*
$a < b$	a is less than (smaller than) b
$a \leq b$	a is less than or equal to b
$a > b$	a is greater than (larger than) b
$a \geq b$	a is greater than or equal to b

The inequalities $a < b$ and $a > b$ are called "strict" inequalities, and $a \leq b$ and $a \geq b$ are called "nonstrict" inequalities.

EXAMPLE 1

INEQUALITIES BETWEEN NUMBERS

a. $3 \leq 5$ **b.** $6 > -2$ **c.** $-10 < -5$

$\uparrow$

-10 is less than (smaller than) -5

Throughout this book are many **Practice Problems**—short questions designed to check your understanding of a topic before moving on to new material. Full solutions are given at the end of the section. Solve the following Practice Problem and then check your answer.

Practice Problem 1   Which number is smaller: $\dfrac{1}{100}$ or $-1{,}000{,}000$?

➤ Solution on page 13

A *double* inequality, such as $a < x < b$, means that *both* the inequalities $a < x$ and $x < b$ hold. The inequality $a < x < b$ can be interpreted graphically as "x is between a and b."

$a < x < b$

Sets and Intervals

Braces { } are read "the set of all" and a vertical bar | is read "such that."

EXAMPLE 2

INTERPRETING SETS

⌐ The set of all

a. $\{ x \mid x > 3 \}$ means "the set of all x such that x is greater than 3."

└ Such that

b. $\{ x \mid -2 < x < 5 \}$ means "the set of all x such that x is between -2 and 5."

Practice Problem 2

a. Write in set notation "the set of all x such that x is greater than or equal to -7."

b. Express in words: $\{ x \mid x < -1 \}$. ➤ Solutions on page 13

The set $\{ x \mid 2 \leq x \leq 5 \}$ can be expressed in *interval* notation by enclosing the endpoints 2 and 5 in square brackets: [2, 5]. The *square* brackets indicate that the endpoints are *in*cluded. The set $\{ x \mid 2 < x < 5 \}$ can be written (2, 5). The *parentheses* indicate that the endpoints 2 and 5 are *ex*cluded. An interval is *closed* if it includes both endpoints, and *open* if it includes neither endpoint. The four types of intervals are shown below: a *solid* dot ● on the graph indicates that the point is *in*cluded in the interval; a *hollow* dot ○ indicates that the point is *ex*cluded.

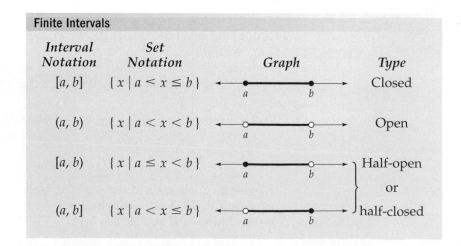

Finite Intervals

Interval Notation	Set Notation	Graph	Type
$[a, b]$	$\{ x \mid a \leq x \leq b \}$		Closed
(a, b)	$\{ x \mid a < x < b \}$		Open
$[a, b)$	$\{ x \mid a \leq x < b \}$		Half-open or half-closed
$(a, b]$	$\{ x \mid a < x \leq b \}$		

An interval may extend infinitely far to the right (indicated by the symbol ∞ for "infinity") or infinitely far to the left (indicated by $-\infty$ for "negative infinity"). Note that ∞ and $-\infty$ are not numbers, but are merely symbols to indicate that the interval extends endlessly in one direction or the other. The infinite intervals in the next box are said to be *closed* or *open* depending on whether they *include* or *exclude* their single endpoint.

Infinite Intervals

Interval Notation	Set Notation	Graph	Type
$[a, \infty)$	$\{\,x \mid x \geq a\,\}$		Closed
(a, ∞)	$\{\,x \mid x > a\,\}$		Open
$(-\infty, a]$	$\{\,x \mid x \leq a\,\}$		Closed
$(-\infty, a)$	$\{\,x \mid x < a\,\}$		Open

The interval $(-\infty, \infty)$ extends infinitely far in *both* directions (meaning the entire real line) and is also denoted by $\mathbb{R}$ (the set of all real numbers).

$$\mathbb{R} = (-\infty, \infty)$$

Cartesian Plane

Two real lines or *axes,* one horizontal and one vertical, intersecting at their zero points, define the *Cartesian plane.** The axes divide the plane into four *quadrants,* I through IV, as shown on the following page.

Any point in the Cartesian plane can be specified uniquely by an ordered pair of numbers (x, y); x, called the *abscissa* or *x-coordinate,* is the number on the horizontal axis corresponding to the point; y, called the *ordinate* or *y-coordinate,* is the number on the vertical axis corresponding to the point.

---

* So named because it was originated by the French philosopher and mathematician René Descartes (1596–1650). Following the custom of the day, Descartes signed his scholarly papers with his Latin name Cartesius, hence "Cartesian" plane.

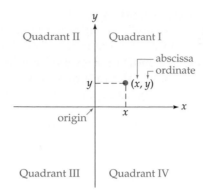

The Cartesian plane

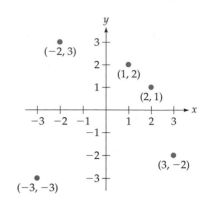

The Cartesian plane with several points. Order matters: (1, 2) is not the same as (2, 1)

Lines and Slopes

The symbol Δ (read "delta," the Greek letter D) means "the change in." For any two points (x_1, y_1) and (x_2, y_2) we define

$$\Delta x = x_2 - x_1 \quad \text{The change in } x \text{ is the difference in the } x\text{-coordinates}$$

$$\Delta y = y_2 - y_1 \quad \text{The change in } y \text{ is the difference in the } y\text{-coordinates}$$

Any two distinct points determine a line. A nonvertical line has a *slope* that measures the steepness of the line, and is defined as *the change in y divided by the change in x* for any two points on the line.

Slope of Line Through (x_1, y_1) and (x_2, y_2)

$$m = \frac{\Delta y}{\Delta x} = \frac{y_2 - y_1}{x_2 - x_1}$$

Slope is the change in y over the change in x ($x_2 \neq x_1$)

The changes Δy and Δx are often called, respectively, the "rise" and the "run," with the understanding that a negative "rise" means a "fall." Slope is then "rise over run."

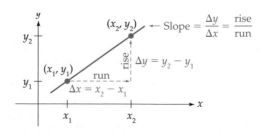

EXAMPLE 3

FINDING SLOPES AND GRAPHING LINES

Find the slope of the line through each pair of points, and graph the line.

a. $(1, 3), (2, 5)$ **b.** $(2, 4), (3, 1)$

c. $(-1, 3), (2, 3)$ **d.** $(2, -1), (2, 3)$

Solution

We use the slope formula $m = \dfrac{y_2 - y_1}{x_2 - x_1}$ for each pair $(x_1, y_1), (x_2, y_2)$.

a. For $(1, 3)$ and $(2, 5)$ the slope is
$$\frac{5 - 3}{2 - 1} = \frac{2}{1} = 2.$$

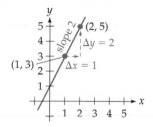

b. For $(2, 4)$ and $(3, 1)$ the slope is $\dfrac{1 - 4}{3 - 2} = \dfrac{-3}{1} = -3.$

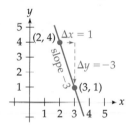

c. For $(-1, 3)$ and $(2, 3)$ the slope is $\dfrac{3 - 3}{2 - (-1)} = \dfrac{0}{3} = 0.$

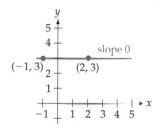

d. For $(2, -1)$ and $(2, 3)$ the slope is *undefined*:
$$\frac{3 - (-1)}{2 - 2} = \frac{4}{0}.$$

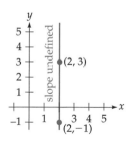

If $\Delta x = 1$, as in Examples 3a and 3b, then the slope is just the "rise," giving

$$\text{Slope} = \begin{pmatrix} \text{Amount that the line rises} \\ \text{when } x \text{ increases by 1} \end{pmatrix}$$

Practice Problem 3 A company president is considering four different business strategies, called S_1, S_2, S_3, and S_4, each with different projected future profits. The graph on the right shows the annual projected profit for the first few years for each of the strategies. Which strategy yields:

a. The highest projected profit in year 1?

b. The highest projected profit in the long run?   ➤ **Solutions on page 13**

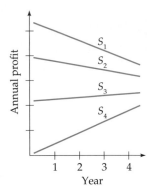

Equations of Lines

The point where a nonvertical line crosses the y-axis is called the *y-intercept* of the line. The y-intercept can be given either as the y-coordinate b or as the point $(0, b)$. Such a line can be expressed very simply in terms of its slope and y-intercept, representing the points by variable coordinates (or "variables") x and y.

Slope-Intercept Form of a Line

$$y = mx + b \qquad \begin{array}{l} m = \text{slope} \\ b = y\text{-intercept} \end{array}$$

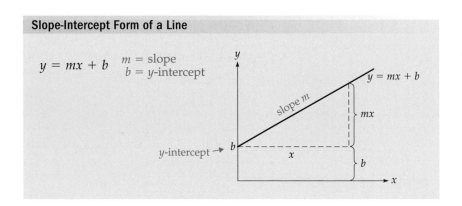

EXAMPLE 4

USING THE SLOPE-INTERCEPT FORM

Find an equation of the line with slope -2 and y-intercept 4.

Solution

$$y = -2x + 4$$

$y = mx + b$ with
$m = -2$ and $b = 4$

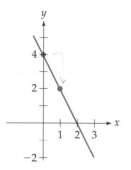

We graph the line by first plotting the y-intercept $(0, 4)$. Using the slope $m = -2$, we plot another point 1 unit over and 2 units *down* from the y-intercept. We then draw the line through these two points, as shown on the left.

Graphing Calculator Exploration

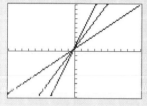

$y_1 = x$, $y_2 = 2x$, and $y_3 = 3x$
on $[-10, 10]$ by $[-10, 10]$

a. Use a graphing calculator to graph the lines $y_1 = x$, $y_2 = 2x$, and $y_3 = 3x$ simultaneously on the window $[-10, 10]$ by $[-10, 10]$. How do the graphs change as the coefficient of x increases from 1 to 2 to 3?

b. Predict what the graph of $y = 0.5x$ would look like. What about $y = -2x$? Check your predictions by graphing them.

c. Describe the graph of the line $y = mx$ for any number m.

Point-Slope Form of a Line

$$y - y_1 = m(x - x_1)$$

$(x_1, y_1) =$ point on the line
$m =$ slope

This form comes directly from the slope formula $m = \dfrac{y_2 - y_1}{x_2 - x_1}$ by replacing x_2 by x, y_2 by y, and multiplying each side by $(x - x_1)$.

EXAMPLE 5

USING THE POINT-SLOPE FORM

Find an equation of the line through $(6, -2)$ with slope $-\frac{1}{2}$.

Solution

$$y - (-2) = -\frac{1}{2}(x - 6)$$

$y-y_1 = m(x-x_1)$ with
$y_1 = -2$, $m = -\frac{1}{2}$, and $x_1 = 6$

$$y + 2 = -\frac{1}{2}x + 3$$

Eliminating parentheses

$$y = -\frac{1}{2}x + 1$$

Subtracting 2 from each side

Alternatively, we could have found this equation using $y = mx + b$, replacing m by the given slope $-\frac{1}{2}$, and then substituting the given $x = 6$ and $y = -2$ to evaluate b.

EXAMPLE 6

FINDING AN EQUATION FOR A LINE THROUGH TWO POINTS

Find an equation for the line through the points (4, 1) and (7, −2).

Solution

The slope is not given, so we calculate it from the two points.

$$m = \frac{-2 - 1}{7 - 4} = \frac{-3}{3} = -1 \qquad m = \frac{y_2 - y_1}{x_2 - x_1} \text{ with } (4, 1) \text{ and } (7, -2)$$

Then we use the point-slope formula with this slope and either of the two points.

$$y - 1 = -1(x - 4)$$

$y - y_1 = m(x - x_1)$ with
slope −1 and point (4, 1)

$$y - 1 = -x + 4$$

Eliminating parentheses

$$y = -x + 5$$

Adding 1 to each side

Practice Problem 4

Find the slope-intercept form of the line through the points (2, 1) and (4, 7).

➤ Solution on page 13

Vertical and horizontal lines have particularly simple equations: a variable equaling a constant.

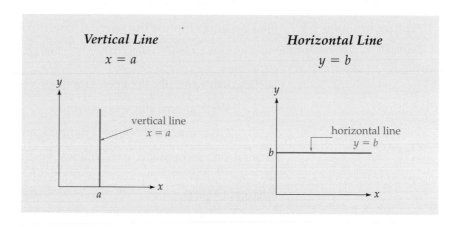

EXAMPLE 7

GRAPHING VERTICAL AND HORIZONTAL LINES

Graph the lines $x = 2$ and $y = 6$.

Solution

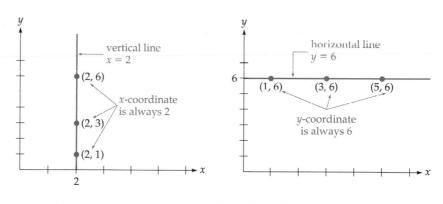

EXAMPLE 8

FINDING EQUATIONS OF VERTICAL AND HORIZONTAL LINES

a. Find an equation for the *vertical* line through the point (3, 2).
b. Find an equation for the *horizontal* line through the point (3, 2).

Solution

a. Vertical line $x = 3$ $x = a$, with a being the
                                                            x-coordinate from (3, 2)

b. Horizontal line $y = 2$ $y = b$, with b being the
                                                            y-coordinate from (3, 2)

Practice Problem 5 Find an equation for the vertical line through the point $(-2, 10)$.

➤ Solution on page 13

Distinguish carefully between slopes of vertical and horizontal lines:

Vertical line: slope is *undefined.*

Horizontal line: slope *is* defined, and is *zero.*

There is one form that covers *all* lines, vertical and nonvertical.

General Linear Equation	
$$ax + by = c$$	For constants a, b, c, with a and b not both zero

Any equation that can be written in this form is called a *linear equation,* and the variables are said to *depend linearly* on each other.

EXAMPLE 9

FINDING THE SLOPE AND THE *y*-INTERCEPT FROM A LINEAR EQUATION

Find the slope and y-intercept of the line $2x + 3y = 12$.

Solution We write the line in slope-intercept form. Solving for y:

$$3y = -2x + 12$$

Subtracting $2x$ from both sides of $2x + 3y = 12$

$$y = -\frac{2}{3}x + 4$$

Dividing each side by 3 gives the slope-intercept form $y = mx + b$

Therefore, the slope is $-\frac{2}{3}$ and the y-intercept is $(0, 4)$.

Practice Problem 6 Find the slope and y-intercept of the line $x - \dfrac{y}{3} = 2$.

➤ Solution on page 13

1.1 Section Summary

An *interval* is a set of real numbers corresponding to a section of the real line. The interval is *closed* if it contains all of its endpoints, and *open* if it contains none of its endpoints.

The nonvertical line through two points (x_1, y_1) and (x_2, y_2) has slope

$$m = \frac{\Delta y}{\Delta x} = \frac{y_2 - y_1}{x_2 - x_1} \qquad x_1 \neq x_2$$

There are five equations or "forms" for lines:

$$y = mx + b \qquad\qquad \begin{array}{l}\text{Slope-intercept form}\\ m = \text{slope}, \ b = y\text{-intercept}\end{array}$$

$$y - y_1 = m(x - x_1) \qquad\qquad \begin{array}{l}\text{Point-slope form}\\ (x_1, y_1) = \text{point}, \ m = \text{slope}\end{array}$$

$$x = a \qquad\qquad \begin{array}{l}\text{Vertical line (slope undefined)}\\ a = x\text{-intercept}\end{array}$$

$$y = b \qquad\qquad \begin{array}{l}\text{Horizontal line (slope zero)}\\ b = y\text{-intercept}\end{array}$$

$$ax + by = c \qquad\qquad \text{General linear equation}$$

➤ **Solutions to Practice Problems**

1. $-1,000,000$ [the negative sign makes it less than (to the left of) the positive number $\frac{1}{100}$]

2. a. $\{ x \mid x \geq -7 \}$

   b. The set of all x such that x is less than -1

3. a. S_1

   b. S_4

4. $m = \dfrac{7 - 1}{4 - 2} = \dfrac{6}{2} = 3$ From points $(2, 1)$ and $(4, 7)$

$\quad y - 1 = 3(x - 2)$ Using the point-slope form with $(x_1, y_1) = (2, 1)$

$\quad y - 1 = 3x - 6$

$\qquad y = 3x - 5$

5. $x = -2$

6. $x - \dfrac{y}{3} = 2$

$\quad -\dfrac{y}{3} = -x + 2$ Subtracting x from each side

$\qquad y = 3x - 6$ Multiplying each side by -3

Slope is $m = 3$ and y-intercept is $(0, -6)$.

1.1 Exercises

Write each interval in set notation and graph it on the real line.

1. $[0, 6)$ **2.** $(-3, 5]$ **3.** $(-\infty, 2]$ **4.** $[7, \infty)$

5. Given the equation $y = 5x - 12$, how will y change if x:
  a. Increases by 3 units?
  b. Decreases by 2 units?

6. Given the equation $y = -2x + 7$, how will y change if x:
  a. Increases by 5 units?
  b. Decreases by 4 units?

Find the slope (if it is defined) of the line determined by each pair of points.

7. $(2, 3)$ and $(4, -1)$ **8.** $(3, -1)$ and $(5, 7)$

9. $(-4, 0)$ and $(2, 2)$ **10.** $(-1, 4)$ and $(5, 1)$

11. $(0, -1)$ and $(4, -1)$ **12.** $(-2, \frac{1}{2})$ and $(5, \frac{1}{2})$

13. $(2, -1)$ and $(2, 5)$ **14.** $(6, -4)$ and $(6, -3)$

For each equation, find the slope m and y-intercept $(0, b)$ (when they exist) and draw the graph.

15. $y = 3x - 4$ **16.** $y = 2x$

17. $y = -\frac{1}{2}x$ **18.** $y = -\frac{1}{3}x + 2$

19. $y = 4$ **20.** $y = -3$

21. $x = 4$ **22.** $x = -3$

23. $2x - 3y = 12$ **24.** $3x + 2y = 18$

25. $x + y = 0$ **26.** $x = 2y + 4$

27. $x - y = 0$ **28.** $y = \frac{2}{3}(x - 3)$

29. $y = \dfrac{x + 2}{3}$ **30.** $\dfrac{x}{2} + \dfrac{y}{3} = 1$

31. $\dfrac{2x}{3} - y = 1$ **32.** $\dfrac{x + 1}{2} + \dfrac{y + 1}{2} = 1$

Use a graphing calculator to graph each line. [*Note:* Your graph will depend on the viewing window you choose. Begin with a "standard" window such as $[-10, 10]$ by $[-10, 10]$, and choose a larger window (or "zoom out") if the line does not appear.]

33. $y = 2x - 8$ **34.** $y = 3x - 6$

35. $y = 7 - 3x$ **36.** $y = 5 - 2x$

37. $y = 50 - x$ **38.** $y = x - 40$

Write an equation of the line satisfying the following conditions. If possible, write your answer in the form $y = mx + b$.

39. Slope -2.25 and y-intercept 3

40. Slope $\frac{2}{3}$ and y-intercept -8

41. Slope 5 and passing through the point $(-1, -2)$

42. Slope -1 and passing through the point $(4, 3)$

43. Horizontal and passing through the point $(1.5, -4)$

44. Horizontal and passing through the point $(\frac{1}{2}, \frac{3}{4})$

45. Vertical and passing through the point $(1.5, -4)$

46. Vertical and passing through the point $(\frac{1}{2}, \frac{3}{4})$

47. Passing through the points $(5, 3)$ and $(7, -1)$

48. Passing through the points $(3, -1)$ and $(6, 0)$

49. Passing through the points $(1, -1)$ and $(5, -1)$

50. Passing through the points $(2, 0)$ and $(2, -4)$

Write an equation of the form $y = mx + b$ for each line in the following graphs. [*Hint:* Either find the slope and y-intercept or use any two points on the line.]

51.

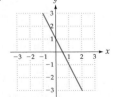

52.

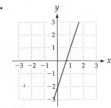

53.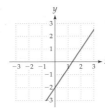

54.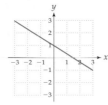

Write equations for the lines determining the four sides of each figure.

55.

56.

57. Show that $y - y_1 = m(x - x_1)$ simplifies to $y = mx + b$ if the point (x_1, y_1) is the y-intercept $(0, b)$.

58. Show that the linear equation $\dfrac{x}{a} + \dfrac{y}{b} = 1$ has x-intercept $(a, 0)$ and y-intercept $(0, b)$. (The x-intercept is the point where the line crosses the x-axis.)

59. Find the x-intercept $(a, 0)$ where the line $y = mx + b$ crosses the x-axis. Under what condition on m will a single x-intercept exist?

60. **i.** Show that the general linear equation $ax + by = c$ with $b \neq 0$ can be written as $y = -\dfrac{a}{b}x + \dfrac{c}{b}$ which is the equation of a line in slope-intercept form.

   ii. Show that the general linear equation $ax + by = c$ with $b = 0$ but $a \neq 0$ can be written as $x = \dfrac{c}{a}$ which is the equation of a vertical line.

[*Note:* Since these steps are reversible, parts (i) and (ii) together show that the general linear equation $ax + by = c$ (for a and b not both zero) includes vertical and nonvertical lines.]

61. **a.** Graph the lines $y_1 = -x$, $y_2 = -2x$, and $y_3 = -3x$ on the window $[-5, 5]$ by $[-5, 5]$ (using the *negation* key $\boxed{(-)}$, not the *subtraction* key $\boxed{-}$). Observe how the coefficient of x changes the slope of the line.

   b. Predict what the line $y = -9x$ would look like, and then check your prediction by graphing it.

62. **a.** Graph the lines $y_1 = x + 2$, $y_2 = x + 1$, $y_3 = x$, $y_4 = x - 1$, and $y_5 = x - 2$ on the window $[-5, 5]$ by $[-5, 5]$. Observe how the constant changes the position of the line.

   b. Predict what the lines $y = x + 4$ and $y = x - 4$ would look like, and then check your prediction by graphing them.

APPLIED EXERCISES

63. **BUSINESS: Energy Usage** A utility considers demand for electricity "low" if it is below 8 mkW (million kilowatts), "average" if it is at least 8 mkW but below 20 mkW, "high" if it is at least 20 mkW but below 40 mkW, and "critical" if it is 40 mkW or more. Express these demand levels in interval notation. [*Hint:* The interval for "low" is $[0, 8)$.]

64. **GENERAL: Grades** If a grade of 90 through 100 is an A, at least 80 but less than 90 is a B, at least 70 but less than 80 a C, at least 60 but less than 70 a D, and below 60 an F, write these

grade levels in interval form (ignoring rounding). [*Hint:* F would be $[0, 60)$.]

65. **ATHLETICS: Mile Run** Read the Application Preview on pages 1–2.

   a. Use the regression line $y = -0.356x + 257.44$ to predict the world record in the year 2010. [*Hint:* If x represents years after 1900, what value of x corresponds to the year 2010? The resulting y will be in seconds, and should be converted to minutes and seconds.]

(continues)

b. According to this formula, when will the record be 3 minutes 30 seconds? [*Hint:* Set the formula equal to 210 seconds and solve. What year corresponds to this x-value?]

66. ATHLETICS: Mile Run Read the Application Preview on pages 1–2. Evaluate the regression line $y = -0.356x + 257.44$ at $x = 720$ and at $x = 722$ (corresponding to the years 2620 and 2622). Does the formula give reasonable times for the mile record in these years? [*Moral:* Linear trends may not continue indefinitely.]

67. BUSINESS: Corporate Profit A company's profit increased linearly from $6 million at the end of year 1 to $14 million at the end of year 3.

  a. Use the two (year, profit) data points (1, 6) and (3, 14) to find the linear relationship $y = mx + b$ between $x =$ year and $y =$ profit.

  b. Find the company's profit at the end of 2 years.

  c. Predict the company's profit at the end of 5 years.

68. ECONOMICS: Per Capita Personal Income In the short run, per capita personal income (PCPI) in the United States grows approximately linearly. In 1990 PCPI was 19.1, and in 2000 it had grown to 29.7 (both in thousands of dollars). (*Source:* Bureau of Economic Analysis)

  a. Use the two given (year, PCPI) data points (0, 19.1) and (10, 29.7) to find the linear relationship $y = mx + b$ between $x =$ years since 1990 and $y =$ PCPI.

  b. Use your linear relationship to predict PCPI in 2010.

69. GENERAL: Temperature On the Fahrenheit temperature scale, water freezes at 32° and boils at 212°. On the Celsius (centigrade) scale, water freezes at 0° and boils at 100°.

  a. Use the two (Celsius, Fahrenheit) data points (0, 32) and (100, 212) to find the linear relationship $y = mx + b$ between $x =$ Celsius temperature and $y =$ Fahrenheit temperature.

  b. Find the Fahrenheit temperature that corresponds to 20° Celsius.

70. ECOLOGY: Waste Disposal The amount of municipal solid waste generated per person per day in the United States has increased approximately linearly, from 2.7 pounds in 1960 to 4.6 pounds in 2000.

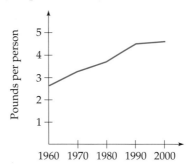

Municipal Solid Waste Generated Per Capita Per Day
(*Source*: U.S. Environmental Protection Agency)

  a. Use the two (year, pound) data points (0, 2.7) and (40, 4.6) to find the linear relationship $y = mx + b$ between $x =$ years since 1960 and $y =$ per capita waste.

  b. Use your formula to predict the amount in the year 2010.

71–72. BUSINESS: Straight-Line Depreciation Straight-line depreciation is a method for estimating the value of an asset (such as a piece of machinery) as it loses value ("depreciates") through use. Given the original *price* of an asset, its *useful lifetime*, and its *scrap value* (its value at the end of its useful lifetime), the value of the asset after t years is given by the formula:

$$\text{Value} = (\text{price}) - \left(\frac{(\text{price}) - (\text{scrap value})}{(\text{useful lifetime})}\right) \cdot t$$

$$\text{for} \quad 0 \leq t \leq (\text{useful lifetime})$$

71. a. A farmer buys a harvester for $50,000 and estimates its useful life to be 20 years, after which its scrap value will be $6000. Use the formula above to find a formula for the value V of the harvester after t years, for $0 \leq t \leq 20$.

  b. Use your formula to find the value of the harvester after 5 years.

  c. Graph the function found in part (a) on a graphing calculator on the window [0, 20] by [0, 50,000]. [*Hint:* Use x instead of t.]

72. a. A newspaper buys a printing press for $800,000 and estimates its useful life to be 20 years, after which its scrap value will be $60,000. Use the formula above Exercise 71 to find a formula for the value V of the press after t years, for $0 \le t \le 20$.
b. Use your formula to find the value of the press after 10 years.
c. Graph the function found in part (a) on a graphing calculator on the window $[0, 20]$ by $[0, 800,000]$. [*Hint:* Use x instead of t.]

73–74. ENVIRONMENTAL SCIENCE: Beverton-Holt Recruitment Curve Some organisms exhibit a density-dependent mortality from one generation to the next. Let $R > 1$ be the net reproductive rate (that is, the number of surviving offspring per parent), let $x > 0$ be the density of parents and y be the density of surviving offspring. The *Beverton-Holt recruitment curve* is

$$y = \frac{Rx}{1 + \left(\dfrac{R-1}{K}\right)x}$$

where $K > 0$ is the *carrying capacity* of the organism's environment. Notice that if $x = K$, then $y = K$.

73. Show that if $x < K$, then $x < y < K$. Explain what this means about the population size over successive generations if the initial population is smaller than the carrying capacity of the environment.

74. Show that if $x > K$, then $K < y < x$. Explain what this means about the population size over successive generations if the initial population is larger than the carrying capacity of the environment.

75. SOCIAL SCIENCE: Age at First Marriage Americans are marrying later and later. Based on data from the U.S. Bureau of the Census for the years 1980 to 2000, the median age at first marriage for men is $y_1 = 24.8 + 0.13x$, and for women it is $y_2 = 22.1 + 0.17x$, where x is the number of years since 1980.

a. Graph these lines on the window $[0, 30]$ by $[20, 30]$.
b. Use these lines to predict the median marriage ages for men and women in the year

2010. [*Hint:* Which x-value corresponds to 2010? Then use TRACE, EVALUATE, or TABLE.]
c. Predict the median marriage ages for men and women in the year 2020.

76. SOCIAL SCIENCE: Equal Pay for Equal Work Women's pay has often lagged behind men's, although Title VII of the Civil Rights Act requires equal pay for equal work. Based on data from 1980–2000, women's annual earnings as a percent of men's can be approximated by the formula $y = 0.97x + 59.5$, where x is the number of years since 1980. (For example, $x = 10$ gives $y = 69.2$, so in 1990 women's wages were about 69.2% of men's wages.)

a. Graph this line on the window $[0, 30]$ by $[0, 100]$.
b. Use this line to predict the percentage in the year 2005. [*Hint:* Which x-value corresponds to 2005? Then use TRACE, EVALUATE, or TABLE.]
c. Predict the percentage in the year 2010. (*Source:* U.S. Department of Labor—Women's Bureau)

77. GENERAL: Smoking and Education According to a recent study,[*] the probability that a smoker will quit smoking increases with the smoker's educational level. The probability (expressed as a percent) that a smoker with x years of education will quit is approximately $y = 0.831x^2 - 18.1x + 137.3$ (for $10 \le x \le 16$).

a. Graph this curve on the window $[10, 16]$ by $[0, 100]$.
b. Find the probability that a high school graduate smoker ($x = 12$) will quit.
c. Find the probability that a college graduate smoker ($x = 16$) will quit.

78. ENVIRONMENTAL SCIENCE: Wind Energy The use of wind power is growing rapidly after a slow start, especially in Europe, where it is seen as an efficient and renewable source of energy. Global wind power generating

---

[*]William Sander, "Schooling and Quitting Smoking," *The Review of Economics and Statistics* LXXVII(1):191–199, February 1995.

capacity for the years 1980 to 2000 is given approximately by $y = 30.5x^2 - 112x + 250$ megawatts, where x is the number of years after 1980. (One megawatt would supply the electrical needs of approximately 100 homes.)

a. Graph this curve on the window [0, 30] by [0, 2500].

b. Use this curve to predict the global wind power generating capacity in the year 2005. [*Hint:* Which x-value corresponds to 2005? Then use TRACE, EVALUATE, or TABLE.]

c. Predict the global wind power generating capacity in the year 2010.
(*Source:* Worldwatch Institute)

79–80. GENERAL: Life Expectancy The following tables give the life expectancy (years of life expected) for a newborn child born in the indicated year (Exercise 79 is for males, Exercise 80 for females). For each exercise:

a. Enter the data into a graphing calculator and make a plot of the resulting points, with Years Since 1960 on the x-axis and Life Expectancy on the y-axis.

b. Use the graphing calculator to find the linear regression line for these points. Enter the resulting function as y_1, which then estimates life expectancy based on the year of birth. Graph the points together with the regression line.

c. Use your line y_1 to estimate the life expectancy of a child born in the year 2025. (This might be your child or grandchild.) [*Hint:* What x-value corresponds to 2025?]

79.

Birth Year	(years since 1960)	Life Expectancy (male)
1960	0	66.6
1970	10	67.1
1980	20	70.0
1990	30	71.8
2000	40	74.4

80.

Birth Year	(years since 1960)	Life Expectancy (female)
1960	0	73.1
1970	10	74.7
1980	20	77.5
1990	30	78.8
2000	40	79.6

1.2 EXPONENTS

Introduction

Not all variables are related linearly. In this section we will discuss exponents, which will enable us to express many nonlinear relationships.

Positive Integer Exponents

Numbers may be expressed with exponents, as in $2^3 = 2 \cdot 2 \cdot 2 = 8$. More generally, for any positive integer n, x^n means the product of n x's.

$$x^n = \overbrace{x \cdot x \cdots x}^{n}$$

The number being raised to the power is called the *base* and the power is the *exponent:*

$$x^n \overset{\frown}{} \text{Exponent or power}$$
$$\underset{\text{Base}}{}$$

There are several *properties of exponents* for simplifying expressions. The first three are known, respectively, as the addition, subtraction, and multiplication properties of exponents.

Properties of Exponents

$x^m \cdot x^n = x^{m+n}$	To *multiply* powers of the same base, *add* the exponents
$\dfrac{x^m}{x^n} = x^{m-n}$	To *divide* powers of the same base, *subtract* the exponents (top exponent minus bottom exponent)
$(x^m)^n = x^{m \cdot n}$	To raise a power to a power, *multiply* the powers
$(xy)^n = x^n \cdot y^n$	To raise a product to a power, raise *each factor* to the power
$\left(\dfrac{x}{y}\right)^n = \dfrac{x^n}{y^n}$	To raise a fraction to a power, raise the numerator *and* denominator to the power

---

EXAMPLE 1

SIMPLIFYING EXPONENTS

a. $x^2 \cdot x^3 = x^5$ $\overset{\frown}{2 + 3}$

b. $\dfrac{x^5}{x^3} = x^2$ $\overset{\frown}{5 - 3}$

c. $(x^2)^3 = x^6$ $\overset{\frown}{2 \cdot 3}$

d. $\dfrac{[(x^2)^3]^4}{x^5 \cdot x^7 \cdot x} = \dfrac{x^{24}}{x^{13}} = x^{11}$ $\overset{\frown}{2 \cdot 3 \cdot 4}$ $\overset{\frown}{24 - 13}$ $\underset{5 + 7 + 1}{}$

e. $(2w)^3 = 2^3 w^3 = 8w^3$

f. $\left(\dfrac{x}{3}\right)^4 = \dfrac{x^4}{3^4} = \dfrac{x^4}{81}$

Practice Problem 1   Simplify: **a.** $\dfrac{x^5 \cdot x}{x^2}$ **b.** $[(x^3)^2]^2$ ➤ Solutions on page 27

Remember: For exponents in the form $x^2 \cdot x^3 = x^5$, *add* exponents.

For exponents in the form $(x^2)^3 = x^6$, *multiply* exponents.

Graphing Calculator Exploration

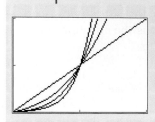

a. Use a graphing calculator to graph $y_1 = x$, $y_2 = x^2$, $y_3 = x^3$, and $y_4 = x^4$ on the viewing window $[0, 2]$ by $[0, 2]$. Use TRACE to identify which curve goes with which power.

b. Which curve is highest for values of x between 0 and 1? Which is lowest?

c. Which curve is highest for values of x greater than 1? Which is lowest?

d. Predict what the curve $y = x^5$ would look like. Check your prediction by graphing it.

e. Predict which of these curves will be positive when x is negative. Check your prediction by changing the viewing window to $[-2, 2]$ by $[-2, 2]$.

Zero and Negative Exponents

For any number x other than zero, we define

$x^0 = 1$	x to the power 0 is 1
$x^{-1} = \dfrac{1}{x}$	x to the power -1 is one over x
$x^{-2} = \dfrac{1}{x^2}$	x to the power -2 is one over x squared
$x^{-n} = \dfrac{1}{x^n}$	x to a negative power is one over x to the positive power

EXAMPLE 2

SIMPLIFYING ZERO AND NEGATIVE EXPONENTS

a. $5^0 = 1$

b. $7^{-1} = \dfrac{1}{7}$

c. $3^{-2} = \dfrac{1}{3^2} = \dfrac{1}{9}$

d. $(-2)^{-3} = \dfrac{1}{(-2)^3} = \dfrac{1}{-8} = -\dfrac{1}{8}$

e. 0^0 and 0^{-3} are undefined.

Practice Problem 2   Evaluate: **a.** 2^0 **b.** 2^{-4} ➤ Solutions on page 28

The definitions of x^0 and x^{-n} are motivated by the following calculations.

$$1 = \frac{x^2}{x^2} = x^{2-2} = x^0$$

The subtraction property of exponents leads to $x^0 = 1$

$$\frac{1}{x^n} = \frac{x^0}{x^n} = x^{0-n} = x^{-n}$$

$x^0 = 1$ and the subtraction property of exponents lead to $x^{-n} = \frac{1}{x^n}$

A fraction to a negative power means *division* by the fraction, so we "invert and multiply."

$$\left(\frac{x}{y}\right)^{-1} = \frac{1}{\frac{x}{y}} = 1 \cdot \frac{y}{x} = \frac{y}{x}$$

—— Reciprocal of the original fraction

Therefore, for $x \neq 0$ and $y \neq 0$,

$$\left(\frac{x}{y}\right)^{-1} = \frac{y}{x}$$

A fraction to the power -1 is the reciprocal of the fraction

$$\left(\frac{x}{y}\right)^{-n} = \left(\frac{y}{x}\right)^{n}$$

A fraction to the negative power is the reciprocal of the fraction to the positive power

EXAMPLE 3

SIMPLIFYING FRACTIONS TO NEGATIVE EXPONENTS

a. $\left(\frac{3}{2}\right)^{-1} = \frac{2}{3}$ **b.** $\left(\frac{1}{2}\right)^{-3} = \left(\frac{2}{1}\right)^{3} = \frac{2^3}{1^3} = 8$

↑
Reciprocal of $\frac{3}{2}$

Practice Problem 3   Simplify: $\left(\frac{2}{3}\right)^{-2}$ ➤ Solution on page 28

Roots and Fractional Exponents

We may take the square root of any *nonnegative* number, and the cube root of *any* number.

EXAMPLE 4

EVALUATING ROOTS

a. $\sqrt{9} = 3$ **b.** $\sqrt{-9}$ is undefined.   Square roots of negative numbers are not defined

c. $\sqrt[3]{8} = 2$ **d.** $\sqrt[3]{-8} = -2$ Cube roots of negative numbers *are* defined

e. $\sqrt[3]{\dfrac{27}{8}} = \dfrac{\sqrt[3]{27}}{\sqrt[3]{8}} = \dfrac{3}{2}$

There are *two* square roots of 9, namely 3 and -3, but the radical sign $\sqrt{}$ means just the *positive* one (the "principal" square root).

$\sqrt[n]{a}$ means the principal *n*th root of *a*.   Principal means the positive root if there are two

---

In general, we may take *odd* roots of *any* number, but *even* roots only if the number is positive or zero.

EXAMPLE 5

EVALUATING ROOTS OF POSITIVE AND NEGATIVE NUMBERS

Odd roots of negative numbers *are* defined

a. $\sqrt[4]{81} = 3$ **b.** $\sqrt[5]{-32} = -2$ Since $(-2)^5 = -32$

Graphing Calculator Exploration

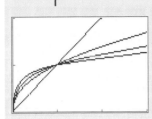

a. Use a graphing calculator to graph $y_1 = x$, $y_2 = \sqrt{x}$, $y_3 = \sqrt[3]{x}$, and $y_4 = \sqrt[4]{x}$ simultaneously on the viewing window $[0, 3]$ by $[0, 2]$. Use TRACE to identify which curve goes with which root.

b. Which curve is highest for values of x between 0 and 1? Which is lowest?

c. Which curve is highest for values of x greater than 1? Which is lowest?

d. Predict what the curve $y = \sqrt[7]{x}$ would look like. Check your prediction by graphing it.

e. Which of these roots are defined for *negative* values of x? Check your answer by changing the window to $[-3, 3]$ by $[-2, 2]$ and using TRACE where x is negative.

Fractional Exponents

Fractional exponents are defined as follows:

$x^{\frac{1}{2}} = \sqrt{x}$	Power $\frac{1}{2}$ means the principal square root
$x^{\frac{1}{3}} = \sqrt[3]{x}$	Power $\frac{1}{3}$ means the cube root
$x^{\frac{1}{n}} = \sqrt[n]{x}$	Power $\frac{1}{n}$ means the principal nth root (for a positive integer n)

EXAMPLE 6

EVALUATING FRACTIONAL EXPONENTS

a. $9^{\frac{1}{2}} = \sqrt{9} = 3$ 　　　　b. $125^{\frac{1}{3}} = \sqrt[3]{125} = 5$

c. $81^{\frac{1}{4}} = \sqrt[4]{81} = 3$ 　　　d. $(-32)^{\frac{1}{5}} = \sqrt[5]{-32} = -2$

e. $\left(-\dfrac{27}{8}\right)^{\frac{1}{3}} = \sqrt[3]{-\dfrac{27}{8}} = -\dfrac{\sqrt[3]{27}}{\sqrt[3]{8}} = -\dfrac{3}{2}$

Practice Problem 4　　Evaluate: **a.** $(-27)^{\frac{1}{3}}$　**b.** $\left(\dfrac{16}{81}\right)^{\frac{1}{4}}$　　　　➤ Solutions on page 28

The definition of $x^{\frac{1}{2}}$ is motivated by the multiplication property of exponents:

$$\left(x^{\frac{1}{2}}\right)^2 = x^{\frac{1}{2}\cdot 2} = x^1 = x$$

Taking square roots of each side of $\left(x^{\frac{1}{2}}\right)^2 = x$ gives

$$x^{\frac{1}{2}} = \sqrt{x}$$

x to the half power means the square root of *x*

To define $x^{\frac{m}{n}}$ for positive integers *m* and *n*, the exponent $\frac{m}{n}$ must be fully reduced (for example, $\frac{4}{6}$ must be reduced to $\frac{2}{3}$). Then

$$x^{\frac{m}{n}} = \left(x^{\frac{1}{n}}\right)^m = \left(x^m\right)^{\frac{1}{n}}$$

Since in both cases the exponents multiply to $\frac{m}{n}$

Therefore we define:

Fractional Exponents

$$x^{\frac{m}{n}} = \left(\sqrt[n]{x}\right)^m = \sqrt[n]{x^m}$$

$x^{m/n}$ means the *m*th power of the *n*th root, or equivalently, the *n*th root of the *m*th power

---

Both expressions, $\left(\sqrt[n]{x}\right)^m$ and $\sqrt[n]{x^m}$, will give the same answer. In either case the numerator determines the power and the denominator determines the root.

Power over root

EXAMPLE 7

EVALUATING FRACTIONAL EXPONENTS

a. $8^{2/3} = \sqrt[3]{8^2} = \sqrt[3]{64} = 4$

First the power, then the root

b. $8^{2/3} = \left(\sqrt[3]{8}\right)^2 = (2)^2 = 4$

First the root, then the power

(same)

c. $25^{3/2} = \left(\sqrt{25}\right)^3 = (5)^3 = 125$

d. $\left(\dfrac{-27}{8}\right)^{2/3} = \left(\sqrt[3]{\dfrac{-27}{8}}\right)^2 = \left(\dfrac{-3}{2}\right)^2 = \dfrac{9}{4}$

Practice Problem 5   Evaluate: **a.** $16^{3/2}$ **b.** $(-8)^{2/3}$ ➤ Solutions on page 28

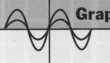

Graphing Calculator Exploration

a. Use a graphing calculator to evaluate $25^{3/2}$. [On some calculators, press 25^(3 ÷ 2).] Your answer should agree with Example 7c on page 24.

b. Evaluate $(-8)^{2/3}$. Use the (−) key for negation, and parentheses around the exponent. Your answer should be 4. If you get an "error," try evaluating the expression as $[(-8)^{1/3}]^2$ or $[(-8)^2]^{1/3}$. Whichever way works, remember it for evaluating negative numbers to fractional powers in the future.

EXAMPLE 8   **EVALUATING NEGATIVE FRACTIONAL EXPONENTS**

a. $8^{-2/3} = \dfrac{1}{8^{2/3}} = \dfrac{1}{(\sqrt[3]{8})^2} = \dfrac{1}{2^2} = \dfrac{1}{4}$

A negative exponent means the reciprocal of the number to the positive exponent, which is then evaluated as before

b. $\left(\dfrac{9}{4}\right)^{-3/2} = \left(\dfrac{4}{9}\right)^{3/2} = \left(\sqrt{\dfrac{4}{9}}\right)^3 = \left(\dfrac{2}{3}\right)^3 = \dfrac{8}{27}$

Interpreting the power 3/2
Reciprocal to the positive exponent
Negative exponent

Practice Problem 6   Evaluate: **a.** $25^{-3/2}$ **b.** $\left(\dfrac{1}{4}\right)^{-1/2}$ **c.** $5^{1.3}$ [*Hint:* Use a calculator.]

➤ Solutions on page 28

Avoiding Pitfalls in Simplifying

The square root of a product is equal to the product of the square roots:

$$\sqrt{a \cdot b} = \sqrt{a} \cdot \sqrt{b}$$

However, the corresponding statement for *sums* is *not* true:

$$\sqrt{a + b} \quad \text{is } not \text{ equal to} \quad \sqrt{a} + \sqrt{b}$$

For example,

$$\underbrace{\sqrt{9 + 16}}_{\sqrt{25}} \neq \underbrace{\sqrt{9}}_{3} + \underbrace{\sqrt{16}}_{4}$$

The two sides are not equal: one is 5 and the other is 7.

Therefore, do not "simplify" $\sqrt{x^2 + 9}$ into $x + 3$. The expression $\sqrt{x^2 + 9}$ *cannot be simplified*. Similarly,

$$(x + y)^2 \quad \text{is } not \text{ equal to} \quad x^2 + y^2$$

The expression $(x + y)^2$ means $(x + y)$ times itself:

$$(x + y)^2 = (x + y)(x + y) = x^2 + xy + yx + y^2 = x^2 + 2xy + y^2$$

This result is worth remembering, since we will use it frequently in Chapter 2.

$(x + y)^2 = x^2 + 2xy + y^2$	$(x + y)^2$ is the first number squared plus twice the product of the numbers plus the second number squared

Learning Curves in Airplane Production

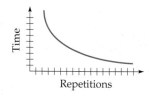

Time

Repetitions

It is a truism that the more you practice a task, the faster you can do it. Successive repetitions generally take less time, following a "learning curve" like that on the left. Learning curves are used in industrial production. For example, it took 150,000 work-hours to build the first Boeing 707 airliner, while later planes ($n = 2, 3, \ldots, 300$) took less time.*

$$\begin{pmatrix} \text{Time to build} \\ \text{plane number } n \end{pmatrix} = 150\, n^{-0.322} \quad \text{thousand work-hours}$$

The time for the 10th Boeing 707 is found by substituting $n = 10$:

$$\begin{pmatrix} \text{Time to build} \\ \text{plane 10} \end{pmatrix} = 150(10)^{-0.322}$$

$150n^{-0.322}$
with $n = 10$

$$\approx 71.46 \text{ thousand work-hours} \qquad \text{Using a calculator}$$

* A work-hour is the amount of work that a person can do in 1 hour. For further information on learning curves in industrial production, see J. M. Dutton et al., "The History of Progress Functions as a Managerial Technology," *Business History Review* 58(1984):204–233.

This shows that building the 10th Boeing 707 took about 71,460 work-hours, which is less than half of the 150,000 work-hours needed for the first. For the 100th 707:

$$\left(\begin{matrix}\text{Time to build}\\ \text{plane 100}\end{matrix}\right) = 150(10)^{-0.322}$$

150$n^{-0.322}$
with $n = 10$

$$\approx 34.05 \text{ thousand work-hours}$$

or about 34,050 work-hours, which is less than the half time needed to build the 10th. Such learning curves are used for determining the cost of a contract to build several planes.

Notice that the learning curve graphed on the previous page decreases less steeply as the number of repetitions increases. This means that while construction time continues to decrease, it does so more slowly for later planes. This behavior, called "diminishing returns," is typical of learning curves.

1.2 Section Summary

We defined zero, negative, and fractional exponents as follows:

$$x^0 - 1 \qquad\qquad \text{for } x \ne 0$$

$$x^{-n} = \frac{1}{x^n} \qquad\qquad \text{for } x \ne 0$$

$$x^{\frac{m}{n}} = \left(\sqrt[n]{x}\right)^m = \sqrt[n]{x^m} \quad m > 0, \quad n > 0, \quad \frac{m}{n} \text{ fully reduced}$$

With these definitions, the following properties of exponents hold for *all* exponents, whether integral or fractional, positive or negative.

$$x^m \cdot x^n = x^{m+n} \qquad (x^m)^n = x^{m \cdot n} \qquad \left(\frac{x}{y}\right)^n = \frac{x^n}{y^n}$$

$$\frac{x^m}{x^n} = x^{m-n} \qquad (xy)^n = x^n \cdot y^n$$

➤ **Solutions to Practice Problems**

1. a. $\dfrac{x^5 \cdot x}{x^2} = \dfrac{x^6}{x^2} = x^4$

   b. $[(x^3)^2]^2 = x^{3 \cdot 2 \cdot 2} = x^{12}$

2. a. $2^0 = 1$

b. $2^{-4} = \dfrac{1}{2^4} = \dfrac{1}{16}$

3. $\left(\dfrac{2}{3}\right)^{-2} = \left(\dfrac{3}{2}\right)^{2} = \dfrac{9}{4}$

4. a. $(-27)^{1/3} = \sqrt[3]{-27} = -3$

b. $\left(\dfrac{16}{81}\right)^{1/4} = \sqrt[4]{\dfrac{16}{81}} = \dfrac{\sqrt[4]{16}}{\sqrt[4]{81}} = \dfrac{2}{3}$

5. a. $16^{3/2} = (\sqrt{16})^3 = 4^3 = 64$

b. $(-8)^{2/3} = (\sqrt[3]{-8})^2 = (-2)^2 = 4$

6. a. $25^{-3/2} = \dfrac{1}{25^{3/2}} = \dfrac{1}{(\sqrt{25})^3} = \dfrac{1}{5^3} = \dfrac{1}{125}$

b. $\left(\dfrac{1}{4}\right)^{-1/2} = \left(\dfrac{4}{1}\right)^{1/2} = \sqrt{4} = 2$

c. $5^{1.3} \approx 8.103$

1.2 Exercises

Evaluate each expression *without* using a calculator.

1. $(2^2 \cdot 2)^2$

2. $(5^2 \cdot 4)^2$

3. 2^{-4}

4. 3^{-3}

5. $\left(\dfrac{1}{2}\right)^{-3}$

6. $\left(\dfrac{1}{3}\right)^{-2}$

7. $\left(\dfrac{5}{8}\right)^{-1}$

8. $\left(\dfrac{3}{4}\right)^{-1}$

9. $4^{-2} \cdot 2^{-1}$

10. $3^{-2} \cdot 9^{-1}$

11. $\left(\dfrac{3}{2}\right)^{-3}$

12. $\left(\dfrac{2}{3}\right)^{-3}$

13. $\left(\dfrac{1}{3}\right)^{-2} - \left(\dfrac{1}{2}\right)^{-3}$

14. $\left(\dfrac{1}{3}\right)^{-2} - \left(\dfrac{1}{2}\right)^{-2}$

15. $\left[\left(\dfrac{2}{3}\right)^{-2}\right]^{-1}$

16. $\left[\left(\dfrac{2}{5}\right)^{-2}\right]^{-1}$

17. $25^{1/2}$

18. $36^{1/2}$

19. $25^{3/2}$

20. $16^{3/2}$

21. $16^{3/4}$

22. $27^{2/3}$

23. $(-8)^{2/3}$

24. $(-27)^{2/3}$

25. $(-8)^{5/3}$

26. $(-27)^{5/3}$

27. $\left(\dfrac{25}{36}\right)^{3/2}$

28. $\left(\dfrac{16}{25}\right)^{3/2}$

29. $\left(\dfrac{27}{125}\right)^{2/3}$

30. $\left(\dfrac{125}{8}\right)^{2/3}$

31. $\left(\dfrac{1}{32}\right)^{2/5}$

32. $\left(\dfrac{1}{32}\right)^{3/5}$

33. $4^{-1/2}$

34. $9^{-1/2}$

35. $4^{-3/2}$

36. $9^{-3/2}$

37. $8^{-2/3}$

38. $16^{-3/4}$

39. $(-8)^{-1/3}$

40. $(-27)^{-1/3}$

41. $(-8)^{-2/3}$

42. $(-27)^{-2/3}$

43. $\left(\dfrac{25}{16}\right)^{-1/2}$

44. $\left(\dfrac{16}{9}\right)^{-1/2}$

45. $\left(\dfrac{25}{16}\right)^{-3/2}$

46. $\left(\dfrac{16}{9}\right)^{-3/2}$

47. $\left(-\dfrac{1}{27}\right)^{-5/3}$

48. $\left(-\dfrac{1}{8}\right)^{-5/3}$

Use a calculator to evaluate each expression. Round answers to two decimal places.

49. $7^{0.39}$ **50.** $5^{0.47}$ **51.** $8^{2.7}$ **52.** $5^{3.9}$

Use a graphing calculator to evaluate each expression.

53. $(-8)^{7/3}$ **54.** $(-8)^{5/3}$ **55.** $\left[\left(\frac{5}{2}\right)^{-1}\right]^{-2}$

56. $\left[\left(\frac{3}{2}\right)^{-2}\right]^{-1}$ **57.** $[(4)^{-1}]^{0.5}$ **58.** $[(0.25)^{-1}]^{0.5}$

59. $(0.4^{-7})^{-1/7}$ **60.** $[(0.5^{-1})^{-2}]^{-3}$

61. $[(0.1)^{0.1}]^{0.1}$ **62.** $\left(1+\frac{1}{1000}\right)^{1000}$

63. $\left(1-\frac{1}{1000}\right)^{-1000}$ **64.** $(1+10^{-6})^{10^{6}}$

Simplify.

65. $(x^3 \cdot x^2)^2$ **66.** $(x^4 \cdot x^3)^2$ **67.** $[z^2(z \cdot z^2)^2 z]^3$

68. $[z(z^3 \cdot z)^2 z^2]^2$ **69.** $[(x^2)^2]^2$ **70.** $[(x^3)^3]^3$

71. $\frac{(ww^2)^3}{w^3 w}$ **72.** $\frac{(ww^3)^2}{w^3 w^2}$ **73.** $\frac{(5xy^4)^2}{25x^3 y^3}$

74. $\frac{(4x^3 y)^2}{8x^2 y^3}$ **75.** $\frac{(9xy^3 z)^2}{3(xyz)^2}$ **76.** $\frac{(5x^2 y^3 z)^2}{5(xyz)^2}$

77. $\frac{(2u^2 vw^3)^2}{4(uw^2)^2}$ **78.** $\frac{(u^3 vw^2)^2}{9(u^2 w)^2}$

APPLIED EXERCISES

79–80. ALLOMETRY: Dinosaurs The study of size and shape is called "allometry," and many allometric relationships involve exponents that are fractions or decimals. For example, the body measurements of most four-legged animals, from mice to elephants, obey (approximately) the following power law:

$$\left(\begin{array}{c}\text{Average body}\\ \text{thickness}\end{array}\right) = 0.4 \text{ (hip-to-shoulder length)}^{3/2}$$

where body thickness is measured vertically and all measurements are in feet. Assuming that this same relationship held for dinosaurs, find the average body thickness of the following dinosaurs, whose hip-to-shoulder length can be measured from their skeletons:

79. Diplodocus, whose hip-to-shoulder length was 16 feet.

80. Triceratops, whose hip-to-shoulder length was 14 feet.

81–82. BUSINESS: The Rule of .6 Many chemical and refining companies use "the rule of point six" to estimate the cost of new equipment. According to this rule, if a piece of equipment (such as a storage tank) originally cost C dollars, then the cost of similar equipment that is x times as large will be approximately $x^{0.6}C$ dollars. For example, if the original equipment cost C dollars, then new equipment with twice the capacity of the old equipment $(x = 2)$ would cost $2^{0.6}C = 1.516C$ dollars—that is, about 1.5 times as much. Therefore, to increase capacity by 100% costs only about 50% more.*

81. Use the rule of .6 to find how costs change if a company wants to quadruple $(x = 4)$ its capacity.

82. Use the rule of .6 to find how costs change if a company wants to triple $(x = 3)$ its capacity.

83–84. BUSINESS: The Rule of .6 (*continuation*) Use a graphing calculator to graph $y = x^{0.6}$, expressing y, the cost multiple for larger equipment, in terms of x, the size multiple. Use the viewing window [0, 5] by [0, 3].

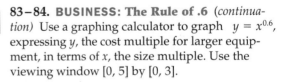

83. By how much can a company multiply its capacity for twice the money? That is, find the value of x that satisfies $x^{0.6} = 2$.

(*continues*)

*Although the rule of .6 is only a rough "rule of thumb," it can be somewhat justified on the basis that the equipment of such industries consists mainly of containers, and the cost of a container depends on its surface area (square units), which increases more slowly than its capacity (cubic units).

[*Hint:* Either use TRACE or find where $y_1 = x^{0.6}$ INTERSECTs $y_2 = 2$.]

84. Does the curve rise more steeply or less steeply as x increases? What does this mean about how rapidly cost increases as equipment size increases?

85–86. ALLOMETRY: Heart Rate It is well known that the hearts of smaller animals beat faster than the hearts of larger animals. The actual relationship is approximately

$$(\text{Heart rate}) = 250(\text{weight})^{-1/4}$$

where the heart rate is in beats per minute and the weight is in pounds. Use this relationship to estimate the heart rate of:

85. A 16-pound dog.

86. A 625-pound grizzly bear.

87–88. ALLOMETRY: Heart Rate (*continuation*) Use a graphing calculator to graph $y = 250x^{-0.25}$, which expresses y, heartbeats per minute, in terms of x, the animal's weight. Use the viewing window [0, 200] by [0, 150].

87. Notice that the curve decreases less steeply for larger values of x. Explain what this means about how rapidly heart rate decreases as body weight increases.

88. Evaluate this formula at your own weight, x, to find your predicted heart rate. Then take your pulse and see if the numbers (roughly) agree.

89–90. BUSINESS AND PSYCHOLOGY: Learning Curves in Airplane Production Recall (pages 26–27) that the learning curve for the production of Boeing 707 airplanes is $150n^{-0.322}$ (thousand work-hours). Find how many work-hours it took to build:

89. The 50th Boeing 707.

90. The 250th Boeing 707.

91. GENERAL: Richter Scale The Richter scale (developed by Charles Richter in 1935) is widely used to measure the strength of earthquakes. Every increase of 1 on the Richter scale corresponds to a 10-fold increase in ground motion. Therefore, an increase on the Richter scale from A to B means that ground motion increases by a factor of 10^{B-A} (for $B > A$).

Find the increase in ground motion between the following earthquakes:

a. The 1994 Northridge, California, earthquake, measuring 6.8 on the Richter scale, and the 1906 San Francisco earthquake, measuring 8.3. (The San Francisco earthquake resulted in 500 deaths and a 3-day fire that destroyed 4 square miles of San Francisco.)

b. The 1995 Kobe (Japan) earthquake, measuring 7.2 on the Richter scale, and the 1933 Miyagi earthquake, measuring 8.1. (The Miyagi earthquake caused a 90-foot-high tsunami, or "tidal wave," that killed 3064 people. The death toll from the Kobe earthquake was more than 5000.)

92. GENERAL: Richter Scale (*continuation*) Every increase of 1 on the Richter scale corresponds to an approximately *30-fold* increase in *energy released*. Therefore, an increase on the Richter scale from A to B means that the energy released increases by a factor of 30^{B-A} (for $B > A$).

a. Find the increase in *energy released* between the earthquakes in Exercise 91a.

b. Find the increase in *energy released* between the earthquakes in Exercise 91b.

93–94. GENERAL: Waterfalls Water falling from a waterfall that is x feet high will hit the ground with speed $\frac{60}{11}x^{0.5}$ miles per hour (neglecting air resistance).

93. Find the speed of the water at the bottom of the highest waterfall in the world, Angel Falls in Venezuela (3281 feet high).

94. Find the speed of the water at the bottom of the highest waterfall in the United States, Ribbon Falls in Yosemite, California (1650 feet high).

95–96. ENVIRONMENTAL SCIENCE: Biodiversity It is well known that larger land areas can support larger numbers of species. According to one study,* multiplying the land area by a factor of x multiplies the number of species by a factor of $x^{0.239}$. Use a graphing calculator to graph $y = x^{0.239}$. Use the viewing window [0, 100] by [0, 4].

*Rober H. MacArthur and Edward O. Wilson, *The Theory of Island Biogeography* (Princeton University Press, 1967).

95. Find the multiple x for the land area that leads to *double* the number of species. That is, find the value of x such that $x^{0.239} = 2$. [*Hint:* Either use TRACE or find where $y_1 = x^{0.239}$ INTERSECTs $y_2 = 2$.]

96. Find the multiple x for the land area that leads to triple the number of species. That is, find the value of x such that $x^{0.239} = 3$. [*Hint:* Either use TRACE or find where $y_1 = x^{0.239}$ INTERSECTs $y_2 = 3$.]

97. BUSINESS: Learning Curves A manufacturer of supercomputers finds that the number of work-hours required to build the 1st, the 10th, the 20th, and the 30th supercomputers are as follows:

Supercomputer Number	Work-Hours Required
1	3200
10	1900
20	1400
30	1300

a. Enter these numbers into a graphing calculator and make a plot of the resulting points (Supercomputer Number on the x-axis and Work-Hours Required on the y-axis).

b. Have the calculator find the power regression formula for these data, fitting a curve of the form $y = ax^b$ to the points. Enter the results in y_1. Plot the points together with the regression curve. Observe that the curve fits the points rather well.

c. Evaluate y_1 at $x = 50$ to predict the number of work-hours required to build the 50th supercomputer.

98. GENERAL: Paper Stacking Suppose that you take an ordinary piece of paper (about $\frac{1}{250}$ of an inch thick), cut it in half, stack the two halves, cut them in two and stack the pieces, and repeat this cutting and stacking operation many times. Each operation doubles the height of the stack, so that after a total of 25 such operations, the stack would be $2^{25} \cdot \frac{1}{250} \cdot \frac{1}{12} \cdot \frac{1}{5280}$ miles high.

a. Evaluate this height.

b. Use a graphing calculator to make a TABLE showing the height of the stack when the number of operations is 15 or more. [*Hint:* Use $y_1 = 2^x/(250 \cdot 12 \cdot 5280)$ for values of x beginning at 15.] For what number of operations is the stack over 1 mile high? over 10 miles high? over 100 miles high?

1.3 **FUNCTIONS**

Introduction

In the previous section we saw that the time required to build a Boeing 707 airliner will vary, depending on the number that have already been built. Mathematical relationships such as this, in which one number depends on another, are called *functions*, and are central to the study of calculus. In this section we define and give some applications of functions.

Functions

A *function** is a rule or procedure for finding, from a given number, a new number. If the function is denoted by f and the given number by

---

*In this chapter the word "function" will mean *function of one variable*. In Chapter 7 we will discuss functions of more than one variable.

x, then the resulting number is written $f(x)$ (read "*f* of *x*") and is called *the value of the function f at x*. The set of numbers *x* for which a function *f* is defined is called the *domain* of *f*, and the set of all resulting function values $f(x)$ is called the *range* of *f*. For any *x* in the domain, $f(x)$ must be a *single* number.

Function

A *function f* is a rule that assigns to each number *x* in a set a number $f(x)$. The set of all allowable values of *x* is called the *domain*, and the set of all values $f(x)$ for *x* in the domain is called the *range*.

For example, recording the temperature at a given location throughout a particular day would define a *temperature* function:

$$f(x) = \begin{pmatrix} \text{Temperature at} \\ \text{time } x \text{ hours} \end{pmatrix} \qquad \text{Domain would be } [0, 24)$$

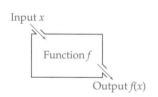

Input *x*

Function *f*

Output $f(x)$

A function *f* may be thought of as a numerical procedure or "machine" that takes an "input" number *x* and produces an "output" number $f(x)$, as shown on the left. The permissible input numbers form the domain, and the resulting output numbers form the range.

We will be mostly concerned with functions that are defined by *formulas* for calculating $f(x)$ from *x*. If the domain of such a function is not stated, then it is always taken to be the *largest* set of numbers for which the function is defined, called the *natural domain* of the function. To *graph* a function *f*, we plot all points (x, y) such that *x* is in the domain and $y = f(x)$. We call *x* the *independent variable* and *y* the *dependent variable*, since *y depends on* (is calculated from) *x*. The domain and range can be illustrated graphically.

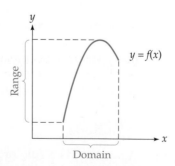

The domain of a function $y = f(x)$ is the set of all allowable *x*-values, and the range is the set of all corresponding *y*-values.

Practice Problem **1**   Find the domain and range of the function graphed below.

➤ Solution on page 45

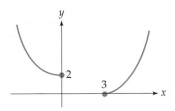

EXAMPLE **1**

FINDING THE DOMAIN AND RANGE OF A RATIONAL FUNCTION

For the function $f(x) = \dfrac{1}{x - 1}$, find:

a. $f(5)$ **b.** the domain   **c.** the range

Solution

a. $f(5) = \dfrac{1}{5 - 1} = \dfrac{1}{4}$ $f(x) = \dfrac{1}{x - 1}$ with $x = 5$

b. Domain $= \{x \mid x \neq 1\}$ $f(x) = \dfrac{1}{x - 1}$ is defined for all x except $x = 1$.

c. The graph of the function (from a graphing calculator) is shown on the right. From it, and realizing that the curve continues upward and downward (as may be verified by zooming out), it is clear that *every* y value is taken except for $y = 0$ (since the curve does not touch the *x*-axis). Therefore:

$$\text{Range} = \{y \mid y \neq 0\}$$

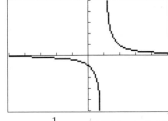

$f(x) = \dfrac{1}{x - 1}$ on $[-5, 5]$ by $[-5, 5]$

May also be written $\{z \mid z \neq 0\}$ or with any other letter

The range could also be found by solving $y = \dfrac{1}{x - 1}$ for x, giving

$x = \dfrac{1}{y} + 1$, which again shows that y can take any value except 0.

EXAMPLE 2

FINDING THE DOMAIN AND RANGE OF A POLYNOMIAL

For $f(x) = 2x^2 + 4x - 5$, determine:

a. $f(-3)$ **b.** the domain   **c.** the range

Solution

a. $f(-3) = 2(-3)^2 + 4 \cdot (-3) - 5$

$\quad\quad = 18 - 12 - 5 = 1$

b. Domain $= \mathbb{R}$

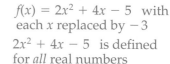
$f(x) = 2x^2 + 4x - 5$ with each x replaced by -3

$2x^2 + 4x - 5$ is defined for *all* real numbers

c. From the graph of $f(x) = 2x^2 + 4x - 5$ at the right, the lowest y value is -7 (as can be found from TRACE or MINIMUM), and all higher y values are taken (since the curve is a parabola opening upward). Therefore:

$$\text{Range} = \{\, y \mid y \geq -7 \,\}$$

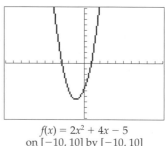

$f(x) = 2x^2 + 4x - 5$
on $[-10, 10]$ by $[-10, 10]$

Any letters may be used for defining a function or describing the domain and the range.

Practice Problem 2

For $g(z) = \sqrt{z - 2}$, determine:

a. $g(27)$ **b.** the domain   **c.** the range

➤ Solutions on page 45

For each x in the domain of a function there must be a *single* number $y = f(x)$, so the graph of a function cannot have two points (x, y) with the same x value but different y values. This leads to the following *graphical* test for functions.

Vertical Line Test for Functions

A curve in the Cartesian plane is the graph of a *function* if and only if no vertical line intersects the curve at more than one point.

EXAMPLE 3

USING THE VERTICAL LINE TEST

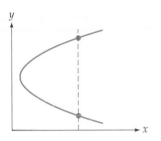

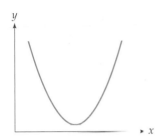

This is *not* the graph of a function of x because there is a vertical line (shown dashed) that intersects the curve twice.

This *is* the graph of a function of x because no vertical line intersects the curve more than once.

A graph that has two or more points (x, y) with the same x-value but different y-values, such as the one on the left above, defines a *relation* rather than a function. We will be concerned exclusively with *functions*, and so we will use the terms "function," "graph," and "curve" interchangeably.

Linear Function

A *linear function* is a function that can be expressed in the form

$$f(x) = mx + b$$

with constants m and b. Its graph is a line with slope m and y-intercept b.

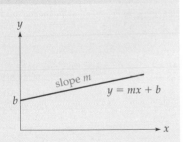

EXAMPLE 4

FINDING A COMPANY'S COST FUNCTION

An electronics company manufactures pocket calculators at a cost of $9 each, and the company's fixed costs (such as rent) amount to $400 per day. Find a function $C(x)$ that gives the total cost of producing x pocket calculators in a day.

Solution

Each calculator costs $9 to produce, so x calculators will cost $9x$ dollars, to which we must add the fixed costs of $400.

$$C(x) \quad = \quad 9x \quad + \quad 400$$

Total   Unit   Number   Fixed
cost   cost   of units   cost

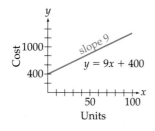

The graph of $C(x) = 9x + 400$ is a line with slope 9 and y-intercept 400, as shown on the left. Notice that the *slope* is the same as the *rate of change* of the cost (increasing at the rate of $9 per additional calculator), which is also the company's *marginal cost* (the cost of producing one more calculator is $9). The *slope*, the *rate of change*, and the *marginal cost* are always the same, as we will see in Chapter 2.

Practice Problem 3

A trucking company will deliver furniture for a charge of $25 plus 5% of the purchase price of the furniture. Find a function $D(x)$ that gives the delivery charge for a piece of furniture that costs x dollars.

➤ Solution on page 45

A mathematical description of a real-world situation is called a *mathematical model*. For example, the cost function $C(x) = 9x + 400$ from the previous example is a mathematical model for the cost of manufacturing calculators. In this model, x, the number of calculators, should take only whole-number values (0, 1, 2, 3, . . .), and the graph should consist of discrete dots rather than a continuous curve. Instead, we will find it easier to let x take *continuous* values, and round up or down as necessary at the end.

Quadratic Function

A *quadratic function* is a function that can be expressed in the form

$$f(x) = ax^2 + bx + c$$

with constants ("coefficients") $a \neq 0$, b, and c. Its graph is called a *parabola*.

The condition $a \neq 0$ keeps the function from becoming $f(x) = bx + c$, which would be linear. Many familiar curves are parabolas.

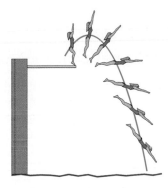

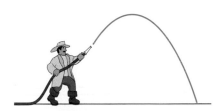

The center of gravity of a
diver describes a parabola.

A stream of water from a hose
takes the shape of a parabola.

The parabola $f(x) = ax^2 + bx + c$ opens *upward* if the constant a is *positive* and opens *downward* if the constant a is *negative*. The *vertex* of a parabola is its "central" point. The vertex is the *lowest* point on the parabola if it opens *up*, and the *highest* point if it opens *down*.

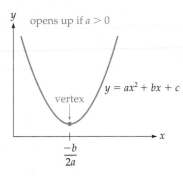

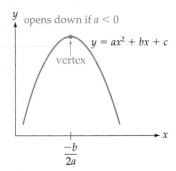

Graphing Calculator Exploration

a. Graph the parabolas $y_1 = x^2$, $y_2 = 2x^2$, and $y_3 = 4x^2$ on the graphing window $[-10, 10]$ by $[-10, 10]$. Use TRACE to identify which curve goes with which formula. How does the shape of the parabola change when the coefficient of x^2 increases?

b. Graph $y_4 = -x^2$. What did the negative sign do to the parabola?

c. Predict the shape of the parabolas $y_5 = -2x^2$ and $y_6 = \frac{1}{3}x^2$. Then check your predictions by graphing the functions.

The x-coordinate of the vertex of a parabola may be found by the following formula, which will be derived in Exercise 57 on page 197.

Vertex Formula for a Parabola

The vertex of the parabola $f(x) = ax^2 + bx + c$ has x-coordinate

$$x = \frac{-b}{2a}$$

EXAMPLE 5 **GRAPHING A QUADRATIC FUNCTION**

Graph the quadratic function $f(x) = 2x^2 - 40x + 104$.

Solution

Graphing using a graphing calculator is largely a matter of finding an appropriate viewing window, as the following three unsatisfactory windows show.

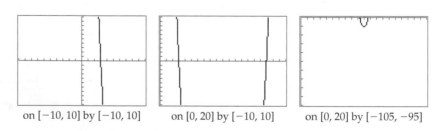

on $[-10, 10]$ by $[-10, 10]$ on $[0, 20]$ by $[-10, 10]$ on $[0, 20]$ by $[-105, -95]$

To find an appropriate viewing window, we use the vertex formula:

$$x = \frac{-b}{2a} = \frac{-(-40)}{2 \cdot 2} = \frac{40}{4} = 10 \qquad \begin{array}{l} x\text{-coordinate of the vertex, from} \\ x = \frac{-b}{2a} \text{ with } a = 2 \text{ and } b = -40 \end{array}$$

We move a few units, say 5, to either side of $x = 10$, making the x-window $[5, 15]$. Using the calculator to EVALUATE the given function at $x = 10$ (or evaluating by hand) gives $y(10) = -96$. Since the parabola opens upward (the coefficient of x^2 is positive), the curve rises up from its vertex, so we select a y-interval from -96 upward, say $[-96, -70]$. Graphing the function on the window $[5, 15]$ by

$[-96, -70]$ gives the following result. (Some other graphing windows are just as good.)

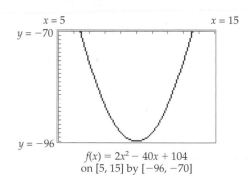

$$f(x) = 2x^2 - 40x + 104$$
on [5, 15] by [−96, −70]

Solving Quadratic Equations

A value of x that solves an equation $f(x) = 0$ is called a *root* of the equation, or a *zero* of the function, or an *x*-intercept of the graph of $y = f(x)$. The roots of a quadratic equation can often be found by factoring.

EXAMPLE 6   **SOLVING A QUADRATIC EQUATION BY FACTORING**

Solve $2x^2 - 4x = 6$.

Solution

$$2x^2 - 4x - 6 = 0$$ Subtracting 6 from each side to get zero on the right

$$2(x^2 - 2x - 3) = 0$$ Factoring out a 2

$$2(x - 3) \cdot (x + 1) = 0$$ Factoring $x^2 - 2x - 3$

   Equals 0   Equals 0   Finding *x*-values that
   at $x = 3$ at $x = -1$ make each factor zero

$$x = 3, \quad x = -1$$ Solutions

Graphing Calculator Exploration

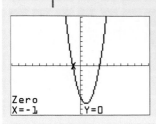

Zero
X=-1   Y=0

Find the solutions to the equation in Example 6 by graphing the function $f(x) = 2x^2 - 4x - 6$ and using ZERO or TRACE to find where the curve crosses the x-axis. Your answers should agree with those found in Example 6.

Practice Problem 4

Solve by factoring or graphing: $9x - 3x^2 = -30$

➤ Solution on page 45

Quadratic equations can often be solved by the "Quadratic Formula." A derivation of this formula is given on page 44.

Quadratic Formula

The solutions to $ax^2 + bx + c = 0$ are

$$x = \frac{-b \pm \sqrt{b^2 - 4ac}}{2a}$$

The "plus or minus" $\pm$ sign means calculate *both* ways, first using the $+$ sign and then using the $-$ sign

---

In a business, it is often important to find a company's *break-even points*, the numbers of units of production where a company's costs are equal to its revenue.

EXAMPLE 7

FINDING BREAK-EVEN POINTS

A company that installs automobile compact disc (CD) players finds that if it installs x CD players per day, then its costs will be $C(x) = 120x + 4800$ and its revenue will be $R(x) = -2x^2 + 400x$ (both in dollars). Find the company's break-even points. (Note: In Section 3.4 we will see how such cost and revenue functions are found.)

Solution

$$120x + 4800 = -2x^2 + 400x \qquad \text{Setting } C(x) = R(x)$$

$$2x^2 - 280x + 4800 = 0 \qquad \begin{array}{l}\text{Combining all terms} \\ \text{on one side}\end{array}$$

$$x = \frac{280 \pm \sqrt{(-280)^2 - 4 \cdot 2 \cdot 4800}}{2 \cdot 2} \qquad \begin{array}{l}\text{Quadratic Formula with} \\ a = 2, \ \ b = -280, \ \text{and} \\ c = 4800\end{array}$$

$$= \frac{280 \pm \sqrt{40{,}000}}{4} = \frac{280 \pm 200}{4}$$

$$= \frac{480}{4} \text{ or } \frac{80}{4} = 120 \text{ or } 20 \qquad \begin{array}{l}\text{Working out the for-} \\ \text{mula on a calculator}\end{array}$$

The company will break even when it installs either 20 or 120 CD players.

Although it is important for a company to know where its break-even points are, most companies want to do better than break even—they want to maximize their profit. Profit is defined as *revenue minus cost* (since profit is what is left over after subtracting expenses from income).

Profit

$$\text{Profit} = \text{Revenue} - \text{Cost}$$

EXAMPLE 8

MAXIMIZING PROFIT

For the CD installer whose daily revenue and cost functions were given in Example 7, find the number of units that maximizes profit, and the maximum profit.

Solution

The profit function is the revenue function minus the cost function.

$$P(x) = \underbrace{-2x^2 + 400x}_{R(x)} - \underbrace{(120x + 4800)}_{C(x)} \qquad \begin{array}{l}P(x) = R(x) - C(x) \ \text{ with} \\ R(x) = -2x^2 + 400x \ \text{ and} \\ C(x) = 120x + 4800\end{array}$$

$$= -2x^2 + 280x - 4800 \qquad \text{Simplifying}$$

Since this function represents a parabola opening downward (because of the -2), it is maximized at its vertex, which is found using the vertex formula.

$$x = \frac{-280}{2(-2)} = \frac{-280}{-4} = 70 \qquad x = \frac{-b}{2a} \text{ with}$$
$$a = -2 \text{ and } b = 280$$

Thus, profit is maximized when 70 CD players are installed. For the maximum profit, we substitute $x = 70$ into the profit function:

$$P(70) = -2(70)^2 + 280 \cdot 70 - 4800 \qquad \begin{array}{l} P(x) = -2x^2 + 280x - 4800 \\ \text{with } x = 70 \end{array}$$

$$= 5000 \qquad \text{Multiplying and combining}$$

Therefore, the company will maximize its profit when it installs 70 CD players per day. Its maximum profit will be $5000 per day.

Why doesn't a company make more profit the more it sells? Because to increase its sales it must lower its prices, which eventually leads to lower profits. The relationship among the cost, revenue, and profit functions can be seen graphically as follows.

Spreadsheet Exploration

The following spreadsheet* shows the graphs of the functions $R(x)$, $C(x)$, and $P(x)$ from Examples 7 and 8. The values for x were made by entering 0 in A3 and then copying the formula =A3+10 for A4 into cells A5 through A18. The values for $R(x)$ were found by entering the formula =-2*A3*A3+400*A3 in B3 and then copying it into cells B4 through B18. Then $C(x)$ was similarly found by starting with C3 being =120*A3+4800. $P(x)$, the difference $R(x) - C(x)$, was found with D3 being =B3-C3. The chart was made by plotting the values for the three columns corresponding to the revenue, cost, and profit.

*See the Preface for how to obtain this and other Excel spreadsheets.

	A	B	C	D	E	F	G	H	I	J
1	Number of units	Revenue	Cost	Profit						
2	x	R(x)	C(x)	P(x)						
3	0	0	4800	-4800						
4	10	3800	6000	-2200						
5	20	7200	7200	0						
6	30	10200	8400	1800						
7	40	12800	9600	3200						
8	50	15000	10800	4200						
9	60	16800	12000	4800						
10	70	18200	13200	5000						
11	80	19200	14400	4800						
12	90	19800	15600	4200						
13	100	20000	16800	3200						
14	110	19800	18000	1800						
15	120	19200	19200	0						
16	130	18200	20400	-2200						
17	140	16800	21600	-4800						
18	150	15000	22800	-7800						

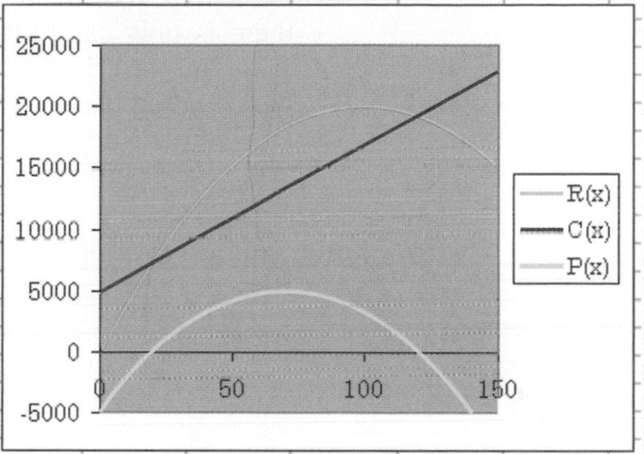

Notice that the break-even points from Example 7 (at $x = 20$ and $x = 120$) correspond to a profit of zero, and that the maximum profit (at $x = 70$) occurs halfway between the two break-even points.

Not all quadratic equations have (real) solutions.

$f(x) = \dfrac{1}{2}x^2 - 3x + 5$

EXAMPLE 9

USING THE QUADRATIC FORMULA

Solve $\dfrac{1}{2}x^2 - 3x + 5 = 0$.

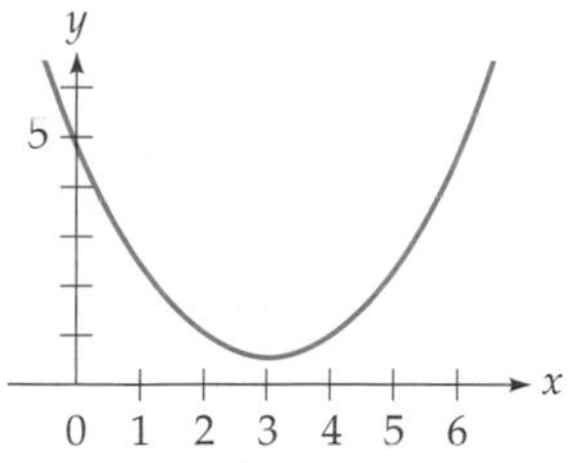

Solution The Quadratic Formula with $a = \frac{1}{2}$, $b = -3$, and $c = 5$ gives $x = \dfrac{3 \pm \sqrt{9 - 4(\frac{1}{2})(5)}}{2(\frac{1}{2})} = \dfrac{3 \pm \sqrt{9 - 10}}{1} = 3 \pm \sqrt{-1}$ Undefined

Therefore, the equation $\frac{1}{2}x^2 - 3x + 5 = 0$ *has no real solutions* (because of the undefined $\sqrt{-1}$). The geometrical reason that there are no solutions can be seen in the graph on the left: The curve never reaches the x-axis, so the function never equals zero.

The quantity $b^2 - 4ac$, whose square root appears in the Quadratic Formula, is called the *discriminant*. If the discriminant is *positive* (as in Example 7), the equation $ax^2 + bx + c = 0$ has *two* solutions (since the square root is added and subtracted). If the discriminant is *zero*, there is only *one* root (since adding and subtracting zero gives the same answer). If the discriminant is *negative* (as in Example 9), then the equation has *no* real roots. Therefore, the discriminant being positive, zero, or negative corresponds to the parabola meeting the *x*-axis at 2, 1, or 0 points, as shown below.

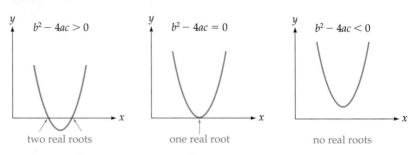

Derivation of the Quadratic Formula

$$ax^2 + bx + c = 0 \qquad \text{The quadratic set equal to zero}$$

$$ax^2 + bx = -c \qquad \text{Subtracting } c$$

$$4a^2x^2 + 4abx = -4ac \qquad \text{Multiplying by } 4a$$

$$4a^2x^2 + 4abx + b^2 = b^2 - 4ac \qquad \text{Adding } b^2$$

$$(2ax + b)^2 = b^2 - 4ac \qquad \text{Since } 4a^2x^2 + 4abx + b^2 = (2ax + b)^2$$

$$2ax + b = \pm\sqrt{b^2 - 4ac} \qquad \text{Taking square roots}$$

$$2ax = -b \pm \sqrt{b^2 - 4ac} \qquad \text{Subtracting } b$$

$$x = \frac{-b \pm \sqrt{b^2 - 4ac}}{2a} \qquad \begin{array}{l}\text{Dividing by } 2a \text{ gives}\\ \text{the Quadratic Formula}\end{array}$$

1.3 Section Summary

In this section we defined and gave examples of *functions*, and saw how to find their domains and ranges. The most important characteristic of a function f is that for any given "input" number x in the domain, there is exactly one "output" number $f(x)$. This requirement is stated geometrically in the *vertical line test*, that no vertical line can intersect the graph of a function at more than one point. We then defined *linear functions* (whose graphs are lines) and *quadratic functions* (whose graphs are parabolas). We solved quadratic equations by factoring, graphing, and using the Quadratic Formula. We maximized and minimized quadratic functions using the vertex formula.

➤ **Solutions to Practice Problems**

1. Domain. $\{\, x \mid x \le 0 \text{ or } x \ge 3 \,\}$; Range: $\{\, y \mid y \ge 0 \,\}$

2. a. $g(27) = \sqrt{27 - 2} = \sqrt{25} = 5$

    b. Domain: $\{\, z \mid z \ge 2 \,\}$

    c. Range : $\{\, y \mid y \ge 0 \,\}$

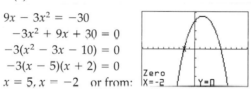

$y_1 = \sqrt{x - 2}$ on $[-1, 10]$ by $[-1, 10]$

3. $D(x) = 25 + 0.05x$

4. $9x - 3x^2 = -30$
$-3x^2 + 9x + 30 = 0$
$-3(x^2 - 3x - 10) = 0$
$-3(x - 5)(x + 2) = 0$
$x = 5, x = -2$ or from:

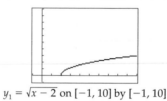

1.3   Exercises

Determine whether each graph defines a function of x.

1.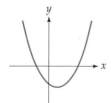

2.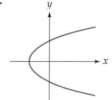

3.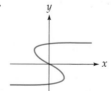

4.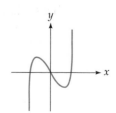

5.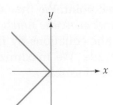

6.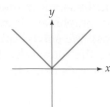

7.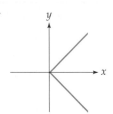

8.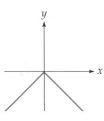

Find the domain and range of each function graphed below.

9.

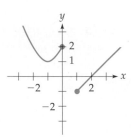

10.

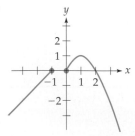

11–22. For each function:

a. Evaluate the given expression.
b. Find the domain of the function.
c. Find the range.
   [*Hint:* Use a graphing calculator.]

11. $f(x) = \sqrt{x - 1}$; find $f(10)$

12. $f(x) = \sqrt{x - 4}$; find $f(40)$

(See instructions in previous column.)

13. $h(z) = \dfrac{1}{z + 4}$; find $h(-5)$

14. $h(z) = \dfrac{1}{z + 7}$; find $h(-8)$

15. $h(x) = x^{1/4}$; find $h(81)$

16. $h(x) = x^{1/6}$; find $h(64)$

17. $f(x) = x^{2/3}$; find $f(-8)$

[*Hint for Exercises 17 and 18:* You may need to enter $x^{m/n}$ as $(x^m)^{1/n}$ or as $(x^{1/n})^m$, as discussed on page 25.]

18. $f(x) = x^{4/5}$; find $f(-32)$

19. $f(x) = \sqrt{4 - x^2}$; find $f(0)$

20. $f(x) = \dfrac{1}{\sqrt{x}}$; find $f(4)$

21. $f(x) = \sqrt{-x}$; find $f(-25)$

22. $f(x) = -\sqrt{-x}$; find $f(-100)$

23–30. Graph each function "by hand." [*Note:* Even if you have a graphing calculator, it is important to be able to sketch simple curves by finding a few important points.]

23. $f(x) = 3x - 2$ **24.** $f(x) = 2x - 3$

25. $f(x) = -x + 1$

26. $f(x) = -3x + 5$

27. $f(x) = 2x^2 + 4x - 16$

28. $f(x) = 3x^2 - 6x - 9$

29. $f(x) = -3x^2 + 6x + 9$

30. $f(x) = -2x^2 + 4x + 16$

31–34. For each quadratic function:

a. Find the vertex using the vertex formula.
b. Graph the function on an appropriate viewing window. (Answers may differ.)

31. $f(x) = x^2 - 40x + 500$

32. $f(x) = x^2 + 40x + 500$

33. $f(x) = -x^2 - 80x - 1800$

34. $f(x) = -x^2 + 80x - 1800$

35–52. Solve each equation by factoring or the Quadratic Formula, as appropriate.

35. $x^2 - 6x - 7 = 0$ **36.** $x^2 - x - 20 = 0$

37. $x^2 + 2x = 15$ **38.** $x^2 - 3x = 54$

39. $2x^2 + 40 = 18x$ **40.** $3x^2 + 18 = 15x$

41. $5x^2 - 50x = 0$ **42.** $3x^2 - 36x = 0$

43. $2x^2 - 50 = 0$ **44.** $3x^2 - 27 = 0$

45. $4x^2 + 24x + 40 = 4$ **46.** $3x^2 - 6x + 9 = 6$

47. $-4x^2 + 12x = 8$ **48.** $-3x^2 + 6x = -24$

49. $2x^2 - 12x + 20 = 0$ **50.** $2x^2 - 8x + 10 = 0$

51. $3x^2 + 12 = 0$ **52.** $5x^2 + 20 = 0$

53–62. Solve each equation using a graphing calculator. [*Hint:* Begin with the viewing window $[-10, 10]$ by $[-10, 10]$ or another of your choice (see Useful Hint in Graphing Calculator Terminology following the Preface) and use ZERO, SOLVE, or TRACE and ZOOM IN.] (In Exercises 61 and 62, round answers to two decimal places.)

53. $x^2 - x - 20 = 0$ **54.** $x^2 + 2x - 15 = 0$

55. $2x^2 + 40 = 18x$ **56.** $3x^2 + 18 = 15x$

57. $4x^2 + 24x + 45 = 9$ **58.** $3x^2 - 6x + 5 = 2$

59. $3x^2 + 7x + 12 = 0$ **60.** $5x^2 + 14x + 20 = 0$

61. $2x^2 + 3x - 6 = 0$ **62.** $3x^2 + 5x - 7 = 0$

 63. Use your graphing calculator to graph the following four equations simultaneously on the viewing window $[-10, 10]$ by $[-10, 10]$:

$$y_1 = 2x + 6$$
$$y_2 = 2x + 2$$
$$y_3 = 2x - 2$$
$$y_4 = 2x - 6$$

a. What do the lines have in common and how do they differ?

b. Write the equation of another line with the same slope that lies 2 units below the lowest line. Then check your answer by graphing it with the others.

64. Use your graphing calculator to graph the following four equations simultaneously on the viewing window $[-10, 10]$ by $[-10, 10]$:

$$y_1 = 3x + 4$$
$$y_2 = 1x + 4$$
$$y_3 = -1x + 4 \quad \text{(Use } \boxed{(-)} \text{ to get } -1x.\text{)}$$
$$y_4 = -3x + 4$$

a. What do the lines have in common and how do they differ?

b. Write the equation of a line through this y-intercept with slope $\frac{1}{2}$. Then check your answer by graphing it with the others.

APPLIED EXERCISES

65. BUSINESS: Cost Functions A lumberyard will deliver wood for $4 per board foot plus a delivery charge of $20. Find a function $C(x)$ for the cost of having x board feet of lumber delivered.

66. BUSINESS: Cost Functions A company manufactures bicycles at a cost of $55 each. If the company's fixed costs are $900, express the company's costs as a linear function of x, the number of bicycles produced.

67. BUSINESS: Salary An employee's weekly salary is $500 plus $15 per hour of overtime.

Find a function $P(x)$ giving his pay for a week in which he worked x hours of overtime.

68. BUSINESS: Salary A sales clerk's weekly salary is $300 plus 2% of her total week's sales. Find a function $P(x)$ for her pay for a week in which she sold x dollars of merchandise.

69. GENERAL: Water Pressure At a depth of d feet underwater, the water pressure is $p(d) = 0.45d + 15$ pounds per square inch. Find the pressure at:

a. The bottom of a 6-foot-deep swimming pool.

b. The maximum ocean depth of 35,000 feet.

70. GENERAL: Boiling Point At higher altitudes, water boils at lower temperatures. This is why at high altitudes foods must be boiled for longer times—the lower boiling point imparts less heat to the food. At an altitude of h thousand feet above sea level, water boils at a temperature of $B(h) = -1.8h + 212$ degrees Fahrenheit. Find the altitude at which water boils at 98.6 degrees Fahrenheit. (Your answer will show that at a high enough altitude, water boils at normal body temperature. This is why airplane cabins must be pressurized—at high enough altitudes one's blood would boil.)

71–72. GENERAL: Stopping Distance According to data from the National Transportation Safety Board, a car traveling at speed v miles per hour should be able to come to a full stop in a distance of

$$D(v) = 0.055v^2 + 1.1v \quad \text{feet}$$

Find the stopping distance required for a car traveling at:

71. 40 mph.

72. 60 mph.

73. BIOMEDICAL: Cell Growth The number of cells in a culture after t days is given by $N(t) = 200 + 50t^2$. Find the size of the culture after:

a. 2 days.
b. 10 days.

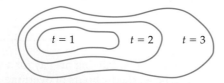

74. GENERAL: Juggling If you toss a ball h feet straight up, it will return to your hand after $T(h) = 0.5\sqrt{h}$ seconds. This leads to the *juggler's dilemma*: Juggling more balls means tossing them higher. However, the square root in the above formula means that tossing them twice as high does not gain twice as much time, but only $\sqrt{2} \approx 1.4$ times as much time. Because of this, there is a limit to the number of balls that a person can juggle, which seems to be about ten. Use this formula to find:

a. How long will a ball spend in the air if it is tossed to a height of 4 feet? 8 feet?
b. How high must it be tossed to spend 2 seconds in the air? 3 seconds in the air?

75. GENERAL: Impact Velocity If a marble is dropped from a height of x feet, it will hit the ground with velocity $v(x) = \frac{60}{11}\sqrt{x}$ miles per hour (neglecting air resistance). Use this formula to find the velocity with which a marble will strike the ground if it is dropped from the top of the tallest building in the United States, the 1454-foot Sears Tower in Chicago.

76. GENERAL: Tsunamis The speed of a tsunami (popularly known as a tidal wave, although it has nothing whatever to do with tides) depends on the depth of the water through which it is traveling. At a depth of d feet, the speed of a tsunami will be $s(d) = 3.86\sqrt{d}$ miles per hour. Find the speed of a tsunami in the Pacific basin where the average depth is 15,000 feet.

77–78. GENERAL: Impact Time of a Projectile If an object is thrown upward so that its height (in feet) above the ground t seconds after it is thrown is given by the function $h(t)$ below, find when the object hits the ground. That is, find the positive value of t such that $h(t) = 0$. Give the answer correct to two decimal places. [*Hint:* Enter the function in terms of x rather than t. Use the ZERO operation, or TRACE and ZOOM IN, or similar operations.]

77. $h(t) = -16t^2 + 45t + 5$

78. $h(t) = -16t^2 + 40t + 4$

79. BUSINESS: Break-Even Points and Maximum Profit A company that produces tracking devices for computer disk drives finds that if it produces x devices per week, its costs will be $C(x) = 180x + 16,000$ and its revenue will be $R(x) = -2x^2 + 660x$ (both in dollars).

a. Find the company's break-even points.
b. Find the number of devices that will maximize profit, and the maximum profit.

80. BUSINESS: Break-Even Points and Maximum Profit City and Country Cycles finds that if it sells x racing bicycles per month, its costs will be $C(x) = 420x + 72,000$ and its revenue will be $R(x) = -3x^2 + 1800x$ (both in dollars).

a. Find the store's break-even points.
b. Find the number of bicycles that will maximize profit, and the maximum profit.

81. **BUSINESS: Break-Even Points and Maximum Profit** A sporting goods store finds that if it sells x exercise machines per day, its costs will be $C(x) = 100x + 3200$ and its revenue will be $R(x) = -2x^2 + 300x$ (both in dollars).

a. Find the store's break-even points.
b. Find the number of sales that will maximize profit, and the maximum profit.

82. **BUSINESS: Break-Even Points and Maximum Profit** A company that installs car alarm systems finds that if it installs x systems per week, its costs will be $C(x) = 210x + 72,000$ and its revenue will be $R(x) = -3x^2 + 1230x$ (both in dollars).

a. Find the company's break-even points.
b. Find the number of installations that will maximize profit, and the maximum profit.

83. **BIOMEDICAL: Muscle Contraction** The fundamental equation of muscle contraction is of the form $(w + a)(v + b) = c$, where w is the weight placed on the muscle, v is the velocity of contraction of the muscle, and a, b, and c are constants that depend upon the muscle and the units of measurement. Solve this equation for v as a function of w, a, b, and c.

84. **GENERAL: Longevity** According to insurance data, when a person reaches age 65, the probability of living for another x decades is approximated by the function $f(x) = -0.077x^2 - 0.057x + 1$ (for $0 \le x \le 3$). Find the probability that such a person will live for another:

a. one decade
b. two decades
c. three decades

85. **BUSINESS: Sales** The following table gives a company's annual sales (in millions of units) at the ends of its first through fourth years.

Year	Sales (millions)
1	3.8
2	3.6
3	3.7
4	4.0

a. Enter the numbers from the table into a graphing calculator and make a plot of the resulting points (Year on the x-axis and Sales on the y-axis).
b. Have your calculator find the quadratic (parabolic) regression formula for these data. Then enter the result in y_1, which gives a formula for sales each year. Plot the points together with the regression line.
c. Predict the sales at the end of year 5 by evaluating $y_1(5)$.

1.4 FUNCTIONS, CONTINUED

Introduction

In this section we will define other useful types of functions and an important operation, the *composition* of two functions.

Polynomial Functions

A *polynomial function* (or simply a *polynomial*) is a function that can be written in the form

$$f(x) = a_n x^n + a_{n-1} x^{n-1} + \cdots + a_2 x^2 + a_1 x + a_0$$

where n is a nonnegative integer and $a_0, a_1, \ldots, a_n$ are (real) numbers, called *coefficients*. The *domain* of a polynomial is $\mathbb{R}$, the set of all (real) numbers. The *degree* of a polynomial is the highest power of the variable. The following are polynomials.

$$f(x) = 2x^8 - 3x^7 + 4x^5 - 5$$
A polynomial of degree 8 (since the highest power of x is 8)

$$f(x) = -4x^2 - \tfrac{1}{3}x + 19$$
A polynomial of degree 2 (a quadratic function)

$$f(x) = x - 1$$
A polynomial of degree 1 (a linear function)

$$f(x) = 6$$
A polynomial of degree 0 (a constant function)

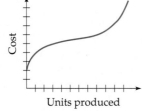

A cost function may increase at different rates at different production levels.

Polynomials are used to model many situations in which change occurs at different rates. For example, the polynomial in the graph on the left might represent the total cost of manufacturing x units of a product. At first, costs rise quite steeply as a result of high start-up expenses, then they rise more slowly as the economies of mass production come into play, and finally they rise more steeply as new production facilities need to be built.

Polynomial equations can often be solved by factoring (just as with quadratic equations).

EXAMPLE 1

SOLVING A POLYNOMIAL EQUATION BY FACTORING

Solve $3x^4 - 6x^3 = 24x^2$

Solution

$$3x^4 - 6x^3 - 24x^2 = 0$$
Rewritten with all the terms on the left side

$$3x^2(x^2 - 2x - 8) = 0$$
Factoring out $3x^2$

$$3x^2 \; (x - 4) \; (x + 2) = 0$$
Factoring further

Finding the zeros of each factor

Equals zero at $x = 0$ Equals zero at $x = 4$ Equals zero at $x = -2$

$$x = 0, \quad x = 4, \quad x = -2$$
Solutions

As in this example, if a positive power of x can be factored out of a polynomial, then $x = 0$ is one of the roots.

Practice Problem 1 Solve $2x^3 - 4x^2 = 48x$ ➤ Solution on page 64

Rational Functions

The word "ratio" means fraction or quotient, and the quotient of two polynomials is called a *rational function*. The following are rational functions.

$$f(x) = \frac{4x^3 + 3x^2}{x^2 - 2x + 1} \qquad g(x) = \frac{1}{x^2 + 1}$$

A rational function is a polynomial over a polynomial

The domain of a rational function is the set of all numbers for which the denominator is not zero. For example, the domain of the function on the left above is $\{x \mid x \neq 1\}$ (since $x = 1$ makes the denominator zero), and the domain of the function on the right is $\mathbb{R}$ (since $x^2 + 1$ is never zero).

Practice Problem 2 What is the domain of the rational function $f(x) = \dfrac{18}{(x + 2)(x - 4)}$?

➤ Solution on page 64

Simplifying a rational function by canceling a common factor from the numerator and the denominator can change the domain of the function, so that the "simplified" and "original" versions may not be equal (since they have different domains). For example, the rational function on the left below is not defined at $x = 1$, whereas the simplified version on the right *is* defined at $x = 1$, so that the two functions are technically not equal.

$$\underbrace{\frac{x^2 - 1}{x - 1}}_{} = \underbrace{\frac{(x + 1)(x - 1)}{x - 1}}_{} \neq \underbrace{x + 1}_{}$$

Not defined at $x = 1$, so the domain is $\{x \mid x \neq 1\}$ Is defined at $x - 1$, so the domain is $\mathbb{R}$

However, the functions *are* equal at every x-value *except* $x = 1$, and the graphs (shown on the following page) are the same except that the rational function omits the point at $x = 1$. We will return to this technical issue when we discuss limits in Section 2.1.

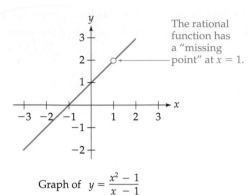

Graph of $y = \dfrac{x^2 - 1}{x - 1}$

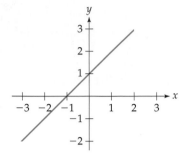

Graph of $y = x + 1$

Exponential Functions

A function in which the independent variable appears in the exponent, such as $f(x) = 2^x$, is called an exponential function.

EXAMPLE 2

GRAPHING AN EXPONENTIAL FUNCTION

Graph the exponential function $f(x) = 2^x$.

Solution

This function is defined for *all* real numbers, so its domain is $\mathbb{R}$. Values of the function are shown in the table on the left below, and plotting these points and drawing a smooth curve through them gives the curve on the right below.

x	$y = 2^x$
3	$2^3 = 8$
2	$2^2 = 4$
1	$2^1 = 2$
0	$2^0 = 1$
-1	$2^{-1} = \frac{1}{2}$
-2	$2^{-2} = \frac{1}{4}$
-3	$2^{-3} = \frac{1}{8}$

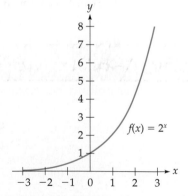

$f(x) = 2^x$ has domain $\mathbb{R}$ and range $\{ y \mid y > 0 \}$

Exponential functions are often used to model population growth and decline.

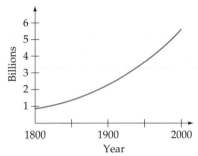

World population since the year 1800 can
be approximated by an exponential function.

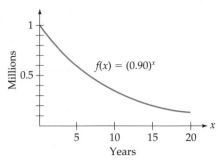

A population of 1 million that declines by 10%
each year is modeled by an exponential function.

In mathematics the letter e is used to represent a constant whose value is approximately 2.718. The exponential function $f(x) = e^x$ will be very important beginning in Chapter 4. Another important function is the logarithmic function to the base e, written $f(x) = \ln x$. These functions are graphed below.

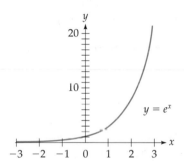

The exponential function has
domain $\mathbb{R}$ and range $\{y \mid y > 0\}$.

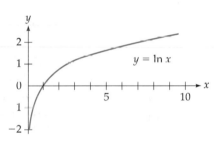

The natural logarithm function has
domain $\{x \mid x > 0\}$ and range $\mathbb{R}$.

Graphing Calculator Exploration

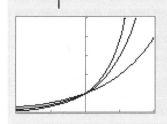

Graph the functions $y_1 = 2^x$, $y_2 = 3^x$, and $y_3 = 4^x$ on the window $[-2, 2]$ by $[0, 5]$.

a. Which function rises most steeply?

b. Between which two curves would e^x lie? Check your prediction by graphing $y_4 = e^x$. [*Hint:* On some calculators, e^x is obtained by pressing [2nd] [ln] [x].]

Piecewise Linear Functions

The rule for calculating the values of a function may be given in several parts. If each part is linear, the function is called a *piecewise linear function*, and its graph consists of "pieces" of straight lines.

EXAMPLE 3

GRAPHING A PIECEWISE LINEAR FUNCTION

Graph $f(x) = \begin{cases} 5 - 2x & \text{if } x \geq 2 \\ x + 3 & \text{if } x < 2 \end{cases}$

This notation means: Use the top formula for $x \geq 2$ and the bottom formula for $x < 2$

Solution We graph one "piece" at a time.

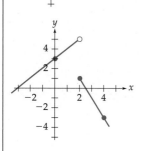

Step 1: To graph the first part, $f(x) = 5 - 2x$ if $x \geq 2$, we use the "endpoint" $x = 2$ and also $x = 4$ (or any other x-value satisfying $x \geq 2$). The points are (2, 1) and (4, −3), with the y-coordinates calculated from $f(x) = 5 - 2x$. Draw the line through these two points, but only for $x \geq 2$ (from $x = 2$ to the *right*), as shown on the left.

Step 2: For the second part, $f(x) = x + 3$ if $x < 2$, the restriction $x < 2$ means that the line ends just *before* $x = 2$. We mark this "missing point" (2, 5) by an "open circle" ○ to indicate that it is *not* included in the graph (the y-coordinate comes from $f(x) = x + 3$). For a second point, choose $x = 0$ (or any other $x < 2$), giving (0, 3). Draw the line through these two points, but only for $x < 2$ (to the *left* of $x = 2$), completing the graph of the function.

An important piecewise linear function is the *absolute value* function.

EXAMPLE 4

THE ABSOLUTE VALUE FUNCTION

The absolute value function is $f(x) = |x|$ defined as

$$f(x) = \begin{cases} x & \text{if } x \geq 0 \\ -x & \text{if } x < 0 \end{cases}$$

The second line, for *negative x*, attaches a *second* negative sign to make the result *positive*

For example, when applied to either 3 or −3, the function gives *positive* 3:

$$f(3) = 3$$

Using the top formula (since $3 \geq 0$)

$$f(-3) = -(-3) = 3$$

Using the bottom formula (since $-3 < 0$)

To graph the absolute value function, we may proceed as in Example 3, or simply observe that for $x \geq 0$, the function gives $y = x$ (a half-line from the origin with slope 1), and for $x < 0$, it gives $y = -x$ (a half-line on the other side of the origin with slope -1), as shown in the following graph.

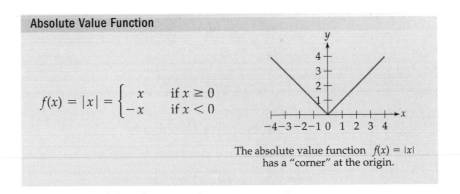

Absolute Value Function

$$f(x) = |x| = \begin{cases} x & \text{if } x \geq 0 \\ -x & \text{if } x < 0 \end{cases}$$

The absolute value function $f(x) = |x|$ has a "corner" at the origin.

Examples 3 and 4 show that the "pieces" of a piecewise linear function may or may not be connected.

EXAMPLE **5**

GRAPHING AN INCOME TAX FUNCTION

Federal income taxes are "progressive," meaning that they take a higher percentage of higher incomes. For example, the 2001 federal income tax for a single taxpayer whose taxable income was less than $136,750 was determined by a three-part rule: 15% of income up to $27,050, plus 27.5% of any amount over $27,050 up to $65,550, plus 30.5% of any amount over $65,550 up to $136,750. For an income of x dollars, the tax $f(x)$ may be expressed as follows:

$$f(x) = \begin{cases} 0.15x & \text{if } 0 \leq x \leq 27{,}050 \\ 4057.50 + 0.275(x - 27{,}050) & \text{if } 27{,}050 < x \leq 65{,}550 \\ 14{,}645 + 0.305(x - 65{,}550) & \text{if } 65{,}550 < x \leq 136{,}750 \end{cases}$$

Graphing this by the same technique as before leads to the graph shown on the following page. The slopes 0.15, 0.275, and 0.305 are called the *marginal tax rates*.

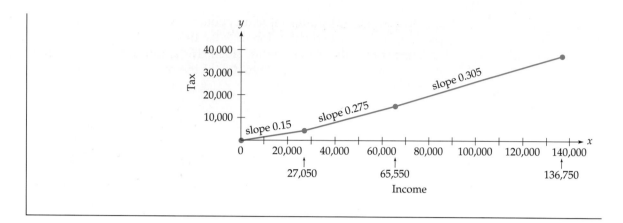

Composite Functions

Just as we substitute a *number* into a function, we may substitute a *function* into a function. For two functions f and g, evaluating f at $g(x)$ gives $f(g(x))$, called the *composition of f with g evaluated at x.*

Composite Functions

The *composition* of f with g evaluated at x is $f(g(x))$.

The *domain* of $f(g(x))$ is the set of all numbers x in the domain of g such that $g(x)$ is in the domain of f. If we think of the functions f and g as "numerical machines," then the composition $f(g(x))$ may be thought of as a *combined* machine in which the output of g is connected to the input of f.

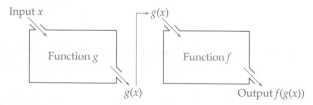

A "machine" for generating the composition of f with g.
A number x is fed into the function g, and the output
$g(x)$ is then fed into the function f, resulting in $f(g(x))$.

*The composition $f(g(x))$ may also be written $(f \circ g)(x)$, although we will not use this notation.

EXAMPLE **6**

FINDING A COMPOSITE FUNCTION

If $f(x) = x^7$ and $g(x) = x^3 - 2x$, find the composition $f(g(x))$.

Solution

$$f(g(x)) \quad = \quad \underbrace{[g(x)]^7}_{\substack{f(x) = x^7 \text{ with } x \\ \text{replaced by } g(x)}} \quad = \quad \underbrace{(x^3 - 2x)^7}_{\substack{\text{Using} \\ g(x) = x^3 - 2x}}$$

EXAMPLE **7**

FINDING BOTH COMPOSITE FUNCTIONS

If $f(x) = \dfrac{x + 8}{x - 1}$ and $g(x) = \sqrt{x}$, find $f(g(x))$ and $g(f(x))$.

Solution

$$f(g(x)) = \frac{g(x) + 8}{g(x) - 1} = \frac{\sqrt{x} + 8}{\sqrt{x} - 1} \qquad \begin{array}{l} f(x) = \dfrac{x + 8}{x - 1} \text{ with } x \\ \text{replaced by } g(x) = \sqrt{x} \end{array}$$

$$g(f(x)) = \sqrt{f(x)} = \sqrt{\frac{x + 8}{x - 1}} \qquad \begin{array}{l} g(x) = \sqrt{x} \text{ with } x \\ \text{replaced by } f(x) = \dfrac{x + 8}{x - 1} \end{array}$$

The order of composition is important: $f(g(x))$ is not the same as $g(f(x))$. To show this, we evaluate the above $f(g(x))$ and $g(f(x))$ at $x = 4$:

$$f(g(4)) = \frac{\sqrt{4} + 8}{\sqrt{4} - 1} = \frac{2 + 8}{2 - 1} = \frac{10}{1} = 10 \qquad f(g(x)) = \frac{\sqrt{x} + 8}{\sqrt{x} - 1} \text{ at } x = 4$$

Different answers

$$g(f(4)) = \sqrt{\frac{4 + 8}{4 - 1}} = \sqrt{\frac{12}{3}} = \sqrt{4} = 2 \qquad g(f(x)) = \sqrt{\frac{x + 8}{x - 1}} \text{ at } x = 4$$

Practice Problem 3 If $f(x) = x^2 + 1$ and $g(x) = \sqrt[3]{x}$, find: **a.** $f(g(x))$, **b.** $g(f(x))$.

➤ Solutions on page 64

EXAMPLE 8   **PREDICTING WATER USAGE**

A planning commission estimates that if a city's population is p thousand people, its daily water usage will be $W(p) = 30p^{1.2}$ thousand gallons. The commission further predicts that the population in t years will be $p(t) = 60 + 2t$ thousand people. Express the water usage W as a function of t, the number of years from now, and find the water usage 10 years from now.

Solution

Water usage W as a function of t is the *composition* of $W(p)$ with $p(t)$:

$$W(p(t)) = 30[p(t)]^{1.2} = 30(60 + 2t)^{1.2} \qquad \begin{array}{l} W = 30p^{1.2} \text{ with } p \\ \text{replaced by } p(t) = 60 + 2t \end{array}$$

To find water usage in 10 years, we evaluate $W(p(t))$ at $t = 10$:

$$W(p(10)) = 30(60 + 2 \cdot 10)^{1.2} \qquad 30(60 + 2t)^{1.2} \text{ with } t = 10$$
$$= 30(80)^{1.2} \approx 5765 \qquad \text{Using a calculator}$$

Thousand gallons

Therefore, in 10 years the city will need about 5,765,000 gallons of water per day.

Shifts of Graphs

Sometimes the graph of a composite function is just a horizontal or vertical shift of an original graph. This occurs when one of the functions is simply the addition or subtraction of a constant. The following diagram shows the graph of $y = x^2$ in the center, along with horizontal shifts (where the *variable* is increased or decreased by a constant) and vertical shifts (where the *function* is increased or decreased by a constant).

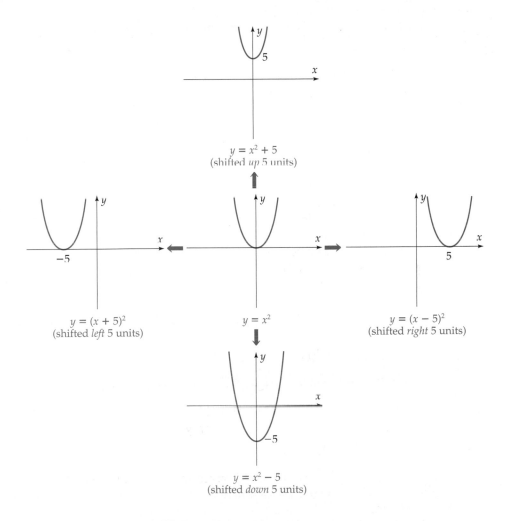

$y = x^2 + 5$
(shifted *up* 5 units)

$y = (x + 5)^2$
(shifted *left* 5 units)

$y = x^2$

$y = (x - 5)^2$
(shifted *right* 5 units)

$y = x^2 - 5$
(shifted *down* 5 units)

In general, if the addition or subtraction is done to the *x-value*, then the shift is *horizontal*, whereas if it is done to the *function*, then the shift is *vertical*. That is, for any function $y = f(x)$ and positive numbers a and b:

The graph of	is the graph of $y = f(x)$ shifted
$y = f(x + a)$	*left* by a units
$y = f(x - a)$	*right* by a units
$y = f(x) + b$	*up* by b units
$y = f(x) - b$	*down* by b units

Of course, a graph can be shifted both horizontally and vertically:

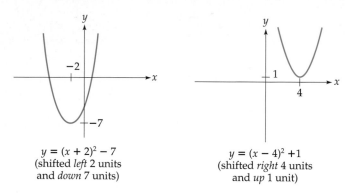

$y = (x + 2)^2 - 7$
(shifted *left* 2 units
and *down* 7 units)

$y = (x - 4)^2 + 1$
(shifted *right* 4 units
and *up* 1 unit)

Such double shifts can be applied to *any* function $y = f(x)$: the graph of $y = f(x + a) + b$ is shifted *left a* units and *up b* units (with the understanding that a *negative a* or *b* means that the direction is reversed).

Be careful: Remember that adding a *positive* number to x means a *left shift*.

Graphing Calculator Exploration

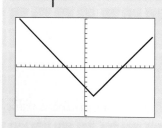

The absolute value function $y = |x|$ may be graphed on some graphing calculator as $y_1 = \text{ABS}(x)$.

a. Graph $y_1 = \text{ABS}(x - 2) - 6$ and observe that the absolute value function is shifted *right* 2 units and *down* 6 units. (The graph shown is drawn using ZOOM ZSquare.)

b. Predict the shift of $y_1 = \text{ABS}(x + 4) + 2$ and then verify your prediction by graphing the function on your calculator.

Given a function $f(x)$, to find an algebraic expression for the "shifted" function $f(x + h)$ we simply replace each occurrence of x by $x + h$.

| EXAMPLE 9 | **FINDING $f(x + h)$ FROM $f(x)$** |

If $f(x) = x^2 - 5x$, find $f(x + h)$.

Solution

$$f(x + h) = \underbrace{(x + h)^2} - \underbrace{5(x + h)}$$

$f(x) = x^2 - 5x$ with each x
replaced by $x + h$

$$= x^2 + 2xh + h^2 - 5x - 5h$$

Expanding

Difference Quotients

The quantity $\dfrac{f(x + h) - f(x)}{h}$ will be very important in Chapter 2 when we begin studying calculus. It is called the *difference quotient*, since it is a quotient whose numerator is a difference. It gives the slope (rise over run) between the points in the curve $y = f(x)$ at x and at $x + h$.

EXAMPLE 10

FINDING A DIFFERENCE QUOTIENT

If $f(x) = x^2 - 4x + 1$, find and simplify $\dfrac{f(x + h) - f(x)}{h}$ $(h \neq 0)$

Solution

$$\frac{f(x + h) - f(x)}{h} = \frac{\overbrace{(x + h)^2 - 4(x + h) + 1}^{f(x+h)} - \overbrace{(x^2 - 4x + 1)}^{f(x)}}{h}$$

$$= \frac{x^2 + 2xh + h^2 - 4x - 4h + 1 - x^2 + 4x - 1}{h}$$ Expanding

$$= \frac{x^2 + 2xh + h^2 - 4x - 4h + 1 - x^2 + 4x - 1}{h}$$ Canceling

$$= \frac{2xh + h^2 - 4h}{h} = \frac{h(2x + h - 4)}{h}$$ Factoring an h from the top

$$= \frac{h(2x + h - 4)}{h} = 2x + h - 4$$ Canceling h from top and bottom (since $h \neq 0$)

Practice Problem **4**   If $f(x) = 3x^2 - 2x + 1$, find and simplify $\dfrac{f(x + h) - f(x)}{h}$.

➤ Solution on page 64

EXAMPLE **11**   **FINDING A DIFFERENCE QUOTIENT**

If $f(x) = \dfrac{1}{x}$, find and simplify $\dfrac{f(x + h) - f(x)}{h}$ $(h \neq 0)$

Solution

$$\frac{f(x + h) - f(x)}{h} = \frac{\overbrace{\dfrac{1}{x + h}}^{f(x+h)} - \overbrace{\dfrac{1}{x}}^{f(x)}}{h}$$

$$= \frac{1}{h}\left(\frac{1}{x + h} - \frac{1}{x}\right) \qquad \text{Multiplying by } 1/h \text{ instead of dividing by } h$$

$$= \frac{1}{h}\left(\frac{x}{(x + h)x} - \frac{x + h}{(x + h)x}\right) \qquad \text{Using the common denominator } (x + h)x$$

$$= \frac{1}{h} \cdot \frac{\overbrace{x - (x + h)}^{-h}}{(x + h)x} = \frac{1}{h} \cdot \frac{-h}{(x + h)x} \qquad \text{Subtracting fractions, and simplifying}$$

$$= \frac{1}{\not{h}} \cdot \frac{\overset{-1}{\not{-h}}}{(x + h)x} = \frac{-1}{(x + h)x} \qquad \text{Canceling } h \ (h \neq 0)$$

1.4 **Section Summary**

We have introduced a variety of functions: polynomials (which include linear and quadratic functions), rational functions, exponential functions, and piecewise linear functions. Examples of these are shown on the following page. You should be able to identify these basic types of functions from their algebraic forms. We also added constants to perform horizontal and vertical *shifts* of graphs of functions, and combined functions by using the "output" of one as the "input" of the other, resulting in *composite* functions.

A Gallery of Functions

POLYNOMIALS

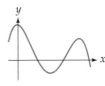

Linear function
$f(x) = mx + b$

Quadratic functions
$f(x) = ax^2 + bx + c$

$f(x) = ax^4 + bx^3 + cx^2 + dx + e$

RATIONAL FUNCTIONS

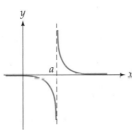

$f(x) = \dfrac{1}{x - a}$

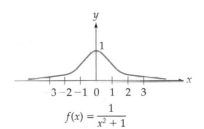

$f(x) = \dfrac{1}{x^2 + 1}$

EXPONENTIAL FUNCTIONS

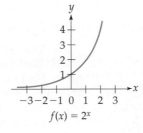

$f(x) = 2^x$

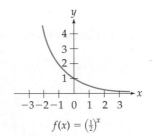

$f(x) = \left(\tfrac{1}{2}\right)^x$

PIECEWISE LINEAR FUNCTIONS

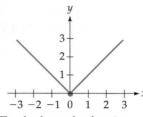

The absolute value function
$f(x) = |x|$

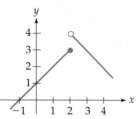

$$f(x) = \begin{cases} x + 1 & \text{if } x \le 2 \\ 6 - x & \text{if } x > 2 \end{cases}$$

➤ **Solutions to Practice Problems**

1. $2x^3 - 4x^2 - 48x = 0$

$2x(x^2 - 2x - 24) = 0$

$2x(x + 4)(x - 6) = 0$

$x = 0, \quad x = -4, \quad x = 6$

2. $\{ x \mid x \ne -2, x \ne 4 \}$

3. **a.** $f(g(x)) = [g(x)]^2 + 1 = \left(\sqrt[3]{x} \right)^2 + 1 \quad \text{or} \quad x^{2/3} + 1$

    b. $g(f(x)) = \sqrt[3]{f(x)} = \sqrt[3]{x^2 + 1} \quad \text{or} \quad (x^2 + 1)^{1/3}$

4. $\dfrac{f(x + h) - f(x)}{h}$

$$= \frac{3(x + h)^2 - 2(x + h) + 1 - (3x^2 - 2x + 1)}{h}$$

$$= \frac{3x^2 + 6xh + 3h^2 - 2x - 2h + 1 - 3x^2 + 2x - 1}{h}$$

$$= \frac{h(6x + 3h - 2)}{h} = 6x + 3h - 2$$

1.4 **Exercises**

Find the domain and range of each function graphed below.

1.

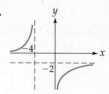

2.

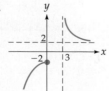

For each function in Exercises 3–10:

a. Evaluate the given expression.

b. Find the domain of the function.

c. Find the range. [*Hint:* Use a graphing calculator. You may have to ignore some false lines on the graph. Graphing in "dot mode" will also eliminate false lines.]

(See instructions on previous page.)

3. $f(x) = \dfrac{1}{x + 4}$; find $f(-3)$

4. $f(x) = \dfrac{1}{(x - 1)^2}$; find $f(-1)$

5. $f(x) = \dfrac{x^2}{x - 1}$; find $f(-1)$

6. $f(x) = \dfrac{x^2}{x + 2}$; find $f(2)$

7. $f(x) = \dfrac{12}{x(x + 4)}$; find $f(2)$

8. $f(x) = \dfrac{16}{x(x - 4)}$; find $f(-4)$

9. $g(x) = 4^x$; find $g\left(-\frac{1}{2}\right)$

10. $g(x) = 8^x$; find $g\left(-\frac{1}{3}\right)$

Solve each equation by factoring.

11. $x^5 + 2x^4 - 3x^3 = 0$ **12.** $x^6 - x^5 - 6x^4 = 0$

13. $5x^3 - 20x = 0$ **14.** $2x^5 - 50x^3 = 0$

15. $2x^3 + 18x = 12x^2$ **16.** $3x^4 + 12x^2 = 12x^3$

17. $6x^5 - 30x^4$ **18.** $5x^4 = 20x^3$

19. $3x^{5/2} - 6x^{3/2} = 9x^{1/2}$

[*Hint for Exercises 19 and 20:* First factor out a fractional power.]

20. $2x^{7/2} + 8x^{5/2} = 24x^{3/2}$

Solve each equation using a graphing calculator.

21. $x^3 - 2x^2 - 8x = 0$ **22.** $x^3 + 2x^2 - 8x = 0$

23. $x^5 - 2x^4 - 3x^3 = 0$ **24.** $x^6 + x^5 - 6x^4 = 0$

25. $2x^3 = 12x^2 - 18x$

26. $3x^4 + 12x^3 + 12x^2 = 0$

27. $6x^5 + 30x^4 = 0$ **28.** $5x^4 + 20x^3 = 0$

29. $2x^{5/2} + 4x^{3/2} = 6x^{1/2}$

30. $3x^{7/2} - 12x^{5/2} = 36x^{3/2}$

31. $x^5 - x^4 - 5x^3 = 0$ **32.** $x^6 + 2x^5 - 5x^4 = 0$

(For Exercises 31 and 32, round answers to two decimal places.)

Graph each function.

33. $f(x) = 3^x$ **34.** $f(x) = \left(\frac{1}{3}\right)^x$

35. $f(x) = \begin{cases} 2x - 7 & \text{if } x \geq 4 \\ 2 - x & \text{if } x < 4 \end{cases}$

36. $f(x) = \begin{cases} 8 - 2x & \text{if } x \geq 2 \\ x + 2 & \text{if } x < 2 \end{cases}$

37. $f(x) = |x - 3| - 3$

38. $f(x) = |x + 2| - 2$

Identify each function in Exercises 39–52 as a polynomial, a rational function, an exponential function, a piecewise linear function, or none of these. (Do not graph them; just identify their types.)

39. $f(x) = x^5$ **40.** $f(x) = 4^x$ **41.** $f(x) = 5^x$

42. $f(x) = x^4$ **43.** $f(x) = x + 2$

44. $f(x) = \begin{cases} 3x - 1 & \text{if } x \geq 2 \\ 1 - x & \text{if } x < 2 \end{cases}$

45. $f(x) = \dfrac{1}{x + 2}$ **46.** $f(x) = x^2 + 9$

47. $f(x) = \begin{cases} x - 2 & \text{if } x < 3 \\ 7 - 4x & \text{if } x > 3 \end{cases}$

48. $f(x) = \dfrac{x}{x^2 + 9}$ **49.** $f(x) = 3x^2 - 2x$

50. $f(x) = x^3 - x^{2/3}$ **51.** $f(x) = x^2 + x^{1/2}$

52. $f(x) = 5$

53. For the functions $y_1 = \left(\frac{1}{3}\right)^x$, $y_2 = \left(\frac{1}{2}\right)^x$, $y_3 = 2^x$, and $y_4 = 3^x$:

  a. Predict which curve will be the highest for large values of x.

  b. Predict which curve will be the lowest for large values of x.

  c. Check your predictions by graphing the functions on the window $[-3, 3]$ by $[0, 5]$.

  d. From your graph, what is the common y-intercept? Why do all such exponential functions meet at this point?

54. Graph the parabola $y_1 = 1 - x^2$ and the semicircle $y_2 = \sqrt{1 - x^2}$ on the window $[-1, 1]$ by $[0, 1]$. (You may want to adjust the window to make the semicircle look more like a semicircle.) Use TRACE to determine which is the "inside" curve (the parabola or the semicircle) and which is the "outside" curve. These graphs show that when you graph a parabola, you should draw the curve near the vertex to be slightly more "pointed" than a circular curve.

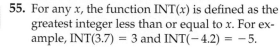 **55.** For any x, the function INT(x) is defined as the greatest integer less than or equal to x. For example, INT(3.7) = 3 and INT(-4.2) = -5.

    a. Use a graphing calculator to graph the function $y_1 = $ INT(x). (You may need to graph it in DOT mode to eliminate false connecting lines.)

    b. From your graph, what are the domain and range of this function?

56. a. Use a graphing calculator to graph the function $y_1 = 2$ INT(x). [See the previous exercise for a definition of INT(x).]

    b. From your graph, what are the domain and range of this function?

For each pair of functions $f(x)$ and $g(x)$, find
a. $f(g(x))$ and **b.** $g(f(x))$.

57. $f(x) = x^5$; $g(x) = 7x - 1$

58. $f(x) = x^8$; $g(x) = 2x + 5$

59. $f(x) = \dfrac{1}{x}$; $g(x) = x^2 + 1$

60. $f(x) = \sqrt{x}$; $g(x) = x^3 - 1$

61. $f(x) = x^3 - x^2$; $g(x) = \sqrt{x} - 1$

62. $f(x) = x - \sqrt{x}$; $g(x) = x^2 + 1$

63. $f(x) = \dfrac{x^3 - 1}{x^3 + 1}$; $g(x) = x^2 - x$

64. $f(x) = \dfrac{x^4 + 1}{x^4 - 1}$; $g(x) = x^3 + x$

65. a. Find the composition $f(g(x))$ of the two linear functions $f(x) = ax + b$ and $g(x) = cx + d$ (for constants a, b, c, and d).

    b. Is the composition of two linear functions always a linear function?

66. a. Is the composition of two quadratic functions always a quadratic function? [*Hint:*

Find the composition of $f(x) = x^2$ and $g(x) = x^2$.]

    b. Is the composition of two polynomials always a polynomial?

For each function in Exercises 67–70, find and simplify $f(x + h)$.

67. $f(x) = 5x^2$ **68.** $f(x) = 3x^2$

69. $f(x) = 2x^2 - 5x + 1$ **70.** $f(x) = 3x^2 - 5x + 2$

For each function in Exercises 71–82, find and simplify $\dfrac{f(x + h) - f(x)}{h}$. (Assume $h \neq 0$.)

71. $f(x) = 5x^2$ **72.** $f(x) = 3x^2$

73. $f(x) = 2x^2 - 5x + 1$ **74.** $f(x) = 3x^2 - 5x + 2$

75. $f(x) = 7x^2 - 3x + 2$ **76.** $f(x) = 4x^2 - 5x + 3$

77. $f(x) = x^3$
    [*Hint:* Use $(x + h)^3 = x^3 + 3x^2h + 3xh^2 + h^3$.]

78. $f(x) = x^4$
    [*Hint:* Use $(x + h)^4 = x^4 + 4x^3h + 6x^2h^2 + 4xh^3 + h^4$.]

79. $f(x) = \dfrac{2}{x}$ **80.** $f(x) = \dfrac{3}{x}$

81. $f(x) = \dfrac{1}{x^2}$ **82.** $f(x) = \sqrt{x}$

[*Hint for Exercise 82:* Multiply top and bottom of the fraction by $\sqrt{x + h} + \sqrt{x}$.]

83. Find, rounding to five decimal places:

    a. $\left(1 + \dfrac{1}{100}\right)^{100}$

    b. $\left(1 + \dfrac{1}{10{,}000}\right)^{10{,}000}$

    c. $\left(1 + \dfrac{1}{1{,}000{,}000}\right)^{1{,}000{,}000}$

    d. Do the resulting numbers seem to be approaching a limiting value? Estimate the limiting value to five decimal places. The number that you have approximated is denoted e, and will be used extensively in Chapter 4.

84. Use the TABLE feature of your graphing calculator to evaluate $\left(1 + \dfrac{1}{x}\right)^{x}$ for values of x such as 100, 10,000, 1,000,000, and higher values. Do the resulting numbers seem to be approaching a limiting value? Estimate the limiting value to five decimal places. The number that you have approximated is denoted e, and will be used extensively in Chapter 4.

85. How will the graph of $y = (x + 3)^3 + 6$ differ from the graph of $y = x^3$? Check by graphing both functions together.

86. How will the graph of $y = -(x - 4)^2 + 8$ differ from the graph of $y = -x^2$? Check by graphing both functions together.

APPLIED EXERCISES

87–88. GENERAL: World Population The world population (in millions) since the year 1700 is approximated by the exponential function $P(x) = 522(1.0053)^x$, where x is the number of years since 1700 (for $0 < x < 200$). Using a calculator, estimate the world population in the year

87. 1750 **88.** 1800

89. ECONOMICS: Income Tax The following function expresses an income tax that is 10% for incomes below $5000, and otherwise is $500 plus 30% of income in excess of $5000.

$$f(x) = \begin{cases} 0.10x & \text{if } 0 \le x < 5000 \\ 500 + 0.30(x - 5000) & \text{if } x \ge 5000 \end{cases}$$

  a. Calculate the tax on an income of $3000.
  b. Calculate the tax on an income of $5000.
  c. Calculate the tax on an income of $10,000.
  d. Graph the function.

90. ECONOMICS: Income Tax The following function expresses an income tax that is 15% for incomes below $6000, and otherwise is $900 plus 40% of income in excess of $6000.

$$f(x) = \begin{cases} 0.15x & \text{if } 0 \le x < 6000 \\ 900 + 0.40(x - 6000) & \text{if } x \ge 6000 \end{cases}$$

  a. Calculate the tax on an income of $3000.
  b. Calculate the tax on an income of $6000.
  c. Calculate the tax on an income of $10,000.
  d. Graph the function.

91–92. GENERAL: Dog Years The usual estimate that each human-year corresponds to 7 dog-years is not very accurate for young dogs, since they quickly reach adulthood. Exercises 91 and 92 give more accurate formulas for converting human-years x into dog-years. For each conversion formula:

a. Find the number of dog-years corresponding to the following amounts of human time: 8 months, 1 year and 4 months, 4 years, 10 years.
b. Graph the function.

91. The following function expresses dog-years as $10\frac{1}{2}$ dog-years per human-year for the first 2 years and then 4 dog-years per human-year for each year thereafter.

$$f(x) = \begin{cases} 10.5x & \text{if } 0 \le x \le 2 \\ 21 + 4(x - 2) & \text{if } x > 2 \end{cases}$$

92. The following function expresses dog-years as 15 dog-years per human-year for the first year, 9 dog-years per human-year for the second year, and then 4 dog-years per human-year for each year thereafter.

$$f(x) = \begin{cases} 15x & \text{if } 0 \le x \le 1 \\ 15 + 9(x - 1) & \text{if } 1 < x < 2 \\ 24 + 4(x - 2) & \text{if } x > 2 \end{cases}$$

93. BUSINESS: Insurance Reserves An insurance company keeps reserves (money to pay claims) of $R(v) = 2v^{0.3}$, where v is the value of all of its policies, and the value of its policies is predicted to be $v(t) = 60 + 3t$, where t is the number of years from now. (Both R and v are in millions of dollars.) Express the reserves R as a function of t, and evaluate the function at $t = 10$.

94. **BUSINESS: Research Expenditures**
An electronics company's research budget is
$R(p) = 3p^{0.25}$, where p is the company's profit,
and the profit is predicted to be $p(t) = 55 + 4t$,
where t is the number of years from now.
(Both R and p are in millions of dollars.) Express the research expenditure R as a function
of t, and evaluate the function at $t = 5$.

95. **BIOMEDICAL: Cell Division** One leukemic
cell in an otherwise healthy mouse will divide
into two cells every 12 hours, so that after x
days the number of leukemic cells will be
$f(x) = 4^x$.
   a. Find the approximate number of leukemic
   cells after 10 days.
   b. If the mouse will die when its body has a
   billion leukemic cells, will it survive beyond day 15?

96. **GENERAL: Mobile Telephones** The use of
cellular mobile telephones has been growing
rapidly, in both developed and developing
countries. The number of "cell" phones worldwide (in millions) is approximated by the exponential function $f(x) = 11(1.52)^x$, where x
is the number of years since 1990. Predict the
number of cell phones in the year 2005.
Source: Worldwatch Institute.

97. **GENERAL: Oil Imports and Fuel Efficiency**
The average fuel efficiency of all cars in America, called the *fleet mpg,* is 21.6 mpg. The
amount of crude oil that would be saved if the
fleet mpg were increased to a value of x is

$$S(x) = 3208 - \frac{69{,}300}{x} \quad \begin{matrix} \text{million barrels} \\ \text{annually} \end{matrix}$$

   a. Graph the function $S(x)$ on the window
   [21.6, 40] by [0, 2000].
   b. For what fleet mpg x will the savings
   reach 720 million barrels, which is the
   amount annually imported from the

Middle East OPEC nations? [*Hint:* Either
ZOOM IN around the point at $y = 720$
or, better, use INTERSECT with
$y_2 = 720$.] [*Note:* New cars average 29
mpg, but fleet mpg always lags. Manufac-
turers have already built and tested proto-
type cars getting from 78 to 138 mpg.]
(*Sources:* Federal Highway Administration
and the Rocky Mountain Institute.)

98. **GENERAL: Long-Term Trends in Higher
Education** The table below gives the percentage of college graduates in the United States
from 1940 to 2000.

	Years Since 1940	Percentage of College Graduates
1940	0	4
1960	20	9
1980	40	18
2000	60	26

   a. Enter these data into a graphing calculator
   and make a plot of the resulting points
   (Years since 1940 on the x-axis and Percentage on the y-axis).
   b. Have your calculator find the cubic (third-
   order polynomial) regression formula for
   these data. Then enter the result as y_1,
   which gives a formula for the percentage
   for each year. Plot the points together with
   the regression curve.
   c. Use the regression formula to predict the
   percentage of college graduates in the year
   2010 ($x = 70$).
   d. Use the regression formula to predict the
   percentage of college graduates in the year
   2060 ($x = 120$). The answer shows that
   polynomial predictions, if extended too far,
   give nonsensical results.

Chapter Summary with Hints and Suggestions

Reading the text and doing the exercises in this chapter have helped you to master the following skills, which are listed by section (in case you need to review them) and are keyed to particular Review Exercises. Answers for all Review Exercises are given at the back of the book, and full solutions can be found in the Student Solutions Manual.

1.1 Real Numbers, Inequalities, and Lines

- Translate an interval into set notation and graph it on the real line.
  (Review Exercises 1–4.)

$$[a, b] \quad (a, b) \quad [a, b) \quad (a, b]$$

$$(-\infty, b] \quad (-\infty, b) \quad [a, \infty) \quad (a, \infty) \quad (-\infty, \infty)$$

- Express given information in interval form.
  (Review Exercises 5–6.)

- Find an equation for a line that satisfies certain conditions.   *(Review Exercises 7–12.)*

$$m = \frac{y_2 - y_1}{x_2 - x_1} \qquad y = mx + b$$

$$y - y_1 = m(x - x_1) \qquad x = a \qquad y = b$$

$$ax + by = c$$

- Find an equation of a line from its graph.
  (Review Exercises 13–14.)

- Use straight-line depreciation to find the value of an asset.   *(Review Exercises 15–16.)*

- Use real-world data to find a regression line and make a prediction.
  (Review Exercises 17–18.)

1.2 Exponents

- Evaluate negative and fractional exponents without a calculator.   *(Review Exercises 19–26.)*

$$x^0 = 1 \qquad x^{-n} = \frac{1}{x^n} \qquad x^{m/n} = \sqrt[n]{x^m} = \left(\sqrt[n]{x}\right)^m$$

- Evaluate an exponential expression using a calculator.   *(Review Exercises 27–28.)*

1.3 Functions

- Evaluate and find the domain and range of a function.   *(Review Exercises 29–32.)*

  A function f is a rule that assigns to each number x in a set (the domain) a (single) number $f(x)$. The range is the set of all resulting values $f(x)$.

- Use the vertical line test to see if a graph defines a function.   *(Review Exercises 33–34.)*

- Graph a linear function: $f(x) = mx + b$
  (Review Exercises 35–36.)

- Graph a quadratic function:
  $f(x) = ax^2 + bx + c$
  (Review Exercises 37–38.)

- Solve a quadratic equation by factoring and by the Quadratic Formula.
  (Review Exercises 39–42.)

Vertex	x-intercepts
$x = \dfrac{-b}{2a}$	$x = \dfrac{-b \pm \sqrt{b^2 - 4ac}}{2a}$

- Use a graphing calculator to graph a quadratic function. *(Review Exercises 43–44.)*

- Construct a linear function from a word problem or from real-life data, and then use the function in an application.
  (Review Exercises 45–48.)

- For given cost and revenue functions, find the break-even points and maximum profit.
  (Review Exercises 49–50.)

1.4 Functions, Continued

- Evaluate and find the domain and range of a more complicated function. *(Review Exercises 51–54.)*

- Solve a polynomial equation by factoring. *(Review Exercises 55–58.)*

- Graph an exponential function. *(Review Exercise 59.)*

- Graph a "shifted" function. *(Review Exercise 60.)*

- Graph a piecewise linear function. *(Review Exercises 61–62.)*

- Given two functions, find their composition. *(Review Exercises 63–66.)*

$$f(g(x)) \qquad g(f(x))$$

- For a given function $f(x)$, find and simplify the difference quotient $\dfrac{f(x + h) - f(x)}{h}$. *(Review Exercises 67–68.)*

- Solve an applied problem involving the composition of functions. *(Review Exercise 69.)*

- Solve a polynomial equation. *(Review Exercises 70–71.)*

- Fit a curve to real-life data and make a prediction. *(Review Exercise 72.)*

Hints and Suggestions

- *(Overview)* In reviewing this chapter, notice the difference between *geometric* objects (points, curves, etc.) and *analytic* objects (numbers, functions, etc.), and the connections between them. Descartes first made this connection: by drawing axes, he saw that points could be specified by numerical coordinates, and so *curves* could be specified by *equations* governing their coordinates. This idea connected geometry to algebra, previously distinct subjects. You should be able to express geometric objects analytically, and vice versa. For example, given a *graph* of a line, you should be able to find an *equation* for it, and given a quadratic *function,* you should be able to *graph* it.

- A graphing calculator or a computer with appropriate software can help you to *explore* a concept more fully (for example, seeing how a curve changes as a coefficient or exponent changes), and also to *solve* a problem (for example, eliminating the point-plotting aspect of graphing, or finding a regression line).

- If you don't have a graphing calculator, you should have a scientific or business calculator to carry out calculations, especially in later chapters.

- The Practice Problems help you to check your mastery of the skills presented. Complete solutions are given at the end of each section.

- The Student Solutions Manual, available separately from your bookstore, provides fully worked-out solutions to selected exercises.

- **Practice for Test:** Review Exercises 1, 9, 11, 13, 15, 17, 19, 31, 33, 35, 37, 43, 47, 49, 55, 61, 65, 67, and 71.

Review Exercises for Chapter 1

Practice test exercise numbers are in green.

1.1 Real Numbers, Inequalities, and Lines

Write each interval in set notation and graph it on the real line.

1. $(2, 5]$ **2.** $[-2, 0)$ **3.** $[100, \infty)$ **4.** $(-\infty, 6]$

5. GENERAL: Wind Speed The United States Coast Guard defines a "hurricane" as winds of at least 74 mph, a "storm" as winds of at least 55 mph but less than 74 mph, a "gale" as winds of at least 38 mph but less than 55 mph, and a "small craft warning" as winds of at least 21 mph but less than 38 mph. Express each of these wind conditions in interval form. [*Hint:* A small craft warning is [21, 38).]

6. State in interval form:
   a. The set of all positive numbers.
   b. The set of all negative numbers.
   c. The set of all nonnegative numbers.
   d. The set of all nonpositive numbers.

Write an equation of the line satisfying each of the following conditions. If possible, write your answer in the form $y = mx + b$.

7. Slope 2 and passing through the point $(1, -3)$

8. Slope -3 and passing through the point $(-1, 6)$

9. Vertical and passing through the point $(2, 3)$

10. Horizontal and passing through the point $(2, 3)$

11. Passing through the points $(-1, 3)$ and $(2, -3)$

12. Passing through the points $(1, -2)$ and $(3, 4)$

Write an equation of the form $y = mx + b$ for each line graphed below.

13.

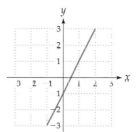

14.

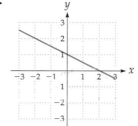

15. BUSINESS: Straight-Line Depreciation A contractor buys a backhoe for $25,000 and estimates its useful life to be 8 years, after which its scrap value will be $1000.

   a. Use straight-line depreciation to find a formula for the value V of the backhoe after t years, for $0 \le t \le 8$.
   b. Use your formula to find the value of the backhoe after 4 years.

16. BUSINESS: Straight-Line Depreciation A trucking company buys a satellite communication system for $78,000 and estimates its useful life to be 15 years, after which its scrap value will be $3000.

a. Use straight-line depreciation to find a formula for the value V of the system after t years, for $0 \le t \le 15$.

b. Use your formula to find the value of the system after 8 years.

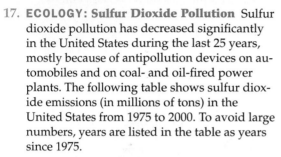

17. ECOLOGY: Sulfur Dioxide Pollution Sulfur dioxide pollution has decreased significantly in the United States during the last 25 years, mostly because of antipollution devices on automobiles and on coal- and oil-fired power plants. The following table shows sulfur dioxide emissions (in millions of tons) in the United States from 1975 to 2000. To avoid large numbers, years are listed in the table as years since 1975.

	Years Since 1975	Sulfur Dioxide Emissions
1975	0	26.0
1980	5	23.5
1985	10	21.6
1990	15	19.3
1995	20	18.2
2000	25	18.1

Sources: Worldwatch Institute and U.S. Environmental Protection Agency

a. Enter the table numbers into a graphing calculator and make a plot of the resulting points (Years Since 1975 on the x-axis and Sulfur Dioxide Emissions on the y-axis).

b. Have your calculator find the linear regression formula for these data. Then enter the result in y_1, which gives a formula for sulfur dioxide emissions in each year. Plot the points together with the regression line. How well does the line fit the data?

c. Use your formula to predict the sulfur dioxide pollution in the year 2010 (assuming that the past trend continues).

18. SOCIAL SCIENCE: Gap Between Rich and Poor During the last few decades, the richest 20% of the world's people have been growing richer, while the poorest 20% have been growing poorer. The probable consequences of this growing gap are not only social instability but also environmental decline, since the richest consume more wastefully while the poorest must cut down rainforests and overgraze land just to survive. The following table shows the ratio of income of the richest 20% to the poorest 20% from 1960 to 2000. To avoid large numbers, years are listed in the table as years since 1960.

	Years Since 1960	Ratio of Richest to Poorest
1960	0	30 to 1
1970	10	32 to 1
1980	20	45 to 1
1990	30	60 to 1
2000	40	72 to 1

Source: United Nations Development Programme

a. Enter the table numbers into a graphing calculator and make a plot of the resulting points [Years Since 1960 on the x-axis and the larger number in the Ratio column (the 30, 32, etc.) on the y-axis].

b. Have your calculator find the linear regression formula for these data. Then enter the result in y_1, which gives a formula for the ratio of richest to poorest in each year. Plot the points together with the regression line. How well does the line fit the data?

c. Use your formula to predict the ratio in the year 2010 (assuming that the past trend continues).

1.2 Exponents

Evaluate each expression without using a calculator.

19. $\left(\frac{1}{6}\right)^{-2}$

20. $\left(\frac{4}{3}\right)^{-1}$

21. $64^{1/2}$

22. $1000^{1/3}$

23. $81^{-3/4}$

24. $100^{-3/2}$

25. $\left(-\frac{8}{27}\right)^{-2/3}$

26. $\left(\frac{9}{16}\right)^{-3/2}$

Use a calculator to evaluate each expression. Round answers to two decimal places.

27. $3^{2.4}$

28. $12^{1.9}$

1.3   Functions

For each function in Exercises 29–32:

a. Evaluate the given expression.
b. Find the domain.
c. Find the range.

29. $f(x) = \sqrt{x - 7}$; find $f(11)$

30. $g(t) = \dfrac{1}{t + 3}$; find $g(-1)$

31. $h(w) = w^{-3/4}$; find $h(16)$

32. $w(z) = z^{-4/3}$; find $w(8)$

Determine whether each graph defines a function of x.

33.

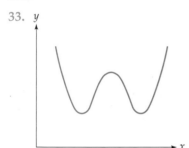

34.

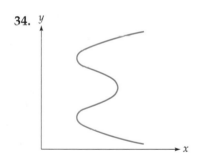

Graph each function.

35. $f(x) = 4x - 8$

36. $f(x) = 6 - 2x$

37. $f(x) = -2x^2 - 4x + 6$

38. $f(x) = 3x^2 - 6x$

Solve each equation by **a.** factoring and **b.** the Quadratic Formula.

39. $3x^2 + 9x = 0$ **40.** $2x^2 - 8x - 10 = 0$

41. $3x^2 + 3x + 5 = 11$

42. $4x^2 - 2 = 2$

For each quadratic function in Exercises 43–44:

a. Find the vertex using the vertex formula.
b. Graph the function on an appropriate viewing window. (Answers may vary.)

43. $f(x) = x^2 - 10x - 25$

44. $f(x) = x^2 + 14x - 15$

45. BUSINESS: Car Rentals A rental company rents cars for $45 per day and $0.12 per mile. Find a function $C(x)$ for the cost of a rented car driven for x miles in a day.

46. BUSINESS: Simple Interest If money is borrowed for a short period of time, generally less than a year, the interest is often calculated as *simple* interest, according to the formula Interest $= P \cdot r \cdot t$, where P is the principal, r is the rate (expressed as a decimal), and t is the time (in years). Find a function $I(t)$ for the interest charged on a loan of $10,000 at an interest rate of 8% for t years. Simplify your answer.

47. GENERAL: Air Temperature The air temperature decreases by about 1 degree Fahrenheit for each 300 feet of altitude. Find a function $T(x)$ for the temperature at an altitude of x feet if the sea level temperature is 70°.

48. ECOLOGY: Carbon Dioxide Pollution The burning of fossil fuels (such as oil and coal) added 27 billion tons of carbon dioxide to the atmosphere during 2000, and this annual amount is growing by 0.58 billion tons per year. Find a function $C(t)$ for the amount of carbon dioxide added during the year t years after 2000, and use the formula to find how soon this annual amount will reach 30 billion tons. [*Note:* Carbon dioxide traps solar heat, increasing the earth's temperature, and may lead to flooding of lowland areas by melting the polar ice.] *Source:* U.S. Environmental Protection Agency.

49. BUSINESS: Break-Even Points and Maximum Profit A store that installs satellite TV receivers finds that if it installs x receivers per week, its costs will be $C(x) = 80x + 1950$ and its revenue will be $R(x) = -2x^2 + 240x$ (both in dollars).

a. Find the store's break-even points.
b. Find the number of receivers the store should install to maximize profit, and the maximum profit.

50. BUSINESS: Break-Even Points and Maximum Profit An air conditioner outlet finds that if it sells x air conditioners per month, its costs will be $C(x) = 220x + 202{,}500$ and its revenue will be $R(x) = -3x^2 + 2020x$ (both in dollars).

a. Find the outlet's break-even points.
b. Find the number of air conditioners the outlet should sell to maximize profit, and the maximum profit.

1.4 Functions, Continued

For each function in Exercises 51–54:

a. Evaluate the given expression.
b. Find the domain.
c. Find the range.

51. $f(x) = \dfrac{3}{x(x - 2)}$; find $f(-1)$

52. $f(x) = \dfrac{16}{x(x + 4)}$; find $f(-8)$

53. $g(x) = 9^x$; find $g(\frac{3}{2})$

54. $g(x) = 8^x$; find $g(\frac{5}{3})$

Solve each equation by factoring.

55. $5x^4 + 10x^3 = 15x^2$ **56.** $4x^5 + 8x^4 = 32x^3$

57. $2x^{5/2} - 8x^{3/2} = 10x^{1/2}$

58. $3x^{5/2} + 3x^{3/2} = 18x^{1/2}$

Graph each function.

59. $f(x) = 4^x$ **60.** $f(x) = (x - 2)^2 - 4$

61. $f(x) = \begin{cases} 3x - 7 & \text{if } x \geq 2 \\ -x - 1 & \text{if } x < 2 \end{cases}$

(If you use a graphing calculator for Exercises 61 and 62, be sure to indicate any missing points.)

62. $f(x) = \begin{cases} 6 - 2x & \text{if } x > 2 \\ 2x - 1 & \text{if } x \leq 2 \end{cases}$

For each pair of functions $f(x)$ and $g(x)$, find
a. $f(g(x))$, **b.** $g(f(x))$.

63. $f(x) = x^2 + 1$; $g(x) = \dfrac{1}{x}$

64. $f(x) = \sqrt{x}$; $g(x) = 5x - 4$

65. $f(x) = \dfrac{x + 1}{x - 1}$; $g(x) = x^3$

66. $f(x) = 2^x$; $g(x) = x^2$

For each function, find and simplify the difference quotient $\dfrac{f(x + h) - f(x)}{h}$.

67. $f(x) = 2x^2 - 3x + 1$ **68.** $f(x) = \dfrac{5}{x}$

69. BUSINESS: Advertising Budget A company's advertising budget is $A(p) = 2p^{0.15}$, where p is the company's profit, and the profit is predicted to be $p(t) = 18 + 2t$, where t is the number of years from now. (Both A and p are in millions of dollars.) Express the advertising budget A as a function of t, and evaluate the function at $t = 4$.

70. a. Solve the equation $x^4 - 2x^3 - 3x^2 = 0$ by factoring.
b. Use a graphing calculator to graph $y = x^4 - 2x^3 - 3x^2$ and find the x-intercepts of the graph. Be sure that you understand why your answers to parts (a) and (b) agree.

71. a. Solve the equation $x^3 + 2x^2 - 3x = 0$ by factoring.
b. Use a graphing calculator to graph $y = x^3 + 2x^2 - 3x$ and find the

x-intercepts of the graph. Be sure that you understand why your answers to parts (a) and (b) agree.

72. **BUSINESS: Revenue** The following table gives a company's annual revenue (in millions of dollars) from its overseas operations during its first five years.

Year	Revenue
1	2.0
2	1.8
3	1.9
4	2.1
5	2.8

a. Enter the table numbers into a graphing calculator and make a plot of the resulting points (Years on the x-axis and Revenue on the y-axis). What kind of curve do the points suggest?

b. Have your calculator fit such a curve to the data. Then enter the result in y_1, which gives a formula for the annual revenue each year. Plot the points together with the regression curve.

c. Use your formula to predict the revenue in years 6 and 7.

2 Derivatives and Their Uses

2.1 LIMITS AND CONTINUITY

Introduction

This chapter introduces the *derivative*, one of the most important concepts in all of calculus. We begin by discussing two preliminary topics, limits and continuity, both of which will be treated intuitively rather than formally, and will be useful when we define the *derivative* in the next section.

Limits

The word "limit" is used in everyday conversation to describe the *ultimate* behavior of something, as in the "limit of one's endurance" or the "limit of one's patience." In mathematics, the word "limit" has a similar but more precise meaning.

The notation $x \to 3$ (read "x approaches 3") means that x takes values *arbitrarily close to 3 without ever reaching 3*. For example, the numbers 2.9, 2.99, 2.999, . . . approach 3 (from the left), and the numbers 3.1, 3.01, 3.001, . . . approach 3 (from the right). We can even let x approach 3 by alternating sides, taking values such as 2.9, 3.01, 2.999, 3.0001, ... We emphasize that $x \to 3$ means that x takes values closer and closer to 3 *but never equaling 3.*

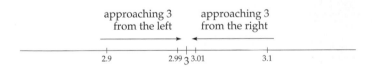

Given a function $f(x)$, if x approaching 3 causes the function to take values arbitrarily close to, say, 10, then we call 10 the *limit* of the function, and write

$$\lim_{x \to 3} f(x) = 10 \qquad \text{Limit of } f(x) \text{ as } x \text{ approaches 3 is 10}$$

The limit of a function may not exist (as we will see later), but if the limit *does* exist, it must be a *single* number. Limits can be defined

Application Preview

Temperature, Superconductivity, and Limits

It has long been known that there is a coldest possible temperature, called absolute zero, the temperature of an object if all heat could be removed from it. On the Fahrenheit temperature scale, absolute zero is 460 degrees below zero. On the "absolute" or "Kelvin" temperature scale (named after the nineteenth-century scientist Lord Kelvin), absolute zero temperature is assigned the value 0. Absolute zero is a temperature that can be *approached* but never actually *reached*. At temperatures approaching absolute zero, some metals become increasingly able to conduct electricity, with efficiencies approaching 100%, a state called *superconductivity*. The graph below gives the electrical conductivity of aluminum, showing the remarkable fact that as temperature decreases to absolute zero, aluminum becomes *superconducting*—its conductivity approaches a limit of 100% (see Exercise 81 on page 95).

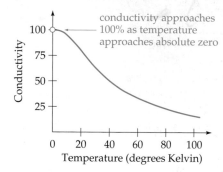

Conductivity of aluminum as a function of temperature

If $f(t)$ stands for the percent conductivity of aluminum at temperature t degrees Kelvin, then the fact that conductivity approaches 100 as temperature t decreases to zero may be written

$$\lim_{t \to 0^+} f(t) = 100 \qquad \text{Limit as } t \to 0^+ \text{ of conductivity is 100}$$

This is an example of *limits*, which are discussed in this section. Superconductivity has many commercial applications, from high-speed "mag-lev" trains that "float" on magnetic fields above the tracks to supercomputers and medical imaging devices.

formally,* but we will treat them intuitively, using the following definition.

Limit of a Function

$$\lim_{x \to c} f(x) \qquad \text{Limit of } f(x) \text{ as } x \text{ approaches } c$$

is the number that $f(x)$ approaches as x approaches c $(x \neq c)$.

Finding Limits

Limits may be found from tables of values of $f(x)$ for x near c.

EXAMPLE 1

FINDING A LIMIT BY TABLES

Use tables to find $\lim_{x \to 3} (2x + 4)$.   Limit of $2x + 4$ as x approaches 3

Solution

We choose x-values approaching 3, and calculate the resulting values of $f(x) = 2x + 4$. We make two tables, for $x < 3$ and for $x > 3$.

	x	$2x + 4$
Numbers approaching 3 but *less* than 3 ↓	2.9	9.8
	2.99	9.98
	2.999	9.998

	x	$2x + 4$
Numbers approaching 3 but *greater* than 3 ↓	3.1	10.2
	3.01	10.02
	3.001	10.002

Reading down these columns, the function values seem to approach 10.

Choosing x-values even closer to 3 (such as 2.9999 and 3.0001) would result in values of $2x + 4$ *even closer to 10, so the limit is 10:*

$$\lim_{x \to 3} (2x + 4) = 10$$

---

* Formally, limits are defined as follows: $\lim_{x \to c} f(x) = L$ if for every number $\epsilon > 0$
there is a number $\delta > 0$ such that $|f(x) - L| < \epsilon$ whenever $0 < |x - c| < \delta$.

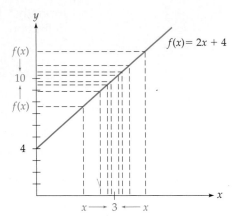

This limit may be seen graphically: As $x \rightarrow 3$ (on the x-axis), $f(x)$ approaches 10 (on the y-axis).

Graph showing $\lim\limits_{x \to 3} (2x + 4) = 10$

The correct limit in Example 1 could have been found simply by *evaluating* the function at $x = 3$:

$$f(3) = 2 \cdot 3 + 4 = 10$$

$f(x) = 2x + 4$ evaluated at $x = 3$ gives the correct limit, 10

However, finding limits by this technique of *direct substitution* is not always possible, as the next example shows.

EXAMPLE 2

FINDING A LIMIT BY TABLES

Find $\lim\limits_{x \to 0} (1 + x)^{1/x}$ correct to three decimal places.

Solution

The function values in the following tables were found using a calculator.

x	$(1 + x)^{1/x}$
0.1	2.594
0.01	2.705
0.001	2.717
0.0001	2.718

x	$(1 + x)^{1/x}$
-0.1	2.868
-0.01	2.732
-0.001	2.720
-0.0001	2.718

To use a graphing calculator to find these numbers, enter the function as $(1 + x)^\wedge(1/x)$, and use the TABLE feature.

$$\lim\limits_{x \to 0} (1 + x)^{1/x} \approx 2.718$$

From the agreement in the last columns of the tables

Practice Problem 1 Evaluate $(1 + x)^{1/x}$ at $x = 0.000001$ and $x = -0.000001$. Based on the results, give a better approximation for $\lim\limits_{x \to 0} (1 + x)^{1/x}$.

➤ Solution on page 91

The actual value of this particular limit is the number $e \approx 2.71828$, which will be very important in our later work. Notice that $(1 + x)^{1/x}$ cannot be evaluated *at* $x = 0$ since the exponent would be $1/0$, which is undefined. This limit *requires* the limit process.

Which limits can be evaluated by direct substitution (as in Example 1) and which cannot (as in Example 2)? The answer comes from the following "Rules of Limits."

Rules of Limits

For any constants a and c, and any positive integer n:

1. $\lim\limits_{x \to c} a = a$

The limit of a constant is just the constant

2. $\lim\limits_{x \to c} x^n = c^n$

The limit of a power is the power of the limit

3. $\lim\limits_{x \to c} \sqrt[n]{x} = \sqrt[n]{c}$ ($c > 0$ if n is even)

The limit of a root is the root of the limit

4. If $\lim\limits_{x \to c} f(x)$ and $\lim\limits_{x \to c} g(x)$ both exist, then

  a. $\lim\limits_{x \to c} [f(x) + g(x)] = \lim\limits_{x \to c} f(x) + \lim\limits_{x \to c} g(x)$

The limit of a sum is the sum of the limits

  b. $\lim\limits_{x \to c} [f(x) - g(x)] = \lim\limits_{x \to c} f(x) - \lim\limits_{x \to c} g(x)$

The limit of a difference is the difference of the limits

  c. $\lim\limits_{x \to c} [f(x) \cdot g(x)] = [\lim\limits_{x \to c} f(x)] \cdot [\lim\limits_{x \to c} g(x)]$

The limit of a product is the product of the limits

  d. $\lim\limits_{x \to c} \dfrac{f(x)}{g(x)} = \dfrac{\lim\limits_{x \to c} f(x)}{\lim\limits_{x \to c} g(x)}$ (if $\lim\limits_{x \to c} g(x) \neq 0$)

The limit of a quotient is the quotient of the limits

These rules, which may be proved from the definition of limit, can be summarized as follows.

Summary of Rules of Limits

For functions composed of additions, subtractions, multiplications, divisions, powers, and roots, limits may be evaluated by direct substitution, provided that the resulting expression is defined.

$$\lim\limits_{x \to c} f(x) = f(c) \qquad \text{Limit evaluated by direct substitution}$$

EXAMPLE 3

FINDING LIMITS BY DIRECT SUBSTITUTION

a. $\lim\limits_{x \to 4} \sqrt{x} = \sqrt{4} = 2$

Direct substitution of $x = 4$ using Rule 3 or the Summary

b. $\lim\limits_{x \to 6} \dfrac{x^2}{x + 3} = \dfrac{6^2}{6 + 3} = \dfrac{36}{9} = 4$

Direct substitution of $x = 6$ (Rules 4, 2, and 1 or the Summary)

Practice Problem 2

Find $\lim\limits_{x \to 3} (2x^2 - 4x + 1)$.

➤ Solution on page 91

If direct substitution into a quotient gives the undefined expression $\dfrac{0}{0}$, factoring, simplifying, and *then* using direct substitution may help.*

EXAMPLE 4

FINDING A LIMIT BY SIMPLIFYING

Find $\lim\limits_{x \to 1} \dfrac{x^2 - 1}{x - 1}$.

Solution

Direct substitution of $x = 1$ into $\dfrac{x^2 - 1}{x - 1}$ gives $\dfrac{0}{0}$, which is undefined. But simplifying the fraction gives

$$\lim\limits_{x \to 1} \frac{x^2 - 1}{x - 1} = \lim\limits_{x \to 1} \underbrace{\frac{(x + 1)(x - 1)}{x - 1}}_{\substack{\text{Factoring the} \\ \text{numerator}}} = \lim\limits_{x \to 1} \underbrace{\frac{(x + 1)(x - 1)}{x - 1}}_{\substack{\text{Canceling the} \\ (x - 1)\text{'s} \\ (\text{since } x \neq 1)}} = \underbrace{\lim\limits_{x \to 1} (x + 1) = 2}_{\substack{\text{Now use} \\ \text{direct} \\ \text{substitution}}}$$

Therefore, the limit *does* exist and equals 2.

---

* Recall from the definition on page 79 that the limit and the result from direct substitution are not always the same, so the limit may still exist even if direct substitution gives an undefined result.

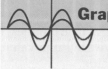

Graphing Calculator Exploration

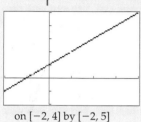

on [−2, 4] by [−2, 5]

The graph of $f(x) = \dfrac{x^2 - 1}{x - 1}$ (from Example 4) is shown on the left.

a. Can you explain why the graph appears to be a straight line?

b. From your knowledge of rational functions, should the graph really be a (complete) line? [*Hint:* See pages 51–52.] Can you see that a point is indeed missing from the graph?*

c. Use the graph on the left (where each tick mark is 1 unit) or enter the function and use TRACE on your calculator to verify that the limit is 2 as x approaches 1.

*To have your calculator show the point as missing, choose a window such that the x-value in question is midway between XMIN and XMAX, or see the Useful Hint in Graphing Calculator Terminology in the Preface.

Practice Problem 3   Find $\displaystyle\lim_{x \to 5} \frac{2x^2 - 10x}{x - 5}$

➤ Solution on page 91

One-Sided Limits

In a *one-sided limit,* the variable approaches the number *from one side only.* For example, the limit as x approaches 3 *from the left,* denoted $x \to 3^-$, means the limit using only x-values to the *left* of 3, such as 2.9, 2.99, 2.999, ... The limit as x approaches 3 *from the right,* denoted $x \to 3^+$, means the limit using only x-values to the *right* of 3, such as 3.1, 3.01, 3.001, ... The limits from the left and from the right (sometimes called left and right limits) are exactly what we found in the two tables of Examples 1 and 2 (see pages 79–80). In the Application Preview on page 77, we used the notation $t \to 0^+$ to indicate that the temperature *decreased* to zero. In general:

Left and Right Limits
$\displaystyle\lim_{x \to c^-} f(x)$ means the limit of $f(x)$ $\hfill$ Limit from the *left*
as $x \to c$ but with $x < c$
$\displaystyle\lim_{x \to c^+} f(x)$ means the limit of $f(x)$ $\hfill$ Limit from the *right*
as $x \to c$ but with $x > c$

The Rules of Limits on page 81 also hold for one-sided limits. The limit that we defined earlier (page 79) is sometimes called the *two-sided* limit to distinguish it from one-sided limits. As in Examples 1 and 2, if the left and right limits both exist and have the same value, then the (two-sided) limit exists and has that same value.

$\lim\limits_{x \to c} f(x) = L$ if and only if *both* one-sided limits $\lim\limits_{x \to c^-} f(x)$ and

$\lim\limits_{x \to c^+} f(x)$ exist and equal the same number L.

EXAMPLE **5**

FINDING ONE-SIDED LIMITS

For the piecewise linear function $f(x) = \begin{cases} x + 1 & \text{if } x \le 3 \\ 8 - 2x & \text{if } x > 3 \end{cases}$ graphed on the left, find the following limits or state that they do not exist.

a. $\lim\limits_{x \to 3^-} f(x)$ **b.** $\lim\limits_{x \to 3^+} f(x)$ **c.** $\lim\limits_{x \to 3} f(x)$

We give two solutions, one using the graph and one using the expression for the function. Both methods are important.

Solution   (Using the graph)

a. $\lim\limits_{x \to 3^-} f(x) = 4$

Approaching 3 *from the left* means using the line on the *left* of $x = 3$, which approaches height 4

b. $\lim\limits_{x \to 3^+} f(x) = 2$

Approaching 3 *from the right* means using the line on the *right* of $x = 3$, which approaches height 2

c. $\lim\limits_{x \to 3} f(x)$ does not exist

The two one-sided limits both exist, but they have different values (4 and 2), so *the limit does not exist*

Solution $\left(\text{Using } f(x) = \begin{cases} x + 1 & \text{if } x \le 3 \\ 8 - 2x & \text{if } x > 3 \end{cases} \right)$

a. For $x \to 3^-$ we have $x < 3$, so $f(x)$ is given by the *upper* line of the function:

$$\lim_{x \to 3^-} f(x) = \lim_{x \to 3^-} (x + 1) = 3 + 1 = 4$$

b. For $x \to 3^+$ we have $x > 3$, so $f(x)$ is given by the *lower* line of the function:

$$\lim_{x \to 3^+} f(x) = \lim_{x \to 3^+} (8 - 2x) = 8 - 2 \cdot 3 = 2$$

c. $\lim_{x \to 3} f(x)$ does not exist since although the two one-sided limits exist, they are not equal.

Notice that the two methods found the same answers.

Practice Problem 4   Explain the difference among $x \to 3^-$, $x \to -3$, and $x \to -3^-$.

➤ Solution on page 91

Infinite Limits

We may use the symbols ∞ (infinity) and $-\infty$ (negative infinity) to indicate that the values of a function become arbitrarily large or arbitrarily small.* The following example shows some of the possibilities. The dashed lines on the graphs, where the function values approach ∞ or $-\infty$, are called *vertical asymptotes*.

EXAMPLE 6   **FINDING LIMITS INVOLVING $\pm\infty$**

For each function graphed below, use the limit notation with ∞ and $-\infty$ to describe its behavior as x approaches the vertical asymptote from the left, from the right, and from both sides.

a.   **b.**

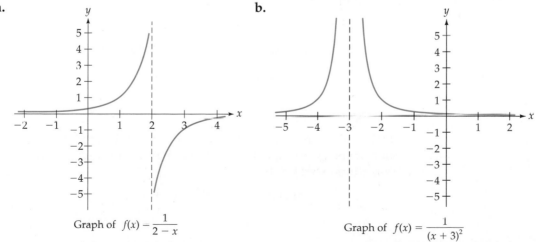

Graph of $f(x) = \dfrac{1}{2 - x}$ Graph of $f(x) = \dfrac{1}{(x + 3)^2}$

* Limits as $x \to \infty$ and as $x \to -\infty$ will be discussed on pages 459–461.

Solution

a. $\lim\limits_{x \to 2^-} f(x) = \infty$ The curve rises arbitrarily high as x approaches 2 from the left

      $\lim\limits_{x \to 2^+} f(x) = -\infty$ The curve falls arbitrarily low as x approaches 2 from the right

      $\lim\limits_{x \to 2} f(x)$ does not exist   The two one-sided limits do not agree

b. $\lim\limits_{x \to -3^-} f(x) = \infty$ The curve rises arbitrarily high as x approaches -3 from the left

      $\lim\limits_{x \to -3^+} f(x) = \infty$ The curve rises arbitrarily high as x approaches -3 from the right

      $\lim\limits_{x \to -3} f(x) = \infty$ The two one-sided limits agree

Be careful! To say that a limit exists means that the limit is a *number*, and since ∞ and $-\infty$ are not numbers, a statement such as $\lim\limits_{x \to c} f(x) = \infty$ means that *the limit does not exist*. The limit statement $\lim\limits_{x \to c} f(x) = \infty$ goes further to explain *why* the limit does not exist: the function values become too large to approach any limit.

Limits of Functions of Two Variables

Some limits involve two variables, with only one variable approaching a limit.

EXAMPLE 7

FINDING A LIMIT OF A FUNCTION OF TWO VARIABLES

Find $\lim\limits_{h \to 0} (x^2 + xh + h^2)$.

Solution

Only h is approaching zero, so x remains unchanged. Since the function involves only powers of h, we may evaluate the limit by direct substitution of $h = 0$:

$$\lim\limits_{h \to 0} (x^2 + xh + h^2) = x^2 + x \cdot 0 + 0^2 = x^2$$

                                              0 0

Practice Problem 5   Find $\lim_{h \to 0} (3x^2 + 5xh + 1)$.   ➤ Solution on page 91

Continuity

Intuitively, a function is said to be *continuous at c* if its graph passes through the point at $x = c$ *without a "hole" or a "jump."* For example, the first function below is *continuous* at c (it has no hole or jump at $x = c$), while the second and third are *discontinuous* at c.

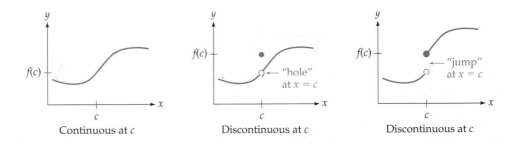

Continuous at c Discontinuous at c Discontinuous at c

In other words, a function is *continuous at c* if the curve *approaches the point at* $x = c$, which may be stated in terms of limits:

$$\lim_{x \to c} f(x) = f(c)$$

This equation means that the quantities on both sides must exist and be *equal*, which we make explicit as follows:

Continuity

A function f is continuous at c if the following three conditions hold:

1. $f(c)$ is defined   Function is *defined* at c

2. $\lim_{x \to c} f(x)$ exists   Left and right limits exist and agree

3. $\lim_{x \to c} f(x) = f(c)$ Limit and value *at c* agree

f is *discontinuous* at c if one or more of these conditions *fails* to hold.

Condition 3, which is just the statement that the expressions in conditions 1 and 2 are equal to each other, may by itself be taken as the definition of continuity.

EXAMPLE 8 FINDING DISCONTINUITIES FROM A GRAPH

Each function below is *discontinuous at c* for the indicated reason.

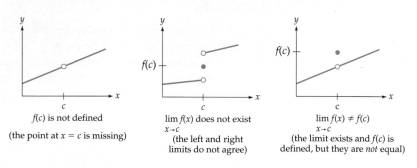

$f(c)$ is not defined	$\lim_{x \to c} f(x)$ does not exist	$\lim_{x \to c} f(x) \neq f(c)$
(the point at $x = c$ is missing)	(the left and right limits do not agree)	(the limit exists and $f(c)$ is defined, but they are *not* equal)

Practice Problem 6

For each graph below, determine whether the function is continuous at c. If it is *not* continuous, indicate the *first* of the three conditions in the definition of continuity (on the previous page) that is violated.

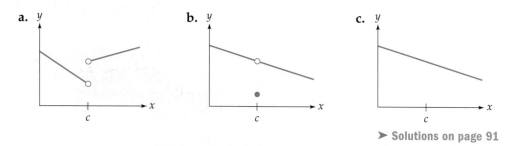

a.   b.   c.

➤ Solutions on page 91

A function is continuous on an *open interval* (a, b) if it is continuous at each point of the interval. A function is continuous on a *closed interval* $[a, b]$ if it is continuous on the open interval (a, b) and has "one-sided continuity" at the endpoints: $\lim_{x \to a^+} f(x) = f(a)$ and $\lim_{x \to b^-} f(x) = f(b)$. A function that is continuous on the entire real line $(-\infty, \infty)$ is said to be *continuous everywhere*, or simply *continuous*.

Which Functions Are Continuous?

Which functions are continuous? *Linear* and *quadratic* functions are continuous, since their graphs are, respectively, straight lines and parabolas, with no holes or jumps. Similarly, *exponential* and *logarithmic* functions are continuous, as may be seen from graphs on page 53. These and other continuous functions can be combined as follows to give other continuous functions.

If functions f and g are continuous at c, then the following are also continuous at c:

1. $f \pm g$	Sums and differences of continuous functions are continuous
2. $a \cdot f$ [for any constant a]	Constant multiples of continuous functions are continuous
3. $f \cdot g$	Products of continuous functions are continuous
4. f/g [if $g(c) \neq 0$]	Quotients of continuous functions are continuous
5. $f(g(x))$ [for f continuous at $g(c)$]	Compositions of continuous functions are continuous

These statements, which can be proved from the Rules of Limits, show that the following types of functions are continuous:

Every polynomial function is continuous.

Every rational function is continuous except where the denominator is zero.

EXAMPLE 9

DETERMINING CONTINUITY

Determine whether each function is continuous.

a. $f(x) = x^3 - 3x^2 - x + 3$ **b.** $f(x) = \dfrac{1}{(x+1)^2}$ **c.** $f(x) = e^{x^2 - 1}$

Solution

The first function is continuous since it is a polynomial. The second (a rational function) is discontinuous at $x = -1$, where the denominator is zero. The rational function is continuous at all *other* values. The third function is continuous since it is the composition of the exponential function e^x and the polynomial $x^2 - 1$.

Calculator-drawn graphs of the functions in Example 9 are shown below. The polynomial (on the left) and the exponential function (on the right) are continuous, although you can't really tell from such graphs. The rational function (in the middle) exhibits the discontinuity at $x = -1$.

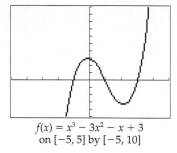

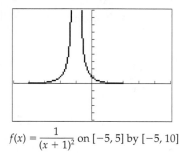

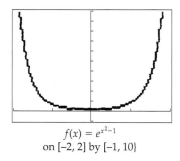

$f(x) = x^3 - 3x^2 - x + 3$
on $[-5, 5]$ by $[-5, 10]$

$f(x) = \dfrac{1}{(x + 1)^2}$ on $[-5, 5]$ by $[-5, 10]$

$f(x) = e^{x^2-1}$
on $[-2, 2]$ by $[-1, 10\}$

2.1 Section Summary

The limit $\lim\limits_{x \to c} f(x)$ is the number (if it exists) that $f(x)$ approaches as x approaches $c\,(x \neq c)$. Left and right limits are found by letting x approach c from one side only. If the left or right limit (or both) do not exist, or do not agree, then the limit does not exist. Limits can often be found from tables of function values for x near c, or by direct substitution of $x = c$ (provided that the function consists of additions, subtractions, multiplications, divisions, roots, and powers, and also provided that the resulting expression is defined). Sometimes it helps to simplify before direct substitution.

Continuity can be understood *geometrically* and *analytically*. Geometrically, a function is continuous at c if its graph has neither a hole nor a jump at $x = c$ (which is equivalent to the curve's passing through the point *at c*). Analytically, a function f is continuous at c if both sides of the following equation are defined and equal:

$$\lim_{x \to c} f(x) = f(c)$$

This equation may be interpreted as saying that continuity is equivalent to being able to *move the limit operation through the function*:

$$\lim_{x \to c} f(x) = f(\overbrace{\lim_{x \to c} x}^{c}) = f(c)$$

A polynomial is continuous everywhere, and a rational function is continuous everywhere except where the denominator is zero.

Solutions to Practice Problems

1. $\lim\limits_{x \to 0} (1 + x)^{1/x} \approx 2.71828$

2. $\lim\limits_{x \to 3} (2x^2 - 4x + 1) = 2 \cdot 3^2 - 4 \cdot 3 + 1$

$$= 18 - 12 + 1 = 7$$

3. $\lim\limits_{x \to 5} \dfrac{2x^2 - 10x}{x - 5} - \lim\limits_{x \to 5} \dfrac{2x(x - 5)}{x - 5}$

$$= \lim\limits_{x \to 5} \frac{2x(x - 5)}{x - 5} = \lim\limits_{x \to 5} 2x = 10$$

4. $x \to 3^-$ means: x approaches (positive) 3 *from the left.*
   $x \to -3$ means: x approaches -3 (the ordinary two-sided limit).
   $x \to -3^-$ means: x approaches -3 *from the left.*

5. $\lim\limits_{h \to 0} (3x^2 + 5xh + 1) = 3x^2 + 5x \cdot 0 + 1 \qquad$ Using direct substitution

$$= 3x^2 + 1$$

6. a. Discontinuous, $f(c)$ is not defined

   b. Discontinuous, $\lim\limits_{x \to c} f(x) \neq f(c)$

   c. Continuous

2.1 Exercises

Complete the tables and use them to find the given limit. Round calculations to three decimal places. A graphing calculator with a TABLE feature will be very helpful.

1. $\lim\limits_{x \to 2} (5x - 7)$

x	5x − 7
1.9	
1.99	
1.999	

x	5x − 7
2.1	
2.01	
2.001	

2. $\lim\limits_{x \to 4} (2x + 1)$

x	2x + 1
3.9	
3.99	
3.999	

x	2x + 1
4.1	
4.01	
4.001	

3. $\lim\limits_{x \to 1} \dfrac{x^3 - 1}{x - 1}$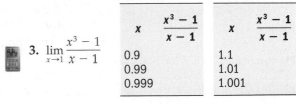

x	$\dfrac{x^3 - 1}{x - 1}$
0.9	
0.99	
0.999	

x	$\dfrac{x^3 - 1}{x - 1}$
1.1	
1.01	
1.001	

4. $\lim\limits_{x \to 1} \dfrac{x^4 - 1}{x - 1}$

x	$\dfrac{x^4 - 1}{x - 1}$
0.9	
0.99	
0.999	

x	$\dfrac{x^4 - 1}{x - 1}$
1.1	
1.01	
1.001	

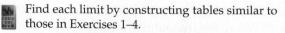

Find each limit by constructing tables similar to those in Exercises 1–4.

5. $\lim\limits_{x \to 0} (1 + 2x)^{1/x}$

6. $\lim\limits_{x \to 0} (1 - x)^{1/x}$

7. $\lim\limits_{x \to 2} \dfrac{\frac{1}{x} - \frac{1}{2}}{x - 2}$

8. $\lim\limits_{x \to 1} \dfrac{\sqrt{x} - 1}{x - 1}$

Find each limit by graphing the function and using TRACE or TABLE to examine the graph near the indicated x-value.

9. $\lim\limits_{x \to 1} \dfrac{\frac{1}{x} - 1}{1 - x}$

Use window $[0, 2]$ by $[0, 5]$.

10. $\lim\limits_{x \to 1.5} \dfrac{2x^2 - 4.5}{x - 1.5}$

Use window $[0, 3]$ by $[0, 10]$.

11. $\lim\limits_{x \to 4} \dfrac{x^{1.5} - 4x^{0.5}}{x^{1.5} - 2x}$

12. $\lim\limits_{x \to 1} \dfrac{x - 1}{x - \sqrt{x}}$

[*Hint:* Choose a window whose x-values are centered at the limiting x-value.]

Find the following limits *without* using a graphing calculator or making tables.

13. $\lim\limits_{x \to 3} (4x^2 - 10x + 2)$

14. $\lim\limits_{x \to 7} \dfrac{x^2 - x}{2x - 7}$

15. $\lim\limits_{x \to 5} \dfrac{3x^2 - 5x}{7x - 10}$

16. $\lim\limits_{t \to 3} \sqrt[3]{t^2 + t - 4}$

17. $\lim\limits_{x \to 3} \sqrt{2}$

18. $\lim\limits_{q \to 9} \dfrac{8 + 2\sqrt{q}}{8 - 2\sqrt{q}}$

19. $\lim\limits_{t \to 25} [(t + 5)t^{-1/2}]$

20. $\lim\limits_{s \to 4} (s^{3/2} - 3s^{1/2})$

21. $\lim\limits_{h \to 0} (5x^3 + 2x^2h - xh^2)$

22. $\lim\limits_{h \to 0} (2x^2 + 4xh + h^2)$

23. $\lim\limits_{x \to 2} \dfrac{x^2 - 4}{x - 2}$

24. $\lim\limits_{x \to 1} \dfrac{x - 1}{x^2 + x - 2}$

25. $\lim\limits_{x \to -3} \dfrac{x + 3}{x^2 + 8x + 15}$

26. $\lim\limits_{x \to -4} \dfrac{x^2 + 9x + 20}{x + 4}$

27. $\lim\limits_{x \to -1} \dfrac{3x^3 - 3x^2 - 6x}{x^2 + x}$

28. $\lim\limits_{x \to 0} \dfrac{x^2 - x}{x^2 + x}$

29. $\lim\limits_{h \to 0} \dfrac{2xh - 3h^2}{h}$

30. $\lim\limits_{h \to 0} \dfrac{5x^4h - 9xh^2}{h}$

31. $\lim\limits_{h \to 0} \dfrac{4x^2h + xh^2 - h^3}{h}$

32. $\lim\limits_{h \to 0} \dfrac{x^2h - xh^2 + h^3}{h}$

For each piecewise linear function $f(x)$ graphed in Exercises 33–36, find:

a. $\lim\limits_{x \to 2^-} f(x)$ **b.** $\lim\limits_{x \to 2^+} f(x)$ **c.** $\lim\limits_{x \to 2} f(x)$

33.

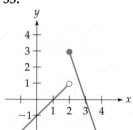

34.

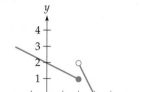

35.

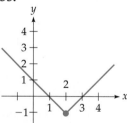

36.

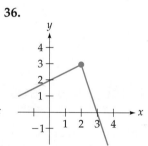

For each piecewise linear function, find:

a. $\lim\limits_{x \to 4^-} f(x)$ **b.** $\lim\limits_{x \to 4^+} f(x)$ **c.** $\lim\limits_{x \to 4} f(x)$

37. $f(x) = \begin{cases} 3 - x & \text{if } x \le 4 \\ 10 - 2x & \text{if } x > 4 \end{cases}$

38. $f(x) = \begin{cases} 5 - x & \text{if } x < 4 \\ 2x - 5 & \text{if } x \ge 4 \end{cases}$

39. $f(x) = \begin{cases} 2 - x & \text{if } x \le 4 \\ x - 6 & \text{if } x > 4 \end{cases}$

40. $f(x) = \begin{cases} 2 - x & \text{if } x < 4 \\ 2x - 10 & \text{if } x \ge 4 \end{cases}$

For each function, find:

a. $\lim\limits_{x \to 0^-} f(x)$ **b.** $\lim\limits_{x \to 0^+} f(x)$ **c.** $\lim\limits_{x \to 0} f(x)$

41. $f(x) = |x|$ **42.** $f(x) = -|x|$

43. $f(x) = \dfrac{|x|}{x}$ **44.** $f(x) = -\dfrac{|x|}{x}$

For each function, use the limit notation with ∞ and $-\infty$ to describe its behavior as x approaches the vertical asymptote from the left, from the right, and from both sides, as in Example 6 on pages 85–86.

45.

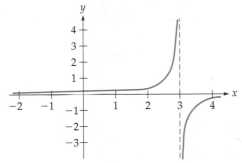

46.

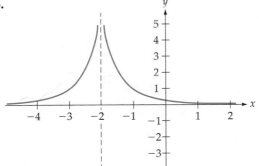

47.

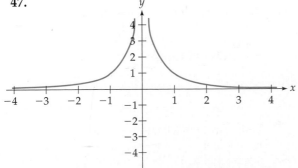

48.

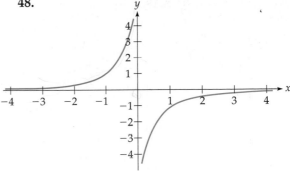

For each function, use the limit notation with ∞ and $-\infty$ to describe its behavior as x approaches the number where the function is undefined from the left, from the right, and from both sides, as in Example 6 on pages 85–86. (A table of values may be helpful.)

49. $f(x) = \dfrac{1}{x + 2}$ **50.** $f(x) = \dfrac{1}{x - 3}$

51. $f(x) = \dfrac{1}{(x - 3)^2}$ **52.** $f(x) = -\dfrac{1}{(x + 2)^2}$

Determine whether each of the following functions is continuous or discontinuous at c. If it is discontinuous, indicate the *first* of the three conditions in the definition of continuity (page 87) that is violated.

53.

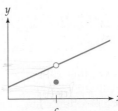

54.

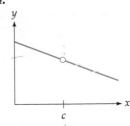

55.

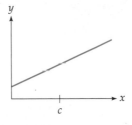

56.

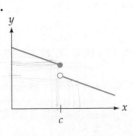

57.

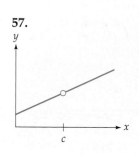

58.

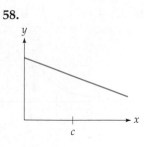

59.

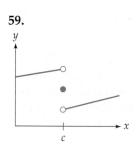

60.

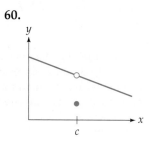

For each piecewise linear function:

a. Draw its graph (by hand or using a graphing calculator).

b. Find the limits as x approaches 3 from the left and from the right.

c. Is it continuous at $x = 3$? If not, indicate the first of the three conditions in the definition of continuity (page 87) that is violated.

61. $f(x) = \begin{cases} x & \text{if } x \le 3 \\ 6 - x & \text{if } x > 3 \end{cases}$

62. $f(x) = \begin{cases} 5 - x & \text{if } x \le 3 \\ x - 2 & \text{if } x > 3 \end{cases}$

63. $f(x) = \begin{cases} x & \text{if } x \le 3 \\ 7 - x & \text{if } x > 3 \end{cases}$

64. $f(x) = \begin{cases} 5 - x & \text{if } x \le 3 \\ x - 1 & \text{if } x > 3 \end{cases}$

Determine whether each function is continuous or discontinuous. If discontinuous, state where it is discontinuous.

65. $f(x) = 7x - 5$

66. $f(x) = 5x^3 - 6x^2 + 2x - 4$

67. $f(x) = \dfrac{x + 1}{x - 1}$

68. $f(x) = \dfrac{x^3}{(x + 7)(x - 2)}$

69. $f(x) = \dfrac{12}{5x^3 - 5x}$

70. $f(x) = \dfrac{x + 2}{x^4 - 3x^3 - 4x^2}$

71. $f(x) = \begin{cases} 3 - x & \text{if } x \le 4 \\ 10 - 2x & \text{if } x > 4 \end{cases}$
[*Hint:* See Exercise 37.]

72. $f(x) = \begin{cases} 5 - x & \text{if } x < 4 \\ 2x - 5 & \text{if } x \ge 4 \end{cases}$
[*Hint:* See Exercise 38.]

73. $f(x) = \begin{cases} 2 - x & \text{if } x \le 4 \\ x - 6 & \text{if } x > 4 \end{cases}$
[*Hint:* See Exercise 39.]

74. $f(x) = \begin{cases} 2 - x & \text{if } x < 4 \\ 2x - 10 & \text{if } x \ge 4 \end{cases}$
[*Hint:* See Exercise 40.]

75. $f(x) = |x|$
[*Hint:* See Exercise 41.]

76. $f(x) = \dfrac{|x|}{x}$
[*Hint:* See Exercise 43.]

77. By canceling the common factor, $\dfrac{(x - 1)(x + 2)}{x - 1}$ simplifies to $x + 2$. At $x = 1$, however, the function $\dfrac{(x - 1)(x + 2)}{x - 1}$ is *dis*continuous (since it is undefined where the denominator is zero), whereas $x + 2$ is *continuous*. Are these two functions, one obtained from the other by simplification, equal to each other? Explain.

APPLIED EXERCISES

78. GENERAL: Relativity According to Einstein's special theory of relativity, under certain conditions a 1-foot-long object moving with velocity v will appear to an observer to have length $\sqrt{1 - (v/c)^2}$, in which c is a constant equal to the speed of light. Find the limiting value of the apparent length as the velocity of the object approaches the speed of light by finding

$$\lim_{v \to c^-} \sqrt{1 - \left(\frac{v}{c}\right)^2}$$

79. BUSINESS: Interest Compounded Continuously If you deposit $1 into a bank account paying 10% interest compounded continuously (see Section 1.2), a year later its value will be

$$\lim_{x \to 0} \left(1 + \frac{x}{10}\right)^{1/x}$$

Find the limit by making a TABLE of values correct to two decimal places, thereby finding the value of the deposit in dollars and cents.

80. BUSINESS: Interest Compounded Continuously If you deposit $1 into a bank ac-

count paying 5% interest compounded continuously (see Section 1.2), a year later its value will be

$$\lim_{x \to 0} \left(1 + \frac{x}{20}\right)^{1/x}$$

Find the limit by making a TABLE of values correct to two decimal places, thereby finding the value of the deposit in dollars and cents.

81. GENERAL: Superconductivity The conductivity of aluminum at temperatures near absolute zero is approximated by the function

$$f(x) = \frac{100}{1 + .001x^2},$$ which expresses the conductivity as a percent. Find the limit of this conductivity percent as the temperature x approaches 0 (absolute zero). (See the Application Preview on page 77.)

2.2   RATES OF CHANGE, SLOPES, AND DERIVATIVES

Introduction

In this section we will define the *derivative*, one of the two most important concepts in all of calculus, which measures the rate of change of a function or, equivalently, the slope of a curve.* We begin by discussing *rates of change*.

Average and Instantaneous Rate of Change

We often speak in terms of *rates of change* to express how one quantity changes with another. For example, in the morning the temperature might be "rising at the rate of 3 degrees per hour" and in the evening it might be "falling at the rate of 2 degrees per hour." In some situations, the rate of change of temperature is extremely important. For example, the manufacture of computer chips involves heating and cooling silicon very gradually. For simplicity, suppose that in one process the temperature of the silicon at time x hours is $f(x) = x^2$ degrees. We shall calculate the *average rate of change of temperature* over various time intervals, taking the change in temperature divided by the change in time. For example, at time 1 the temperature is $1^2 = 1$ degrees, and at time 3 the temperature is $3^2 = 9$ degrees, so the

*The second most important concept in calculus is the definite integral, which will be defined in Section 5.3.

temperature went up by $9 - 1 = 8$ degrees in 2 hours, for an average rate of:

$$\begin{pmatrix} \text{Average rate} \\ \text{of change} \\ \text{from 1 to 3} \end{pmatrix} = \frac{3^2 - 1^2}{2} = \frac{9 - 1}{2} = \frac{8}{2} = 4 \qquad \begin{array}{l} \text{Average rate} \\ \text{of change} \\ \text{over 2 hours} \end{array}$$

Similarly,

degrees per hour

$$\begin{pmatrix} \text{Average rate} \\ \text{of change} \\ \text{from 1 to 2} \end{pmatrix} = \frac{2^2 - 1^2}{1} = \frac{4 - 1}{1} = 3 \qquad \begin{array}{l} \text{Average rate} \\ \text{of change} \\ \text{over 1 hour} \end{array}$$

$$\begin{pmatrix} \text{Average rate} \\ \text{of change} \\ \text{from 1 to 1.5} \end{pmatrix} = \frac{1.5^2 - 1^2}{0.5} = \frac{2.25 - 1}{0.5} = \frac{1.25}{0.5} = 2.5 \qquad \begin{array}{l} \text{Average rate} \\ \text{of change} \\ \text{over 0.5 hour} \end{array}$$

$$\begin{pmatrix} \text{Average rate} \\ \text{of change} \\ \text{from 1 to 1.1} \end{pmatrix} = \frac{1.1^2 - 1^2}{0.1} = \frac{1.21 - 1}{0.1} = \frac{0.21}{0.1} = 2.1 \qquad \begin{array}{l} \text{Average rate} \\ \text{of change} \\ \text{over 0.1 hour} \end{array}$$

We see that rate of change of temperature over shorter and shorter time intervals is decreasing from 4 to 3 to 2.5 to 2.1, numbers that seem to be approaching 2 degrees per hour. To verify that this is indeed true, we generalize the process. We have been finding the average over a time interval from 1 to a slightly later time that we now call $1 + h$, where h is a small positive number. Notice that in each step we calculated the following expression for successively smaller values of h.

$$\begin{pmatrix} \text{Average rate} \\ \text{of change} \\ \text{from 1 to } 1 + h \end{pmatrix} = \frac{(1 + h)^2 - 1^2}{h} \qquad \begin{array}{l} \text{We used } h = 2, \ h = 1, \\ h = 0.5, \text{ and } h = 0.1 \end{array}$$

If we now use our limit notation to let h approach zero, the amount of time will shrink to an instant, giving what is called the *instantaneous rate of change*:

$$\begin{pmatrix} \text{Instantaneous} \\ \text{rate of change} \\ \text{at time 1} \end{pmatrix} = \lim_{h \to 0} \frac{(1 + h)^2 - (1)^2}{h} \qquad \begin{array}{l} \text{Taking the limit as} \\ h \text{ approaches zero} \end{array}$$

For simplicity we have been using the function $f(x) = x^2$ and the time $x = 1$, but the same procedure applies to *any* function f and number x, leading to the following general definition:

Average and Instantaneous Rate of Change

The *average* rate of change of a function f between x and $x + h$ is defined as

$$\frac{f(x + h) - f(x)}{h} \qquad \begin{array}{l} \text{Difference quotient gives} \\ \text{the } average \text{ rate of change} \end{array}$$

> The *instantaneous* rate of change of a function f at the number x is defined as
>
> $$\lim_{h \to 0} \frac{f(x + h) - f(x)}{h}$$
>
> Taking the limit makes it *instantaneous*

The fraction is just the difference quotient that we introduced on page 61; the numerator is the change in the *function* between two x-values, and the denominator is the change between the two x-values: $(x + h) - (x) = h$. We may use the second formula to check our guess that the average rate of change of temperature is indeed approaching 2 degrees per hour.

EXAMPLE 1

FINDING AN INSTANTANEOUS RATE OF CHANGE

Find the instantaneous rate of change of the temperature function $f(x) = x^2$ at time $x = 1$.

Solution

$$\lim_{h \to 0} \frac{f(x + h) - f(x)}{h}$$

Formula for the instantaneous rate of change of f at x

$$= \lim_{h \to 0} \frac{f(1 + h) - f(1)}{h}$$

Substituting $x = 1$

$$= \lim_{h \to 0} \frac{(1 + h)^2 - (1)^2}{h}$$

$f(x) = x^2$ gives $f(1 + h) = (1 + h)^2$ and $f(1) = 1^2$

$$= \lim_{h \to 0} \frac{1 + 2h + h^2 - 1}{h}$$

Expanding $(1 + h)^2 = 1 + 2h + h^2$

$$= \lim_{h \to 0} \frac{\cancel{1} + 2h + h^2 - \cancel{1}}{h}$$

Simplifying

$$= \lim_{h \to 0} \frac{h(2 + h)}{h} = \lim_{h \to 0} \frac{\cancel{h}(2 + h)}{\cancel{h}}$$

Factoring out h and canceling (since $h \neq 0$)

$$= \lim_{h \to 0} (2 + h) = 2$$

Evaluating the limit by direct substitution

Since $f(x)$ gives the temperature at time x hours, this means that *the instantaneous rate of change of temperature at time $x = 1$ hour is 2 degrees per hour* (just as we had guessed earlier).

In this Example, f gave the temperature (degrees) at time x (hours), so the units of the rate of change were *degrees per hour*. In general, for a function f, the units of the instantaneous rate of change are *function units per x unit*.

Practice Problem 1 If $f(x)$ gives the population of a city in year x, what are the units of the instantaneous rate of change?

➤ Solution on page 108

Secant and Tangent Lines

We may "see" the average and instantaneous rates of change on the graph of a function. First, some terminology. A *secant line* to a curve is a line that passes through two points of the curve. A *tangent line* is a line that passes through a point of the curve and matches exactly the steepness of the curve at that point.*

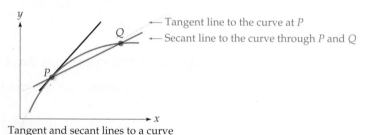

Tangent and secant lines to a curve

If the curve is the graph of a function f and the points P and Q have x-coordinates x and $x + h$, respectively (and so y-coordinates $f(x)$ and $f(x + h)$, respectively), then the graph of the curve and the secant line takes the form shown below.

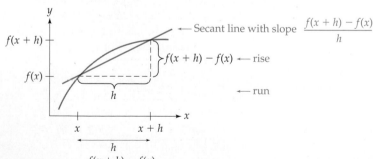

The slope $\dfrac{f(x + h) - f(x)}{h}$ of the secant line is the average rate of change of the function.

---

* The word "secant" comes from the Latin *secare*, "to cut," suggesting that the secant line "cuts" the curve at two points. The word "tangent" comes from the Latin *tangere*, "to touch," suggesting that the tangent line just "touches" the curve at one point.

Observe that the *slope* (rise over run) of the secant line is the difference quotient $\frac{f(x + h) - f(x)}{h}$, exactly the same as our earlier definition of the average rate of change of the function between x and $x + h$.

Furthermore, the two points where the secant line meets the curve are separated by a distance h along the x-axis. Letting $h \to 0$ forces the second point to approach the first, causing the secant line to approach the tangent line, the slope of which is then $\lim\limits_{h \to 0} \frac{f(x + h) - f(x)}{h}$. But this is exactly the same as our earlier definition of the *instantaneous rate of change of the function at x.*

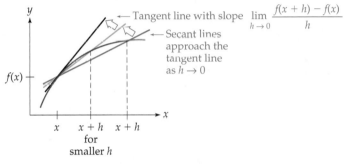

The slope $\lim\limits_{h \to 0} \dfrac{f(x + h) - f(x)}{h}$ of the *tangent* line is the *instantaneous* rate of change of the function.

The last two diagrams have shown that the slope of the *secant* line is the *average* rate of change of the function, and the slope of the *tangent* lines is the *instantaneous* rate of change of the function. The fact that rates of change are related to slopes should come as no surprise: if a function has a large rate of change, its graph must be rising rapidly, giving it a large slope. Similarly, if a function has only a small rate of change, its graph will be rising slowly, giving it a small slope.

To summarize: For a function f and its graph:

$$\begin{pmatrix} \textit{Average rate} \\ \textit{of change of } f \\ \textit{between } x \textit{ and } x + h \end{pmatrix} = \begin{pmatrix} \textit{Slope of the secant} \\ \textit{line through the} \\ \textit{points at } x \textit{ and } x + h \end{pmatrix} = \frac{f(x + h) - f(x)}{h} \qquad \textit{Without the limit}$$

$$\begin{pmatrix} \textit{Instantaneous} \\ \textit{rate of change} \\ \textit{of } f \textit{ at } x \end{pmatrix} = \begin{pmatrix} \textit{Slope of the} \\ \textit{tangent line} \\ \textit{at } x \end{pmatrix} = \lim_{h \to 0} \frac{f(x + h) - f(x)}{h} \qquad \textit{With the limit}$$

In Example 1 we found the instantaneous rate of change of a function. We now use the same formula to find the slope of the tangent line to the graph of a function.

EXAMPLE 2

FINDING THE SLOPE OF A TANGENT LINE

Find the slope of the tangent line to $f(x) = \dfrac{1}{x}$ at $x = 2$.

Solution

$$\lim_{h \to 0} \frac{f(x + h) - f(x)}{h}$$ — Formula for the slope of the tangent line at x

$$= \lim_{h \to 0} \frac{f(2 + h) - f(2)}{h}$$ — Substituting $x = 2$

$$= \lim_{h \to 0} \frac{\dfrac{1}{2 + h} - \dfrac{1}{2}}{h}$$ — Using the function $f(x) = \dfrac{1}{x}$

$$= \lim_{h \to 0} \frac{1}{h}\left(\frac{1}{2 + h} - \frac{1}{2}\right)$$ — Since multiplying by $1/h$ is equivalent to dividing by h

$$= \lim_{h \to 0} \frac{1}{h} \cdot \frac{2 - (2 + h)}{(2 + h) \cdot 2}$$ — Combining fractions using the common denominator $(2 + h) \cdot 2$

$$= \lim_{h \to 0} \frac{1}{h} \cdot \frac{-h}{(2 + h) \cdot 2}$$ — Simplifying the numerator

$$= \lim_{h \to 0} \frac{1}{\cancel{h}} \cdot \frac{\overset{-1}{\cancel{-h}}}{(2 + h) \cdot 2} = \lim_{h \to 0} \frac{-1}{(2 + h) \cdot 2}$$ — Cancelling

$$= \frac{-1}{(2 + 0) \cdot 2} = \frac{-1}{4} = -\frac{1}{4}$$ — Evaluating the limit by direct substitution of $h = 0$

Therefore, the curve $y = \dfrac{1}{x}$ has slope $-\dfrac{1}{4}$ at $x = 2$, as shown in the graph on the left.

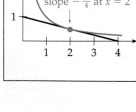

$y = \dfrac{1}{x}$

Tangent line has slope $-\dfrac{1}{4}$ at $x = 2$

EXAMPLE 3

FINDING A TANGENT LINE

Use the result of the preceding example to find the equation of the tangent line to $f(x) = \dfrac{1}{x}$ at $x = 2$.

Solution

From Example 2 we know that the *slope* of the tangent line is $m = -\dfrac{1}{4}$. The point on the curve at $x = 2$ is $(2, \frac{1}{2})$, the y-coordinate coming from $y = f(2) = \dfrac{1}{2}$. The point-slope formula then gives:

$$y - \tfrac{1}{2} = -\tfrac{1}{4}(x - 2)$$

$y - y_1 = m(x - x_1)$ with $m = -\tfrac{1}{4}$, $x_1 = 2$, and $y_1 = \tfrac{1}{2}$

$$y - \tfrac{1}{2} = -\tfrac{1}{4}x + \tfrac{1}{2}$$

Multiplying out

$$y = 1 - \tfrac{1}{4}x$$

Simplifying

Equation of the tangent line

Graphing Calculator Exploration

On a graphing calculator, graph the curve $y = 1/x$ together with the line $y = 1 - x/4$ using the window [0, 4] by [0, 4] to see that the line is indeed the tangent line to the curve at $x = 2$. (For a way to have your graphing calculator find the *equation* for the tangent line, see Exercises 47–48 on page 110.)

The first of the following graphs shows the curve $y = \tfrac{1}{x}$ along with the tangent line that we found at $x = 2$. The next two graphs show the results of successively "zooming in" near the point of tangency, showing that the curve seems to straighten out and almost *become its own tangent line* on a smaller and smaller viewing window. For this reason, the tangent line is called *the best linear approximation to the curve* near the point of tangency.

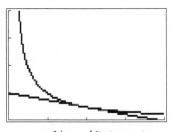

$y = 1/x$ and its tangent line at $x = 2$ on [0, 4] by [0, 4]

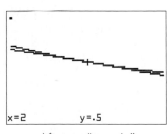

After one "zoom in" near $(2, 1/2)$

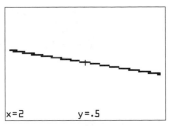

After a second "zoom in" centered at $(2, 1/2)$

Graphing Calculator Exploration

Use a graphing calculator to graph $y = x^3 - 2x^2 - 3x + 4$ (or any function of your choice). Then "zoom in" a few times around a point to see the curve straighten out and almost *become* its own tangent line.

The Derivative

In Examples 1 and 2 we found the instantaneous rate of change of a function and the slope of a tangent line of a curve *at a particular number or point.* It is much more efficient to carry out the same calculation but keeping x as a *variable,* obtaining a new function that gives the instantaneous rate of change or the slope of the tangent line at *any* value of x. This new function is denoted with a *prime,* $f'(x)$ (read: "f prime of x"), and is called *the derivative of f at x.*

Derivative

For a function f, the *derivative of f at x* is defined as

$$f'(x) = \lim_{h \to 0} \frac{f(x + h) - f(x)}{h} \qquad \text{Limit of the difference quotient}$$

(provided that the limit exists). The derivative $f'(x)$ gives the instantaneous rate of change of f at x and also the slope of the graph of f at x.

In general, the units of the derivative are *function units* per x *unit.*

EXAMPLE 4

FINDING THE DERIVATIVE OF A FUNCTION FROM THE DEFINITION

Find the derivative of $f(x) = x^2 - 7x + 150$.

Solution

$$f'(x) = \lim_{h \to 0} \frac{f(x + h) - f(x)}{h} \qquad \text{Definition of the derivative}$$

$$= \lim_{h \to 0} \frac{\overbrace{(x + h)^2 - 7(x + h) + 150}^{f(x+h)} - \overbrace{(x^2 - 7x + 150)}^{f(x)}}{h} \qquad \text{Using } f(x) = x^2 - 7x + 150$$

$$= \lim_{h \to 0} \frac{x^2 + 2xh + h^2 - 7x - 7h + 150 - x^2 + 7x - 150}{h} \qquad \text{Expanding and simplifying}$$

$$= \lim_{h \to 0} \frac{2xh + h^2 - 7h}{h} = \lim_{h \to 0} \frac{h(2x + h - 7)}{h} \qquad \text{Simplifying (since } h \neq 0)$$

$$= \lim_{h \to 0} (2x + h - 7) = 2x - 7 \qquad \text{Evaluating the limit by direct substitution}$$

Therefore, the derivative of $f(x) = x^2 - 7x + 150$ is $f'(x) = 2x - 7$.

The operation of calculating derivatives should be thought of as an operation on *functions*, taking one function [such as $f(x) = x^2 - 7x + 150$] and giving another $[f'(x) = 2x - 7]$. The resulting function is called "the derivative" because it is *derived* from the first, and the process of obtaining it is called "differentiation." If the derivative is defined at x, then the original function is said to be *differentiable* at x.

EXAMPLE 5

USING A DERIVATIVE IN AN APPLICATION

Refining crude oil into various products, such as gasoline, heating oil, and plastics, requires heating and cooling the oil at different rates. Suppose that the temperature of the oil at time x hours is $f(x) = x^2 - 7x + 150$ degrees Fahrenheit (for $0 \le x \le 8$). Find the instantaneous rate of change of temperature at times $x = 6$ hours and $x = 2$ hours and interpret the results.

Solution

The instantaneous rate of change means the derivative, so ordinarily we would now take the derivative of the temperature function $f(x) = x^2 - 7x + 150$. However, this is just what we did in the previous example, so we will use the result that the derivative is $f'(x) = 2x - 7$, evaluating this at $x = 6$ and at $x = 2$ and interpreting the results.

$$f'(6) = 2 \cdot 6 - 7 = 5$$
Evaluating $f'(x) = 2x - 7$ at $x = 6$

Interpretation: After 6 hours, the temperature is increasing at the rate of 5 degrees per hour.

$$f'(2) = 2 \cdot 2 - 7 = -3$$
Evaluating $f(x) = 2x - 7$ at $x = 2$

Interpretation: After 2 hours, the temperature is *decreasing* at the rate of 3 degrees per hour.

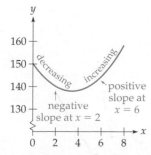

A derivative that is *positive* means that the original quantity is *increasing*, and a derivative that is *negative* means that the quantity is *decreasing*.

The graph of $f(x) = x^2 - 7x + 150$ on the left shows that the slope of the curve is indeed negative at $x = 2$ and positive at $x = 6$.

EXAMPLE 6

FINDING A DERIVATIVE FROM THE DEFINITION

Find the derivative of $f(x) = x^3$.

Solution

In our solution we will use the expansion $(x + h)^3 = x^3 + 3x^2h + 3xh^2 + h^3$ (found by multiplying together three copies of $(x + h)$).

$$f'(x) = \lim_{h \to 0} \frac{f(x + h) - f(x)}{h}$$ Definition of $f'(x)$

$$= \lim_{h \to 0} \frac{(x + h)^3 - x^3}{h}$$ Using $f(x) = x^3$

$$= \lim_{h \to 0} \frac{x^3 + 3x^2h + 3xh^2 + h^3 - x^3}{h}$$ Using the expansion of $(x + h)^3$

$$= \lim_{h \to 0} \frac{x^3 + 3x^2h + 3xh^2 + h^3 - x^3}{h} = \lim_{h \to 0} \frac{h(3x^2 + 3xh + h^2)}{h}$$ Cancelling and then factoring out an h

$$= \lim_{h \to 0} \frac{h(3x^2 + 3xh + h^2)}{h} = \lim_{h \to 0} 3x^2 + 3xh + h^2$$ Cancelling again

$$= 3x^2$$ Evaluating the limit by direct substitution

Therefore, the derivative of $f(x) = x^3$ is $f'(x) = 3x^2$.

Graphing Calculator Exploration

Some advanced graphing calculators have computer algebra systems that can simplify algebraic expressions. For example, the Texas Instruments *TI-89* calculator will find and simplify the difference quotient for the function $f(x) = x^3$ that we found in the preceding Example as follows:

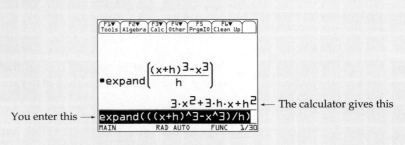

You enter this ⟶ expand(((x+h)^3-x^3)/h)

The calculator gives this ⟵ $3 \cdot x^2 + 3 \cdot h \cdot x + h^2$

The calculator finds the (simplified) difference quotient as $3x^2 + 3xh + h^2$, and taking the limit (using direct substitution of $h = 0$) gives the derivative $f'(x) = 3x^2$, just as we found above.

Leibniz's Notation for the Derivative

Calculus was developed by Isaac Newton (1642–1727) and Gottfried Wilhelm Leibniz (1646–1716) in two different countries, so there naturally developed two different notations for the derivative. Newton denoted derivatives by a dot over the function, $\dot{f}$, a notation that has been largely replaced by our "prime" notation. Leibniz wrote the derivative of $f(x)$ by writing $\dfrac{d}{dx}$ in front of the function: $\dfrac{d}{dx} f(x)$. In Leibniz's notation, the fact that the derivative of x^3 is $3x^2$ is written

$$\frac{d}{dx} x^3 = 3x^2 \qquad \text{The derivative of } x^3 \text{ is } 3x^2$$

The following table shows equivalent expressions in the two notations.

Prime Notation		*Leibniz's Notation*	
$f'(x)$	$=$	$\dfrac{d}{dx} f(x)$	Prime and $\frac{d}{dx}$ both mean the derivative
y'	$=$	$\dfrac{dy}{dx}$	For y a function of x

Each notation has its own advantages, and we will use both.* Leibniz's notation comes from writing the definition of the derivative as:

$$\frac{dy}{dx} = \lim_{\Delta x \to 0} \frac{f(x + \Delta x) - f(x)}{\Delta x} \qquad \text{Definition of the derivative (page 102)} \\ \text{with the change in } x \text{ written as } \Delta x$$

or

$$\frac{dy}{dx} = \lim_{\Delta x \to 0} \frac{\Delta y}{\Delta x} \qquad \begin{array}{l} f(x + \Delta x) - f(x) \text{ is the change in } y, \\ \text{and so can be written } \Delta y \end{array}$$

* Other notations for the derivative are $Df(x)$ and $D_x f(x)$, but we will not use them.

It is as if the limit turns Δ (read "Delta," the Greek letter D) into d, changing the $\dfrac{\Delta y}{\Delta x}$ into $\dfrac{dy}{dx}$. That is, Leibniz's notation reminds us that the derivative $\dfrac{dy}{dx}$ is the limit of the slope $\dfrac{\Delta y}{\Delta x}$.

Some functions are *not differentiable* (the derivative does not exist) at certain x-values. For example, the following diagram shows a function that has a "corner point" at $x = 1$. At this point the slope (and therefore the derivative) cannot be defined, so the function is not differentiable at $x = 1$. Other nondifferentiable functions will be discussed in Section 2.7.

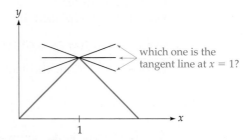

Since the tangent line cannot be uniquely defined at $x = 1$, the slope, and therefore the derivative, is undefined at $x = 1$.

The following diagram shows the geometric relationship between a function (upper graph) and its derivative (lower graph). Observe carefully how the *slope of f* is shown by the *sign of f'*.

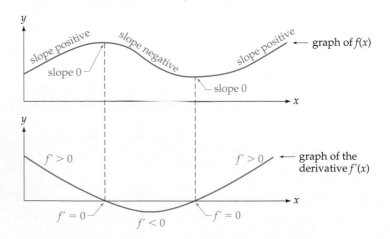

Practice Problem 2   The graph shows a function and its derivative. Which is the original function (#1 or #2) and which is its derivative? [*Hint:* Which curve has *slope* zero where the other has *value* zero?]

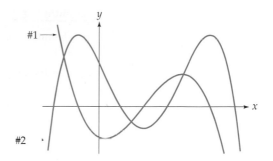

➤ Solution on page 108

2.2 Section Summary

The derivative of the function f at x is

$$f'(x) = \lim_{h \to 0} \frac{f(x + h) - f(x)}{h} \qquad \text{Provided that the limit exists}$$

Some students remember the steps as **DESL** (pronounced "diesel"): write the **D**efinition, **E**xpress the numerator in terms of the function, **S**implify, and take the **L**imit.

The derivative $f'(x)$ gives both the *slope of the graph* of the function at x and the *instantaneous rate of change of the function* at x. In other words, "derivative," "slope," and "instantaneous rate of change" are merely the mathematical, the geometric, and the analytic versions of the same idea.

The derivative gives the rate of change *at a particular instant,* not an actual change over a period of time. Instantaneous rates of changes are like the speeds on an automobile speedometer—a reading of 50 mph at one moment does not mean that you will travel exactly 50 miles in the next hour, since the actual distance depends upon your speed during the entire hour. The derivative, however, may be interpreted as the *approximate* change resulting from a 1-unit increase in the independent variable. For example, if your speedometer reads 50 mph, then you may say that you will travel *about* 50 miles during the next hour, meaning that this will be true provided that your speed remains steady throughout the hour. (In Chapter 5 we will see how to calculate actual changes from rates of change that do not stay constant.)

➤ **Solutions to Practice Problems**

1. The units are people per year, measuring the rate of growth of the population.

2. #2 is the original function and #1 is its derivative.

2.2 Exercises

By imagining tangent lines at points P_1, P_2, and P_3, state whether the slopes are positive, zero, or negative at these points.

1.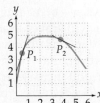

2.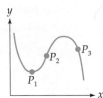

3.

4.

Use the tangent lines shown at points P_1 and P_2 to find the slopes of the curve at these points.

5.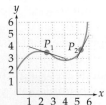

6.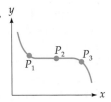

In Exercises 7 and 8, use the graph of each function $f(x)$ to make a rough sketch of the derivative $f'(x)$ showing where $f'(x)$ is positive, negative, and zero. (Omit scale on y-axis.)

7.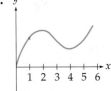

8.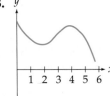

Find the average rate of change of the given function between the following pairs of x-values. [*Hint:* See page 96.]

   a. $x = 1$ and $x = 3$
   b. $x = 1$ and $x = 2$
   c. $x = 1$ and $x = 1.5$
   d. $x = 1$ and $x = 1.1$
   e. $x = 1$ and $x = 1.01$
   f. What number do your answers seem to be approaching?

9. $f(x) = x^2 + x$ **10.** $f(x) = 2x^2 + 5$

Find the average rate of change of the given function between the following pairs of x-values. [*Hint:* See page 96.]

   a. $x = 2$ and $x = 4$
   b. $x = 2$ and $x = 3$
   c. $x = 2$ and $x = 2.5$
   d. $x = 2$ and $x = 2.1$
   e. $x = 2$ and $x = 2.01$
   f. What number do your answers seem to be approaching?

11. $f(x) = 2x^2 + x - 2$ **12.** $f(x) = x^2 + 2x - 1$

Find the average rate of change of the given function between the following pairs of x-values. [*Hint:* See page 96.]

   a. $x = 3$ and $x = 5$
   b. $x = 3$ and $x = 4$
   c. $x = 3$ and $x = 3.5$
   d. $x = 3$ and $x = 3.1$
   e. $x = 3$ and $x = 3.01$
   f. What number do your answers seem to be approaching?

13. $f(x) = 5x + 1$ **14.** $f(x) = 7x - 2$

Find the average rate of change of the given function between the following pairs of x-values. [*Hint:* See page 96.]

   a. $x = 4$ and $x = 6$
   b. $x = 4$ and $x = 5$
   c. $x = 4$ and $x = 4.5$
   d. $x = 4$ and $x = 4.1$
   e. $x = 4$ and $x = 4.01$
   f. What number do your answers seem to be approaching?

15. $f(x) = \sqrt{x}$ **16.** $f(x) = \dfrac{4}{x}$

Use the formula on page 97 to find the instantaneous rate of change of the function at the given x-value. If you did the related problem in Exercises 9–16, compare your answers. [*Hint:* See Example 1.]

17. $f(x) = x^2 + x$ at $x = 1$

18. $f(x) = x^2 + 2x - 1$ at $x = 2$

19. $f(x) = 5x + 1$ at $x = 3$

20. $f(x) = \dfrac{4}{x}$ at $x = 4$

Use the formula on page 97 to find the slope of the tangent line to the curve at the given x-value. If you did the related problem in Exercises 9–16, compare your answers. [*Hint:* See Example 2.]

21. $f(x) = 2x^2 + x - 2$ at $x = 2$

22. $f(x) = 2x^2 + 5$ at $x = 1$

23. $f(x) = \sqrt{x}$ at $x = 4$

24. $f(x) = 7x - 2$ at $x = 3$

Find $f'(x)$ by using the definition of the derivative.

25. $f(x) = x^2 - 3x + 5$

26. $f(x) = 2x^2 - 5x + 1$

27. $f(x) = 1 - x^2$ **28.** $f(x) = \frac{1}{2}x^2 + 1$

29. $f(x) = 9x - 2$ **30.** $f(x) = -3x + 5$

31. $f(x) = \dfrac{x}{2}$ **32.** $f(x) = 0.01x + 0.05$

33. $f(x) = 4$ **34.** $f(x) = \pi$

35. $f(x) = ax^2 + bx + c$
   (a, b, and c are constants)

36. $f(x) = (x + a)^2$
   (a is a constant.) [*Hint:* First expand $(x + a)^2$.]

37. $f(x) = x^5$
   [*Hint:* Use $(x + h)^5 =$
   $x^5 + 5x^4h + 10x^3h^2 + 10x^2h^3 + 5xh^4 + h^5$]

38. $f(x) = x^4$
   [*Hint:* Use $(x + h)^4 =$
   $x^4 + 4x^3h + 6x^2h^2 + 4xh^3 + h^4$]

39. $f(x) = \dfrac{2}{x}$ **40.** $f(x) = \dfrac{1}{x^2}$

41. $f(x) = \sqrt{x}$ **42.** $f(x) = \dfrac{1}{\sqrt{x}}$

[*Hint:* Multiply the numerator and denominator of the difference quotient by $(\sqrt{x + h} + \sqrt{x})$ and then simplify.]

[*Hint:* Multiply the numerator and denominator of the difference quotient by $(\sqrt{x} + \sqrt{x + h})$ and then simplify.]

43. $f(x) = x^3 + x^2$ **44.** $f(x) = \dfrac{1}{2x}$

45. a. Find the equation for the tangent line to the curve $f(x) = x^2 - 3x + 5$ at $x = 2$, writing the equation in slope-intercept form. [*Hint:* Use your answer to Exercise 25.]

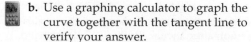 **b.** Use a graphing calculator to graph the curve together with the tangent line to verify your answer.

46. a. Find the equation for the tangent line to the curve $f(x) = 2x^2 - 5x + 1$ at $x = 2$, writing the equation in slope-intercept form.
[*Hint:* Use your answer to Exercise 26.]

b. Use a graphing calculator to graph the curve together with the tangent line to verify your answer.

47. a. Graph the function $f(x) = x^2 - 3x + 5$ on the window $[-10, 10]$ by $[-10, 10]$. Then use the DRAW menu to graph the TANGENT line at $x = 2$. Your screen should also show the *equation* of the tangent line. (If you did Exercise 45, this equation for the tangent line should agree with the one you found there.)

b. Add to your graph the tangent line at $x = 1$, and the tangent lines at any other x-values that you choose.

48. a. Graph the function $f(x) = 2x^2 - 5x + 1$ on the window $[-10, 10]$ by $[-10, 10]$.

Then use the DRAW menu to graph the TANGENT line at $x = 2$. Your screen should also show the *equation* of the tangent line. (If you did Exercise 46, this equation for the tangent line should agree with the one you found there.)

b. Add to your graph the tangent line at $x = 0$, and the tangent lines at any other x-values that you choose.

For each function in Exercises 49–54:

a. Find $f'(x)$ using the definition of the derivative.

b. Explain, by considering the original function, why the derivative is a constant.

49. $f(x) = 3x - 4$ **50.** $f(x) = 2x - 9$

51. $f(x) = 5$ **52.** $f(x) = 12$

53. $f(x) = mx + b$
(m and b are constants)

54. $f(x) = b$
(b is a constant)

APPLIED EXERCISES

55. BUSINESS: Temperature The temperature in an industrial pasteurization tank is $f(x) = x^2 - 8x + 110$ degrees centigrade after x minutes (for $0 \le x \le 12$).

a. Find $f'(x)$ by using the definition of the derivative.

b. Use your answer to part (a) to find the instantaneous rate of change of the temperature after 2 minutes. Be sure to interpret the sign of your answer.

c. Use your answer to part (a) to find the instantaneous rate of change after 5 minutes.

56. GENERAL: Population The population of a town is $f(x) = 3x^2 - 12x + 200$ people after x weeks (for $0 \le x \le 20$).

a. Find $f'(x)$ by using the definition of the derivative.

b. Use your answer to part (a) to find the instantaneous rate of change of the population after 1 week. Be sure to interpret the sign of your answer. *(continues)*

c. Use your answer to part (a) to find the instantaneous rate of change of the population after 5 weeks.

57. BEHAVIORAL SCIENCE: Learning Theory In a psychology experiment, a person could memorize x words in $f(x) = 2x^2 - x$ seconds (for $0 \le x \le 10$).

a. Find $f'(x)$ by using the definition of the derivative.

b. Evaluate your answer at $x = 5$ and interpret it as an instantaneous rate of change in the proper units.

58. BUSINESS: Advertising An automobile dealership finds that the number of cars that it sells on day x of an advertising campaign is $S(x) = -x^2 + 10x$ (for $0 \le x \le 7$).

a. Find $S'(x)$ by using the definition of the derivative.

b. Use your answer to part (a) to find the instantaneous rate of change on day $x = 3$. *(continues)*

c. Use your answer to part (a) to find the instantaneous rate of change on day $x = 6$.

Be sure to interpret the signs of your answers.

59. **BIOMEDICAL: Temperature** The temperature of a patient in a hospital on day x of an illness is given $T(x) = -x^2 + 5x + 100$ degrees Fahrenheit (for $1 < x < 5$).

a. Find $T'(x)$ by using the definition of the derivative.
b. Use your answer to part (a) to find the instantaneous rate of change of temperature on day 2.
c. Use your answer to part (a) to find the instantaneous rate of change of temperature on day 3. *(continues)*

d. What do your answers to parts (b) and (c) tell you about the patient's health on those two days?

60. **BIOMEDICAL: Bacteria** The number of bacteria in a culture x hours after treatment with an antibiotic is given by
$f(x) = -x^2 + 12x + 1000$ (for $1 < x < 30$).

a. Find $f'(x)$ by using the definition of the derivative.
b. Use your answer to part (a) to find the instantaneous rate of change after 2 hours.
c. Use your answer to part (a) to find the instantaneous rate of change after 20 hours.

Be sure to interpret the signs of your answers.

SOME DIFFERENTIATION FORMULAS

Introduction

In Section 2.2 we defined the *derivative* of a function and used it to calculate instantaneous rates of change and slopes. Even for a function as simple as $f(x) = x^2$, however, calculating the derivative from the definition was rather involved. Calculus would be of limited usefulness if all derivatives had to be calculated in this way.

In this section we will learn several *rules of differentiation* that will simplify finding the derivatives of many useful functions. The rules are derived from the definition of the derivative, which is why we studied the definition first. We will also learn another important use for differentiation: calculating "marginals" (marginal revenue, marginal cost, and marginal profit), which are used extensively in business and economics.

Derivative of a Constant

The first rule of differentiation shows how to differentiate a constant function.

For any constant c,

$$\frac{d}{dx} c = 0$$

Derivative of a constant is zero

| EXAMPLE **1** | **DIFFERENTIATING A CONSTANT** |

$$\frac{d}{dx}7 = 0$$

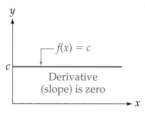

A constant function
(a horizontal line) has
derivative (slope) zero.

This rule is obvious geometrically. The graph of a constant function $f(x) = c$ is the horizontal line $y = c$. Since the slope of a horizontal line is zero, the derivative of $f(x) = c$ is zero.

This rule follows immediately from the definition of the derivative. The constant function $f(x) = c$ has the same value c for *any* value of x, so, in particular, $f(x + h) = c$ and $f(x) = c$. Substituting these into the definition of the derivative gives

$$f'(x) = \lim_{h \to 0} \frac{\overbrace{f(x + h)}^{c} - \overbrace{f(x)}^{c}}{h} = \lim_{h \to 0} \frac{c - c}{h} = \lim_{h \to 0} \frac{\overbrace{0}^{0}}{h} = \lim_{h \to 0} 0 = 0$$

Therefore, the derivative of a constant function $f(x) = c$ is $f'(x) = 0$.

Power Rule

One of the most useful differentiation formulas in all of calculus is called the *Power Rule*. It tells how to differentiate powers such as x^7 or x^{100}.

Power Rule

For any constant exponent n,

$$\frac{d}{dx}x^n = n \cdot x^{n-1}$$

To differentiate x^n, bring down the exponent as a mutiplier and then decrease the exponent by 1

A derivation of the Power Rule for positive integer exponents is given at the end of this section. We will use the Power Rule for *all* real numbers n, since more general proofs will be given later (see pages 139, 256–257, and 314).

| EXAMPLE 2 | **USING THE POWER RULE** |

a. $\dfrac{d}{dx}x^7 = 7x^{7-1} = 7x^6$

Bring down ⌝ ⌜ Decrease the
the exponent ⌝ ⌜ exponent by 1

b. $\dfrac{d}{dx}x^{100} = 100x^{100\ 1} = 100x^{99}$

c. $\dfrac{d}{dx}x^{-2} = -2x^{-2\ 1} = -2x^{-3}$

The Power Rule holds for negative exponents

d. $\dfrac{d}{dx}\sqrt{x} = \dfrac{d}{dx}x^{\frac{1}{2}} = \dfrac{1}{2}x^{\frac{1}{2}-1} = \dfrac{1}{2}x^{-\frac{1}{2}}$

And for fractional exponents

e. $\dfrac{d}{dx}x^1 = 1x^{1-1} = x^0 = 1$

This last result is used so frequently that it should be remembered separately.

$$\dfrac{d}{dx}x = 1 \qquad \text{The derivative of } x \text{ is } 1$$

From now on we will skip the middle step in these examples, differentiating powers in one step:

$$\dfrac{d}{dx}x^{50} = 50x^{49}$$

$$\dfrac{d}{dx}x^{2/3} = \dfrac{2}{3}x^{-1/3} \qquad \tfrac{2}{3}-1$$

| Practice Problem 1 | Find |

a. $\dfrac{d}{dx}x^2$ **b.** $\dfrac{d}{dx}x^{-5}$ **c.** $\dfrac{d}{dx}\sqrt[4]{x}$ ➤ Solutions on page 123

Constant Multiple Rule

The Power Rule shows how to differentiate a power such as x^3. The *Constant Multiple Rule* extends this result to functions such as $5x^3$, a constant *times* a function. Briefly, to differentiate a constant times a function, we simply "carry along" the constant and differentiate the function.

Constant Multiple Rule

For any constant c,

$$\frac{d}{dx}[c \cdot f(x)] = c \cdot f'(x)$$

The derivative of a constant times a function is the constant times the derivative of the function

---

(provided, of course, that the derivative $f'(x)$ exists). A derivation of this rule is given at the end of this section.

EXAMPLE 3

USING THE CONSTANT MULTIPLE RULE

a. $\dfrac{d}{dx} 5x^3 = 5 \cdot 3x^2 = 15x^2$ **b.** $\dfrac{d}{dx} 3x^{-4} = 3(-4)x^{-5} = -12x^{-5}$

Carry along the constant

Derivative of x^3

Again we will skip the middle step, bringing down the exponent and immediately multiplying it by the number in front of the x.

EXAMPLE 4

CALCULATING DERIVATIVES MORE QUICKLY

a. $\dfrac{d}{dx} 8x^{-1/2} = -4x^{-3/2}$ **b.** $\dfrac{d}{dx} 7x = 7 \cdot 1 = 7$

$8\left(-\frac{1}{2}\right)$

Derivative of x

This last example, showing that the derivative of $7x$ is just 7, leads to a very useful general rule.

For any constant c,

$$\frac{d}{dx}(cx) = c$$

The derivative of a constant times x is just the constant

EXAMPLE 5

FINDING DERIVATIVES INVOLVING CONSTANTS

a. $\dfrac{d}{dx}(7x) = 7$ Using $\dfrac{d}{dx}(cx) = c$

b. $\dfrac{d}{dx}7 = 0$ For a constant alone, the derivative is zero

c. $\dfrac{d}{dx}(7x^2) = 7 \cdot 2x = 14x$ But for a constant times a function, the derivative is the constant times the derivative of the function

Sum Rule

The *Sum Rule* extends differentiation to sums of functions. Briefly, to differentiate a *sum* of two functions, just differentiate the functions separately and add the results.

Sum Rule

$$\frac{d}{dx}[f(x) + g(x)] = f'(x) + g'(x)$$

The derivative of a sum is the sum of the derivatives

(provided, of course, that both the derivatives $f'(x)$ and $g'(x)$ exist). A derivation of the Sum Rule is given at the end of this section. For example, the derivative of the sum

is just the sum of the derivatives

$$x^3 + x^5 \qquad \text{Sum of } x^3 \text{ and } x^5$$
$$\downarrow \quad \downarrow$$
$$3x^2 + 5x^4 \qquad \begin{array}{l}\text{Differentiating each}\\ \text{separately and adding}\end{array}$$

A similar rule holds for the *difference* of two functions,

$$\frac{d}{dx}[f(x) - g(x)] = f'(x) - g'(x)$$

The derivative of a difference is the difference of the derivatives

(provided that $f'(x)$ and $g'(x)$ exist). These two rules may be combined:

Sum-Difference Rule

$$\frac{d}{dx}[f(x) \pm g(x)] = f'(x) \pm g'(x)$$

Use both upper signs or both lower signs

Similar rules hold for sums and differences of any finite number of terms. Using these rules, we may differentiate any polynomial or, more generally, functions with variables raised to *any* constant powers.

EXAMPLE 6

USING THE SUM-DIFFERENCE RULE

a. $\dfrac{d}{dx}(x^3 - x^5) = 3x^2 - 5x^4$

Derivatives taken separately

b. $\dfrac{d}{dx}(5x^{-2} - 6x^{1/3} + 4) = -10x^{-3} - 2x^{-2/3}$

The constant 4 has derivative 0

Leibniz's Notation and Evaluation of Derivatives

Leibniz's derivative notation, $\dfrac{d}{dx}$, is often read "the derivative with respect to x" to emphasize that the independent variable is x. To differentiate a function of some *other* variable, the x in $\dfrac{d}{dx}$ is replaced by the other variable. For example:

Function	*Derivative*	
$f(t)$	$\dfrac{d}{dt}f(t)$	Use $\dfrac{d}{dt}$ for the derivative with respect to t
w^3	$\dfrac{d}{dw}w^3$	Use $\dfrac{d}{dw}$ for the derivative with respect to w

The following two notations both mean the derivative *evaluated* at $x = 2$.

Derivative

$f'(2)$

Evaluated at $x = 2$

Derivative

$\dfrac{df}{dx}\Big|_{x=2}$

Evaluated at $x = 2$

Bar | means "evaluated at"

Be careful! Both notations mean *first* differentiate and *then* evaluate.

EXAMPLE 7

EVALUATING A DERIVATIVE

If $f(x) = x^4$, find $f'(2)$.

Solution

$$f'(x) = 4x^3 \qquad \text{First differentiate}$$
$$f'(2) = 4 \cdot 2^3 = 4 \cdot 8 = 32 \qquad \text{Then evaluate}$$

Practice Problem 2   If $f(x) = x^3$, find $\left. \dfrac{df}{dx} \right|_{x=-1}$ ▶ Solution on page 123

Derivatives in Business and Economics: Marginals

There is another interpretation for the derivative, one that is particularly important in business and economics. Suppose that a company has calculated its revenue, cost, and profit functions, as defined below.

$$R(x) = \begin{pmatrix} \text{Total revenue (income)} \\ \text{from selling } x \text{ units} \end{pmatrix} \qquad \text{Revenue function}$$

$$C(x) = \begin{pmatrix} \text{Total cost of} \\ \text{producing } x \text{ units} \end{pmatrix} \qquad \text{Cost function}$$

$$P(x) = \begin{pmatrix} \text{Total profit from producing} \\ \text{and selling } x \text{ units} \end{pmatrix} \qquad \text{Profit function}$$

The term "marginal cost" means the additional cost of producing one more unit, $C(x + 1) - C(x)$, which may be written $\dfrac{C(x + 1) - C(x)}{1}$, which is just the difference quotient $\dfrac{C(x + h) - C(x)}{h}$ with $h = 1$. If many units are being produced, then $h = 1$ is a relatively small number compared with x, so this difference quotient may be approximated by its limit as $h \to 0$, that is by the *derivative* of the cost function. In view of this approximation, in calculus the marginal cost is *defined* to be the derivative of the cost function:

$$MC(x) = C'(x) \qquad \text{Marginal cost is the derivative of cost}$$

The marginal *revenue* function $MR(x)$ and the marginal *profit* function $MP(x)$ are similarly defined as the derivatives of the revenue and cost functions.

$$MR(x) = R'(x) \qquad \text{Marginal revenue is the derivative of revenue}$$

$$MP(x) = P'(x) \qquad \text{Marginal profit is the derivative of profit}$$

All of this can be summarized very briefly: "marginal" means "derivative of." We now have three interpretations for the derivative: *slopes, instantaneous rates of change,* and *marginals.*

EXAMPLE 8

FINDING AND INTERPRETING MARGINAL COST

A company manufactures cordless telephones and finds that its cost function (the total cost of manufacturing x telephones) is

$$C(x) = 400\sqrt{x} + 500 \qquad \text{Cost function}$$

dollars, where x is the number of telephones produced.

a. Find the marginal cost function $MC(x)$.

b. Find the marginal cost when 100 telephones have been produced, and interpret your answer.

Solution

a. The marginal cost function is the derivative of the cost function $C(x) = 400x^{1/2} + 500$:

$$MC(x) = 200x^{-1/2} = \frac{200}{\sqrt{x}} \qquad \text{Derivative of } C(x)$$

b. To find the marginal cost when 100 telephones have been produced, we evaluate the marginal cost function at $x = 100$:

$$MC(100) = \frac{200}{\sqrt{100}} = \frac{200}{10} = \$20 \qquad MC(x) = \frac{200}{\sqrt{x}} \text{ evaluated at } x = 100$$

Interpretation: When 100 telephones have been produced, the marginal cost is $20, meaning that to produce one more telephone costs about $20.

EXAMPLE 9

FINDING A LEARNING RATE

A psychology researcher finds that the number of names that a person can memorize in x minutes is approximately $f(x) = 6\sqrt[3]{x^2}$. Find the instantaneous rate of change of this function after 8 minutes and interpret your answer.

Solution

$$f(x) = 6x^{2/3} \qquad \text{\small $6\sqrt[3]{x^2}$ in exponential form}$$

$$f'(x) = 6 \cdot \frac{2}{3} x^{-1/3} = 4x^{-1/3} \qquad \text{\small The instantaneous rate of change is $f'(x)$}$$

$$f'(8) = 4(8)^{-1/3} = 4\left(\frac{1}{\sqrt[3]{8}}\right) = 4\left(\frac{1}{2}\right) = 2 \qquad \text{\small Evaluating at $x = 8$}$$

Interpretation: After 8 minutes the person can memorize about two additional names per minute.

Functions as Single Objects

You may have noticed that calculus requires a more abstract point of view than precalculus mathematics. In earlier courses you looked at functions and graphs as collections of individual points, to be plotted one at a time. Now, however, we are operating on *whole functions* all at once (for example, differentiating the function x^3 to obtain the function $3x^2$). In calculus, the basic objects of interest are *functions*, and a function should be thought of as a *single* object.

This is in keeping with a trend toward increasing abstraction as you learn mathematics. You first studied single numbers, then points (pairs of numbers), then functions (collections of points), and now collections of functions (polynomials, differentiable functions, and so on). Each stage has been a generalization of the previous stage as you have reached higher levels of sophistication. This process of generalization or "chunking" of knowledge enables you to express ideas of wider applicability and power.

Derivatives on a Graphing Calculator

Graphing calculators have an operation called NDERIV (or something similar), standing for *numerical derivative*, which gives an *approximation* of the derivative of a function. Most do so by evaluating the *symmetric difference quotient,* $\dfrac{f(x + h) - f(x - h)}{2h}$ for a small value of h, such as

$h = 0.001$.   The numerator represents the change in the function when x changes by $2h$ (from $x - h$ to $x + h$), and the denominator divides by this change in x. Geometrically, the symmetric difference quotient gives the slope of the secant line through two points on the curve h units on either side of the point at x. While NDERIV usually approximates the derivative quite closely, it sometimes gives erroneous results, as we will see in later sections. For this reason, using a graphing calculator effectively requires an understanding of both the calculus that underlies it and the technology that limits it.

Graphing Calculator Exploration

a. Use a graphing calculator to graph $y_1 = x^3 - x^2 - 6x + 3$ on $[-5, 5]$ by $[-10, 10]$.

b. Define y_2 as the derivative of y_1 (using NDERIV) and graph both functions.

c. Observe that where y_1 is horizontal, the *value* of y_2 is zero; where y_1 slopes *upward*, y_2 is *positive*; and where y_1 slopes *downward*, y_2 is *negative*. Would you be able to use these observations to identify which curve is the original function and which is the derivative?

d. Now check your answer to Example 9 as follows: Redefine y_1 as $y_1 = 6x^{2/3}$, reset the window to $[0, 10]$ by $[-10, 30]$, GRAPH y_1 and y_2, and EVALUATE y_2 at $x = 8$. Your answer should agree with that of Example 9.

2.3   Section Summary

Our development of calculus has followed two quite different lines—one technical (the *rules* of derivatives) and the other conceptual (the *meaning* of derivatives).

On the conceptual side, derivatives have three meanings:

- Instantaneous rates of change
- Slopes of curves
- Marginals

The fact that the derivative represents all three of these ideas simultaneously is one of the reasons that calculus is so useful.

On the technical side, although we have learned several differentiation rules, we really know how to differentiate only one kind of function, *x to a constant power:*

$$\frac{d}{dx} x^n = nx^{n-1}$$

The other rules,

$$\frac{d}{dx} [c \cdot f(x)] = c \cdot f'(x)$$

and

$$\frac{d}{dx} [f(x) \pm g(x)] = f'(x) \pm g'(x)$$

simply extend the Power Rule to sums, differences, and constant multiples of such powers. Therefore, any function to be differentiated must first be expressed in terms of powers. This is why we reviewed exponential notation so carefully in Chapter 1.

Verification of the Power Rule for Positive Integer Exponents

Multiplying $(x + h)$ times itself repeatedly gives

$$(x + h)^2 = x^2 + 2xh + h^2$$

$$(x + h)^3 = x^3 + 3x^2h + 3xh^2 + h^3$$

and in general, for any positive integer n,

$$(x + h)^n = x^n + nx^{n-1}h + \tfrac{1}{2}n(n - 1)x^{n-2}h^2 + \cdots + nxh^{n-1} + h^n$$

$$= x^n + nx^{n-1}h + h^2[\tfrac{1}{2}n(n - 1)x^{n-2} + \cdots + nxh^{n-3} + h^{n-2}] \quad \text{Factoring out } h^2$$

$$= x^n + nx^{n-1}h + h^2 \cdot P \qquad \qquad \begin{array}{l} P \text{ stands for the polynomial} \\ \text{in the square bracket above} \end{array}$$

The resulting formula

$$(x + h)^n = x^n + nx^{n-1}h + h^2 \cdot P$$

will be useful in the following verification. To prove the Power Rule for any positive integer n, we use the definition of the derivative to differentiate $f(x) = x^n$.

$$f'(x) = \lim_{h \to 0} \frac{f(x + h) - f(x)}{h}$$

Definition of the derivative

$$= \lim_{h \to 0} \frac{(x + h)^n - x^n}{h}$$

Since $f(x + h) = (x + h)^n$ and $f(x) = x^n$

$$= \lim_{h \to 0} \frac{x^n + nx^{n-1}h + h^2 \cdot P - x^n}{h}$$

Expanding, using the formula derived earlier

$$= \lim_{h \to 0} \frac{nx^{n-1}h + h^2 \cdot P}{h}$$

Canceling the x^n and the $-x^n$

$$= \lim_{h \to 0} \frac{h(nx^{n-1} + h \cdot P)}{h}$$

Factoring out an h

$$= \lim_{h \to 0} (nx^{n-1} + h \cdot P)$$

Canceling the h (since $h \neq 0$)

$$= nx^{n-1}$$

Evaluating the limit by direct substitution

This shows that for any positive integer n, the derivative of x^n is nx^{n-1}.

Verification of the Constant Multiple Rule

For a constant c and a function f, let $g(x) = c \cdot f(x)$. If $f'(x)$ exists, we may calculate the derivative $g'(x)$ as follows:

$$g'(x) = \lim_{h \to 0} \frac{g(x + h) - g(x)}{h}$$

Definition of the derivative

$$= \lim_{h \to 0} \frac{c \cdot f(x + h) - c \cdot f(x)}{h}$$

Since $g(x + h) = c \cdot f(x + h)$ and $g(x) = c \cdot f(x)$

$$= \lim_{h \to 0} \frac{c \cdot [f(x + h) - f(x)]}{h}$$

Factoring out the c

$$= c \cdot \lim_{h \to 0} \frac{f(x + h) - f(x)}{h}$$

Taking c outside the limit leaves just the definition of the derivative $f'(x)$.

$$= c \cdot f'(x)$$

Constant Multiple Rule

This shows that the derivative of a constant times a function, $c \cdot f(x)$, is the constant times the derivative of the function, $c \cdot f'(x)$.

Verification of the Sum Rule

For two functions f and g, let their sum be $s = f + g$. If $f'(x)$ and $g'(x)$ exist, we may calculate $s'(x)$ as follows:

$$s'(x) = \lim_{h \to 0} \frac{s(x+h) - s(x)}{h}$$

Definition of the derivative

$$= \lim_{h \to 0} \frac{[f(x+h) + g(x+h)] - [f(x) + g(x)]}{h}$$

Since $s(x+h) = f(x+h) + g(x+h)$ and $s(x) = f(x) + g(x)$

$$= \lim_{h \to 0} \frac{f(x+h) + g(x+h) - f(x) - g(x)}{h}$$

Eliminating the brackets

$$- \lim_{h \to 0} \frac{f(x+h) - f(x) + g(x+h) - g(x)}{h}$$

Rearranging the numerator

$$= \lim_{h \to 0} \left[\frac{f(x+h) - f(x)}{h} + \frac{g(x+h) - g(x)}{h} \right]$$

Separating the fraction into two parts

$$= \underbrace{\lim_{h \to 0} \frac{f(x+h) - f(x)}{h}}_{f'(x)} + \underbrace{\lim_{h \to 0} \frac{g(x+h) - g(x)}{h}}_{g'(x)}$$

Using Limit Rule 4a on page 81

Recognizing the definition of the derivatives of f and g

$$= f'(x) + g'(x)$$

Sum Rule

This shows that the derivative of a sum $f(x) + g(x)$ is the sum of the derivatives $f'(x) + g'(x)$.

➤ **Solutions to Practice Problems**

1. a. $\dfrac{d}{dx} x^2 = 2x^{2-1} = 2x$

b. $\dfrac{d}{dx} x^{-5} = -5x^{-5-1} = -5x^{-6}$

c. $\dfrac{d}{dx} \sqrt[4]{x} = \dfrac{d}{dx} x^{1/4} = \dfrac{1}{4} x^{(1/4)-1} = \dfrac{1}{4} x^{-3/4}$

2. $\dfrac{df}{dx} = 3x^2$

$\left. \dfrac{df}{dx} \right|_{x=-1} = 3(-1)^2 = 3$

2.3 Exercises

Find the derivative of each function.

1. $f(x) = x^4$ **2.** $f(x) = x^5$ **3.** $f(x) = x^{500}$

4. $f(x) = x^{1000}$ **5.** $f(x) = x^{1/2}$ **6.** $f(x) = x^{1/3}$

7. $g(x) = \frac{1}{2}x^4$ **8.** $f(x) = \frac{1}{3}x^9$

9. $g(w) = 6\sqrt[3]{w}$ **10.** $g(w) = 12\sqrt{w}$

11. $h(x) = \dfrac{3}{x^2}$ **12.** $h(x) = \dfrac{4}{x^3}$

13. $f(x) = 4x^2 - 3x + 2$

14. $f(x) = 3x^2 - 5x + 4$

15. $f(x) = \dfrac{1}{x^{1/2}}$ **16.** $f(x) = \dfrac{1}{x^{2/3}}$

17. $f(x) = \dfrac{6}{\sqrt[3]{x}}$ **18.** $f(x) = \dfrac{4}{\sqrt{x}}$

19. $f(r) = \pi r^2$ **20.** $f(r) = \dfrac{4}{3}\pi r^3$

21. $f(x) = \dfrac{1}{6}x^3 + \dfrac{1}{2}x^2 + x + 1$

22. $f(x) = \dfrac{1}{24}x^4 + \dfrac{1}{6}x^3 + \dfrac{1}{2}x^2 + x + 1$

23. $g(x) = \sqrt{x} - \dfrac{1}{x}$ **24.** $g(x) = \sqrt[3]{x} - \dfrac{1}{x}$

25. $h(x) = 6\sqrt[3]{x^2} - \dfrac{12}{\sqrt[3]{x}}$ **26.** $h(x) = 8\sqrt{x^3} - \dfrac{8}{\sqrt[4]{x}}$

27. $f(x) = \dfrac{10}{\sqrt{x}} - 9\sqrt[3]{x^5} + 17$

28. $f(x) = \dfrac{9}{\sqrt[3]{x}} - 16\sqrt{x^5} - 14$

29. $f(x) = \dfrac{x^2 + x^3}{x}$ **30.** $f(x) = x^2(x + 1)$

31. a. Find the derivative of $f(x) = 2$.
   b. Interpret your answer in terms of slope.
   c. Interpret your answer in terms of instantaneous rate of change.

32. a. Find the derivative of $f(x) = 3x$.

b. Interpret your answer in terms of slope.
c. Interpret your answer in terms of instantaneous rate of change.

Find the indicated derivatives.

33. If $f(x) = x^5$, find $f'(-2)$.

34. If $f(x) = x^4$, find $f'(-3)$.

35. If $f(x) = 6\sqrt[3]{x^2} - \dfrac{48}{\sqrt[3]{x}}$, find $f'(8)$.

36. If $f(x) = 12\sqrt[3]{x^2} + \dfrac{48}{\sqrt[3]{x}}$, find $f'(8)$.

37. If $f(x) = x^3$, find $\left.\dfrac{df}{dx}\right|_{x=-3}$

38. If $f(x) = x^4$, find $\left.\dfrac{df}{dx}\right|_{x=-2}$

39. If $f(x) = \dfrac{16}{\sqrt{x}} + 8\sqrt{x}$, find $\left.\dfrac{df}{dx}\right|_{x=4}$

40. If $f(x) = \dfrac{54}{\sqrt{x}} + 12\sqrt{x}$, find $\left.\dfrac{df}{dx}\right|_{x=9}$

41. Use a graphing calculator to verify that the derivative of a constant is zero, as follows. Define y_1 to be a constant (such as $y_1 = 5$) and then use NDERIV to define y_2 to be the derivative of y_1. Then graph the two functions together on an appropriate window and use TRACE to observe that the derivative y_2 is zero (graphed as a line along the x-axis), showing that the derivative of a constant is zero.

42. Use a graphing calculator to verify that the derivative of a linear function is a constant, as follows. Define y_1 to be a linear function (such as $y_1 = 3x - 4$) and then use NDERIV to define y_2 to be the derivative of y_1. Then graph the two functions together on an appropriate window and observe that the derivative y_2 is a constant (graphed as a horizontal line, such as $y_2 = 3$), verifying that the derivative of $y_1 = mx + b$ is $y_2 = m$.

APPLIED EXERCISES

43. BUSINESS: Marginal Profit An electronics company finds that its total profit from selling x computer chips is $P(x) = 0.02x^{3/2} - 3000$ dollars.

  a. Find the company's marginal profit function.

  b. Find the marginal profit when 10,000 units have been sold, and interpret your answer.

44. BUSINESS: Marginal Cost A steel mill finds that its cost function is

$$C(x) = 8000 \sqrt{x} - 6000 \sqrt[3]{x}$$

dollars, where x is the (daily) production of steel (in tons).

  a. Find the marginal cost function.

  b. Find the marginal cost when 64 tons of steel are produced.

45. BUSINESS: Marginal Profit (*43 continued*) Use a calculator to find the actual profit from the 10,001st computer chip, $P(10,001) - P(10,000)$, by evaluating the expression

$$\overbrace{[0.02(10,001)^{3/2} - 3000]}^{P(10,001)} - \overbrace{[0.02(10,000)^{3/2} - 3000]}^{P(10,000)}$$

Is your answer close to the answer of $3 found in Exercise 43(b)? Which way of finding the marginal profit was easier, using calculus (Exercise 43) or carrying out the calculation in this exercise?

46. BUSINESS: Marginal Cost (*44 continued*) Use a calculator to find the actual cost of the 65th ton of steel, $C(65) - C(64)$, by evaluating the expression:

$$\overbrace{(8000 \sqrt{65} - 6000 \sqrt[3]{65})}^{C(65)} - \overbrace{(8000 \sqrt{64} - 6000 \sqrt[3]{64})}^{C(64)}$$

Is your answer close to the answer of $375 found in Exercise 44(b)? Which way of finding the marginal cost was easier, using calculus (Exercise 44) or carrying out the calculation in this exercise?

47. GENERAL: Population A company that makes games for teenage children forecasts that the teenage population in the United States x years from now will be

$$P(x) = 12,000,000 - 12,000x + 600x^2 + 100x^3$$

Find the rate of change of the teenage population:

  a. x years from now.

  b. 1 year from now and interpret your answer.

  c. 10 years from now and interpret your answer.

48. BIOMEDICAL: Flu Epidemic The number of people newly infected on day t of a flu epidemic is $f(t) = 13t^2 - t^3$ (for $0 \le t \le 13$). Find the instantaneous rate of change of this number on

  a. day 5 and interpret your answer.

  b. day 10 and interpret your answer.

49. BUSINESS: Advertising It has been estimated that the number of people who will see a newspaper advertisement that has run for x consecutive days is of the form $N(x) = T - \frac{1}{2}T/x$ for $x \ge 1$, where T is the total readership of the newspaper. If a newspaper has a circulation of 400,000, an ad that runs for x days will be seen by

$$N(x) = 400,000 - \frac{200,000}{x}$$

people. Find how fast this number of potential customers is growing when this ad has run for 5 days.

50. ENVIRONMENTAL SCIENCE: Pollution An electrical generating plant burns high-sulfur oil, and the amount of sulfur dioxide pollution x miles downwind of the plant is $f(x) = 108x^{-2}$ parts per million (ppm). Find the instantaneous rate of change of the pollution level 2 miles from the source. Interpret your answer in the proper units.

51. BIOMEDICAL: Blood Flow Nitroglycerin is often prescribed to enlarge blood vessels that have become too constricted. If the cross-sectional area of a blood vessel t hours after nitroglycerin is administered is $A(t) = 0.01t^2$ square centimeters (for $1 \le t \le 5$), find the instantaneous rate of change of the cross-sectional area 4 hours after the administration of nitroglycerin.

52. **GENERAL: Hailstones** Hailstones are frozen raindrops that increase in size as long as the updrafts keep them in the clouds. The weight of a typical hailstone that remains in a cloud for t minutes is $W(t) = 0.05t^3$ ounces. Find the instantaneous rate of change of the weight after 2 minutes.

53. **PSYCHOLOGY: Learning Rates** A language school has found that its students can memorize $p(t) = 24\sqrt{t}$ phrases in t hours of class (for $1 \le t \le 10$). Find the instantaneous rate of change of this quantity after 4 hours of class.

54. **ENVIRONMENTAL SCIENCE: Water Quality** Downstream from a waste treatment plant the amount of dissolved oxygen in the water usually decreases for some distance (due to bacteria consuming the oxygen) and then increases (due to natural purification). A graph of the dissolved oxygen at various distances downstream looks like the curve below (known as the "oxygen sag"). The amount of dissolved oxygen is usually taken as a measure of the health of the river.

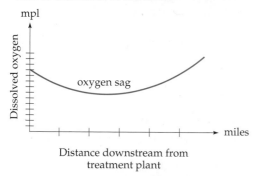

Distance downstream from treatment plant

Suppose that the amount of dissolved oxygen x miles downstream is $D(x) = 0.2x^2 - 2x + 10$ mpl (milligrams per liter) for $0 \le x \le 20$. Use this formula to find the instantaneous rate of change of the dissolved oxygen:

a. 1 mile downstream.
b. 10 miles downstream.

Interpret the signs of your answers.

55–56: **ECONOMICS: Marginal Utility** Generally, the more you have of something, the less valuable each additional unit becomes. For example, a dollar is less valuable to a millionaire than to a beggar. Economists define a person's "utility function" $U(x)$ for a product as the "perceived value" of having x units of that product. The *derivative* of $U(x)$ is called the *marginal utility function*, $MU(x) = U'(x)$. Suppose that a person's utility function for money is given by the function below. That is, $U(x)$ is the utility (perceived value) of x dollars.

a. Find the marginal utility function $MU(x)$.
b. Find $MU(1)$, the marginal utility of the first dollar.
c. Find $MU(1,000,000)$, the marginal utility of the millionth dollar.

55. $U(x) = 100\sqrt{x}$ 56. $U(x) = 12\sqrt[3]{x}$

57. **GENERAL: Smoking and Education** According to a recent study,* the probability that a smoker will quit smoking increases with the smoker's educational level. The probability (expressed as a percent) that a smoker with x years of education will quit is approximated by the equation $f(x) = 0.831x^2 - 18.1x + 137.3$ (for $10 \le x \le 16$).

a. Find $f(12)$ and $f'(12)$ and interpret these numbers. [*Hint:* $x = 12$ corresponds to a high school graduate.]
b. Find $f(16)$ and $f'(16)$ and interpret these numbers. [*Hint:* $x = 16$ corresponds to a college graduate.]

58. **BIOMEDICAL: Lung Cancer** Asbestos has been found to be a potent cause of lung cancer. According to one study of asbestos workers, the number of lung cancer cases in the group depended on the number t of years of exposure to asbestos according to the function $N(t) = 0.00437t^{3.2}$.

a. Graph this function on the window $[0, 15]$ by $[-10, 30]$.
b. Find $N(10)$ and $N'(10)$ and interpret these numbers.

* William Sander, "Schooling and Quitting Smoking," *The Review of Economics and Statistics* LXXVII(1):191–199, February 1995.

59–60. GENERAL: College Tuition The following tables give the annual college tuition costs (in dollars) for a year at a private (Exercise 59) or public (Exercise 60) college for the academic years ending in 1970–2000. To avoid large numbers, years are listed as years since 1970.

a. Enter these numbers into your graphing calculator and make a plot of the resulting points (Years Since 1970 on the *x*-axis and Tuition on the *y*-axis).
b. Have your calculator find the quadratic regression formula for these data. Then enter the result in function y_1. Plot the points together with the regression curve. Observe that the curve fits the points quite well.
c. Predict the tuition in the year 2010 by evaluating y_1 at $x = 40$ (years since 1970).

d. Define y_2 to be the derivative of y_1 (using NDERIV).
e. Predict the rate of change of tuition in the year 2010 by evaluating y_2 at $x = 40$. State your answer in the proper units.

59.

Years Since 1970	Tuition (private college)
0	1533
10	3130
20	8147
30	15,518

60.

Years Since 1970	Tuition (public college)
0	323
10	583
20	1356
30	3362

Sources: U.S. Department of Education, The College Board

THE PRODUCT AND QUOTIENT RULES

Introduction

In the previous section we learned how to differentiate the sum and difference of two functions—we simply take the sum or difference of the derivatives. In this section we learn how to differentiate the *product* and *quotient* of two functions. Unfortunately, we do not simply take the product or quotient of the derivatives. Matters are a little more complicated.

Product Rule

To differentiate the product of two functions, $f(x) \cdot g(x)$, we use the *Product Rule.*

Product Rule

$$\frac{d}{dx}[f(x) \cdot g(x)] = f'(x) \cdot g(x) + f(x) \cdot g'(x)$$

The derivative of a product is the derivative of the first times the second plus the first times the derivative of the second (provided, of course, that the derivatives $f'(x)$ and $g'(x)$ both exist). The formula is clearer if we write the functions simply as f and g.

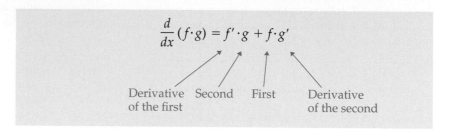

$$\frac{d}{dx}(f \cdot g) = f' \cdot g + f \cdot g'$$

Derivative of the first   Second   First   Derivative of the second

A derivation of the Product Rule is given at the end of this section.

EXAMPLE 1

USING THE PRODUCT RULE

Use the Product Rule to calculate $\dfrac{d}{dx}(x^3 \cdot x^5)$.

Solution

$$\frac{d}{dx}(x^3 \cdot x^5) = 3x^2 \cdot x^5 + x^3 \cdot 5x^4 = 3x^7 + 5x^7 = 8x^7$$

Derivative of the first   Second   First   Derivative of the second

We may check this answer by simplifying the original product, $x^3 \cdot x^5 = x^8$, and then differentiating:

$$\frac{d}{dx}(\underbrace{x^3 \cdot x^5}_{x^8}) = \frac{d}{dx}x^8 = 8x^7$$

Agrees with above answer

Notice that the derivative of a product is *not* the product of the derivatives: $(f \cdot g)' \neq f' \cdot g'$. For $x^3 \cdot x^5$ the product of the derivatives would be $3x^2 \cdot 5x^4 = 15x^6$, which is *not* the correct answer $8x^7$ that we found above. The Product Rule shows the correct way to differentiate a product.

EXAMPLE 2

USING THE PRODUCT RULE

Use the Product Rule to find $\dfrac{d}{dx}[(x^2 - x + 2)(x^3 + 3)]$.

Solution

$$\frac{d}{dx}[(x^2 - x + 2)(x^3 + 3)]$$

$$= (2x - 1)(x^3 + 3) + (x^2 - x + 2)(3x^2)$$

<u>Derivative</u>   <u>Derivative</u>
of $x^2 - x + 2$ of $x^3 + 3$

$$= 2x^4 + 6x - x^3 - 3 + 3x^4 - 3x^3 + 6x^2 \qquad \text{Multiplying out}$$

$$= 5x^4 - 4x^3 + 6x^2 + 6x - 3 \qquad\qquad \text{Simplifying}$$

Practice Problem 1   Use the Product Rule to find $\dfrac{d}{dx}[x^3(x^2 - x)]$.

➤ Solution on page 140

Quotient Rule

The *Quotient Rule* shows how to differentiate a quotient of two functions.

Quotient Rule

$$\frac{d}{dx}\left(\frac{f(x)}{g(x)}\right) = \frac{g(x)\cdot f'(x) - g'(x)\cdot f(x)}{[g(x)]^2}$$

⟵ The bottom times the
derivative of the top,
minus the derivative of
the bottom times the top

The bottom squared

(provided that the derivatives $f'(x)$ and $g'(x)$ both exist and that $g(x) \neq 0$). A derivation of the Quotient Rule is given at the end of this section.

The Quotient Rule looks less formidable if we write the functions simply as f and g,

$$\frac{d}{dx}\left(\frac{f}{g}\right) = \frac{g\cdot f' - g'\cdot f}{g^2}$$

or even as

$$\frac{d}{dx}\left(\frac{\text{top}}{\text{bottom}}\right) = \frac{(\text{bottom})\cdot\left(\dfrac{d}{dx}\text{top}\right) - \left(\dfrac{d}{dx}\text{bottom}\right)\cdot(\text{top})}{(\text{bottom})^2}$$

EXAMPLE 3

USING THE QUOTIENT RULE

Use the Quotient Rule to find $\dfrac{d}{dx}\left(\dfrac{x^9}{x^3}\right)$.

Solution

$$\frac{d}{dx}\left(\frac{x^9}{x^3}\right) = \frac{\overset{\text{Bottom}}{(x^3)}\overset{\substack{\text{Derivative}\\\text{of the top}}}{(9x^8)} - \overset{\substack{\text{Derivative}\\\text{of the bottom}}}{(3x^2)}\overset{\text{Top}}{(x^9)}}{\underset{\text{Bottom squared}}{(x^3)^2}} = \frac{9x^{11} - 3x^{11}}{x^6} = \frac{6x^{11}}{x^6} = 6x^5$$

We may check this answer by simplifying the original quotient and then differentiating:

$$\frac{d}{dx}\left(\frac{x^9}{x^3}\right) = \frac{d}{dx}\,x^6 = 6x^5 \qquad\qquad \text{Agrees with above answer}$$

$$\underset{x^6}{}$$

Notice that the derivative of a quotient is *not* the quotient of the derivatives:

$$\left(\frac{f}{g}\right)' \quad \text{is } not \text{ equal to} \quad \frac{f'}{g'}$$

For the quotient $\dfrac{x^9}{x^3}$, taking the quotient of the derivatives would give $\dfrac{9x^8}{3x^2} = 3x^6$, which is *not* the correct answer $6x^5$ that we found above. The Quotient Rule shows the correct way to differentiate a quotient.

EXAMPLE 4

USING THE QUOTIENT RULE

Find $\dfrac{d}{dx}\left(\dfrac{x^2}{x+1}\right)$.

Solution   The Quotient Rule gives

$$\frac{d}{dx}\left(\frac{x^2}{x+1}\right) = \frac{(x+1)(2x) - (1)(x^2)}{(x+1)^2} = \frac{2x^2 + 2x - x^2}{(x+1)^2} = \frac{x^2 + 2x}{(x+1)^2}$$

Bottom — Derivative of the top — Derivative of the bottom — Top — Bottom squared

---

Practice Problem 2   Find $\dfrac{d}{dx}\left(\dfrac{2x^2}{x^2+1}\right)$.

➤ Solution on page 140

Not every quotient requires the Quotient Rule. Some are simple enough to be differentiated by the Power Rule.

EXAMPLE 5

DIFFERENTIATING A QUOTIENT BY THE POWER RULE

Find the derivative of $y = \dfrac{5}{x^2}$

Solution

$$\frac{d}{dx}\left(\frac{5}{x^2}\right) = \frac{d}{dx}(5x^{-2}) = -10x^{-3} = -\frac{10}{x^3}$$

Differentiated by the power rule

In this example we rewrote the expression before and after the differentiation:

Begin	**Rewrite**	**Differentiate**	**Rewrite**
$y = \dfrac{5}{x^2}$	$y = 5x^{-2}$	$\dfrac{dy}{dx} = -10x^{-3}$	$\dfrac{dy}{dx} = -\dfrac{10}{x^3}$

This way is often much easier than using the Quotient Rule if the numerator or denominator is a constant.

EXAMPLE 6

FINDING THE COST OF CLEANER WATER

Practically every city must purify its drinking water and treat its wastewater. The cost of the treatment rises steeply for higher degrees of purity. If the cost of purifying a gallon of water to a purity of x percent is

$$C(x) = \frac{2}{100 - x} \qquad \text{for } 80 < x < 100$$

dollars, find the rate of change of the purification costs when the purity is:

a. 90%   **b.** 98%

Solution

The rate of change of cost is the *derivative* of the cost function:

Derivative of 2
Derivative of $100 - x$

$$C'(x) = \frac{d}{dx}\left(\frac{2}{100 - x}\right) = \frac{(100 - x)(0) - (-1)(2)}{(100 - x)^2} \qquad \text{Differentiating by the Quotient Rule}$$

$$= \frac{0 + 2}{(100 - x)^2} = \frac{2}{(100 - x)^2} \qquad \begin{array}{l}\text{Simplifying (the}\\ \text{derivative is}\\ \text{undefined at}\\ x = 100)\end{array}$$

a. For 90% purity we evaluate at $x = 90$:

$$C'(90) = \frac{2}{(100 - 90)^2} = \frac{2}{10^2} = \frac{2}{100} = 0.02 \qquad \begin{array}{l}C'(x) = \dfrac{2}{(100 - x)^2}\\[4pt] \text{evaluated at } x = 90\end{array}$$

Interpretation: At 90% purity, the rate of change of the cost is 0.02 dollar, meaning that the costs increase by about *2 cents for each additional percentage of purity.*

b. For 98% purity we evaluate $C'(x)$ at $x = 98$:

$$C'(98) = \frac{2}{(100 - 98)^2} = \frac{2}{2^2} = \frac{2}{4} = \frac{1}{2} = 0.50 \qquad \begin{array}{l}C'(x) = \dfrac{2}{(100 - x)^2}\\[4pt] \text{evaluated at } x = 98\end{array}$$

Interpretation: At 98% purity, the rate of change of the cost is 0.50 dollar, meaning that the costs increase by about *50 cents for each additional percentage of purity.*

Notice that an extra percentage of purity above the 98% level is 25 times as costly as an extra percentage above the 90% purity level.

Graphing Calculator Exploration

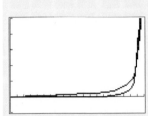

a. On a graphing calculator, enter the cost function from Example 6 as $y_1 = \dfrac{2}{(100 - x)}$. Then use NDERIV to define y_2 to be the derivative of y_1.

b. Graph both y_1 and y_2 on the window [80, 100] by [−1, 5]. Your graph should resemble the one on the left (but you may have an additional "false" vertical line on the right).

c. Verify the results of Example 6 by evaluating y_2 at $x = 90$ and at $x = 98$.

d. Evaluate y_2 at $x = 100$, giving (supposedly) the derivative of y_1 at $x = 100$. However, in Example 6 we saw that the derivative of y_1 is *undefined* at $x = 100$. Your calculator is giving you a "false value" for the derivative, resulting from NDERIV's use of a symmetric difference quotient (see pages 119–120) and a positive value for h. Therefore, to use your calculator effectively, you must also understand calculus.

Marginal Average Cost

It is often useful to calculate not just the *total* cost of producing x units of some product, but also the *average cost per unit*, denoted $AC(x)$, which is found by dividing the total cost $C(x)$ by the number of units x.

$$AC(x) = \frac{C(x)}{x}$$
Average cost per unit is total cost divided by the number of units

The derivative of the average cost function is called the *marginal average cost, MAC.**

$$MAC(x) = \frac{d}{dx}\left[\frac{C(x)}{x}\right]$$
Marginal average cost is the derivative of average cost

---

* The marginal average cost function is sometimes denoted $\overline{C}'(x)$, with similar notations used for marginal average revenue and marginal average profit.

Marginal average revenue *MAR*, and marginal average profit *MAP*, are defined similarly as the derivatives of average revenue per unit, $\frac{R(x)}{x}$, and average profit per unit, $\frac{P(x)}{x}$.

$$MAR(x) = \frac{d}{dx}\left[\frac{R(x)}{x}\right]$$

Marginal average revenue is the derivative of average revenue $\frac{R(x)}{x}$

$$MAP(x) = \frac{d}{dx}\left[\frac{P(x)}{x}\right]$$

Marginal average profit is the derivative of average profit $\frac{P(x)}{x}$

EXAMPLE **7**

FINDING AND INTERPRETING MARGINAL AVERAGE COST

It costs a book publisher $12 to produce each book, and fixed costs are $1500. Therefore, the company's cost function is

$$C(x) = 12x + 1500 \qquad \text{Total cost of producing } x \text{ books}$$

a. Find the average cost function.

b. Find the marginal average cost function.

c. Find the marginal average cost at $x = 100$ and interpret your answer.

Solution

a. The average cost function is

$$AC(x) = \underbrace{\frac{12x + 1500}{x}}_{\substack{\text{Total cost divided} \\ \text{by number of units}}} = \underbrace{12 + \frac{1500}{x}}_{\text{Simplifying}} = \underbrace{12 + 1500x^{-1}}_{\text{In power form}}$$

b. The *marginal* average cost is the derivative of average cost. We could use the Quotient Rule on the first expression above, but it is easier to use the Power Rule on the last expression:

$$MAC(x) = \frac{d}{dx}(12 + 1500x^{-1}) = -1500x^{-2} = -\frac{1500}{x^2}$$

c. Evaluating at $x = 100$:

$$MAC(100) = -\frac{1500}{100^2} = -\frac{1500}{10,000} = -0.15 \qquad -\frac{1500}{x^2} \text{ at } x = 100$$

Interpretation: When 100 books have been produced, the average cost per book is decreasing (because of the negative sign) by about *15 cents per additional book produced.* This reflects the fact that while *total* costs rise when you produce more, the *average cost per unit* decreases, because of the economies of mass production.

Graphing Calculator Exploration

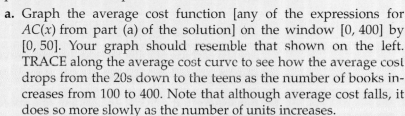

Use a graphing calculator to investigate further the effects of mass production in Example 7.

a. Graph the average cost function [any of the expressions for $AC(x)$ from part (a) of the solution] on the window $[0, 400]$ by $[0, 50]$. Your graph should resemble that shown on the left. TRACE along the average cost curve to see how the average cost drops from the 20s down to the teens as the number of books increases from 100 to 400. Note that although average cost falls, it does so more slowly as the number of units increases.

b. To see exactly how rapidly the average cost declines, graph the *marginal* average cost function on the window $[0, 400]$ by $[-1, 1]$. TRACE along this curve to see how the marginal average cost (which is negative since costs are decreasing) approaches zero as the number of units increases (the law of diminishing returns).

EXAMPLE 8

FINDING TIME SAVED BY SPEEDING

A certain mathematics professor drives 25 miles to his office every day, mostly on highways. If he drives at constant speed v miles per hour, his travel time (distance divided by speed) is

$$T(v) = \frac{25}{v}$$

hours. Find $T'(55)$ and interpret this number.

Solution Since $T(v) = \dfrac{25}{v}$ is a quotient, we could differentiate it by the Quotient Rule. However, it is easier to write $\dfrac{25}{v}$ as a *power*,

$$T(v) = 25v^{-1}$$

and differentiate using the Power Rule:

$$T'(v) = -25v^{-2}$$

This gives the rate of change of the travel time with respect to driving speed. $T'(v)$ is negative, showing that as speed increases, travel time *decreases*. Evaluating this at speed $v = 55$ gives

$$T'(55) = -25(55)^{-2} = \frac{-25}{(55)^2} \approx -0.00826 \qquad \text{Using a calculator}$$

This number, the rate of change of travel time with respect to driving speed, means that when driving at 55 miles per hour, you save only 0.00826 hour for each extra mile per hour of speed. Multiplying by 60 gives the saving in *minutes:*

$$(-0.00826)(60) \approx -0.50 = -\frac{1}{2}$$

That is, each extra mile per hour of speed saves only about half a minute, or 30 seconds. For example, speeding by 10 mph would save only about $\frac{1}{2} \cdot 10 = 5$ minutes. One must then decide whether this slight savings in time is worth the risk of an accident or a speeding ticket.

2.4 Section Summary

The following is a list of the differentiation formulas that we have learned so far. The letters c and n stand for constants, and f and g stand for differentiable functions of x.

$$\frac{d}{dx}c = 0$$

$$\frac{d}{dx}x^n = nx^{n-1} \qquad\qquad \text{special case: } \frac{d}{dx}x = 1$$

$$\frac{d}{dx}(c \cdot f) = c \cdot f' \qquad\qquad \text{special case: } \frac{d}{dx}(cx) = c$$

$$\frac{d}{dx}(f \pm g) = f' \pm g'$$

$$\frac{d}{dx}(f \cdot g) = f' \cdot g + f \cdot g'$$

$$\frac{d}{dx}\left(\frac{f}{g}\right) = \frac{g \cdot f' - g' \cdot f}{g^2} \qquad g \neq 0$$

These formulas are used extensively throughout calculus, and you are not yet ready to proceed to the next section until you have mastered them.

Verification of the Differentiation Formulas

We conclude this section with derivations of the Product and Quotient Rules, and the Power Rule in the case of *negative* integer exponents. First, however, we need to establish a preliminary result about an arbitrary function g:

$$\text{If } g'(x) \text{ exists, then } \lim_{h \to 0} g(x + h) = g(x).$$

We begin with $\lim_{h \to 0} g(x + h)$ and show that it is equal to $g(x)$:

$$\lim_{h \to 0} g(x + h) = \lim_{h \to 0} [g(x + h) - g(x) + g(x)] \qquad \text{Subtracting and adding } g(x)$$

$$= \lim_{h \to 0} \left[\frac{g(x + h) - g(x)}{h} \cdot h + g(x) \right] \qquad \text{Dividing and multiplying by } h$$

$$= \lim_{h \to 0} \left[\frac{g(x + h) - g(x)}{h} \cdot h \right] + \lim_{h \to 0} g(x) \qquad \begin{array}{l}\text{The limit of a sum is the sum} \\ \text{of the limits}\end{array}$$

$$= \underbrace{\lim_{h \to 0} \frac{g(x + h) - g(x)}{h}}_{g'(x)} \cdot \underbrace{\lim_{h \to 0} h}_{0} + g(x) \qquad \begin{array}{l}\text{The limit of a product is} \\ \text{the product of the limits}\end{array}$$

$$= g'(x) \cdot 0 + g(x) \qquad \begin{array}{l}\text{Since the first limit above is} \\ \text{the definition of } g'(x) \text{ and} \\ \text{the second limit is zero}\end{array}$$

$$= g(x) \qquad \text{Simplifying}$$

This proves the result that if $g'(x)$ exists, then $\lim_{h \to 0} g(x + h) = g(x)$.

Replacing $x + h$ by a new variable y, this equation becomes

$$\lim_{y \to x} g(y) = g(x) \qquad\qquad y = x + h, \text{ so } h \to 0 \text{ implies } y \to x$$

According to the definition of continuity on page 87 (but with different letters), this equation means that the function g is *continuous* at x. Therefore, the result that we have shown can be stated simply:

If a function is *differentiable* at x, then it is *continuous* at x.

Or, even more briefly:

> **Differentiability implies continuity.**

Verification of the Product Rule

For two functions f and g, let their product be $p(x) = f(x) \cdot g(x)$. If $f'(x)$ and $g'(x)$ exist, we may calculate $p'(x)$ as follows.

$$p'(x) = \lim_{h \to 0} \frac{p(x+h) - p(x)}{h}$$ Definition of the derivative

$$= \lim_{h \to 0} \frac{f(x+h)g(x+h) - f(x)g(x)}{h}$$ $p(x+h) = f(x+h) \cdot g(x+h)$ and $p(x) = f(x) \cdot g(x)$

$$= \lim_{h \to 0} \frac{f(x+h)g(x+h) - f(x)g(x+h) + f(x)g(x+h) - f(x)g(x)}{h}$$ Subtracting and adding $f(x)g(x+h)$

$$= \lim_{h \to 0} \left[\frac{f(x+h)g(x+h) - f(x)g(x+h)}{h} + \frac{f(x)g(x+h) - f(x)g(x)}{h} \right]$$ Separating the fraction into two parts

$$= \lim_{h \to 0} \frac{[f(x+h) - f(x)]g(x+h)}{h} + \lim_{h \to 0} \frac{f(x)[g(x+h) - g(x)]}{h}$$ Using Limit Rule 4a on page 81 and factoring

$$= \lim_{h \to 0} \frac{[f(x+h) - f(x)]}{h} \underbrace{\lim_{h \to 0} g(x+h)} + f(x) \lim_{h \to 0} \frac{[g(x+h) - g(x)]}{h}$$ Using Limit Rule 4c on page 81

$$\underbrace{}_{f'(x)} \quad \underbrace{}_{g(x)} \quad \underbrace{}_{f(x)} \quad \underbrace{}_{g'(x)}$$ Recognizing the definitions of $f'(x)$ and $g'(x)$

$$= f'(x) \cdot g(x) + f(x) \cdot g'(x)$$ Product Rule

Verification of the Quotient Rule

For two functions f and g with $g(x) \neq 0$, let the quotient be $q(x) = \dfrac{f(x)}{g(x)}$. If $f'(x)$ and $g'(x)$ exist, we may calculate $q'(x)$ as follows.

$$q'(x) = \lim_{h \to 0} \frac{q(x+h) - q(x)}{h}$$ Definition of the derivative

$$= \lim_{h \to 0} \frac{\dfrac{f(x+h)}{g(x+h)} - \dfrac{f(x)}{g(x)}}{h}$$ $q(x+h) = \dfrac{f(x+h)}{g(x+h)}$ and $q(x) = \dfrac{f(x)}{g(x)}$

$$= \lim_{h \to 0} \frac{1}{h} \left[\frac{f(x + h)}{g(x + h)} - \frac{f(x)}{g(x)} \right]$$

Since dividing by h is equivalent to multiplying by $1/h$

$$= \lim_{h \to 0} \left[\frac{1}{h} \cdot \frac{g(x)f(x + h) - g(x + h)f(x)}{g(x + h)g(x)} \right]$$

Subtracting the fractions, using the common denominator $g(x + h)g(x)$

$$= \lim_{h \to 0} \left[\frac{1}{h} \cdot \frac{g(x)f(x + h) - g(x)f(x) - [g(x + h)f(x) - g(x)f(x)]}{g(x + h)g(x)} \right]$$

Subtracting and adding $g(x)f(x)$

$$= \lim_{h \to 0} \left[\frac{1}{g(x + h)g(x)} \cdot \frac{g(x)[f(x + h) - f(x)] - [g(x + h) - g(x)]f(x)}{h} \right]$$

Factoring in the numerator; switching the denominators

$$= \lim_{h \to 0} \left[\frac{1}{g(x + h)g(x)} \left(g(x) \lim_{h \to 0} \frac{f(x + h) - f(x)}{h} - \lim_{h \to 0} \frac{g(x + h) - g(x)}{h} f(x) \right) \right]$$

Using Limit Rules 4b and 4c on page 81

$$\underbrace{\phantom{\frac{1}{g(x+h)g(x)}}}_{\substack{\text{Approaches} \\ g(x)}} \qquad \underbrace{\phantom{g(x) \lim \frac{f(x+h)-f(x)}{h}}}_{f'(x)} \qquad \underbrace{\phantom{\lim \frac{g(x+h)-g(x)}{h}}}_{g'(x)}$$

$$= \frac{1}{[g(x)]^2} [g(x)f'(x) - g'(x)f(x)]$$

Using Limit Rules 1 and 4d on page 81

$$= \frac{g(x)f'(x) - g'(x)f(x)}{[g(x)]^2}$$

Quotient Rule

Verification of the Power Rule for Negative Integer Exponents

On pages 121–122 we proved the Power Rule for *positive* integer exponents. Using the Quotient Rule, we may now prove the Power Rule for *negative* integer exponents. Any negative integer n may be written as $n = -p$, where p is a *positive* integer. Then

$$\frac{d}{dx} x^n = \frac{d}{dx} \left(\frac{1}{x^p} \right)$$

Since $x^n = x^{-p} = \dfrac{1}{x^p}$

$$= \frac{x^p \cdot 0 - px^{p-1} \cdot 1}{x^{2p}}$$

Using the Quotient Rule, with $\dfrac{d}{dx} 1 = 0$ and $\dfrac{d}{dx} x^p = px^{p-1}$

$$= \frac{-px^{p-1}}{x^{2p}}$$

Simplifying

$$= -px^{\underbrace{p - 1 - 2p}_{-p-1}} = -px^{\overbrace{-p}^{n} \overbrace{-1}^{n-1}}$$

Subtracting exponents and simplifying

Since $-p = n$

$$= nx^{n-1}$$

Power Rule

This proves the Power Rule, $\dfrac{d}{dx} x^n = nx^{n-1}$, for negative integer exponents n.

➤ **Solutions to Practice Problems**

1. $\dfrac{d}{dx}[x^3(x^2 - x)] = 3x^2(x^2 - x) + x^3(2x - 1)$

$\qquad\qquad\qquad\quad = 3x^4 - 3x^3 + 2x^4 - x^3$

$\qquad\qquad\qquad\quad = 5x^4 - 4x^3$

2. $\dfrac{d}{dx}\left(\dfrac{2x^2}{x^2 + 1}\right) = \dfrac{(x^2 + 1)4x - 2x\cdot 2x^2}{(x^2 + 1)^2}$

$\qquad\qquad\qquad = \dfrac{4x^3 + 4x - 4x^3}{(x^2 + 1)^2} = \dfrac{4x}{(x^2 + 1)^2}$

2.4 Exercises

Find the derivative of each function in two ways:

   a. Using the *Product* Rule.
   b. Multiplying out the function and using the *Power* Rule.

Your answers to parts (a) and (b) should agree.

1. $x^4 \cdot x^6$ **2.** $x^7 \cdot x^2$

3. $x^4(x^5 + 1)$ **4.** $x^5(x^4 + 1)$

Find the derivative of each function by using the Product Rule.

5. $f(x) = x^2(x^3 + 1)$ **6.** $f(x) = x^3(x^2 + 1)$

7. $f(x) = x(5x^2 - 1)$ **8.** $f(x) = 2x(x^4 + 1)$

9. $f(x) = (x^2 + 1)(x^2 - 1)$

10. $f(x) = (x^3 - 1)(x^3 + 1)$

11. $f(x) = (x^2 + x)(3x + 1)$

12. $f(x) = (x^2 + 2x)(2x + 1)$

13. $f(x) = (\sqrt{x} - 1)(\sqrt{x} + 1)$

14. $f(x) = (\sqrt{x} + 2)(\sqrt{x} - 2)$

15. $f(t) = 6t^{4/3}(3t^{2/3} + 1)$

16. $f(t) = 4t^{3/2}(2t^{1/2} - 1)$

17. $f(z) = (z^4 + z^2 + 1)(z^3 - z)$

18. $f(z) = (\sqrt[4]{z} + \sqrt{z})(\sqrt[4]{z} - \sqrt{z})$

Find the derivative of each function in two ways:

   a. Using the *Quotient* rule.
   b. Simplifying the original function and using the *Power* Rule.

Your answers to parts (a) and (b) should agree.

19. $\dfrac{x^8}{x^2}$ **20.** $\dfrac{x^9}{x^3}$

21. $\dfrac{1}{x^3}$ **22.** $\dfrac{1}{x^4}$

Find the derivative of each function by using the Quotient Rule.

23. $f(x) = \dfrac{x^4 + 1}{x^3}$ **24.** $f(x) = \dfrac{x^5 - 1}{x^2}$

25. $f(x) = \dfrac{x + 1}{x - 1}$ **26.** $f(x) = \dfrac{x - 1}{x + 1}$

27. $f(t) = \dfrac{t^2 - 1}{t^2 + 1}$ **28.** $f(t) = \dfrac{t^2 + 1}{t^2 - 1}$

29. $f(s) = \dfrac{s^3 - 1}{s + 1}$ **30.** $f(s) = \dfrac{s^3 + 1}{s - 1}$

31. $f(x) = \dfrac{x^4 + x^2 + 1}{x^2 + 1}$

32. $f(x) = \dfrac{x^5 + x^3 + x}{x^3 + x}$

Differentiate each function by rewriting before and after differentiating, as on page 131.

Begin	Rewrite	Differentiate	Rewrite

33. $y = \dfrac{3}{x}$

34. $y = \dfrac{x^2}{4}$

35. $y - \dfrac{3x^4}{8}$

36. $y = \dfrac{3}{2x^2}$

37. PRODUCT RULE FOR THREE FUNCTIONS
Show that if f, g, and h are differentiable functions of x, then

$$\frac{d}{dx}(f \cdot g \cdot h) = f' \cdot g \cdot h + f \cdot g' \cdot h + f \cdot g \cdot h'$$

[*Hint:* Write the function as $f \cdot (g \cdot h)$ and apply the product rule twice.]

38. Derive the Quotient Rule from the Product Rule as follows.

a. Define the quotient to be a single function,

$$Q(x) = \frac{f(x)}{g(x)}.$$

b. Multiply both sides by $g(x)$ to obtain the equation $Q(x) \cdot g(x) = f(x)$.

c. Differentiate each side, using the Product Rule on the left side.   (*continues*)

d. Solve the resulting formula for the derivative $Q'(x)$.

e. Replace $Q(x)$ by $\dfrac{f(x)}{g(x)}$ and show that the resulting formula for $Q'(x)$ is the same as the Quotient Rule.

Note that in this derivation when we differentiated $Q(x)$ we *assumed* that the derivative of the quotient exists, whereas in the derivation on pages 138–139 we *proved* that the derivative exists.

39. Find a formula for $\dfrac{d}{dx}[f(x)]^2$ by writing it as $\dfrac{d}{dx}[f(x)f(x)]$ and using the Product Rule. Be sure to simplify your answer.

40. Find a formula for $\dfrac{d}{dx}[f(x)]^{-1}$ by writing it as $\dfrac{d}{dx}\left[\dfrac{1}{f(x)}\right]$ and using the Quotient Rule. Be sure to simplify your answer.

Find the derivative of each function.

41. $(x^3 + 2)\dfrac{x^2 + 1}{x + 1}$

42. $(x^5 + 1)\dfrac{x^3 + 2}{x + 1}$

43. $\dfrac{(x^2 + 3)(x^3 + 1)}{x^2 + 2}$

44. $\dfrac{(x^3 + 2)(x^2 + 2)}{x^3 + 1}$

45. $\dfrac{\sqrt{x} - 1}{\sqrt{x} + 1}$

46. $\dfrac{\sqrt{x} + 1}{\sqrt{x} - 1}$

APPLIED EXERCISES

47. ECONOMICS: Marginal Average Revenue
Use the Quotient Rule to find a general expression for the marginal average revenue. That is, calculate $\dfrac{d}{dx}\left[\dfrac{R(x)}{x}\right]$ and simplify your answer.

48. ECONOMICS: Marginal Average Profit
Use the Quotient Rule to find a general expression for the marginal average profit. That is, calculate $\dfrac{d}{dx}\left[\dfrac{P(x)}{x}\right]$ and simplify your answer.

49. ENVIRONMENTAL SCIENCE: Water Purification If the cost of purifying a gallon of water to a purity of x percent is

$$C(x) = \frac{100}{100 - x} \text{ cents} \quad \text{for } 50 \le x < 100$$

a. Find the instantaneous rate of change of the cost with respect to purity.

b. Evaluate this rate of change for a purity of 95% and interpret your answer.

c. Evaluate this rate of change for a purity of 98% and interpret your answer.

50. **BUSINESS: Marginal Average Cost** A toy company can produce plastic trucks at a cost of $8 each, while fixed costs are $1200 per day. Therefore, the company's cost function is $C(x) = 8x + 1200$.

    a. Find the average cost function
    $$AC(x) = \frac{C(x)}{x}.$$

    b. Find the marginal average cost function $MAC(x)$.

    c. Evaluate $MAC(x)$ at $x = 200$ and interpret your answer.

51. **ENVIRONMENTAL SCIENCE: Water Purification** (49 continued)

    a. Use a graphing calculator to graph the cost function $C(x)$ from Exercise 49 on the window [50, 100] by [0, 20]. TRACE along the curve to see how rapidly costs increase for purity (x-coordinate) increasing from 50 to near 100.

    b. To check your answers to Exercise 49, use the "dy/dx" or SLOPE feature of your calculator to find the slope of the cost curve at $x = 95$ and at $x = 98$. The resulting rates of change of the cost should agree with your answers to Exercise 49(b) and (c). Note that further purification becomes increasingly expensive at higher purity levels.

52. **BUSINESS: Marginal Average Cost** (50 continued)

    a. Graph the average cost function $AC(x)$ that you found in Exercise 50(a) on the window [0, 400] by [0, 50]. TRACE along the average cost curve to see how the average cost falls from the 20s down to the teens as the number of trucks increases. Note that although average cost falls, it does so more slowly as the number of units increases.

    b. To check your answer to Exercise 50, use the "dy/dx" or SLOPE feature of your calculator to find the slope of the average cost curve at $x = 200$. This slope gives the rate of change of the cost, which should agree with your answer to Exercise 50(c). Find the slope (rate of change) for other x-values to see that the rate of change of average cost tends toward zero (the law of diminishing returns).

53. **BIOMEDICAL: Beverton-Holt Recruitment Curve** Some organisms exhibit a density-dependent mortality from one generation to the next. Let $R > 1$ be the net reproductive rate (that is, the number of surviving offspring per parent), let $x > 0$ be the density of parents, and y be the density of surviving offspring. The *Beverton-Holt recruitment curve* is

$$y = \frac{Rx}{1 + \left(\dfrac{R-1}{K}\right)x}$$

where $K > 0$ is the *carrying capacity* of the organism's environment. Show that $\dfrac{dy}{dx} > 0$, and interpret this as a statement about the parents and the offspring.

54. **BIOMEDICAL: Murrell's Rest Allowance** Work-rest cycles for workers performing tasks that expend more than 5 kilocalories per minute (kcal/min) are often based on Murrell's formula

$$R(w) = \frac{w-5}{w-1.5} \qquad \text{for } w \geq 5$$

for the number of minutes $R(w)$ of rest for each minute of work expending w kcal/min. Show that $R'(w) > 0$ for $w \geq 5$ and interpret this fact as a statement about the additional amount of rest required for more strenuous tasks.

55. **BUSINESS: Marginal Average Profit** A company's profit function is $P(x) = 12x - 1800$ dollars where x is the number sold.

    a. Find the average profit function $AP(x) = P(x)/x$.

    b. Find the marginal average profit function $MAP(x)$.

    c. Evaluate $MAP(x)$ at $x = 300$ and interpret your answer.

56. **BUSINESS: Sales** The number of bottles of whiskey that a store will sell in a month at a price of p dollars per bottle is

$$N(p) = \frac{2250}{p+7} \qquad (p \geq 5)$$

Find the rate of change of this quantity when the price is $8 and interpret your answer.

57. GENERAL: Body Temperature If a person's temperature after x hours of strenuous exercise is $T(x) = x^3(4 - x^2) + 98.6$ degrees Fahrenheit (for $0 \le x \le 2$), find the rate of change of the temperature after 1 hour.

58. BUSINESS: CD Sales After x months, monthly sales of a compact disc are predicted to be $S(x) = x^2(8 - x^3)$ thousand (for $0 \le x \le 2$). Find the rate of change of the sales after 1 month.

59. GENERAL: Body Temperature (*57 continued*)

   a. Graph the temperature function $T(x)$ given in Exercise 57 on the window [0, 2] by [90, 110]. TRACE along the temperature curve to see how the temperature rises and then falls as time increases.

   b. To check your answer to Exercise 57, use the "dy/dx" or SLOPE feature of your calculator to find the slope (rate of change) of the curve at $x = 1$. Your answer should agree with your answer to Exercise 57.

   c. TRACE along the temperature curve to estimate the maximum temperature.

60. BUSINESS: CD Sales (*58 continued*)

   a. Graph the sales function $S(x)$ given in Exercise 58 on the window [0, 2] by [0, 12]. TRACE along the sales curve to see how the sales rise and then fall as x, the number of months, increases.

   b. To check your answer to Exercise 58, use the "dy/dx" or SLOPE feature of your calculator to find the slope (rate of change) of the curve at $x = 1$. Your answer should agree with your answer to Exercise 58.

   c. TRACE along the curve to estimate the maximum sales.

61. ECONOMICS: National Debt The following table gives the national debt (the amount of money that the federal government has borrowed from, and therefore owes to, its people), in millions of dollars, for the years 1970 to 2000. To avoid large numbers, years are listed in the table as years since 1970.

	Years Since 1970	National Debt (millions of dollars)
1970	0	370,100
1975	5	533,200
1980	10	907,700
1985	15	1,823,100
1990	20	3,233,300
1995	25	4,974,000
2000	30	5,674,200

Source: U.S. Treasury Department

   a. Enter these numbers into your graphing calculator and make a plot of the resulting points (Years Since 1970 on the x-axis and National Debt on the y-axis).

   b. Have your calculator find the quadratic regression formula for these data. Then enter the result in function y_1. Plot the points together with the regression curve. Observe that the curve fits the points quite well.

   c. The population of the United States (in millions) for these years is approximated by the linear function $y = 2.378x + 204.1$, where x is the number of years since 1970. Enter this function into your calculator in y_2.

   d. Use your calculator to define the function y_3 to be $y_1 \div y_2$, the national debt divided by the number of people, giving the "per capita national debt." Graph the function on the window [0, 40] by [0, 35,000] to show the per capita national debt for the years 1970 to 2010.

   e. TRACE along the curve y_3. The y-values represent the amount of money that the federal government owes to each one of its citizens in year x after 1970 (if the debt were equally distributed). Observe how rapidly the amount grows.

   f. Define function y_4 to be the derivative of y_3 (using the NDERIV feature of your calculator) and graph y_4 on the window [0, 40] by [0, 1500]. Use TABLE to evaluate both y_3 and y_4 at $x = 40$ and interpret your answers.

62–63. PITFALLS OF NDERIV ON A GRAPH-ING CALCULATOR

a. Find the derivative (by hand) of each function below, and observe that the derivative is undefined at $x = 0$.

b. Find the derivative of each function below by using NDERIV on a graphing calculator and evaluate the derivative at $x = 0$. If your cal-culator gives you an answer, this is a "false value" for the derivative, since in part (a) you showed that the derivative is undefined at $x = 0$. [For an explanation, see the Graphing Calculator Exploration part (d) on page 133.]

62. $y = \dfrac{1}{x}$ **63.** $y = \dfrac{1}{x^2}$

2.5 HIGHER-ORDER DERIVATIVES

Introduction

We have seen that from one function we can calculate a new function, the *derivative* of the original function. This new function, however, can itself be differentiated, giving what is called the *second derivative* of the original function. Differentiating again gives the *third derivative* of the original function, and so on. In this section we will calculate and interpret such higher-order derivatives.

Calculating Higher-Order Derivatives

EXAMPLE 1

FINDING HIGHER DERIVATIVES OF A POLYNOMIAL

From $f(x) = x^3 - 6x^2 + 2x - 7$ we may calculate

$$f'(x) = 3x^2 - 12x + 2 \qquad \text{"First" derivative of } f$$

Differentiating again gives

$$f''(x) = 6x - 12 \qquad \begin{array}{l}\text{Second derivative of } f,\\ \text{read "} f \text{ double prime"}\end{array}$$

and a third time:

$$f'''(x) = 6 \qquad \begin{array}{l}\text{Third derivative of } f,\\ \text{read "} f \text{ triple prime"}\end{array}$$

and a fourth time:

$$f''''(x) = 0 \qquad \begin{array}{l}\text{Fourth derivative of } f,\\ \text{read "} f \text{ quadruple prime"}\end{array}$$

All further derivatives of this function will, of course, be zero.

We also denote derivatives by replacing the primes by the number of differentiations in *parentheses*. For example, the fourth derivative may be denoted $f^{(4)}(x)$.

Practice Problem 1

If $f(x) = x^3 - x^2 + x - 1$, find:

a. $f'(x)$ **b.** $f''(x)$ **c.** $f'''(x)$ **d.** $f^{(4)}(x)$

➤ Solutions on page 154

While Example 1 showed that a polynomial can be differentiated "down to zero," the same is not true for all functions.

EXAMPLE 2

FINDING HIGHER DERIVATIVES OF A RATIONAL FUNCTION

Find the first five derivatives of $f(x) = \dfrac{1}{x}$.

Solution

$f(x) = x^{-1}$	$f(x)$ in power form
$f'(x) = -x^{-2}$	First derivative
$f''(x) = 2x^{-3}$	Second derivative
$f'''(x) = -6x^{-4}$	Third derivative
$f^{(4)}(x) = 24x^{-5}$	Fourth derivative
$f^{(5)}(x) = -120x^{-4}$	Fifth derivative

Clearly, we will never get to zero no matter how many times we differentiate.

Practice Problem 2

If $f(x) = 16x^{-1/2}$, find:

a. $f'(x)$ **b.** $f''(x)$

➤ Solutions on page 154

In Leibniz's notation, the second derivative $\dfrac{d}{dx}\dfrac{df}{dx}$ is written $\dfrac{d^2f}{dx^2}$.
The superscript goes after the d in the numerator and after the dx in the denominator. The following table shows equivalent statements in the two notations.

Prime Notation		*Leibniz's Notation*	
$f''(x)$	$=$	$\dfrac{d^2}{dx^2}f(x)$	Second derivative
y''	$=$	$\dfrac{d^2y}{dx^2}$	
$f'''(x)$	$=$	$\dfrac{d^3}{dx^3}f(x)$	Third derivative
y'''	$=$	$\dfrac{d^3y}{dx^3}$	
$f^{(n)}(x)$	$=$	$\dfrac{d^n}{dx^n}f(x)$	nth derivative
$y^{(n)}$	$=$	$\dfrac{d^ny}{dx^n}$	

Calculating higher derivatives merely requires repeated use of the same differentiation rules that we have been using.

EXAMPLE 3

FINDING A SECOND DERIVATIVE USING THE QUOTIENT RULE

Find $\dfrac{d^2}{dx^2}\left(\dfrac{x^2+1}{x}\right)$.

Solution

$$\frac{d}{dx}\left(\frac{x^2+1}{x}\right) = \frac{x(2x)-(x^2+1)}{x^2}$$ First derivative, using the Quotient Rule

$$= \frac{2x^2-x^2-1}{x^2} = \frac{x^2-1}{x^2}$$ Simplifying

Differentiating this answer gives the *second* derivative of the original:

$$\frac{d}{dx}\left(\frac{x^2-1}{x^2}\right) = \frac{x^2(2x)-2x(x^2-1)}{x^4}$$ Second derivative (derivative of the derivative)

$$= \frac{2x^3-2x^3+2x}{x^4} = \frac{2x}{x^4} = \frac{2}{x^3}$$ Simplifying

Answer: $\dfrac{d^2}{dx^2}\left(\dfrac{x^2+1}{x}\right) = \dfrac{2}{x^3}$

The function in this example was a quotient, so it was perhaps natural to use the Quotient Rule. It is easier, however, to simplify the original function first,

$$\frac{x^2+1}{x} = \frac{x^2}{x} + \frac{1}{x} = x + x^{-1}$$

and then differentiate by the Power Rule. So, the first derivative of $x + x^{-1}$ is $1 - x^{-2}$, and differentiating again gives $2x^{-3}$, agreeing with the answer found by the Quotient Rule. *Moral:* **Always simplify before differentiating.**

Practice Problem 3   Find $f''(x)$ if $f(x) = \dfrac{x+1}{x}$.

➤ Solution on page 154

EXAMPLE 4

EVALUATING A SECOND DERIVATIVE

If $f(x) = \dfrac{1}{\sqrt{x}}$, find $f''\left(\dfrac{1}{4}\right)$.

First differentiate, then evaluate

Solution

$$f(x) = x^{-1/2}$$ $f(x)$ in power form

$$f'(x) = -\frac{1}{2}x^{-3/2}$$ Differentiating once

$$f''(x) = \frac{3}{4}x^{-5/2}$$ Differentiating again

$$f''\left(\frac{1}{4}\right) = \frac{3}{4}\left(\frac{1}{4}\right)^{-5/2} = \frac{3}{4}(4)^{5/2}$$ Evaluating $f''(x)$ at $\frac{1}{4}$

$$= \frac{3}{4}(\sqrt{4})^5 = \frac{3}{4}(2)^5 = \frac{3}{4}(32) = 24$$

Practice Problem 4   Find $\dfrac{d^2}{dx^2}(x^4 + x^3 + 1)\Big|_{x=-1}$.

➤ Solution on page 154

Velocity and Acceleration

There is another important interpretation for the derivative, one that also gives a meaning to the *second* derivative. Imagine that you are driving along a straight road, and let $s(t)$ stand for your distance (in miles) from your starting point after t hours of driving. Then the derivative $s'(t)$ gives the instantaneous rate of change of distance with respect to time (miles per hour). However, "miles per hour" means speed or velocity, so the derivative of the *distance* function $s(t)$ is just the *velocity* function $v(t)$, giving your velocity at any time t.

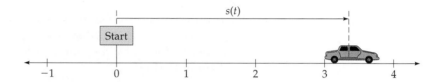

In general, for an object moving along a straight line, with distance measured from some fixed point, measured positively in one direction and negatively in the other (sometimes called "directed distance"),

$$\text{if} \qquad s(t) = \begin{pmatrix} \text{Distance} \\ \text{at time } t \end{pmatrix}$$

$$\text{then} \qquad s'(t) = \begin{pmatrix} \text{Velocity} \\ \text{at time } t \end{pmatrix}$$

Letting $v(t)$ stand for the velocity at time t, we may state this simply as:

$$v(t) = s'(t) \qquad \text{The velocity function is the derivative of the distance function}$$

The units of velocity come directly from the distance and time units of $s(t)$. For example, if distance is measured in feet and time in seconds, then the velocity is in *feet per second*, whereas if distance is in miles and time is in hours, then velocity is in *miles per hour*.

In everyday speech, the word "accelerating" means "speeding up." That is, acceleration means the rate of increase of speed, and since rates of increase are just derivatives, *acceleration is the derivative of velocity*. Since velocity is itself the derivative of distance, acceleration is the *second* derivative of distance. Letting $a(t)$ stand for the acceleration at time t, we have:

Distance, Velocity, and Acceleration

$$s(t) = \begin{pmatrix} \text{Distance} \\ \text{at time } t \end{pmatrix}$$

$v(t) = s'(t)$ Velocity is the derivative of distance

$a(t) = v'(t) = s''(t)$ Acceleration is the derivative of velocity, and the second derivative of distance

Therefore, we now have an interpretation for the *second* derivative: If $s(t)$ represents distance, then the *first* derivative represents velocity, and the *second* derivative represents *acceleration*. (In physics, there is even an interpretation for the third derivative, which gives the rate of change of acceleration: It is called the "jerk," since it is related to motion being "jerky."*)

EXAMPLE 5

FINDING AND INTERPRETING VELOCITY AND ACCELERATION

A delivery truck is driving along a straight road, and after t hours its distance (in miles) east of its starting point is

$$s(t) = 24t^2 - 4t^3 \qquad \text{for } 0 \leq t \leq 6$$

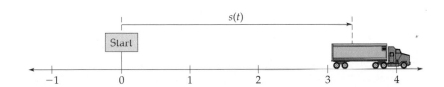

a. Find the velocity of the truck after 2 hours.
b. Find the velocity of the truck after 5 hours.
c. Find the acceleration of the truck after 1 hour.

Solution

a. To find velocity, we differentiate distance:

$v(t) = 48t - 12t^2$ Differentiating $s(t) = 24t^2 - 4t^3$

$v(2) = 48 \cdot 2 - 12 \cdot (2)^2$ Evaluating $v(t)$ at $t = 2$

$\quad\quad = 96 - 48 = 48$ miles per hour   Velocity after 2 hours

*See T. R. Sandlin, "The Jerk," *Physics Teacher* 28:36–40, January 1990.

b. At $t = 5$ hours:

$$v(5) = 48 \cdot 5 - 12 \cdot (5)^2 \qquad \text{Evaluating } v(t) \text{ at } t = 5$$
$$= 240 - 300 = -60 \text{ miles per hour} \qquad \text{Velocity after 5 hours}$$

What does the negative sign mean? Since distances are measured *eastward* (according to the original problem), the "positive" direction is east, so a negative velocity means a *westward* velocity. Therefore, at time $t = 5$ the truck is driving *westward at 60 miles per hour* (that is, back toward its starting point).

c. The acceleration is

$$a(t) = 48 - 24t \qquad \text{Differentiating } v(t) = 48t - 12t^2$$
$$a(1) = 48 - 24 = 24 \qquad \text{Acceleration after 1 hour}$$

Therefore, after 1 hour the acceleration of the truck is 24 mi/hr^2 (it is speeding up).

Graphing Calculator Exploration

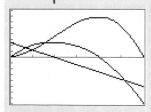

Distance, velocity, and
acceleration on [0, 6]
by [−150, 150].
Which is which?

Use a graphing calculator to graph the distance function $y_1 = 24x^2 - 4x^3$ (use x instead of t), the velocity function $y_2 = 48x - 12x^2$, and the acceleration function $y_3 = 48 - 24x$. (Alternatively, you could define y_2 and y_3 using NDERIV.) Your display should look like the one on the left. By looking at the graph, can you determine which curve represents distance, which represents velocity, and which represents acceleration? [*Hint:* Which curve gives the slope of which other curve?] Check your answer by using TRACE to identify functions 1, 2, and 3.

In Example 5, velocity was in miles per hour (mi/hr), so acceleration, the rate of change of velocity with respect to time, is in miles per hour *per hour,* written mi/hr^2. In general, the units of acceleration are distance/time^2.

Practice Problem 5

A helicopter rises vertically, and after t seconds its height above the ground is $s(t) = 6t^2 - t^3$ feet (for $0 \le t \le 6$).

a. Find its velocity after 2 seconds.

b. Find its velocity after 5 seconds.

c. Find its acceleration after 1 second.

[*Hint:* Distances are measured *upward,* so a negative velocity means downward.]   ➤ Solutions on page 154

Other Interpretations of Second Derivatives

The second derivative has other meanings besides acceleration. In general, second derivatives measure how the rate of change is itself changing. That is, if the first derivative measures the rate of growth, then the second derivative tells whether the growth is speeding up or slowing down.

EXAMPLE 6

PREDICTING POPULATION GROWTH

Demographers predict that t years from now the population of a city will be:

$$P(t) = 2,000,000 + 28,800t^{1/3}$$

Find $P'(8)$ and $P''(8)$ and interpret these numbers.

Solution The derivative is

$$P'(t) = 9600t^{-2/3} \qquad \text{Derivative of } P(t)$$

so

$$P'(8) = 9600(8)^{-2/3} = 9600 \cdot \frac{1}{4} = 2400 \qquad \text{Evaluating at } t = 8$$

Interpretation: Eight years from now the population will be growing at the rate of 2400 people per year.

The second derivative is

$$P''(t) = -6400t^{-5/3} \qquad \text{Derivative of } P'(t) = 9600t^{-2/3}$$
$$P''(8) = -6400(8)^{-5/3} \qquad \text{Evaluating at } t = 8$$

$$= -6400 \cdot \frac{1}{32} = -200$$

The fact that the first derivative is positive and the second derivative (the rate of change of the derivative) is negative means that the growth is continuing but more slowly.

Interpretation: After 8 years, the growth rate is *decreasing* by about 200 people per year each year. In other words, in the following year the population will continue to grow, but at the slower rate of about $2400 - 200 = 2200$ people per year.

Graphing Calculator Exploration

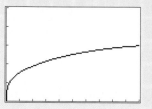

Use a graphing calculator to graph the population function $y_1 = 2,000,000 + 28,800x^{1/3}$ (using x instead of t). Your graph should resemble the one on the left. Can you see from the graph that the first derivative is positive (sloping upward), and that the second derivative is negative (slope decreasing)? You may check these facts numerically using NDERIV.

$y_1 = 2,000,000 + 28,800x^{1/3}$
on [0, 10] by
[2,000,000, 2,100,000]

ECONOMIC GROWTH SLOWED SHARPLY IN THIRD QUARTER

AN ANNUAL RATE OF 2.7%

Signs Point to Temporary Dip, but Strong Gains of Recent Years May Have Ended

---

By LOUIS UCHITELLE

The pace of economic activity slowed sharply in the summer quarter, the Commerce Department reported yester-

Statements about first and second derivatives occur frequently in everyday life, and may even make the front page of the newspaper. For example, the newspaper headline on the left (*New York Times*, October 28, 2000) says that the United States economy grew (the first derivative is positive) but more slowly (the second derivative is negative), following a curve like that shown below.

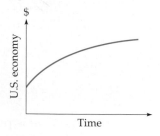

If y represents the size of the economy, then

$$\frac{dy}{dx} > 0 \quad \text{and} \quad \frac{d^2y}{dx^2} < 0$$

Practice Problem 6   The following headlines appeared recently in the *New York Times*. For each headline, sketch a curve representing the type of growth described and indicate the correct signs of the first and second derivatives.

a. Consumer Prices Rose in October at a Slower Rate

b. Households Still Shrinking, but Rate is Slower

➤ Solutions on page 155

Practice Problem 7   A recent mathematics publication* included the following statement: "In the fall of 1972 President Nixon announced that the rate of increase of inflation was decreasing. This was the first time a sitting president used the third derivative to advance his case for reelection." Explain why Nixon's announcement involved a third derivative.

➤ Solution on page 155

2.5 **Section Summary**

By simply repeating the process of differentiation we can calculate second, third, and higher derivatives. We also have another interpretation for the derivative, one that gives an interpretation for the second derivative as well. For distance measured along a straight line from some fixed point:

$$\text{If} \qquad s(t) = \textit{distance at time } t$$
$$\text{then} \qquad s'(t) = \textit{velocity at time } t$$
$$\text{and} \qquad s''(t) = \textit{acceleration at time } t.$$

Therefore, whenever you are driving along a straight road, your speedometer is the derivative of your odometer reading.

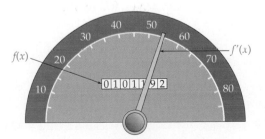

Velocity is the derivative of distance

*Notices of the American Mathematical Society, vol. 43, no. 10 (1996), page 1108.

We now have *four* interpretations for the derivative: instantaneous rate of change, slope, marginals, and velocity. It has been said that science is at its best when it unifies, and the derivative, unifying these four different concepts, is one of the most important ideas in all of science. We also saw that the second derivative, which measures the rate of change of the rate of change, can show whether growth is speeding up or slowing down.

Remember, however, that derivatives measure just what an automobile speedometer measures: the velocity at a particular *instant*. Although this statement may be obvious for velocities, it is easy to forget when dealing with marginals. For example, suppose that the marginal cost for a product is $15 when 100 units have been produced [which may be written $C'(100) = 15$]. Therefore, costs are increasing at the rate of $15 per additional unit, but only at the instant when $x = 100$. Although this may be used to *estimate* future costs (*about* $15 for each additional unit), it does not mean that one additional unit will increase costs by exactly $15, two more by exactly $30, and so on, since the marginal rate usually changes as production increases. A marginal cost is only an *approximate* predictor of future costs.

➤ **Solutions to Practice Problems**

1. a. $f'(x) = 3x^2 - 2x + 1$

   b. $f''(x) = 6x - 2$

   c. $f'''(x) = 6$

   d. $f^{(4)}(x) = 0$

2. a. $f'(x) = -8x^{-3/2}$

   b. $f''(x) = 12x^{-5/2}$

3. $f(x) = \dfrac{x}{x} + \dfrac{1}{x} = 1 + x^{-1}$ Simplifying first

   $f'(x) = -x^{-2}$

   $f''(x) = 2x^{-3}$

4. $\dfrac{d}{dx}(x^4 + x^3 + 1) = 4x^3 + 3x^2$

   $\dfrac{d^2}{dx^2}(x^4 + x^3 + 1) = \dfrac{d}{dx}(4x^3 + 3x^2) = 12x^2 + 6x$

   $(12x^2 + 6x)\,\big|_{x=-1} = 12 - 6 = 6$

5. a. $v(t) = 12t - 3t^2$

   $v(2) = 24 - 12 = 12 = 12$ ft/sec

   b. $v(5) = 60 - 75 = -15$ ft/sec or 15 ft/sec *downward*

   c. $a(t) = 12 - 6t$

   $a(1) = 12 - 6 = 6$ ft/sec^2

6. a. **b.**

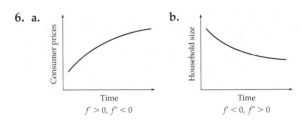

$f' > 0, f'' < 0$ $f' < 0, f'' > 0$

7. Inflation is itself a derivative since it is the rate of change of the consumer price index. Therefore its growth rate is a second derivative, and the slowing of this growth would be a third derivative.

2.5 Exercises

For each function, find:

    a. $f'(x)$ **b.** $f''(x)$ **c.** $f'''(x)$ **d.** $f^{(4)}(x)$

1. $f(x) = x^4 - 2x^3 - 3x^2 + 5x - 7$

2. $f(x) = x^4 - 3x^3 + 2x^2 - 8x + 4$

3. $f(x) = 1 + x + \frac{1}{2}x^2 + \frac{1}{6}x^3 + \frac{1}{24}x^4 + \frac{1}{120}x^5$

4. $f(x) = 1 + x + \frac{1}{2}x^2 + \frac{1}{6}x^3 + \frac{1}{24}x^4$

5. $f(x) = \sqrt{x^5}$ **6.** $f(x) = \sqrt{x^3}$

For each function, find **a.** $f''(x)$ and **b.** $f''(3)$.

7. $f(x) = \dfrac{x-1}{x}$ **8.** $f(x) = \dfrac{x+2}{x}$

9. $f(x) = \dfrac{x+1}{2x}$ **10.** $f(x) = \dfrac{x-2}{4x}$

11. $f(x) = \dfrac{1}{6x^2}$ **12.** $f(x) = \dfrac{1}{12x^3}$

Find the *second* derivative of each function.

13. $f(x) = (x^2 - 2)(x^2 + 3)$

14. $f(x) = (x^2 - 1)(x^2 + 2)$

15. $f(x) = \dfrac{27}{\sqrt[3]{x}}$ **16.** $f(x) = \dfrac{32}{\sqrt[4]{x}}$

17. $f(x) = \dfrac{x}{x-1}$ **18.** $f(x) = \dfrac{x}{x-2}$

Evaluate each expression.

19. $\dfrac{d^2}{dr^2}(\pi r^2)$ **20.** $\dfrac{d^3}{dr^3}\left(\dfrac{4}{3}\pi r^3\right)$

21. $\dfrac{d^2}{dx^2}x^{10}\bigg|_{x=-1}$ **22.** $\dfrac{d^2}{dx^2}x^{11}\bigg|_{x=-1}$

23. $\dfrac{d^3}{dx^3}x^{10}\bigg|_{x=-1}$ **24.** $\dfrac{d^3}{dx^3}x^{11}\bigg|_{x=-1}$

25. $\dfrac{d^2}{dx^2}\sqrt{x^3}\bigg|_{x=1/16}$ **26.** $\dfrac{d^2}{dx^2}\sqrt[3]{x^4}\bigg|_{x-1/27}$

27. GENERAL: Velocity Each of the following three "stories," labeled **a**, **b**, and **c**, matches one of the velocity graphs, labeled (i), (ii), and (iii). For each story, choose the most appropriate graph.

    a. I left my home and drove to meet a friend, but I got stopped for a speeding ticket. Afterward I drove on more slowly.

    b. I started driving but then stopped to look at the map. Realizing that I was going the wrong way, I drove back the other way.

    c. After driving for a while I got into some stop-and-go driving. Once past the tie-up I could speed up again.

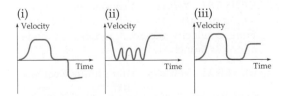

28. BUSINESS: Profit Each of the following three descriptions of a company's profit over time, labeled **a, b,** and **c,** matches one of the graphs, labeled (i), (ii), and (iii). For each description, choose the most appropriate graph.

a. Profits were growing increasingly rapidly.
b. Profits were declining but the rate of decline was slowing.
c. Profits were rising, but more and more slowly.

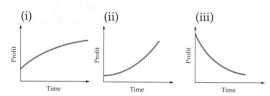

29. Find $\dfrac{d^{100}}{dx^{100}}(x^{99} - 4x^{98} + 3x^{50} + 6)$.

[*Hint:* No calculation is necessary. Think of what happens when an nth-degree polynomial is differentiated $n + 1$ times. For example, try differentiating x^3 four times.]

30. Find a general formula for $\dfrac{d^n}{dx^n} x^{-1}$.

[*Hint:* Calculate the first few derivatives and look for a pattern. You may use the "factorial" notation: $n! = n(n-1) \cdots 1$. For example, $3! = 3 \cdot 2 \cdot 1 = 6$.]

31. Verify the following formula for the *second* derivative of a product, where f and g are differentiable functions of x:

$$\frac{d^2}{dx^2}(f \cdot g) = f'' \cdot g + 2f' \cdot g' + f \cdot g''$$

[*Hint:* Use the Product Rule repeatedly.]

32. Verify the following formula for the *third* derivative of a product, where f and g are differentiable functions of x:

$$\frac{d^3}{dx^3}(f \cdot g) = f''' \cdot g + 3f'' \cdot g' + 3f' \cdot g'' + f \cdot g'''$$

[*Hint:* Differentiate the formula in Exercise 31 by the Product Rule.]

APPLIED EXERCISES

33. GENERAL: Velocity After t hours a freight train is $s(t) = 18t^2 - 2t^3$ miles due north of its starting point (for $0 \le t \le 9$).

a. Find its velocity at time $t = 3$ hours.
b. Find its velocity at time $t = 7$ hours.
c. Find its acceleration at time $t = 1$ hour.

34. GENERAL: Velocity After t hours a passenger train is $s(t) = 24t^2 - 2t^3$ miles due west of its starting point (for $0 \le t \le 12$).

a. Find its velocity at time $t = 4$ hours.
b. Find its velocity at time $t = 10$ hours.
c. Find its acceleration at time $t = 1$ hour.

35. GENERAL: Velocity A rocket can rise to a height of $h(t) = t^3 + 0.5t^2$ feet in t seconds. Find its velocity and acceleration 10 seconds after it is launched.

36. GENERAL: Velocity After t hours a car is a distance $s(t) = 60t + \dfrac{100}{t+3}$ miles from its starting point. Find the velocity after 2 hours.

37. GENERAL: Impact Velocity A penny dropped from a building will fall a distance $s(t) = 16t^2$ feet in t seconds (neglecting air resistance).

a. With what velocity will it hit the ground if it does so 5 seconds after it is dropped? (This is called the *impact velocity*.)
b. Find the acceleration at any time t. (This number is called the *acceleration due to gravity*.)

38. GENERAL: Impact Velocity If a marble is dropped from the top of the Sears Tower in Chicago, its height above the ground t seconds after it is dropped will be $s(t) = 1454 - 16t^2$ feet (neglecting air resistance).

a. How long will it take to reach the ground? [*Hint:* Find when the height equals zero.]
b. Use your answer to part (a) to find the velocity with which it will strike the ground.

39. GENERAL: Maximum Height If a bullet from a 9-millimeter pistol is fired straight up from the ground, its height t seconds after it is fired will be $s(t) = -16t^2 + 1280t$ feet (neglecting air resistance) for $0 \le t \le 80$.

a. Find the velocity function.

b. Find the time t when the bullet will be at its maximum height. [*Hint:* At its maximum height the bullet is moving neither up nor down, and has velocity zero. Therefore, find the time when the velocity $v(t)$ equals zero.]

c. Find the maximum height the bullet will reach. [*Hint:* Use the time found in part (b) together with the height function $s(t)$.]

40. BIOMEDICAL: Fever The temperature of a patient t hours after taking a fever reducing medicine is $T(t) = 98 + 8/\sqrt{t}$ degrees Fahrenheit. Find $T(2)$, $T'(2)$, and $T''(2)$, and interpret these numbers.

41. ECONOMICS: National Debt The national debt of a South American country t years from now is predicted to be $D(t) = 65 + 9t^{4/3}$ billion dollars. Find $D'(8)$ and $D''(8)$ and interpret your answers.

42. ENVIRONMENTAL SCIENCE: Earth Temperature If the average temperature of the earth t years from now is predicted to be $T(t) = 65 - 4/t$ (for $t \ge 8$), find $T'(10)$ and $T''(10)$ and interpret your answers.

43–44. ENVIRONMENTAL SCIENCE: Sea Level The burning of fossil fuels (such as coal and oil) generates carbon dioxide, which traps heat in the atmosphere, thereby increasing the temperature of the earth (the "greenhouse effect").

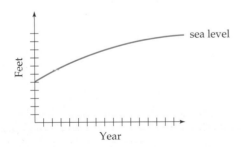

The higher temperature in turn melts the polar icecaps, raising the sea level. The precise results are very difficult to predict, but suppose the following:

43. In t years the average sea level on the East Coast of the United States will be $L(t) = 30 - 4t^{-1/2}$ feet (for $t > 2$). Find $L'(4)$ and $L''(4)$ and interpret your answers.

44. In t years the average sea level on the West Coast of the United States will be $L(t) = 35 - 8t^{-1/2}$ feet (for $t > 2$). Find $L'(4)$ and $L''(4)$ and interpret your answers.

45. BUSINESS: Profit The annual profit of the Digitronics company x years from now is predicted to be $P(x) = 5.27x^{0.3} - 0.463x^{1.52}$ million dollars (for $0 \le x \le 8$). Evaluate the profit function and its first and second derivatives at $x = 3$ and interpret your answers. [*Hint:* Enter the given function in y_1, define y_2 to be the derivative of y_1 (using NDERIV), and define y_3 to be the derivative of y_2. Then evaluate each at the stated x-value.]

46. GENERAL: Population The population of a country x years from now is predicted to be $P(x) = 3.17x^{1.3} + 0.192x^{0.74}$ million people (for $0 \le x \le 10$). Evaluate the population function and its first and second derivatives at $x = 3.75$ and interpret your answers. [See the hint in Exercise 45.]

47. GENERAL: Windchill Index The windchill index (revised in 2001) for a temperature of 32 degrees Fahrenheit and wind speed x miles per hour is

$$W(x) = 55.628 - 22.07x^{0.16}$$

a. Graph the windchill index on a graphing calculator using the window [0, 50] by [0, 40]. Then find the windchill index for wind speeds of $x = 15$ and $x = 30$ mph.

b. Notice from your graph that the windchill index has first derivative negative and second derivative positive. What does this mean about how successive 1-mph increases in wind speed affect the windchill index?

c. Verify your answer to part (b) by defining y_2 to be the derivative of y_1 (using NDERIV), evaluating it at $x = 15$ and $x = 30$, and interpreting your answers.

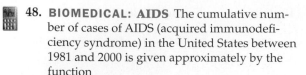

48. BIOMEDICAL: AIDS The cumulative number of cases of AIDS (acquired immunodeficiency syndrome) in the United States between 1981 and 2000 is given approximately by the function

$$f(x) = -0.0182x^4 + 0.526x^3 - 1.3x^2 + 1.3x + 5.4$$

in thousands of cases, where x is the number of years since 1980.

a. Graph this function on your graphing calculator on the window $[1, 20]$ by $[0, 800]$. Notice that at some time in the 1990s the rate of growth began to slow.

b. Find when the AIDS epidemic began to slow. [*Hint:* Find where the second deriva-

tive of $f(x)$ is zero, and then convert the x-value to a year.]

(*Source:* Centers for Disease Control)

Find the second derivative of each function.

49. $(x^2 - x + 1)(x^3 - 1)$

50. $(x^3 + x - 1)(x^3 + 1)$

51. $\dfrac{x}{x^2 + 1}$

52. $\dfrac{x}{x^2 - 1}$

53. $\dfrac{2x - 1}{2x + 1}$

54. $\dfrac{3x + 1}{3x - 1}$

2.6 THE CHAIN RULE AND THE GENERALIZED POWER RULE

Introduction

In this section we will learn the last of the general rules of differentiation, the *Chain Rule* for differentiating composite functions. We will then prove a very useful special case of it, the *Generalized Power Rule* for differentiating powers of functions. We begin by reviewing composite functions.

Composite Functions

As we saw on page 56, *composite* functions are simply functions of functions: The composition of f with g evaluated at x is $f(g(x))$.

EXAMPLE 1

FINDING A COMPOSITE FUNCTION

For $f(x) = x^2$ and $g(x) = 4 - x$, find $f(g(x))$.

Solution

$$f(g(x)) = (4 - x)^2 \qquad \begin{array}{l} f(x) = x^2 \text{ with } x \\ \text{replaced by } g(x) = 4 - x \end{array}$$

Practice Problem 1 For the same $f(x) = x^2$ and $g(x) = 4 - x$, find $g(f(x))$.

➤ Solution on page 168

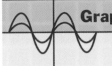

Graphing Calculator Exploration

Use a graphing calculator to verify that the compositions $f(g(x))$ and $g(f(x))$ above are different.

a. Enter $y_1 = x^2$ and $y_2 = 4 - x$.

b. Then define y_3 and y_4 to be the compositions in the two orders [on some calculators this is done by defining $y_3 = y_1(y_2)$ and $y_4 = y_2(y_1)$].

c. Graph y_3 and y_4 (but turn "off" y_1 and y_2) on the standard window and notice that the graphs are very different.

Besides building compositions out of simpler functions, we can also *de*compose functions into compositions of simpler functions.

EXAMPLE 2

DECOMPOSING A COMPOSITE FUNCTION

Find functions $f(x)$ and $g(x)$ such that $(x^2 + 1)^5$ is the composition $f(g(x))$.

Solution

Think of $(x^2 + 1)^5$ as an inside function $x^2 + 1$ followed by an outside operation $(\quad)^5$. We match the "inside" and "outside" parts of $(x^2 + 1)^5$ and $f(g(x))$.

$$
\overset{\text{Outside function}}{(x^2 + 1)^5} \; = \; f(g(x))
$$

Inside function

Therefore, $(x^2 + 1)^5$ can be written as $f(g(x))$ with $\begin{cases} f(x) = x^5 \\ g(x) = x^2 + 1 \end{cases}$

(Other answers are possible.)

Note that expressing a function as a composition involves thinking of the function in terms of "blocks," an inside block that starts a calculation and an outside block that completes it.

Practice Problem 2   Find $f(x)$ and $g(x)$ such that $\sqrt{x^5 - 7x + 1}$ is the composition $f(g(x))$.

➤ Solution on page 168

The Chain Rule

If we were asked to differentiate the function $(x^2 - 5x + 1)^{10}$, we could first multiply together ten copies of $x^2 - 5x + 1$ (certainly a time-consuming, tedious, and error-prone process), and then differentiate the resulting polynomial. There is, however, a much easier way, using the *Chain Rule*, which shows how to differentiate a composite function of the form $f(g(x))$.

Chain Rule

$$\frac{d}{dx} f(g(x)) = f'(g(x)) \cdot g'(x)$$

To differentiate $f(g(x))$, differentiate $f(x)$, then replace each x by $g(x)$, and finally multiply by the derivative of $g(x)$.

(provided that the derivatives on the right-hand side of the equation exist). The name comes from thinking of compositions as "chains" of functions. A verification of the Chain Rule is given at the end of this section.

EXAMPLE 3

DIFFERENTIATING USING THE CHAIN RULE

Use the Chain Rule to find $\dfrac{d}{dx}(x^2 - 5x + 1)^{10}$.

Solution

$(x^2 - 5x + 1)^{10}$ is $f(g(x))$ with $\begin{cases} f(x) = x^{10} & \text{Outside function} \\ g(x) = x^2 - 5x + 1 & \text{Inside function} \end{cases}$

Since $f'(x) = 10x^9$, we have

$$f'(g(x)) = 10(g(x))^9 \qquad \begin{array}{l} f'(x) = 10x^9 \ \text{with } x \\ \text{replaced by} \ g(x) \end{array}$$

$$= 10(x^2 - 5x + 1)^9 \qquad \text{Using } g(x) = x^2 - 5x + 1$$

Substituting this last expression into the Chain Rule gives:

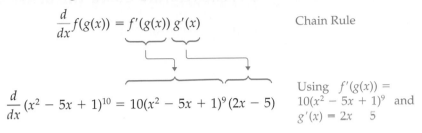

$$\frac{d}{dx} f(g(x)) = f'(g(x))\, g'(x) \qquad \text{Chain Rule}$$

$$\frac{d}{dx}(x^2 - 5x + 1)^{10} = 10(x^2 - 5x + 1)^9 (2x - 5)$$

Using $f'(g(x)) = 10(x^2 - 5x + 1)^9$ and $g'(x) = 2x - 5$

This result says that to differentiate $(x^2 - 5x + 1)^{10}$, we bring down the exponent 10, reduce the exponent to 9 (steps familiar from the Power Rule), and finally multiply by the derivative of the inside function.

$$\frac{d}{dx}(x^2 - 5x + 1)^{10} = 10(x^2 - 5x + 1)^9 (2x - 5)$$

| Inside function | Bring down the power n | Power $n-1$ | Derivative of the inside function |

Generalized Power Rule

Example 3 suggests a general rule for differentiating a function to a power.

Generalized Power Rule

$$\frac{d}{dx}[g(x)]^n = n \cdot [g(x)]^{n-1} \cdot g'(x)$$

To differentiate a function to a power, bring down the power as a multiplier, reduce the exponent by 1, and then multiply by the derivative of the inside function

(provided, of course, that the derivative $g'(x)$ exists). The Generalized Power Rule follows from the Chain Rule by reasoning similar to that of Example 3: The derivative of $f(x) = x^n$ is $f'(x) = nx^{n-1}$, so

$$\frac{d}{dx} f(g(x)) = f'(g(x))\, g'(x) \qquad \text{Chain Rule}$$

gives

$$\frac{d}{dx}[g(x)]^n = n[g(x)]^{n-1} g'(x) \qquad \text{Generalized Power Rule}$$

EXAMPLE 4

DIFFERENTIATING USING THE GENERALIZED POWER RULE

Find $\dfrac{d}{dx}\sqrt{x^4 - 3x^3 - 4}$.

Solution

$$\frac{d}{dx}(x^4 - 3x^3 - 4)^{1/2} = \frac{1}{2}(x^4 - 3x^3 - 4)^{-1/2}(4x^3 - 9x^2)$$

Inside function   Bring down the n Power $n-1$ Derivative of the inside function

Think of the Generalized Power Rule "from the outside in." That is, first bring down the outer exponent and reduce the exponent by 1, and only then multiply by the derivative of the inside function.

Be careful! It is the *original function* (not the differentiated function) that is raised to the power $n - 1$. Only at the end do you multiply by the derivative of the inside function.

EXAMPLE 5

SIMPLIFYING AND DIFFERENTIATING

Find $\dfrac{d}{dx}\left(\dfrac{1}{x^2 + 1}\right)^3$.

Solution   Writing the function as $(x^2 + 1)^{-3}$ gives

$$\frac{d}{dx}(x^2 + 1)^{-3} = -3(x^2 + 1)^{-4}(2x) = -6x(x^2 + 1)^{-4} = -\frac{6x}{(x^2 + 1)^4}$$

Inside function   Bring down the n Power $n-1$ Derivative of the inside function

Practice Problem 3   Find $\dfrac{d}{dx}(x^3 - x)^{-1/2}$.   ➤ Solution on page 168

EXAMPLE 6

FINDING THE GROWTH RATE OF AN OIL SLICK

An oil tanker hits a reef, and after t days the radius of the oil slick is $r(t) = \sqrt{4t + 1}$ miles. How fast is the radius of the oil slick expanding after 2 days?

Solution   To find the rate of change of the radius, we differentiate:

$$\frac{d}{dt}(4t + 1)^{1/2} = \frac{1}{2}(4t + 1)^{-1/2}(4) = 2(4t + 1)^{-1/2}$$

└──── Derivative of $4t + 1$

At $t = 2$ this is

$$2(4 \cdot 2 + 1)^{-1/2} = 2 \cdot 9^{-1/2} = 2 \cdot \frac{1}{3} = \frac{2}{3}$$

Interpretation: After 2 days the radius of the oil slick is growing at the rate of $\frac{2}{3}$ of a mile per day.

Graphing Calculator Exploration

a. On a graphing calculator, enter the radius of the oil slick as $y_1 = \sqrt{4x + 1}$. Then use NDERIV to define y_2 to be the derivative of y_1. Graph both y_1 and y_2 on the window $[0, 6]$ by $[0, 5]$.

b. Verify your answer to Example 6 by evaluating y_2 at $x = 2$. Do you get $\frac{2}{3}$?

c. Notice from your graph that the (derivative) function y_2 is *decreasing*, meaning that the radius is growing more *slowly*. Estimate when the radius will be growing by only $\frac{1}{2}$ mile per day. [*Hint:* One way is to use TRACE to follow along the curve y_2 to the x-value (number of days) when the y-coordinate is 0.5, zooming in if necessary. Your answer should be between 3 and 4 days. You can also use INTERSECT with $y_3 = 0.5$.

Some problems require the Generalized Power Rule in combination with another differentiation rule, such as the Product or Quotient Rule.

EXAMPLE 7

DIFFERENTIATING USING TWO RULES

Find $\dfrac{d}{dx}[(5x - 2)^4(9x + 2)^7]$.

Solution Since this is a product of powers, $(5x - 2)^4$ times $(9x + 2)^7$, we use the Product Rule together with the Generalized Power Rule.

$$\dfrac{d}{dx}[(5x - 2)^4(9x + 2)^7] = \underbrace{4(5x - 2)^3(5)}(9x + 2)^7 + (5x - 2)^4\underbrace{[7(9x + 2)^6(9)]}$$

$$\text{Derivative of} \qquad\qquad\qquad\qquad \text{Derivative of}$$
$$(5x - 2)^4 \qquad\qquad\qquad\qquad\qquad (9x + 2)^7$$

$$= 20\underset{4 \cdot 5}{(5x - 2)^3}(9x + 2)^7 + 63\underset{7 \cdot 9}{(5x - 2)^4}(9x + 2)^6 \qquad \text{Simplifying}$$

EXAMPLE 8

DIFFERENTIATING USING TWO RULES

Find $\dfrac{d}{dx}\left(\dfrac{x}{x + 1}\right)^4$.

Solution Since the function is a quotient raised to a power, we use the Quotient Rule together with the Generalized Power Rule. Working from the outside in, we obtain

$$\dfrac{d}{dx}\left(\dfrac{x}{x + 1}\right)^4 = 4\left(\dfrac{x}{x + 1}\right)^3 \underbrace{\dfrac{(x + 1)(1) - (1)(x)}{(x + 1)^2}}$$

$$\text{Derivative of the}$$
$$\text{inside function } \dfrac{x}{x + 1}$$

$$= 4\left(\dfrac{x}{x + 1}\right)^3\dfrac{x + 1 - x}{(x + 1)^2} = 4\left(\dfrac{x}{x + 1}\right)^3\dfrac{1}{(x + 1)^2} \qquad \text{Simplifying}$$

$$= 4\dfrac{x^3}{(x + 1)^3}\dfrac{1}{(x + 1)^2} = \dfrac{4x^3}{(x + 1)^5} \qquad \text{Simplifying further}$$

EXAMPLE 9

DIFFERENTIATING USING A RULE TWICE

Find $\dfrac{d}{dz}[z^2 + (z^2 - 1)^3]^5$.

Solution   Since this is a function to a power, where the inside function also contains a function to a power, we must use the Generalized Power Rule *twice*.

$$\frac{d}{dz}[z^2 + (z^2 - 1)^3]^5 = 5[z^2 + (z^2 - 1)^3]^4\underbrace{[2z + 3(z^2 - 1)^2(2z)]}_{\substack{\text{Derivative of} \\ z^2 + (z^2 - 1)^3}}$$

$$= 5[z^2 + (z^2 - 1)^3]^4[2z + \underset{\underset{3 \cdot 2}{\uparrow}}{6}z(z^2 - 1)^2]$$ Simplifying

Chain Rule in Leibniz's Notation

A composition may be written in two parts:

$y = f(g(x))$ is equivalent to $y = f(u)$ and $u = g(x)$

The derivatives of these last two functions are:

$$\frac{dy}{du} = f'(u) \quad \text{and} \quad \frac{du}{dx} = g'(x)$$

The Chain Rule

$$\frac{d}{dx}\underbrace{f(g(x))}_{y} = f'(g(x)) \cdot g'(x)$$ Chain Rule

$$\underbrace{\phantom{\frac{d}{dx}f(g(x))}}_{\dfrac{dy}{dx}} \quad \underbrace{}_{\dfrac{dy}{du}} \underbrace{}_{\dfrac{du}{dx}}$$

with the indicated substitutions then becomes:

Chain Rule in Leibniz's Notation

For $y = f(u)$ with $u = g(x)$,

$$\frac{dy}{dx} = \frac{dy}{du} \cdot \frac{du}{dx}$$

In this form the Chain Rule is easy to remember, since it looks as if the *du* in the numerator and the denominator cancel:

$$\frac{dy}{dx} = \frac{dy}{du} \cdot \frac{du}{dx}$$

However, since derivatives are not really fractions (they are *limits* of fractions), this is only a convenient device for remembering the Chain Rule.

The Product Rule (page 127) showed that the derivative of a product is not the product of the derivatives. We now see where the product of the derivatives *does* appear: it appears in the Chain Rule, when differentiating composite functions. In other words, the product of the derivatives comes not from *products* but from *compositions* of functions.

A Simple Example of the Chain Rule

The derivation of the Chain Rule is rather technical, but we can show the basic idea in a simple example. Suppose that your company produces steel, and you want to calculate your company's total revenue in dollars per year. You would take the revenue from a ton of steel (dollars per ton) and multiply by your company's output (tons per year). In symbols:

$$\frac{\$}{\text{year}} = \frac{\$}{\text{ton}} \cdot \frac{\text{ton}}{\text{year}} \qquad \text{Note that ``ton'' cancels}$$

If we were to express these rates as derivatives, the equation above would become the Chain Rule.

2.6 Section Summary

To differentiate a composite function (a function of a function), we have the Chain Rule:

$$\frac{d}{dx}f(g(x)) = f'(g(x)) \cdot g'(x)$$

or, in Leibniz's notation, writing $y = f(g(x))$ as $y = f(u)$ and $u = g(x)$,

$$\frac{dy}{dx} = \frac{dy}{du} \cdot \frac{du}{dx} \qquad \text{The derivative of a composite function is the product of the derivatives}$$

To differentiate a function to a power, $[f(x)]^n$, we have the *Generalized Power Rule* (a special case of the Chain Rule when the "outer" function is a power):

$$\frac{d}{dx}[f(x)]^n = n \cdot [f(x)]^{n-1} \cdot f'(x)$$

For now, the Generalized Power Rule is more useful than the Chain Rule, but in Chapter 5 we will make important use of the Chain Rule.

Verification of the Chain Rule

Let $f(x)$ and $g(x)$ be differentiable functions. We define k by

$$k = g(x + h) - g(x)$$

or, equivalently,

$$g(x + h) = g(x) + k$$

Then

$$\lim_{h \to 0} \underbrace{[g(x + h) - g(x)]}_{k} = \lim_{h \to 0} k = 0$$

showing that $h \to 0$ implies $k \to 0$ (see pages 137–138). With these relations we may calculate the derivative of the composition $f(g(x))$.

$\dfrac{d}{dx} f(g(x)) = \lim\limits_{h \to 0} \dfrac{f(g(x + h)) - f(g(x))}{h}$ Definition of the derivative of $f(g(x))$

$= \lim\limits_{h \to 0} \left[\dfrac{f(g(x + h)) - f(g(x))}{g(x + h) - g(x)} \cdot \dfrac{g(x + h) - g(x)}{h} \right]$ Dividing and multiplying by $g(x + h) - g(x)$

$= \lim\limits_{h \to 0} \dfrac{f(g(x + h)) - f(g(x))}{g(x + h) - g(x)} \lim\limits_{h \to 0} \dfrac{g(x + h) - g(x)}{h}$ The limit of a product is the product of the limits (Limit Rule 4c on page 81)

$= \underbrace{\lim\limits_{k \to 0} \dfrac{f(g(x) + k) - f(g(x))}{k}}_{f'(g(x))} \underbrace{\lim\limits_{h \to 0} \dfrac{g(x + h) - g(x)}{h}}_{g'(x)}$ Using the relations $g(x + h) = g(x) + k$ and $k = g(x + h) - g(x)$ and that $h \to 0$ implies $k \to 0$

$= f'(g(x)) \cdot g'(x)$ Chain Rule

The last step comes from recognizing the first limit as the definition of the derivative f' at $g(x)$, and the second limit as the definition of the derivative $g'(x)$. This verifies the Chain Rule,

$$\frac{d}{dx} f(g(x)) = f'(g(x)) \cdot g'(x)$$

Strictly speaking, this verification requires an additional assumption, that the denominator $g(x + h) - g(x)$ is never zero. This assumption can be avoided by a more technical argument, which we omit.

➤ **Solutions to Practice Problems**

1. $g(f(x)) = 4 - f(x) = 4 - x^2$
2. $f(x) = \sqrt{x}, \quad g(x) = x^5 - 7x + 1$
3. $\dfrac{d}{dx}(x^3 - x)^{-1/2} = -\dfrac{1}{2}(x^3 - x)^{-3/2}(3x^2 - 1)$

2.6 Exercises

Find functions f and g such that the given function is the composition $f(g(x))$.

1. $\sqrt{x^2 - 3x + 1}$ **2.** $(5x^2 - x + 2)^4$

3. $(x^2 - x)^{-3}$ **4.** $\dfrac{1}{x^2 + x}$ **5.** $\dfrac{x^3 + 1}{x^3 - 1}$

6. $\dfrac{\sqrt{x} - 1}{\sqrt{x} + 1}$ **7.** $\left(\dfrac{x + 1}{x - 1}\right)^4$ **8.** $\sqrt{\dfrac{x - 1}{x + 1}}$

9. $\sqrt{x^2 - 9} + 5$ **10.** $\sqrt[3]{x^3 + 8} - 5$

Use the Generalized Power Rule to find the derivative of each function.

11. $f(x) = (x^2 + 1)^3$ **12.** $f(x) = (x^3 + 1)^4$

13. $h(z) = (3z^2 - 5z + 2)^4$

14. $h(z) = (5z^2 + 3z - 1)^3$

15. $f(x) = \sqrt{x^4 - 5x + 1}$ **16.** $f(x) = \sqrt{x^6 + 3x - 1}$

17. $w(z) = \sqrt[3]{9z - 1}$ **18.** $w(z) = \sqrt[5]{10z - 4}$

19. $y = (4 - x^2)^4$ **20.** $y = (1 - x)^{50}$

21. $y = \left(\dfrac{1}{w^3 - 1}\right)^4$ **22.** $y = \left(\dfrac{1}{w^4 + 1}\right)^5$

23. $y = x^4 + (1 - x)^4$

24. $f(x) = (x^2 + 4)^3 - (x^2 + 4)^2$

25. $f(x) = \dfrac{1}{\sqrt[3]{(9x + 1)^2}}$ **26.** $f(x) = \dfrac{1}{\sqrt[3]{(3x - 1)^2}}$

27. $f(x) = [(x^2 + 1)^3 + x]^3$

28. $f(x) = [(x^3 + 1)^2 - x]^4$

29. $f(x) = 3x^2(2x + 1)^5$ **30.** $f(x) = 2x(x^3 - 1)^4$

31. $f(x) = (2x + 1)^3(2x - 1)^4$

32. $f(x) = (2x - 1)^3(2x + 1)^4$

33. $f(x) = \left(\dfrac{x + 1}{x - 1}\right)^3$ **34.** $f(x) = \left(\dfrac{x - 1}{x + 1}\right)^5$

35. $f(x) = x^2\sqrt{1 + x^2}$ **36.** $f(x) = x^2\sqrt{x^2 - 1}$

37. $f(x) = \sqrt{1 + \sqrt{x}}$ **38.** $f(x) = \sqrt[3]{1 + \sqrt[3]{x}}$

39. Find the derivative of $(x^2 + 1)^2$ in two ways:
   a. By the Generalized Power Rule.
   b. By "squaring out" the original expression and then differentiating.
   Your answers should agree.

40. Find the derivative of $\dfrac{1}{x^2}$ in three ways:
   a. By the Quotient Rule.
   b. By writing $\dfrac{1}{x^2}$ as $(x^2)^{-1}$ and using the Generalized Power Rule
   c. By writing $\dfrac{1}{x^2}$ as x^{-2} and using the (ordinary) Power Rule.
   Your answers should agree.

41. Find the derivative of $\dfrac{1}{3x + 1}$ in two ways:

   a. By the Quotient Rule.

   b. By writing the function as $(3x + 1)^{-1}$ and using the Generalized Power Rule.

Your answers should agree. Which way was easier? Remember this for the future.

42. Find an expression for the derivative of the composition of three functions, $\dfrac{d}{dx}f(g(h(x)))$.

[*Hint:* Use the Chain Rule twice.]

43. Suppose that $L(x)$ is a function such that $L'(x) = \dfrac{1}{x}$. Use the Chain Rule to show that

the derivative of the composite function

$$L(g(x)) \text{ is } \frac{d}{dx}L(g(x)) = \frac{g'(x)}{g(x)}.$$

44. Suppose that $E(x)$ is a function such that $E'(x) = E(x)$. Use the Chain Rule to show that the derivative of the composite function

$$E(g(x)) \text{ is } \frac{d}{dx}E(g(x)) = E(g(x)) \cdot g'(x).$$

Find the *second* derivative of each function.

45. $f(x) = (x^2 + 1)^{10}$ **46.** $f(x) = (x^3 - 1)^5$

APPLIED EXERCISES

47. BUSINESS: Cost A company's cost function is $C(x) = \sqrt{4x^2 + 900}$ dollars, where x is the number of units. Find the marginal cost function and evaluate it at $x = 20$.

48. BUSINESS: Cost *(continuation)* Graph the cost function $y_1 = \sqrt{4x^2 + 900}$ on the viewing window $[0, 30]$ by $[-10, 70]$. Then use NDERIV to define y_2 as the derivative of y_1. Verify the answer to Exercise 47 by evaluating the marginal cost function y_2 at $x = 20$.

49. BUSINESS: Cost *(continuation)* Find the number x of units at which the marginal cost is 1.75. [*Hint:* TRACE along the marginal cost function y_2 to find where the y-coordinate is 1.75, giving your answer as the x-coordinate rounded to the nearest whole number.]

50. SOCIOLOGY: Educational Status A study* estimated how a person's social status (rated on a scale where 100 indicates the status of a college graduate) depended on years of education. Based on this study, with e years of education, a person's status is $S(e) = 0.22(e + 4)^{2.1}$. Find $S'(12)$ and interpret your answer.

*Robert L. Hamblin, "Mathematical Experimentation and Sociological Theory: A Critical Analysis," *Sociometry* 34:423–452, 1971.

51. SOCIOLOGY: Income Status A study* estimated how a person's social status (rated on a scale where 100 indicates the status of a college graduate) depended upon income. Based on this study, with an income of i thousand dollars, a person's status is $S(i) = 17.5(i - 1)^{0.53}$. Find $S'(25)$ and interpret your answer.

52. ECONOMICS: Compound Interest If $1000 is deposited in a bank paying $r\%$ interest compounded annually, 5 years later its value will be

$$V(r) = 1000(1 + 0.01r)^5 \quad \text{dollars}$$

Find $V'(6)$ and interpret your answer. [*Hint:* $r = 6$ corresponds to 6% interest.]

53. BIOMEDICAL: Drug Sensitivity The strength of a patient's reaction to a dose of x milligrams of a certain drug is $R(x) = 4x\sqrt{11 + 0.5x}$ for $0 \le x \le 140$. The derivative $R'(x)$ is called the *sensitivity* to the drug. Find $R'(50)$, the sensitivity to a dose of 50 mg.

54. GENERAL: Population The population of a city x years from now is predicted to be $P(x) = \sqrt[4]{x^2 + 1}$ million people for $1 \le x \le 5$. Find when the population will be growing at the rate of a quarter of a million people per year. [*Hint:* On a graphing calculator, enter the given population function in y_1, use NDERIV to define y_2 to be the derivative of y_1, and graph both on the window [1, 5] by [0, 3]. Then TRACE along y_2 to find the x-coordinate (rounded to the nearest tenth of a unit) at which the y-coordinate is 0.25. You may have to ZOOM IN to find the correct x-value.]

55. BIOMEDICAL: Drug Sensitivity (*53 continued*) For the reaction function given in Exercise 53, find the dose at which the sensitivity is 25. [*Hint:* On a graphing calculator, enter the reaction function $y_1 = 4x\sqrt{11 + 0.5x}$, and use NDERIV to define y_2 to be the derivative of y_1. Then graph y_2 on the window [0, 140] by [0, 50] and TRACE along y_2 to find the x-coordinate (rounded to the nearest whole number) at which the y-coordinate is 25. You may have to ZOOM IN to find the correct x-value.]

56. BIOMEDICAL: Blood Flow It follows from Poiseuille's Law that blood flowing through certain arteries will encounter a resistance of $R(x) = 0.25(1 + x)^4$, where x is the distance (in meters) from the heart. Find the instantaneous rate of change of the resistance at:

a. 0 meters.　　**b.** 1 meter.

57. ENVIRONMENTAL SCIENCE: Pollution The carbon monoxide level in a city is predicted to be $0.02x^{3/2} + 1$ ppm (parts per million), where x is the population in thousands. In t years the population of the city is predicted to be $x(t) = 12 + 2t$ thousand people. Therefore, in t years the carbon monoxide level will be

$$P(t) = 0.02(12 + 2t)^{3/2} + 1$$

ppm. Find $P'(2)$, the rate at which carbon monoxide pollution will be increasing in 2 years.

58. PSYCHOLOGY: Learning After p practice sessions, a subject could perform a task in $T(p) = 36(p + 1)^{-1/3}$ minutes for $0 \le p \le 10$. Find $T'(7)$ and interpret your answer.

59. GENERAL: Greenhouse Gases and Global Warming The burning of oil and other "fossil fuels" is increasing the concentration of "greenhouse gases" in the atmosphere. The chief greenhouse gas is carbon dioxide (CO_2), and the table below gives the atmospheric CO_2 concentration in parts per million (ppm) for every five years since 1970. To avoid large numbers, years are listed in the table as years since 1970.

	Years Since 1970	CO$_2$ (ppm)
1970	0	325.3
1975	5	331.0
1980	10	338.5
1985	15	345.7
1990	20	354.0
1995	25	362.0
2000	30	370.0

Source: Scripps Institute of Oceanography

a. Enter the table numbers into your graphing calculator and make a plot of the resulting points (Years Since 1970 on the x-axis and CO_2 on the y-axis).

b. Have your calculator find the linear regression formula for these data. Then enter the result as y_1, which gives a formula for CO_2 concentrations for each year. Plot the points together with the regression line. Observe that the line fits the points quite well.

c. Define $y_2 = 0.024x + 51.3$. This function is an estimate of the average global temperature y_2 (in degrees Fahrenheit) for any CO_2 concentration x (which is the "output" of function y_1). (This function was found similarly by another linear regression on temperature and CO_2 data.)

d. Define y_3 to be the *composition* of y_2 and y_1 [on some calculators, by defining $y_3 = y_2(y_1)$.] Therefore, y_3 gives the average global temperature for any value of x (Years Since 1970).

e. Turn "off" the point plots [part (a)] and the functions y_1 and y_2. Then graph y_3 on the window [0, 100] by [55, 65]. Find the slope of this linear function (using NDERIV or dy/dx), giving the predicted temperature rise per year.　　(*continues*)

f. Use the slope (degrees per year) from part (e) to estimate how long it will take for temperatures to increase by 1.8 degrees. (It has been estimated that a 1.8-degree temperature increase could raise the sea level by 1 foot, inundating low-lying areas of the Mississippi delta and the food-producing river floodplains of Africa and Asia, and seriously disrupting world food supplies.)*

*The function y_1 that you found, describing how CO_2 levels increase with time, is widely accepted as accurate. The given function y_2 connecting temperature to CO_2 concentrations is less well established, since factors other than CO_2 affect temperature.

2.7 NONDIFFERENTIABLE FUNCTIONS

Introduction

In spite of all of the rules of differentiation, there are nondifferentiable functions—functions that cannot be differentiated at certain values. We begin this section by exhibiting such a function (the absolute value function) and showing that it is not differentiable at $x = 0$. We will then discuss general geometric conditions for a function to be nondifferentiable. Knowing where a function is not differentiable is important for understanding graphs and for interpreting answers from a graphing calculator.

Absolute Value Function

On page 55 we defined the absolute value function,

$$f(x) = |x| = \begin{cases} x & \text{if } x \geq 0 \\ -x & \text{if } x < 0 \end{cases}$$

Although the absolute value function is *defined* for *all* values of x, we will show that it is *not* differentiable at $x = 0$.

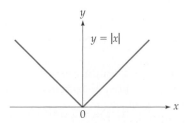

The graph of the absolute value function $f(x) = |x|$ has a "corner" at the origin.

EXAMPLE 1

SHOWING NONDIFFERENTIABILITY

Show that $f(x) = |x|$ is not differentiable at $x = 0$.

Solution   We have no "rules" for differentiating the absolute value function, so we must use the definition of the derivative:

$$f'(x) = \lim_{h \to 0} \frac{f(x + h) - f(x)}{h}$$

provided that this limit exists. It is this provision, which until now we have steadfastly ignored, that will be important in this example. We will show that this limit, and hence the derivative, does not exist at $x = 0$.

For $x = 0$ the definition becomes

$$\lim_{h \to 0} \frac{f(0 + h) - f(0)}{h} = \lim_{h \to 0} \frac{f(h) - f(0)}{h} = \lim_{h \to 0} \frac{|h| - |0|}{h} = \lim_{h \to 0} \frac{|h|}{h}$$

$$\underbrace{\qquad}_{\substack{\text{Using} \\ f(x) = |x|}}$$

Now h can approach 0 through *positive* numbers such as 0.01, 0.001, 0.0001, ... (denoted $h \to 0^+$), or through *negative* numbers such as $-0.01, -0.001, -0.0001, ...$ (denoted $h \to 0^-$).

For *positive* h:

$$\lim_{h \to 0^+} \frac{|h|}{h} = \lim_{h \to 0^+} \frac{h}{h} = \lim_{h \to 0^+} 1 = 1$$

$$\begin{array}{cc} \text{Since } |h| = h & \dfrac{h}{h} = 1 \\ \text{for } h > 0 & \end{array}$$

For *negative* h:

$$\lim_{h \to 0^-} \frac{|h|}{h} = \lim_{h \to 0^-} \frac{-h}{h} = \lim_{h \to 0^-} (-1) = -1$$

$$\begin{array}{cc} \text{For } h < 0, |h| = -h & \dfrac{-h}{h} = -1 \\ \text{(the negative sign} & \\ \text{makes the negative} & \\ h \text{ positive)} & \end{array}$$

The limit as $h \to 0$ must be a *single* number, the same regardless of how h approaches 0. Since the right and left limits do not agree (the first is $+1$ and the second is -1), the limit $\lim_{h \to 0} \dfrac{|h|}{h}$ does not exist, so *the derivative does not exist.* This is what we wanted to show—that the absolute value function is not differentiable at $x = 0$.

Geometric Explanation of Nondifferentiability

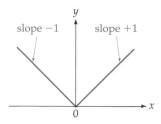

slope −1   slope +1

We can give a geometric and intuitive reason why the absolute value function is not differentiable at $x = 0$. Its graph consists of two straight lines with slopes +1 and −1 that meet in a corner at the origin. To the right of the origin the slope is +1 and to the left of the origin the slope is −1, but *at* the origin the two conflicting slopes make it impossible to define a *single* slope. Therefore, the slope (and hence the derivative) is undefined at $x = 0$.

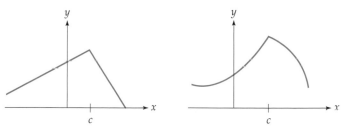

Graphing Calculator Exploration

If your graphing calculator has an operation like ABS or something similar for the absolute value function, graph $y_1 = \text{ABS}(x)$ on the window $[-2, 2]$ by $[-1, 2]$. Use NDERIV to "find" the derivative of y_1 at $x = 0$. Your calculator may give a "false value" such as 0, resulting from its *approximating* the derivative by the symmetric difference quotient (see pages 119–120). The correct answer is that the derivative is *undefined* at $x = 0$, as we just showed. This is why it is important to understand nondifferentiability—your calculator may give a misleading answer.

Other Nondifferentiable Functions

For the same reason, at any "corner point" of a graph, where two different slopes conflict, the function will not be differentiable.

Each of the functions graphed here has a "corner point"
at $x = c$, and so is not differentiable at $x = c$.

There are other reasons, besides a corner point, why a function may not be differentiable. If a curve has a vertical tangent line at a point, the slope will not be defined at that x-value, since the slope of a vertical line is undefined.

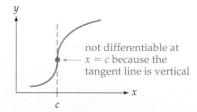

not differentiable at
$x = c$ because the
tangent line is vertical

We showed on pages 137–138 that if a function is differentiable, then it is continuous. Therefore, if a function is discontinuous (has a "jump") at some point, then it will not be differentiable at that *x*-value.

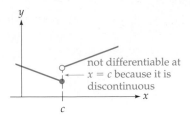

not differentiable at *x* = *c* because it is discontinuous

Therefore, if a function *f* satisfies *any* of the following conditions:

1. *f* has a corner point at *x* = *c*,
2. *f* has a vertical tangent at *x* = *c*,
3. *f* is discontinuous at *x* = *c*,

then *f* will not be differentiable at *c*. In Chapter 3, when we use calculus for graphing, it will be important to remember these three conditions that make the derivative fail to exist.

Practice Problem

For the function graphed below, find the *x*-values at which the derivative is undefined.

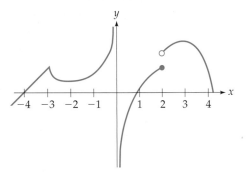

➤ Solution on page 175

All differentiable functions are continuous (see pages 137–138), but *not* all continuous functions are differentiable—for example, *f*(*x*) = |*x*|. These facts are shown in the following diagram

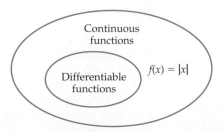

Spreadsheet Exploration

Another function that is not differentiable is $f(x) = x^{2/3}$. The following spreadsheet* calculates values of the difference quotient $\frac{f(x+h) - f(x)}{h}$ at $x = 0$ for this function. Since $f(0) = 0$, the difference quotient at $x = 0$ simplifies to:

$$\frac{f(x+h) - f(x)}{h} = \frac{f(0+h) - f(0)}{h} = \frac{f(h)}{h} = \frac{h^{2/3}}{h} = h^{-1/3}.$$

For example, cell **B5** evaluates $h^{-1/3}$ at $h = \frac{1}{1000}$ obtaining $\left(\frac{1}{1000}\right)^{-1/3} = 1000^{1/3} = \sqrt[3]{1000} = 10$. Column **B** evaluates this different quotient for the *positive* values of h in column **A**, while column **E** evaluates it for the corresponding negative values of h in column **D**.

	B5	▼	=	=A5^(-1/3)	
	A	B	C	D	E
1	h	(f(0+h)-f(0))/h		h	(f(0+h)-f(0))/h
2	1.0000000	1.0000000		-1.0000000	-1.0000000
3	0.1000000	2.1544347		-0.1000000	-2.1544347
4	0.0100000	4.6415888		-0.0100000	-4.6415888
5	0.0010000	10.0000000		-0.0010000	-10.0000000
6	0.0001000	21.5443469		-0.0001000	-21.5443469
7	0.0000100	46.4158883		-0.0000100	-46.4158883
8	0.0000010	100.0000000		-0.0000010	-100.0000000
9	0.0000001	215.4434690		-0.0000001	-215.4434690

Notice that the values in column **B** are becoming arbitrarily large, while the values in column **E** are becoming arbitrarily small, so the difference quotient does not approach a limit as $h \to 0$. This shows that the derivative of $f(x) = x^{2/3}$ at 0 does not exist, so the function $f(x) = x^{2/3}$ is *not* differentiable at $x = 0$.

➤ **Solution to Practice Problem**

$x = -3$, $x = 0$, and $x = 2$

* See the Preface for how to obtain this and other Excel spreadsheets.

2.7 Exercises

For each function graphed below, find the x-values at which the derivative does not exist.

1.

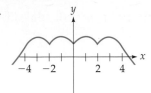

2.

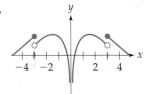

3.

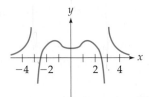

4.

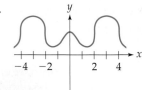

Use the definition of the derivative to show that the following functions are not differentiable at $x = 0$.

5. $f(x) = |2x|$ **6.** $f(x) = |3x|$

[*Hint for Exercises 5 and 6:* Modify the calculations on page 172 to apply to these functions.]

7. $f(x) = x^{2/5}$ **8.** $f(x) = x^{4/5}$

 Use NDERIV to "find" the derivative of each function at $x = 0$. Is the result correct?

9. $f(x) = \dfrac{1}{x}$ **10.** $f(x) = \dfrac{1}{x^2}$

 11. a. Show that the definition of the derivative applied to the function $f(x) = \sqrt{x}$ at $x = 0$ gives $f'(0) = \lim\limits_{h \to 0} \dfrac{\sqrt{h}}{h}$.

b. Use a calculator to evaluate the difference quotient $\dfrac{\sqrt{h}}{h}$ for the following values of h: 0.1, 0.001, and 0.00001. [*Hint:* Enter the calculation into your calculator with h replaced by 0.1, and then change the value of h by inserting zeros.]

c. From your answers to part (b), does the limit exist? Does the derivative of $f(x) = \sqrt{x}$ at $x = 0$ exist?

d. Graph $f(x) = \sqrt{x}$ on the window $[0, 1]$ by $[0, 1]$. Do you see why the slope at $x = 0$ does not exist?

12. a. Show that the definition of the derivative applied to the function $f(x) = \sqrt[3]{x}$ at $x = 0$ gives $f'(0) = \lim\limits_{h \to 0} \dfrac{\sqrt[3]{h}}{h}$.

b. Use a calculator to evaluate the difference quotient $\dfrac{\sqrt[3]{h}}{h}$ for the following values of h: 0.1, 0.0001, and 0.0000001. [*Hint:* Enter the calculation into your calculator with h replaced by 0.1, and then change the value of h by inserting zeros.]

c. From your answers to part (b), does the limit exist? Does the derivative of $f(x) = \sqrt[3]{x}$ at $x = 0$ exist?

d. Graph $f(x) = \sqrt[3]{x}$ on the window $[-1, 1]$ by $[-1, 1]$. Do you see why the slope at $x = 0$ does not exist?

Chapter Summary with Hints and Suggestions

Reading the text and doing the exercises in this chapter have helped you to master the following skills, which are listed by section (in case you need to review them) and are keyed to particular Review Exercises. Answers for all Review Exercises are given at the back of the book, and full solutions can be found in the Student Solutions Manual.

2.1 Limits and Continuity

- Find the limit of a function from tables. *(Review Exercises 1–2.)*

- Find left and right limits. *(Review Exercises 3–4.)*

- Find the limit of a function by direct substitution. *(Review Exercises 5–12.)*

- Determine whether a function is continuous or discontinuous. *(Review Exercises 13–20.)*

2.2 Rates of Change, Slopes, and Derivatives

- Find the derivative of a function from the *definition* of the derivative. *(Review Exercises 21–24.)*

$$f'(x) = \lim_{h \to 0} \frac{f(x + h) - f(x)}{h}$$

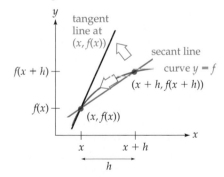

2.3 Some Differentiation Formulas

- Find the derivative of a function using the rules of differentiation. *(Review Exercises 25–30.)*

$$\frac{d}{dx} c = 0 \qquad \frac{d}{dx} x^n = nx^{n-1}$$

$$\frac{d}{dx}(c \cdot f) = c \cdot f' \qquad \frac{d}{dx}(f \pm g) = f' \pm g'$$

- Calculate and interpret a company's marginal cost. *(Review Exercise 31.)*

$$MC(x) = C'(x) \qquad MR(x) = R'(x)$$

$$MP(x) = P'(x)$$

- Find and interpret the derivative of a learning curve. *(Review Exercise 32.)*

- Find and interpret the derivative of an area or volume formula. *(Review Exercises 33–34.)*

2.4 The Product and Quotient Rules

- Find the derivative of a function using the Product Rule or Quotient Rule. *(Review Exercises 35–45.)*

$$\frac{d}{dx}(f \cdot g) = f' \cdot g + f \cdot g'$$

$$\frac{d}{dx}\left(\frac{f}{g}\right) = \frac{g \cdot f' - g' \cdot f}{g^2}$$

- Use differentiation to solve an applied problem and interpret the answer. *(Review Exercises 46–48.)*

2.5 Higher-Order Derivatives

- Calculate the second derivative of a function. *(Review Exercises 49–58.)*

- Find and interpret the first and second derivatives in an applied problem. *(Review Exercise 59.)*

■ Find the velocity and acceleration of a rocket.
(*Review Exercise 60.*)

$$v(t) = s'(t) \qquad a(t) = v'(t) = s''(t)$$

■ Find the maximum height of a projectile.
(*Review Exercise 61.*)

■ Find a company's profit function, then calculate and interpret the first and second derivatives. (*Review Exercise 62.*)

2.6 The Chain Rule and the Generalized Power Rule

■ Find the derivative of a function using the Generalized Power Rule.
(*Review Exercises 63–72.*)

$$\frac{d}{dx} f^n = n \cdot f^{n-1} \cdot f'$$

$$\frac{d}{dx} f(g(x)) = f'(g(x)) \cdot g'(x)$$

$$\frac{dy}{dx} = \frac{dy}{du} \cdot \frac{du}{dx}$$

■ Find the derivative of a function using *two* differentiation rules. (*Review Exercises 73–84.*)

■ Find the *second* derivative of a function using the Generalized Power Rule.
(*Review Exercises 85–88.*)

■ Find the derivative of a function in several different ways. (*Review Exercises 89–90.*)

■ Use the Generalized Power Rule to find the derivative in an applied problem and interpret the answer. (*Review Exercises 91–92.*)

■ Compare the profit from one unit to the marginal profit found by differentiation.
(*Review Exercise 93.*)

■ Find where the marginal profit equals a given number. (*Review Exercise 94.*)

■ Use the Generalized Power Rule to solve an applied problem and interpret the answer.
(*Review Exercises 95–96.*)

2.7 Nondifferentiable Functions

■ See from a graph where the derivative is undefined. (*Review Exercises 97–100.*)

$$f' \text{ is undefined at } \begin{cases} \text{corner points} \\ \text{vertical tangents} \\ \text{discontinuities} \end{cases}$$

■ Prove that a function is not differentiable at a given value. (*Review Exercises 101–102.*)

Hints and Suggestions

■ (*Overview*) This chapter introduced one of the most important concepts in all of calculus, the *derivative*. First we defined it (using limits), then we developed several "rules of differentiation" to simplify its calculation.

■ Remember the four interpretations of the derivative—*slopes, instantaneous rates of change, marginals,* and *velocities.*

■ The *second* derivative gives the rate of change of the rate of change, and acceleration.

■ Graphing calculators help to find limits, graph curves and their tangent lines, and calculate derivatives (using NDERIV) and second derivatives (using NDERIV twice). NDERIV, however, provides only an *approximation* to the derivative, and therefore sometimes gives a misleading result.

■ The units of the derivative are important in applied problems. For example, if $f(x)$ gives the temperature in degrees at time x hours, then the derivative $f'(x)$ is in *degrees per hour*. In general, the units of the derivative $f'(x)$ are "*f*-units" per "*x*-unit."

■ **Practice for Test:** Review Exercises 1, 2, 3, 4, 7, 9, 11, 15, 19, 20, 21, 25, 31, 37, 43, 47, 49, 57, 59, 62, 67, 71, 77, 83, 91, 94, 97.

Review Exercises for Chapter 2

Practice test exercise numbers are in green.

2.1 Limits and Continuity

For each limit, complete the limit table and find the limit (or state that the limit does not exist). Round calculations to three decimal places.

1. $\lim\limits_{x \to 2} (4x + 2)$

x	4x + 2
1.9	
1.99	
1.999	

x	4x + 2
2.1	
2.01	
2.001	

2. $\lim\limits_{x \to 0} \dfrac{\sqrt{x + 1} - 1}{x}$

x	$\dfrac{\sqrt{x+1}-1}{x}$
−0.1	
−0.01	
−0.001	

x	$\dfrac{\sqrt{x+1}-1}{x}$
0.1	
0.01	
0.001	

For each piecewise linear function, find:

  a. $\lim\limits_{x \to 5^-} f(x)$ **b.** $\lim\limits_{x \to 5^+} f(x)$ **c.** $\lim\limits_{x \to 5} f(x)$

3. $f(x) = \begin{cases} 2x - 7 & \text{if } x \le 5 \\ 3 - x & \text{if } x > 5 \end{cases}$

4. $f(x) = \begin{cases} 4 - x & \text{if } x < 5 \\ 2x - 11 & \text{if } x \ge 5 \end{cases}$

Find the following limits (*without* using limit tables).

5. $\lim\limits_{x \to 4} \sqrt{x^2 + x + 5}$ **6.** $\lim\limits_{x \to 0} \pi$

7. $\lim\limits_{s \to 16} \left(\dfrac{1}{2} s - s^{1/2} \right)$ **8.** $\lim\limits_{r \to 8} \dfrac{r}{r^2 - 30\sqrt[3]{r}}$

9. $\lim\limits_{x \to 1} \dfrac{x^2 - x}{x^2 - 1}$ **10.** $\lim\limits_{x \to -1} \dfrac{3x^3 - 3x}{2x^2 + 2x}$

11. $\lim\limits_{h \to 0} \dfrac{2x^2 h - xh^2}{h}$ **12.** $\lim\limits_{h \to 0} \dfrac{6xh^2 - x^2 h}{h}$

For each function, state whether it is continuous or discontinuous. If it is discontinuous, state the values of x at which it is discontinuous.

13. $f(x) = 2x + 5$ **14.** $f(x) = x^2 - 1$

15. $f(x) = \dfrac{1}{x + 1}$ **16.** $f(x) = \dfrac{1}{x^2 + 1}$

17. $f(x) = \dfrac{x - 1}{x^2 + x}$ **18.** $f(x) = \dfrac{1}{|x| - 3}$

19. $f(x) = \begin{cases} 2x - 7 & \text{if } x \le 5 \\ 3 - x & \text{if } x > 5 \end{cases}$

    [*Hint:* See Exercise 3.]

20. $f(x) = \begin{cases} 4 - x & \text{if } x < 5 \\ 2x - 11 & \text{if } x \ge 5 \end{cases}$

    [*Hint:* See Exercise 4.]

2.2 Rates of Change, Slopes, and Derivatives

Find the derivative of each function using the definition of the derivative (that is, as you did in Section 2.2).

21. $f(x) = 2x^2 + 3x - 1$

22. $f(x) = 3x^2 + 2x - 3$

(See instructions on previous page.)

23. $f(x) = \dfrac{3}{x}$ **24.** $f(x) = 4\sqrt{x}$

2.3 Some Differentiation Formulas

Find the derivative of each function.

25. $f(x) = 6\sqrt[3]{x^5} - \dfrac{4}{\sqrt{x}} + 1$

26. $f(x) = 4\sqrt{x^5} - \dfrac{6}{\sqrt[3]{x}} + 1$

Evaluate each expression.

27. If $f(x) = \dfrac{1}{x^2}$, find $f'\left(\dfrac{1}{2}\right)$.

28. If $f(x) = \dfrac{1}{x}$, find $f'\left(\dfrac{1}{3}\right)$.

29. If $f(x) = 12\sqrt[3]{x}$, find $f'(8)$.

30. If $f(x) = 6\sqrt[3]{x}$, find $f'(-8)$.

31. BUSINESS: Marginal Cost A company's
cost function is $C(x) = 20 + 3x + \dfrac{54}{\sqrt{x}}$ dollars
where x is the number of units (for $5 \le x \le 20$).
  a. Find the marginal cost function.
  b. Find the marginal cost at $x = 9$ and interpret your answer.

32. Learning Curves in Industry From pages
26–27 the learning curve for building Boeing
707 airplanes is $f(x) = 150x^{-0.322}$, where $f(x)$
is the time (in thousands of hours) that it took
to build the xth Boeing 707. Find the instantaneous rate of change of this production time
for the tenth plane, and interpret your answer.

33. GENERAL: Geometry The formula for the
area of a circle is $A = \pi r^2$, where r is the radius of the circle and π is a constant.

  a. Show that the derivative of the area formula is $2\pi r$, the formula for the circumference of a circle.

  b. Give an explanation for this in terms of rates of change.

34. GENERAL: Geometry The formula for the
volume of a sphere is $V = \frac{4}{3}\pi r^3$, where r is
the radius of the sphere and π is a constant.

  a. Show that the derivative of the volume formula is $4\pi r^2$, the formula for the surface area of a sphere.

  b. Give an explanation for this in terms of rates of change.

2.4 The Product and Quotient Rules

Find the derivative of each function.

35. $f(x) = 2x(5x^3 + 3)$

36. $f(x) = x^2(3x^3 - 1)$

37. $f(x) = (x^2 + 5)(x^2 - 5)$

38. $f(x) = (x^2 + 3)(x^2 - 3)$

39. $y = (x^4 + x^2 + 1)(x^5 - x^3 + x)$

40. $y = (x^5 + x^3 + x)(x^4 - x^2 + 1)$

41. $y = \dfrac{x - 1}{x + 1}$ **42.** $y = \dfrac{x + 1}{x - 1}$

43. $y = \dfrac{x^5 + 1}{x^5 - 1}$ **44.** $y = \dfrac{x^6 - 1}{x^6 + 1}$

45. Find the derivative of $f(x) = \dfrac{2x + 1}{x}$ in
three different ways, and check that the
answers agree:
  a. By the Quotient Rule
  b. By writing the function in the form
  $f(x) = (2x + 1)(x^{-1})$ and using the Product
  Rule.
  c. By thinking of another way, which is the
  easiest of all.

46. BUSINESS: Sales The manager of an electronics store estimates that the number of cassette tapes that a store will sell at a price of x dollars is

$$S(x) = \frac{2250}{x + 9}$$

Find the rate of change of this quantity when the price is \$6 per tape, and interpret your answer.

47. **BUSINESS: Marginal Average Profit** A company's profit function is $P(x) = 6x - 200$ dollars, where x is the number of units.

   a. Find the average profit function.
   b. Find the marginal average profit function.
   c. Evaluate the marginal average profit function at $x = 10$ and interpret your answer.

48. **BUSINESS: Marginal Average Cost** A company's cost function is $C(x) = 5x + 100$ dollars, where x is the number of units.

   a. Find the average cost function.
   b. Find the marginal average cost function.
   c. Evaluate the marginal average cost function at $x = 20$ and interpret your answer.

2.5 Higher-Order Derivatives

Find the *second* derivative of each function.

49. $f(x) = 12\sqrt{x^3} - 9\sqrt[3]{x}$

50. $f(x) = 18\sqrt[3]{x^2} - 4\sqrt{x^3}$

51. $f(x) = \dfrac{1}{3x^2}$ 52. $f(x) = \dfrac{1}{2x^3}$

Evaluate each expression.

53. If $f(x) = \dfrac{2}{x^3}$, find $f''(-1)$.

54. If $f(x) = \dfrac{3}{x^4}$, find $f''(-1)$.

55. $\left.\dfrac{d^2}{dx^2} x^6\right|_{x=-2}$ 56. $\left.\dfrac{d^2}{dx^2} x^{-2}\right|_{x=-2}$

57. $\left.\dfrac{d^2}{dx^2} \sqrt{x^5}\right|_{x=16}$ 58. $\left.\dfrac{d^2}{dx^2} \sqrt{x^7}\right|_{x=4}$

59. **GENERAL: Population** The population of a city t years from now is predicted to be

$P(t) = 0.25t^3 - 3t^2 + 5t + 200$ thousand people. Find $P(10)$, $P'(10)$, and $P''(10)$ and interpret your answers.

60. **GENERAL: Velocity** A rocket rises $s(t) = 8t^{5/2}$ feet in t seconds. Find its velocity and acceleration after 25 seconds.

61. **GENERAL: Velocity** The fastest baseball pitch on record (thrown by Lynn Nolan Ryan of the California Angels on August 20, 1974) was clocked at 100.9 miles per hour (148 feet per second).

   a. If this pitch had been thrown straight up, its height after t seconds would have been $s(t) = -16t^2 + 148t + 5$ feet. Find the maximum height the ball would have reached.
   b. Verify your answer to part (a) by graphing the height function $y_1 = -16x^2 + 148x + 5$ on the window [0, 10] by [0, 400]. Then TRACE along the curve to find its highest point (or use the MAXIMUM feature of your calculator).

62. **BUSINESS: Profit** The table below gives a company's annual profit at the end of the year for years 1 through 6.

Year	Annual Profit (millions of dollars)
1	3.4
2	3.1
3	3.0
4	3.1
5	4.1
6	5.2

   a. Enter the data in the table into your graphing calculator and make a plot of the resulting points (Year on the x-axis and Annual Profit on the y-axis).
   b. Have your calculator find the quadratic regression formula for these data. Then enter the result in y_1, giving the annual profit for year x. Plot the points together with the regression curve (on an appropriate window). Observe that the curve fits the points quite well.
   c. Predict the annual profit at the end of year 7 by evaluating $y_1(7)$.   (*continues*)

d. Define y_2 to be the derivative of y_1 (using NDERIV). Find $y_2(7)$ and interpret this number.

e. Define y_3 to be the derivative of y_2 (so y_3 is the *second* derivative of y_1). Find and interpret $y_3(7)$.

f. Graph all three functions and the points on the window $[0, 7]$ by $[-1, 7]$. Can you explain why y_3 appears to be a horizontal straight line?

2.6 The Chain Rule and the Generalized Power Rule

Find the derivative of each function.

63. $h(z) = (4z^2 - 3z + 1)^3$

64. $h(z) = (3z^2 - 5z - 1)^4$

65. $g(x) = (100 - x)^5$ **66.** $g(x) = (1000 - x)^4$

67. $f(x) = \sqrt{x^2 - x + 2}$

68. $f(x) = \sqrt{x^2 - 5x - 1}$

69. $w(z) = \sqrt[3]{6z - 1}$ **70.** $w(z) = \sqrt[3]{3z + 1}$

71. $h(x) = \dfrac{1}{\sqrt[5]{(5x + 1)^2}}$ **72.** $h(x) = \dfrac{1}{\sqrt[5]{(10x + 1)^3}}$

73. $g(x) = x^2(2x - 1)^4$ **74.** $g(x) = 5x(x^3 - 2)^4$

75. $y = x^3\sqrt[3]{x^3 + 1}$ **76.** $y = x^4\sqrt{x^2 + 1}$

77. $f(x) = [(2x^2 + 1)^4 + x^4]^3$

78. $f(x) = [(3x^2 - 1)^3 + x^3]^2$

79. $f(x) = \sqrt{(x^2 + 1)^4 - x^4}$

80. $f(x) = \sqrt{(x^3 + 1)^2 + x^2}$

81. $f(x) = (3x + 1)^4(4x + 1)^3$

82. $f(x) = (x^2 + 1)^3(x^2 - 1)^4$

83. $f(x) = \left(\dfrac{x + 5}{x}\right)^4$ **84.** $f(x) = \left(\dfrac{x + 4}{x}\right)^5$

Find the *second* derivative of each function.

85. $h(w) = (2w^2 - 4)^5$ **86.** $h(w) = (3w^2 + 1)^4$

87. $g(z) = z^3(z + 1)^3$ **88.** $g(z) = z^4(z + 1)^4$

89. Find the derivative of $(x^3 - 1)^2$ in two ways:

a. By the Generalized Power Rule.

b. By "squaring out" the original expression and then differentiating.

Your answers should agree.

90. Find the derivative of $g(x) = \dfrac{1}{x^3 + 1}$ in two ways:

a. By the Quotient Rule.

b. By the Generalized Power Rule.

Your answers should agree.

91. BUSINESS: Marginal Profit A company's profit from producing x tons of polyurethane is $P(x) = \sqrt{x^3 - 3x + 34}$ thousand dollars (for $0 \le x \le 10$). Find $P'(5)$ and interpret your answer.

92. GENERAL: Compound Interest If $500 is deposited in an account earning interest at r percent annually, after 3 years its value will be $V(r) = 500(1 + 0.01r)^3$ dollars. Find $V'(8)$ and interpret your answer.

93. BUSINESS: Marginal Profit (*91 continued*) Using a graphing calculator, graph the profit function from Exercise 91 on the window $[0, 10]$ by $[0, 30]$.

a. Evaluate the profit function at 4, 5, and 6 and calculate the following actual costs:

$P(5) - P(4)$ (actual cost of the fifth ton)

$P(6) - P(5)$ (actual cost of the sixth ton)

Compare these results with the "instantaneous" marginal profit at $x = 5$ found in Exercise 91.

b. Define another function to be the derivative of the previously entered profit function (using NDERIV) and graph both together. Find (to the nearest tenth of a unit) the x-value where the marginal profit reaches 4.

94. On a graphing calculator, graph the cost function $C(x) = \sqrt[3]{x^2 + 25x + 8}$ as y_1 on the window $[0, 20]$ by $[-2, 10]$. Use NDERIV to define y_2 to be its derivative, graphing both. TRACE along the derivative to find the x-value (to the nearest unit) where the marginal cost is 0.25.

95. BIOMEDICAL: Blood Flow Blood flowing through an artery encounters a resistance of $R(x) = 0.25(0.01x + 1)^4$, where x is the distance (in centimeters) from the heart. Find the instantaneous rate of change of the resistance 100 centimeters from the heart.

96. GENERAL: Survival Rate Suppose that for a group of 10,000 people, the number who survive to age x is $N(x) = 1000\sqrt{100 - x}$. Find $N'(96)$ and interpret your answer.

2.7 Nondifferentiable Functions

For each function graphed below, find the values of x at which the derivative does not exist.

97.

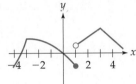

98.

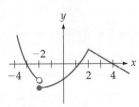

99.

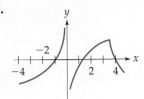

100.

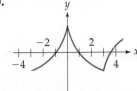

Use the definition of the derivative to show that the following functions are not differentiable at $x = 0$.

101. $f(x) = |5x|$ **102.** $f(x) = x^{3/5}$

3 Further Applications of Derivatives

Application Preview

Stevens' Law of Psychophysics

Suppose that you are given two weights and asked to judge how much heavier one is than the other. If one weight is actually *twice* as heavy as the other, most people will judge the heavier weight as being *less* than twice as heavy. This is one of the oldest problems in experimental psychology—how sensation (perceived weight) varies with stimulus (actual weight). Similar experiments can be performed for perceived brightness of a light compared with actual brightness, perceived effort compared with actual work, and so on. The results will vary somewhat from person to person, but the following diagram shows some typical stimulus-response curves (in arbitrary units).

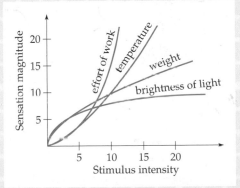

Notice, for example, that perceived effort increases more rapidly than actual work, which suggests that a 10% increase in an employee's work should be rewarded with a *greater* than 10% increase in pay.

Such stimulus-response curves were studied by the psychologist S. S. Stevens* at Harvard, who expressed them as power functions.

$$\text{Response} = a(\text{stimulus})^b$$

or

$$f(x) = ax^b \qquad \text{for constants } a \text{ and } b$$

In this chapter we will see that calculus can be very helpful for graphing such functions, such as showing whether they "curl upward" like the work and temperature curves or "curl downward" like the weight and brightness curves (see page 213).

*See S. S. Stevens, "On the Psychophysical Law," *Psychological Review* 64:153–181, 1957.

3.1 GRAPHING USING THE FIRST DERIVATIVE

Introduction

In this chapter we will put derivatives to two major uses: graphing and optimization. Graphing involves using calculus to find the most important points on a curve, then sketching the curve either by hand or by using a graphing calculator. Optimization means finding the largest or smallest values of a function (for example, maximizing profit or minimizing risk). We begin with graphing, since it will form the basis for optimization later in the chapter.

We saw in Chapter 2 that the derivative of a function gives the slope of its graph.

If $f' > 0$ on an interval, then f is *increasing* on that interval.

If $f' < 0$ on an interval, then f is *decreasing* on that interval.

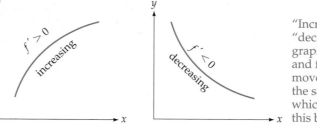

"Increasing" and "decreasing" on a graph mean rising and falling as you move *from left to right*, the same direction in which you are reading this book.

Relative Extreme Points and Critical Numbers

On a graph, a *relative maximum point* is a point that is at least as *high* as the neighboring points of the curve on either side, and a *relative minimum point* is a point that is at least as *low* as the neighboring points on either side.

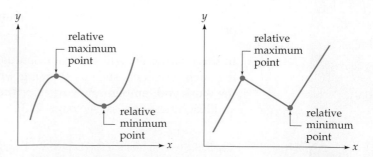

The word "relative" means that although these points may not be the highest and lowest on the *entire* curve, they are the highest and lowest *relative to points nearby*. (Later we will use the terms "absolute maximum" and "absolute minimum" to mean the highest and lowest points on the entire curve.) A curve may have any number of relative maximum and minimum points (collectively, "relative extreme points"), even none. For a function f, the relative extreme points may be defined more formally in terms of the values of f.

> f has a *relative maximum value* at c if $f(c) \geq f(x)$ for all values of x near* c.

> f has a *relative minimum value* at c if $f(c) \leq f(x)$ for all values of x near* c.

In the first of the two graphs below, the relative extreme points occur where the slope is *zero* (where the tangent line is horizontal), and in the second graph they occur where the slope is *undefined* (at corner points). The x-coordinates of such points are called *critical numbers*.

Critical Number

A *critical number* of a function f is an x-value in the domain of f at which either

$$f'(x) = 0$$

or

$$f'(x) \text{ is undefined}$$

Derivative is zero or undefined

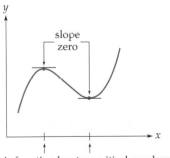

This function has two critical numbers (where the derivative is zero).

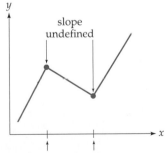

This function also has two critical numbers (but where the derivative is undefined).

* "Near c" means in some open interval containing c.

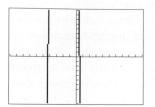

A "useless" graph of
$f(x) = x^3 - 12x^2 - 60x + 36$
on the standard window
$[-10, 10]$ by $[-10, 10]$

Graphing Functions

We graph a function by finding its critical numbers, making a "sign diagram" for the derivative to show the intervals of increase and decrease and the relative extreme points, and then drawing the curve on a graphing calculator or "by hand." Obtaining a reasonable graph even with a graphing calculator requires more than just pushing buttons, as shown in the graph of the function $f(x) = x^3 - 12x^2 - 60x + 36$ on the left. We will improve on this graph in the following example.

EXAMPLE 1

GRAPHING A FUNCTION

Graph the function $f(x) = x^3 - 12x^2 - 60x + 36$.

Solution

Step 1: Find critical numbers.

$$f'(x) = 3x^2 - 24x - 60 \qquad \text{Derivative of } f(x) = x^3 - 12x^2 - 60x + 36$$

$$= 3(x^2 - 8x - 20) = 3(x - 10)(x + 2) = 0 \qquad \text{Factoring and setting equal to zero}$$

Zero at   Zero at
$x = 10$ $x = -2$

The derivative is zero at $x = 10$ and at $x = -2$, and there are no numbers at which the derivative is undefined (it is a polynomial), so the critical numbers (CNs) are

$$\text{CN} \begin{cases} x = 10 \\ x = -2 \end{cases} \qquad \text{Both are in the domain of the orginal function}$$

Step 2: Make a sign diagram for the derivative f'. A sign diagram for f' begins with a copy of the x-axis with the critical numbers written below it and the behavior of f' indicated above it.

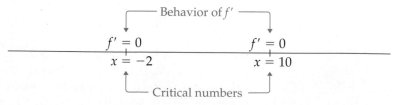

Since f' is continuous, it can change sign only at critical numbers, so f' must keep the same sign between consecutive critical numbers. We determine the sign of f' in each interval by choosing a "test point" in each interval and substituting it into f'. It is easiest to use the factored form: $f'(x) = 3(x - 10)(x + 2)$. For example, in the first interval, $f'(-3) = 3(-3 - 10)(-3 + 2) = 3(\text{negative})(\text{negative}) = (\text{positive})$. We indicate the sign of f' (the slope of f) by arrows: $\nearrow$ for positive slope, $\rightarrow$ for zero slope, and $\searrow$ for negative slope.

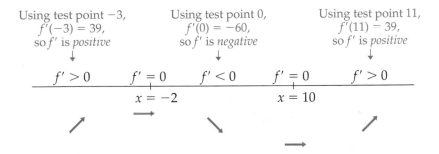

The sign diagram shows that to the left of $x = -2$ the function increases, then between $x = -2$ and $x = 10$ it decreases, and then to the right of $x = 10$ it increases again. Therefore, the open intervals of increase are $(-\infty, -2)$ and $(10, \infty)$, and the open interval of decrease is $(-2, 10)$.

Arrows $\nearrow \rightarrow \searrow$ indicate a relative *maximum* point, and arrows $\searrow \rightarrow \nearrow$ indicate a relative *minimum* point. We then list these points under the critical numbers.

$f' > 0$	$f' = 0$	$f' < 0$	$f' = 0$	$f' > 0$
	$x = -2$		$x = 10$	
$\nearrow$	$\rightarrow$	$\searrow$	$\rightarrow$	$\nearrow$
	rel max $(-2, 100)$		rel min $(10, -764)$	

The y-coordinates of the points were found by evaluating the *original* function at the x-coordinate: $f(-2) = 100$ and $f(10) = -764$. [*Hint:* Use 📟 with TABLE or EVALUATE.]

***Step 3:* Sketch the graph.** The arrows $\nearrow \rightarrow \searrow \rightarrow \nearrow$ show the general shape of the curve: going up, level, down, level, and up again. The critical numbers show that we want to include x-values from before $x = -2$ to after $x = 10$, suggesting an interval such as $[-10, 20]$. The y-coordinates show that we want to go from above 100 to below -764, suggesting an interval of y-values such as $[-800, 200]$.

Using a graphing calculator, you then would graph on the window $[-10, 20]$ by $[-800, 200]$ (or some other reasonable window), as shown below.

By hand, you would plot the relative maximum point (with a "cap"⁀) and the relative minimum point (with a "cup" ‿), and then draw an "up-down-up" curve through them.

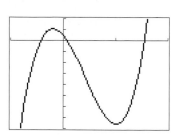

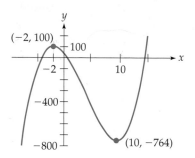

Two important observations:

1. Even with a graphing calculator, you still need calculus to find the relative extreme points that determine an appropriate window.

2. Given a calculator-drawn graph such as that on the left above, you should be able to make a hand-drawn graph such as that on the right, including numbers on the axes and coordinates of important points (using TRACE or TABLE if necessary).

Graphing Calculator Exploration

Do you see where the "useless" graph shown on page 188 fits into the "useful" graph on the left above? Check this by graphing $y_1 = x^3 - 12x^2 - 60x + 36$ first on the window $[-10, 20]$ by $[-800, 200]$ (obtaining the graph shown above) and then on the standard window $[-10, 10]$ by $[-10, 10]$.

Practice Problem 1   Find the critical numbers of $f(x) = x^3 - 12x + 8$.

➤ Solution on page 195

In Example 1 there were no critical numbers at which the derivative was *undefined* (it was a polynomial). For an example of a function with a critical number where the derivative is *undefined*, think of the absolute value function $f(x) = |x|$. The graph has a "corner point"

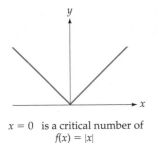

$x = 0$ is a critical number of
$f(x) = |x|$

at the origin (as shown on the left), and so the derivative is undefined at the critical number $x = 0$.

First-Derivative Test for Relative Extreme Values

The graphical idea from the sign diagram, that ↗ → ↘ (up, level, and down) indicates a relative maximum and ↘ → ↗ (down, level, and up) indicates a relative minimum, can be stated more formally in terms of the derivative.

First-Derivative Test

If a function f has a critical number c, then at $x = c$ the function has

a *relative maximum* if $f' > 0$ just before c and $f' < 0$ just after c.

a *relative minimum* if $f' < 0$ just before c and $f' > 0$ just after c.

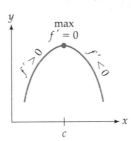

If f' is positive then negative,
then f has a relative *maximum* at c.

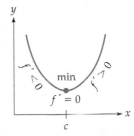

If f' is negative then positive,
then f has a relative *minimum* at c.

If the derivative has the *same* sign on both sides of c, then the function has *neither* a relative maximum nor a relative minimum at $x = c$.

[*Hint:* Use 🔲 with TABLE or EVALUATE.]

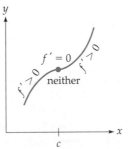

f' is positive on both sides, so
f has *neither* at $x = c$.

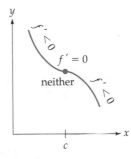

f' is negative on both sides, so
f has *neither* at $x = c$.

The diagrams below show that the first-derivative test applies even at critical numbers where the derivative is *undefined* (abbreviated: f' und).

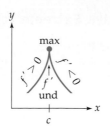

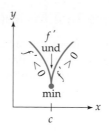

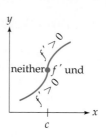

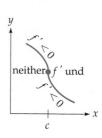

f' is positive then negative, so f has a relative *maximum* at $x = c$.

f' is negative then positive, so f has a relative *minimum* at $x = c$.

f' is positive on both sides, so f has *neither* at $x = c$.

f' is negative on both sides, so f has *neither* at $x = c$.

EXAMPLE 2

GRAPHING A FUNCTION

Graph $f(x) = -x^4 + 4x^3 - 20$.

Solution

$$f'(x) = -4x^3 + 12x^2 \qquad \text{Differentiating}$$

$$= -4x^2(x - 3) = 0 \qquad \text{Factoring and setting equal to zero}$$

Critical numbers:

$$\text{CN} \begin{cases} x = 0 \\ x = 3 \end{cases} \qquad \begin{array}{l} \text{From} \quad -4x^2 = 0 \\ \text{From} \quad (x - 3) = 0 \end{array}$$

We make a sign diagram for the derivative:

$f' = 0$		$f' = 0$	Behavior of f'
$x = 0$		$x = 3$	Critical numbers

We determine the sign of $f'(x) = -4x^2(x - 3)$ using test points in each interval, and then add arrows.

$f' > 0$	$f' = 0$	$f' > 0$	$f' = 0$	$f' < 0$
	$x = 0$		$x = 3$	

Finally, we interpret the arrows to describe the behavior of the functions, which we state at the bottom.

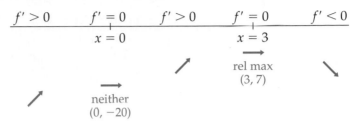

The open intervals of increase are $(-\infty, 0)$ and $(0, 3)$, and the open interval of decrease is $(3, \infty)$.

| **Using a graphing calculator,** we would choose a window such as $[-2, 5]$ by $[-30, 10]$ to include the points $(0, -20)$ and $(3, 7)$, and graph the function. | **By hand,** we would plot $(0, -20)$ and $(3, 7)$, and join them by a curve that goes, according to the arrows, "up-level-up-level-down." |

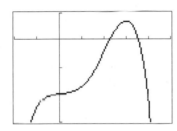

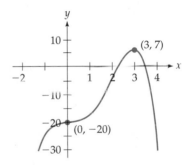

With a graphing calculator, an incomplete sign diagram may be enough to find an appropriate viewing window.

EXAMPLE **3**   **GRAPHING A RATIONAL FUNCTION**

Graph the rational function $f(x) = \dfrac{1}{x^2 - 4x}$.

Solution

$$f(x) = \frac{1}{x^2 - 4x} = \frac{1}{x(x - 4)}$$
   Original function with factored denominator

The denominator is zero at $x = 0$ and at $x = 4$, so the function is *undefined* at these numbers. The derivative is

$$f'(x) = \frac{(x^2 - 4x)\cdot 0 - (2x - 4)\cdot 1}{(x^2 - 4x)^2} \qquad \text{Using the quotient rule on } \frac{1}{x^2 - 4x}$$

$$= \frac{-(2x - 4)}{(x^2 - 4x)^2} = \frac{-2(x - 2)}{(x^2 - 4x)^2} \qquad \begin{array}{l}\text{Simplifying, and factoring}\\\text{the numerator}\end{array}$$

Zero at $x = 2$

The derivative is zero at $x = 2$ and undefined (abbreviated: f' und) at $x = 0$ and at $x = 4$ (as was the original function). We list all of these on a sign diagram along with the signs of f' (from test points).

$$f' > 0 \quad f' \text{ und} \quad f' > 0 \quad f' = 0 \quad f' < 0 \quad f' \text{ und} \quad f' < 0$$

| $x = 0$ | | $x = 2$ | | $x = 4$ |

$(2, -\tfrac{1}{4})$

y-coordinate from

$f(x) = \dfrac{1}{x^2 - 4x}$ at $x = 2$

For the viewing window, we might choose x-values $[-2, 6]$ (to include the x-values in the sign diagram) and y-values $[-2, 2]$ (narrow, so that the $-\frac{1}{4}$ will be distinct from the x-axis). The graph is shown below (you may need to ignore some false vertical lines). A hand-drawn graph, which you should be able to make from the calculator graph, is next to it.

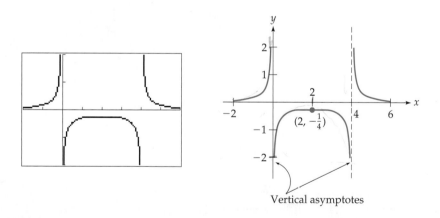

Vertical asymptotes

Notice in the previous example that where the function is undefined, the curve has a *vertical asymptote*. This is true of any (simplified) rational function, since the one-sided limits will be ∞ or $-\infty$ as x approaches one of these values.

Graphing Calculator Exploration

Graph the function $f(x) = \dfrac{1}{x^2 - 4x}$ on the standard window $[-10, 10]$ by $[-10, 10]$. Is the graph as clear as when graphed on the window we chose in Example 3? Choosing a good window is very important.

Practice Problem 2

Explain why, for a rational function, the function and its derivative will be undefined at the same x-values. Are such x-values critical numbers? ➤ Solution on page 196

3.1 Section Summary

To sketch a graph:

- Find the domain (by excluding any x-values at which the function is undefined).
- Find the critical numbers (where f' is zero or undefined, but where f is defined).
- List all of these on a sign diagram for the derivative, indicating the behavior of f' on each interval. Add arrows and relative extreme points.
- Finally, sketch the curve. If you use a graphing calculator, the sign diagram will suggest an appropriate window. If you graph by hand, your sign diagram will show you the shape of the curve.

If there are no relative extreme points, choose a few x-values, including any of special interest. On a graphing calculator, use these x-values to determine an x-interval, and use TABLE or TRACE to find a y-interval; then graph the function on the resulting window. By hand, plot the points corresponding to the chosen x-values and draw an appropriate curve through the points using your sign diagram.

➤ **Solutions to Practice Problems**

1. $f'(x) = 3x^2 - 12 = 3(x^2 - 4) = 3(x + 2)(x - 2)$

CN $\begin{cases} x = -2 \\ x = 2 \end{cases}$

2. Because the denominator of f' will be the square of the denominator of f (from the Quotient Rule). Such x values are *not* critical numbers because they are not in the domain of the original function (see the definition on page 187).

page 187

3.1 **Exercises**

For the functions graphed in Exercises 1 and 2:

a. Find the intervals on which the derivative is positive.

b. Find the intervals on which the derivative is negative.

1. 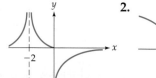 **2.**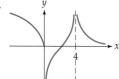

3. Which of the numbers 1, 2, 3, 4, 5, and 6 are critical numbers of the function graphed below?

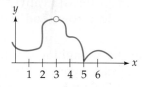

4. The first column in parts a through d shows the graphs of four functions, and the second column shows the graphs of their derivatives, but not necessarily in the same order. Write below each derivative the correct function from which it came.

a.

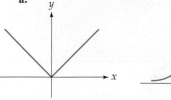

function f_1

a) derivative of function: _____

b.

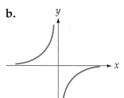

function f_2

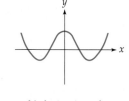

b) derivative of function: _____

c.

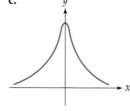

function f_3

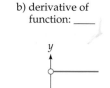

c) derivative of function: _____

d.

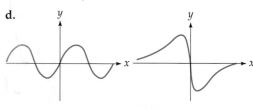

function f_4

d) derivative of function: _____

Find the critical numbers of each function.

5. $f(x) = x^3 - 48x$

6. $f(x) = x^3 - 27x$

7. $f(x) = x^4 + 4x^3 - 8x^2 + 1$

8. $f(x) = x^4 + 4x^3 - 20x^2 - 12$

9. $f(x) = (2x - 6)^4$

10. $f(x) = (x^2 + 6x - 7)^2$

11. $f(x) = 3x + 5$

12. $f(x) = 4x - 12$

13. $f(x) = x^3 - 2x^2 + x + 11$

14. $f(x) = x^3 - x^2 + 15$

Sketch the graph of each function "by hand" after making a sign diagram for the derivative and finding all open intervals of increase and decrease.

15. $f(x) = x^4 + 4x^3 - 8x^2 + 64$

16. $f(x) = x^4 - 4x^3 - 8x^2 + 64$

17. $f(x) = -x^4 + 4x^3 - 4x^2 + 1$

18. $f(x) = -x^4 - 4x^3 - 4x^2 + 1$

19. $f(x) = 3x^4 - 8x^3 + 6x^2$

20. $f(x) = 3x^4 + 8x^3 + 6x^2$

21. $f(x) = (x - 1)^6$

22. $f(x) = (x - 1)^5$

23. $f(x) = (x^2 - 4)^2$

24. $f(x) = (x^2 - 2x - 8)^2$

25. $f(x) = x^2(x - 4)^2$

26. $f(x) = x(x - 4)^3$

27. $f(x) = x^2(x - 5)^3$

28. $f(x) = x^3(x - 5)^2$

Graph each function using a graphing calculator by first making a sign diagram for the derivative. For Exercises 29–32, also make a sketch from the screen, showing numbers on the axes and coordinates of relative extreme points. (Answers may vary depending on the window chosen.)

29. $f(x) = x^3 - 300x$ **30.** $f(x) = x^3 - 243x$

31. $f(x) = x^4 - 50x^2 - 25$

32. $f(x) = x^4 - 72x^2 - 4$

33. $f(x) = x^5 - 5x^4 + 5x^3 - 23$

34. $f(x) = x^5 + 5x^4 + 5x^3 - 2$

35. $f(x) = 0.01x^5 - 0.05x$

36. $f(x) = 0.02x^5 - 0.1x$

37. $f(x) = x^3 - 2x^2 + x + 11$

38. $f(x) = x^3 - x^2 + 12$

39. $f(x) = \sqrt{400 - x^2}$

40. $f(x) = \sqrt{2x - x^2}$

For each function, find all critical numbers and graph the function on an appropriate window using a graphing calculator. For Exercises 41 and 42, also make a sketch from the screen, showing numbers on the axes and coordinates of relative extreme points. You may need to ignore some false lines. (Answers may vary depending on the window chosen.)

41. $f(x) = \dfrac{1}{x^2 - 2x - 8}$

42. $f(x) = \dfrac{1}{x^2 - 2x - 3}$

43. $f(x) = \dfrac{8}{x^2 + 4}$ This curve is called the Witch of Agnesi.

44. $f(x) = \dfrac{4x}{x^2 + 4}$ This curve is called Newton's serpentine.

45. $f(x) = \dfrac{x^2}{x^2 + 1}$ **46.** $f(x) = \dfrac{1}{x^2 + 1}$

47. $f(x) = \dfrac{x^2}{x - 3}$ **48.** $f(x) = \dfrac{x^2 - 3x - 1}{x - 5}$

49. $f(x) = \dfrac{10x^2}{x^2 - 5}$ **50.** $f(x) = \dfrac{x^2 + 4}{x}$

51. $f(x) = \dfrac{2x^2}{x^4 + 1}$ **52.** $f(x) = \dfrac{x^4 - 2x^2 + 1}{x^4 + 2}$

Graph each function on an appropriate window using a graphing calculator. (Answers may vary depending on the window chosen.)

53. $f(x) = \dfrac{1}{x - 5}$ **54.** $f(x) = \dfrac{1}{2 + x}$

55. $f(x) = \dfrac{1}{0.1 + 2^{-0.5x}}$ **56.** $f(x) = \dfrac{3x}{\sqrt{x^2 + 1}}$

57. Derive the formula $x = \dfrac{-b}{2a}$ for the x-coordinate of the vertex of parabola $y = ax^2 + bx + c$. [*Hint:* The slope is zero at the vertex, so finding the vertex means finding the critical number.]

58. Derive the formula $x = -b$ for the x-coordinate of the vertex of parabola $y = a(x + b)^2 + c$.

[*Hint:* The slope is zero at the vertex, so finding the vertex means finding the critical number.]

APPLIED EXERCISES

59. BIOMEDICAL: Bacterial Growth A population of bacteria grows to size $p(x) =$ $x^3 - 9x^2 + 24x + 10$ after x hours (for $x \geq 0$). Graph this population curve (based on, if you wish, a calculator graph), showing the coordinates of the relative extreme points.

60. BEHAVIORAL SCIENCE: Learning Curves A learning curve is a function $L(x)$ that gives the amount of time that a person requires to learn x pieces of information. Many learning curves take the form $L(x) = (x - a)^n + b$ (for $x \geq 0$), where a, b, and n are constants. Graph the learning curve $L(x) = (x - 2)^3 + 8$ (based on, if you wish, a calculator graph), showing the coordinates of all corresponding points to critical numbers.

61. BUSINESS: Marginal and Average Cost A company's cost function is

$$C(x) = x^2 + 2x + 4$$

dollars, where x is the number of units.

a. Enter the cost function in y_1 on a graphing calculator.

b. Define y_2 to be the *marginal* cost function by defining y_2 to be the derivative of y_1 (using NDERIV).

c. Define y_3 to be the company's *average* cost function, $AC(x) = \dfrac{C(x)}{x}$, by defining $y_3 = \dfrac{y_1}{x}$.

d. Turn off the function y_1 so that it will not be graphed, but graph the marginal cost function y_2 and the average cost function y_3 on the window [0, 10] by [0, 10]. Observe that the marginal cost function pierces the average cost function at its minimum point (use TRACE to see which curve is which function).

e. To see that the final sentence of part (d) is true in general, change the coefficients in the cost function $C(x)$, or change the cost function to a cubic or some other function [so that $C(x)/x$ has a minimum]. Again turn off the cost function and graph the other two to see that the marginal cost function pierces the average cost function at its minimum. We will return to this observation later.

62. GENERAL: Airplane Flight Path A plane is to take off and reach a level cruising altitude of 5 miles after a horizontal distance of 100 miles, as shown in the diagram below. Find a polynomial flight path of the form $f(x) = ax^3 + bx^2 + cx + d$ by following steps i to iv to determine the constants a, b, c, and d.

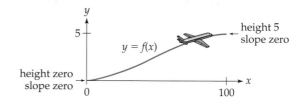

i. Use the fact that the plane is on the ground at $x = 0$ [that is, $f(0) = 0$] to determine the value of d.

ii. Use the fact that the path is horizontal at $x = 0$ [that is, $f'(0) = 0$] to determine the value of c.

iii. Use the fact that at $x = 100$ the height is 5 and the path is horizontal to determine the values of a and b. State the function $f(x)$ that you have determined.

iv. Use a graphing calculator to graph your function on the window [0, 100] by [0, 10] to verify its shape.

63. GENERAL: Drug Interception Suppose that the cost of a border patrol that intercepts x percent of the illegal drugs crossing a state border is

$$C(x) = \frac{600}{100 - x} \quad \text{million dollars (for } x < 100\text{).}$$

a. Graph this function on [0, 100] by [0, 100].
b. Observe that the curve is at first rather flat, but then rises increasingly steeply as x nears 100. Predict what the graph of the derivative would look like.
c. Check your prediction by defining y_2 to be the derivative of y_1 (using NDERIV) and graphing both y_1 and y_2.

64. GENERAL: Aspirin Clinical studies have shown that the analgesic (pain-relieving) effect of aspirin is approximately $f(x) = \dfrac{100x^2}{x^2 + 0.02}$ where $f(x)$ is the percentage of pain relief from x grams of aspirin.

a. Graph this "dose-response" curve for doses up to 1 gram, that is, on the window [0, 1] by [0, 100].

b. TRACE along the curve to see that the curve is very close to its maximum height of 100% or 1 by the time x reaches 0.65. This means that there is very little added effect in going above 650 milligrams, the amount of aspirin in two regular tablets, notwithstanding the aspirin companies' promotion of "extra strength" tablets.

[*Note:* Aspirin's dose-response curve is extremely unusual in that it levels off quite early, and, even more unusual, aspirin's effect in protecting against heart attacks even *decreases* as the dosage x increases above about 80 milligrams.]

3.2 GRAPHING USING THE FIRST AND SECOND DERIVATIVES

Introduction

In the previous section we used the first derivative to find the function's slope and relative extreme points and to draw its graph. In this section we will use the *second* derivative to find the "curl" or *concavity* of the curve, and to define the important concept of *inflection point*. The second derivative also gives us a very useful way to distinguish between maximum and minimum points of a curve.

Concavity and Inflection Points

A curve that curls upward is said to be *concave up*, and a curve that curls downward is said to be *concave down*. A point where the concavity *changes* (from up to down or down to up) is called an *inflection point*.

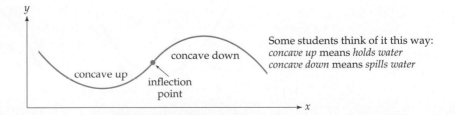

Concavity shows how a curve *curls* or *bends* away from straightness.

A straight line (with any slope) has *no concavity*.

However, bending the two ends *upward* makes it *concave up*,

and bending the two ends *downward* makes it *concave down*.

As these pictures show, a curve that is concave *up* lies *above* its tangent, whereas a curve that is concave *down* lies *below* its tangent (except, of course, at the point of tangency).

Practice Problem 1

For each of the following curves, label the parts that are concave up and the parts that are concave down. Then find all inflection points.

a.

b.

➤ Solutions on page 211

How can we use calculus to determine concavity? The key is the second derivative. The second derivative, being the derivative of the derivative, gives the rate of change of the slope, showing whether the slope is increasing or decreasing. That is, $f'' > 0$ means that the slope is increasing, and so the curve must be *concave up* (as in the diagram on the left below). Similarly, $f'' < 0$ means that the slope is decreasing, and so the curve must be *concave down* (as on the right below).

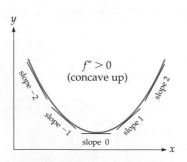

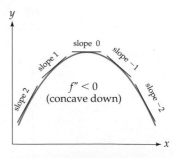

$f'' > 0$ means that the slope is increasing, so f is *concave up*.

$f'' < 0$ means that the slope is decreasing, so f is *concave down*.

Since an inflection point is where the concavity changes, the second derivative must be negative on one side and positive on the other. Therefore, *at* an inflection point, f'' must be either zero or undefined. All of this may be summarized as follows.

Concavity and Inflection Points

$f'' > 0$ on an interval means that f is *concave up* (curls upward) on that interval.

$f'' < 0$ on an interval means that f is *concave down* (curls downward) on that interval.

An *inflection point* is where the concavity *changes* (f'' must be zero or undefined).

Graphing Calculator Exploration

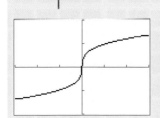

a. Use a graphing calculator to graph $y_1 = \sqrt[3]{x}$ on the window $[-3, 3]$ by $[-2, 2]$. Observe where the curve is concave up and where it is concave down.

b. Use NDERIV to define y_2 to be the derivative of y_1, and y_3 to be the derivative of y_2. Graph y_1 and y_3 (but turn off y_2 so that it will not be graphed).

c. Verify that y_3 (the second derivative) is positive where y_1 is concave up, and negative where y_1 is concave down.

d. Now change y_1 to $y_1 = \dfrac{x^2 + 2}{x^2 + 1}$ and observe that where this curve is concave up or down agrees with where y_3 is positive or negative. According to y_3, how many inflection points does y_1 have? Can you see them on y_1?

To find inflection points, we make a sign diagram for the *second* derivative to show where the concavity changes (where f'' changes sign). An example will make the method clear.

EXAMPLE **1**

GRAPHING AND INTERPRETING A COMPANY'S ANNUAL PROFIT FUNCTION

A company's annual profit after x years is $f(x) = x^3 - 9x^2 + 24x$ million dollars (for $x \geq 0$). Graph this function, showing all relative extreme points and inflection points. Interpret the inflection points.

Solution

$$f'(x) = 3x^2 - 18x + 24 \qquad \text{Differentiating}$$
$$= 3(x^2 - 6x + 8) = 3(x - 2)(x - 4) \qquad \text{Factoring}$$

The critical numbers are $x = 2$ and $x = 4$, and the sign diagram for f' (found in the usual way) is

$f' > 0$	$f' = 0$	$f' < 0$	$f' = 0$	$f' > 0$
	$x = 2$		$x = 4$	

↗ rel max (2, 20) ↘ rel min (4, 16) ↗

To find the inflection points, we calculate the second derivative:

$$f''(x) = 6x - 18 = 6(x - 3) \qquad \text{Differentiating } f'(x) = 3x^2 - 18x + 24$$

This is zero at $x = 3$, which we enter on a sign diagram for the *second* derivative.

$f'' = 0$
$x = 3$

← Behavior of f''
← Where f'' is zero or undefined

We use test points to determine the sign of $f''(x) = 6(x - 3)$ on either side of $x = 3$, just as we did for the first derivative.

$f''(2) = 6(2 - 3) < 0$ ⌐ $f''(4) = 6(4 - 3) > 0$ ⌐

$f'' < 0$	$f'' = 0$	$f'' > 0$
	$x = 3$	

con dn   con up

IP (3,18)

Concave down, concave up (so concavity *does* change)

IP means inflection point. The 18 comes from substituting $x = 3$ into $f(x) = x^3 - 9x^2 + 24x$

Using a graphing calculator, we would choose an x-interval such as $[0, 6]$ (to include the x-values on the sign diagrams) and a y-interval such as $[0, 30]$ (to include the origin and the y-values on the sign diagrams), and graph the function.

By hand, we would plot the relative maximum ($\frown$), minimum ($\smile$), and inflection point and sketch the curve according to the sign diagrams, being sure to show the concavity changing at the inflection point.

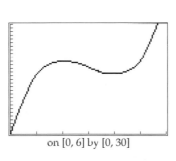

on $[0, 6]$ by $[0, 30]$

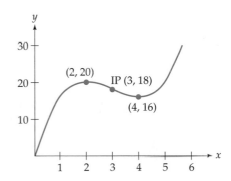

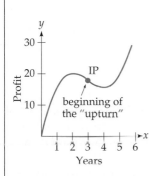

Interpretation of the inflection point: Observe what the graph shows— that the company's profit increased (up to year 2), then decreased (up to year 4), and then increased again. The inflection point at $x = 3$ is where the profit *first began to show signs of improvement.* It marks the end of the period of increasingly steep decline and the first sign of an "upturn," where a clever investor might begin to "buy in."

At an inflection point, the concavity (that is, the sign of f'') must *actually change.* For example, a second derivative sign diagram such as

$$\underline{\quad f'' < 0 \qquad \overset{f'' = 0}{\underset{x = 3}{+}} \qquad f'' < 0 \quad}$$

con dn   con dn

sign of f'' (concavity) does *not* change

would mean that there is *not* an inflection point at $x = 3$, since the concavity is the same on both sides. For there to be an inflection point, the signs of f'' on the two sides *must be different.*

Practice Problem 2   For each curve on the following page, is there an inflection point? [*Hint:* Does the concavity change?]

a.

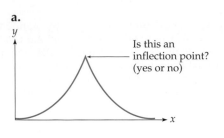

Is this an
inflection point?
(yes or no)

b.

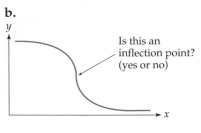

Is this an
inflection point?
(yes or no)

➤ Solutions on page 211

Inflection Points in Everyday Life

Inflection points have important interpretations besides the "first sign of the upturn" in Example 1.*

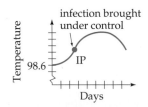

infection brought under control

IP

98.6

Temperature

Days

The graph on the left shows a person's temperature during an illness, and the inflection point is where the temperature ends its steepest rise and begins to moderate—that is, the point at which the illness is *first brought under control.*

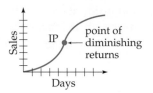

Sales

IP

point of
diminishing
returns

Days

The graph on the left gives a company's total sales after x days of an advertising campaign. The inflection *point gives the point of diminishing returns,* beyond which advertising will still bring additional sales, but at a slower rate.

Distinguish carefully between slope and concavity: *Slope* measures *steepness,* whereas *concavity* measures *curl.* All combinations of slope and concavity are possible. A graph may be

Increasing and concave *up* ($f' > 0, f'' > 0$), such as

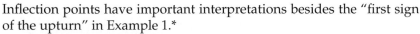

Increasing and concave *down* ($f' > 0, f'' < 0$), such as

---

* For an interesting application of concavity to economics, see Harry M. Markowitz, "The Utility of Wealth," *Journal of Applied Political Economy*, 60(2):151–158, 1952. In this article, "concave" means "concave up" and "convex" means "concave down." Markowitz won the Nobel Memorial Prize in Economics in 1990.

Decreasing and concave *up* $(f' < 0, f'' > 0)$, such as

Decreasing and concave *down* $(f' < 0, f'' < 0)$, such as

The four quarters of a circle illustrate all four possibilities, as shown at the left.

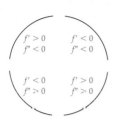

EXAMPLE 2

GRAPHING A FRACTIONAL POWER FUNCTION

Graph $f(x) = 18x^{1/3}$.

Solution The derivative is

$$f'(x) = 6x^{-2/3} = \frac{6}{\sqrt[3]{x^2}}$$

Undefined at $x = 0$ (zero denominator)

The sign diagram for f' is

$$\underline{\quad f' > 0 \qquad f' \text{ und} \qquad f' > 0 \quad}$$
$$\qquad\qquad x = 0 \qquad \nearrow$$
$$\qquad\qquad \text{neither}$$
$$\nearrow \qquad\quad (0, 0)$$

f' is undefined at $x = 0$ and positive on either side (using test points)

The *second* derivative is

$$f''(x) = -4x^{-5/3} = \frac{-4}{\sqrt[3]{x^5}}$$

Also undefined at $x = 0$

The sign diagram for f'' is

$$\underline{\quad f'' > 0 \qquad f'' \text{ und} \qquad f'' < 0 \quad}$$
$$\text{con up} \qquad x = 0 \qquad \text{con dn}$$
$$\qquad\qquad \text{IP } (0, 0)$$

Concavity is different on either side of $x = 0$ (using test points), so there is an inflection point at $x = 0$

Based on this information, we may graph the function with a graphing calculator or by hand:

Using a graphing calculator, we experiment with viewing windows centered at the inflection point $(0, 0)$ until we find one that shows the curve effectively. The graph on $[-2, 2]$ by $[-20, 20]$ is shown below. Many other windows would be just as good.

By hand, we use the sign diagrams to draw the curve to the *left* of $x = 0$ as *increasing* and concave *up,* and to the right of $x = 0$ as increasing and concave *down,* with the two parts meeting at the origin. The scale comes from calculating the points $(1, 18)$ and $(-1, -18)$.

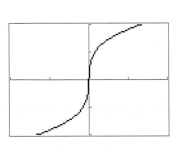

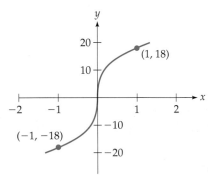

The fact that the derivative is *undefined* at $x = 0$ is shown in the graph by the *vertical tangent* at the origin (since the slope of a vertical line is undefined). This function is the stimulus-response curve for brightness of light (for $x \geq 0$; see page 185), and the vertical tangent indicates the disproportionately large effect of small increases in dim light.

EXAMPLE **3**

GRAPHING USING A GRAPHING CALCULATOR

Use a graphing calculator to graph $f(x) = 36\sqrt[3]{(x - 1)^2}$.

Solution Using a standard window $[-10, 10]$ by $[-10, 10]$ gives the graph on the left, which is useless.

Instead, we begin by differentiating $f(x) = 36(x - 1)^{2/3}$ and making its sign diagram.

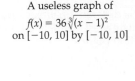

A useless graph of
$f(x) = 36\sqrt[3]{(x - 1)^2}$
on $[-10, 10]$ by $[-10, 10]$

$$f'(x) = 24(x - 1)^{-1/3} = \frac{24}{\sqrt[3]{x - 1}}$$

Undefined at $x = 1$

$f' < 0$ f' und $f' > 0$

$x = 1$

↘   ↗

rel min
$(1, 0)$

Using test points on either side of $x = 1$.

The y-coordinate in $(1, 0)$ comes from evaluating the original function at $x = 1$

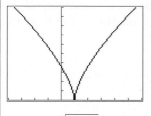

$y = 36 \sqrt[3]{(x-1)^2}$ on
$[-4, 6]$ by $[0, 100]$

From the sign diagram we choose an x-interval centered at the critical number $x = 1$, such as $[-4, 6]$. After some experimentation, we choose the y-interval $[0, 100]$ (beginning at $x = 0$ because the minimum point has y-coordinate 0) so that the curve fills the screen. A sharp point on a graph such as the one at $x = 1$ is called a *cusp*, where the function is not differentiable.

Graphing Calculator Exploration

a. In the graph at the very beginning of Example 3, why didn't the curve reach the x-axis, as the later graph shows that it does? [*Hint:* It has to do with pixels.]

b. Try graphing the same function $36\sqrt[3]{(x-1)^2}$ but entered in exponential form as $y_1 = 36(x-1)^\wedge(2/3)$. If your calculator displays only half of the curve, try separating the exponent $2/3$ into two parts, such as $36((x-1)^\wedge 2)^\wedge(1/3)$ or instead as $36((x-1)^\wedge(1/3))^\wedge 2$. You may need to make similar adjustments in the future depending on your calculator.

Second-Derivative Test

Determining whether a twice-differentiable function has a relative maximum or minimum at a critical number is merely a question of concavity: concave *up* means a relative *minimum*, and concave *down* means a relative *maximum*.

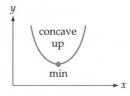

Concave *up* at a critical number: relative *minimum*.

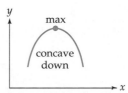

Concave *down* at a critical number: relative *maximum*.

Since the second derivative determines concavity, we have the following *second-derivative test*, which will be very useful in the next two sections.

Second-Derivative Test for Relative Extreme Points

If $x = c$ is a critical number of f at which f'' is defined, then

$f''(c) > 0$ means that f has a relative *minimum* at $x = c$.

$f''(c) < 0$ means that f has a relative *maximum* at $x = c$.

To use the second-derivative test, first find all critical numbers, substituting each into the second derivative and determining the sign of the result: A *positive* result means a *minimum* at the critical number, and a *negative* result means a *maximum*. (If the second derivative is zero, then the test is inconclusive, and you should use the first-derivative test (page 191) or make a sign diagram for f'.)

EXAMPLE 4

USING THE SECOND-DERIVATIVE TEST

Use the second-derivative test to find all relative extreme points of $f(x) = x^3 - 9x^2 + 24x$.

Solution

$$f'(x) = 3x^2 - 18x + 24 \qquad \text{The derivative}$$
$$= 3(x^2 - 6x + 8) = 3(x - 2)(x - 4) \qquad \text{Factoring}$$

$$\text{CN} \begin{cases} x = 2 \\ x = 4 \end{cases} \qquad \text{Critical numbers}$$

We substitute each critical number into $f''(x) = 6x - 18$.

$$f''(2) = 6 \cdot 2 - 18 = -6 \quad \text{(negative)} \qquad \begin{array}{l} f''(x) = 6x - 18 \\ \text{at } x = 2 \end{array}$$

Therefore, f has a relative *maximum* at $x = 2$.

$$f''(4) = 6 \cdot 4 - 18 = 6 \quad \text{(positive)} \qquad \begin{array}{l} f''(x) = 6x - 18 \\ \text{at } x = 4 \end{array}$$

Therefore, f has a relative *minimum* at $x = 4$.

The relative maximum at $x = 2$ and the relative minimum at $x = 4$ are exactly what we found before when we graphed this function on pages 202–203.

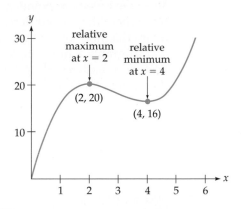

Be careful! The second-derivative test tells us *nothing* if the second derivative is zero at a critical number. This is shown by the three functions below. Each function has a critical number $x = 0$ at which f'' is zero (as you may check), but one has a maximum, one has a minimum, and one has neither.

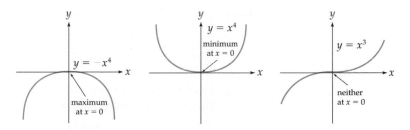

Three functions showing that a critical number where $f'' = 0$
may be a maximum, a minimum, or neither.

3.2 **Section Summary**

The main developments of this section were the use of the second derivative to determine the *concavity* or *curl* of a function, and finding and interpreting *inflection points.*

$f'' > 0$ on an interval means that f is concave up on that interval.

$f'' < 0$ on an interval means that f is concave down on that interval.

To locate inflection points (points at which the concavity changes), we find where the second derivative is zero or undefined, and then make a sign diagram for f'' to see whether the concavity *actually changes*.

To graph a function, we use the *first*-derivative sign diagram to find *slope* and *relative extreme points,* and the *second*-derivative sign diagram to find *concavity* and *inflection points.* Then we graph the function, either by hand or on a graphing calculator (using the relative extreme points and inflection points to determine the viewing window).

The following steps may be useful when graphing a function by hand.

Curve Sketching

1. Find the domain.
2. Find the first derivative.
3. Set the derivative equal to zero and solve for x. Also find the x-values at which the derivative is undefined.
4. Find the y-value for each x-value found in step 3 that is in the domain of the function.
5. Make a sign diagram for the first derivative, marking points found in steps 3 and 4 as "max," "min," or "neither."
6. Find the second derivative.
7. Set the second derivative equal to zero and solve for x. Also find the x-values at which the second derivative is undefined.
8. Find the y-value for each x-value found in step 7 that is in the domain of the function.
9. Make a sign diagram for the second derivative, marking points found in steps 7 and 8 as inflection points *if* the second derivative changes sign there.
10. From the sign diagrams, sketch the graph, labeling all maximum, minimum, and inflection points.

The second-derivative test shows whether a function has a relative maximum or minimum at a critical value (provided that f'' is defined and not zero):

$$f'' > 0 \quad \text{means a relative minimum.}$$

$$f'' < 0 \quad \text{means a relative maximum.}$$

The second-derivative test should be thought of as a simple application of concavity.

➤ **Solutions to Practice Problems**

1. a.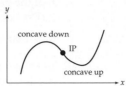

b.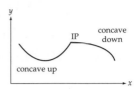

2. a. No (concave up on both sides)

   b. Yes (concave down then concave up)

3.2 **Exercises**

For each graph, which of the numbered points are inflection points?

1.

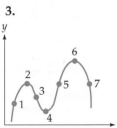

2.

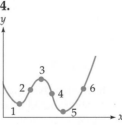

3.

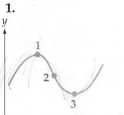

4.

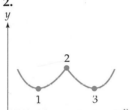

5.

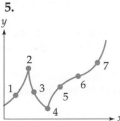

6.

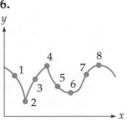

For each function:
a. Make a sign diagram for the first derivative.
b. Make a sign diagram for the second derivative.

c. Sketch the graph by hand, showing all relative extreme points and inflection points.

7. $f(x) = x^3 + 3x^2 - 9x + 5$

8. $f(x) = x^3 - 3x^2 - 9x + 7$

9. $f(x) = x^3 - 3x^2 + 3x + 4$

10. $f(x) = x^3 + 3x^2 + 3x + 6$

11. $f(x) = x^4 - 8x^3 + 18x^2 + 2$

12. $f(x) = x^4 + 8x^3 + 18x^2 + 8$

13. $f(x) = 5x^4 - x^5$ **14.** $f(x) = (x - 2)^3 + 2$

15. $f(x) = (2x + 4)^5$ **16.** $f(x) = (3x - 6)^6 + 1$

17. $f(x) = x(x - 3)^2$ **18.** $f(x) = x^3(x - 4)$

19. $f(x) = x^{3/5}$ **20.** $f(x) = x^{1/5}$

21. $f(x) = \sqrt[5]{x^4} + 2$ **22.** $f(x) = \sqrt[5]{x^2} - 1$

23. $f(x) = \sqrt[4]{x^3}$ **24.** $f(x) = \sqrt{x^5}$

25. $f(x) = \sqrt[3]{(x - 1)^2}$ **26.** $f(x) = \sqrt[5]{x + 2} + 3$

Graph each function using a graphing calculator by first making sign diagrams for the first and second derivatives. Make a sketch from the screen, showing the coordinates of all relative extreme points and inflection points. Graphs may vary depending on the window chosen.

27. $f(x) = x^3 - 18x^2 + 60x + 20$

28. $f(x) = x^3 - 300x$

29. $f(x) = x^4 - 16x^3$ **30.** $f(x) = x^4 - 20x^3$

31. $f(x) = x^3 - 9x^2 - 48x + 48$

32. $f(x) = x^3 - 6x^2 - 63x + 42$

33. $f(x) = x^3 - 2x^2 + x + 5$

34. $f(x) = x^3 + 2x^2 + x - 4$

35. $f(x) = 36\sqrt[3]{x - 1}$ **36.** $f(x) = 45\sqrt[3]{(x - 1)^2}$

Graph each function using a graphing calculator by first making a sign diagram for just the first derivative. Make a sketch from the screen, showing the coordinates of all relative extreme points and inflection points. Graphs may vary depending on the window chosen.

37. $f(x) = x^{1/2}$ **38.** $f(x) = x^{3/2}$

39. $f(x) = x^{-1/2}$ **40.** $f(x) = x^{-3/2}$

41. $f(x) = 9x^{2/3} - 6x$ **42.** $f(x) = 30x^{1/3} - 10x$

43. $f(x) = 8x - 10x^{4/5}$ **44.** $f(x) = 6x - 10x^{3/5}$

45. $f(x) = 3x^{2/3} - x^2$ **46.** $f(x) = 3x^{4/3} - 2x^2$

47. CONCAVITY OF A PARABOLA Show that the quadratic function $f(x) = ax^2 + bx + c$ is concave up if $a > 0$ and is concave down if $a < 0$. Therefore, the rule that a parabola opens up if $a > 0$ and down if $a < 0$ is merely an application of concavity. [*Hint:* Find the second derivative.]

48. INFLECTION POINT OF A CUBIC Show that the general "cubic" (third degree) function $f(x) = ax^3 + bx^2 + cx + d$ (with $a \neq 0$) has an inflection point at $x = \dfrac{-b}{3a}$.

49. INFLECTION POINTS Explain why, at an inflection point, a curve must cross its tangent line (assuming that the tangent line exists).

50. INFLECTION POINTS For a twice-differentiable function, explain why the slope must have a relative maximum or minimum value at an inflection point. [*Hint:* Use the fact that the concavity changes at an inflection point, and then interpret concavity in terms of increasing and decreasing slope.]

51–52. FINDING INFLECTION POINTS Use a graphing calculator to estimate the x-coordinates of the inflection points of each function, rounding your answers to two decimal places. [*Hint:* Graph the second derivative, either calculating it directly or using NDERIV twice, and see where it crosses the x-axis.]

51. $f(x) = x^5 - 2x^3 + 3x + 4$

52. $f(x) = x^5 - 3x^3 + 6x + 2$

APPLIED EXERCISES

53. BUSINESS: Revenue A company's annual revenue after x years is
$$f(x) = x^3 - 9x^2 + 15x + 25$$
thousand dollars (for $x \geq 0$).
a. Make sign diagrams for the first and second derivatives.
b. Sketch the graph of the revenue function, showing all relative extreme points and inflection points.

54. BUSINESS SALES A company's weekly sales (in thousands) after x weeks are given by $f(x) = -x^4 + 4x + 70$ (for $0 \leq x \leq 3$).
a. Make sign diagrams for the first and second derivatives.

b. Sketch the graph of the sales function, showing all relative extreme points and inflection points.

55. GENERAL: Temperature The temperature in a refining tower is $f(x) = x^4 - 4x + 112$ degrees Fahrenheit after x hours (for $x \geq 0$).
a. Make sign diagrams for the first and second derivatives.
b. Sketch the graph of the temperature function, showing all relative extreme points and inflection points.

56. BIOMEDICAL: Dosage Curve The dose-response curve for x grams of a drug is $f(x) = 8(x - 1)^3 + 8$ (for $x \geq 0$).

a. Make sign diagrams for the first and second derivatives.

b. Sketch the graph of the response function, showing all relative extreme points and inflection points.

57. PSYCHOLOGY: Stimulus and Response
Sketch the graph of the brightness response curve $f(x) = x^{2/5}$ for $x \geq 0$, showing all relative extreme points and inflection points.

58. PSYCHOLOGY: Stimulus and Response
Sketch the graph of the loudness response curve $f(x) = x^{4/5}$ for $x \geq 0$, showing all relative extreme points and inflection points.

59–60. SOCIOLOGY: Status Sociologists have estimated how a person's "status" in society (as perceived by others) depends on the person's income and education level. One estimate is that status S depends on income i according to the formula $S(i) = 16\sqrt{i}$ (for $i \geq 0$), and that status depends upon education level e according to the formula $S(e) = \frac{1}{4}e^2$ (for $e \geq 0$).

59. a. Sketch the graph of the function $S(i)$ above.
b. Is the curve concave up or down? What does this signify about the rate at which status increases at higher income levels?

60. a. Sketch the graph of the function $S(e)$ above.
b. Is the curve concave up or down? What does this signify about the rate at which status increases at higher education levels?

61–62. PSYCHOLOGY: Response Curves The data in the tables below were collected from two stimulus-response experiments. The table in Exercise 61 gives weight (in pounds, actual and perceived), and the table in Exercise 62 gives temperature (degrees Celsius above room temperature, actual and perceived).

61.

Actual Weight	Perceived Weight
1	2
2	3
5	6
10	8
20	15

62.

Actual Temperature	Perceived Temperature
2	1
5	3
10	9
12	12
15	16

a. Enter the data from the table you are using into a graphing calculator and make a plot of the resulting points (Actual on the x-axis and Perceived on the y-axis).
b. Have your calculator find the power regression formula (of the form ax^b) for these data. Then enter the results in y_1, which gives a stimulus-response curve. Plot the points together with the power regression curve. Observe that the curve fits the points quite well.
c. Is the stimulus-response curve concave up or down?
d. Use your stimulus-response curve to predict the *perceived* value if the *actual* value were 25.
e. Use your stimulus-response curve to predict the *actual* value if the *perceived* value were 8.

63. GENERAL: Airplane Flight Path In Exercise 62 on page 198 it was found that the flight path that satisfies the conditions in the diagram shown below was $y = -0.00001x^3 + 0.0015x^2$. Find the inflection point of this path, and explain why it represents the point of steepest ascent of the airplane.

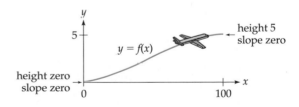

3.3 OPTIMIZATION

Introduction

Many problems consist of "optimizing" a function—that is, finding its maximum or minimum value. For example, you might want to maximize your profit, or to minimize the time required to do a task. If you could express your happiness as a function, you would want to maximize it.* One of the principal uses of calculus is that it provides a very general technique for optimizing functions.

We will concentrate on *applications* of optimization. Accordingly, we will optimize continuous functions that are defined on closed intervals, or functions that have only one critical number in their domain. Most applications fall into these two categories, and the wide range of examples and exercises in this and the following sections will demonstrate the power of these techniques.

Absolute Extreme Values

The *absolute maximum value* of a function is the *largest* value of the function on its domain. Similarly, the *absolute minimum value* of a function is the *smallest* value of the function on its domain. An absolute *extreme* value is a value that is either the absolute maximum or the absolute minimum value of the function. (This use of the word "absolute" has nothing to do with its use in the *absolute value function* defined on page 55.) The maximum and minimum values of the function correspond to the highest and lowest points on its graph.

For a given function, both absolute extreme values may exist, or one or both may fail to exist, as the following graphs show.

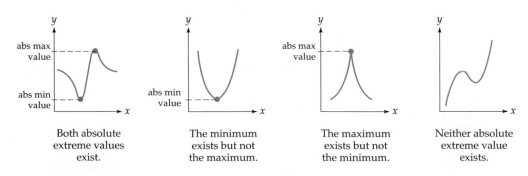

Both absolute extreme values exist.

The minimum exists but not the maximum.

The maximum exists but not the minimum.

Neither absolute extreme value exists.

---

* Expressing happiness in numbers has an honorable past: Plato (*Republic* IX, 587) calculated that a king is exactly 729 ($= 3^6$) times as happy as a tyrant.

When will both extreme values exist? For a *continuous* function (one whose graph is a single unbroken curve) defined on a *closed* interval (that is, including both endpoints), the absolute maximum and minimum values are *guaranteed* to exist. We know from graphing functions that maximum and minimum values can occur only at critical numbers, unless they occur at endpoints, where the curve is "cut off" from rising or falling further. To summarize:

Optimizing Continuous Functions on Closed Intervals

A continuous function f on a closed interval $[a, b]$ has both an absolute maximum value and an absolute minimum value. To find them:

1. Find all critical numbers of f in $[a, b]$.
2. Evaluate f at the critical numbers and at the endpoints a and b.

The largest and smallest values found in step 2 will be the absolute maximum and minimum values of f on $[a, b]$.

Simply stated, to find the absolute extreme values, we need to consider only *critical numbers and endpoints.*

EXAMPLE 1

OPTIMIZING A CONTINUOUS FUNCTION ON A CLOSED INTERVAL

Find the absolute extreme values of $f(x) = x^3 - 9x^2 + 15x$ on $[0, 3]$.

Solution

The function is continuous (it is a polynomial) and the interval is closed, so both extreme values exist. First we find the critical numbers.

$$f'(x) = 3x^2 - 18x + 15 \qquad \text{The derivative}$$
$$= 3(x^2 - 6x + 5) = 3(x - 1)(x - 5) \qquad \text{Factoring}$$

$$\text{CN} \begin{cases} x - 1 \\ x = 5 \end{cases} \quad \leftarrow \text{Not in the given domain} \\ \qquad\qquad\quad [0, 3], \text{ so we eliminate it}$$

We evaluate f at the remaining critical numbers and at the endpoints (EP).

CN: $x = 1$ $f(1) = 1 - 9 + 15 \qquad = 7$ ← Largest function value

EP $\begin{cases} x = 0 & f(0) = 0 - 0 + 0 \qquad\qquad = 0 \\ x = 3 & f(3) = 27 - 9 \cdot 9 + 15 \cdot 3 = -9 \end{cases}$ ← Smallest function value

The largest (7) and the smallest (-9) of the resulting values of the function are the absolute extreme values of f on $[0, 3]$:

Maximum value of f is   7 (occurring at $x = 1$).

Minimum value of f is -9 (occurring at $x = 3$).

The graph of the function shows these absolute extreme values, one occurring at a critical number ($x = 1$), and the other occurring at an endpoint ($x = 3$).

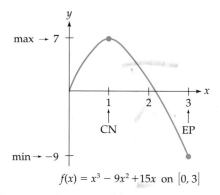

$f(x) = x^3 - 9x^2 + 15x$ on $[0, 3]$

In other problems, both extreme values might occur at critical numbers or both might occur at endpoints.

Notice how calculus helped in this example. The absolute extreme values could have occurred at *any* x-values in $[0, 3]$. Calculus reduced this infinite list of possibilities (*all* numbers between 0 and 3, not just integers) to a mere three numbers (0, 1, and 3). We then had only to "test" these numbers by substituting them into the function to find which numbers made f largest and smallest.

Second-Derivative Test for Absolute Extreme Values

Earlier we used the second-derivative test to find *relative* maximum and minimum values. However, if a continuous function has only one critical number in its domain, then the second-derivative test can be used to find *absolute* extreme values (since without a second critical number the function must continue to increase or decrease away from the relative extreme point). Some students refer to this as "the only critical point in town" test.

Applications of Optimization

If a timber forest is allowed to grow for t years, the value of the timber increases in proportion to the square root of t, while maintenance costs are proportional to t. Therefore, the value of the forest after t years is of the form

$$V(t) = a\sqrt{t} - bt \qquad \qquad a \text{ and } b \text{ are constants}$$

EXAMPLE **2**   **OPTIMIZING THE VALUE OF A TIMBER FOREST**

The value of a timber forest after t years is $V(t) = 96\sqrt{t} - 6t$ thousand dollars (for $t > 0$). Find when its value is maximized.

Solution

$$V(t) = 96t^{1/2} - 6t \qquad \qquad V(t) \text{ in exponential form}$$

$$V'(t) = 48t^{-1/2} - 6 = 0 \qquad \text{Setting the derivative equal to zero}$$

$$48t^{-1/2} = 6 \qquad \qquad \text{Adding 6 to each side of } 48t^{-1/2} - 6 = 0$$

$$t^{-1/2} = \frac{6}{48} = \frac{1}{8} \qquad \qquad \text{Dividing by 48}$$

$$\frac{1}{\sqrt{t}} = \frac{1}{8} \qquad \qquad \text{Expressing } t^{-1/2} \text{ in radical form}$$

$$\sqrt{t} = 8 \qquad \qquad \text{Inverting both sides}$$

$$t = 64 \qquad \qquad \text{Squaring both sides}$$

Since there is a *single* critical number ($t = 0$, which makes the derivative undefined, is not in the domain), we may use the second-derivative test. The second derivative is

$$V''(t) = -24t^{-3/2} = -\frac{24}{\sqrt{t^3}} \qquad \text{Differentiating } V'(t) = 48t^{-1/2} - 6$$

At $t = 64$ this is clearly negative, so by the second-derivative test $V(t)$ is *maximized* at $t = 64$. Finally, we state the answer clearly:

The value of the forest is maximized after 64 years. The maximum value is $V(64) = 96\sqrt{64} - 6 \cdot 64 = 384$ thousand dollars, or $384,000.

Maximizing Profit

The famous economist John Maynard Keynes said, "The engine that drives Enterprise is Profit." Many management problems consist of maximizing profit, and require *constructing* the profit function before maximizing it. Such problems have three economic ingredients. The first is that profit is defined as *revenue minus cost:*

$$\text{Profit} = \text{Revenue} - \text{Cost}$$

The second ingredient is that revenue is *price times quantity*. For example, if a company sells 100 toasters for $25 each, the revenue will obviously be $25 \cdot 100 = \$2500$.

$$\text{Revenue} = \left(\begin{matrix} \text{Unit} \\ \text{price} \end{matrix}\right) \cdot (\text{Quantity})$$

The third economic ingredient reflects the fact that, in general, price and quantity are inversely related: increasing the price decreases sales, whereas decreasing the price increases sales. To put this another way, "flooding the market" with a product drives the price down, whereas creating a shortage drives the price up. If the relationship between the price p and the quantity x that consumers will buy at that price is expressed as a function $p(x)$, it is called the *price function.*[*]

> ### Price Function
>
> $p(x)$ gives the price p at which consumers will buy exactly x units of the product.

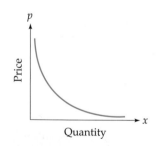

The price function $p(x)$ shows the inverse relation between price p and quantity x.

The price function, relating price and quantity, may be linear or curved (as shown on the left), but it will always be a *decreasing* function.

In actual practice, price functions are very difficult to determine, requiring extensive (and expensive) market research. In this section we will be given the price function. In the next section we will see how to do without price functions, at least in simple cases.

EXAMPLE 3

MAXIMIZING A COMPANY'S PROFIT

It costs the American Automobile Company $8000 to produce each automobile, and fixed costs (rent and other expenses that do not depend on the amount of production) are $20,000 per week. The company's price function is $p(x) = 22{,}000 - 70x$, where p is the price at which exactly x cars will be sold.

* We will use *lowercase p* for price and *capital P* for profit.

a. How many cars should be produced each week to maximize profit?

b. For what price should they be sold?

c. What is the company's maximum profit?

Solution **Revenue** is price times quantity, $R = p \cdot x$:

$$R = p \cdot x = (22{,}000 - 70x)x = 22{,}000x - 70x^2$$

Replacing p by the price function
$p = 22{,}000 - 70x$

$\underbrace{\phantom{(22{,}000 - 70x)}}_{p(x)}$ Revenue function $R(x)$

Cost is the cost per car ($8000) times the number of cars (x) plus the fixed cost ($20,000):

$$C(x) = 8000x + 20{,}000 \qquad \text{(Unit cost)·(Quantity) + (Fixed costs)}$$

Profit is revenue minus cost:

$$P(x) = \underbrace{(22{,}000x - 70x^2)}_{R(x)} - \underbrace{(8000x + 20{,}000)}_{C(x)}$$

$$= -70x^2 + 14{,}000x - 20{,}000$$

Profit function
(after simplification)

a. We maximize the profit by setting its derivative equal to zero:

$$P'(x) = -140x + 14{,}000 = 0$$

Differentiating
$P = -70x^2 + 14{,}000x - 20{,}000$

$$-140x = -14{,}000$$

Solving

$$x = \frac{-14{,}000}{-140} = 100$$

Only one critical number

$$P''(x) = -140$$

From $P'(x) = -140x + 14{,}000$

The second derivative is negative, so the profit is maximized at the critical number. (If the second derivative had involved x, we would have substituted the critical number $x = 100$.) Since x is the number of cars, the company should produce 100 cars per week (the time period stated in the problem).

b. The selling price p is found from the price function:

$$p = 22{,}000 - 70 \cdot 100 = \$15{,}000$$

$p(x) = 22{,}000 - 70x$
evaluated at $x = 100$

c. The maximum profit is found from the profit function:

$$P(100) = -70(100)^2 + 14{,}000(100) - 20{,}000$$

$P(x) = -70x^2 + 14{,}000x - 20{,}000$
evaluated at $x = 100$

$$= \$680{,}000$$

Finally, state the answer clearly in words.

The company should make 100 cars per week and sell them for $15,000 each. The maximum profit will be $680,000.

Actually, automobile dealers seem to prefer prices like $14,999 as if $1 makes a difference.

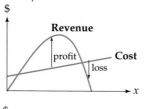

Graphs of the Revenue, Cost, and Profit Functions

The graphs of the revenue and cost functions are shown on the left. At x-values where revenue is above cost, there is a profit, and where the cost is above the revenue, there is a loss.

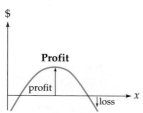

The height of the profit function at any x is the amount by which the revenue is above the cost in the graph on the left. Since profit equals revenue minus cost, we may differentiate each side of $P(x) = R(x) - C(x)$, obtaining

$$P'(x) = R'(x) - C'(x)$$

This shows that setting $P'(x) = 0$ (which we do to maximize profit) is equivalent to setting $R'(x) - C'(x) = 0$, which is equivalent to $R'(x) = C'(x)$. This last equation may be expressed in marginals, $MR = MC$, which is a classic economic criterion for maximum profit.

Classic Economic Criterion for Maximum Profit

At maximum profit:

$$\begin{pmatrix} \text{Marginal} \\ \text{revenue} \end{pmatrix} = \begin{pmatrix} \text{Marginal} \\ \text{cost} \end{pmatrix}$$

EXAMPLE 4

MAXIMIZING THE AREA OF AN ENCLOSURE

A farmer has 1000 feet of fence and wants to build a rectangular enclosure along a straight wall. If the side along the wall needs no fence, find the dimensions that make the enclosure as large as possible. Also find the maximum area.

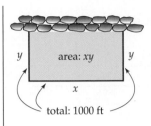

Solution   The largest enclosure means, of course, the largest area.

We let variables stand for the length and width:

$$x = \text{length (parallel to wall)}$$
$$y = \text{width (perpendicular to wall)}$$

The problem becomes

Maximize $A = xy$	Area is length times width
subject to $x + 2y = 1000$	One x side and two y sides from 1000 feet of fence

We must express the area $A = xy$ in terms of one variable. We use

$x = 1000 - 2y$	Solving $x + 2y = 1000$ for x
$A = xy = \underbrace{(1000 - 2y)y}_{x} = 1000y - 2y^2$	Substituting $x = 1000 - 2y$ into $A = xy$
$A' = 1000 - 4y = 0$	Maximizing $A = 1000y - 2y^2$ by setting the derivative equal to zero
$y = 250$	Solving $1000 - 4y = 0$ for y

Since $A'' = -4$,   the second-derivative test shows that the area is indeed *maximized* when $y = 250$.   The length x is

$$x = 1000 - 2 \cdot 250 = 500 \quad \text{Evaluating } x = 1000 - 2y \text{ at } y = 250$$

Length (parallel to the wall) is 500 feet, width (perpendicular to the wall) is 250 feet, and area (length times width) is 125,000 square feet.

Spreadsheet Exploration

In the preceding example you might think that it does not matter how the fence is laid out as long as all 1000 feet are used. To see that the area enclosed really *does* change, and that the maximum occurs at $y = 250$,   the following spreadsheet* calculates the area $A(y) = 1000y - 2y^2$ for y-values from 245 to 255. Notice that the area is largest for $y = 250$,   and that for each change of 1 in the y-value the change in the area is smallest for widths closest to 250. This

* See the Preface for how to obtain this and other Excel spreadsheets.

verifies *numerically* that the derivative (rate of change) becomes zero as y approaches 250.

B8 ▼	=	=1000*A8-2*A8^2
	A	**B**
1	Width (perpendicular to wall)	Area Enclosed
2		
3	245	124950
4	246	124968
5	247	124982
6	248	124992
7	249	124998
8	250	125000
9	251	124998
10	252	124992
11	253	124982
12	254	124968
13	255	124950

Based on this spreadsheet, how much area will the farmer lose if he mistakenly makes the width 249 or 251 feet instead of 250 (and therefore the length 502 or 498 feet)? Is this loss significant based on an area of 125,000 square feet? This is characteristic of maximization problems where the slope is zero—begin *near* the maximizing value is essentially as good as being *at* it.

EXAMPLE 5

MAXIMIZING THE VOLUME OF A BOX

An open-top box is to be made from a square sheet of metal 12 inches on each side by cutting a square from each corner and folding up the sides, as in the diagram below. Find the volume of the largest box that can be made in this way.

Square sheet

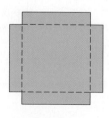

Corners removed

Side flaps folded up to make open-top box.

Solution   Let x = the length of the side of the square cut from each corner.

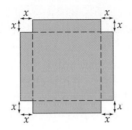

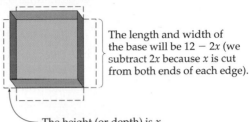

The length and width of the base will be $12 - 2x$ (we subtract $2x$ because x is cut from both ends of each edge).

The 12" by 12" square with four x by x corners removed.

The height (or depth) is x, the size of the edge folded up.

Therefore, the volume is

$$V(x) = (12 - 2x)(12 - 2x)x \qquad \text{(length)·(width)·(height)}$$

Since x is a length, $x > 0$,   and since x inches are cut from *both* sides of each 12-inch edge, we must have $2x < 12$,   so $x < 6$.   The problem becomes

$$\text{Maximize} \quad V(x) = (12 - 2x)(12 - 2x)x \qquad \text{on } 0 < x < 6$$
$$V(x) = (144 - 48x + 4x^2)x \qquad \text{Multiplying out}$$
$$= 4x^3 - 48x^2 + 144x \qquad \text{Multiplying out}$$
$$V'(x) = 12x^2 - 96x + 144 \qquad \text{Differentiating}$$
$$= 12(x^2 - 8x + 12) \qquad \text{Factoring}$$
$$= 12(x - 2)(x - 6) \qquad \text{Factoring}$$

$$\text{CN} \begin{cases} x = 2 \\ x = \cancel{6} \qquad \leftarrow \text{Not in the domain, so we eliminate it} \end{cases}$$

The second derivative is $V''(x) = 24x - 96$,   which at $x = 2$ is

$$V''(2) = 48 - 96 < 0$$

Therefore, the volume is *maximized* at $x = 2$.

Maximum volume is 128 cubic inches.   From $V(x)$ evaluated at $x = 2$

Graphing Calculator Exploration

Do the previous example (and then modify it) on a graphing calculator as follows:

a. Enter the volume as $y_1 = (12 - 2x)(12 - 2x)x$ and graph it on $[0, 6]$ by $[0, 150]$.

b. Use MAXIMUM to maximize y_1. Your answer should agree with that found above. (While this may seem faster than doing the problem "by hand," you would still need to have found the volume function and the domain, which was most of the work.)

c. What if the beginning size were not a 12-inch by 12-inch square, but a standard 8.5-inch by 11-inch sheet of paper? Go back to y_1 and replace the two 12s by 8.5 and 11 and find the x and the maximum volume. (*Answer:* volume about 66 cubic inches)

d. What about a 3 by 5 card?

Parts (c) and (d), which would be more difficult to solve by hand, show how useful a graphing calculator is for solving related problems after one has been analyzed using calculus.

3.3 Section Summary

There is no single, all-purpose procedure for solving word problems. You must think about the problem, draw a picture if possible, and express the quantity to be maximized or minimized in terms of some appropriate variable. You can become good at it only with practice.

We have two procedures for optimizing continuous functions on intervals:

1. If the function has only one critical number in the interval, we find it and use the second-derivative test to show whether the function is maximized or minimized there.

2. If the interval is closed, we evaluate the function at all critical numbers and endpoints in the interval; the largest and smallest resulting values will be the maximum and minimum values of the function.

For functions satisfying neither of these two conditions, graph the function. The maximum (or minimum) value will be the y-coordinate of the highest (or lowest) point on the graph.

3.3 Exercises

Find (*without* using a calculator) the absolute extreme values of each function on the given interval.

1. $f(x) = x^3 - 6x^2 + 9x + 8$ on $[-1, 2]$

2. $f(x) = x^3 - 6x^2 + 22$ on $[-2, 2]$

3. $f(x) = x^3 - 12x$ on $[-3, 3]$

4. $f(x) = x^3 - 27x$ on $[-2, 2]$

5. $f(x) = x^4 + 4x^3 + 4x^2$ on $[-2, 1]$

6. $f(x) = x^4 - 4x^3 + 4x^2$ on $[0, 3]$

7. $f(x) = 2x^5 - 5x^4$ on $[-1, 3]$

8. $f(x) = 4x^5 - 5x^4$ on $[0, 2]$

9. $f(x) = 3x^2 - x^3$ on $[0, 5]$

10. $f(x) = 6x^2 - x^3$ on $[0, 5]$

11. $f(x) = 5 - x$ on $[0, 5]$

12. $f(x) = x(100 - x)$ on $[0, 100]$

13. $f(x) = (x^2 - 1)^2$ on $[-1, 1]$

14. $f(x) = \sqrt[3]{x^2}$ on $[-1, 8]$

15. $f(x) = \dfrac{x}{x^2 + 1}$ on $[-3, 3]$

16. $f(x) = \dfrac{1}{x^2 + 1}$ on $[-3, 3]$

17. Find the number in the interval $[0, 3]$ such that the number minus its square is:

   a. As large as possible.
   b. As small as possible.

18. Find the number in the interval $[\frac{1}{3}, 2]$ such that the sum of the number and its reciprocal is:

   a. As large as possible.
   b. As small as possible.

 19. ONE FUNCTION, DIFFERENT DOMAINS

   a. Graph the function $y_1 = x^3 - 15x^2 + 63x$ on the window $[0, 10]$ by $[0, 130]$. By visual inspection of this function on this domain, where do the absolute maximum and minimum values occur: both at critical numbers, both at endpoints, or one at a critical number and one at an endpoint?

   b. Now change the domain to $[0, 8]$ and answer the same question.

   c. Now change the domain to $[2, 8]$ and answer the same question.

   d. Can you find a domain such that the minimum occurs at a critical number and the maximum at an endpoint?

 20. EXISTENCE OF EXTREME VALUES

   a. Graph the function $y_1 = 5/x$ on the window $[1, 10]$ by $[0, 10]$. By visual inspection, does the function have an absolute maximum value on this domain? An absolute minimum value?

   b. Now change the x window to $[0, 10]$ and answer the same questions. (We cannot now call $[0, 10]$ the domain, since the function is not defined at 0.)

   c. Based on the screen display, answer the same questions for the domain $(0, \infty)$.

   d. Is there a domain on which this function has an absolute maximum but no absolute minimum?

   e. It was stated in the box on page 215 that a continuous function on a closed interval will always have an absolute maximum and minimum value. Is this claim violated by the function $f(x) = 5/x$ on $[0, 10]$ [part (b)]? Is it violated by $f(x) = 5/x$ on $(0, \infty)$ [part (c)]?

APPLIED EXERCISES

21. BIOMEDICAL: Pollen Count The average pollen count in New York City on day x of the pollen season is $P(x) = 8x - 0.2x^2$ (for $0 < x < 40$). On which day is the pollen count highest?

22. GENERAL: Fuel Economy The fuel economy (in miles per gallon) of an average American compact car is $E(x) = -0.015x^2 + 1.14x + 8.3$, where x is the driving speed (in miles per hour, $20 \le x \le 60$). At what speed is fuel economy greatest?

23. GENERAL: Fuel Economy The fuel economy (in miles per gallon) of an average American midsized car is $E(x) = -0.01x^2 + 0.62x + 10.4$, where x is the driving speed (in miles per hour, $20 \le x \le 60$). At what speed is fuel economy greatest?

24. GENERAL: Water Power The proportion of a river's energy that can be obtained from an undershot waterwheel is $E(x) = 2x^3 - 4x^2 + 2x$, where x is the speed of the waterwheel relative to the speed of the river. Find the maximum value of this function on the interval $[0, 1]$, thereby showing that only about 30% of a river's energy can be captured. Your answer should agree with the old millwright's rule that the speed of the wheel should be about one-third of the speed of the river.

25. GENERAL: Timber Value The value of a timber forest after t years is $V(t) = 480\sqrt{t} - 40t$ (for $0 \le t \le 50$). Find when its value is maximized.

26. GENERAL: Longevity and Exercise A recent study of the exercise habits of 17,000 Harvard alumni found that the death rate (deaths per 10,000 person-years) was approximately $R(x) = 5x^2 - 35x + 104$, where x is the weekly amount of exercise in thousands of calories ($0 \le x \le 4$). Find the exercise level that minimizes the death rate.

27. ENVIRONMENTAL SCIENCE: Pollution Two chemical factories are discharging toxic waste into a large lake, and the pollution level at a point x miles from factory A toward factory B is $P(x) = 3x^2 - 72x + 576$ parts per million (for $0 \le x \le 50$). Find where the pollution is the least.

28. BUSINESS: Maximum Profit City Cycles Incorporated finds that it costs $70 to manufacture each bicycle, and fixed costs are $100 per day. The price function is $p(x) = 270 - 10x$, where p is the price (in dollars) at which exactly x bicycles will be sold. Find the quantity City Cycles should produce and the price it should charge to maximize profit. Also find the maximum profit.

29. BUSINESS: Maximum Profit Country Motorbikes Incorporated finds that it costs $200 to produce each motorbike, and that fixed costs are $1500 per day. The price function is $p(x) = 600 - 5x$, where p is the price (in dollars) at which exactly x motorbikes will be sold. Find the quantity Country Motorbikes should produce and the price it should charge to maximize profit. Also find the maximum profit.

30. BUSINESS: Maximum Profit A retired potter can produce china pitchers at a cost of $5 each. She estimates her price function to be $p = 17 - 0.5x$, where p is the price at which exactly x pitchers will be sold per week. Find the number of pitchers that she should produce and the price that she should charge in order to maximize profit. Also find the maximum profit.

31. GENERAL: Parking Lot Design A company wants to build a parking lot along the side of one of its buildings using 800 feet of fence. If the side along the building needs no fence, what are the dimensions of the largest possible parking lot?

32. GENERAL: Area A farmer wants to make two identical rectangular enclosures along a straight river, as in the diagram shown below. If he has 600 yards of fence, and if the sides along the river need no fence, what should be the dimensions of each enclosure if the total area is to be maximized?

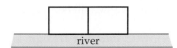

33. **GENERAL: Area** A farmer wants to make three identical rectangular enclosures along a straight river, as in the diagram shown below. If he has 1200 yards of fence, and if the sides along the river need no fence, what should be the dimensions of each enclosure if the total area is to be maximized?

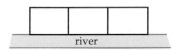

34. **GENERAL: Area** What is the area of the largest rectangle whose perimeter is 100 feet?

35. **GENERAL: Package Design** An open-top box is to be made from a square piece of cardboard that measures 18 inches by 18 inches by removing a square from each corner and folding up the sides. What are the dimensions and volume of the largest box that can be made in this way?

36. **GENERAL: Gutter Design** A long gutter is to be made from a 12-inch-wide strip of metal by folding up the two edges. How much of each edge should be folded up in order to maximize the capacity of the gutter? [*Hint:* Maximizing the capacity means maximizing the cross-sectional area, shown below.]

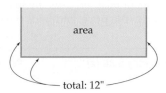

total: 12"

37. **GENERAL: Maximizing a Product** Find the two numbers whose sum is 50 and whose product is a maximum.

38. **GENERAL: Maximizing Area** Show that the largest rectangle with a given perimeter is a square.

39. **BIOMEDICAL: Coughing** When you cough, you are using a high-speed stream of air to clear your trachea (windpipe). During a cough your trachea contracts, forcing the air to move faster, but also increasing the friction. If a trachea contracts from a normal radius of 3 centimeters to a radius of r centimeters, the velocity of the airstream is $V(r) = c(3 - r)r^2$, where c is a positive constant depending on the length and the elasticity of the trachea. Find the radius r that maximizes this velocity. (X-ray pictures verify that the trachea does indeed contract to this radius.)

40. **GENERAL: "Efishency"** At what speed should a fish swim upstream so as to reach its destination with the least expenditure of energy? The energy depends on the friction of the fish through the water and on the duration of the trip. If the fish swims with velocity v, the energy has been found experimentally to be proportional to v^k (for constant $k > 2$) times the duration of the trip. A distance of s miles against a current of speed c requires time $\frac{s}{v - c}$ (distance divided by speed). The energy required is then proportional to $\frac{v^k s}{v - c}$. For $k = 3$, minimizing energy is equivalent to minimizing

$$E(v) = \frac{v^3}{v - c}$$

Find the speed v with which the fish should swim in order to minimize its expenditure $E(v)$. (Your answer will depend on c, the speed of the current.)

41. **GENERAL: Athletic Fields** A running track consists of a rectangle with a semicircle at each end, as shown below. If the perimeter is to be exactly 440 yards, find the dimensions (x and r) that maximize the area of the rectangle. [*Hint:* The perimeter is $2x + 2\pi r$.]

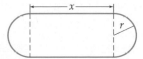

42. GENERAL: Window Design A Norman window consists of a rectangle topped by a semicircle, as shown below. If the perimeter is to be 18 feet, find the dimensions (x and r) that maximize the area of the window. [*Hint:* The perimeter is $2x + 2r + \pi r$.]

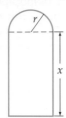

43–44. GENERAL: Maximizing Capacity of a Computer Disk Personal computers store information on disks by magnetically writing data on concentric circular "tracks" on the disk. The capacity of the disk depends on the radius x of the innermost track according to the formulas given below (in which x is in inches and y is in megabytes). Enter the formula into a graphing calculator and use MAXIMUM to find the inner radius x that maximizes the disk's capacity y. Also find the maximum capacity. (Computers actually use an inner radius slightly larger than the optimum value, which reduces the capacity slightly.)

43. $y = 0.988x(2.25 - x)$ for $0 \le x \le 2.25$ (For 5.25-inch high-density double-sided disks)

44. $y = 1.06x(1.68 - x)$ for $0 \le x \le 1.68$ (For 3.5-inch double-density double-sided disks)

45. BIOMEDICAL: Bacterial Growth A chemical reagent is introduced into a bacterial population, and t hours later the number of bacteria (in thousands) is $N(t) = 1000 + 15t^2 - t^3$ (for $0 \le t \le 15$).

a. When will the population be the largest, and how large will it be?
b. When will the population be growing at the fastest rate, and how fast? (What word applies to such a point?)

46. GENERAL: Value of a Pulpwood Forest The value of a pulpwood forest after growing for

x years is predicted to be $V(t) = 400x^{0.4} - 40x$ thousand dollars (for $0 \le x \le 25$). Use a graphing calculator with MAXIMUM to find when the value will be maximized, and what the maximum value will be.

47–48. GENERAL: Package Design Use a graphing calculator (as explained on page 224) to find the side of the square removed and the volume of the box described in Example 5 (pages 222–223) if the square piece of metal is replaced by a:

47. 5 by 7 card (5 inches by 7 inches)

48. 6 by 8 card (6 inches by 8 inches)

You might try constructing such a box.

49. BUSINESS: Maximum Revenue A restaurant manager keeps records of the number of bottles of wine he sells per hour at different prices, and finds the following data.

Quantity Sold (x)	Price (p)
10	10
8	16
6	28

a. Enter these numbers into your graphing calculator and find the linear regression line giving price as a function of quantity sold, thereby finding the price function $p(x)$.
b. Find the company's revenue function. [*Hint:* Revenue is price times quantity.]
c. Find the quantity and price that maximize revenue.

50. BUSINESS: Maximum Revenue The manager of a campus bookstore keeps records of the number of baseball caps she sells per week as she varies the price, and finds the following data.

Quantity Sold (x)	Price (p)
15	14
12	17
9	23

a. Enter these numbers into your graphing calculator and find the linear regression line giving price as a function of quantity sold, thereby finding the price function $p(x)$.

b. Find the store's revenue function. [*Hint:* Revenue is price times quantity.]

c. Find the quantity and price that maximize revenue.

3.4 FURTHER APPLICATIONS OF OPTIMIZATION

Introduction

In this section we continue to solve optimization problems. In particular, we will see how to maximize a company's profit if we are not given the price function, provided that we are given information describing how price changes will affect sales. We will also see that sometimes x should be chosen as something *other* than the quantity sold.

EXAMPLE 1 **FINDING PRICE AND QUANTITY FUNCTIONS**

A store can sell 20 bicycles per week at a price of $400 each. The manager estimates that for each $10 price reduction she can sell two more bicycles per week. The bicycles cost the store $200 each. If x stands for *the number of $10 price reductions*, express the price p and the quantity q as functions of x.

Solution Let

$$x = \text{the number of \$10 price reductions}$$

For example, $x = 4$ means that the price is reduced by $40 (four $10 price reductions). Therefore, in general, if there are x $10 price reductions from the original $400 price, then the price $p(x)$ is

$$p(x) = 400 - 10x \qquad\qquad \text{Price}$$

Original price — Less x $10 price reductions

The quantity sold $q(x)$ will be

$$q(x) = 20 + 2x \qquad\qquad \text{Quantity}$$

Original quantity — Plus two for each price reduction

We will return to this example and maximize the store's profit after a practice problem.

Practice Problem

A computer manufacturer can sell 1500 personal computers per month at a price of $3000 each. The manager estimates that for each $200 price reduction he will sell 300 more each month. If x stands for *the number of $200 price reductions*, express the price p and the quantity q as functions of x. ➤ Solution on page 236

EXAMPLE 2

MAXIMIZING PROFIT (*Continuation of Example 1*)

Using the information in Example 1, find the price of the bicycles and the quantity that maximize profit. Also find the maximum profit.

Solution   In Example 1 we found

$$p(x) = 400 - 10x \qquad \text{Price}$$
$$q(x) = 20 + 2x \qquad \text{Quantity sold at that price}$$

Revenue is price times quantity, $p(x) \cdot q(x)$:

$$R(x) = (400 - 10x)(20 + 2x) \qquad p(x)q(x)$$
$$= 8000 + 600x - 20x^2 \qquad \text{Multiplying out and simplifying}$$

The cost function is unit cost times quantity:

$$C(x) = 200\,(20 + 2x) = 4000 + 400x \qquad \begin{array}{l}\text{If there were a fixed cost}\\\text{we would add it}\end{array}$$

Unit Quantity
cost $q(x)$

Profit is revenue minus cost:

$$P(x) = \underbrace{(8000 + 600x - 20x^2)}_{R(x)} - \underbrace{(4000 + 400x)}_{C(x)}$$

$$= 4000 + 200x - 20x^2 \qquad \text{Simplifying}$$

We maximize profit by setting the derivative equal to zero:

$$200 - 40x = 0 \qquad \text{Differentiating } P = 4000 + 200x - 20x^2$$

The critical number is $x = 5$. The second derivative, $P''(x) = -40$, shows that the profit is *maximized* at $x = 5$. Since $x = 5$ is the number of $10 price reductions, the original price of $400 should be lowered by $50 ($10 five times), from $400 to $350. The quantity sold is found from the quantity function:

$$q(5) = 20 + 2 \cdot 5 = 30 \qquad q(x) = 20 + 2x \ \text{ at } \ x = 5$$

Finally, we state the answer clearly.

Sell the bicycles for $350 each.

Quantity sold: 30 per week.

Maximum profit: $4500.   From $P(x) = 4000 + 200x - 20x^2$ at $x = 5$

Exercise 25 will show how a graphing calculator enables you to modify the problem (such as changing the cost per bicycle) and then immediately recalculate the new answer.

Choosing Variables

Notice that in Example 2 we did not choose x to be the quantity sold, but instead to be *the number of $10 price reductions*. (Therefore, a negative x would have meant a price *increase*.) We chose this x because from it we could easily calculate both the new price and the new quantity. Other choices for x are also possible, but in situations where a price change will make one quantity rise and another fall, it is often easiest to choose x to be the *number of such changes*.

EXAMPLE 3 **MAXIMIZING HARVEST SIZE**

An orange grower finds that if he plants 80 orange trees per acre, each tree will yield 60 bushels of oranges. He estimates that for each additional tree that he plants per acre, the yield of each tree will decrease by 2 bushels. How many trees should he plant per acre to maximize his harvest?

Solution   We take x equal to the number of "changes"—that is, let

$$x = \text{the number of added trees per acre}$$

With x extra trees per acre,

Trees per acre: $80 + x$ Original 80 plus x more

Yield per tree: $60 - 2x$ Original yield less 2 per extra tree

Therefore, the total yield per acre will be

$$Y(x) = \underbrace{(60 - 2x)}_{\substack{\text{Yield} \\ \text{per tree}}}\underbrace{(80 + x)}_{\substack{\text{Trees} \\ \text{per acre}}} = 4800 - 100x - 2x^2$$

We maximize this by setting the derivative equal to zero:

$$-100 - 4x = 0 \qquad \text{Differentiating} \quad Y = 4800 - 100x - 2x^2$$

$$x = -25 \qquad \text{Negative!}$$

The number of *added* trees is negative, meaning that the grower should plant 25 *fewer* trees per acre. The second derivative, $Y''(x) = -4$, shows that the yield is indeed maximized at $x = -25$. Therefore:

$$\text{Plant 55 trees per acre.} \qquad 80 - 25 = 55$$

Earlier problems involved maximizing areas and volumes using only a fixed amount of material (such as a fixed length of fence). Instead, we could minimize the amount of materials for a fixed area or volume.

EXAMPLE 4 **MINIMIZING PACKAGE MATERIALS**

A moving company wishes to design an open-top box with a square base whose volume is exactly 32 cubic feet. Find the dimensions of the box requiring the least amount of materials.

Solution   The base is square, so we define

$$x = \text{length of side of base}$$
$$y = \text{height}$$

The volume (length·width·height) is $x·x·y$ or x^2y, which (according to the problem) must equal 32 cubic feet:

$$x^2y = 32$$

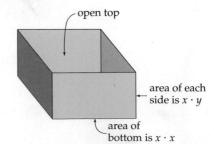

The box consists of a bottom (area x^2) and four sides (each of area xy). Minimizing the amount of materials means minimizing the surface area of the bottom and four sides:

$$A = x^2 + 4xy \qquad \binom{\text{Area of}}{\text{bottom}} + \binom{\text{Area of}}{\text{four sides}}$$

As usual, we must express this area in terms of just *one* variable, so we use the volume requirement to express y in terms of x:

$$y = \frac{32}{x^2} \qquad \text{Solving} \quad x^2 y = 32 \quad \text{for } y$$

The area function becomes

$$A = x^2 + 4x\frac{32}{x^2} \qquad A = x^2 + 4xy \text{ with } y \text{ replaced by } \frac{32}{x^2}$$

$$= x^2 + \frac{128}{x} \qquad \text{Simplifying}$$

$$= x^2 + 128x^{-1} \qquad \text{Writing } \frac{1}{x} \text{ as } x^{-1}$$

We minimize this by finding the critical number:

$$A'(x) = 2x - 128x^{-2} \qquad \text{Differentiating} \quad A = x^2 + 128x^{-1}$$

$$2x - \frac{128}{x^2} = 0 \qquad \text{Setting the derivative equal to zero}$$

$$2x^3 - 128 = 0 \qquad \text{Multiplying by } x^2 \text{ (since } x > 0)$$

$$x^3 = 64 \qquad \text{Adding 128 and then dividing by 2}$$

$$x = 4 \qquad \text{Taking cube roots}$$

The second derivative

$$A''(x) = 2 + 256x^{-3} = 2 + \frac{256}{x^3} \qquad \text{From} \quad A'(x) = 2x - 128x^{-2}$$

is positive at $x = 4$, so the area is minimized. Therefore, the dimensions are

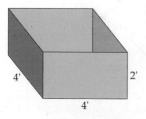

Base: 4 feet on each side

Height: 2 feet

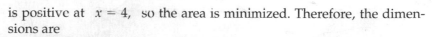

Height from $y = 32/x^2$ at $x = 4$

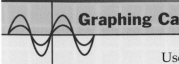

Graphing Calculator Exploration

Use a graphing calculator to solve the previous example as follows:

a. Enter the area function to be minimized as $y_1 = x^2 + 128/x$.

b. Graph y_1 for x-values $[0, 10]$, using TABLE or TRACE to find where y_1 seems to "bottom out" to determine an appropriate y-interval.

c. Graph the function on the window determined in part (b) and then use MINIMUM to find the minimum value. Your answer should agree with that found above. (Your calculator may give an inexact answer that needs to be rounded.)

Notice that either way required first finding the function to be minimized.

Minimizing the Cost of Materials

How would this problem have changed if the material for the bottom of the box had been more costly than the material for the sides? If, for example, the material for the sides cost $2 per square foot and the material for the base, needing greater strength, cost $4 per square foot, then instead of simply minimizing the surface area, we would minimize *total cost*:

$$\text{Cost} = \begin{pmatrix} \text{Area of} \\ \text{bottom} \end{pmatrix}\begin{pmatrix} \text{Cost of bottom} \\ \text{per square foot} \end{pmatrix} + \begin{pmatrix} \text{Area of} \\ \text{sides} \end{pmatrix}\begin{pmatrix} \text{Cost of sides} \\ \text{per square foot} \end{pmatrix}$$

Since the areas would be just as before, this cost would be

$$\text{Cost} = (x^2)(4) + (4xy)(2) = 4x^2 + 8xy$$

From here on we would proceed just as before, eliminating the y (using the volume relationship $x^2y = 32$) and then setting the derivative equal to zero.

Maximizing Tax Revenue

Governments raise money by collecting taxes. If a sales tax or an import tax is too high, trade will be discouraged and tax revenues will fall. If, on the other hand, the tax rate is too low, trade may flourish but tax revenues will again fall. Economists often want to determine the tax rate that maximizes revenue for the government. To do this, they must first predict the relationship between the tax on an item and the total sales of the item.

Suppose, for example, that the relationship between the tax rate t on an item and its total sales S is

$$S(t) = 9 - 20\sqrt{t} \qquad \begin{array}{l} t = \text{tax rate } (0 \le t \le 0.20) \\ S = \text{total sales (millions of dollars)} \end{array}$$

If the tax rate is $t = 0$ (0%), then the total sales will be

$$S(0) = 9 - 20\sqrt{0} = 9 \qquad\qquad \text{\$9 million}$$

If the tax rate is raised to $t = 0.16$ (16%), then sales will be

$$S(0.16) = 9 - 20\sqrt{0.16}$$
$$= 9 - (20)(0.4) = 9 - 8 = 1 \qquad \text{\$1 million}$$

That is, raising the tax rate from 0% to 16% will discourage \$8 million worth of sales. The graph of $S(t)$ on the left shows how total sales decrease as the tax rate increases. With such information (which may be found from historical data), one can find the tax rate that maximizes revenue.

EXAMPLE **5**

MAXIMIZING TAX REVENUE

If economists predict that the relationship between the tax rate t on an item and the total sales S of that item (in millions of dollars) is

$$S(t) = 9 - 20\sqrt{t} \qquad\qquad \text{For } 0 \le t \le 0.20$$

find the tax rate that maximizes revenue to the government.

Solution

The government's revenue R is the tax rate t times the total sales $S(t) = 9 - 20\sqrt{t}$:

$$R(t) = \underbrace{t(9 - 20t^{1/2})}_{S(t)} = 9t - 20t^{3/2}$$

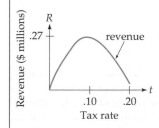

The graph of this function is shown on the left. To maximize it, we set its derivative equal to zero:

$$9 - 30t^{1/2} = 0 \qquad\qquad \text{Derivative of } 9t - 20t^{3/2}$$

$$9 = 30t^{1/2} \qquad\qquad \text{Adding } 30t^{1/2} \text{ to each side}$$

$$t^{1/2} = \frac{9}{30} = 0.3 \qquad\qquad \begin{array}{l}\text{Switching sides and}\\ \text{dividing by 30}\end{array}$$

$$t = 0.09 \qquad\qquad \text{Squaring both sides}$$

This gives a tax rate of $t = 9\%$. The second derivative,

$$R''(t) = -30 \cdot \frac{1}{2} t^{-1/2} = -\frac{15}{\sqrt{t}} \qquad \text{From } R' = 9 - 30t^{1/2}$$

is negative at $t = 0.09$, showing that the revenue is maximized. Therefore:

A tax rate of 9% maximizes revenue for the government.

Graphing Calculator Exploration

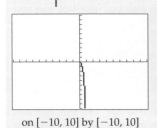

on $[-10, 10]$ by $[-10, 10]$

The graph of the function from Example 5, $y_1 = 9x - 20x^{3/2}$ (written in x instead of t for ease of entry) is shown on the left on the standard window $[-10, 10]$ by $[-10, 10]$. This might lead you to believe, erroneously, that the function is maximized at the endpoint $(0, 0)$.

a. Why does this graph not look like the graph at the bottom of the previous page? [*Hint:* Look at the scale.]

b. Can you find a window on which your graphing calculator will show a graph like the one on the left?

This example illustrates one of the pitfalls of graphing calculators—the part of the curve where the "action" takes place may be entirely hidden in one pixel. Calculus, on the other hand, will *always* find the critical value, no matter where it is, and then a graphing calculator can be used to confirm your answer by showing the graph on an appropriate window.

➤ **Solution to Practice Problem**

Price: $p(x) = 3000 - 200x$
Quantity: $q(x) = 1500 + 300x$

3.4 Exercises

1. **BUSINESS: Maximum Profit** An automobile dealer can sell 12 cars per day at a price of $15,000. He estimates that for each $300 price reduction he can sell two more cars per day. If each car costs him $12,000, and fixed costs are $1000, what price should he charge to maximize his profit? How many cars will he sell at this price? [*Hint:* Let x = the number of $300 price reductions.]

2. **BUSINESS: Maximum Profit** An automobile dealer can sell four cars per day at a price of $12,000. She estimates that for each $200 price reduction she can sell two more cars per day. If each car costs her $10,000, and her fixed costs are $1000, what price should she charge to maximize her profit? How many cars will she sell at this price? [*Hint:* Let $x =$ the number of $200 price reductions.]

3. **BUSINESS: Maximum Revenue** An airline finds that if it prices a cross-country ticket at $200, it will sell 300 tickets per day. It estimates that each $10 price reduction will result in 30 more tickets sold per day. Find the ticket price (and the number of tickets sold) that will maximize the airline's revenue.

4. **ECONOMICS: Oil Prices** An oil-producing country can sell 1 million barrels of oil a day at a price of $25 per barrel. If each $1 price increase will result in a sales decrease of 50,000 barrels per day, what price will maximize the country's revenue? How many barrels will it sell at that price?

5. **BUSINESS: Maximum Revenue** Rent-A-Reck Incorporated finds that it can rent 60 cars if it charges $80 for a weekend. It estimates that for each $5 price increase it will rent three fewer cars. What price should it charge to maximize its revenue? How many cars will it rent at this price?

6. **GENERAL: Maximum Yield** A peach grower finds that if he plants 40 trees per acre, each tree will yield 60 bushels of peaches. He also estimates that for each additional tree that he plants per acre, the yield of each tree will decrease by 2 bushels. How many trees should he plant per acre to maximize his harvest?

7. **GENERAL: Maximum Yield** An apple grower finds that if she plants 20 trees per acre, each tree will yield 90 bushels of apples. She also estimates that for each additional tree that she plants per acre, the yield of each tree will decrease by 3 bushels. How many trees should she plant per acre to maximize her harvest?

8. **GENERAL: Fencing** A farmer has 1200 feet of fence and wishes to build two identical rectan-

gular enclosures, as in the diagram. What should be the dimensions of each enclosure if the total area is to be a maximum?

9. **GENERAL: Minimum Materials** An open-top box with a square base is to have a volume of 4 cubic feet. Find the dimensions of the box that can be made with the smallest amount of material.

10. **GENERAL: Minimum Materials** An open-top box with a square base is to have a volume of 108 cubic inches. Find the dimensions of the box that can be made with the smallest amount of material.

11. **GENERAL: Largest Postal Package** The U.S. Postal Service will accept a package if its length plus its girth (the distance all the way around) does not exceed 84 inches. Find the dimensions and volume of the largest package with a square base that can be mailed.

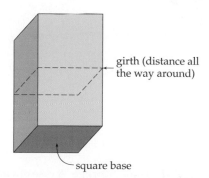

girth (distance all the way around)

square base

12. **GENERAL: Fencing** A homeowner wants to build, along his driveway, a garden surrounded by a fence. If the garden is to be 800 square feet, and the fence along the driveway costs $6 per foot whereas on the other three sides it costs only $2 per foot, find the

dimensions that will minimize the cost. Also find the minimum cost.

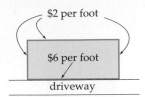

$2 per foot

$6 per foot

driveway

13. GENERAL: Fencing A homeowner wants to build, along her driveway, a garden surrounded by a fence. If the garden is to be 5000 square feet, and the fence along the driveway costs $6 per foot whereas on the other three sides it costs only $2 per foot, find the dimensions that will minimize the cost. Also find the minimum cost. (See the diagram above.)

14–15. ECONOMICS: Tax Revenue Suppose that the relationship between the tax rate t on imported shoes and the total sales S (in millions of dollars) is given by the function below. Find the tax rate t that maximizes revenue for the government.

14. $S(t) = 4 - 6\sqrt[3]{t}$ **15.** $S(t) = 8 - 15\sqrt[3]{t}$

16. BIOMEDICAL: Drug Concentration If the amount of a drug in a person's blood after t hours is $f(t) = t/(t^2 + 9)$, when will the drug concentration be the greatest?

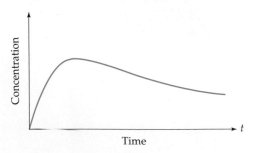

17. GENERAL: Wine Storage A case of vintage wine appreciates in value each year, but there is also an annual storage charge. The value of the wine after t years is $V(t) = 2000 + 96\sqrt{t} - 12t$ dollars (for $0 \le t \le 25$). Find the storage time that will maximize the value of the wine.

18. GENERAL: Bus Shelter Design A bus stop shelter, consisting of two square sides, a back,

and a roof, as shown below, is to have volume 1024 cubic feet. What are the dimensions that require the least amount of materials?

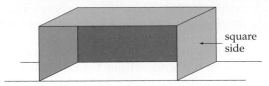

square side

19. GENERAL: Area Show that the rectangle of fixed area whose perimeter is a minimum is a square.

20. POLITICAL SCIENCE: Campaign Expenses A politician estimates that by campaigning in a county for x days, she will gain $2x$ (thousand) votes, but her campaign expenses will be $5x^2 + 500$ dollars. She wants to campaign for the number of days that maximizes the number of votes per dollar, $f(x) = \dfrac{2x}{5x^2 + 500}$ For how many days should she campaign?

21. GENERAL: Page Design A page of 96 square inches is to have margins of 1 inch on either side and $1\frac{1}{2}$ inches at the top and bottom, as in the diagram. Find the dimensions of the page that maximize the print area.

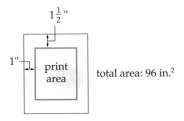

$1\frac{1}{2}$"

1"

print area

total area: 96 in.^{2}

22. BIOMEDICAL: Contagion If an epidemic spreads through a town at a rate that is proportional to the number of infected people and to the number of uninfected people, then the rate is $R(x) = cx(p - x)$, where x is the number of infected people and c and p (the population) are positive constants. Show that the rate $R(x)$ is greatest when half of the population is infected.

23. BIOMEDICAL: Contagion If an epidemic spreads through a town at a rate that is proportional to the number of uninfected people and to the square of the number of infected people, then the rate is $R(x) = cx^2(p - x)$, where x is the number of infected people and c and p (the population) are positive constants. Show that the rate $R(x)$ is greatest when two-thirds of the population is infected.

24. BUSINESS: Maximizing Profit An electronics store can sell 35 cellular telephones per week at a price of $200. The manager estimates that for each $20 price reduction she can sell 9 more per week. The telephones cost the store $100 each, and fixed costs are $700 per week.

a. If x is the number of $20 price reductions, find the price $p(x)$ and enter it in y_1. Then enter the quantity function $q(x)$ in y_2.

b. Make y_3 the revenue function by defining $y_3 = y_1 y_2$ (price times quantity).

c. Make y_4 the cost function by defining y_4 as unit cost times y_2 plus fixed costs.

d. Make y_5 the profit function by defining $y_5 = y_3 - y_4$ (revenue minus cost).

e. Turn off y_1, y_2, y_3, and y_4 and graph the profit function y_5 for x-values $[-10, 10]$, using TABLE or TRACE to find an appropriate y-interval. Then use MAXIMUM to maximize it.

f. Use EVALUATE to find the price and the quantity for this maximum profit.

25. BUSINESS: Exploring a Profit Maximization Problem Use a graphing calculator to further explore Example 2 (pages 230–231) as follows:

a. Enter the price function $y_1 = 400 - 10x$ and the quantity function $y_2 = 20 + 2x$ into your graphing calculator.

b. Make y_3 the revenue function by defining $y_3 = y_1 y_2$ (price times quantity).

c. Make y_4 the cost function by defining $y_4 = 200 y_2$ (unit cost times quantity).

d. Make y_5 the profit function by defining $y_5 = y_3 - y_4$ (revenue minus cost).

e. Turn off y_1, y_2, y_3, and y_4 and graph the profit function y_5 on the window $[0, 10]$ by $[0, 10{,}000]$ and then use MAXIMUM to maximize it. Your answer should agree with that found in Example 2.

Now change the problem!

f. What if the store finds that it can buy the bicycles from another wholesaler for $150 instead of $200? In y_4, change the 200 to 150. Then graph the profit y_5 (you may have to turn off y_4 again) and maximize it. Find the new price and quantity by evaluating y_1 and y_2 (using EVALUATE) at the new x-value.

g. What if cycling becomes more popular and the manager estimates that she can sell 30 instead of 20 bicycles per week at the original $400 price? Go back to y_2 and change 20 to 30 (keeping the change made earlier) and graph and find the price and quantity that maximize profit now.

Notice how flexible this setup is for changing any of the numbers.

3.5 OPTIMIZING LOT SIZE AND HARVEST SIZE

Introduction

In this section we discuss two important applications of optimization, one economic and one ecological. The first concerns the most efficient way for a business to order merchandise (or for a manufacturer to produce merchandise), and the second concerns the preservation of animal populations that are harvested by people. Either of these applications may be read independently of the other.

Minimizing Inventory Costs

A business encounters two kinds of costs in maintaining inventory: storage costs (warehouse and insurance costs for merchandise not yet sold) and reorder costs (delivery and bookkeeping costs for each order). For example, if a furniture store expects to sell 250 sofas in a year, it could order all 250 at once (incurring high storage costs), or it could order them in many small lots, say 50 orders of five each, spaced throughout the year (incurring high reorder costs). Obviously, the best order size (or "lot" size) is the one that minimizes the total of storage plus reorder costs.

EXAMPLE **1** **MINIMIZING INVENTORY COSTS**

A furniture showroom expects to sell 250 sofas a year. Each sofa costs the store $300, and there is a fixed charge of $500 per order. If it costs $100 to store a sofa for a year, how large should each order be and how often should orders be placed to minimize inventory costs?

Solution   Let

$$x = \text{lot size} \qquad x \text{ is the number of sofas in each order}$$

Storage Costs: If the sofas sell steadily throughout the year, and if the store reorders x more whenever the stock runs out, then its inventory during the year looks like the following graph.

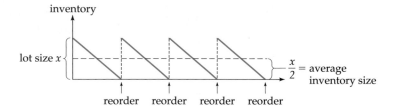

Notice that the inventory level varies from the lot size x down to zero, with an average inventory of $x/2$ sofas throughout the year. Because it costs $100 to store a sofa for a year, the total (annual) storage costs are

$$\begin{pmatrix} \text{Storage} \\ \text{costs} \end{pmatrix} = \begin{pmatrix} \text{Storage} \\ \text{per item} \end{pmatrix} \cdot \begin{pmatrix} \text{Average num-} \\ \text{ber of items} \end{pmatrix}$$

$$= \quad 100 \cdot \frac{x}{2} \quad = 50x$$

Reorder Costs: Each sofa costs $300, so an order of lot size x costs $300x$, plus the fixed order charge of $500:

$$\left(\begin{array}{c}\text{Cost} \\ \text{per order}\end{array}\right) = 300x + 500$$

The yearly supply of 250 sofas, with x sofas in each order, requires $\dfrac{250}{x}$ orders. (For example, 250 sofas at 5 per order require $\dfrac{250}{5} = 50$ orders.) Therefore, the yearly reorder costs are

$$\left(\begin{array}{c}\text{Reorder} \\ \text{costs}\end{array}\right) = \left(\begin{array}{c}\text{Cost} \\ \text{per order}\end{array}\right) \cdot \left(\begin{array}{c}\text{Number} \\ \text{of orders}\end{array}\right)$$

$$= (300x + 500) \cdot \left(\frac{250}{x}\right)$$

Total Cost: $C(x)$ is storage costs plus reorder costs:

$$C(x) = \left(\begin{array}{c}\text{Storage} \\ \text{costs}\end{array}\right) + \left(\begin{array}{c}\text{Reorder} \\ \text{costs}\end{array}\right)$$

$$= 100\frac{x}{2} + (300x + 500)\left(\frac{250}{x}\right) \qquad \text{Using the storage and reorder costs found earlier}$$

$$= 50x + 75{,}000 + 125{,}000x^{-1} \qquad \text{Simplifying}$$

To minimize $C(x)$, we differentiate:

$$C'(x) = 50 - 125{,}000x^{-2} = 50 - \frac{125{,}000}{x^2} \qquad \begin{array}{l}\text{Differentiating} \\ C = 50x + 75{,}000 + \\ 125{,}000x^{-1}\end{array}$$

$$50 - \frac{125{,}000}{x^2} = 0 \qquad \begin{array}{l}\text{Setting the deriva-} \\ \text{tive equal to zero}\end{array}$$

$$50x^2 = 125{,}000 \qquad \begin{array}{l}\text{Multiplying by } x^2 \\ \text{and adding 125,000} \\ \text{to each side}\end{array}$$

$$x^2 = \frac{125{,}000}{50} = 2500 \qquad \begin{array}{l}\text{Dividing each} \\ \text{side by 50}\end{array}$$

$$x = 50 \qquad \begin{array}{l}\text{Taking square roots} \\ (x > 0) \text{ gives} \\ \text{lot size 50}\end{array}$$

$$C''(x) = 25{,}000x^{-3} = 250{,}000\,\frac{1}{x^3} \qquad \begin{array}{l}C'' \text{ is positive, so} \\ C \text{ is minimized} \\ \text{at } x = 50\end{array}$$

At 50 sofas per order, the yearly 250 will require $\frac{250}{50} = 5$ orders. Therefore:

Lot size is 50 sofas, with orders placed five times a year.

Graphing Calculator Exploration

a. Verify the answer to the previous example by graphing the total cost function $y_1 = 50x + 75,000 + 125,000/x$ on the window [0, 200] by [0, 150,000] and using MINIMUM.

b. Notice that the curve is rather flat to the right of $x = 50$. From this observation, if you cannot order exactly 50 at a time, would it be better to order somewhat more than 50 or somewhat less than 50?

Modifications and Assumptions

If the number of orders per year is not a whole number, say 7.5 orders per year, we just interpret it as 15 orders in 2 years, and handle it accordingly.

We made two major assumptions in Example 1. We assumed that there was a steady demand, and that orders were equally spaced throughout the year. These are reasonable assumptions for many products, while for seasonal products such as bathing suits or fur coats, separate calculations can be done for the "on" and "off" seasons.

Production Runs

Similar analysis applies to manufacturing. For example, if a book publisher can estimate the yearly demand for a book, she may print the yearly total all at once, incurring high storage costs, or she may print them in several smaller runs throughout the year, incurring setup costs for each run. Here the setup costs for each printing run play the role of the reorder costs for a store.

EXAMPLE 2

MINIMIZING INVENTORY COSTS FOR A PUBLISHER

A publisher estimates the annual demand for a book to be 4000 copies. Each book costs $8 to print, and setup costs are $1000 for each printing. If storage costs are $2 per book per year, find how many books should be printed per run and how many printings will be needed if costs are to be minimized.

Solution   Let

$$x = \text{the number of books in each run}$$

Storage Costs: As before, an average of $\frac{x}{2}$ books are stored throughout the year, at a cost of $2 each, so annual storage costs are

$$\begin{pmatrix} \text{Storage} \\ \text{costs} \end{pmatrix} = \left(\frac{x}{2}\right) \cdot 2 = x$$

Production Costs: The cost per run is

$$\begin{pmatrix} \text{Costs} \\ \text{per run} \end{pmatrix} = 8x + 1000 \qquad \begin{array}{l} x \text{ books at \$8 each, plus} \\ \text{\$1000 setup costs} \end{array}$$

The 4000 books at x books per run will require $\dfrac{4000}{x}$ runs. Therefore, production costs are

$$\begin{pmatrix} \text{Production} \\ \text{costs} \end{pmatrix} = (8x + 1000)\left(\frac{4000}{x}\right) \qquad \begin{array}{l} \text{Cost per run times} \\ \text{number of runs} \end{array}$$

Total Cost: The total cost is storage costs plus production costs:

$$C(x) = x + (8x + 1000)\left(\frac{4000}{x}\right) \qquad \text{Storage + production}$$

$$= x + 32{,}000 + 4{,}000{,}000x^{-1} \qquad \text{Multiplying out}$$

We differentiate, set the derivative equal to zero, and solve, just as before (omitting the details), obtaining $x = 2000$. The second-derivative test will show that costs are minimized at 2000 books per run. The 4000 books require $\frac{4000}{2000} = 2$ printings. Therefore, the publisher should:

Print 2000 books per run, with two printings.

Maximum Sustainable Yield

The next application involves industries such as fishing, in which a naturally occurring animal population is "harvested." Taking too large a harvest will kill off the animal population (like the bowhead whale, hunted almost to extinction in the nineteenth century). We want to find the "maximum sustainable yield" that may be taken, year after year, while preserving a stable animal population.

For some animals one can determine a *reproduction function, f(p)*, which gives the expected population a year from now if the present population is p.

Reproduction Function

A reproduction function $f(p)$ gives the population a year from now if the current population is p.

For example, the reproduction function $f(p) = -\frac{1}{4}p^2 + 3p$ (where p and $f(p)$ are measured in thousands) means that if the population is now $p = 6$ (thousand), then a year from now the population will be

$$f(6) = -\frac{1}{4}6^2 + 3 \cdot 6 = -9 + 18 = 9 \qquad \text{(thousand)}$$

Therefore, during the year the population will increase from 6000 to 9000.

If, on the other hand, the present population is $p = 10$ (thousand), a year later the population will be

$$f(10) = -\frac{1}{4} \cdot 10^2 + 3 \cdot 10 = -25 + 30 = 5 \qquad \text{(thousand)}$$

That is, during the year the population will decline from 10,000 to 5000 (perhaps because of inadequate food to support such a large population). In actual practice, reproduction functions are very difficult to calculate, but can sometimes be estimated by analyzing previous population and harvest data.*

Suppose that we have a reproduction function f and a current population of size p, which will therefore grow to size $f(p)$ next year. The *amount of growth* in the population during that year is

$$\begin{pmatrix} \text{Amount} \\ \text{of growth} \end{pmatrix} = f(p) - p$$

Next year's population ⟶ $f(p)$
Current population ⟶ p

Harvesting this amount removes only the *growth*, returning the population to its former size p. The population will then repeat this growth, and taking the same harvest $f(p) - p$ will cause this situation to repeat itself year after year. The quantity $f(p) - p$ is called the sustainable yield.

Sustainable Yield

For reproduction function $f(p)$, the sustainable yield is

$$Y(p) = f(p) - p$$

---

* For more information, see J. Blower, L. Cook, and J. Bishop, *Estimating the Size of Animal Populations* (London: George Allen and Unwin Ltd., 1981).

We want the population size p that maximizes the sustainable yield $Y(p)$. To maximize $Y(p)$, we set its derivative equal to zero:

$$Y'(p) = f'(p) - 1 = 0 \qquad \text{Derivative of } Y = f(p) - p$$
$$f'(p) = 1 \qquad \text{Solving for } f'(p)$$

For a given reproduction function $f(p)$, we find the maximum sustainable yield by solving this equation (provided that the second-derivative test gives $Y''(p) = f''(p) < 0$).

Maximum Sustainable Yield

For reproduction function $f(p)$, the population p that results in the maximum sustainable yield is the solution to

$$f'(p) = 1$$

(provided that $f''(p) < 0$). The maximum sustainable yield is then

$$Y(p) = f(p) - p$$

Once we calculate the population p that gives the maximum sustainable yield, we wait until the population reaches this size and then harvest, year after year, an amount $Y(p)$.

Note that to find the maximum sustainable yield, we set the derivative $f'(p)$ equal to 1, not 0. This is because we are maximizing not the reproductive function $f(p)$ but rather the yield function $Y(p) = f(p) - p$.

EXAMPLE 3

FINDING MAXIMUM SUSTAINABLE YIELD

The reproduction function for the American lobster in an East Coast fishing area is $f(p) = -0.02p^2 + 2p$ (where p and $f(p)$ are in thousands). Find the population p that gives the maximum sustainable yield and find the size of the yield.

Solution We set the derivative of the reproduction function equal to 1:

$$f'(p) = -0.04p + 2 = 1 \qquad \text{Differentiating } f(p) = -0.02p^2 + 2p$$
$$-0.04p = -1 \qquad \text{Subtracting 2 from each side}$$
$$p = \frac{-1}{-0.04} = 25 \qquad \text{Dividing by } -0.04$$

The second derivative is $f''(p) = -0.04$, which is negative, showing that $p = 25$ (thousand) is the population that gives the maximum sustainable yield. The actual yield is found from the yield function $Y(p) = f(p) - p$:

$$Y(p) = \underbrace{-0.02p^2 + 2p}_{f(p)} - p = -0.02p^2 + p$$

$$Y(25) = -0.02(25)^2 + 25 \qquad \text{Evaluating at } p = 25$$

$$= -12.5 + 25 = 12.5 \qquad \text{(thousand)}$$

The population size for the maximum sustainable yield is 25,000, and the yield is 12,500 lobsters. 25,000 from $p = 25$

Graphing Calculator Exploration

Solve the previous example on a graphing calculator as follows:

a. Enter the reproduction function as $y_1 = -0.02x^2 + 2x$.

b. Define y_2 to be the derivative of y_1 (using NDERIV).

c. Define $y_3 = 1$.

d. Turn off y_1 and graph y_2 and y_3 on the window $[0, 40]$ by $[0, 2]$.

e. Use INTERSECT to find where y_2 and y_3 meet, thereby solving $f' = 1$. (You should find $x = 25$, as above.)

f. Find the yield by evaluating $y_1 - x$ at the x-value found in part (e).

In this problem, solving "by hand" was probably easier, but this graphing calculator method may be preferable if the reproduction function is more complicated.

3.5 Exercises

Lot Size

1. A supermarket expects to sell 4000 boxes of sugar in a year. Each box costs $2, and there is a fixed delivery charge of $20 per order. If it costs $1 to store a box for a year, what is the order size and how many times a year should the orders be placed to minimize inventory costs?

2. A supermarket expects to sell 5000 boxes of rice in a year. Each box costs $2, and there is a fixed delivery charge of $50 per order. If it costs $2 to store a box for a year, what is the order size and how many times a year should the orders be placed to minimize inventory costs?

3. A liquor warehouse expects to sell 10,000 bottles of scotch whiskey in a year. Each bottle costs $12, plus a fixed charge of $125 per order. If it costs $10 to store a bottle for a year, how many bottles should be ordered at a time and how many orders should the warehouse place in a year to minimize inventory costs?

4. A wine warehouse expects to sell 30,000 bottles of wine in a year. Each bottle costs $9, plus a fixed charge of $200 per order. If it costs $3 to store a bottle for a year, how many bottles should be ordered at a time and how many orders should the warehouse place in a year to minimize inventory costs?

5. An automobile dealer expects to sell 800 cars a year. The cars cost $9000 each plus a fixed charge of $1000 per delivery. If it costs $1000 to store a car for a year, find the order size and the number of orders that minimize inventory costs.

6. An automobile dealer expects to sell 400 cars a year. The cars cost $11,000 each plus a fixed charge of $500 per delivery. If it costs $1000 to store a car for a year, find the order size and the number of orders that minimize inventory costs.

Production Runs

7. A toy manufacturer estimates the demand for a game to be 2000 per year. Each game costs $3 to manufacture, plus setup costs of $500 for each production run. If a game can be stored for a year for a cost of $2, how many should be manufactured at a time and how many production runs should there be to minimize costs?

8. A toy manufacturer estimates the demand for a doll to be 10,000 per year. Each doll costs $5 to manufacture, plus setup costs of $800 for each production run. If it costs $4 to store a doll for a year, how many should be manufactured at a time and how many production runs should there be to minimize costs?

9. A producer of audio tapes estimates the yearly demand for a tape to be 1,000,000. It costs $800 to set up the machinery for the tape, plus $10 for each tape produced. If it costs the company $1 to store a tape for a year, how many should be produced at a time and how many production runs will be needed to minimize costs?

10. A compact disc manufacturer estimates the yearly demand for a CD to be 10,000. It costs $400 to set the machinery for the CD, plus $3 for each CD produced. If it costs the company $2 to store a CD for a year, how many should be burned at a time and how many production runs will be needed to minimize costs?

Maximum Sustainable Yield

11. Marine ecologists estimate the reproduction curve for swordfish in the Georges Bank fishing grounds to be $f(p) = -0.01p^2 + 5p$, where p and $f(p)$ are in hundreds. Find the population that gives the maximum sustainable yield, and the size of the yield.

12. The reproduction function for the Hudson Bay lynx is estimated to be $f(p) = -0.02p^2 + 5p$, where p and $f(p)$ are in thousands. Find the population that gives the maximum sustainable yield, and the size of the yield.

13. The reproduction function for the Antarctic blue whale is estimated to be $f(p) = -0.0004p^2 + 1.06p$, where p and $f(p)$ are in thousands. Find the population that gives the maximum sustainable yield, and the size of the yield.

14. The reproduction function for the Canadian snowshoe hare is estimated to be $f(p) = -0.025p^2 + 4p$, where p and $f(p)$ are in thousands. Find the population that gives the maximum sustainable yield, and the size of the yield.

15. A conservation commission estimates the reproduction function for rainbow trout in a large lake to be $f(p) = 50\sqrt{p}$, where p and $f(p)$ are in thousands. Find the population that gives the maximum sustainable yield, and the size of the yield.

16. The reproduction function for oysters in a large bay is $f(p) = 30\sqrt[3]{p^2}$, where p and $f(p)$ are in pounds. Find the size of the population that gives the maximum sustainable yield, and the size of the yield.

17. Suppose that the reproduction function for Pacific salmon in a northwest fishing area is $f(p) = 24\sqrt[3]{p} - 9\sqrt{p}$ (for $0 \le p < 50$), where p and $f(p)$ are in thousands. Use a graphing calculator to find the population that gives the maximum sustainable yield, and the size of the yield. [*Hint:* Follow the steps in the Graphing Calculator Exploration on page 246 using this reproduction function and x-interval [0, 50].]

18. **BUSINESS: Exploring a Lot Size Problem**
Use a graphing calculator to explore Example 1 (pages 240–241) as follows:

a. Enter the total cost function, unsimplified, as $y_1 = 100(x/2) + (300x + 500)(250/x)$.

b. Graph y_1 on the window [0, 200] by [0, 150,000] and use MINIMUM to minimize it. Your answer should agree with the $x = 50$ found in Example 1.

c. Suppose that business improves, and the showroom expects to sell 350 per year instead of 250. In y_1, change the 250 to 350 and minimize it.

d. Suppose that a modest recession decreases sales (so change the 350 back to 250), and that inflation has driven the cost of storage up to $125 (so change the 100 to 125), and minimize y_1.

IMPLICIT DIFFERENTIATION AND RELATED RATES

Introduction

A function written in the form $y = f(x)$ is said to be defined *explicitly*, meaning that y is defined by a rule or *formula $f(x)$ in x alone*. A function may instead be defined *implicitly*, meaning that y is defined by an *equation in x and y*, such as $x^2 + y^2 = 25$. In this section we will see how to differentiate such *implicit functions* when ordinary "explicit" differentiation is difficult or impossible. We will then use implicit differentiation to find rates of change.

Implicit Differentiation

The equation $x^2 + y^2 = 25$ defines a circle. While a circle is not the graph of a function (it violates the vertical line test, see page 34), the top half by itself defines a function, as does the bottom half by itself. To find these two functions, we solve $x^2 + y^2 = 25$ for y:

$$y^2 = 25 - x^2$$

Subtracting x^2 from each side of $x^2 + y^2 = 25$

$$y = \pm\sqrt{25 - x^2}$$

Plus or minus since when squared either one gives $25 - x^2$

The *positive* square root defines the top half of the circle (where y is positive), and the *negative* square root defines the bottom half (where y is negative). The equation $x^2 + y^2 = 25$ defines *both* functions at the same time.

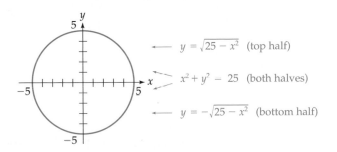

$y = \sqrt{25 - x^2}$ (top half)

$x^2 + y^2 - 25$ (both halves)

$y = -\sqrt{25 - x^2}$ (bottom half)

To find the slope anywhere on the circle, we could differentiate the "top" and "bottom" functions separately. However, it is easier to find both answers at once by differentiating *implicitly*, that is by differentiating both sides of the equation $x^2 + y^2 = 25$ with respect to x. Remember, however, that y is a *function* of x, so differentiating y^2 means differentiating a *function* squared, which requires the Generalized Power Rule:

$$\frac{d}{dx} y^n = n \cdot y^{n-1} \frac{dy}{dx}$$

EXAMPLE 1

DIFFERENTIATING IMPLICITLY

Use implicit differentiation to find $\dfrac{dy}{dx}$ when $x^2 + y^2 - 25$.

Solution

We differentiate both sides of the equation with respect to x:

$$\underbrace{\frac{d}{dx}x^2}_{\downarrow} + \underbrace{\frac{d}{dx} y^2}_{\downarrow} = \underbrace{\frac{d}{dx}25}_{\downarrow} \qquad \text{Differentiating } x^2 + y^2 = 25$$

$$2x + 2y\frac{dy}{dx} = 0 \qquad \text{Using the Generalized Power Rule on } y^2$$

Solving for $\dfrac{dy}{dx}$:

$$2y\frac{dy}{dx} = -2x \qquad\qquad \text{Subtracting } 2x$$

$$\frac{dy}{dx} = -\frac{x}{y} \qquad\qquad \begin{array}{l}\text{Canceling the 2's}\\ \text{and dividing by } y\end{array}$$

Therefore, $\dfrac{dy}{dx} = -\dfrac{x}{y}$ when x and y are related by $x^2 + y^2 = 25$.

Notice that the formula for $\dfrac{dy}{dx}$ involves both x and y. Implicit differentiation enables us to find derivatives that would otherwise be difficult or impossible to calculate, but at a "cost"—the result may depend on both x and y.

Remember that x and y play different roles: x is the *independent* variable, and y is a *function*. Therefore, we must include a $\dfrac{dy}{dx}$ (from the Generalized Power Rule) when differentiating y^n, but not when differentiating x^n since $\dfrac{dx}{dx} = 1$.

EXAMPLE 2

EVALUATING AN IMPLICIT DERIVATIVE *(Continuation of Example 1)*

Find the slope of the circle $x^2 + y^2 = 25$ at the points $(3, 4)$ and $(3, -4)$.

Solution

We simply evaluate the derivative $\dfrac{dy}{dx} = -\dfrac{x}{y}$ (found in Example 1) at the given points.

At $(3, 4)$: $\dfrac{dy}{dx} = -\dfrac{3}{4}$ $\quad\leftarrow x$
$\quad\leftarrow y$

At $(3, -4)$: $\dfrac{dy}{dx} = -\dfrac{3}{-4} = \dfrac{3}{4}$

Note that the negative sign in $\dfrac{dy}{dx} = -\dfrac{x}{y}$ gives the slope the correct sign: negative at $(3, 4)$ and positive at $(3, -4)$.

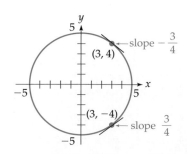

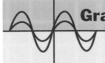

Graphing Calculator Exploration

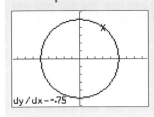

a. Graph the entire circle by graphing $y_1 = \sqrt{25 - x^2}$ and $y_2 = -\sqrt{25 - x^2}$. You may have to adjust the window to make the circle look "circular."

b. Verify the answer to Example 2 by finding the derivatives of y_1 and y_2 at $x = 3$.

c. Can you find the derivative at $x = 5$? Why not?

Derivatives should be evaluated only at points on the curve, so we evaluate $\dfrac{dy}{dx}$ only at x- and y-values *satisfying the original equation.* (It is easy to check that $x = 3$ and $y = \pm 4$ *do* satisfy $x^2 + y^2 = 25$.) Evaluating at a point not on the curve, such as (2, 3), would give a meaningless result.

The following are typical "pieces" that might appear in implicit differentiation problems.

EXAMPLE 3

FINDING DERIVATIVES—IMPLICIT AND EXPLICIT

a. $\dfrac{d}{dx} y^3 = 3y^2 \dfrac{dy}{dx}$ $\qquad$ Differentiating y^3, so include $\dfrac{dy}{dx}$

b. $\dfrac{d}{dx} x^3 = 3x^2$ $\qquad$ Differentiating x^3, so no $\dfrac{dx}{dx}$

c. $\dfrac{d}{dx} (x^3 y^5) = \underbrace{3x^2}_{\frac{d}{dx} x^3} \cdot y^5 + x^3 \cdot \underbrace{5y^4 \dfrac{dy}{dx}}_{\frac{d}{dx} y^5}$ $\qquad$ Using the Product Rule

$\qquad\qquad\quad = 3x^2 y^5 + 5x^3 y^4 \dfrac{dy}{dx}$ $\qquad$ Try to do problems such as this in one step, putting the constants in front from the start

Practice Problem Find:

a. $\dfrac{d}{dx} x^4$ **b.** $\dfrac{d}{dx} y^2$ **c.** $\dfrac{d}{dx} (x^2 y^3)$ $\qquad$ ➤ Solutions on page 257

In general, finding $\dfrac{dy}{dx}$ from an equation that defines y implicitly involves three steps:

1. Differentiate both sides of the equation *with respect to x*.

2. Collect all terms involving $\dfrac{dy}{dx}$ on one side, and all others on the other side.

3. Factor out the $\dfrac{dy}{dx}$ and solve for it by dividing.

EXAMPLE 4

FINDING AND EVALUATING AN IMPLICIT DERIVATIVE

For $y^4 + x^4 - 2x^2y^2 = 9$, find $\dfrac{dy}{dx}$ and evaluate it at $x = 2, y = 1$.

Solution

$$4y^3 \frac{dy}{dx} + 4x^3 - 4xy^2 - 4x^2y \frac{dy}{dx} = 0$$

Differentiating with respect to x, putting constants first

$$4y^3 \frac{dy}{dx} - 4x^2y \frac{dy}{dx} = -4x^3 + 4xy^2$$

Collecting dy/dx terms on the left, others on the right

$$(4y^3 - 4x^2y) \frac{dy}{dx} = -4x^3 + 4xy^2$$

Factoring out $\dfrac{dy}{dx}$

$$\frac{dy}{dx} = \frac{-4x^3 + 4xy^2}{4y^3 - 4x^2y}$$

Dividing by $4y^3 - 4x^2y$ to solve for dy/dx

$$= \frac{-x^3 + xy^2}{y^3 - x^2y}$$

Dividing by 4

$$\frac{dy}{dx} = \frac{-(2)^3 + (2)(1)^2}{(1)^3 - (2)^2(1)} = \frac{-6}{-3} = 2$$

Evaluating at $x = 2, y = 1$

Note that in the example above the given point *is* on the curve, since $x = 2$ and $y = 1$ satisfy the original equation:

$$1^4 + 2^4 - 2 \cdot 2^2 \cdot 1 = 1 + 16 - 8 = 9$$

In economics, a *demand equation* is the relationship between the price p of an item and the quantity x that consumers will demand at that price. (All prices are in dollars unless otherwise stated.)

EXAMPLE **5**

FINDING AND INTERPRETING AN IMPLICIT DERIVATIVE

For the demand equation $x = \sqrt{1900 - p^3}$, use implicit differentiation to find dp/dx. Then evaluate it at $x = 30$, $p = 10$ and interpret your answer.

Solution

$$x^2 = 1900 - p^3$$
Simplifying by squaring both sides of $x = \sqrt{1900 - p^3}$

$$2x = -3p^2 \frac{dp}{dx}$$
Differentiating both sides with respect to x

$$\frac{dp}{dx} = -\frac{2x}{3p^2}$$
Solving for $\frac{dp}{dx}$

$$\frac{dp}{dx} = -\frac{60}{300} = -0.2$$
Evaluating at $p = 10$ and $x = 30$

Interpretation: $dp/dx = -0.2$ says that the rate of change of price with respect to quantity is -0.2, so that increasing the quantity by 1 means decreasing the price by 0.20 (or 20 cents). Therefore, each 20-cent price decrease brings approximately one more sale (at the given values of x and p).

Our first step of squaring both sides of the original equation was not necessary, but it made the differentiation easier by avoiding the generalized Power Rule.

Notice that this particular demand function $x = \sqrt{1900 - p^3}$ can be solved *explicitly* for p:

$$x^2 = 1900 - p^3$$
Squaring

$$p^3 = 1900 - x^2$$
Adding p^3 and subtracting x^2

$$p = (1900 - x^2)^{1/3}$$
Taking cube roots

We can differentiate this *explicitly* with respect to x:

$$p' = \frac{1}{3}(1900 - x^2)^{-2/3}(-2x)$$
Using the Generalized Power Rule

$$= -\frac{2}{3}x(1900 - x^2)^{-2/3}$$
Simplifying

Evaluating at the given values of $x = 30$ and $p = 10$ gives

$$p' = -\frac{2}{3} \cdot 30(1900 - 30^2)^{-2/3}$$
Substituting $x = 30$

$$= -20(1000)^{-2/3} = -\frac{20}{100} = -0.2$$

This agrees with the answer by implicit differentiation. Which way was easier?

Related Rates

Sometimes *both* variables in an equation will be functions of a *third* variable, usually *t* for time. For example, for a seasonal product such as fur coats, the price *p* and weekly sales *x* will be related by a demand equation, and both price *p* and quantity *x* will depend on the time of year. Differentiating both sides of the demand equation with respect to time *t* will give an equation relating the derivatives dp/dt and dx/dt. Such "related rates" equations show how fast one quantity is changing relative to another. First, an "everyday" example.

EXAMPLE 6

FINDING RELATED RATES

A pebble thrown into a pond causes circular ripples to radiate outward. If the radius of the outer ripple is growing by 2 feet per second, how fast is the area of its circle growing at the moment when the radius is 10 feet?

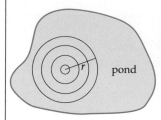
pond

Solution　The formula for the area of a circle is $A = \pi r^2$. Both the area *A* and the radius *r* of the circle increase with time, so both are functions of *t*. We are told that the radius is increasing by 2 feet per second $(dr/dt = 2)$, and we want to know how fast the area is changing (dA/dt). To find the relationship between dA/dt and dr/dt, we differentiate both sides of $A = \pi r^2$ with respect to *t*.

$$\frac{dA}{dt} = 2\pi r \cdot \frac{dr}{dt}$$ Writing the 2 before the π

$$\frac{dA}{dt} = 2\pi \cdot \underbrace{10}_{r} \cdot \underbrace{2}_{\frac{dr}{dt}} = \underbrace{40\pi \approx 125.6}_{\substack{\text{Using} \\ \pi \approx 3.14}}$$ Substituting $r = 10$ and $\frac{dr}{dt} = 2$

Therefore, at the moment when the radius is 10 feet, the area of the circle is growing at the rate of about 126 square feet per second.

We should be ready to interpret any *rate* as a derivative, just as we interpreted the radius growing by 2 feet per second as $dr/dt = 2$.

EXAMPLE **7**

USING RELATED RATES TO FIND PROFIT GROWTH

A boat yard's total profit from selling x outboard motors is $P = -x^2 + 1000x - 2000$. If the outboards are selling at the rate of 20 per week, how fast is the profit changing when 400 motors have been sold?

Solution   Profit P and quantity x both change with time, so both are functions of t. We differentiate both sides of $P = -x^2 + 1000x - 2000$ with respect to t and then substitute the given data.

$$\frac{dP}{dt} = -2x\frac{dx}{dt} + 1000\frac{dx}{dt} \qquad \text{Differentiating with respect to } t$$

$$= -2(400)20 + 1000 \cdot 20$$

$$\underbrace{}_{\substack{x \quad dx/dt}} \quad \underbrace{}_{dx/dt} \qquad \begin{array}{l}\text{Substituting } x - 400 \text{ (number sold)} \\ \text{and } dx/dt = 20 \text{ (sales per week)}\end{array}$$

$$= -16{,}000 + 20{,}000 = 4000$$

Therefore, the company's profits are growing at the rate of $4000 per week.

EXAMPLE **8**

USING RELATED RATES TO PREDICT POLLUTION

A study of urban pollution predicts that sulfur oxide emissions in a city will be $S = 2 + 20x + 0.1x^2$ tons, where x is the population (in thousands). The population of the city t years from now is expected to be $x = 800 + 20\sqrt{t}$ thousand people. Find how rapidly sulfur oxide pollution will be increasing 4 years from now.

Solution

Finding the rate of increase of pollution means finding $\dfrac{dS}{dt}$.

$$\frac{dS}{dt} = 20\frac{dx}{dt} + 0.2x\frac{dx}{dt} \qquad \begin{array}{l}S = 2 + 20x + 0.1x^2 \\ \text{differentiated with respect to } t \\ (x \text{ is also a function of } t)\end{array}$$

We then find dx/dt from the other given equation:

$$\frac{dx}{dt} = 10t^{-1/2} \qquad \begin{array}{l}x = 80 + 20t^{1/2} \text{ differentiated} \\ \text{with respect to } t\end{array}$$

$$= 10 \cdot 4^{-1/2} = 10 \cdot \frac{1}{2} = 5 \qquad \begin{array}{l}\text{Substituting the given } t = 4 \\ \text{gives } dx/dt = 5\end{array}$$

$\dfrac{dS}{dt}$ then becomes

$$\underset{dx/dt}{\underbrace{\dfrac{dS}{dt} = 20 \cdot 5}} + 0.2\underset{x}{\underbrace{(800 + 20\sqrt{4})}}\underset{dx/dt}{\underbrace{5}}$$

$\dfrac{dS}{dt} = 20\,\dfrac{dx}{dt} + 0.2x\,\dfrac{dx}{dt}$
with $dx/dt = 5$ and
$x = 800 + 20t^{1/2}$ at $t = 4$

$$= 100 + 0.2(840)5 = 100 + 840 = 940$$

Therefore, in 4 years the sulfur oxide emissions will be increasing at the rate of 940 tons per year.

Graphing Calculator Exploration

Verify the answer to the preceding example on a graphing calculator as follows:

a. Define $y_1 = 2 + 20x + 0.1x^2$ (the sulfur oxide function).

b. Define $y_2 = 800 + 20\sqrt{x}$ (the population function, using x for ease of entry).

c. Define $y_3 = y_1(y_2)$ (the composition of y_1 and y_2, giving pollution in year x).

d. Define y_4 to be the derivative of y_3 (using NDERIV, giving rate of change of pollution).

e. Graph these on the window $[0, 10]$ by $[0, 1500]$. (Which function does *not* appear on the screen?)

f. Evaluate y_4 at $x = 4$ to verify the answer to Example 8.

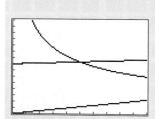

Verification of the Power Rule for Rational Powers

On page 112 we stated the Power Rule for differentiation:

$$\dfrac{d}{dx}x^n = nx^{n-1}$$

Although we proved it only for *integer* powers, we have been using the Power Rule for *all* constant powers n. Using implicit differentiation, we may now prove the Power Rule for *rational* powers. (Recall that a rational number is of the form p/q, where p and q are integers with $q \neq 0$.) Let $y = x^n$ for a rational exponent $n = p/q$, and let x be a number at which $x^{p/q}$ is differentiable. Then

$$y = x^n = x^{p/q}$$ Since $n = p/q$

$$y^q = x^p$$ Raising each side to the power q

$$q y^{q-1} \frac{dy}{dx} = p x^{p-1}$$ Differentiating each side implicitly with respect to x

$$\frac{dy}{dx} = \frac{p x^{p-1}}{q y^{q-1}}$$ Dividing each side by $q y^{q-1}$

$$= \frac{p x^{p-1}}{q \left(x^{\frac{p}{q}} \right)^{q-1}} = \frac{p}{q} \frac{x^{p-1}}{x^{p-\frac{p}{q}}}$$ Using $y = x^{p/q}$ and multiplying out the exponents in the denominator

$$= \frac{p}{q} x^{\overbrace{p-1-\left(p - \frac{p}{q} \right)}^{-1+\frac{p}{q}}} = \frac{p}{q} x^{\frac{p}{q}-1} = n x^{n-1}$$ Subtracting powers, simplifying, and replacing p/q by n (twice)

This is what we wanted to show, that the derivative of $y = x^n$ is $dy/dx = n x^{n-1}$ for any rational exponent $n = p/q$. This proves the Power Rule for rational exponents.

3.6 Section Summary

An equation in x and y may define one or more functions $y = f(x)$, which we may need to differentiate. Instead of solving the equation for y, which may be difficult or impossible, we can differentiate *implicitly*, differentiating both sides of the original equation with respect to x (writing a dy/dx or y' whenever we differentiate y) and solving for the derivative dy/dx. The derivative at any point of the curve may then be found by substituting the coordinates of that point.

Implicit differentiation is especially useful when several variables in an equation depend on an underlying variable, usually t for time. Differentiating the equation implicitly with respect to this underlying variable gives an equation involving the rates of change of the original variables. Numbers may then be substituted into this "related rate equation" to find a particular rate of change.

▶ **Solution to Practice Problem**

a. $\dfrac{d}{dx} x^4 = 4x^3$ **b.** $\dfrac{d}{dx} y^2 = 2y \dfrac{dy}{dx}$ **c.** $\dfrac{d}{dx} (x^2 y^3) = 2xy^3 + 3x^2 y^2 \dfrac{dy}{dx}$

3.6 EXERCISES

For each equation, use implicit differentiation to find dy/dx.

1. $y^3 - x^2 = 4$ **2.** $y^2 = x^4$

3. $x^3 = y^2 - 2$ **4.** $x^2 + y^2 = 1$

5. $y^4 - x^3 = 2x$ **6.** $y^2 = 4x + 1$

7. $(x + 1)^2 + (y + 1)^2 = 18$

8. $xy = 12$ **9.** $x^2y = 8$

10. $x^2y + xy^2 = 4$ **11.** $xy - x = 9$

12. $x^3 + 2xy^2 + y^3 = 1$ **13.** $x(y - 1)^2 = 6$

14. $(x - 1)(y - 1) = 25$

15. $y^3 - y^2 + y - 1 = x$

16. $x^2 + y^2 = xy + 4$ **17.** $\dfrac{1}{x} + \dfrac{1}{y} = 2$

18. $\sqrt[3]{x} + \sqrt[3]{y} = 2$ **19.** $x^3 = (y - 2)^2 + 1$

20. $\sqrt{xy} = x + 1$

For each equation, find dy/dx evaluated at the given values.

21. $y^2 - x^3 = 1$ at $x = 2, y = 3$

22. $x^2 + y^2 = 25$ at $x = -3, y = 4$

23. $y^2 = 6x - 5$ at $x = 1, y = -1$

24. $xy = 12$ at $x = 6, y = 2$

25. $x^2y + y^2x = 0$ at $x = -2, y = 2$

26. $y^2 + y + 1 = x$ at $x = 1, y = -1$

27. $x^2 + y^2 = xy + 7$ at $x = 3, y = 2$

28. $\sqrt[3]{x} + \sqrt[3]{y} = 3$ at $x = 1, y = 8$

For each demand equation, use implicit differentiation to find dp/dx.

29. $p^2 + p + 2x = 100$

30. $p^3 + p + 6x = 50$

31. $12p^2 + 4p + 1 = x$

32. $8p^2 + 2p + 100 = x$

33. $xp^3 = 36$

34. $xp^2 = 96$

35. $(p + 5)(x + 2) = 120$

36. $(p - 1)(x + 5) = 24$

APPLIED EXERCISES ON IMPLICIT DIFFERENTIATION

37. BUSINESS: Demand Equation A company's demand equation is $x = \sqrt{68 - p^2}$, where p is the price in dollars. Find dp/dx when $p = 2$ and interpret your answer.

38. BUSINESS: Demand Equation A company's demand equation is $x = \sqrt{650 - p^2}$, where p is the price in dollars. Find dp/dx when $p = 5$ and interpret your answer.

39. BUSINESS: Sales If a company spends r million dollars on research, its sales will be s million dollars, where r and s are related by $s^2 = r^3 - 55$.

a. Find ds/dr by implicit differentiation and evaluate it at $r = 4$, $s = 3$. [*Hint:* Differentiate the equation with respect to r.]

b. Find dr/ds by implicit differentiation and evaluate it at $r = 4$, $s = 3$. [*Hint:* Differentiate the original equation with respect to s.]

c. Interpret your answers to parts (a) and (b) as rates of change.

40. BIOMEDICAL: Muscle Contraction When a muscle lifts a load, it does so according to the "fundamental equation of muscle contraction" $(L + m)(V + n) = k$, where L is the load that the muscle is lifting, V is the velocity of contraction of the muscle, and m, n, and k are constants. Use implicit differentiation to find dV/dL.

EXERCISES ON RELATED RATES

In each equation, x and y are functions of t. Differentiate with respect to t to find a relation between dx/dt and dy/dt.

41. $x^3 + y^2 = 1$ **42.** $x^5 - y^3 = 1$

43. $x^2y = 80$

44. $xy^2 = 96$

45. $3x^2 - 7xy = 12$

46. $2x^3 - 5xy = 14$

47. $x^2 + xy = y^2$

48. $x^3 - xy = y^3$

APPLIED EXERCISES ON RELATED RATES

49. GENERAL: Snowballs A large snowball is melting so that its radius is decreasing at the rate of 2 inches per hour. How fast is the volume decreasing at the moment when the radius is 3 inches? [*Hint:* The volume of a sphere of radius r is $V = \frac{4}{3}\pi r^3$.]

50. GENERAL: Hailstones A hailstone (a small sphere of ice) is forming in the clouds so that its radius is growing at the rate of 1 millimeter per minute. How fast is its volume growing at the moment when the radius is 2 millimeters? [*Hint:* The volume of a sphere of radius r is $V = \frac{4}{3}\pi r^3$.]

51. BIOMEDICAL: Tumors The radius of a spherical tumor is growing by $\frac{1}{2}$ centimeter per week. Find how rapidly the volume is increasing at the moment when the radius is 4 centimeters. [*Hint:* The volume of a sphere of radius r is $V = \frac{4}{3}\pi r^3$.]

52. BUSINESS: Profit A company's profit from selling x units of an item is $P = 1000x - \frac{1}{2}x^2$ dollars. If sales are growing at the rate of 20 per day, find how rapidly profit is growing (in dollars per day) when 600 units have been sold.

53. BUSINESS: Revenue A company's revenue from selling x units of an item is given as $R = 1000x - x^2$ dollars. If sales are increasing at the rate of 80 per day, find how rapidly revenue is growing (in dollars per day) when 400 units have been sold.

54. SOCIAL SCIENCE: Accidents The number of traffic accidents per year in a city of population p is predicted to be $T = 0.002p^{3/2}$. If the population is growing by 500 people a year, find the rate at which traffic accidents will be rising when the population is $p = 40,000$.

55. SOCIAL SCIENCE: Welfare The number of welfare cases in a city of population p is expected to be $W = 0.003p^{4/3}$. If the population is growing by 1000 people per year, find the rate at which the number of welfare cases will be increasing when the population is $p = 1,000,000$.

56. GENERAL: Rockets A rocket fired straight up is being tracked by a radar station 3 miles from the launching pad. If the rocket is traveling at 2 miles per second, how fast is the distance between the rocket and the tracking station changing at the moment when the rocket is 4 miles up? [*Hint:* The distance D in the illustration satisfies $D^2 = 9 + y^2$. To find the value of D, solve $D^2 = 9 + 4^2$.]

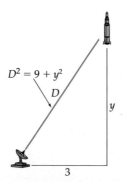

$D^2 = 9 + y^2$

D

y

3

57. BIOMEDICAL: Poiseuille's Law Blood flowing through an artery flows faster in the center of the artery and more slowly near the sides (because of friction). The speed of the blood is $V = c(R^2 - r^2)$ millimeters (mm) per second, where R is the radius of the artery, r is the distance of the blood from the center of the artery, and c is a constant. Suppose that arteriosclerosis is narrowing the artery at the rate of $dR/dt = -0.01$ mm per year.

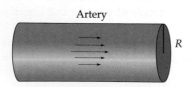

Artery

R

Find the rate at which blood flow is being reduced in an artery whose radius is $R = 0.05$ mm with $C = 500$. [*Hint:* Find dV/dt, considering r to be a constant. The units of dV/dt will be mm per second per year.]

58. IMPLICIT AND EXPLICIT DIFFERENTIA-TION The equation $x^2 + 4y^2 = 100$ describes an ellipse.

a. Use implicit differentiation to find its slope at the points $(8, 3)$ and $(8, -3)$.

b. Solve the equation for y, obtaining *two* functions, and differentiate both to find the slopes at $x = 8$. [Answers should agree with part (a).]

c. Use a graphing calculator to graph the two functions found in part (b) on an appropriate window. Then use NDERIV to find the derivatives at (or near) $x = 8$. [Answers should agree with parts (a) and (b).]

Notice that differentiating implicitly was easier than solving for y and then differentiating.

59. RELATED RATES: Speeding A traffic patrol helicopter is stationary a quarter of a mile directly above a highway, as shown in the diagram below. Its radar detects a car whose line-of-sight distance from the helicopter is half a mile and is increasing at the rate of 57 mph. Is the car exceeding the highway's speed limit of 60 mph?

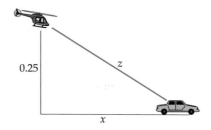

0.25

z

x

60. RELATED RATES: Speeding (*59 continued*)

In Exercise 59 you found that the car's speed was $\dfrac{(0.5)(57)}{\sqrt{(0.5)^2 - (0.25)^2}}$ mph. Enter this expression into a graphing calculator and then replace both occurrences of the line-of-sight distance 0.5 by 0.4 and calculate the new speed of the car. What if the line-of-sight distance were 0.3 mile?

Chapter Summary with Hints and Suggestions

Reading the text and doing the exercises in this chapter have helped you to master the following skills, which are listed by section (in case you need to review them) and are keyed to particular Review Exercises. Answers for all Review Exercises are given at the back of the book, and full solutions can be found in the Student Solutions Manual.

3.1 Graphing Using the First Derivative

3.2 Graphing Using the First and Second Derivatives

- Graph a polynomial, showing all relative extreme points and inflection points. (*Review Exercises 1–8.*)

  f' gives slope, f'' gives concavity

- Graph a fractional power function, showing all relative extreme points and inflection points. (*Review Exercises 9–12.*)

- Graph a rational function, showing all relative extreme points. (*Review Exercises 13–18.*)

3.3 Optimization

- Find the absolute extreme values of a given function on a given interval. (*Review Exercises 19–28.*)

- Show that the marginal cost function pierces the average cost function where the average cost is minimized. (*Review Exercises 29–30.*)

- Maximize the efficiency of a tugboat or a flying bird. (*Review Exercises 31, 38.*)

3.4 Further Applications of Optimization

- Solve a geometric optimization problem. (*Review Exercises 32–37.*)

- Maximize profit for a company. (*Review Exercise 39.*)

- Maximize revenue from an orchard. (*Review Exercise 40.*)

- Minimize the cost of a power cable. (*Review Exercise 41.*)

- Maximize tax revenue to the government. (*Review Exercise 42.*)

- Minimize the materials used in a container. (*Review Exercises 43–45.*)

3.5 Optimizing Lot Size and Harvest Size

- Find the lot size that minimizes production costs or inventory costs. (*Review Exercises 46–47.*)

- Find the population size that allows the maximum sustainable yield. (*Review Exercises 48–49.*)

3.6 Implicit Differentiation and Related Rates

- Find a derivative by implicit differentiation. (*Review Exercises 50–53.*)

- Find a derivative by implicit differentiation and evaluate it. (*Review Exercises 54–57.*)

- Solve a geometric related rates problem. (*Review Exercise 58.*)

- Use related rates to find the growth in profit or revenue. (*Review Exercises 59–60.*)

Hints and Suggestions

- Do not confuse *relative* and *absolute* extremes: a function may have several *relative* maximum points (high points compared to their neighbors), but it can have at most one *absolute* maximum value (the largest value of the function on its entire domain). *Relative* extremes are used in graphing, and *absolute* extremes are used in optimization.

- Graphing calculators can be very helpful for graphing functions. However, you must first find a window that shows the interesting parts of the curve (relative extreme points and inflection points), and that is where calculus is essential.

- We have two procedures for optimizing continuous functions. Both begin by finding all critical numbers of the function in the domain. Then:

   1. If the function has only one critical number, find the sign of the second derivative there: a *positive* sign means an absolute *minimum*, and a *negative* sign means an absolute *maximum*.

   2. If the interval is closed, the maximum and minimum values may be found by evaluating the function at all critical numbers in the interval and endpoints—the largest and

smallest resulting values are the maximum and minimum values of the function.

A good strategy is this: Find all critical numbers in the domain. If there is only one, use procedure 1 above. Otherwise, try to define the function on a closed interval and use procedure 2. If all else fails, make a sketch of the graph.

- Don't forget to use the second-derivative test in applied problems. Your employer will not be happy if you accidentally *minimize* your company's profits.

- In implicit differentiation problems, remember which is the *function* (usually y) and which is the independent variable (usually x).

- In related rate problems, begin by looking for an equation that relates the variables, and then differentiate it with respect to the underlying variable (usually t).

- **Practice for Test:** Review Exercises 3, 11, 21, 25, 33, 35, 39, 43, 45, 47, 53, 55.

Review Exercises for Chapter 3

Practice test exercise numbers are in green.

3.1 and 3.2 Graphing

Graph each function "by hand," showing all relative extreme points and inflection points.

1. $f(x) = x^3 - 3x^2 - 9x + 12$

2. $f(x) = x^3 + 3x^2 - 9x - 7$

3. $f(x) = x^4 - 4x^3 + 15$

4. $f(x) = x^4 + 4x^3 + 17$

5. $f(x) = x(x + 3)^2$ 6. $f(x) = x(x - 6)^2$

7. $f(x) = x(x - 4)^3$ 8. $f(x) = x(x + 4)^3$

9. $f(x) = \sqrt[7]{x^5} + 1$ 10. $f(x) = \sqrt[7]{x^6} + 1$

11. $f(x) = \sqrt[7]{x^4} + 1$ 12. $f(x) = \sqrt[7]{x^3} + 1$

Graph each function, showing all relative extreme points. If you use a graphing calculator, make a hand-drawn sketch showing all relative extreme points.

13. $f(x) = \dfrac{1}{x^2 - 6x}$ 14. $f(x) = \dfrac{1}{x^2 + 4x}$

15. $f(x) = \dfrac{x^2}{x^4 + 1}$ 16. $f(x) = \dfrac{2x^2}{1 + x^2}$

17. $f(x) = \dfrac{1 - 2x}{x^2}$ 18. $f(x) = \dfrac{1 - x}{x^2}$

3.3 and 3.4 Optimization

Find the absolute extreme values of each function on the given interval.

19. $f(x) = 2x^3 - 6x$ on $[0, 5]$

20. $f(x) = 2x^3 - 24x$ on $[0, 5]$

21. $f(x) = x^4 - 4x^3 - 8x^2 + 64$ on $[-1, 5]$

22. $f(x) = x^4 - 4x^3 + 4x^2 + 1$ on $[0, 10]$

23. $h(x) = (x - 1)^{2/3}$ on $[0, 9]$

24. $f(x) = \sqrt{100 - x^2}$ on $[-10, 10]$

25. $g(w) = (w^2 - 4)^2$ on $[-3, 3]$

26. $g(x) = x(8 - x)$ on $[0, 8]$

27. $f(x) = \dfrac{x}{x^2 + 1}$ on $[-3, 3]$

28. $f(x) = \dfrac{x}{x^2 + 4}$ on $[-4, 4]$

29. BUSINESS: Average and Marginal Cost
For the cost function $C(x) = 10,000 + x^2$, graph the marginal cost function $MC(x)$ and also the average cost function $AC(x) = C(x)/x$ on the same graph, showing that they intersect at the point where the average cost is minimized.

30. Prove that the marginal cost function intersects the average cost function at the point where the average cost is minimized (as shown in the following diagram) by justifying each numbered step in the following series of equations.
At the x-value where the average cost function $AC(x) = \dfrac{C(x)}{x}$ is minimized, we must have

$$0 \overset{①}{=} \frac{d}{dx} AC(x) \overset{②}{=} \frac{xC'(x) - C(x)}{x^2}$$

$$\overset{③}{=} \frac{1}{x}\left[\frac{xC'(x) - C(x)}{x}\right]$$

$$\overset{④}{=} \frac{1}{x}\left[C'(x) - \frac{C(x)}{x}\right] \overset{⑤}{=} \frac{1}{x}[MC(x) - AC(x)].$$

⑥ Finally, explain why the equality of the first and last expressions in the series of equations proves the result.

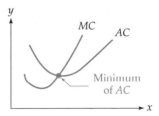

The marginal cost function pierces the average cost function at its minimum.

31. GENERAL: Fuel Efficiency At what speed should a tugboat travel upstream so as to use the least amount of fuel to reach its destination? If the tugboat's speed through the water is v, and if the speed of the current (relative to the land) is c, then the energy used is proportional to $E(v) = \dfrac{v^2}{v - c}$. Find the velocity v that minimizes the energy $E(v)$. Your answer will depend upon c, the speed of the current.

32. GENERAL: Fencing A homeowner wants to enclose three adjacent rectangular pens of equal size along a straight wall, as in the following diagram. If the side along the wall needs no fence, what is the largest total area that can be enclosed using only 240 feet of fence?

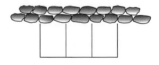

33. GENERAL: Maximum Area A homeowner wants to enclose three adjacent rectangular pens of equal size, as in the diagram below. What is the largest total area that can be enclosed using only 240 feet of fence?

34. GENERAL: Unicorns To celebrate the acquisition of Styria in 1261, Ottokar II sent hunters into the Bohemian woods to capture a unicorn. To display the unicorn at court, the king built a rectangular cage. The material for three sides of the cage cost 3 ducats per running cubit, while the fourth was to be gilded and cost 51 ducats per running cubit. In 1261 it was well known that a happy unicorn requires an area of 2025 square cubits. Find the dimensions that would keep the unicorn happy at the lowest cost.

35. GENERAL: Box Design An open-top box with a square base is to have a volume of exactly 500 cubic inches. Find the dimensions of the box that can be made with the smallest amount of materials.

36. GENERAL: Packaging Find the dimensions of the cylindrical tin can with volume 16π cubic inches that can be made from the least amount of tin. (*Note:* $16\pi \approx 50$ cubic inches.)

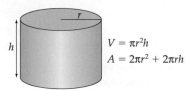

$$V = \pi r^2 h$$
$$A = 2\pi r^2 + 2\pi r h$$

37. GENERAL: Packaging Find the dimensions of the open-top cylindrical tin can with volume 8π cubic inches that can be made from the least amount of tin. (*Note:* $8\pi \approx 25$ cubic inches.)

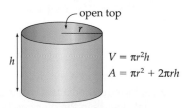

$$V = \pi r^2 h$$
$$A = \pi r^2 + 2\pi r h$$

38. GENERAL: Bird Flight Let v be the flying speed of a bird, and let w be its weight. The power P that the bird must maintain during flight is $P = \dfrac{aw^2}{v} + bv^3$, where a and b are positive constants depending on the shape of

the bird and the density of the air. Find the speed v that minimizes the power P.

39. BUSINESS: Maximum Profit A computer dealer can sell 12 personal computers per week at a price of $2000 each. He estimates that each $400 price decrease will result in three more sales per week. If the computers cost him $1200 each, what price should he charge to maximize his profit? How many will he sell at that price?

40. GENERAL: Farming A peach tree will yield 100 pounds of peaches now, which will sell for 40 cents a pound. Each week that the farmer waits will increase the yield by 10 pounds, but the selling price will decrease by 2 cents per pound. How long should the farmer wait to pick the fruit in order to maximize her revenue?

41. GENERAL: Minimum Cost A cable is to connect a power plant to an island that is 1 mile offshore and 3 miles downshore from the power plant. It costs $5000 per mile to lay a cable underwater and $3000 per mile to lay it underground. If the cost of laying the cable is to be minimized, find the distance x downshore from the island where the cable should meet the land.

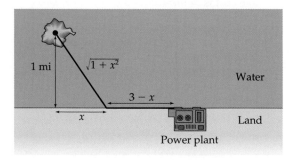

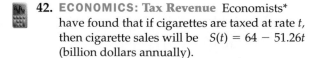

42. ECONOMICS: Tax Revenue Economists[*] have found that if cigarettes are taxed at rate t, then cigarette sales will be $S(t) = 64 - 51.26t$ (billion dollars annually).

---

[*]Michael Grossman, Gary Becker (winner of the 1992 Nobel Memorial Prize in Economics), Frank Chaloupka, and Kevin Murphy.

a. Use a graphing calculator to find the tax rate t that maximizes revenue to the government.

b. Multiply this tax rate times $2.02 (the average price of a pack of cigarettes) to find the actual tax per pack.

43. GENERAL: Packaging A 12-ounce soft drink can has volume 21.66 cubic inches. If the top and bottom are twice as thick as the sides, find the dimensions (radius and height) that minimize the amount of metal used in the can.

44. GENERAL: Box Design A standard 8.5- by 11-inch piece of paper can be made into a box with a lid by cutting x- by x-inch squares from two corners and x- by 5.5-inch rectangles from the other corners and then folding, as shown below. What value of x maximizes the volume of the box, and what is the maximum volume?

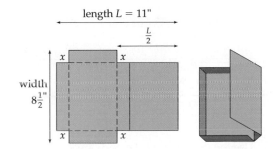

length $L = 11"$

$\frac{L}{2}$

width $8\frac{1}{2}"$

x x

x x

45. GENERAL: Box Design A standard 6- by 8-inch card can be made into a box with a lid by cutting x- by x-inch squares from two corners and x- by 4-inch rectangles from the other corners and then folding. (See the diagram above, but use length 8 and width 6.) What value of x maximizes the volume of the box, and what is the maximum volume?

3.5 Optimizing Lot Size and Harvest Size

46. BUSINESS: Production Runs A wallpaper company estimates the demand for a certain pattern to be 900 rolls per year. It costs $800 to set up the presses to print the pattern, plus $200 to print each roll. If the company can store a roll of wallpaper for a year at a cost of $4, how many rolls should it print at a time and how many printing runs will it need in a year to minimize production costs?

47. BUSINESS: Lot Size A motorcycle shop estimates that it will sell 500 motorbikes in a year. Each bike costs $300, plus a fixed charge of $500 per order. If it costs $200 to store a motorbike for a year, what is the order size and how many orders will be needed in a year to minimize inventory costs?

48. MAXIMUM SUSTAINABLE YIELD The reproduction function for the North American duck is estimated to be $f(p) = -0.02p^2 + 7p$, where p and $f(p)$ are measured in thousands. Find the size of the population that allows the maximum sustainable yield, and also find the size of the yield.

49. MAXIMUM SUSTAINABLE YIELD Ecologists estimate the reproduction function for striped bass in an East Coast fishing ground to be $f(p) = 60\sqrt{p}$, where p and $f(p)$ are measured in thousands. Find the size of the population that allows the maximum sustainable yield, and also the size of the yield.

3.6 Implicit Differentiation and Related Rates

For each equation, use implicit differentiation to find dy/dx.

50. $6x^2 + 8xy + y^2 = 100$

51. $8xy^2 - 8y = 1$

52. $2xy^2 - 3x^2y = 0$ **53.** $\sqrt{x} - \sqrt{y} = 10$

For each equation, find dy/dx evaluated at the given values.

54. $x + y = xy$ at $x = 2, y = 2$

55. $y^3 - y^2 - y = x$ at $x = 2, y = 2$

56. $xy^2 = 81$ at $x = 9, y = 3$

57. $x^2y^2 - xy = 2$ at $x = -1, y = 1$

58. GENERAL: Melting Ice A cube of ice is melting so that each edge is decreasing at the rate of 2 inches per hour. Find how fast the volume of the ice is decreasing at the moment when each edge is 10 inches long.

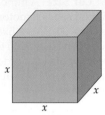

59. BUSINESS: Profit A company's profit from selling x units of a product is $P = 2x^2 - 20x$ dollars. If sales are growing at the rate of 30 per day, find the rate of change of profit when 40 units have been sold.

60. BUSINESS: Revenue A company finds that its revenue from selling x units of a product is $R = x^2 + 500x$ dollars. If sales are increasing at the rate of 50 per month, find the rate of change of revenue when 200 units have been sold.

61. BIOMEDICAL: Medication You swallow a spherical pill whose radius is 0.5 centimeter (cm), and it dissolves in your stomach so that its radius decreases at the rate of 0.1 cm per minute. Find the rate at which the volume is decreasing (the rate at which the medication is being made available to your system) when the radius is

a. 0.5 cm
b. 0.2 cm

Cumulative Review for Chapters 1–3

The following exercises review some of the basic techniques that you learned in Chapters 1–3. Answers to all of these cumulative review exercises are given in the answer section at the back of the book.

1. Find an equation for the line through the points $(-4, 3)$ and $(6, -2)$. Write your answer in the form $y = mx + b$.

2. Simplify $\left(\frac{4}{25}\right)^{-1/2}$

3. Find, correct to three decimal places:
$\lim\limits_{x \to 0} (1 + 3x)^{1/x}$

4. For the function $f(x) = \begin{cases} 4x - 8 & \text{if } x < 3 \\ 7 - 2x & \text{if } x \geq 3 \end{cases}$

  a. Draw its graph.

  b. Find $\lim\limits_{x \to 3^-} f(x)$.

  c. Find $\lim\limits_{x \to 3^+} f(x)$.

  d. Find $\lim\limits_{x \to 3} f(x)$.

  e. Is $f(x)$ continuous or discontinuous, and if it is discontinuous, where?

5. Use the definition of the derivative, $f'(x) = \lim\limits_{h \to 0} \dfrac{f(x + h) - f(x)}{h}$, to find the derivative of
$f(x) = 2x^2 - 5x + 7$.

6. Find the derivative of $f(x) = 8\sqrt{x^3} - \dfrac{3}{x^2} + 5$.

7. Find the derivative of $f(x) = (x^5 - 2)(x^4 + 2)$.

8. Find the derivative of $f(x) = \dfrac{2x - 5}{3x - 2}$.

9. The population of a city x years from now is predicted to be $P(x) = 3600x^{2/3} + 250{,}000$ people. Find $P'(8)$ and $P''(8)$ and interpret your answers.

10. Find $\dfrac{d}{dx}\sqrt{2x^2 - 5}$ and write your answer in radical form.

11. Find $\dfrac{d}{dx}[(3x + 1)^4(4x + 1)^3]$.

12. Find $\dfrac{d}{dx}\left(\dfrac{x - 2}{x + 2}\right)^3$ and simplify your answer.

13. Make sign diagrams for the first and second derivatives and draw the graph of the function $f(x) = x^3 - 12x^2 - 60x + 400$. Show on your graph all relative extreme points and inflection points.

14. Make sign diagrams for the first and second derivatives and draw the graph of the function $f(x) = \sqrt[3]{x^2} - 1$. Show on your graph all relative extreme points and inflection points.

15. A homeowner wishes to use 600 feet of fence to enclose two identical adjacent pens, as in the diagram below. Find the largest total area that can be enclosed.

16. A store can sell 12 telephone answering machines per day at a price of $200 each. The manager estimates that for each $10 price reduction she can sell 2 more per day. The answering machines cost the store $80 each. Find the price and the quantity sold per day to maximize the company's profit.

17. For y defined implicitly by $x^3 + 9xy^2 + 3y = 43$, find $\dfrac{dy}{dx}$ and evaluate it at the point $(1, 2)$.

18. A large spherical balloon is being inflated at the rate of 32 cubic feet per minute. Find how fast the radius is increasing at the moment when the radius is 2 feet.

4 Exponential and Logarithmic Functions

Application Preview

Carbon 14 Dating and the Shroud of Turin

Carbon 14 dating is a method for estimating the age of ancient plants and animals. While a plant is alive, it absorbs carbon dioxide, and with it both "ordinary" carbon and carbon 14, a radioactive form of carbon produced by cosmic rays in the upper atmosphere. When the plant dies, it stops absorbing both types of carbon, and the carbon 14 decays exponentially at a known rate, as shown in the following graph.

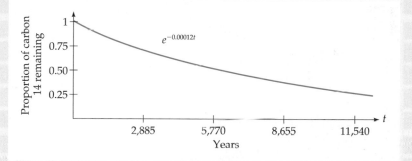

The time since the plant died can then be estimated by comparing the amount of carbon 14 that remains with the amount of ordinary carbon remaining, since they were originally in a known proportion. The comparison technique involves *logarithms*, as explained in this chapter. Animal remains can also be dated in this way, since their diets contain plant matter.

In 1988, the Roman Catholic Church used carbon 14 dating to determine the authenticity of the Shroud of Turin, believed by many to be the burial cloth of Jesus. This was done by estimating the age of the flax plants from which the linen shroud was woven. In Exercise 32 on page 302 you will find that the shroud is only a few hundred years old, and therefore cannot be the burial cloth of Christ.

A few of the technical details of carbon 14 dating are as follows. The amount of carbon 14 remaining in a sample can be found by burning the sample near a Geiger counter to measure the radioactivity. For the shroud, this would have required a handkerchief-sized sample, and so instead a more advanced method was employed, using a tandem accelerator mass spectrometer and requiring only a postage-stamp-sized sample.

**Carbon 14 Dating and
the Shroud of Turin**
(continued)

Carbon 14 dating assumes, however, that the ratio of carbon 14 to ordinary carbon in the atmosphere has remained steady over the centuries. This assumption seemed difficult to verify until the discovery in 1955 of a bristlecone pine tree in the White Mountains of California that was over 2000 years old (according to its growth rings). Each annual ring had absorbed carbon and carbon 14 from the air, and so provided a year-by-year record of their ratio. The analysis showed that the ratio has remained relatively steady over the centuries, and where variations did occur, it enabled scientists to construct a table for making corrections in the original carbon 14 dates. (Incidentally, bristlecone pine trees that are 4900 years old have been found, making them the oldest living things on earth.) For extremely old remains, such as dinosaur fossils, archeologists use longer-lasting radioactive elements, such as potassium 40.

Carbon 14 dating, for which its inventor Willard Libby received a Nobel prize in 1960, has become an invaluable tool in the social and biological sciences.

4.1 EXPONENTIAL FUNCTIONS

Introduction

In this chapter we introduce exponential and logarithmic functions, and apply them to a wide variety of problems. We begin with exponential functions, showing how they are used to model the processes of growth and decay.

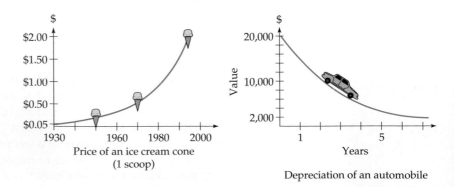

Price of an ice cream cone
(1 scoop)

Depreciation of an automobile

We will also define the very important mathematical constant e.

Exponential Functions

A function that has a variable in an exponent, such as $f(x) = 2^x$, is called an *exponential function.*

$$f(x) = 2^x$$

Exponent

Base

The table below shows some values of the exponential function $f(x) = 2^x$, and its graph (based on these points) is shown on the right.

x	y = 2^x
−3	$2^{-3} = \frac{1}{8}$
−2	$2^{-2} = \frac{1}{4}$
−1	$2^{-1} = \frac{1}{2}$
0	$2^0 = 1$
1	$2^1 = 2$
2	$2^2 = 4$
3	$2^3 = 8$

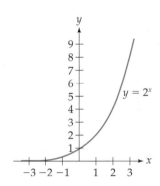

Domain of $y = 2^x$ is $\mathbb{R} = (-\infty, \infty)$ and range is $(0, \infty)$.

Clearly, the exponential function 2^x is quite different from the parabola x^2.

The exponential function $f(x) = \left(\frac{1}{2}\right)^x$ has base $\frac{1}{2}$. The following table shows some of its values, and its graph is shown to the right of the table. Notice that it is the mirror image of the curve $y = 2^x$.

x	y = (1/2)^x
−3	$\left(\frac{1}{2}\right)^{-3} = 8$
−2	$\left(\frac{1}{2}\right)^{-2} = 4$
−1	$\left(\frac{1}{2}\right)^{-1} = 2$
0	$\left(\frac{1}{2}\right)^0 = 1$
1	$\left(\frac{1}{2}\right)^1 = \frac{1}{2}$
2	$\left(\frac{1}{2}\right)^2 = \frac{1}{4}$
3	$\left(\frac{1}{2}\right)^3 = \frac{1}{8}$

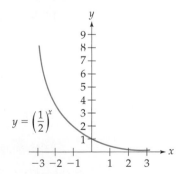

Domain of $y = \left(\frac{1}{2}\right)^x$ is $\mathbb{R} = (-\infty, \infty)$ and range is $(0, \infty)$.

We can define an exponential function $f(x) = a^x$ for any positive base a. We always take the base to be positive, so for the rest of this section *the letter a will stand for a positive constant.*

Exponential functions with bases $a > 1$ are used to model *growth,* as in populations or savings accounts, and exponential functions with bases $a < 1$ are used to model *decay,* as in depreciation. (For base $a = 1$, the graph is a horizontal line, since $1^x = 1$ for all x.)

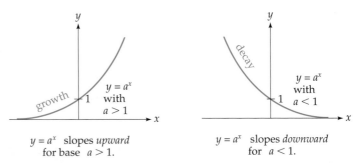

$y = a^x$ slopes *upward*
for base $a > 1$.

$y = a^x$ slopes *downward*
for $a < 1$.

Compound Interest

Money invested at compound interest grows exponentially. (The word "compound" means that the interest is added to the account, earning more interest.) For example, if you invest P dollars (the "principal") at 8% interest compounded annually, then after 1 year you will have P dollars plus 8% of P dollars:

$$\left(\begin{matrix}\text{Value after} \\ \text{1 year}\end{matrix}\right) = P + 0.08P = P \cdot (1 + 0.08)$$

Notice that increasing a quantity by 8% is the same as multiplying it by $(1 + 0.08)$. Therefore, to find the value after n years of compounding, we simply multiply by $(1 + 0.08)$ n times.

$$\left(\begin{matrix}\text{Value after} \\ n \text{ years}\end{matrix}\right) = \overbrace{P \cdot (1 + 0.08) \cdot (1 + 0.08) \cdot \ \ldots \ \cdot (1 + 0.08)}^{n \text{ times}}$$
$$= P \cdot (1 + 0.08)^n$$

The 0.08 can be replaced by any interest rate r (written in decimal form).

Compound Interest

For P dollars invested at interest rate r per period,

$$\left(\begin{matrix}\text{Value after} \\ n \text{ periods}\end{matrix}\right) = P(1 + r)^n$$

Remember, however, that banks always state *annual* interest rates, but the compounding may be done more frequently, such as quarterly or daily. The *r* in the formula above is the interest rate *per compounding period*, and the *n* is the number of *periods*. For example, if a bank offers 8% compounded *quarterly*, the interest must be calculated on a *quarterly* basis: the annual 8% is equivalent to a *quarterly* rate of 2% (one-quarter of 8%), so $r = 0.02$,   and *n* is the number of *quarters*.

EXAMPLE **1**

FINDING VALUE UNDER COMPOUND INTEREST

Find the value of $1000 invested for 2 years at 8% compounded quarterly.

Solution

For quarterly compounding, the interest rate *r* is then one-quarter of the annual 8%, so $r = \frac{1}{4} \cdot 8\% = 2\% = 0.02$.   The number of periods (quarters) in 2 years is $n = 4 \cdot 2 = 8$.   The principal is $P = 1000$. With these numbers, the compound interest formula gives

$$P(1 + r)^n = 1000 \cdot (1 + 0.02)^8 \qquad r = \left(\begin{array}{c}\text{rate per}\\\text{quarter}\end{array}\right) = 2\%$$

$$\underbrace{}_{P} \qquad \overset{\uparrow}{r} \ \overset{\uparrow}{n}$$

$$= 1000 \cdot \underbrace{(1.02)^8}_{\substack{1.17166 \\ \text{(from a calculator)}}} \approx 1171.66 \qquad n = \left(\begin{array}{c}\text{number of}\\\text{quarters}\end{array}\right) = 8$$

Therefore, the value is $1171.66.

We may interpret the formula $P(1 + 0.02)^8$ intuitively as follows: multiplying the principal by $(1 + 0.02)$ means that you keep the original amount (the "1") plus some interest (the 0.02), and the exponent 8 means that this is done a total of 8 times.

Practice Problem **1**

Find the value of $2000 invested for 2 years at 36% compounded monthly. [*Hint:* Convert the 36% to a monthly rate and let *n* be the number of months in 2 years. You will need a calculator.]

➤ Solution on page 283

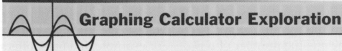

Graphing Calculator Exploration

In the long run, which is more important—principal or interest rate? The following graph shows the value of $1000 at 5% interest, together with the value of a mere $200 at the higher rate of 10% (both compounded annually). The fact that the (initially) lower curve eventually surpasses the higher one illustrates a general fact: the higher interest rate will eventually prevail, regardless of the size of the initial investment.

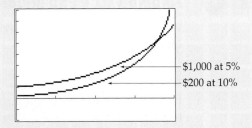

$1,000 at 5%

$200 at 10%

Find how soon the $200 at 10% will surpass the $1000 at 5% by graphing $y_1 = 1000(1 + 0.05)^x$ and $y_2 = 200(1 + 0.10)^x$ on the window [0, 40] by [−2000, 8000] and using INTERSECT. What if the higher rate is only 9%? (You will have to extend your window.)

Present Value

The value to which a sum will grow under compound interest is often called its *future value.* That is, Example 1 showed that the future value of $1000 (at 8% compounded quarterly for 2 years) is $1171.66.

Reversing the order, we can speak of the *present value* of a future payment. Example 1 shows that if a payment of $1171.66 is to be made in 2 years, its *present value* is $1000 (at this interest rate). That is, a promise to pay you $1171.66 in 2 years is worth exactly $1000 now, since $1000 deposited in a bank now would be worth that much in 2 years (at the stated interest rate). To find the *future* value we *multiply P* by $(1 + r)^n$, and therefore to find the *present* value, we simply *divide P* by $(1 + r)^n$.

Present Value

For P dollars, at interest rate r per period for n periods,

$$\left(\begin{matrix}\text{Present} \\ \text{value}\end{matrix}\right) = \frac{P}{(1 + r)^n}$$

EXAMPLE 2

FINDING PRESENT VALUE

Find the present value of $5000 to be paid 8 years from now at 10% interest compounded semiannually.

Solution

The principal is $P = 5000$, semiannual (half-yearly) compounding means that $r = 0.05$ (half of 10%), and the number of interest periods is $n = 16$ (the number of half-years in 8 years). The present value formula gives

$$\frac{P}{(1 + r)^n} = \frac{5000}{(1 + 0.05)^{16}} = \frac{5000}{1.05^{16}} \approx \$2290.56 \qquad \text{Using a calculator}$$

Therefore, the present value of the $5000 is just $2290.56.

Depreciation by a Fixed Percentage

Depreciation by a fixed percentage means that a piece of equipment loses a fixed percentage (say 30%) of its value each year. Losing a percentage of value is like compound interest but with a negative interest rate. Therefore, we use the compound interest formula, $P(1 + r)^n$, but for depreciation r is negative.

EXAMPLE 3

DEPRECIATING AN ASSET

A car worth $15,000 depreciates in value by 40% each year. How much is it worth after 3 years?

Solution

The car loses 40% of its value each year, which is equivalent to an interest rate of *negative* 40%. The compound interest formula $P(1 + r)^n$ with $P = 15,000$, $r = -0.40$, and $n = 3$ gives

$$P(1 + r)^n = 15,000(1 - 0.40)^3 = 15,000(0.60)^3 = \$3240$$

Using a calculator

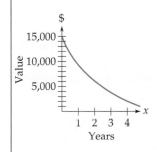

The exponential function $f(x) = 15,000(0.60)^x$, giving the value of the car after x years of depreciation, is graphed on the left.

Practice Problem **2** A printing press, originally worth $50,000, loses 20% of its value each year. What is its value after 4 years? ➤ Solution on page 283

The graph on the previous page shows that depreciation by a fixed percentage is quite different from "straight-line" depreciation. Under straight-line depreciation the same *dollar* value is lost each year, whereas under fixed-percentage depreciation the same *percentage* of value is lost each year, resulting in larger dollar losses in the early years and smaller dollar losses in later years. Depreciation by a fixed percentage (also called the "declining balance" method) is one type of "accelerated" depreciation. The method of depreciation that one uses depends on how one chooses to estimate value, and in practice is often determined by the tax laws.

The Number *e*

Imagine that a bank offers 100% interest, and that you deposit $1 for 1 year. Let us see how the value changes under different types of compounding.

For *annual* compounding, your $1 would in a year grow to $2 (the original dollar plus a dollar interest).

For *quarterly* compounding, we use the compound interest formula with $P = 1$, $n = 4$, and $r = \frac{1}{4} \cdot 100\% = 25\% = 0.25$:

$$P(1 + r)^n = 1(1 + 0.25)^4 = (1.25)^4 \approx 2.44 \qquad \text{Using a calculator}$$

or 2 dollars and 44 cents, an improvement of 44 cents over annual compounding.

For *daily* compounding, the value after a year would be

$$\left(1 + \frac{1}{365}\right)^{365} \approx 2.71 \qquad \begin{array}{l} n = 365 \text{ periods} \\ r = \dfrac{100\%}{365} = \dfrac{1}{365} \end{array}$$

an increase of 27 cents over quarterly compounding. Clearly, if the interest rate, the principal, and the amount of time stay the same, the value increases as the compounding is done more frequently.

In general, if the compounding is done n times a year, the value of the dollar after a year will be

$$\left(\begin{array}{l} \text{Value of \$1 after 1 year at 100\%} \\ \text{interest compounded } n \text{ times a year} \end{array}\right) = \left(1 + \frac{1}{n}\right)^n$$

The following table shows the value of $\left(1 + \dfrac{1}{n}\right)^n$ for various values of n.

Value of \$1 at 100% Interest Compounded n Times a Year for 1 Year

n	$\left(1 + \dfrac{1}{n}\right)^n$	Answer (rounded)	
1	$\left(1 + \dfrac{1}{1}\right)^1 = 2.00000$		Annual compounding
4	$\left(1 + \dfrac{1}{4}\right)^4 \approx 2.44141$		Quarterly compounding
365	$\left(1 + \dfrac{1}{365}\right)^{365} \approx 2.71457$		Daily compounding
10,000	$\left(1 + \dfrac{1}{10,000}\right)^{10,000} \approx 2.71815$		
100,000	$\left(1 + \dfrac{1}{100,000}\right)^{100,000} \approx 2.71827$		
1,000,000	$\left(1 + \dfrac{1}{1,000,000}\right)^{1,000,000} \approx 2.71828$		Answers agree to five decimal places
10,000,000	$\left(1 + \dfrac{1}{10,000,000}\right)^{10,000,000} \approx 2.71828$		

Notice that as n increases, the values in the right-hand column seem to be settling down to a definite value, approximately 2.71828. That is, as n approaches infinity, the limit of $\left(1 + \dfrac{1}{n}\right)^n$ is approximately 2.71828. This particular number is very important in mathematics, and is given the name e (just as 3.14159... is given the name π).

The Constant e

$$e = \lim_{n \to \infty} \left(1 + \frac{1}{n}\right)^n = 2.71828...$$

$n \to \infty$ is read "n approaches infinity." The dots mean that the decimal expansion goes on forever

The same e appears in probability and statistics in the formula for the "bell-shaped" or "normal" curve. Its value has been calculated to many thousands of decimal places, and its value to 15 decimal places is $e \approx 2.718281828459045$.

Continuous Compounding of Interest

This kind of compound interest, the limit as the compounding frequency approaches infinity, is called *continuous* compounding. We have shown that $1 at 100% interest compounded continuously for 1 year would be worth precisely e dollars (about $2.72). The formula for continuous compound interest at other rates is as follows (a justification for it is given at the end of this section).

Continuous Compounding

For P dollars invested at interest rate r compounded continuously,

$$\left(\begin{matrix}\text{Value after}\\ n \text{ years}\end{matrix}\right) = Pe^{rn}$$

EXAMPLE 4 **FINDING VALUE WITH CONTINUOUS COMPOUNDING**

Find the value of $1000 at 8% interest compounded continuously for 20 years.

Solution

We use the formula Pe^{rn} with $P = 1000$, $r = 0.08$, and $n = 20$.

$$Pe^{rn} = \underbrace{1000}_{P} \cdot e^{(\overset{\uparrow}{0.08})(\overset{\uparrow}{20})} = 1000 \cdot \underbrace{e^{1.6}}_{4.95303} \approx \$4953.03$$

$e^{1.6}$ is usually found using the 2nd and LN keys

Present Value with Continuous Compounding

As before, the value that a sum will attain in n years is often called its *future* value, and the current value of a future sum is its *present* value. Under continuous compounding, to find future value we multiply P by e^{rn}, and so to find *present* value we *divide* by e^{rn}.

> **Present Value with Continuous Compounding**
>
> For P dollars and interest rate r compounded continuously for n years,
>
> $$\binom{\text{Present}}{\text{value}} = \frac{P}{e^{rn}} = Pe^{-rn}$$

EXAMPLE **5**

FINDING PRESENT VALUE WITH CONTINUOUS COMPOUNDING

The present value of $5000 to be paid in 10 years, at 7% interest compounded continuously, is

$$\frac{5000}{e^{(0.07)(10)}} = \frac{5000}{e^{0.7}} \approx \$2482.93 \qquad \text{Using a calculator}$$

Intuitive Meaning of Continuous Compounding

Under quarterly compounding, your money is, in a sense, earning interest throughout the quarter, but the interest is not added to your account until the end of the quarter. Under continuous compounding, the interest is added to your account *as it is earned*, with no delay. The extra earnings in continuous compounding come from this "instant crediting" of interest, since then your interest starts earning more interest immediately.

How to Compare Interest Rates

How do you compare different interest rates, such as 16% compounded quarterly and 15.8% compounded continuously? You simply see what each will do for a deposit of $1 for 1 year.

EXAMPLE **6**

COMPARING INTEREST RATES

Which gives a better return: 16% compounded quarterly or 15.8% compounded continuously?

Solution

For 16% compounded quarterly (on $1 for 1 year),

$$1 \cdot \left(1 + \frac{0.16}{4}\right)^4 = (1 + 0.04)^4 \approx 1.170 \qquad \text{Using} \quad P(1 + r)^n$$

For 15.8% compounded continuously, ↙ Better

$$1 \cdot e^{0.158 \cdot 1} = e^{0.158} \approx 1.171$$ Using Pe^{rn}

Therefore, 15.8% compounded continuously is better. (The difference is only a tenth of a cent, but it would be more impressive for principals larger than $1 or periods longer than 1 year.)

The actual percentage increase during 1 year is called the *annual percentage rate* or APR.* For example, the "continuous" return of $1.171, after subtracting the dollar invested, leaves a gain of 0.171 on the original $1, which means an APR of 17.1%. The *stated* rate (here, 15.8%) is called the *nominal* rate of interest. That is, a nominal rate of 15.8% compounded continuously is equivalent to an APR of 17.1%. The 1993 Truth in Savings Act requires that banks always state the annual percentage rate.

Practice Problem 3

From the previous example, what is the annual percentage rate for 16% compounded quarterly? [*Hint:* No calculation is needed.]

➤ Solution on page 283

The Function $y = e^x$

The number e gives us a new exponential function $f(x) = e^x$. This function is used extensively in business, economics, and all areas of science. The table below shows the values of e^x for various values of x. These values lead to the graph of $f(x) = e^x$ shown at the right.

x	$y = e^x$
-3	$e^{-3} \approx 0.05$
-2	$e^{-2} \approx 0.14$
-1	$e^{-1} \approx 0.37$
0	$e^0 = 1$
1	$e^1 \approx 2.72$
2	$e^2 \approx 7.39$
3	$e^3 \approx 20.09$

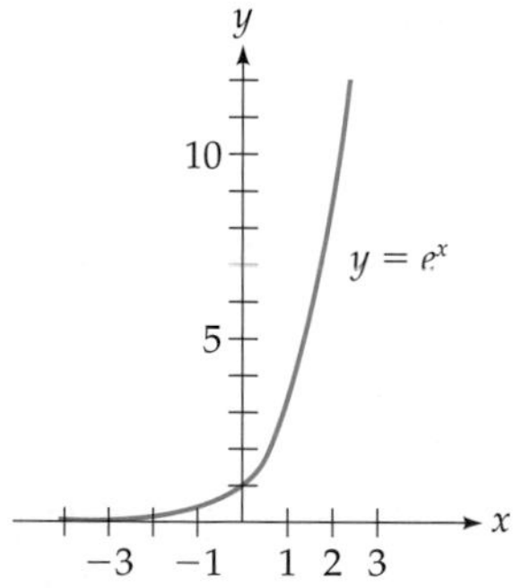

Domain of $y = e^x$ is $\mathbb{R} = (-\infty, \infty)$ and range is $(0, \infty)$.

* Also called the *annual percentage yield* (APY) or the *effective* rate of interest.

Notice that e^x is never zero, and is positive for all values of x, even when x is negative. We restate this important observation as follows:

> e to any power is positive.

The following graph shows the function $f(x) = e^{kx}$ for various values of the constant k. For positive values of k the curve rises, and for negative values of k the curve falls (as you move to the right). For higher values of k the curve rises more steeply. Each curve crosses the y-axis at 1.

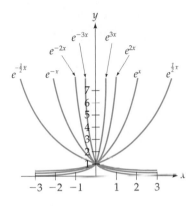

$f(x) = e^{kx}$ for various values of k.

Graphing Calculator Exploration

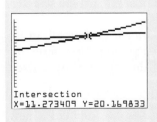

Intersection
X=11.273409 Y=20.169833

The most populous states are California and Texas, with New York third and Florida fourth but gaining. According to data from the Census Bureau, x years after 2000 the population of New York will be $19e^{0.0053x}$ and the population of Florida will be $15.9e^{0.0211x}$ (both in millions).

a. Graph these two functions on the window [0, 20] by [0, 25]. [Use the 2nd and LN keys for entering e to powers.]

b. Use INTERSECT to find the x-value where the curves intersect.

c. From your answer to part (b), in which year is Florida projected to overtake New York as the third largest state? [*Hint: x* is years after 2000.]

Exponential Growth

All exponential growth, whether continuous or discrete, has one common characteristic: the amount of growth is proportional to the size. For example, the interest that a bank account earns is proportional to the size of the account, and the growth of a population is proportional to the size of the population. This is in contrast, for example, to a person's height, which does not increase exponentially. That is, exponential growth occurs in those situations where a quantity grows *in proportion to its size*.

Justification of the Formula for Continuous Compounding

The compound interest formula Pe^{rn} is derived as follows: P dollars invested for n years at interest rate r compounded k times a year yields

$$P\left(1 + \frac{r}{k}\right)^{kn}$$

From the formula on page 272

Define $m = \dfrac{k}{r}$, so that $k = rm$. Replacing k by rm and letting k (and therefore m) approach ∞, this becomes

$$P\left(1 + \frac{r}{rm}\right)^{rmn} = P\left[\underbrace{\left(1 + \frac{1}{m}\right)^{m}}_{\substack{\text{Approaches} \\ e \text{ as } m \to \infty}}\right]^{rn} \to Pe^{rn}$$

Letting $m \to \infty$

The limit shown on the right is the continuous compounding formula Pe^{rn}.

4.1 Section Summary

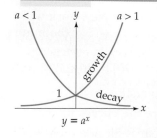

Exponential functions have exponents that involve variables. The exponential functions $f(x) = a^x$ slope *upward* or *downward* depending upon whether the base a (which must be positive) satisfies $a > 1$ or $a < 1$.

The formula $P(1 + r)^n$ gives the value after n periods of an investment of P dollars that increases at rate r per period. This same formula, with a *negative* growth rate r, governs depreciation by a fixed percentage.

By considering a 100% increase not given all at once but divided up into smaller and smaller successive increases, we defined a new constant e:

$$e = \lim_{n \to \infty}\left(1 + \frac{1}{n}\right)^n \approx 2.71828$$

Exponential functions with base e are used extensively in modeling many types of growth, such as the growth of populations and interest that is compounded continuously. For interest compounded *continuously*, the formula is Pe^{rn}.

> ➤ **Solutions to Practice Problems**

1. $2000(1 + 0.03)^{24} = 2000(1.03)^{24} \approx 2000(2.032794) \approx \$4065.59$

2. $50{,}000(1 - 0.20)^4 = 50{,}000(0.8)^4 = 50{,}000(0.4096) = \$20{,}480$

3. 17% (from 1.170)

4.1 Exercises

Use a calculator to evaluate, rounding to three decimal places.

1. a. e^2 **b.** e^{-2} **c.** $e^{1/2}$

2. a. e^3 **b.** e^{-3} **c.** $e^{1/3}$

Express as a power of e.

3. a. $e^5 e^{-2}$ **b.** $\dfrac{e^5}{e^3}$ **c.** $\dfrac{e^5 e^{-1}}{e^{-2} e}$

4. a. $e^{-3} e^5$ **b.** $\dfrac{e^4}{e}$ **c.** $\dfrac{e^2 e}{e^{-2} e^{-1}}$

Graph each function. If you are using a graphing calculator, make a hand-drawn sketch from the screen.

5. $y = 3^x$ **6.** $y = 5^x$

7. $y = \left(\tfrac{1}{3}\right)^x$ **8.** $y = \left(\tfrac{1}{5}\right)^x$

Calculate each value of e^x using a calculator.

9. $e^{1.74}$ **10.** $e^{-0.09}$

11. e^x **Versus** x^n Which curve is eventually higher, x to a power or e^x?

    a. Graph x^2 and e^x on the window $[0, 5]$ by $[0, 20]$. Which curve is higher?

    b. Graph x^3 and e^x on the window $[0, 6]$ by $[0, 200]$. Which curve is higher for large values of x?

    c. Graph x^4 and e^x on the window $[0, 10]$ by $[0, 10{,}000]$. Which curve is higher for large values of x?

    d. Graph x^5 and e^x on the window $[0, 15]$ by $[0, 1{,}000{,}000]$. Which curve is higher for large values of x?

    e. Do you think that e^x will exceed x^6 for large values of x? Based on these observations, can you make a conjecture about e^x and *any* power of x?

12. Linear Versus Exponential Growth

    a. Graph $y_1 = x$ and $y_2 = e^{0.01x}$ in the window $[0, 10]$ by $[0, 10]$. Which curve is higher for x near 10?

    b. Then graph the same curves on the window $[0, 1000]$ by $[0, 1000]$. Which curve is higher for x near 1000?

    A function such as y_1 represents *linear* growth, and y_2 represents *exponential* growth, and the result here is true in general: exponential growth always beats linear growth (eventually, no matter what the constants).*

* The realization that populations grow exponentially while food supplies grow only linearly caused the great nineteenth-century essayist Thomas Carlyle to dub economics the "dismal science." He was commenting not on how interesting economics is, but on the grim conclusions that follow from populations outstripping their food supplies.

APPLIED EXERCISES

13. BUSINESS: Interest Find the value of $1000 deposited in a bank at 10% interest for 8 years compounded:

  a. annually.   **b.** quarterly.
  c. continuously.

14. BUSINESS: Interest Find the value of $1000 deposited in a bank at 12% interest for 8 years compounded:

  a. annually.   **b.** quarterly.
  c. continuously.

15. GENERAL: Interest A loan shark lends you $100 at 2% compound interest per week (this is a *weekly*, not an annual rate).

  a. How much will you owe after 3 years?
  b. In "street" language, the profit on such a loan is known as the "vigorish" or the "vig." Find the shark's vig.

16. GENERAL: Compound Interest In 1626, Peter Minuit purchased Manhattan Island from the native Americans for $24 worth of trinkets and beads. Find what the $24 would be worth in the year 2000 if it had been deposited in a bank paying 5% interest compounded quarterly.

17. GENERAL: Interest Find the error in the ad shown below, which appeared in a New York newspaper. [*Hint:* Check that the nominal rate is equivalent to the effective rate. For daily compounding, some banks use 365 days and some use 360 days in the year. Try it both ways.]

> At T&M Bank, flexibility is the key word. You can choose the length of time and the amount of your deposit, which will earn an annual yield of 9.825% based on a rate of 9.25% compounded daily.

18. GENERAL: Interest Find the error in the following ad, which appeared in a Washington, D.C. newspaper. Assume that the compounding is done daily. [*Hint:* Check that the nominal rate is equivalent to the effective rate. For daily compounding, some banks use 365 days and some use 360 days in the year. Try it both ways.]

> **Your Money's in 7th Heaven** when you get Hans Johnson's "sky high" return.
> No time restrictions or withdrawal penalties! Funds available when you want them.
>
> Current Annual   7% Regular
> **Yield: 7.19%**   **Passbooks**

19. GENERAL: Present Value A rich uncle wants to make you a millionaire. How much money must he deposit in a trust fund paying 8% compounded quarterly at the time of your birth to yield $1,000,000 when you retire at age 60?

20. GENERAL: Present Value If a college education costs $50,000, how large a trust fund (paying 8% compounded quarterly) must be established at a child's birth to ensure sufficient funds at age 18?

21. GENERAL: Consumer Fraud Buying a "vacation timeshare" means buying the right to use a vacation property for a fixed period each year. Suppose that you pay $500 for a vacation timeshare and receive a "money-back guarantee" that at any time the company will buy back your timeshare, or if not, give you a $1000 bond. The deception, however, is that the bond is not redeemable for 45 years. Find the real value of the "guarantee"—that is, find the present value of $1000 in 45 years (assume a 6% interest rate compounded annually).

22. GENERAL: Zero-Coupon Bonds A zero-coupon bond is a bond that makes no payments (coupons) until it matures, at which time it pays its "face value" of $1000. (You buy it for much less than $1000.) Find the selling price of a 10-year zero-coupon bond if the interest rate is 9% compounded annually. [*Hint:* The selling price equals the present value of $1000 to be paid in 10 years at the stated interest rate.]

23. GENERAL: Compound Interest Which is better: 10% interest compounded quarterly or 9.8% compounded continuously?

24. GENERAL: Compound Interest Which is better: 8% interest compounded quarterly or 7.8% compounded continuously?

25. BUSINESS: Depreciation A $15,000 automobile depreciates by 35% per year. Find its value after:

a. 4 years.   **b.** 6 months.

26. BUSINESS: Appreciation A $10,000 painting appreciates in value by 12% each year. Find its value after 16 years.

27. GENERAL: Population According to the United Nations Fund for Population Activities, the population of the world x years after the year 2000 will be $5.89e^{0.0175x}$ billion people (for $0 \leq x \leq 20$). Use this formula to predict the world population in the year 2010.

28. GENERAL: Population The most populous country is China, with a (2000) population of 1.26 billion, growing at the rate of 0.8% a year. The 2000 population of India was 1.02 billion, growing at 1.7% a year. Assuming that these growth rates continue, which population will be larger in the year 2025?

29. GENERAL: Nuclear Meltdown According to the Nuclear Regulatory Commission, the probability of a "severe core meltdown accident" at a nuclear reactor in the United States within the next n years is $1 - (0.9997)^{100n}$. Find the probability of a meltdown:

a. within 25 years.
b. within 40 years.

(The 1986 core meltdown in the Chernobyl reactor in the Soviet Union spread radiation over much of Eastern Europe, leading to an undetermined number of fatalities.)

30. GENERAL: Mosquitoes Female mosquitoes *(Culex pipiens)* feed on blood (only the females drink blood) and then lay several hundred eggs. In this way each mosquito can, on the average, breed another 300 mosquitoes in about 9 days. Find the number of great-grandchildren mosquitoes that will be descended from one female mosquito, assuming that all eggs hatch and mature.

31. ENVIRONMENTAL SCIENCE: Light According to the Bouguer–Lambert Law, the proportion of light that penetrates ordinary seawater to a depth of x feet is $e^{-0.44x}$. Find

the proportion of light that penetrates to a depth of:

a. 3 feet.   **b.** 10 feet.

32–33. BIOMEDICAL: Drug Dosage If a dosage d of a drug is administered to a patient, the amount of the drug remaining in the tissues t hours later will be $f(t) = de^{-kt}$, where k (the "absorption constant") depends on the drug.*

32. For the immunosuppressant cyclosporine, the absorption constant is $k = 0.012$. For a dose of $d = 400$ milligrams, use the previous formula to find the amount of cyclosporine remaining in the tissues after:

a. 24 hours.   **b.** 48 hours.

33. For the cardioregulator digoxin, the absorption constant is $k = 0.018$. For a dose of $d = 2$ milligrams, use the previous formula to find the amount remaining in the tissues after:

a. 24 hours.   **b.** 48 hours.

34. BIOMEDICAL: Bacterial Growth A colony of bacteria in a petri dish doubles in size every hour. At noon the petri dish is just covered with bacteria. At what time was the petri dish:

a. 50% covered? [*Hint:* No calculation needed.]
b. 25% covered?

35. BUSINESS: Advertising A company finds that x days after the conclusion of an advertising campaign the daily sales of a new product are $S(x) = 100 + 800e^{-0.2x}$. Find the daily sales 10 days after the end of the advertising campaign.

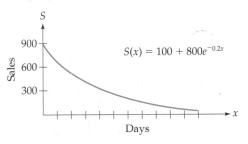

*For further details, see T. R. Harrison, ed., *Principles of Internal Medicine,* 11th ed. (New York: McGraw-Hill, 1987), 342–352.

36. BUSINESS: Quality Control A company finds that the proportion of its light bulbs that will burn continuously for longer than t weeks is $e^{-0.01t}$. Find the proportion of bulbs that burn for longer than 10 weeks.

37. GENERAL: Temperature A covered mug of coffee originally at 200 degrees Fahrenheit, if left for t hours in a room whose temperature is 70 degrees, will cool to a temperature of $70 + 130e^{-1.8t}$ degrees. Find the temperature of the coffee after:

a. 15 minutes.   **b.** half an hour.

38. BEHAVIORAL SCIENCE: Learning In certain experiments the percentage of items that are remembered after t time units is

$$p(t) = 100 \frac{1 + e}{1 + e^{t+1}}$$

Such curves are called "forgetting" curves. Find the percentage remembered after:

a. 0 time units.   **b.** 2 time units.

39. BIOMEDICAL: Epidemics The Reed–Frost model for the spread of an epidemic predicts that the number I of newly infected people is $I = S(1 - e^{-rx})$, where S is the number of susceptible people, r is the effective contact rate, and x is the number of infectious people. Suppose that a school reports an outbreak of measles with $x = 10$ cases, and that the effective contact rate is $r = 0.01$. If the number of susceptibles is $S = 400$, use the Reed–Frost model to estimate how many students will be newly infected during this stage of the epidemic.

40. SOCIAL SCIENCE: Election Cost The cost of winning a seat in the House of Representatives in recent years has been approximately $805e^{0.0625x}$ thousand dollars, where x is the number of years since 2000. Estimate the cost of winning a House seat in the year 2010. (*Source:* Center for Responsive Politics)

41. GENERAL: Rate of Return An investment of $8000 grows to $10,291.73 in 4 years. Find the annual rate of return for annual compounding. [*Hint:* Use $P(1 + r)^n$ and solve for r (rounded).]

42. GENERAL: Rate of Return An investment of $9000 grows to $10,380.65 in 2 years. Find the annual rate of return for quarterly compounding. [*Hint:* Use $P(1 + r)^n$ and solve for r (rounded).]

43. ATHLETICS: Olympic Games When the Olympic Games were held near Mexico City in the summer of 1968, many athletes were concerned that the high elevation would affect their performance. If air pressure decreases exponentially by 0.4% for each 100 feet of altitude, by what percentage did the air pressure decrease in moving from Tokyo (the site of the 1964 Summer Olympics, at altitude 30 feet) to Mexico City (altitude 7347 feet)?

44. BIOMEDICAL: Radioactive Contamination The core meltdown and explosions at the nuclear reactor in Chernobyl in 1986 released large amounts of strontium 90, which decays exponentially at the rate of 2.5% per year. Areas downwind of the reactor will be uninhabitable for 100 years. What percent of the original strontium 90 contamination will still be present after:

a. 50 years?   **b.** 100 years?

45. GENERAL: Population As stated earlier, the most populous state is California, with Texas second but gaining. According to the Census Bureau, x years after 2000 the population of California will be $34e^{0.013x}$ and the population of Texas will be $21e^{0.021x}$ (all in millions).

a. Graph these two functions on a calculator on the window [0, 100] by [0, 100].
b. In which year is Texas projected to overtake California as the most populous state? [*Hint:* Use INTERSECT.]

46. GENERAL: St. Louis Arch The Gateway Arch in St. Louis is built around a mathematical curve called a "catenary." The height of this catenary above the ground at a point x feet from the center line is

$$y = 688 - 31.5(e^{0.01033x} + e^{-0.01033x})$$

a. Graph this curve on a calculator on the window [−400, 400] by [0, 700].
b. Find the height of the Gateway Arch at its highest point, using the fact that the top of the arch is 5 feet higher than the top of the central catenary.

4.2 LOGARITHMIC FUNCTIONS

Introduction

In this section we introduce logarithmic functions, emphasizing the *natural* logarithm function. We then apply natural logarithms to a wide variety of problems, from doubling money under compound interest to drug dosage.

Logarithms

The word "logarithm" (abbreviated "log") means *power* or *exponent*. The number being raised to the power is called the *base* and is written as a subscript. For example, the expression

$$\log_{10} 1000 \qquad \text{Read "log (base 10) of 1000"}$$

Base

means the *exponent* to which we have to raise 10 to get 1000. Since $10^3 = 1000,$ the exponent is 3, so the *logarithm* is 3:

$$\log_{10} 1000 = 3 \qquad \text{Since } 10^3 = 1000$$

We find logarithms by writing them as exponents and then finding the exponent.

EXAMPLE 1

FINDING A LOGARITHM

Evaluate $\log_{10} 100$.

Solution

$$\log_{10} 100 = y \qquad \text{is equivalent to} \qquad 10^y = 100 \qquad y = 2 \text{ works, so 2 is the logarithm}$$

The logarithm y is the exponent that solves

Therefore

$$\log_{10} 100 = 2 \qquad \text{Since } 10^2 = 100$$

EXAMPLE 2

FINDING A LOGARITHM

Evaluate $\log_{10} \dfrac{1}{10}$.

Solution

$$\log_{10} \frac{1}{10} = y \qquad \text{is equivalent to} \qquad 10^y = \frac{1}{10}$$

$y = -1$ works, so -1 is the logarithm

The logarithm y is the exponent that solves

Therefore

$$\log_{10} \frac{1}{10} = -1 \qquad\qquad \text{Since} \quad 10^{-1} = \frac{1}{10}$$

Practice Problem 1   Evaluate $\log_{10} 10{,}000$. ➤ Solution on page 300

Logarithms to the base 10 are called *common* logarithms and are often written without the base, so that $\log 100$ means $\log_{10} 100$. The LOG key on many calculators evaluates common logarithms.

We may find logarithms to other bases as well. Any positive number other than 1 may be used as a base.

EXAMPLE 3   **FINDING A LOGARITHM**

Find $\log_9 3$.

Solution

$$\log_9 3 = y \qquad \text{is equivalent to} \qquad 9^y = 3$$

$y = \frac{1}{2}$ works, so $\frac{1}{2}$ is the logarithm

Therefore

$$\log_9 3 = \frac{1}{2} \qquad\qquad \text{Since} \quad 9^{1/2} = \sqrt{9} = 3$$

In general, for any positive base a other than 1,

Logarithms

$$\log_a x = y \qquad \text{is equivalent to} \qquad a^y = x$$

$\log_a x$ means the exponent of a that gives x

Practice Problem 2

Find:

a. $\log_5 125$ **b.** $\log_8 2$ ➤ Solutions on page 300

Natural Logarithms

The most widely used of all bases is e, the number (approximately 2.718) that we defined on page 277. Logarithms to the base e are called *natural* or *Napierian logarithms*.* The natural logarithm of x is written ln x ("n" for "natural") instead of $\log_e x$, and may be found using the LN key on a calculator.

> **Natural Logarithms**
>
> ln x = logarithm of x to the base e ln x means $\log_e x$

Practice Problem 3

Use a calculator to find   ln 8.34. ➤ Solution on page 300

The following table shows some values of the natural logarithm function $f(x) = \ln x$,   and its graph is shown at the right.

x	$y = \ln x$
0.1	-2.30
0.5	-0.69
1	0
2	0.69
5	1.61
10	2.30
15	2.71

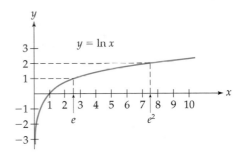

Domain of $y = \ln x$ is $(0, \infty)$
and range is $\mathbb{R} = (-\infty, \infty)$.

The natural logarithm function may be used for modeling growth that continually slows. Notice that   ln x is defined only for $x > 0$.   This is because e to any power is positive.

   The following graph shows logarithm functions for several different bases. Notice that each is defined only for $x > 0$,   and that they all

---

*After John Napier, a 17th-century Scottish mathematician and, incidentally, the inventor of the decimal point.

pass through the point $(1, 0)$, since $a^0 = 1$. We will concentrate on the *natural* logarithm function, since it is the one most used in applications.

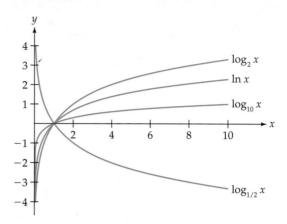

Since logs are exponents, each property of exponents (see page 19) can be restated as a property of logarithms. The first three properties show that some natural logarithms can be found without using a calculator.

Properties of Natural Logarithms

1. $\ln 1 = 0$ The natural log of 1 is 0 (since $e^0 = 1$)

2. $\ln e = 1$ The natural log of e is 1 (since $e^1 = e$)

3. $\ln e^x = x$ The natural log of e to a power is just the power (since $e^x = e^x$)

4. $e^{\ln x} = x$ e raised to the natural log of a number is just the number

The first two properties are special cases of the third, with $x = 0$ and $x = 1$. Since logarithms are exponents, the third property simply says that the exponent of e that gives e^x is x, which is obvious when you think about it. Since $\ln x$ is the power of e that gives x, raising e to that power must give x, which is the fourth property.

EXAMPLE 4

USING THE PROPERTIES OF NATURAL LOGARITHMS

a. $\ln e^7 = 7$

b. $\ln e^{3x} = 3x$ The natural log of e to a power is just the power

c. $\ln \sqrt[3]{e^2} = \ln e^{2/3} = \frac{2}{3}$

d. $e^{\ln 5} = 5$ e raised to the natural log of 5 is just 5

Graphing Calculator Exploration

a. Evaluate $\ln e^{17}$ to verify that the answer is 17. Change 17 to other numbers (positive, negative, or zero) in order to verify that $\ln e^x = x$ for any x.

b. Evaluate $e^{\ln 29}$ to verify that the answer is 29. Change the 29 to another positive number to verify that $e^{\ln x} = x$. What about negative numbers, or zero?

The next four properties enable us to simplify logs of products, quotients, and powers. For any positive numbers M and N:

Properties of Natural Logarithms

5. $\ln (M \cdot N) = \ln M + \ln N$	The log of a product is the sum of the logs
6. $\ln \left(\dfrac{1}{N}\right) = -\ln N$	The log of 1 over a number is the negative of the log of the number
7. $\ln \left(\dfrac{M}{N}\right) = \ln M - \ln N$	The log of a quotient is the difference of the logs
8. $\ln (M^N) = N \ln M$	The log of a number to a power is the power times the log

Property 8 can be stated simply: *logarithms bring down exponents*. A justification for these four properties is given at the end of this section.

EXAMPLE 5

USING THE PROPERTIES OF NATURAL LOGARITHMS

a. $\ln (2 \cdot 3) = \ln 2 + \ln 3$ Property 5

b. $\ln \frac{1}{7} = -\ln 7$ Property 6

c. $\ln \frac{2}{3} = \ln 2 - \ln 3$ Property 7

d. $\ln (2^3) = 3 \ln 2$ $\ln(2^3) = 3 \ln 2$

Properties 1 through 8 are very useful for simplifying functions.

EXAMPLE 6

SIMPLIFYING A FUNCTION

$$f(x) = \ln(2x) - \ln 2$$

$$= \ln 2 + \ln x - \ln 2 \qquad \text{Since} \ \ln(2x) = \ln 2 + \ln x \\ \text{by Property 5}$$

$$= \ln x \qquad \text{Canceling}$$

EXAMPLE 7

SIMPLIFYING A FUNCTION

$$f(x) = \ln\left(\frac{x}{e}\right) + 1$$

$$= \ln x - \ln e + 1 \qquad \text{Since} \ \ln(x/e) = \ln x - \ln e \\ \text{by Property 7}$$

$$= \ln x - 1 + 1 \qquad \text{Since} \ \ln e = 1 \ \text{by Property 2}$$

$$= \ln x \qquad \text{Canceling}$$

EXAMPLE 8

SIMPLIFYING A FUNCTION

$$f(x) = \ln(x^5) - \ln(x^3)$$

$$= 5\ln x - 3\ln x \qquad \text{Bringing down exponents} \\ \text{by Property 8}$$

$$= 2\ln x \qquad \text{Combining}$$

Graphing Calculator Exploration

Some advanced graphing calculators have computer algebra systems that can simplify algebraic expressions. For example, the Texas Instruments *TI-89* graphing calculator simplifies logarithmic expressions, but only if the condition $x > 0$ is included so that the logarithms are defined.

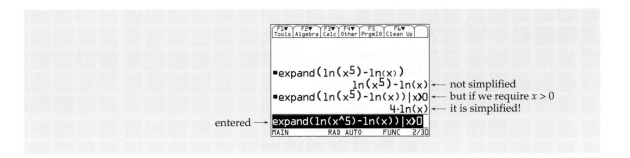

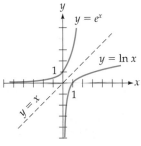

Graphing Calculator Exploration

a. Graph the function $y = \ln x^2$ on the window $[-5, 5]$ by $[-5, 5]$.

b. Change the function to $y = 2 \ln x$ and explain why the two graphs are different. (Doesn't Property 8 say that the two functions should be the same?)

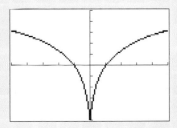

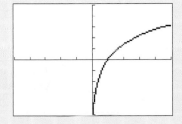

The natural logarithm function $\ln x$ and the exponential function e^x are *inverse functions* in that either one "undoes" or "inverts" the other. (Notationally, e^x raises x up to the exponent, and logarithms bring down exponents, so it is reasonable that they "undo" each other.) This inverse relationship between e^x and $\ln x$ is equivalent to the geometric fact that their graphs are reflections of each other in the diagonal line $y = x$, as shown on the left.

$y = e^x$ and $y = \ln x$
are inverse functions.

Doubling Under Compound Interest

How soon will money invested at compound interest double in value? The solution to this question makes important use of the property $\ln (M^N) = N \ln M$ ("logs bring down exponents").

EXAMPLE 9

FINDING DOUBLING TIME

A sum is invested at 12% interest compounded quarterly. How soon will it double in value?

Solution

We use the formula $P(1 + r)^n$ with (quarterly) interest rate $r = \frac{1}{4} \cdot 12\% = 3\% = 0.03$. Since double P dollars is $2P$ dollars, we want to solve

$$P(1 + 0.03)^n = 2P \qquad n \text{ is the number of quarters}$$

$$\underset{\text{—— } P \text{ dollars doubled}}{}$$

$$1.03^n = 2 \qquad \text{Canceling the } P\text{'s and simplifying}$$

The variable is in the *exponent,* so we take logarithms to bring it down.

$$\ln (1.03^n) = \ln 2 \qquad \text{Taking natural logs of both sides}$$

$$n \ln 1.03 = \ln 2 \qquad \begin{array}{l}\text{Bringing down the exponent} \\ \text{(Property 8 of logarithms)}\end{array}$$

$$n = \frac{\ln 2}{\ln 1.03} \qquad \text{Dividing each side by } \ln 1.03 \text{ to solve for } n$$

$$\approx \frac{0.6931}{0.0296} \approx 23.4 \qquad \begin{array}{l}\text{Evaluating the logs using a calculator} \\ \text{(answer means 23.4 quarters)}\end{array}$$

Since n is in *quarters,* we divide by 4 to convert to years, $\frac{23.4}{4} \approx 5.9$. A sum at 12% compounded quarterly doubles in about 5.9 years.

Of course, for quarterly compounding you will not get the $2P$ dollars until the end of the last quarter, by which time it will be slightly more.

Notice that the principal P canceled out after the first step, showing that *any* sum will double in the same amount of time: 1 dollar will double into 2 dollars in exactly the same time that a million dollars will double into 2 million dollars.

To find how soon the value *triples,* we solve $P(1 + r)^n = 3P$, and similarly for any other multiple. To find how soon the value would *increase by 50%* (that is, become 1.5 times its original value), we would solve

$$P(1 + r)^n = 1.5P \qquad \text{Multiplying by 1.5 increases } P \text{ by 50\%}$$

Practice Problem 4

A sum is invested at 10% interest compounded semiannually (twice a year). How soon will it increase by 60%?

➤ Solution on page 300

EXAMPLE 10 **FINDING TRIPLING TIME WITH CONTINUOUS COMPOUNDING**

A sum is invested at 15% compounded continuously. How soon will it triple?

Solution   The idea is the same as before, but for continuous compounding we use the formula Pe^{rn}:

$$Pe^{0.15n} = 3P \qquad \text{3P since the value triples}$$

$$e^{0.15n} = 3 \qquad \text{Canceling the } P\text{'s}$$

$$\ln e^{0.15n} = \ln 3 \qquad \text{Taking the natural log of both sides}$$

$$\underbrace{\quad}_{0.15n}$$

$$0.15n = \ln 3 \qquad \begin{array}{l}\text{Since the natural log of } e \text{ to a} \\ \text{power is just the power}\end{array}$$

$$n = \frac{\ln 3}{0.15} \qquad \text{Dividing by 0.15 to solve for } n$$

$$\approx \frac{1.0986}{0.15} \approx 7.3 \qquad \text{Using a calculator}$$

A sum at 15% compounded continuously triples in about 7.3 years.

Drug Dosage

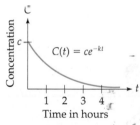

Drug concentration in the blood

The amount of a drug that remains in a person's bloodstream decreases exponentially with time. If the initial concentration is c (milligrams per milliliter of blood), the concentration t hours later will be

$$C(t) = ce^{-kt}$$

where the "absorption constant" k measures how rapidly the drug is absorbed.

Every medicine has a minimum concentration below which it is not effective. When the concentration falls to this level, another dose should be administered. If doses are administered regularly every T hours over a period of time, the concentration will look like the following graph.

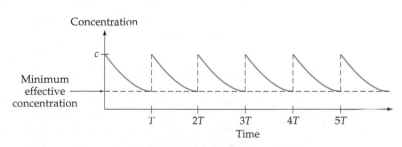

Drug concentration with repeated doses

The problem is to determine the time T at which the dose should be repeated so as to maintain an effective concentration.

EXAMPLE **11**

CALCULATING DRUG DOSAGE

The absorption constant for penicillin is $k = 0.11$, and the minimum effective concentration is 2 milligrams. If the original concentration is $c = 5$ milligrams, find when another dose should be administered in order to maintain an effective concentration.

Solution The concentration formula $C(t) = ce^{-kt}$ with $c = 5$ and $k = 0.11$ is

$$C(t) = 5e^{-0.11t}$$

To find the time when this concentration reaches the minimum effective level of 2, we solve

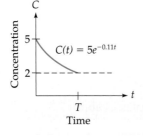

$$5e^{-0.11t} = 2$$

$$e^{-0.11t} = 0.4 \qquad \text{Dividing by 5}$$

$$\underbrace{\ln e^{-0.11t}}_{-0.11t} = \ln 0.4 \qquad \begin{array}{l}\text{Taking natural logs to bring}\\\text{down the exponent}\end{array}$$

$$-0.11t = \ln 0.4$$

$$t = \frac{\ln 0.4}{-0.11} \approx \frac{-0.9163}{-0.11} \approx 8.3 \qquad \begin{array}{l}\text{Solving for } t \text{ and}\\\text{using a calculator}\end{array}$$

The concentration will reach the minimum effective level in 8.3 hours, so another dose should be taken approximately every 8 hours.

The last two examples have led to equations of the form $e^{at} = b$. Such equations occur frequently in applications, and are solved by taking the natural log of each side to bring down the exponent. Such equations may also be solved on a graphing calculator by graphing the function as y_1 and the constant as y_2 and using INTERSECT to find where they meet.

Carbon 14 Dating

All living things absorb small amounts of radioactive carbon 14 from the atmosphere. When they die, the carbon 14 stops being absorbed and decays exponentially into ordinary carbon. Therefore, the proportion of carbon 14 still present in a fossil or other ancient remain can be used to estimate how old it is. The proportion of the original carbon 14 that will be present after t years is

$$\left(\begin{array}{l}\text{Proportion of carbon 14} \\ \text{remaining after } t \text{ years}\end{array}\right) = e^{-0.00012t}$$

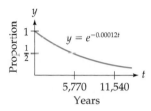

EXAMPLE 12 **DATING BY CARBON 14**

The Dead Sea Scrolls, discovered in a cave near the Dead Sea in what was then Jordan, are among the earliest documents of Western civilization. Estimate the age of the Dead Sea Scrolls if the animal skins on which some were written contain 78% of their original carbon 14.

Solution   The proportion of carbon 14 remaining after t years is $e^{-0.00012t}$. We equate this formula to the actual proportion (expressed as a decimal):

$e^{-0.00012t} = 0.78$	Equating the proportions
$\ln e^{-0.00012t} = \ln 0.78$	Taking natural logs
$-0.00012t = \ln 0.78$	$\ln e^{-0.00012t} = -0.00012t$ by Property 3
$t = \dfrac{\ln 0.78}{-0.00012} \approx \dfrac{-0.24846}{-0.00012} \approx 2071$	Solving for t and using a calculator

Therefore, the Dead Sea Scrolls are approximately 2070 years old.

Graphing Calculator Exploration

Solve Example 12 by graphing $y_1 = e^{-0.00012x}$ and $y_2 = 0.78$ on the window $[0, 10{,}000]$ by $[0, 1]$ and using INTERSECT to find where they meet.

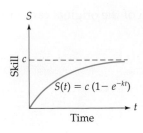

Behavioral Science: Learning Theory

Your ability to do a task generally improves with practice. Frequently, your skill after t units of practice is given by a function of the form

$$S(t) = c(1 - e^{-kt})$$

where c and k are positive constants.

EXAMPLE **13**

ESTIMATING LEARNING TIME

After t weeks of training, your secretary can type

$$S(t) = 100(1 - e^{-0.25t})$$

words per minute. How many weeks will he take to reach 80 words per minute?

Solution We solve for t in the following equation:

$100(1 - e^{-0.25t}) = 80$	Setting $S(t)$ equal to 80
$1 - e^{-0.25t} = 0.80$	Dividing by 100
$-e^{-0.25t} = -0.20$	Subtracting 1
$e^{-0.25t} = 0.20$	Multiplying by -1
$-0.25t = \ln 0.20$	Taking natural logs
$t = \dfrac{\ln 0.20}{-0.25}$	Solving for t
$\approx \dfrac{-1.6094}{-0.25} \approx 6.4$	Using a calculator

He will reach 80 words per minute in about $6\frac{1}{2}$ weeks.

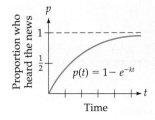

Social Science: Diffusion of Information by Mass Media

When a news bulletin is repeatedly broadcast over radio and television, the proportion of people who hear the bulletin within t hours is

$$p(t) = 1 - e^{-kt}$$

for some constant k.

EXAMPLE   14   **PREDICTING THE SPREAD OF INFORMATION**

A storm warning is broadcast, and the proportion of people who hear the bulletin within t hours of its first broadcast is $p(t) = 1 - e^{-0.30t}$. When will 75% of the people have heard the bulletin?

Solution   Equating the proportions gives $1 - e^{-0.30t} = 0.75$. Solving this equation as in the previous examples (omitting the details) gives $t \approx 4.6$. Therefore, it takes about $4\frac{1}{2}$ hours for 75% of the people to hear the news.

4.2   Section Summary

We define logarithms as follows:

$$\log_a x = y \quad \text{is equivalent to} \quad a^y = x \qquad \text{Logarithm to the base } a$$
$$\ln x = y \quad \text{is equivalent to} \quad e^y = x \qquad \text{Natural logarithm (base } e)$$

Natural logarithms have the following properties, which are also listed inside the back cover.

1. $\ln 1 = 0$

2. $\ln e = 1$

3. $\ln e^x = x$

4. $e^{\ln x} = x$

5. $\ln (M{\cdot}N) = \ln M + \ln N$

6. $\ln \left(\dfrac{1}{N}\right) = -\ln N$

7. $\ln \left(\dfrac{M}{N}\right) = \ln M - \ln N$

8. $\ln (M^N) = N \ln M$

Justification of the Properties of Logarithms

Properties 1–4 were justified on page 290.

Property 5 follows from the addition property of exponents, $e^x {\cdot} e^y = e^{x+y}$ ("the exponent of a product is the sum of the exponents"). Since logs are exponents, this can be restated "the log of a product is the sum of the logs," which is Property 5.

Property 7 follows from the subtraction property of exponents, $e^x/e^y = e^{x-y}$ ("the exponent of a quotient is the difference of the exponents"). This translates into "the log of a quotient is the difference of the logs," which is Property 7.

Property 6 is just a special case of Property 7 (with $M = 1$), along with Property 1.

Property 8 comes from the property of exponents, $(e^x)^y = e^{x \cdot y}$, which says that the exponent y can be "brought down" and multiplied by the x. Since logs are exponents, this says that in $\ln(M^N)$ the exponent N can be brought down and multiplied by the logarithm, which gives $N \ln M$, and this is just Property 8.

Each property of logs is equivalent to a property of exponents:

Logarithmic Property	*Exponential Property*
$\ln 1 = 0$	$e^0 = 1$
$\ln(M \cdot N) = \ln M + \ln N$	$e^x \cdot e^y = e^{x+y}$
$\ln\left(\dfrac{1}{N}\right) = -\ln N$	$\dfrac{1}{e^y} = e^{-y}$
$\ln\left(\dfrac{M}{N}\right) = \ln M - \ln N$	$\dfrac{e^x}{e^y} = e^{x-y}$
$\ln M^N = N \ln M$	$(e^x)^y = e^{xy}$

➤ **Solutions to Practice Problems**

1. $\log_{10} 10{,}000 = 4$ (Since $10^4 = 10{,}000$)

2. **a.** $\log_5 125 = 3$ (Since $5^3 = 125$)

   b. $\log_8 2 = \frac{1}{3}$ (Since $8^{1/3} = \sqrt[3]{8} = 2$)

3. $\ln 8.34 \approx 2.121$ (rounded)

4. $P(1 + 0.05)^n = 1.60 \cdot P$

$$(1.05)^n = 1.6$$
$$\ln(1.05)^n = \ln 1.6$$
$$n \ln 1.05 = \ln 1.6$$
$$n = \frac{\ln 1.6}{\ln 1.05} \approx 9.6 \text{ half-years}$$

In about 4.8 years.

4.2 Exercises

Find each logarithm *without* using a calculator or tables.

1. **a.** $\log_5 25$ **b.** $\log_3 81$ **c.** $\log_3 \frac{1}{3}$

   d. $\log_3 \frac{1}{9}$ **e.** $\log_4 2$ **f.** $\log_4 \frac{1}{2}$

2. **a.** $\log_3 27$ **b.** $\log_2 16$ **c.** $\log_{16} 4$

   d. $\log_4 \frac{1}{4}$ **e.** $\log_2 \frac{1}{8}$ **f.** $\log_9 \frac{1}{3}$

3. **a.** $\ln(e^{10})$ **b.** $\ln \sqrt{e}$ **c.** $\ln \sqrt[3]{e^4}$

   d. $\ln 1$ **e.** $\ln(\ln(e^e))$ **f.** $\ln\left(\frac{1}{e^3}\right)$

4. **a.** $\ln(e^{-5})$ **b.** $\ln e$ **c.** $\ln \sqrt[3]{e}$

   d. $\ln \sqrt{e^5}$ **e.** $\ln\left(\frac{1}{e}\right)$ **f.** $\ln(\ln e)$

Use the properties of natural logarithms to simplify each function.

5. $f(x) = \ln(9x) - \ln 9$

6. $f(x) = \ln\left(\frac{x}{2}\right) + \ln 2$

7. $f(x) = \ln(x^3) - \ln x$

8. $f(x) = \ln(4x) - \ln 4$

9. $f(x) = \ln\left(\frac{x}{4}\right) + \ln 4$

10. $f(x) = \ln(x^3) - 3 \ln x$

11. $f(x) = \ln(e^{5x}) - 2x - \ln 1$

12. $f(x) = \ln(e^{-2x}) + 3x + \ln 1$

13. $f(x) = 8x - e^{\ln x}$

14. $f(x) = e^{\ln x} + \ln(e^{-x})$

 15–16. Find the domain and range of each function:

15. $f(x) = \ln(x^2 - 1)$ 16. $f(x) = \ln(1 - x^2)$

APPLIED EXERCISES

17. **GENERAL: Interest** An investment grows at 24% compounded monthly. How many years will it take to:

    a. double?   **b.** increase by 50%?

18. **GENERAL: Interest** An investment grows at 36% compounded monthly. How many years will it take to:

    a. double?   **b.** increase by 50%?

19. **GENERAL: Interest** A bank offers 7% compounded continuously. How soon will a deposit:

    a. triple?   **b.** increase by 25%?

20. **GENERAL: Interest** A bank offers 6% compounded continuously. How soon will a deposit:

    a. quadruple?   **b.** increase by 75%?

21. **GENERAL: Depreciation** An automobile depreciates by 30% per year. How soon will it be

worth only half its original value? [*Hint:* Depreciation is like interest but at a negative rate *r*.]

22. **GENERAL: Population Growth** During the years 1990 to 2000, the population of Georgia increased by 2.4% annually. Assuming that this trend continues, in how many years will the population double?

23. **ECONOMICS: Energy Output** The world's output of primary energy (petroleum, natural gas, coal, hydroelectricity, and nuclear electricity) is increasing at the rate of 2.4% annually. How long will it take to increase by 50%?

24. **BUSINESS: Appreciation** An art collection appreciates in value by 8% per year. How soon will its value double?

25–26. **BUSINESS: Advertising** After *t* days of advertisements for a new laundry detergent, the proportion of shoppers in a town who have seen the ads is $1 - e^{-0.03t}$. How long must the ads run to reach:

25. 90% of the shoppers?

26. 99% of the shoppers?

27. BEHAVIORAL SCIENCE: Forgetting The proportion of students in a psychology experiment who could remember an eight-digit number correctly for t minutes was $0.9 - 0.2 \ln t$ (for $t > 1$). Find the proportion that remembered the number for 5 minutes.

28. SOCIAL SCIENCE: Diffusion of Information by Mass Media Election returns are broadcast in a town of 1 million people, and the number of people who have heard the news within t hours is $1,000,000(1 - e^{-0.4t})$. How long will it take for 900,000 people to hear the news?

29. BEHAVIORAL SCIENCE: Learning After t weeks of practice, a typing student can type $100(1 - e^{-0.4t})$ words per minute (wpm). How soon will the student type 80 wpm?

30. BIOMEDICAL: Drug Dose If the original concentration of a drug in a patient's bloodstream is c (milligrams per liter), t hours later the concentration will be $C(t) = ce^{-kt}$, where k is the absorption constant. If the original concentration of the asthma medication theophylline is $c = 20$ and the absorption constant is $k = 0.23$, when should the drug be readministered so that the concentration does not fall below the minimum effective concentration of 5?

31–32. GENERAL: Carbon 14 Dating The proportion of carbon 14 still present in a sample after t years is $e^{-0.00012t}$.

31. Estimate the age of the cave paintings discovered in 1994 in the Ardéche region of France if the carbon with which they were drawn contains only 2.3% of its original carbon 14. They are the oldest known paintings in the world.

32. Estimate the age of the Shroud of Turin, believed by many to be the burial cloth of Christ (see the Application Preview on pages 269–270), from the fact that its linen fibers contained 92.3% of their original carbon 14.

33–34. GENERAL: Potassium 40 Dating The radioactive isotope potassium 40 is used to date very old remains. The proportion of potassium 40 that remains after t million years is $e^{-0.00054t}$. Use this function to estimate the age of the following fossils.

33. The most complete skeleton of an early human ancestor ever found was discovered in Kenya in 1984. Use the formula above to estimate the age of the remains if they contained 99.91% of their original potassium 40.

34. DATING OLDER WOMEN Use the formula above to estimate the age of the partial skeleton of *Australopithecus afarensis* (known as "Lucy") that was found in Ethiopia in 1977 if it had 99.82% of its original potassium 40.

35. ENVIRONMENTAL SCIENCE: Radioactive Waste Hospitals use radioactive tracers in many medical tests. After the tracer is used, it must be stored as radioactive waste until its radioactivity has decreased enough for it to be disposed of as ordinary chemical waste. For the radioactive isotope of potassium, the proportion of radioactivity remaining after t days is $e^{-0.05t}$. How soon will the proportion of radioactivity decrease to 0.001 so that it can be disposed of as ordinary chemical waste?

36. ENVIRONMENTAL SCIENCE: Rainforests It has been estimated* that the world's tropical rainforests are disappearing at the rate of 1.8% per year. If this rate continues, how soon will the rainforests be reduced to 50% of their present size? (Rainforests not only generate much of the oxygen that we breathe, but also contain plants with unique medical properties, such as the rosy periwinkle, which has revolutionized the treatment of leukemia.)

37. BIOMEDICAL: Half-Life of a Drug The time required for the amount of a drug in one's body to decrease by half is called the "half-life" of the drug.

    a. For a drug with absorption constant k, derive the following formula for its half-life:

* Paul R. Ehrlich and Edward O. Wilson, "Biodiversity Studies: Science and Policy," *Science* 253:758–762, August 16, 1991.

$$\left(\begin{array}{c}\text{Half-}\\\text{life}\end{array}\right) = \frac{\ln 2}{k}$$

[*Hint:* Solve the equation $e^{-kt} = \frac{1}{2}$ for t and use the properties of logarithms.]

b. Find the half-life of the cardioregulator digoxin if its absorption constant is $k = 0.018$ and time is measured in hours.

38–39. SOCIAL SCIENCE: Education and Income According to a 1992 study,* each additional year of education increases one's income by 16%. Therefore, with x extra years of education, your income will be multiplied by a factor of 1.16^x.

38. How many additional years of education are required to *double* your income? That is, find the x that satisfies $1.16^x = 2$.

39. How many additional years of education are required to increase your income by 50%? That is, find the x that satisfies $1.16^x = 1.5$.

40. BIOMEDICAL: Population Growth The *Gompertz growth curve* models the size $N(t)$ of a population at time $t \geq 0$ as $N(t) = Ke^{-ae^{-bt}}$ where K and b are positive constants. Show that if $N_0 = N(0)$ is the *initial population* at time $t = 0$, then $a = \ln(K/N_0)$.

41. BIOMEDICAL: Heterozygosity In 1931, Sewell Wright showed that the frequency $f(x)$ of heterozygotes (averaged across many populations) after x generations is $f(x) = H_0\left(1 - \frac{1}{2N}\right)^x$ where H_0 is the initial heterozygosity and N is the number of individuals in the population. For a population of size $N = 500$, how many generations are needed to reduce the frequency by 6%?

42. BIOMEDICAL: Ricker Recruitment The population dynamics of many fish (such as salmon) can be described by the *Ricker curve* $y = axe^{-bx}$ for $x \geq 0$ where $a > 1$ and $b > 0$ are constants, x is the size of the parental stock, and y is the number of recruits

(offspring). Determine the size of the equilibrium population for which $y = x$.

Solve the following exercises on a graphing calculator by graphing an appropriate exponential function (using x for ease of entry) together with a constant function and using INTERSECT to find where they meet. You will have to choose an appropriate window.

43. GENERAL: Inflation At 2% inflation, prices increase by 2% compounded annually. How soon will prices

a. double? **b.** triple?

44. GENERAL: Inflation At 5% inflation, prices increase by 5% compounded annually. How soon will prices:

a. double? **b.** triple?

45. GENERAL: Interest A bank account grows at 6% compounded quarterly. How many years will it take to:

a. double? **b.** increase by 50%?

46. GENERAL: Interest A bank account grows at 7% compounded continuously. How many years will it take to:

a. double? **b.** increase by 25%?

47. BUSINESS: Advertising After a sale has been advertised for t days, the proportion of shoppers in a city who have seen the ad is $1 - e^{-0.08t}$. How long must the ad run to reach:

a. 50% of the shoppers?
b. 60% of the shoppers?

48. BIOMEDICAL: Drug Dosage If the original concentration of a drug in a patient's bloodstream is 5 (milligrams per milliliter), and if the absorption constant is 0.15, then t hours later the concentration will be $5e^{-0.15t}$. When should the drug be readministered so that the concentration does not fall below the minimum effective concentration of 2.7?

49. GENERAL: Carbon 14 Dating In 1991 two hikers in the Italian Alps found the frozen but well-preserved body of the most ancient human ever found, dubbed "Iceman." Estimate

* Orley Ashenfelter and Alan Krueger, "Estimate of the Economic Return to Schooling from a New Sample of Twins," *American Economic Review* 84(5):1157–1173, December 1994.

the age of Iceman if his grass cape contained 53% of its original carbon 14. (Use the carbon 14 decay function stated in Exercises 31–32.)

50. **GENERAL: Potassium 40 Dating** Estimate the age of the oldest known dinosaur, a dog-sized creature called *Herrerasaurus* found in Argentina in 1988, if volcanic material found with it contained 88.4% of its original potassium 40. (Use the potassium 40 decay function given in Exercises 33–34.)

51. **ENVIRONMENTAL SCIENCE: Nuclear Waste** More than half a century after the beginning of the nuclear age, not a single country has found a safe or permanent way to dispose of long-lived radioactive waste. Among the most hazardous radioactive waste is irradiated fuel from nuclear power plants, totaling 245,000 tons in 2000 and growing by 11.3% annually. At this rate, how long will it take for this amount to double? *(Source: World-watch Institute)*

52. **SOCIAL SCIENCE: Cell Phone Usage** Between 1990 and 2000 the number of cell phone subscribers worldwide increased by 52% annually. At this rate, what is the doubling time for cell phone subscribers? Express your answer in years and months.

4.3 DIFFERENTIATION OF LOGARITHMIC AND EXPONENTIAL FUNCTIONS

Introduction

You may have noticed that Sections 4.1 and 4.2 contained no calculus. Their purpose was to introduce logarithmic and exponential functions. In this section we differentiate these new functions and use their derivatives for graphing, optimization, and finding rates of change. We emphasize *natural* (base *e*) logs and exponentials, since most applications use these exclusively. Verifications of the differentiation rules are given at the end of the section.

Derivatives of Logarithmic Functions

The rule for differentiating the natural logarithm function is as follows:

Derivative of $\ln x$

$$\frac{d}{dx} \ln x = \frac{1}{x} \qquad \text{The derivative of } \ln x \text{ is 1 over } x$$

EXAMPLE 1 **DIFFERENTIATING A LOGARITHMIC FUNCTION**

Differentiate $f(x) = x^3 \ln x$.

Solution The function is a *product*, x^3 times $\ln x$, so we use the Product Rule.

$$\frac{d}{dx}(x^3 \ln x) = 3x^2 \ln x + x^3 \frac{1}{x} = 3x^2 \ln x + x^2$$

$$\underbrace{\hspace{3cm}}_{} \qquad \text{From } x^3 \frac{1}{x} = x^2$$

Derivative   Second   First   Derivative
of the first   left alone   left alone   of $\ln x$

Practice Problem 1 Differentiate $f(x) = \dfrac{\ln x}{x}$. ➤ Solution on page 316

The preceding rule, together with the Chain Rule, shows how to differentiate the natural logarithm of a *function*. For any differentiable function $f(x)$ that is positive:

Derivative of ln $f(x)$

$$\frac{d}{dx} \ln f(x) = \frac{f'(x)}{f(x)}$$

The derivative of the natural log of a function is the derivative of the function over the function

---

Notice that the right-hand side does not involve logarithms at all.

EXAMPLE 2

DIFFERENTIATING A LOGARITHMIC FUNCTION

$$\frac{d}{dx} \ln(x^2 + 1) = \frac{2x}{x^2 + 1}$$

   ← Derivative of $x^2 + 1$
   ← Original function (without the ln)

As we observed, the answer does not involve logarithms.

Practice Problem 2 Find $\dfrac{d}{dx} \ln(x^3 - 5x + 1)$. ➤ Solution on page 316

EXAMPLE **3**

DIFFERENTIATING A LOGARITHMIC FUNCTION

Find the derivative of $f(x) = \ln(x^4 - 1)^3$.

Solution For this problem we need the rule for differentiating the natural logarithm of a function, together with the Generalized Power Rule [for differentiating $(x^4 - 1)^3$].

$$\frac{d}{dx} \ln(x^4 - 1)^3 = \frac{\frac{d}{dx}(x^4 - 1)^3}{(x^4 - 1)^3} \qquad \text{Using } \frac{d}{dx} \ln f = \frac{f'}{f}$$

$$= \frac{3(x^4 - 1)^2 4x^3}{(x^4 - 1)^3} \qquad \text{Using the Generalized Power Rule}$$

$$= \frac{12x^3}{x^4 - 1} \qquad \text{Dividing top and bottom by } (x^4 - 1)^2$$

Alternative Solution It is easier if we simplify first, using Property 8 of logarithms (see the inside back cover) to bring down the exponent 3:

$$\ln(x^4 - 1)^3 = 3 \ln(x^4 - 1) \qquad \text{Using } \ln(M^N) = N \ln M$$

Now we differentiate the simplified expression:

$$\frac{d}{dx} 3 \ln(x^4 - 1) = 3 \frac{4x^3}{x^4 - 1} = \frac{12x^3}{x^4 - 1} \qquad \text{Same answer as before}$$

Moral: Changing $\ln (\cdot \cdot \cdot)^n$ to $n \ln (\cdot \cdot \cdot)$ simplifies differentiation.

Derivatives of Exponential Functions

The rule for differentiating the exponential function e^x is as follows:

Derivative of e^x

$$\frac{d}{dx} e^x = e^x \qquad \text{The derivative of } e^x \text{ is simply } e^x$$

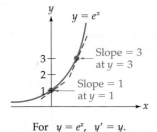

For $y = e^x$, $y' = y$.

This shows the rather surprising fact that e^x is its own derivative. Stated another way, the function e^x is unchanged by the operation of differentiation.

This rule can be interpreted graphically. If $y = e^x$, then $y' = e^x$, so that $y = y'$. This means that on the graph of $y = e^x$, the slope y' always equals the y-coordinate, as shown in the graph on the left.

Graphing Calculator Exploration

a. Define $y_1 = e^x$ and y_2 as the derivative of y_1 (using NDERIV) and graph them together on the window $[-3, 3]$ by $[-1, 10]$.

b. Why does the screen show only one curve?

c. Use TRACE to compare the values of y_1 and y_2 at some chosen x-value. Do the y-values agree *exactly*? If not, explain the slight discrepancy. [*Hint:* Is NDERIV really the derivative?]

EXAMPLE 4

FINDING A DERIVATIVE INVOLVING e^x

Find $\dfrac{d}{dx}\left(\dfrac{e^x}{x}\right)$.

Solution   Since the function is a quotient, we use the Quotient Rule:

$$\overbrace{}^{\text{Derivative of } e^x}$$
$$\overbrace{}^{\text{Derivative of } x}$$

$$\frac{d}{dx}\left(\frac{e^x}{x}\right) = \frac{x \cdot e^x - 1 \cdot e^x}{x^2} = \frac{xe^x - e^x}{x^2}$$

EXAMPLE 5

EVALUATING A DERIVATIVE INVOLVING e^x

If $f(x) = x^2 e^x$, find $f'(1)$.

Solution

$$f'(x) = 2xe^x + x^2 e^x \qquad \text{Using the Product Rule on } x^2 \cdot e^x$$
$$f'(1) = 2(1)e^1 + (1)^2 e^1 \qquad \text{Substituting } x = 1$$
$$= 2e + e = 3e \qquad \text{Simplifying}$$

In these problems we leave our answers in their "exact" forms, leaving e as e. Later, in applied problems, we will approximate our answers using $e \approx 2.718$ or a calculator.

Practice Problem 3   If $f(x) = xe^x$, find $f'(1)$.   ➤ Solution on page 316

The rule for differentiating e^x, together with the Chain Rule, shows how to differentiate $e^{f(x)}$. For any differentiable function $f(x)$:

Derivative of $e^{f(x)}$

$$\frac{d}{dx}e^{f(x)} = e^{f(x)} \cdot f'(x)$$

The derivative of e to a function is e to the function times the derivative of the function

---

That is, to differentiate $e^{f(x)}$ we simply "copy" the original $e^{f(x)}$ and then multiply by the derivative of the exponent.

EXAMPLE 6

DIFFERENTIATING AN EXPONENTIAL FUNCTION

$$\frac{d}{dx}e^{x^4+1} = e^{x^4+1}(4x^3) = 4x^3 e^{x^4+1}$$

Reversing the order

Copied — Derivative of the exponent

EXAMPLE 7

DIFFERENTIATING AN EXPONENTIAL FUNCTION

Find $\dfrac{d}{dx}e^{x^2/2}$.

Solution The exponent $x^2/2$ should first be rewritten as $\frac{1}{2}x^2$, a constant times x to a power, since then its derivative is easily seen to be x.

$$\frac{d}{dx}e^{x^2/2} = \frac{d}{dx}e^{\frac{1}{2}x^2} = e^{\frac{1}{2}x^2}(x) = xe^{\frac{1}{2}x^2} = xe^{x^2/2}$$

Rewriting the exponent in its original form

— Derivative of the exponent

Practice Problem 4 Find $\dfrac{d}{dx}e^{1+x^3/3}$.

➤ Solution on page 316

The formulas for differentiating natural logarithmic and exponential functions are summarized as follows, with $f(x)$ written simply as f.

Logarithmic Formulas	Exponential Formulas	
$\dfrac{d}{dx}\ln x = \dfrac{1}{x}$	$\dfrac{d}{dx}e^x = e^x$	Top formulas apply only to $\ln x$ and e^x
$\dfrac{d}{dx}\ln f = \dfrac{f'}{f}$	$\dfrac{d}{dx}e^f = e^f \cdot f'$	Bottom formulas apply to $\ln$ and e of a *function*

e^x versus x^n

Notice that we do *not* take the derivative of e^x by the Power Rule,

$$\frac{d}{dx}x^n = nx^{n-1}$$

This is because the Power Rule applies to x^n, *a variable to a constant power*, whereas e^x is *a constant to a variable power*. The two types of functions are quite different, as their graphs show.

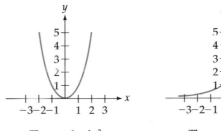

The graph of x^2
(a variable to a
constant power)

The graph of e^x
(a constant to a
variable power)

Each type of function has its own differentiation formula.

$$\frac{d}{dx}x^n = nx^{n-1} \qquad\qquad \frac{d}{dx}e^x = e^x$$

For a variable
x to a constant
power n

For the constant
e to a variable
power x

EXAMPLE 8

DIFFERENTIATING A LOGARITHMIC AND EXPONENTIAL FUNCTION

Find the derivative of $\ln(1 + e^x)$.

Solution

$$\frac{d}{dx}\ln(1 + e^x) = \frac{\frac{d}{dx}(1 + e^x)}{1 + e^x} = \frac{e^x}{1 + e^x}$$

$$\underbrace{\qquad}_{\text{Using } \frac{d}{dx}\ln f = \frac{f'}{f}} \qquad \underbrace{\qquad}_{\substack{\text{Working out} \\ \text{the numerator}}}$$

Functions of the form e^{kx} (for constant k) arise in many applications. The derivative of e^{kx} is as follows.

$$\frac{d}{dx}e^{kx} = e^{kx} \cdot k = ke^{kx} \qquad\qquad \text{Using } \frac{d}{dx}e^f = e^f \cdot f'$$

$$\underset{\text{Derivative of the exponent}}{\uparrow}$$

This result is so useful that we record it as a separate formula.

Derivative of e^{kx}

$$\frac{d}{dx}e^{kx} = ke^{kx} \qquad\qquad\qquad \text{For any constant } k$$

This formula says that the rate of change (the derivative) of e^{kx} is proportional to itself. That is, the function $y = e^{kx}$ satisifies the *differential equation*

$$y' = ky$$

We noted this earlier when we observed that in exponential growth a quantity *grows in proportion to itself* (as in populations and savings accounts).

These differentiation formulas enable us to find instantaneous rates of change of logarithmic and exponential functions. In many applications the variable stands for time, so we use t instead of x.

EXAMPLE 9   **FINDING A RATE OF IMPROVEMENT OF A SKILL**

After t weeks of practice a pole vaulter can vault

$$H(t) = 14(1 - e^{-0.10t})$$

feet. Find the rate of change of the athlete's jumps after

a. 0 weeks (at the beginning of training)   **b.** 12 weeks

Solution

$$H(t) = 14(1 - e^{-0.10t}) = 14 - 14e^{-0.10t} \qquad \text{$H(t)$ multiplied out}$$

We differentiate to find the rate of change

$$H'(t) = \underbrace{-14(-0.10)e^{-0.10t}}_{} = \underbrace{1.4e^{-0.10t}}_{} \qquad \begin{array}{l}\text{Differentiating} \\ 14 - 14e^{-0.10t}\end{array}$$

$$\text{Using } \frac{d}{dt}e^{kt} = ke^{kt} \qquad \text{Simplifying}$$

a. For the rate of change after 0 weeks:

$$H'(0) = 1.4e^{-0.10(0)} = 1.4e^0 = 1.4 \qquad \begin{array}{l}H'(t) = 1.4e^{-0.10t} \\ \text{with } t = 0\end{array}$$

b. After 12 weeks:

$$H'(12) = 1.4e^{-0.10(12)}$$
$$= 1.4e^{-1.2} \approx 1.4(0.30) = 0.42 \qquad \begin{array}{l}H'(t) = 1.4e^{-0.10t} \\ \text{with } t = 12\end{array}$$

Using a calculator

At first the vaults increased by 1.4 feet per week. After 12 weeks, the gain was only 0.42 foot (about 5 inches) per week.

This result is typical of learning a new skill: early improvement is rapid, later improvement is slower.

Maximizing Consumer Expenditure

The amount of a commodity that consumers will buy depends on the price of the commodity. For a commodity whose price is p, let the consumer demand be given by a function $D(p)$. Multiplying the number of units $D(p)$ by the price p gives the total *consumer expenditure* for the commodity.

Consumer Demand and Expenditure

Let $D(p)$ be the consumer demand at price p. Then the consumer expenditure is

$$E(p) = p \cdot D(p)$$

MAXIMIZING CONSUMER EXPENDITURE

If consumer demand for a commodity is $D(p) = 10{,}000e^{-0.02p}$ units per week, where p is the selling price, find the price that maximizes consumer expenditure.

Solution Using the formula above for consumer expenditure,

$$E(p) = p \cdot 10{,}000e^{-0.02p} = 10{,}000pe^{-0.02p} \qquad E(p) = p \cdot D(p)$$

To maximize $E(p)$ we differentiate:

$$E'(p) = \underbrace{10{,}000e^{-0.02p}}_{\substack{\text{Derivative} \\ \text{of } 10{,}000p}} + \underbrace{10{,}000p(-0.02)e^{-0.02p}}_{\substack{\text{Derivative} \\ \text{of } e^{-0.02p}}}$$

Using the Product Rule to differentiate $E(p) = 10{,}000p \cdot e^{-0.02p}$

$$= 10{,}000e^{-0.02p} - 200pe^{-0.02p} \qquad \text{Simplifying}$$

$$= 200e^{-0.02p}(50 - p) \qquad \text{Factoring}$$

$$\text{CN:} \quad p = 50$$

Critical number from $(50 - p)$ (since e to a power is never zero)

We calculate E'' for the second-derivative test:

$$E''(p) = 200(-0.02)e^{-0.02p}(50 - p) + 200e^{-0.02p}(-1)$$

From $E'(p) = 200e^{-0.02p}(50 - p)$ using the Product Rule

$$= -4e^{-0.02p}(50 - p) - 200e^{-0.02p} \qquad \text{Simplifying}$$

At the critical number $p = 50$,

$$E''(50) = -4e^{-0.02(50)}(50 - 50) - 200e^{-0.02(50)} \qquad \text{Substituting } p = 50$$

$$= -200e^{-1} = \frac{-200}{e} \qquad \text{Simplifying}$$

E'' is negative, so the expenditure $E(p)$ is maximized at $p = 50$:

Consumer expenditure is maximized at price $p = \$50$.

Graphing Calculator Exploration

Use a graphing calculator to verify the answer to the previous example by graphing $E(x) = 10{,}000xe^{-0.02x}$ (using x instead of p) on the window $[0, 200]$ by $[0, 200{,}000]$ and using MAXIMUM.

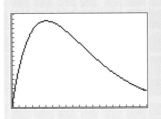

How did we choose the graphing window? We first found the critical number "by hand" (setting the derivative equal to zero and finding, in this example, 50) and chose x-values around it. For the y-values, we evaluated the function at the critical number (using EVALUATE, finding in this example $y \approx 184{,}000$) and chose y-values including it. Notice how graphing calculators and calculus are most effective when used together.

Graphing Logarithmic and Exponential Functions

To graph logarithmic and exponential functions using a graphing calculator, we first find critical points and possible inflection points, and then graph the function on a window including these points. (If graphing "by hand," we would make sign diagrams for the first and second derivatives and then sketch the graph, as in Chapter 3.)

EXAMPLE 11

GRAPHING AN EXPONENTIAL FUNCTION

Graph $f(x) = e^{-x^2/2}$.

Solution As before, we write the function as $f(x) = e^{-\frac{1}{2}x^2}$. The derivative is

$$f'(x) = e^{-\frac{1}{2}x^2}(-x) = -xe^{-\frac{1}{2}x^2} \qquad \text{Using } \frac{d}{dx}e^f = e^f \cdot f'$$

⎿ Derivative of the exponent

CN: $x = 0$ Critical number is 0

    $y = 1$ From $y = e^{-\frac{1}{2}x^2}$ evaluated at $x = 0$

The second derivative is

$$f''(x) = (-1)e^{-\frac{1}{2}x^2} - xe^{-\frac{1}{2}x^2}(-x) \qquad \begin{array}{l}\text{From } f'(x) = -x \cdot e^{-\frac{1}{2}x^2}\\ \text{using the Product Rule}\end{array}$$

$$= -e^{-\frac{1}{2}x^2} + x^2 e^{-\frac{1}{2}x^2} \qquad \text{Simplifying}$$

$$= e^{-\frac{1}{2}x^2}(-1 + x^2) \qquad \text{Factoring}$$

$$= (x^2 - 1)e^{-\frac{1}{2}x^2} \qquad \text{Rearranging}$$

$$= (x + 1)(x - 1)e^{-\frac{1}{2}x^2} \qquad \text{Factoring}$$

$$x = \pm 1 \qquad \text{Where } f'' = 0$$

$$y = e^{-\frac{1}{2}} \approx 0.6 \qquad \begin{array}{l}\text{From } y = e^{-\frac{1}{2}x^2}\\ \text{evaluated at } x = \pm 1\end{array}$$

Based on these values, we choose the graphing window as follows. For the x-values we choose $[-3, 3]$ (to include 0 and ± 1 and beyond), and for the y-values we choose $[-1, 2]$ (to include 1 and 0.6 and above and below). This window gives the following graph. (Many other windows would be just as good, and after seeing the graph you might want to adjust the window.)

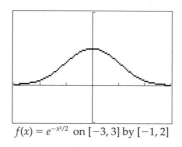

$f(x) = e^{-x^2/2}$ on $[-3, 3]$ by $[-1, 2]$

This function, multiplied by the constant $1/\sqrt{2\pi}$, is the famous "bell-shaped curve" of statistics, used for predictions of many things from the IQs of newborn babies to stock prices.

Verification of the Power Rule for Arbitrary Powers

On pages 121–122 we proved the Power Rule $\dfrac{d}{dx} x^n = nx^{n-1}$ for positive integer exponents, and on pages 256–257 we extended it to rational exponents. We can now show that the Power Rule holds for *all* exponents. We begin by using one of the "inverse" properties of logarithms, $x = e^{\ln x}$, but with x replaced by x^n:

$$f(x) = x^n = e^{\ln x^n} = e^{n \ln x}$$

Using the property that logs bring down exponents

Differentiating:

$$f'(x) = e^{n \ln x}\left(n\frac{1}{x} \right)$$

x^{n-1} Derivative of the exponent

Differentiating $f(x) = e^{n \ln x}$ by the $e^{f(x)}$ formula (page 308)

$$= x^n \cdot \frac{1}{x} \cdot n = nx^{n-1}$$

x^{n-1}

Replacing $e^{n \ln x}$ by x^n, reordering, and combining x's

This is what we wanted to show—that the derivative of $f(x) = x^n$ is $f'(x) = nx^{n-1}$ for *all* exponents n. (The equation $x^n = e^{n \ln x}$ can be taken as a *definition* of x to a power. A definition of logarithms without recourse to exponents will be given in Exercise 75 on page 386).

Verification of the Differentiation Formulas

The formula for the derivative of the natural logarithm function comes from applying the definition of the derivative to $f(x) = \ln x$.

$$f'(x) = \lim_{h \to 0} \frac{f(x + h) - f(x)}{h} = \lim_{h \to 0} \frac{\ln (x + h) - \ln x}{h}$$
Definition of $f'(x)$

$$= \lim_{h \to 0} \frac{1}{h} [\ln (x + h) - \ln x]$$
Dividing by h is equivalent to multiplying by $1/h$

$$= \lim_{h \to 0} \frac{1}{h} \ln \left(\frac{x + h}{x} \right) = \lim_{h \to 0} \frac{1}{h} \ln \left(1 + \frac{h}{x} \right)$$
Using Property 7 of logarithms (inside back cover), and simplifying

$$= \lim_{h \to 0} \frac{1}{x} \frac{x}{h} \ln \left(1 + \frac{h}{x} \right) = \lim_{h \to 0} \frac{1}{x} \ln \left(1 + \frac{h}{x} \right)^{x/h}$$
Dividing and multiplying by x, and then using Property 8 of logarithms

$$= \lim_{n \to \infty} \frac{1}{x} \ln \underbrace{\left(1 + \frac{1}{n} \right)^{n}}_{\substack{\text{Approaches} \\ e \text{ as } n \to \infty}}$$
Defining $n = x/h$, so that $h \to 0$ implies $n \to \infty$ (for $h > 0$)

$$= \frac{1}{x} \ln e = \frac{1}{x}$$
Since $\ln e = 1$ (the same conclusion follows if $h < 0$)

This is the result that we wanted to show—that the derivative of the natural logarithm function $f(x) = \ln x$ is $f'(x) = \frac{1}{x}$.

For the rule to differentiate the natural logarithm of a (positive) *function*, we begin with

$$\frac{d}{dx} f(g(x)) = f'(g(x))g'(x)$$
Chain Rule (from page 160) for differentiable functions f and g

$$\frac{d}{dx} \ln (g(x)) = \frac{1}{g(x)} g'(x) = \frac{g'(x)}{g(x)}$$
Taking $f(x) = \ln x$, so $f'(x) = \frac{1}{x}$

Replacing g by f, this is exactly the formula we wanted to show,

$$\frac{d}{dx} \ln f(x) = \frac{f'(x)}{f(x)}$$

To derive the rule for differentiating e^x we begin with

$$\ln e^x = x$$
Property 3 of natural logarithms

and differentiate both sides:

$$\frac{\frac{d}{dx} e^x}{e^x} = 1$$
Using $\frac{d}{dx} \ln f = \frac{f'}{f}$ on the left side

Multiplying each side by e^x gives

$$\frac{d}{dx}e^x = e^x$$

This is the rule for differentiating e^x. This rule together with the Chain Rule gives the rule for differentiating $e^{f(x)}$, just as before.

➤ **Solutions to Practice Problems**

1. $f'(x) = \dfrac{d}{dx}\left(\dfrac{\ln x}{x}\right) = \dfrac{x\left(\frac{1}{x}\right) - 1\ln x}{x^2}$

    $= \dfrac{1 - \ln x}{x^2}$

    Derivative of e^x
    Derivative of x

2. $\dfrac{d}{dx}\ln(x^3 - 5x + 1) = \dfrac{3x^2 - 5}{x^3 - 5x + 1}$

3. $f'(x) = e^x + xe^x$

    $f'(1) = e^1 + 1e^1 = 2e$ (which is approximately $2 \cdot 2.718 = 5.436$)

4. $\dfrac{d}{dx}e^{1 + x^3/3} = \dfrac{d}{dx}e^{1 + \frac{1}{3}x^3} = e^{1 + \frac{1}{3}x^3}(x^2) = x^2e^{1 + x^3/3}$

4.3 Exercises

Find the derivative of each function.

1. $f(x) = x^2 \ln x$

2. $f(x) = \dfrac{\ln x}{x^3}$

3. $f(x) = \ln x^2$

4. $f(x) = \ln(x^3 + 1)$

5. $f(x) = \ln \sqrt{x}$

6. $f(x) = \sqrt{\ln x}$

7. $f(x) = \ln(x^2 + 1)^3$

8. $f(x) = \ln(x^4 + 1)^2$

9. $f(x) = \ln(-x)$

10. $f(x) = \ln(5x)$

11. $f(x) = \dfrac{e^x}{x^2}$

12. $f(x) = x^3e^x$

13. $f(x) = e^{x^3 + 2x}$

14. $f(x) = 2e^{7x}$

15. $f(x) = e^{x^3/3}$

16. $f(x) = \ln(e^x - 2x)$

17. $f(x) = x - e^{-x}$

18. $f(x) = x \ln x - x$

19. $f(x) = \ln e^{2x}$

20. $f(x) = \ln e^x$

21. $f(x) = e^{1 + e^x}$

22. $f(x) = \ln(e^x + e^{-x})$

23. $f(x) = x^e$

24. $f(x) = ex$

25. $f(x) = e^3$

26. $f(x) = \sqrt{e}$

27. $f(x) = \ln(x^4 + 1) - 4e^{x/2} - x$

28. $f(x) = x^2e^x - 2\ln x + (x^2 + 1)^3$

29. $f(x) = x^2 \ln x - \frac{1}{2}x^2 + e^{x^2} + 5$

30. $f(x) = e^{-2x} - x \ln x + x - 7$

For each function, find the indicated expressions.

31. $f(x) = \dfrac{\ln x}{x^5}$, find **a.** $f'(x)$ **b.** $f'(1)$

32. $f(x) = x^4 \ln x$, find **a.** $f'(x)$ **b.** $f'(1)$

33. $f(x) = \ln(x^4 + 48)$, find **a.** $f'(x)$ **b.** $f'(2)$

34. $f(x) = x^2 \ln x - x^2$, find **a.** $f'(x)$ **b.** $f'(e)$

35. $f(x) = \ln(e^x - 3x)$, find **a.** $f'(x)$ **b.** $f'(0)$

36. $f(x) = \ln(e^x + e^{-x})$, find **a.** $f'(x)$ **b.** $f'(0)$

For each function:

a. Find $f'(x)$.

b. Evaluate the given expression and approximate it to three decimal places.

37. $f(x) = 5x \ln x$, find and approximate $f'(2)$.

38. $f(x) = e^{x^2/2}$, find and approximate $f'(2)$.

39. $f(x) = \dfrac{e^x}{x}$, find and approximate $f'(3)$.

40. $f(x) = \ln(e^x - 1)$, find and approximate $f'(3)$.

Find the *second* derivative of each function.

41. $f(x) = e^{-x^5/5}$ **42.** $f(x) = e^{-x^6/6}$

By calculating the first few derivatives, find a formula for the nth derivative of each function (k is a constant).

43. $f(x) = e^{kx}$ **44.** $f(x) = e^{-kx}$

Use your graphing calculator to graph each function on a window that includes all relative extreme points and inflection points, and give the coordinates of these points (rounded to two decimal places). [*Hint:* Use NDERIV once or twice with ZERO.] (Answers may vary depending on the graphing window chosen.)

45. $f(x) = e^{-2x^2}$ **46.** $f(x) = 1 - e^{-x^2/2}$

47. $f(x) = \ln(1 + x^2)$ **48.** $f(x) = e^x + e^{-x}$

Use your graphing calculator to graph each function on the indicated interval, and give the coordinates of all relative extreme points and inflection points (rounded to two decimal places). [*Hint:* Use NDERIV once or twice together with ZERO.] (Answers may vary depending on the graphing window chosen.)

49. $f(x) = \dfrac{x^2}{e^x}$ for $-1 \le x \le 8$

50. $f(x) = \dfrac{x}{e^x}$ for $-1 \le x \le 5$

51. $f(x) = x \ln|x|$ for $-2 \le x \le 2$

[*Hint for Exercises 51–52:* $|x|$ is sometimes entered as ABS (x).]

52. $f(x) = x^2 \ln|x|$ for $-2 \le x \le 2$

Use implicit differentiation to find dy/dx.

53. $y^2 - ye^x = 12$ **54.** $y^2 - x \ln y = 10$

APPLIED EXERCISES

55. GENERAL: Compound Interest A sum of $1000 at 5% interest compounded continuously will grow to $V(t) = 1000e^{0.05t}$ dollars in t years. Find the rate of growth after:

    a. 0 years (the time of the original deposit).
    b. 10 years.

[*Hint:* The rate of growth means the derivative.]

56. GENERAL: Depreciation A $10,000 automobile depreciates so that its value after t years is $V(t) = 10{,}000e^{-0.35t}$ dollars. Find the rate of change of its value:

    a. when it is new ($t = 0$).
    b. after 2 years.

[*Hint:* The rate of change means the derivative.]

57. GENERAL: Population The world population (in billions) is predicted to be $P(t) = 5.89e^{0.0175t}$, where t is the number of years after 2000. Find the rate of change of the population in the year 2010. [*Hint:* The rate of change means the derivative.]

58. BEHAVIORAL SCIENCE: Ebbinghaus Memory Model According to the

Ebbinghaus model of memory, if one is shown a list of items, the percentage of items that one will remember t time units later is $P(t) = (100 - a)e^{-bt} + a$, where a and b are constants. For $a = 25$ and $b = 0.2$, this function becomes $P(t) = 75e^{-0.2t} + 25$. Find the rate of change of this percentage:

a. at the beginning of the test ($t = 0$).
b. after 3 time units.

[*Hint:* The rate of change means the derivative.]

59. BIOMEDICAL: Drug Dosage A patient receives an injection of 1.2 milligrams of a drug, and the amount remaining in the bloodstream t hours later is $A(t) = 1.2e^{-0.05t}$. Find the rate of change of this amount:

a. just after the injection (at time $t = 0$).
b. after 2 hours.

[*Hint:* The rate of change means the derivative.]

60. GENERAL: Temperature A covered cup of coffee at 200 degrees, if left in a 70-degree room, will cool to $T(t) = 70 + 130e^{-2.5t}$ degrees in t hours. Find the rate of change of the temperature:

a. at time $t = 0$.   **b.** after 1 hour.

61. BUSINESS: Sales The weekly sales (in thousands) of a new product are predicted to be $S(x) = 1000 - 900e^{-0.1x}$ after x weeks. Find the rate of change of sales after

a. 1 week.   **b.** 10 weeks.

62. SOCIAL SCIENCE: Diffusion of Information by Mass Media The number of people in a town of 50,000 who have heard an important news bulletin within t hours of its first broadcast is $N(t) = 50{,}000(1 - e^{-0.4t})$. Find the rate of change of the number of informed people:

a. at time $t = 0$.   **b.** after 8 hours.

63–64. ECONOMICS: Consumer Expenditure If consumer demand for a commodity is given by the function below (where p is the selling price in dollars), find the price that maximizes consumer expenditure.

63. $D(p) = 5000e^{-0.01p}$ **64.** $D(p) = 8000e^{-0.05p}$

65–66. BUSINESS: Maximizing Revenue Each of the following functions is a company's price function, where p is the price (in dollars) at which quantity x (in thousands) will be sold.

a. Find the revenue function $R(x)$. [*Hint:* Revenue is price times quantity, $p \cdot x$.]
b. Find the quantity and price that will maximize revenue.

65. $p = 400\,e^{-0.20x}$ **66.** $p = 4 - \ln x$

67. BIOMEDICAL: Population Growth The *Gompertz growth curve* models the size $N(t)$ of a population at time $t \geq 0$ as $N(t) = Ke^{-ae^{-bt}}$ where K and b are positive constants. Show that $\dfrac{dN}{dt} = bN\ln\left(\dfrac{K}{N}\right)$ and interpret this derivative to make statements about the population growth when $N < K$ and when $N > K$.

68. BIOMEDICAL: Ricker Recruitment The population dynamics of many fish (such as salmon) can be described by the *Ricker curve* $y = axe^{-bx}$ for $x \geq 0$ where $a > 1$ and $b > 0$ are constants, x is the size of the parental stock, and y is the number of recruits (offspring). Determine the size of the parental stock that maximizes the number of recruits.

69. BIOMEDICAL: Reynolds Number An important characteristic of blood flow is the "Reynolds number." As the Reynolds number increases, blood flows less smoothly. For blood flowing through certain arteries, the Reynolds number is

$$R(r) = a \ln r - br$$

where a and b are positive constants and r is the radius of the artery. Find the radius r that maximizes the Reynolds number R. (Your answer will involve the constants a and b.)

70. BIOMEDICAL: Drug Concentration If a drug is injected intramuscularly, the concentration of the drug in the bloodstream after t hours will be

$$A(t) = \frac{c}{b - a}(e^{-at} - e^{-bt})$$

If the constants are $a = 0.4$, $b = 0.6$, and $c = 0.1$, find the time of maximum concentration.

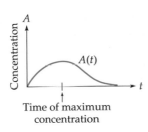

Time of maximum
concentration

71. GENERAL: Temperature A mug of beer chilled to 40 degrees, if left in a 70-degree room, will warm to a temperature of $T(t) = 70 - 30e^{-3.5t}$ degrees in t hours. Enter this temperature function as y_1 (using x for ease of entry), define y_2 as its derivative (using NDERIV), and graph them on the window $[0, 2]$ by $[0, 80]$.

    a. Evaluate y_1 and y_2 at $x = 0.25$ (using EVALUATE) and interpret your answers.

    b. Evaluate y_1 and y_2 at $x = 1$ and interpret your answers.

72. SOCIAL SCIENCE: Diffusion of Information by Mass Media The number of people in a city of 200,000 who have heard a weather bulletin within t hours of its first broadcast is $N(t) = 200,000(1 - e^{-0.5t})$. Enter this function as y_1 (using x for ease of entry), define y_2 as its derivative (using NDERIV), and graph them on the window $[0, 4]$ by $[0, 200,000]$.

    a. Evaluate y_1 and y_2 at $x = 0.5$ (using EVALUATE) and interpret your answers.

    b. Evaluate y_1 and y_2 at $x = 3$ and interpret your answers.

73–74. ATHLETICS. World's Record 100-Meter Run In 1987 Carl Lewis set a new world's record of 9.93 seconds for the 100 meter run. The distance that he ran in the first x seconds was

$$11.274[x - 1.06(1 - e^{-x/1.06})] \text{ meters}$$

for $0 \le x \le 9.93$.* Enter this function as y_1, and define y_2 as its derivative (using NDERIV), so that y_2 gives the velocity after x seconds. Graph them on the window $[0, 9.93]$ by $[0, 100]$.

73. Trace along the velocity curve to verify that Lewis's maximum speed was about 11.27 me-

ters per second. Find how quickly he reached a speed of 10 meters per second, which is 95% of his maximum speed.

74. Define y_3 as the derivative of y_2 (using NDERIV) so that y_3 gives the acceleration after x seconds, and graph y_2 and y_3 on the window $[0, 9.93]$ by $[0, 20]$. Evaluate both y_2 and y_3 at $x = 0.1$ and also at $x = 9.93$ (using EVALUATE). Interpret your answers.

75. ECONOMICS: Consumer Expenditure If consumer demand for a commodity is given by $D(p) = 4000e^{-0.002p}$ (where p is the selling price in dollars), find the price that maximizes consumer expenditure. [*Hint*: Find the critical number of the expenditure function "by hand" and then graph it (using x for ease of entry) on an appropriate window to verify that it is maximized at this critical number.]

76. BUSINESS: Maximizing Revenue An electronics company finds that the price function for mobile telephones is $p(x) = 400e^{-0.005x}$, where p is the price in dollars at which quantity x (in thousands) will be sold. Find the quantity x that maximizes revenue. [*Hint*: Revenue is price times quantity, $p \cdot x$. Find the critical number "by hand" and then graph the revenue function on an appropriate window to verify that it is maximized at this critical number.]

77. BIOMEDICAL: Drug Concentration If a certain drug is injected intramuscularly, the concentration of the drug in the bloodstream after t hours will be $C(t) = 0.75(e^{-0.2t} - e^{-0.6t})$. Find the time of maximum concentration.

78. ATHLETICS: How Fast Do Old Men Slow Down? The fastest times for the marathon (26.2 miles) for male runners aged 35 to 80 are approximated by the function*

$$f(x) = \begin{cases} 106.2e^{0.0063x} & \text{if } x \le 58.2 \\ 850.4e^{0.000614x^2 - 0.0652x} & \text{if } x > 58.2 \end{cases}$$

in minutes, where x is the age of the runner.

(continues)

---

*See W. G. Pritchard, "Mathematical Models of Running," *SIAM Review* 35(3):359–379, September 1993.

*Ray C. Fair, "How Fast Do Old Men Slow Down?," *Review of Economics and Statistics* LXXVI(1):103–118, February 1994.

a. Graph this function on the window [35, 80] by [0, 240].
[*Hint:* On some calculators, enter
$y_1 = (106.2e^{0.0063x})(x \le 58.2) + (850.4e^{0.000614x^2-0.0652x})(x > 58.2).$]

b. Find $f(35)$ and $f'(35)$ and interpret these numbers. [*Hint:* Use NDERIV or dy/dx.]
c. Find $f(80)$ and $f'(80)$ and interpret these numbers.

EXPONENTIAL AND LOGARITHMIC FUNCTIONS TO OTHER BASES

The rules for differentiating exponential functions with (positive) base a are shown below:

Derivatives of a^x and $a^{f(x)}$

$$\frac{d}{dx}a^x = (\ln a)a^x$$

$$\frac{d}{dx}a^{f(x)} = (\ln a)a^{f(x)}f'(x)$$ For a differentiable function f

For example,

$$\frac{d}{dx}2^x = (\ln 2)2^x$$

$$\frac{d}{dx}5^{3x^2+1} = (\ln 5)5^{3x^2+1}(6x) = 6(\ln 5)x\,5^{3x^2+1}$$

These formulas are more complicated than the corresponding base e formulas (page 309), which is why e is called the "natural" base: it makes the derivative formulas simplest. These formulas reduce to the natural (base e) formulas if $a = e$.

Use the formulas above to find the derivative of each function.

79. a. $f(x) = 10^x$ **b.** $f(x) = 3^{x^2+1}$ **c.** $f(x) = 2^{3x}$
d. $f(x) = 5^{3x^2}$ **e.** $f(x) = 2^{4-x}$

80. a. $f(x) = 5^x$ **b.** $f(x) = 2^{x^2-1}$ **c.** $f(x) = 3^{4x}$
d. $f(x) = 9^{5x^2}$ **e.** $f(x) = 10^{1-x}$

The rules for differentiating logarithmic functions with (positive) base a are as follows:

Derivatives of $\log_a x$ and $\log_a f(x)$

$$\frac{d}{dx}\log_a x = \frac{1}{(\ln a)x}$$

$$\frac{d}{dx}\log_a f(x) = \frac{f'(x)}{(\ln a)f(x)} \qquad \text{For a differentiable function } f > 0$$

For example,

$$\frac{d}{dx}\log_5 x = \frac{1}{(\ln 5)x}$$

$$\frac{d}{dx}\log_2 (x^3 + 1) = \frac{3x^2}{(\ln 2)(x^3 + 1)}$$

These formulas are more complicated than the corresponding base e formulas (page 309), and again the simplicity of the base e formulas is why e is called the "natural" base. As before, these formulas reduce to the natural (base e) formulas if $a = e$.

Use the formulas above to find the derivative of each function.

81. a. $\log_2 x$ **b.** $\log_{10}(x^2 - 1)$ **c.** $\log_3 (x^4 - 2x)$ **82. a.** $\log_3 x$ **b.** $\log_2(x^2 + 1)$ **c.** $\log_{10}(x^3 - 4x)$

4.4 TWO APPLICATIONS TO ECONOMICS: RELATIVE RATES AND ELASTICITY OF DEMAND

Introduction

In this section we define *relative rates of change* and see how they are used in economics. (*Relative* rates are not the same as the *related* rates discussed in Section 3.6.) We then define the very important economic concept of *elasticity of demand*.

Relative versus Absolute Rates

The derivative of a function gives its rate of change. For example, if $f(t)$ is the cost of a pair of shoes at time t years, then $f'(t)$ is the rate of change of cost (in dollars per year). That is, $f' = 3$ would mean that the price of shoes is increasing at the rate of $3 per year. Similarly, if $g(t)$ is the price of a new automobile at time t years, then $g' = 300$ would mean that automobile prices are increasing at the rate of $300 per year.

Does this mean that car prices are rising 100 times as fast as shoe prices? In absolute terms, yes. However, this does not take into account the enormous price difference between automobiles and shoes.

Relative Rates of Change

If shoe prices are increasing at the rate of $3 per year, and if the current price of a pair of shoes is $60, the *relative* rate of increase is $\frac{3}{60} = \frac{1}{20} = 0.05$, which means that shoe prices are increasing at the relative rate of 5% per year. Similarly, if the price of an average automobile is $15,000, then an increase of $300 relative to this price is $\frac{300}{15,000} = \frac{1}{50} = 0.02$, for a *relative* rate of 2% per year. Therefore, in a *relative* sense (that is, as a fraction of the current price), car prices are increasing *less* rapidly than shoe prices.

In general, if $f(t)$ is the price of an item at time t, then the rate of change is $f'(t)$, and the relative rate of change is $f'(t)/f(t)$, the derivative divided by the function. We will sometimes call the derivative $f'(x)$ the "absolute" rate of change to distinguish it from the relative rate of change $f'(x)/f(x)$.

Relative rates are often more meaningful than absolute rates. For example, it is easier to grasp the fact that the gross domestic product is growing at the relative rate of 4% a year than that it is growing at the absolute rate of $300,000,000,000 per year.

The expression $f'(x)/f(x)$ is the derivative of the natural logarithm of $f(x)$:

$$\frac{d}{dx} \ln f(x) = \frac{f'(x)}{f(x)}$$

This provides an alternative expression for the relative rate of change, in terms of logarithms.

Relative Rate of Change

$$\left(\begin{array}{c} \text{Relative rate of} \\ \text{change of } f(t) \end{array} \right) = \frac{d}{dt} \ln f(t) = \frac{f'(t)}{f(t)} \qquad \text{For a differentiable function } f > 0$$

We use the variable t since it often stands for time. The middle expression is sometimes called the *logarithmic derivative*, since it is found by first taking the logarithm and then the derivative.

The relative rate of change, being a ratio or a percent, does not depend on the units of the function (whether dollars or euros, pounds or kilos). Therefore, relative rates can be compared between different products, and even between different nations. This is in contrast to absolute rates of change (that is, derivatives), which *do* depend on the units (for example, dollars per year).

EXAMPLE 1

FINDING A RELATIVE RATE OF CHANGE

If the gross domestic product t years from now is predicted to be $G(t) = 8.2e^{\sqrt{t}}$ trillion dollars, find the relative rate of change 25 years from now.

We give two solutions, showing the use of both formulas.

Solution $\left[\text{using the } \dfrac{d}{dt} \ln f(t) \text{ formula} \right]$

$\ln G(t) = \ln 8.2e^{\sqrt{t}}$	Taking natural logs
$= \ln 8.2 + \ln e^{\sqrt{t}}$	Log of a product is the sum of the logs
$= \ln 8.2 + \sqrt{t}$	$\ln e^{\sqrt{t}} = \sqrt{t}$ by Property 3 of logs (see inside back cover)
$= \ln 8.2 + t^{1/2}$	In exponent form

Then we differentiate:

$$\frac{d}{dt}(\ln 8.2 + t^{1/2}) = 0 + \frac{1}{2}t^{-1/2} = \frac{1}{2}t^{-1/2} \qquad \text{\small $\ln 8.2$ is a constant, so its derivative is zero}$$

Finally, we evaluate at the given time $t = 25$:

$$\frac{1}{2}(25)^{-1/2} = \frac{1}{2}\frac{1}{\sqrt{25}} = \frac{1}{2}\frac{1}{5} = \frac{1}{10} = 0.10 \qquad \text{\small $\frac{1}{2}t^{-1/2}$ evaluated at $t = 25$}$$

Therefore, in 25 years the gross domestic product will be increasing at the relative rate of 0.10, or 10%, per year.

Alternative Solution $\left[\text{using the } \dfrac{f'(t)}{f(t)} \text{ formula} \right]$

$$G(t) = 8.2e^{\sqrt{t}} = 8.2e^{t^{1/2}} \qquad \text{\small Writing $G(t)$ with fractional exponents}$$

$$G'(t) = 8.2e^{t^{1/2}}\left(\frac{1}{2}t^{-1/2}\right) \qquad \text{\small Differentiating}$$

⌣
↑
└─── Derivative of the exponent

Therefore, the relative rate of change $\dfrac{G'(t)}{G(t)}$ is

$$\frac{G'(t)}{G(t)} = \frac{8.2e^{t^{1/2}}\left(\frac{1}{2}t^{-1/2}\right)}{8.2e^{t^{1/2}}} = \frac{1}{2}t^{-1/2}$$ Same result as with the first formula

At $t = 25$,

$$\frac{1}{2}(25)^{-1/2} = \frac{1}{2}\frac{1}{\sqrt{25}} = \frac{1}{2}\frac{1}{5} = \frac{1}{10} = 0.10$$ Again the same

Therefore, the relative growth rate is 10%, just as we found before.

Both formulas give the same answer, so you should use the one that is easier to apply in your particular problem. The $\dfrac{d}{dx}\ln f(t)$ formula sometimes allows simplification before the differentiation, whereas the $\dfrac{f'(t)}{f(t)}$ formula often involves simplification afterward.

Practice Problem 1

An investor estimates that if a piece of land is held for t years, it will be worth $f(t) = 300 + t^2$ thousand dollars. Find the relative rate of change at time $t = 10$ years. $\left[\textit{Hint:} \text{ Use the } \dfrac{f'(t)}{f(t)} \text{ formula.} \right]$

➤ Solution on page 331

Graphing Calculator Exploration

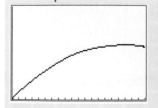

Enter the function from Practice Problem 1 into a graphing calculator as $y_1 = 300 + x^2$, and then turn off the function so that it will not graph. Then graph $y_2 = \dfrac{d}{dx}\ln y_1$ and $y_3 = \dfrac{y_1'}{y_1}$ (using NDERIV) together on the window [0, 20] by [0, 0.1]. Why do you get only one curve for the two functions?

Elasticity of Demand

Farmers are aware of the paradox that an abundant harvest usually brings *lower* total revenue than a poor harvest. The reason is simply that the larger quantities in an abundant harvest result in lower prices, which in turn cause increased demand, but the demand does *not* increase enough to compensate for the lower prices.

Revenue is price times quantity, and when one of these quantities increases, the other generally decreases. The question is whether the increase in one is enough to compensate for the decrease in the other. The concept of *elasticity of demand* was devised to answer this question.

Intuitively, elasticity of demand is a measure of how *responsive* demand is to price changes. Think of "elastic" as meaning "very responsive." If demand is elastic, a small price cut will bring a large increase in demand, so total revenue will rise. On the other hand, if demand is *in*elastic, a price cut will bring only a slight increase in demand, so total revenue will fall.

In general, *elastic* demand means that consumers will purchase *significantly* more or less in response to price changes. *Inelastic* demand means that consumers will buy only *slightly* more or less in response to price changes. (This is the cause of the farmers' difficulties: demand for farm products is inelastic.)

Demand Functions

In general, if the price of an item rises, the demand will fall, and vice versa. If the relationship between the price p of an item and the quantity x that will be sold at that price can be expressed with x as a function of p, $x = D(p)$, that function is called the *demand function*.

Demand Function

The demand function

$$x = D(p)$$

gives the quantity x of an item that will be demanded by consumers at price p.

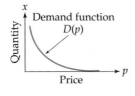

Law of downward-sloping demand

Since, in general, demand falls as prices rise, the slope of the demand function is negative, as shown on the left. This is known as the *law of downward-sloping demand*.

Calculating Elasticity of Demand

For a demand function $D(p)$, let us calculate the relative rate of change of demand divided by the relative rate of change of price. Using the derivative-of-the-logarithm formula,

$$\frac{\left(\begin{array}{c}\text{Relative rate of}\\\text{change of demand}\end{array}\right)}{\left(\begin{array}{c}\text{Relative rate of}\\\text{change of price}\end{array}\right)} = \frac{\dfrac{d}{dp}\ln D(p)}{\dfrac{d}{dp}\ln p} = \frac{\dfrac{D'(p)}{D(p)}}{\dfrac{1}{p}} = \underbrace{\frac{pD'(p)}{D(p)}}_{\text{Simplified}}$$

Because most demand functions are downward-sloping, the derivative $D'(p)$ is generally negative. Economists prefer to work with positive numbers, so the *elasticity of demand* is taken to be the negative of this quantity (in order to make it positive).*

Elasticity of Demand

For a demand function $D(p)$, the elasticity of demand is

$$E(p) = \frac{-p \cdot D'(p)}{D(p)}$$

Demand is *elastic* if $E(p) > 1$ and *inelastic* if $E(p) < 1$.

Elasticity, being composed of *relative* rates of change, does not depend on the units of the demand function. Therefore, elasticities can be compared between different products, and even between different countries.

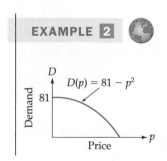

D

$D(p) = 81 - p^2$

81

Demand

Price

p

EXAMPLE 2

FINDING ELASTICITY OF DEMAND FOR COMMUTER BUS SERVICE

A bus line estimates the demand function for its daily commuter tickets to be $D(p) = 81 - p^2$ (in thousands of tickets), where p is the price in dollars $(0 \le p \le 9)$. Find the elasticity of demand when the price is:

a. $3   **b.** $6

*Some economists omit the negative sign.

Solution

$$E(p) = \frac{-pD'(p)}{D(p)} \qquad \text{Definition of elasticity}$$

$$= \frac{-p(-2p)}{81 - p^2} \qquad \begin{array}{l}\text{Substituting} \quad D(p) = 81 - p^2 \\ \text{so} \quad D'(p) = 2p\end{array}$$

$$= \frac{2p^2}{81 - p^2} \qquad \text{Simplifying}$$

a. Evaluating at $p = 3$ gives

$$E(3) = \frac{2(3)^2}{81 - (3)^2} = \frac{18}{81 - 9} = \frac{18}{72} = \frac{1}{4} \qquad E(p) = \frac{2p^2}{81 - p^2} \text{ with } p = 3$$

Interpretation: The elasticity is less than 1, so demand for tickets is *in-elastic* at a price of $3. This means that a small price change (up or down from this level) will cause only a *slight* change in demand. More precisely, elasticity of $\frac{1}{4}$ means that a 1% price change will cause only about a $\frac{1}{4}$% change in demand.

b. At the price of $6, the elasticity of demand is

$$E(6) = \frac{2(6)^2}{81 - (6)^2} = \frac{72}{81 - 36} = \frac{8}{5} = 1.6 \qquad E(p) = \frac{2p^2}{81 - p^2} \text{ with } p = 6$$

Interpretation: The elasticity is greater than 1, so demand is *elastic* at a price of $6. This means that a small change in price (up or down from this level) will cause a relatively *large* change in demand. In particular, an elasticity of 1.6 means that a price change of 1% will cause about a 1.6% change in demand.

The changes in demand are, of course, in the opposite direction from the changes in price. That is, if prices are *raised* by 1% (from the $6 level), demand will *fall* by 1.6%, whereas if prices are *lowered* by 1%, demand will *rise* by 1.6%. In the future we will assume that the *direction* of the change is clear, and say simply that a 1% change in price will cause about a 1.6% change in demand.

Practice Problem 2

For the demand function $D(p) = 90 - p$, find the elasticity of demand $E(p)$ and evaluate it at $p = 30$ and $p = 75$. (Be sure to complete this Practice Problem, as the results will be used shortly.)

➤ Solution on page 331

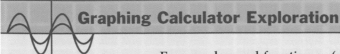

Graphing Calculator Exploration

For any demand function y_1 (written in terms of x), define y_2 as the elasticity function for y_1 by defining $y_2 = -x \cdot y_1'/y_1$ (using NDERIV). Try entering the demand function from Example 2 or Practice Problem 2 in y_1 and then evaluating y_2 at appropriate numbers to check the answers found there.

Using Elasticity to Increase Revenue

In Example 2 we found that at a price of $3, demand is inelastic $(E = \frac{1}{4} < 1)$, and so demand responds only *weakly* to price changes. Therefore, to increase revenue the company should *raise* prices, since the higher prices will drive away only a relatively small number of customers. On the other hand, at a price of $6, demand is elastic $(E = 1.6 > 1)$, and so demand is very responsive to price changes. In this case, to increase revenue the company should *lower* prices, since this will attract more than enough new customers to compensate for the price decrease. In general:

Using Elasticity to Increase Revenue

To increase revenue:

> *Raise* prices if demand is *inelastic* $(E < 1)$.
>
> *Lower* prices if demand is *elastic* $(E > 1)$.

This statement shows why elasticity of demand is important to any company that cuts prices in an attempt to boost revenue, or to any utility that raises prices in order to increase revenue. Elasticity shows whether the strategy will succeed or fail.

Spreadsheet Exploration

The following spreadsheet* is based on the demand function $D(p) = 90 - p$ from Practice Problem 2 on page 327, and shows the

* See the Preface for how to obtain this and other Excel spreadsheets.

elasticity of demand along with the demand and revenue for various prices.

D6	▼	=	=$A6/(90-$A6)
	A	**B**	**C**	**D**
1	Price	Demand	Revenue	Elasticity
2				
3	30	60	1800	0.500
4	35	55	1925	0.636
5	40	50	2000	0.800
6	45	45	2025	1.000
7	50	40	2000	1.250
8	55	35	1925	1.571
9	60	30	1800	2.000
10	65	25	1625	2.600
11	70	20	1400	3.500
12	75	15	1125	5.000

Notice that where elasticity is less than 1 (the first three rows), revenue rises as the price increases, but where elasticity is greater than 1 (the last six rows), revenue falls as the price increases.

The borderline case, $E = 1$ (called *unitary elasticity*), is where revenue cannot be raised, which will be the case if revenue is at its maximum. Therefore, elasticity must be unitary when revenue is maximized.

At maximum revenue, elasticity of demand must equal 1.

We could use this fact as a basis for a new method for maximizing revenue, but instead, we will stick with our earlier (and easier) method of maximizing functions by finding critical numbers.

Practice Problem 3

According to the preceding spreadsheet, at what price is revenue maximized, and what is the elasticity of demand at that price?

➤ Solution on page 331

Verification of the Relationship Between Elasticity and Revenue

We may verify the relationship between elasticity of demand and revenue as follows: revenue is price p times quantity x.

$$R = px = p \cdot D(p) \qquad \text{Using } x = D(p)$$

Differentiating will show how revenue responds to price changes.

$$R'(p) = D(p) + pD'(p) \qquad \text{Using the Product Rule on } p \cdot D(p)$$

$$= D(p)\left[1 + \frac{pD'(p)}{D(p)}\right] \qquad \text{Factoring out } D(p)$$

$$= D(p)\left[1 - \frac{-pD'(p)}{D(p)}\right] \qquad \text{Replacing the plus sign by two minus signs}$$

This is the definition of elasticity $E(p)$

$$= D(p)[1 - E(p)] \qquad \text{Replacing } \frac{-pD'(p)}{D(p)} \text{ by } E(p)$$

If demand is *elastic*, $E > 1$, then the quantity in brackets is negative, and so the derivative $R'(p)$ is negative, showing that revenue *decreases* as price increases. Therefore, to increase revenue, one should *lower* prices. On the other hand, if the demand is *inelastic*, $E < 1$, then the quantity in brackets is positive, and so the derivative $R'(p)$ is positive, showing that revenue *increases* as price increases. In this case, to increase revenue, one should *raise* prices. This proves the statements on page 328 in the box headed "Using Elasticity to Increase Revenue."

Elasticity Is Not the Same as Slope

Do not confuse elasticity of demand with the slope of the demand curve. The two ideas are quite different. For example, in Practice Problem 2 on page 327 we found that the linear demand function $D(p) = 90 - p$ (which has slope -1 all along it) has elasticity $\frac{1}{2}$ at one point and elasticity 5 at another.

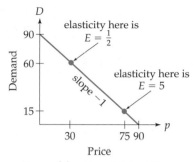

Demand function $D(p) = 90 - p$,
showing the elasticities found in Practice Problem 2

To understand why the elasticity changes, notice that the upper point represents a high demand (vertical axis) and a low price (horizontal axis), compared with the lower point, which represents a low demand and a high price. The demand function having slope -1 means that each \$1 price increase lowers demand by 1 unit, but losing one low-priced sale out of many is less important than losing one high-priced sale out of a few. That is, to see how revenue changes, we must look not at the absolute rates of change of quantity and price, but at their *relative* rates of change, which is what elasticity is all about.

> ➤ **Solutions to Practice Problems**

1. $\dfrac{f'(t)}{f(t)} = \dfrac{2t}{300 + t^2}$

At $t = 10$, $\dfrac{2 \cdot 10}{300 + 10^2} = \dfrac{20}{400} = \dfrac{1}{20} = 0.05$ or 5%

2. $E(p) = \dfrac{-pD'(p)}{D(p)} = \dfrac{-p(-1)}{90 - p} = \dfrac{p}{90 - p}$

At $p = 30$, $E(30) = \dfrac{30}{90 - 30} = \dfrac{30}{60} = \dfrac{1}{2} = 0.5$ (demand is inelastic)

At $p = 75$, $E(75) = \dfrac{75}{90 - 75} = \dfrac{75}{15} = 5$ (demand is elastic)

3. At $p = 45$. Elasticity is 1.

4.4 Exercises

EXERCISES ON RELATIVE RATES

For each function:

a. Find the relative rate of change.

b. Evaluate the relative rate of change at the given value(s) of t.

1. $f(t) = t^2$, $t = 1$ and $t = 10$

2. $f(t) = t^3$, $t = 1$ and $t = 10$

3. $f(t) = 100e^{0.2t}$, $t = 5$

4. $f(t) = 100e^{-0.5t}$, $t = 4$

5. $f(t) = e^{t^2}$, $t = 10$ **6.** $f(t) = e^{t^3}$, $t = 5$

7. $f(t) = e^{-t^2}$, $t = 10$ **8.** $f(t) = e^{-t^3}$, $t = 5$

9. $f(t) = 25\sqrt{t - 1}$, $t = 6$

10. $f(t) = 100\sqrt[3]{t + 2}$, $t = 8$

APPLIED EXERCISES ON RELATIVE RATES

11–12. ECONOMICS: National Debt If the national debt of a country (in trillions of dollars) t years from now is given by the indicated function, find the relative rate of change of the debt 10 years from now.

11. $N(t) = 0.5 + 1.1e^{0.01t}$

12. $N(t) = 0.4 + 1.2e^{0.01t}$

 13–14. GENERAL: Population The population (in millions) of a city t years from now is given by the indicated function.

a. Find the relative rate of change of the population 8 years from now.

b. Will the relative rate of change ever reach 1.5%?

13. $P(t) = 4 + 1.3e^{0.04t}$ **14.** $P(t) = 6 + 1.7e^{0.05t}$

EXERCISES ON ELASTICITY OF DEMAND

For each demand function $D(p)$:

a. Find the elasticity of demand $E(p)$.

b. Determine whether the demand is elastic, inelastic, or unitary elastic at the given price p.

15. $D(p) = 200 - 5p$, $p = 10$

16. $D(p) = 60 - 8p$, $p = 5$

17. $D(p) = 300 - p^2$, $p = 10$

18. $D(p) = 100 - p^2$, $p = 5$

19. $D(p) = \dfrac{300}{p}$, $p = 4$

20. $D(p) = \dfrac{500}{p}$, $p = 2$

21. $D(p) = \sqrt{175 - 3p}$, $p = 50$

22. $D(p) = \sqrt{100 - 2p}$, $p = 20$

23. $D(p) = \dfrac{100}{p^2}$, $p = 40$

24. $D(p) = \dfrac{600}{p^3}$, $p = 25$

25. $D(p) = 4000e^{-0.01p}$, $p = 200$

26. $D(p) = 6000e^{-0.05p}$, $p = 100$

APPLIED EXERCISES ON ELASTICITY OF DEMAND

27. **AUTOMOBILE SALES** An automobile dealer is selling cars at a price of $12,000. The demand function is $D(p) = 2(15 - 0.001p)^2$, where p is the price of a car. Should the dealer raise or lower the price to increase revenue?

28. **LIQUOR SALES** A liquor distributor wants to increase its revenues by discounting its best-selling liquor. If the demand function for this liquor is $D(p) = 60 - 3p$, where p is the price per bottle, and if the current price is $15, will the discount succeed?

29. **CITY BUS REVENUES** The manager of a city bus line estimates the demand function to be $D(p) = 150,000\sqrt{1.75 - p}$, where p is the fare in dollars. The bus line currently charges a fare of $1.25, and it plans to raise the fare to increase its revenues. Will this strategy succeed?

30. **NEWSPAPER SALES** The demand function for a newspaper is $D(p) = 80,000\sqrt{75 - p}$, where p is the price in cents. The publisher currently charges 50 cents, and it plans to raise the price to increase revenues. Will this strategy succeed?

31. **ELECTRICITY RATES** An electrical utility asks the Federal Regulatory Commission for permission to raise rates to increase revenues. The utility's demand function is

$$D(p) = \frac{120}{10 + p}$$

where p is the price (in cents) of a kilowatt-hour of electricity. If the utility currently charges 6 cents per kilowatt-hour, should the commission grant the request?

32. **OIL PRICES** A Middle Eastern oil-producing country estimates that the demand for oil (in millions of barrels) is $D(p) = 28e^{-0.04p}$, where p is the price of a barrel of oil. To raise its revenues, should it raise or lower its price from its current level of $20 per barrel?

33. **OIL PRICES** A European oil-producing country estimates that the demand for its oil (in millions of barrels) is $D(p) = 41e^{-0.06p}$, where

p is the price of a barrel of oil. To raise its revenues, should it raise or lower its price from its current level of $20 per barrel?

34–35. LIQUOR AND BEER The demand functions for distilled spirits and for beer are given below, where p is the retail price and $D(p)$ is the demand in gallons per capita.* For each demand function, find the elasticity of demand for any price p. [*Note:* You will find, in each case, that demand is inelastic. This means that taxation, which acts like a price increase, is an ineffective way of discouraging liquor consumption, but is an effective way of raising revenue.]

34. $D(p) = 3.509p^{-0.859}$ (for distilled spirits)

35. $D(p) = 7.881p^{-0.112}$ (for beer)

36. **CONSTANT ELASTICITY**

   a. Show that for a demand function of the form $D(p) = \dfrac{c}{p^n}$, where c and n are positive constants, the elasticity is constant.
   b. What type of demand function has elasticity equal to 1 for every value of p?

37. **LINEAR ELASTICITY** Show that for a demand function of the form $D(p) = ae^{-cp}$, where a and c are positive constants, the elasticity of demand is $E(p) = cp$.

38–39. ELASTICITY OF SUPPLY A supply function $S(p)$ gives the total amount of a product that producers are willing to supply at a given price p. The *elasticity of supply* is defined as

$$E_s(p) = \frac{p \cdot S'(p)}{S(p)}$$

Elasticity of supply measures the relative increase in supply resulting from a small relative increase in

* Stanley Ornstein and Dominique Hanssens, "Alcohol Control Laws and the Consumption of Distilled Spirits and Beer," *Journal of Consumer Research* 12:200–213, September 1985. Variables in this study other than price have been ignored.

price. It is less useful than elasticity of demand, however, since it is not related to total revenue.

38. Use the preceding formula to find the elasticity of supply for a supply function of the form $S(p) = ae^{cp}$, where a and c are positive constants.

39. Use the preceding formula to find the elasticity of supply for a supply function of the form $S(p) = ap^n$, where a and n are positive constants.

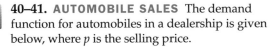 **40–41. AUTOMOBILE SALES** The demand function for automobiles in a dealership is given below, where p is the selling price.

a. Use the method described in the Graphing Calculator Exploration on page 328 to find the elasticity of demand at a price of $12,000.

b. Should the dealer raise or lower the price from this level to increase revenue?

c. Find the price at which elasticity equals 1. [*Hint:* Use INTERSECT.]

40. $D(p) = 3\sqrt{20 - 0.001p}$

41. $D(p) = \dfrac{200}{8 + e^{0.0001p}}$

Chapter Summary with Hints and Suggestions

Reading the text and doing the exercises in this chapter have helped you to master the following skills, which are listed by section (in case you need to review them) and are keyed to particular Review Exercises. Answers for all Review Exercises are given at the back of the book, and full solutions can be found in the Student Solutions Manual.

4.1 Exponential Functions

- Find the value of money invested at compound interest. *(Review Exercise 1.)*

$$P(1 + r)^n \qquad Pe^{rn}$$

- Determine which of two banks gives better interest. *(Review Exercise 2.)*

- Depreciate an asset. *(Review Exercise 3.)*

$$P(1 + r)^n \quad \text{(for depreciation, } r \text{ is negative)}$$

- Determine which of two drugs provides more medication. *(Review Exercise 4.)*

- Predict the world's largest city. *(Review Exercise 5.)*

- Predict computer memory capacity. *(Review Exercise 6.)*

4.2 Logarithmic Functions

- Find doubling (and other) times for compound interest. *(Review Exercises 7–8.)*

$$\ln M^N = N \cdot \ln M \qquad \ln e^x = x$$

- Date a fossil by potassium 40. *(Review Exercises 9–10.)*

- Predict the spread of information, or oil demand. *(Review Exercises 11–12.)*

- Use and analyze the accuracy of the "rule of 72" for compound interest. *(Review Exercises 13–14.)*

- Find the k-multiple time for an investment. (*Review Exercises 15–16.*)

- Find the time required for a given percentage change in a bank deposit or in advertising results. (*Review Exercises 17–18.*)

4.3 Differentiation of Logarithmic and Exponential Functions

- Find the derivative of a logarithmic or exponential function. (*Review Exercises 19–34.*)

$$\frac{d}{dx}\ln x = \frac{1}{x} \qquad \frac{d}{dx}e^x = e^x \qquad \frac{d}{dx}\ln f = \frac{f'}{f}$$

$$\frac{d}{dx}e^f = e^f \cdot f' \qquad \frac{d}{dx}e^{kx} = ke^{kx}$$

- Graph an exponential or a logarithmic function. (*Review Exercises 35–36.*)

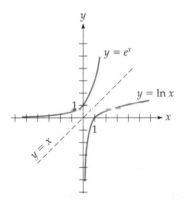

- Find the rate of change of sales, amount of medication, learning, temperature, or diffusion of information. (*Review Exercises 37–41.*)

- Find when a company maximizes its present value. (*Review Exercise 42.*)

- Maximize a company's revenue. (*Review Exercises 43–44.*)

- Maximize consumer expenditure for a product. (*Review Exercise 45.*)

- Graph an exponential or logarithmic function. (*Review Exercises 46–47.*)

- Maximize consumer expenditure or maximize revenue. (*Review Exercises 48–49.*)

4.4 Two Applications to Economics: Relative Rates and Elasticity of Demand

- Find the relative rate of change of a country's gross domestic product. (*Review Exercises 50–51.*)

$$\begin{pmatrix}\text{Relative}\\\text{rate}\end{pmatrix} = \frac{d}{dt}\ln f = \frac{f'}{f}$$

- Find the elasticity of demand for a product, and its consequences. (*Review Exercises 52–54.*)

$$\begin{pmatrix}\text{Elasticity}\\\text{of demand}\end{pmatrix} = \frac{-p \cdot D'(p)}{D(p)}$$

- Find the relative rate of change of population. (*Review Exercise 55.*)

- Find the elasticity of demand for a product, how it affects revenue, and what price gives unitary elasticity. (*Review Exercise 56.*)

Hints and Suggestions

- (*Overview*) Exponential and logarithmic functions should be thought of as just other types of functions, like polynomials, but having their own differentiation rules. In fact, in a sense they are more "natural" than polynomials because they give natural growth rates.

- A graphing calculator helps by drawing graphs of exponential and logarithmic functions, and finding intersection points (for example, where one population overtakes another). It is also useful for checking derivatives and graphically verifying maximum and minimum values.

- Interest rates are always *annual* rates unless stated otherwise. To use $P(1 + r)^n$ for finding value under compound interest, first determine the *compounding period*. Then r is the interest rate *per period*, and n is the *number of periods*.

■ When do you use Pe^{rn} and when do you use $P(1 + r)^n$? Use Pe^{rn} if the word "continuous" occurs, and use $P(1 + r)^n$ if it does not. $P(1 + r)^n$ with negative r gives depreciation. Dividing P by e^{rn} or $(1 + r)^n$ gives present value.

■ The Power Rule $\dfrac{d}{dx}x^n = nx^{n-1}$ is for differentiating a *variable to a constant power*, and $\dfrac{d}{dx}e^x = e^x$ is for differentiating the *constant e to a variable power*.

■ **Practice for Test:** Review Exercises 1, 3, 5, 7, 8, 11, 17, 19, 23, 25, 27, 31, 35, 37, 41, 45, 49, 51, 53, 55.

Review Exercises for Chapter 4

Practice test exercise numbers are in green.

4.1 Exponential Function

1. **GENERAL: Interest** Find the value of $10,000 invested for 8 years at 8% interest if the compounding is done:

    a. quarterly.   **b.** continuously.

2. **GENERAL: Interest** One bank offers 6% compounded quarterly and a second offers 5.98% compounded continuously. Where should you take your money?

3. **BUSINESS: Depreciation** An $800,000 computer loses 20% of its value each year.

    a. Give a formula for its value after t years.
    b. Find its value after 4 years.

4. **BIOMEDICAL: Drug Concentration** If the concentration of a drug in a patient's bloodstream is c (milligrams per milliliter), then t hours later the concentration will be $C(t) = ce^{-kt}$, where k is a constant (the "elimination constant"). Two drugs are being compared: drug A with initial concentration $c = 2$ and elimination constant $k = 0.2$, and drug B with initial concentration $c = 3$ and elimination constant $k = 0.25$. Which drug will have the greater concentration 4 hours later?

5. **GENERAL: Population** The largest city in the world is Tokyo, with São Paulo (Brazil) smaller but growing faster. According to the Census Bureau, x years after 2000 the population of Tokyo will be $27.5e^{0.0034x}$ and the population of São Paulo will be $17.4e^{0.011x}$ (both in millions). Graph both functions on a calculator with the window [0, 100] by [0, 50]. When will São Paulo overtake Tokyo as the world's largest city (assuming that these rates continue to hold)?

6. **GENERAL: Moore's Law of Computer Memory** The amount of information that can be stored on a computer chip can be measured in megabits (a "bit" is a binary digit, 0 or 1, and a "megabit" is a million bits). The first 1-megabit chips became available in 1987, and 4-megabit chips became available in 1990. This quadrupling of capacity every 3 years is expected to continue, so that chip capacity will be $C(t) = 4^{t/3}$ megabits where t is the number of years after 1987. Use this formula (known as Moore's Law, after Gordon Moore, a founder of the Intel Corporation) to predict chip capacity in the year 2008. [*Hint:* What value of t corresponds to 2008?]

4.2 Logarithmic Functions

7. **GENERAL: Interest** Find how soon an investment at 10% interest compounded semiannually will:

    a. double in value.   **b.** increase by 50%.

8. GENERAL: Interest Find how soon an investment at 7% interest compounded continuously will:

a. double in value.   **b.** increase by 50%.

9–10. GENERAL: Fossils In the following exercises, use the fact that the proportion of potassium 40 remaining after t million years is $e^{-0.00054t}$.

9. In 1984 in the Wind River Basin of Wyoming, scientists discovered a fossil of a small, three-toed horse, an ancestor of the modern horse. Estimate the age of this fossil if it contained 97.3% of its original potassium 40.

10. Estimate the age of a skull found in 1959 in Tanzania (dubbed "Nutcracker Man" because of its huge jawbone) that contained 99.9% of its original potassium 40.

11. SOCIAL SCIENCE: Diffusion of Information by Mass Media In a city of a million people, election results broadcast over radio and television will reach $N(t) = 1,000,000(1 - e^{-0.3t})$ people within t hours. Find when the news will have reached 500,000 people.

12. ECONOMICS: Oil Demand The demand for oil in the United States is increasing by 3% per year. Assuming that this rate continues, how soon will demand increase by 50%?

13–14. BUSINESS: Rule of 72 If a sum is invested at interest rate r compounded continuously, the doubling time (the time in which it will double in value) is found by solving the equation $Pe^{rt} = 2P$. The solution (by the usual method of canceling the P and taking logs) is $t = \dfrac{\ln 2}{r} \approx \dfrac{0.69}{r}$.

For *annual* compounding, the doubling time should be somewhat longer, and may be estimated by replacing 69 by 72.

Rule of 72
For r% interest compounded annually, the doubling time is approximately $\dfrac{72}{r}$ years.

For example, to estimate the doubling time for an investment at 8% compounded annually we would

divide 72 by 8, giving $\frac{72}{8} = 9$ years. The 72, however, is only a rough "upward adjustment" of 69, and the rule is most accurate for interest rates around 9%. For each interest rate:

a. Use the rule of 72 to estimate the doubling time for annual compounding.

b. Use the compound interest formula $P(1 + r)^n$ to find the actual doubling time for annual compounding.

13. 6%   (*See instructions above.*)

14. 1% (This shows that for interest rates very different from 9% the rule of 72 is less accurate.)

15. GENERAL: Interest Find a formula for the time required for an investment to grow to k times its original size if it grows at interest rate r compounded annually.

16. GENERAL: Interest Find a formula for the time required for an investment to grow to k times its original size if it grows at interest rate r compounded continuously.

 17. GENERAL: Interest If a bank offers 6.5% interest, in how many years will a deposit increase by 50% if the compounding is done:

a. quarterly?   **b.** continuously?

18. BUSINESS: Advertising After the opening of a new store has been advertised for t days, the proportion of people in a city who have seen the ad is $p(t) = 1 - e^{-0.032t}$. How long must the ad run to reach:

a. 30% of the people?
b. 40% of the people?

4.3 Differentiation of Logarithmic and Exponential Functions

Find the derivative of each function

19. $f(x) = \ln 2x$

20. $f(x) = \ln(x^2 - 1)^2$

21. $f(x) = \ln(1 - x)$

22. $f(x) = \ln \sqrt{x^2 + 1}$

23. $f(x) = \ln \sqrt[3]{x}$

24. $f(x) = \ln e^x$

25. $f(x) = \ln x^2$

26. $f(x) = x \ln x - x$

27. $f(x) = e^{-x^2}$

28. $f(x) = e^{1-x}$

29. $f(x) = \ln e^{x^2}$ **30.** $f(x) = e^{x^2 \ln x - x^2/2}$

31. $f(x) = 5x^2 + 2x \ln x + 1$

32. $f(x) = 2x^3 + 3x \ln x - 1$

33. $f(x) = 2x^3 - 3xe^{2x}$ **34.** $f(x) = 4x - 2x^2e^{2x}$

Graph each function, showing all relative extreme points and inflection points.

35. $f(x) = \ln(x^2 + 4)$ **36.** $f(x) = 16e^{-x^2/8}$

37. BUSINESS: Sales The weekly sales (in thousands) of a new product after x weeks of advertising is $S(x) = 2000 - 1500e^{-0.1x}$. Find the rate of change of sales after:

a. 1 week. **b.** 10 weeks.

38. BIOMEDICAL: Drug Dosage A patient receives an injection of 1.5 milligrams of a drug, and the amount remaining in the bloodstream t hours later is $A(t) = 1.5e^{-0.08t}$. Find the instantaneous rate of change of this amount:

a. immediately after the injection (time $t = 0$).
b. after 5 hours.

39. BEHAVIORAL SCIENCE: Learning In a test of short-term memory, the percent of subjects who remember an eight-digit number for at least t seconds is $P(t) = 100 - 200 \ln (t + 1)$. Find the rate of change of this percent after 5 seconds.

40. GENERAL: Temperature A thermos bottle that is chilled to 35 degrees Fahrenheit and then left in a 70-degree room will warm to a temperature of $T(t) = 70 - 35e^{-0.1t}$ degrees after t hours. Find the rate of change of the temperature:

a. at time $t = 0$. **b.** after 5 hours.

41. SOCIAL SCIENCE: Diffusion of Information by Mass Media The number of people in a town of 30,000 who have heard an important news bulletin within t hours of its first broadcast is $N(t) = 30,000(1 - e^{-0.3t})$. Find the instantaneous rate of change of the number of informed people after:

a. 1 hour. **b.** 8 hours.

42. BUSINESS: Maximizing Present Value A new company is growing so that its value t years from now will be $50t^2$ dollars. Therefore, its present value (at the rate of 8% compounded continuously) is

$$V(t) = 50t^2e^{-0.08t} \quad \text{dollars (for } t > 0)$$

Find the number of years that maximizes the present value.

43–44. BUSINESS: Maximizing Revenue The given function is a company's price function, where x is the quantity (in thousands) that will be sold at price p dollars.

a. Find the revenue function $R(x)$. [*Hint:* Revenue is price times quantity, $p \cdot x$.]
b. Find the quantity and price that will maximize revenue.

43. $p = 200e^{-0.25x}$ **44.** $p = 5 - \ln x$

45. ECONOMICS: Maximizing Consumer Expenditure Consumer demand for a commodity is estimated to be $D(p) = 25,000e^{-0.02p}$ units per month, where p is the selling price in dollars. Find the selling price that maximizes consumer expenditure.

46–47. Use your graphing calculator to graph each function on a window that includes all relative extreme points and inflection points, and give the coordinates of these points (rounded to two decimal places). [*Hint:* Use NDERIV once or twice with ZERO.] (Answers may vary depending on the graphing window chosen.)

46. $f(x) = \dfrac{x^4}{e^x}$ **47.** $f(x) = x^3 \ln |x|$

48. ECONOMICS: Consumer Expenditure If consumer demand for a commodity is $D(p) = 200e^{-0.0013p}$ (where p is the selling price in dollars), find the price that maximizes consumer expenditure.

49. BUSINESS: Maximizing Revenue A manufacturer finds that the price function for autofocus cameras is $p(x) = 20e^{-0.0077x}$, where p is the price in dollars at which quantity x (in thousands) will be sold. Find the quantity x that maximizes revenue.

4.4 Two Applications to Economics: Relative Rates and Elasticity of Demand

50–51. ECONOMICS: Relative Rate of Change
The gross domestic product of a developing country is forecast to be $G(t) = 5 + 2e^{0.01t}$ million dollars t years from now. Find the relative rate of change:

50. 20 years from now.   **51.** 10 years from now.

52. ECONOMICS: Elasticity of Demand
A South American country exports coffee and estimates the demand function to be $D(p) = 63 - 2p^2$. If the country wants to raise revenues to improve its balance of payments, should it raise or lower the price from the present level of $3 per pound?

53. ECONOMICS: Elasticity of Demand
A South African country exports gold and estimates the demand function to be
$D(p) = 400\sqrt{600 - p}$. If the country wants to raise revenues to improve its balance of payments, should it raise or lower the price from the present level of $350 per ounce?

54. ECONOMICS: Elasticity of Demand The demand function for cigarettes is of the form $D(p) = 1.2p^{-0.44}$, where p is the price of a pack of cigarettes and $D(p)$ is the demand measured in packs per day per capita.* Find the elasticity of demand. [*Note:* You will find that demand is inelastic. This means that taxation, which acts like a price increase, is an ineffective way of discouraging smoking, but is an effective way of raising revenue.]

55. GENERAL: Relative Rate of Change The population of a city x years from now is projected to be $P(x) = 3.25 + 0.04x + 0.002x^3$ million people (for $0 \le x \le 10$). Find the relative rate of change 9 years from now.

56. GENERAL: Elasticity of Demand A boat dealer finds that the demand function for outboard motor boats near a large lake is $D(p) = 200 - 20p + p^2 - 0.03p^3$ (for $0 \le p \le 15$), where p is the selling price in thousands of dollars.

a. Use a graphing calculator to find the elasticity of demand at a price of $10,000. [*Hint:* What value of p corresponds to $10,000?]
b. Should the dealer raise or lower the price from this level to increase revenue?
c. Find the price at which elasticity equals 1.

* Jeffrey E. Harris, "Taxing Tar and Nicotine," *American Economic Review* 70(3):300–311.

5 Integration and Its Applications

Application Preview

Cigarette Smoking

Most cigarettes today have filters to absorb some of the toxic material or "tar" in the smoke before it is inhaled. The tobacco near the filter acts like an additional filter, absorbing tar until it is itself smoked, at which time it releases all of its accumulated toxins. A typical cigarette consists of 8 centimeters of tobacco followed by a filter.

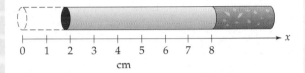

As the cigarette is smoked, tar is typically inhaled at the rate of $r(x) = 300e^{0.025x} - 240e^{0.02x}$ milligrams (mg) of tar per centimeter (cm) of tobacco, where x is the distance along the cigarette.* The amount of tar inhaled from any particular segment of the cigarette is the area under the graph of this function over that interval, which can be calculated by a process called *definite integration*, as explained in this chapter.

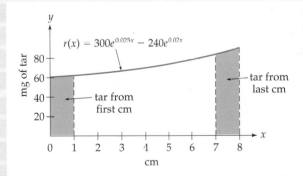

The results of integrating over the first and last centimeter of the cigarette are as follows (omitting the details—see Exercise 100 on page 388). The numbers represent milligrams of tar from the beginning and end of the cigarette.

* The actual rate depends upon the type of cigarette and the proportion of time that it is smoked rather than left to burn in the air. For further information, see Helen Marcus-Roberts and Maynard Thompson (eds), *Modules in Applied Mathematics*, vol 4, *Life Science Models*, Springer-Verlag, 1976, pp. 238–249.

Cigarette Smoking
(continued)

$$\begin{pmatrix} \text{Tar from} \\ \text{first cm} \end{pmatrix} = \int_0^1 (300e^{0.025x} - 240e^{0.2x})\, dx \approx 61$$

Integrate from 0 to 1 for the first centimeter

$$\begin{pmatrix} \text{Tar from} \\ \text{last cm} \end{pmatrix} = \int_7^8 (300e^{0.025x} - 240e^{0.02x})\, dx \approx 83$$

Integrate from 7 to 8 for the last centimeter

Notice that the last centimeter releases significantly more tar (about 36% more) than the first centimeter.

Moral: The Surgeon General has determined that smoking is hazardous to your health and the last puffs are 36% more hazardous than the first.

This is just one of the many applications of integration discussed in this chapter.

5.1 ANTIDERIVATIVES AND INDEFINITE INTEGRALS

Introduction

We have been studying differentiation and its uses. We now consider the reverse process, *anti*differentiation, which, for a given derivative, essentially recovers the original function. Antidifferentiation has many uses. For example, differentiation turns a cost function into a marginal cost function, and so antidifferentiation turns marginal cost back into cost. Later we will use antidifferentiation for other purposes, such as finding areas.

Antiderivatives and Indefinite Integrals

We begin with a simple example of antidifferentiation. Since the derivative of x^2 is $2x$, an *anti*derivative of $2x$ is x^2:

An antiderivative of $2x$ is x^2 Since the derivative of x^2 is $2x$

There are, however, other antiderivatives of $2x$. Each of the following is an antiderivative of $2x$:

$x^2 + 1$ $x^2 - 17$ $x^2 + e$ Since the derivative of each is $2x$

Clearly, we may add *any* constant to x^2 and the derivative will still be $2x$. Therefore, $x^2 + C$ is an antiderivative of $2x$ for *any* constant C.

Furthermore, it can be shown that there are no other antiderivatives of $2x$, and so the *most general* antiderivative of $2x$ is $x^2 + C$. The most general antiderivative of $2x$ is called the *indefinite integral* of $2x$, and is written with the $2x$ between an *integral sign* $\int$ and a *dx*:

$$\int 2x\, dx = x^2 + C$$

The indefinite integral of $2x$ is $x^2 + C$ (because the derivative of $x^2 + C$ is $2x$)

Integral sign $\quad$ Integrand $\quad$ Arbitrary constant

The function to be integrated (here $2x$) is called the *integrand*. The *dx* reminds us that the variable of integration is x. The constant C is called an *arbitrary constant* because it may take any value, positive, negative, or zero.

An indefinite integral (sometimes called simply an *integral*) can always be checked by differentiation: the derivative of the answer must equal the integrand (as is the case with $x^2 + C$ and $2x$).

Indefinite Integral

$$\int f(x)\, dx = g(x) + C \qquad \text{The integral of } f(x) \text{ is } g(x) + C$$

if and only if

$$g'(x) = f(x) \qquad \text{the derivative of } g(x) \text{ is } f(x)$$

Integration Rules

There are several "rules" that simplify integration. The first, which is one of the most useful rules in all of calculus, shows how to integrate x to a constant power. It comes from "reversing" the power rule for differentiation (see page 112).

Power Rule for Integration

$$\int x^n\, dx = \frac{1}{n+1} x^{n+1} + C \qquad (n \neq -1)$$

To integrate x to a power, add 1 to the power and multiply by 1 over the new power

EXAMPLE 1

FINDING AN INDEFINITE INTEGRAL

$$\int x^3 \, dx = \frac{1}{4} x^4 + C \qquad \text{Using the Power Rule with } n = 3$$

$$\uparrow n = 3 \quad \uparrow \quad \underset{n+1 = 4}{\underbrace{\qquad}}$$

$$\frac{1}{n+1} = \frac{1}{4}$$

Differentiating the answer $\frac{1}{4}x^4 + C$ immediately gives the integrand x^3, so the answer is correct.

Practice Problem 1 Find $\int x^2 \, dx$ and check your answer by differentiation.

➤ Solution on page 353

The proof of the Power Rule for Integration consists simply of differentiating the right-hand side.

$$\frac{d}{dx} \left(\frac{1}{n+1} x^{n+1} + C \right) = \frac{1}{n+1} (n+1) x^n = x^n$$

The power $n + 1$ ⟶   The power   Simplified
brought down   decreased by 1

Since the derivative is the integrand x^n, the Power Rule for Integration is correct.

To integrate functions like $\sqrt{x}$ and $\frac{1}{x^2}$, we first express them as powers.

EXAMPLE 2

EXPRESSING AS A POWER BEFORE INTEGRATING

$$\int \sqrt{x} \, dx = \int x^{1/2} \, dx = \frac{2}{3} x^{3/2} + C \qquad \begin{array}{l} \text{Using the Power Rule for} \\ \text{Integration with } \; n = 1/2 \end{array}$$

$$n = \frac{1}{2} \quad\qquad n+1 = \frac{1}{2} + 1 = \frac{3}{2}$$

$$\frac{1}{n+1} = \frac{1}{3/2} = \frac{2}{3}$$

Note that in the answer, the multiple $\frac{2}{3}$ and the exponent $\frac{3}{2}$ are reciprocals of each other. Can you see, from the Power Rule, why this will always be so?

| EXAMPLE 3 | **EXPRESSING AS A POWER BEFORE INTEGRATING** |

$$\int \frac{1}{x^2}\, dx = \int x^{-2}\, dx = \frac{1}{-1} x^{-1} + C = -x^{-1} + C$$

$$\underset{n = -2}{\underbrace{}}$$

$$n + 1 = -1$$

Simplified
answer

| Practice Problem 2 | Find $\displaystyle\int \frac{dx}{x^3}.$ $\left[\textit{Hint: } \text{Equivalent to } \displaystyle\int \frac{1}{x^3}\, dx.\right]$ |

➤ **Solution on page 353**

| EXAMPLE 4 | **INTEGRATING 1** |

$$\int 1\, dx = \int x^0\, dx = \frac{1}{1} x^1 + C = x + C \qquad \text{Using } x^0 = 1$$

$$\underset{n = 0}{\underbrace{}} \quad \underset{n + 1 = 1}{\underbrace{}}$$

This result is so useful that it should be memorized.

$$\int 1\, dx = x + C \qquad \text{The integral of 1 is } x + C$$

| EXAMPLE 5 | **INTEGRATING WITH OTHER VARIABLES** |

a. $\displaystyle\int t^3\, dt = \frac{1}{4} t^4 + C$

Using the Power Rule
($n = 3$) with dt since the
variable is t ("integrating
with respect to t")

b. $\displaystyle\int u^{-4}\, du = \frac{1}{-3} u^{-3} + C = -\frac{1}{3} u^{-3} + C$

Using the Power Rule
($n = -4$) with du since the
variable is u ("integrating
with respect to u")

Practice Problem 3   Find $\int z^{-1/2}\, dz.$ $= \frac{2}{1} z^{1/2} + C = 2z^{1/2} + C$ ➤ Solution on page 353

Notice what happens if we try to integrate x^{-1}:

$$\int x^{-1}\, dx = \frac{1}{0} x^0 + C \qquad \text{Undefined because of the } \tfrac{1}{0}$$

$n = -1$

$n + 1 = 0$

The Power Rule for Integration fails for the exponent -1 because it leads to the undefined expression $\frac{1}{0}$. For this reason, the Power Rule for Integration includes the restriction "$n \neq -1$." It can integrate any power of x *except* x^{-1}.

The Constant Multiple and Sum Rules for differentiation (pages 114 and 115) lead immediately to analogous rules for simplifying integrals. The first rule says that the sum of two functions may be integrated one at a time.

Sum Rule for Integration

$$\int [f(x) + g(x)]\, dx = \int f(x)\, dx + \int g(x)\, dx \qquad \begin{array}{l}\text{The integral of a}\\\text{sum is the sum of}\\\text{the integrals}\end{array}$$

EXAMPLE 6   **USING THE SUM RULE**

$$\int (x^2 + x^3)\, dx = \int x^2\, dx + \int x^3\, dx \qquad \begin{array}{l}\text{Using the Sum Rule to break}\\\text{the integral into two integrals}\end{array}$$

$$= \frac{1}{3} x^3 + \frac{1}{4} x^4 + C \qquad \begin{array}{l}\text{Using the Power Rule on each}\\\text{(one "}+ C\text{" is enough)}\end{array}$$

The second rule says that a constant may be moved across the integral sign.

Constant Multiple Rule for Integration

For any constant k,

$$\int k \cdot f(x)\, dx = k \int f(x)\, dx$$

The integral of a constant times a function is the constant times the integral of the function

EXAMPLE 7 **USING THE CONSTANT MULTIPLE RULE**

$$\int 6x^2\, dx = 6 \int x^2\, dx$$

Using the Constant Multiple Rule to move the 6 across the integral sign

$$= 6 \cdot \frac{1}{3}x^3 + C$$

Using the Power Rule with $n-2$

$$= 2x^3 + C$$

Simplifying

EXAMPLE 8 **INTEGRATING A CONSTANT**

$$\int 7\, dx = 7 \int 1\, dx = 7x + C$$

Moving the constant outside

The integral of 1 is x (plus C)

This leads to a very useful general rule. For any constant k,

Integral of a Constant

$$\int k\, dx = kx + C$$

The integral of a constant is the constant times x (plus C)

The Sum Rule can be extended to integrate the sum or difference of *any* number of terms, writing only one $+ C$ at the end (since any number of arbitrary constants can be added together to give just one).

EXAMPLE 9

INTEGRATING A SUM OF POWERS

$$\int (6x^2 - 3x^{-2} + 5)\, dx$$

$$= 6 \int x^2\, dx - 3 \int x^{-2}\, dx + 5 \int 1\, dx \qquad \text{Breaking up the integral and moving constants outside}$$

$$= 6 \cdot \frac{1}{3} x^3 - 3 \cdot \frac{1}{-1} x^{-1} + 5x + C \qquad \text{Integrating each separately}$$

From integrating the 1

$$= 2x^3 + 3x^{-1} + 5x + C \qquad \text{Simplifying}$$

EXAMPLE 10

REWRITING BEFORE INTEGRATING

$$\int \left(\frac{3\sqrt{x}}{2} - \frac{2}{\sqrt{x}} \right) dx = \int \left(\frac{3}{2} x^{1/2} - 2x^{-1/2} \right) dx \qquad \text{Writing as powers of } x$$

$$= \frac{3}{2} \cdot \frac{2}{3} x^{3/2} - 2 \cdot \frac{2}{1} x^{1/2} + C \qquad \text{Integrating each term separately}$$

$$= x^{3/2} - 4x^{1/2} + C \qquad \text{Simplifying}$$

Practice Problem 4   Find $\int \left(\sqrt[3]{w} - \frac{4}{w^3} \right) dw$. ▶ Solution on page 353

Some integrals are so simple that they can be integrated "at sight."

EXAMPLE 11

INTEGRATING "AT SIGHT"

a. $\int 4x^3\, dx = x^4 + C$
By remembering that $4x^3$ is the derivative of x^4

b. $\int 7x^6\, dx = x^7 + C$
By remembering that $7x^6$ is the derivative of x^7

Practice Problem 5

Integrate "at sight" by noticing that each integrand is of the form nx^{n-1} and integrating to x^n without working through the Power Rule.

a. $\int 5x^4\, dx$ b. $\int 3x^2\, dx$

$= x^5 + c$ $= x^3 + C$

➤ Solution on page 353

Algebraic Simplification of Integrals

Sometimes an integrand needs to be multiplied out or otherwise rewritten before it can be integrated.

EXAMPLE 12

EXPANDING BEFORE INTEGRATING

Find $\int x^2(x + 6)^2\, dx$.

Solution

$$\int x^2(x + 6)^2\, dx = \int x^2\underbrace{(x^2 + 12x + 36)}_{(x + 6)^2}\, dx \qquad \text{"Squaring out" the } (x + 6)^2$$

$$= \int (x^4 + 12x^3 + 36x^2)\, dx \qquad \text{Multiplying out}$$

$$= \frac{1}{5}x^5 + 12\cdot\frac{1}{4}x^4 + 36\cdot\frac{1}{3}x^3 + C \qquad \text{Integrating each term separately}$$

$$= \frac{1}{5}x^5 + 3x^4 + 12x^3 + C \qquad \text{Simplifying}$$

Practice Problem 6

Find $\int \dfrac{6t^2 - t}{t}\, dt. = 6t - t$ [*Hint:* First simplify the integrand.]

➤ Solution on page 353

Since differentiation turns a cost function into a marginal cost function, integration turns a marginal cost function back into a cost function. To evaluate the constant, however, we need the fixed costs.

EXAMPLE 13

RECOVERING COST FROM MARGINAL COST

A company's marginal cost function is $MC(x) = 6\sqrt{x}$ and the fixed cost is $1000. Find the cost function.

Solution

We integrate the marginal cost to find the cost function.

$$C(x) = \int MC(x)\,dx = \int 6\sqrt{x}\,dx = 6\int x^{1/2}\,dx \qquad \text{Integrating}$$

$$= 6\cdot\frac{2}{3}x^{3/2} + K = 4x^{3/2} + K \qquad \begin{array}{l}\text{From the Power Rule} \\ \text{(using } K \text{ to avoid} \\ \text{confusion with } C \text{ for} \\ \text{cost)}\end{array}$$

It remains to find the value of the constant K. The cost function evaluated at $x = 0$ always gives the fixed cost (because when nothing is produced, only the fixed cost remains):

$$\underbrace{C(0) = K}_{\text{Fixed cost}} \qquad \begin{array}{l}\text{Evaluating } C(x) = 4x^{3/2} + K \\ \text{at } x = 0 \text{ gives } K\end{array}$$

Therefore, K equals the fixed cost, which is given as 1000, so $K = 1000$. The completed cost function is obtained by replacing K by 1000.

$$C(x) = 4x^{3/2} + 1000 \qquad \begin{array}{l}C(x) = 4x^{3/2} + K \\ \text{with } K = 1000\end{array}$$

Finding the cost function involved two steps:

a. Integrating the marginal cost to find the cost function
b. Using the fixed cost to evaluate the arbitrary constant

Integration, being the reverse of differentiation, can recover *any* quantity from its rate of change (together with one fixed value). For example, given the present size of an economy and its rate of growth, we can integrate to find the size of the economy at any time in the future.

EXAMPLE 14

RECOVERING A QUANTITY FROM ITS RATE OF CHANGE

The gross domestic product (GDP) of a country is $78 billion and growing at the rate of $4.4t^{-1/3}$ billion dollars per year after t years. Find a formula for GDP after t years. Then use your formula to find the GDP after 8 years.

Solution

$$G(t) = \int 4.4t^{-1/3}\,dt = 4.4 \int t^{-1/3}\,dt \qquad \text{Integrating the rate of change}$$

$$= 4.4 \left(\frac{3}{2}\right) t^{2/3} + C = 6.6t^{2/3} + C \qquad \begin{array}{l}\text{Using the Power Rule and}\\ \text{then simplifying}\end{array}$$

As before, evaluating $G(t) = 6.6t^{2/3} + C$ at $t = 0$ gives $G(0) = C$, which shows that the constant C is the GDP at time $t = 0$, which is given as $78 billion. Therefore

$$G(t) = 6.6t^{2/3} + 78 \qquad\qquad G(t) = 6.6t^{2/3} + C \text{ with } C = 78$$

For the GDP after 8 years, we evaluate $G(t)$ at $t = 8$:

$$G(8) = 6.6\cdot 8^{2/3} + 78 = 6.6\underset{4}{(\underbrace{4})} + 78 = 104.4 \qquad \begin{array}{l}G(t) = 6.6t^{2/3} + 78\\ \text{at } t = 8\end{array}$$

Therefore, the GDP after 8 years is $104.4 billion.

Notice that the $+C$, which at first may have seemed like a pointless formality, enabled us to give the cost function its correct fixed cost in Example 13 and to give the GDP its correct current value in Example 14.

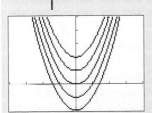

Graphing Calculator Exploration

A graphing calculator can show the *geometric* meaning of the arbitrary constant. In $\int 2x\,dx = x^2 + C$, the solution $x^2 + C$ is actually a whole collection of functions, a different curve for each value of C. Examine these curves for various values of C as follows:

a. Graph the five functions $x^2 - 2$, $x^2 - 1$, x^2, $x^2 + 1$, and $x^2 + 2$ on the graphing window $[-4, 4]$ by $[-2, 5]$. Use TRACE to see how the constant shifts the curve vertically.

b. Predict what the curves $x^2 + 4$ and $x^2 - 4$ (the solutions with $C = \pm 4$) would look like. Then check your predictions by graphing them.

c. Find the slope (using NDERIV or dy/dx) of several of the curves at a particular x-value and check that in each case the slope is twice the x-value. This verifies that the derivative of each curve is $2x$, so each is an integral of $2x$.

Integrals as Continuous Sums

We have seen that differentiation is a kind of "breaking down," changing total cost into marginal (per unit) cost. Integration, the reverse process, is therefore a kind of "adding up," recovering the total from its parts. For example, integration recovers the total cost from marginal (per unit) cost, and integration adds up the growth of an economy over several years to give its total size.

To put this another way, ordinary addition is for summing "chunks," such as $2 plus $3 equals $5, whereas integration is for summing *continuous* change, such as the slow but steady growth of an economy over time. In other words:

> Integration is continuous summation.

5.1 ## Section Summary

In this section we have discussed both the *techniques* of integration and the *meanings* or *uses* of integration.

On the *technical* side, we defined indefinite integration as the reverse process of differentiation. The Power Rule enables us to integrate powers.

$$\int x^n \, dx = \frac{1}{n+1} x^{n+1} + C \qquad n \neq -1$$

(Increasing the exponent by 1 should seem quite reasonable: Differentiation lowers the exponent by 1, so integration, the reverse process, should raise it by 1.) The Sum and Constant Multiple Rules extended integration to more complicated functions such as polynomials. Sometimes algebra is necessary to express a function in power form. *Reminder:* Don't forget the C.

As for the *uses* of integration, it recovers cost, revenue, and profit from their marginals, and in general it recovers any quantity from its rate of change (together with a particular value). The inverse relationship between integration and differentiation is shown in the following diagram.

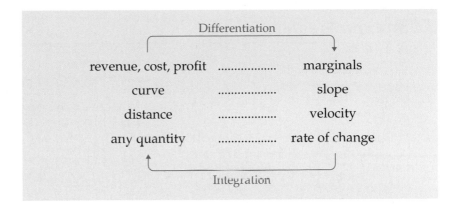

> ## Solutions to Practice Problems

1. $\displaystyle\int x^2\, dx = \frac{1}{3}x^3 + C$ $\qquad$ Check: $\dfrac{d}{dx}\left(\dfrac{1}{3}x^3 + C\right) = x^2$

2. $\displaystyle\int \frac{1}{x^3}\, dx = \int x^{-3}\, dx = -\tfrac{1}{2}x^{-2} + C$

3. $\displaystyle\int z^{-1/2}\, dz = \frac{1}{1/2}z^{1/2} + C = 2z^{1/2} + C$

4. $\displaystyle\int (w^{1/3} - 4w^{-3})\, dw = \int w^{1/3}\, dw - 4\int w^{-3}\, dw$

$$= \tfrac{3}{4}w^{4/3} - 4\left(-\tfrac{1}{2}\right)w^{-2} + C$$

$$= \tfrac{3}{4}w^{4/3} + 2w^{-2} + C$$

5. a. $\displaystyle\int 5x^4\, dx = x^5 + C$ $\qquad$ **b.** $\displaystyle\int 3x^2\, dx = x^3 + C$

6. $\displaystyle\int \frac{6t^2 - t}{t}\, dt = \int \frac{t(6t - 1)}{t}\, dt$

$$= \int (6t - 1)\, dt$$

$$= 6\int t\, dt - \int 1\, dt = 6\cdot\tfrac{1}{2}t^2 - t + C$$

$$= 3t^2 - t + C$$

5.1 Exercises

Find each indefinite integral.

1. $\int x^4\, dx$

2. $\int x^7\, dx$

3. $\int x^{2/3}\, dx$

4. $\int x^{3/2}\, dx$

5. $\int \sqrt{u}\, du$

6. $\int \sqrt[3]{u}\, du$

7. $\int \dfrac{dw}{w^4}$

8. $\int \dfrac{dw}{w^2}$

9. $\int \dfrac{dz}{\sqrt{z}}$

10. $\int \dfrac{dz}{\sqrt[3]{z}}$

11. $\int 6x^5\, dx$

12. $\int 9x^8\, dx$

13. $\int (8x^3 - 3x^2 + 2)\, dx$

14. $\int (12x^3 + 3x^2 - 5)\, dx$

15. $\int \left(6\sqrt{x} + \dfrac{1}{\sqrt[3]{x}} \right) dx$

16. $\int \left(3\sqrt{x} + \dfrac{1}{\sqrt{x}} \right) dx$

17. $\int \left(16\sqrt[3]{x^5} - \dfrac{16}{\sqrt[3]{x^5}} \right) dx$

18. $\int \left(14\sqrt[4]{x^3} - \dfrac{3}{\sqrt[4]{x^3}} \right) dx$

19. $\int \left(10\sqrt[3]{t^2} + \dfrac{1}{\sqrt[3]{t^2}} \right) dt$

20. $\int \left(21\sqrt{t^5} + \dfrac{6}{\sqrt{t^5}} \right) dt$

21. $\int (x - 1)^2\, dx$

22. $\int (x + 2)^2\, dx$

23. $\int (1 + 10w)\, \sqrt{w}\, dw$

24. $\int (1 - 7w)\, \sqrt[3]{w}\, dw$

25. $\int \dfrac{6x^3 - 6x^2 + x}{x}\, dx$

26. $\int \dfrac{4x^4 + 4x^2 - x}{x}\, dx$

27. $\int (x - 2)(x + 4)\, dx$

28. $\int (x + 5)(x - 3)\, dx$

29. $\int (r - 1)(r + 1)\, dr$

30. $\int (3s + 1)(3s - 1)\, ds$

31. $\int \dfrac{x^2 - 1}{x + 1}\, dx$

32. $\int \dfrac{x^2 - 1}{x - 1}\, dx$

33. $\int (t + 1)^3\, dt$

34. $\int (t - 1)^3\, dt$

35. Find:

  a. $\int \dfrac{1}{x^3}\, dx$

  b. $\dfrac{\int 1\, dx}{\int x^3\, dx}$

  Notice that the answers are not the same, showing that you do not integrate a fraction by integrating the numerator and denominator separately.

36. Evaluate

  a. $\int x\, dx$

  b. $x \int 1\, dx$

  Notice that the two answers are not the same, showing that a *variable* cannot be moved across the integral sign.

37. a. Verify that $\int x^2\, dx = \frac{1}{3}x^3 + C$.

  b. Graph the five functions $\frac{1}{3}x^3 - 2, \frac{1}{3}x^3 - 1,$ $\frac{1}{3}x^3, \frac{1}{3}x^3 + 1,$ and $\frac{1}{3}x^3 + 2$ (the solutions for five different values of C) on the graphing window $[-3, 3]$ by $[-5, 5]$. Use TRACE to see how the constant shifts the curve vertically.

  c. Find the slopes (using NDERIV or dy/dx) of several of the curves at a particular x-value and check that in each case the slope is the square of the x-value. This verifies that the derivative of each curve is x^2, and so each is an integral of x^2.

38. a. Graph the five functions $\ln x - 2,$ $\ln x - 1, \ \ln x, \ \ln x + 1,$ and $\ln x + 2$ on the graphing window $[0, 4]$ by $[-3, 3]$.

(continues)

b. Find the slope (using NDERIV or dy/dx) of several of the curves at a particular x-value and check that in each case the slope is the reciprocal of the x-value. This suggests that the derivative of each function is $1/x$.

c. Based on part (b), conjecture what is the indefinite integral of the function $1/x$ (for $x > 0$).

APPLIED EXERCISES

39. BUSINESS: Cost A company's marginal cost function is $MC - 20x^{3/2} - 15x^{2/3} + 1$, where x is the number of units, and fixed costs are $4000. Find the cost function.

40. BUSINESS: Cost A company's marginal cost function is $MC = 21x^{4/3} - 6x^{1/2} + 50$, where x is the number of units, and fixed costs are $3000. Find the cost function.

41. BUSINESS: Revenue A company's marginal revenue function is $MR = 12\sqrt[3]{x} + 3\sqrt{x}$, where x is the number of units. Find the revenue function. (Evaluate C so that revenue is zero when nothing is produced.)

42. BUSINESS: Revenue A company's marginal revenue function is $MR - 15\sqrt{x} + 4\sqrt[3]{x}$, where x is the number of units. Find the revenue function. (Evaluate C so that revenue is zero when nothing is produced.)

43. GENERAL: Velocity A Porsche 928 can accelerate from a standing start to a speed of $v(t) = -0.09t^2 + 8t$ feet per second after t seconds (for $0 \le t < 35$).

a. Find a formula for the distance that it will travel from its starting point in the first t seconds. [*Hint:* Integrate velocity to find distance, and then use the fact that distance is 0 at time $t = 0$.]

b. Use the formula that you found in part (a) to find the distance that the car will travel in the first 10 seconds.

44. GENERAL: Velocity A BMW 733i can accelerate from a standing start to a speed of $v(t) = -0.09t^2 + 6t$ feet per second after t seconds (for $0 \le t < 40$).

a. Find a formula for the distance that it will travel from its starting point in the first t seconds. [*Hint:* Integrate velocity to find

distance, and then use the fact that distance is 0 at time $t = 0$.]

b. Use the formula that you found in part (a) to find the distance that the car will travel in the first 10 seconds.

45. GENERAL: Learning A person can memorize words at the rate of $3/\sqrt{t}$ words per minute.

a. Find a formula for the total number of words that can be memorized in t minutes. [*Hint:* Evaluate C so that 0 words have been memorized at time $t = 0$.]

b. Use the formula that you found in part (a) to find the total number of words that can be memorized in 25 minutes.

46. BIOMEDICAL: Temperature A patient's temperature is 108 degrees Fahrenheit and is changing at the rate of $t^2 - 4t$ degrees per hour, where t is the number of hours since taking a fever-reducing medication $(0 \le t \le 3)$.

a. Find a formula for the patient's temperature after t hours. [*Hint:* Evaluate the constant C so that the temperature is 108 at time $t = 0$.]

b. Use the formula that you found in part (a) to find the patient's temperature after 3 hours.

47. ENVIRONMENTAL SCIENCE: Pollution A chemical plant is adding pollution to a lake at the rate of $40\sqrt{t^3}$ tons per year, where t is the number of years that the plant has been in operation.

a. Find a formula for the total amount of pollution that will enter the lake in the first t years of the plant's operation. [*Hint:* Evaluate C so that no pollution has been added at time $t = 0$.] (*continues*)

b. Use the formula that you found in part (a) to find how much pollution will enter the lake in the first 4 years of the plant's operation.

c. If all life in the lake will cease when 400 tons of pollution have entered the lake, will the lake "live" beyond 4 years?

48. BUSINESS: Appreciation A $20,000 art collection is increasing in value at the rate of $300\sqrt{t}$ dollars per year after t years.

a. Find a formula for its value after t years. [*Hint:* Evaluate C so that its value at time $t = 0$ is $20,000.]

b. Use the formula that you found in part (a) to find its value after 25 years.

REVIEW EXERCISE

This exercise will be important in the next section.

49. Find $\dfrac{d}{dx}\ln(-x)$.

5.2 INTEGRATION USING LOGARITHMIC AND EXPONENTIAL FUNCTIONS

Introduction

In the previous section we defined integration as the reverse of differentiation, and we introduced several integration formulas. In this section we develop integration formulas involving logarithmic and exponential functions. One of these formulas will answer a question that we could not answer earlier—namely, how to integrate x^{-1}, the only power not covered by the Power Rule.

The Integral $\displaystyle\int e^{ax}\,dx$

On page 310 we saw that to *differentiate* e^{ax}, we *multiply* by a to get ae^{ax}. Therefore, to *integrate* e^{ax}, the reverse process, we must *divide* by a. That is, for any $a \neq 0$:

$$\int e^{ax}\,dx = \frac{1}{a}e^{ax} + C$$

The integral of e to a constant times x is 1 over the constant times the original function (plus C)

EXAMPLE 1

INTEGRATING AN EXPONENTIAL FUNCTION

$$\int e^{2x}\, dx = \frac{1}{2} e^{2x} + C \qquad\qquad \text{Using the formula with } a = 2$$

$$\underset{a=2}{\uparrow} \quad \underset{\substack{1\\[-2pt]\overline{a}}}{\smile} \quad \underset{\text{Original function}}{\uparrow}$$

As always, we may check the answer by differentiation.

$$\frac{d}{dx}\left(\frac{1}{2}e^{2x} + C\right) = \frac{1}{2}\cdot 2\cdot e^{2x} = e^{2x} \qquad\qquad \text{Using } \frac{d}{dx}e^{ax} = ae^{ax}$$

The result is the integrand e^{2x}, so the integration is correct.

The proof of this rule consists simply of differentiating the right-hand side.

$$\frac{d}{dx}\left(\frac{1}{a}e^{ax} + C\right) = \frac{1}{a}ae^{ax} = e^{ax} \qquad\qquad \text{Using } \frac{d}{dx}e^{ax} = ae^{ax}$$

The result is the integrand, so the integration formula is correct.

EXAMPLE 2

INTEGRATING EXPONENTIAL FUNCTIONS

a. $\displaystyle\int e^{\frac{1}{2}x}\, dx = 2e^{\frac{1}{2}x} + C$ $\qquad\qquad a = \dfrac{1}{2}\ \ \text{so}\ \ \dfrac{1}{a} = \dfrac{1}{\frac{1}{2}} = 2$

b. $\displaystyle\int 6e^{-3x}\, dx = 6\int e^{-3x}\, dx = 6\left(-\frac{1}{3}\right)\cdot e^{-3x} + C = -2e^{-3x} + C$

$$\underset{a=-3}{\uparrow} \qquad \underset{\frac{1}{a}\,=\,-\frac{1}{3}}{\uparrow}$$

c. $\displaystyle\int e^{x}\, dx = 1e^{x} + C = e^{x} + C$ $\qquad\qquad a = 1,\ \ \text{so}\ \ \dfrac{1}{a} = 1$

Each of these answers may be checked by differentiation. The last one says that e^{x} is the integral of itself, just as e^{x} is the derivative of itself.

$$\int e^x \, dx = e^x + C \qquad \text{The integral of } e^x \text{ is } e^x \text{ (plus } C\text{)}$$

(handwritten:) $e^{\frac{1}{3}x}$ a. $3e^{4x} + C$
b. $3e^{\frac{1}{3}x} = 3e^{\frac{x}{3}} + C$

Practice Problem 1 Find: **a.** $\displaystyle\int 12e^{4x} \, dx$ **b.** $\displaystyle\int e^{x/3} \, dx$

[*Hint:* For part (b), write $e^{x/3}$ as $e^{\frac{1}{3}x}$, then integrate, and rewrite again.]

➤ Solutions on page 366

When using these new integration formulas to solve applied problems, be careful to evaluate the constant C correctly. It will not always be equal to the initial value of the function.

EXAMPLE 3   **FINDING TOTAL FLU CASES FROM THE RATE OF CHANGE**

An influenza epidemic hits a large city and spreads at the rate of $12e^{0.2t}$ new cases per day, where t is the number of days since the epidemic began. The epidemic began with 4 cases.

a. Find a formula for the total number of flu cases in the first t days of the epidemic.

b. Use your formula to find the number of cases during the first 30 days.

Solution

a. To find the total number of cases, we integrate the growth rate $12e^{0.2t}$.

$$\underbrace{f(t)}_{\substack{\text{Total cases} \\ \text{in first } t \text{ days}}} = \int 12e^{0.2t} \, dt = 12 \int e^{0.2t} \, dt \qquad \text{Taking out the constant}$$

$$= 12\,\frac{1}{0.2}\,e^{0.2t} + C = 60e^{0.2t} + C \qquad \text{Using the } \int e^{ax}\,dx \text{ formula}$$

$$\uparrow 5$$

Evaluating $f(t)$ at $t = 0$ must give the initial number of cases:

$$f(0) = 60e^{0.2(0)} + C = 60 + C$$

$\underbrace{}$ Initial number of cases $\underbrace{}$ $e^0 = 1$

$f(t) = 60e^{0.2t} + C$ evaluated at $t = 0$ (the beginning of the epidemic)

This initial number must equal the given initial number, 4:

$$60 + C = 4$$ Initial number from the formula set equal to the given initial number

$$C = -56$$ Solving for C

Replacing C by -56 gives the total number of flu cases within t days:

$$f(t) \quad = \quad 60e^{0.2t} - 56$$

$\underbrace{}$ Number of cases in first t days $\underbrace{\phantom{60e^{0.2t} - 56}}$ Answer to Part (a)

$f(t) = 60e^{0.2t} + C$ with $C = -56$

b. To find the number within 30 days, we evaluate at $t = 30$:

$$f(30) = 60e^{0.2(30)} - 56 \qquad f(t) = 60e^{0.2t} + C \text{ at } t = 30$$
$$= 60e^6 - 56 \approx 24{,}150 \qquad \text{Using a calculator}$$

Therefore, within 30 days the epidemic will have grown to more than 24,000 cases.

Evaluating the Constant C

Notice that we did *not* simply replace the constant C by the initial number of cases, 4. (This would have given the wrong initial number of cases.) Instead, we evaluated the function at the initial time $t = 0$ and set it equal to the initial number of cases:

$$60e^0 + C \quad = \quad 4$$

$\underbrace{}$ $f(t)$ evaluated at $t = 0$ $\underset{\smile}{}$ Given initial value

We then solved to find the correct value of C, $C = -56$, which we then substituted into the formula.

In general, to evaluate the constant C:

1. Evaluate the integral at the given number (usually $t = 0$) and set the result equal to the stated initial value.

2. Solve for C.

3. Write the answer with C replaced by its correct value.

The Integral $\displaystyle\int \frac{1}{x} \, dx$

The differentiation formula $\dfrac{d}{dx} \ln x = \dfrac{1}{x}$ can be read "backward" as an integration formula:

$$\int \frac{1}{x} \, dx = \ln x + C \qquad \text{The integral of 1 over } x \text{ is the natural log of } x \text{ (plus } C\text{)}$$

This formula, however, is restricted to $x > 0$, for only then is $\ln x$ defined. For $x < 0$ we can differentiate $\ln(-x)$, giving

$$\frac{d}{dx} \ln(-x) = \frac{-1}{-x} = \frac{1}{x} \qquad \text{Using } \frac{d}{dx} \ln f = \frac{f'}{f} \text{ and simplifying}$$

This result says that for $x < 0$, the integral of $1/x$ is $\ln(-x)$. The negative sign in $\ln(-x)$ serves only to make the already negative x positive, and this could be accomplished just as well with absolute value bars.

$$\int \frac{1}{x} \, dx = \ln |x| + C \qquad \text{The integral of 1 over } x \text{ is the natural logarithm of the absolute value of } x$$

This formula holds for negative *and* positive values of x, since in both cases $\ln |x|$ is defined. The integral can be written in three different ways, all of which have the same answer.

$$\int \frac{1}{x} \, dx \;=\; \int \frac{dx}{x} \;=\; \int x^{-1} \, dx \;=\; \ln |x| + C$$

EXAMPLE 4

INTEGRATING USING THE LN RULE

$$\int \frac{5}{2x} \, dx \;=\; \int \frac{5}{2} \frac{1}{x} \, dx \;=\; \frac{5}{2} \int \frac{1}{x} \, dx \;=\; \frac{5}{2} \ln |x| + C$$

$\uparrow$

Taking out the constant   Using the ln formula

EXAMPLE 5

INTEGRATING NEGATIVE POWERS

$$\int (x^{-1} + x^{-2})\, dx = \int x^{-1}\, dx + \int x^{-2}\, dx = \ln |x| - x^{-1} + C$$

From the ↗
natural log
formula

From the
Power Rule
with $n = -2$

Practice Problem 2

Find $\int \dfrac{3}{4x}\, dx.$ $= \int \dfrac{3}{4} \dfrac{1}{x}\, dx = \dfrac{3}{4} \int \dfrac{1}{x}\, dx$

$= \dfrac{3}{4} \ln |x| + c$

➤ Solution on page 366

EXAMPLE 6

FINDING TOTAL SALES FROM THE SALES RATE

An electronics dealer estimates that during month t of a sale, a discontinued computer will sell at a rate of approximately $25/t$ per month, where $t = 1$ corresponds to the beginning of the sale, at which time none have been sold. Find a formula for the total number of computers that will be sold up to month t. Will the store's inventory of 64 computers be sold by month $t = 12$?

Solution

To find the total sales, we integrate the sales rate $\dfrac{25}{t}$:

$$S(t) = \int \frac{25}{t}\, dt = 25 \int \frac{1}{t}\, dt = 25 \ln t + C \qquad \text{Omitting absolute values, since } t > 0$$

Total sales in
first t months

To evaluate C, we evaluate at the given starting time $t = 1$:

$$S(1) = 25 \ln 1 + C = C$$

Initial number
of sales
0

The initial number of sales must be zero, giving $C = 0$, and substituting this into the sales function gives

$$S(t) = 25 \ln t \qquad\qquad S(t) = 25 \ln t + C \text{ with } C = 0$$

Total number sold up to month t

To find the number sold up to month 12, we evaluate at $t = 12$:

$$S(12) = 25 \ln 12 \approx 62 \qquad \text{Using a calculator}$$

Therefore, all but two of the 64 computers will be sold by month 12.

In this example the initial time (the beginning of the sale) was given as $t = 1$ rather than the more usual $t = 0$. The initial time will be clear from the problem.

Consumption of Natural Resources

Just as the world population grows exponentially, so does the world's annual consumption of natural resources. We can estimate the total consumption at any time in the future by integrating the rate of consumption, and from this predict when the known reserves will be exhausted.

EXAMPLE **7**   **FINDING A FORMULA FOR TOTAL CONSUMPTION FROM THE RATE**

The annual world consumption of tin is predicted* to be $0.24e^{0.01t}$ million metric tons per year, where t is the number of years since 2000. Find a formula for the total tin consumption within t years of 2000 and estimate when the known world reserves of 6 million metric tons will be exhausted.

Solution

To find the total consumption, we integrate the rate $0.24e^{0.01t}$.

$$\underbrace{C(t)}_{\substack{\text{Total tin consumed} \\ \text{in first } t \text{ years} \\ \text{after 2000}}} = \int 0.24e^{0.01t}\, dt = 0.24 \int e^{0.01t}\, dt \qquad \begin{array}{l}\text{Taking out} \\ \text{the constant}\end{array}$$

$$= 0.24\, \underbrace{\frac{1}{0.01}}_{100}\, e^{0.01t} + C = 24e^{0.01t} + C \qquad \begin{array}{l}\text{Using the} \\ \int e^{ax}\, dx \;\; \text{formula}\end{array}$$

---

*World Resources Institute and U.S. Geological Survey. Data for several exercises in this chapter come from these sources.

The total consumed in the first zero years must be zero, so $C(0) = 0$.

$$C(0) = 24e^0 + C = 24 + C \qquad\qquad C(t) = 24e^{0.01t} + C \text{ with } t = 0$$

$\underbrace{}$
0

$\underbrace{24 + C}$
Must equal zero

From $24 + C = 0$ we find $C = -24$. We substitute this into the formula for $C(t)$:

$$C(t) = \underbrace{24e^{0.01t} - 24} \qquad\qquad C(t) = 24e^{0.01t} + C \text{ with } C = -24$$

Formula for total consumption within first t years since 2000

To predict when the total world reserves of 6 million metric tons will be exhausted, we set this function equal to 6 and solve for t:

$$24e^{0.01t} - 24 = 6 \qquad\qquad C(t) \text{ set equal to 6}$$

$$24e^{0.01t} = 30 \qquad\qquad \text{Adding 24}$$

$$e^{0.01t} = \frac{30}{24} = 1.25 \qquad\qquad \text{Dividing by 24}$$

$$\ln e^{0.01t} = \ln 1.25 \qquad\qquad \begin{array}{l}\text{Taking natural logs}\\\text{and using } \ln e^x = x\end{array}$$

$\underbrace{0.01t} \quad \underbrace{0.223}$

$$t \approx \frac{0.223}{0.01} = 22.3 \qquad\qquad \text{Solving } 0.01t = 0.223 \text{ for } t$$

Therefore, the known world supply of tin will be exhausted in about 22 years after 2000, which means in about the year 2022 (assuming that consumption continues at the predicted rate).

Spreadsheet Exploration

Since integration is continuous summation, instead of integrating we could add up the annual tin consumption for each year to see when the total will reach the known reserves of 6 million metric tons. The following spreadsheet* shows this annual consumption

* See the Preface for how to obtain this and other Excel spreadsheets.

for each year (using the formula $0.24e^{0.01t}$ from the preceding example) in column B continuing into column F, with the cumulative totals shown in column C continuing into column G.

	B2	▼	= =0.24*EXP(0.01*$A2)				
	A	B	C	D	E	F	G
1	Year	Annual Consumption	Cumulative Consumption		Year	Annual Consumption	Cumulative Consumption
2	1	0.24241204	0.24241204		16	0.28164261	4.18511691
3	2	0.24484832	0.48726036		17	0.28447316	4.46959007
4	3	0.24730909	0.73456945		18	0.28733217	4.75692224
5	4	0.24979459	0.98436404		19	0.29021990	5.04714215
6	5	0.25230506	1.23666910		20	0.29313666	5.34027881
7	6	0.25484077	1.49150987		21	0.29608273	5.63636154
8	7	0.25740196	1.74891183		22	0.29905842	5.93541996
9	8	0.25998890	2.00890073		23	0.30206400	6.23748396
10	9	0.26260183	2.27150256		24	0.30509980	6.54258376
11	10	0.26524102	2.53674358		25	0.30816610	6.85074986
12	11	0.26790674	2.80465032		26	0.31126322	7.16201308
13	12	0.27059924	3.07524956		27	0.31439147	7.47640454
14	13	0.27331881	3.34856837		28	0.31755115	7.79395570
15	14	0.27606571	3.62463408		29	0.32074260	8.11469830
16	15	0.27884022	3.90347430		30	0.32396611	8.43866441

Since the cumulative consumption is less than 6 million metric tons in year 22 and more in year 23, we conclude that the known world reserves of tin will be exhausted sometime in 2022, in agreement with Example 7.

The integration method of Example 7 has two advantages over addition: it is often easier (especially for a large number of years) and it includes changes in the amount *during* each year rather than just at the end of each year.

Incidentally, tin is used mainly for coating steel (a "tin" can is actually a steel can with a thin protective coating of tin to prevent rust). The predicted unavailability of tin is already causing major changes in the food-packaging industry.

Power Rule for Integration, Revisited

Our new integration formula shows how to integrate x^{-1}, the only power not covered by the Power Rule. Therefore, we can now write one "combined" formula for the integral $\int x^n \, dx$ for *any power n*.

Integrals of Powers of x

$$\int x^n \, dx = \begin{cases} \dfrac{1}{n+1}x^{n+1} + C & \text{if } n \neq -1 \\[2ex] \ln|x| + C & \text{if } n = -1 \end{cases}$$

Use the Power Rule if n is *other* than -1

Use the ln formula if n equals -1

It is a curious fact that every power of x integrates to another power of x, with the single exception of x^{-1}, which integrates to an entirely different kind of function, the natural logarithm.

5.2 Section Summary

We have three integration formulas:

$$\int x^n \, dx = \frac{1}{n+1}x^{n+1} + C \qquad\qquad n \neq -1$$

$$\int e^{ax} \, dx = \frac{1}{a}e^{ax} + C \qquad\qquad a \neq 0$$

$$\int \frac{1}{x} \, dx = \int \frac{dx}{x} = \int x^{-1} \, dx = \ln|x| + C$$

What is the difference between the bottom formula with and without the absolute value bars?

$$\int \frac{1}{x} \, dx = \ln|x| + C \qquad \text{versus} \qquad \int \frac{1}{x} \, dx = \ln x + C$$

Both are correct, but the second holds only for *positive* x-values (so that $\ln x$ is defined; see page 289), while the first holds for *negative* as well as positive x-values. We will sometimes use the second when we know that x is positive.

To evaluate C in an application we set the function (evaluated at the given number) equal to the stated initial value and solve for C.

➤ **Solutions to Practice Problems**

1. a. $\displaystyle\int 12e^{4x}\,dx = 12\int e^{4x}\,dx = 12\cdot\tfrac{1}{4}e^{4x} + C$

$$= 3e^{4x} + C$$

b. $\displaystyle\int e^{x/3}\,dx = \int e^{\frac{1}{3}x}\,dx = 3e^{\frac{1}{3}x} + C = 3e^{x/3} + C$

2. $\displaystyle\int \frac{3}{4x}\,dx = \frac{3}{4}\int \frac{1}{x}\,dx = \frac{3}{4}\ln|x| + C$

5.2   Exercises

Find each indefinite integral.

1. $\displaystyle\int e^{3x}\,dx$

2. $\displaystyle\int e^{4x}\,dx$

3. $\displaystyle\int e^{x/4}\,dx$

4. $\displaystyle\int e^{x/3}\,dx$

5. $\displaystyle\int e^{0.05x}\,dx$

6. $\displaystyle\int e^{0.02x}\,dx$

7. $\displaystyle\int e^{-2y}\,dy$

8. $\displaystyle\int e^{-3y}\,dy$

9. $\displaystyle\int e^{-0.5x}\,dx$

10. $\displaystyle\int e^{-0.4x}\,dx$

11. $\displaystyle\int 6e^{2x/3}\,dx$

12. $\displaystyle\int 24e^{-2u/3}\,du$

13. $\displaystyle\int -5x^{-1}\,dx$

14. $\displaystyle\int -\frac{1}{2}x^{-1}\,dx$

15. $\displaystyle\int \frac{3\,dx}{x}$

16. $\displaystyle\int \frac{dx}{2x}$

17. $\displaystyle\int \frac{3}{2v}\,dv$

18. $\displaystyle\int \frac{2}{3v}\,dv$

19. $\displaystyle\int \left(e^{3x} - \frac{3}{x}\right)dx$

20. $\displaystyle\int \left(e^{2x} - \frac{2}{x}\right)dx$

21. $\displaystyle\int (3e^{0.5t} - 2t^{-1})\,dt$ **22.** $\displaystyle\int (5e^{0.5t} - 4t^{-1})\,dt$

23. $\displaystyle\int (x^2 + x + 1 + x^{-1} + x^{-2})\,dx$

24. $\displaystyle\int (x^{-2} - x^{-1} + 1 - x + x^2)\,dx$

25. $\displaystyle\int (5e^{0.02t} - 2e^{0.01t})\,dt$

26. $\displaystyle\int (3e^{0.05t} - 2e^{0.04t})\,dt$

APPLIED EXERCISES

27–28. BIOMEDICAL: Epidemics A flu epidemic hits a college community, beginning with five cases on day $t = 0$. The rate of growth of the epidemic (new cases per day) is given by the following function $r(t)$, where t is the number of days since the epidemic began.

a. Find a formula for the total number of cases of flu in the first t days.

b. Use your answer to part (a) to find the total number of cases in the first 20 days.

27. $r(t) = 18e^{0.05t}$ **28.** $r(t) = 20e^{0.04t}$

29. BUSINESS: Sales In an effort to reduce its inventory, a music store runs a sale on its least popular compact discs (CDs). The sales rate (CDs sold per day) on day t of the sale is predicted to be $50/t$ (for $t \geq 1$), where $t = 1$ corresponds to the beginning of the sale, at which time none of the inventory of 200 CDs had been sold.

 a. Find a formula for the total number of CDs sold up to day t.

 b. Will the store have sold its inventory of 200 CDs by day $t = 30$?

30. BUSINESS: Sales In an effort to reduce its inventory, a book store runs a sale on its least popular mathematics books. The sales rate (books sold per day) on day t of the sale is predicted to be $60/t$ (for $t \geq 1$), where $t = 1$ corresponds to the beginning of the sale, at which time none of the inventory of 350 books had been sold.

 a. Find a formula for the number of books sold up to day t.

 b. Will the store have sold its inventory of 350 books by day $t = 30$?

31. GENERAL: Consumption of Natural Resources World consumption of silver is running at the rate of $17e^{0.02t}$ thousand metric tons per year, where t is measured in years and $t = 0$ corresponds to 2000.

 a. Find a formula for the total amount of silver that will be consumed within t years of 2000.

 b. When will the known world resources of 420 thousand metric tons of silver be exhausted? (Silver is used extensively in photography.)

32. GENERAL: Consumption of Natural Resources World consumption of copper is running at the rate of $15e^{0.04t}$ million metric tons per year, where t is measured in years and $t = 0$ corresponds to 2000.

 a. Find a formula for the total amount of copper that will be used within t years of 2000.

 b. When will the known world resources of 650 million metric tons of copper be exhausted?

33. GENERAL: Cost of Maintaining a Home The cost of maintaining a home generally increases as the home becomes older. Suppose that the rate of cost (dollars per year) for a home that is x years old is $200e^{0.4x}$.

 a. Find a formula for the total maintenance cost during the first x years. (Total maintenance should be zero at $x = 0$.)

 b. Use your answer to part (a) to find the total maintenance cost during the first 5 years.

34. BIOMEDICAL: Cell Growth A culture of bacteria is growing at the rate of $20e^{0.8t}$ cells per day, where t is the number of days since the culture was started. Suppose that the culture began with 50 cells.

 a. Find a formula for the total number of cells in the culture after t days.

 b. If the culture is to be stopped when the population reaches 500, when will this occur?

35. GENERAL: Freezing of Ice An ice cube tray filled with tap water is placed in the freezer, and the temperature of the water is changing at the rate of $-12e^{-0.2t}$ degrees Fahrenheit per hour after t hours. The original temperature of the tap water was 70 degrees.

 a. Find a formula for the temperature of water that has been in the freezer for t hours.

 b. When will the ice be ready? (Water freezes at 32 degrees.)

36. SOCIAL SCIENCE: Divorces The number of divorces per year in the United States is approximately $1.14e^{0.01t}$ million, where t is measured in years and $t = 0$ corresponds to 2000. Find a formula for the total number of divorces expected within t years of 2000.

37. BUSINESS: Total Savings A factory installs new equipment that is expected to generate savings at the rate of $800e^{-0.2t}$ dollars per year, where t is the number of years that the equipment has been in operation.

 a. Find a formula for the total savings that the equipment will generate during its first t years.

 b. If the equipment originally cost $2000, when will it "pay for itself"?

38. BUSINESS: Total Savings A company installs a new computer that is expected to generate savings at the rate of $20{,}000e^{-0.02t}$ dollars per year, where t is the number of years that the computer has been in operation.

a. Find a formula for the total savings that the computer will generate during its first t years.

b. If the computer originally cost $250,000, when will it "pay for itself"?

39. BUSINESS: Value of an Investment A real estate investment, originally worth $5000, grows continuously at the rate of $400e^{0.05t}$ dollars per year, where t is the number of years since the investment was made.

a. Find a formula for the value of the investment after t years.

b. Use your formula to find the value of the investment after 10 years.

40. BUSINESS: Value of an Investment A biotechnology investment, originally worth $20,000, grows continuously at the rate of $1000e^{0.10t}$ dollars per year, where t is the number of years since the investment was made.

a. Find a formula for the value of the investment after t years.

b. Use your formula to find the value of the investment after 7 years.

Find each indefinite integral [*Hint:* Use some algebra first.]

41. $\displaystyle\int \frac{(x+1)^2}{x}\,dx$ **42.** $\displaystyle\int \frac{(x-1)^2}{x}\,dx$

43. $\displaystyle\int \frac{(t-1)(t+3)}{t^2}\,dt$ **44.** $\displaystyle\int \frac{(t+2)(t-4)}{t^2}\,dt$

45. $\displaystyle\int \frac{(x-2)^3}{x}\,dx$ **46.** $\displaystyle\int \frac{(x+2)^3}{x}\,dx$

47. GENERAL: Consumption of Natural Resources World consumption of lead is running at the rate of $5.6e^{0.01t}$ million metric tons per year, where t is measured in years and $t = 0$ corresponds to 2000.

a. Find a formula for the total amount of lead that will be consumed within t years of 2000.

b. Use a graphing calculator to find when the world's known resources of 140 million metric tons of lead will be exhausted. [*Hint:* Use INTERSECT.] Lead has many uses, from batteries to shields against radioactivity.

48. GENERAL: Total Savings A homeowner installs a solar water heater that is expected to generate savings at the rate of $70e^{0.03t}$ dollars per year, where t is the number of years since it was installed.

a. Find a formula for the total savings within the first t years of operation.

b. Use a graphing calculator to find when the heater will "pay for itself" if it cost $800. [*Hint:* Use INTERSECT.]

5.3 DEFINITE INTEGRALS AND AREAS

Introduction

We begin this section by calculating areas under curves, leading to a definition of the *definite integral* of a function. The *Fundamental Theorem of Integral Calculus* then provides an easier way to calculate definite integrals using *indefinite* integrals. Finally, we will illustrate the wide variety of applications of definite integrals.

Area Under a Curve

The diagram below shows a continuous nonnegative function. We want to calculate the *area* under the curve and above the x-axis between the vertical lines $x = a$ and $x = b$.

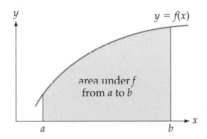

We begin by *approximating* the area by rectangles. In the first graph on the left below, the area under the curve is approximated by five rectangles with equal bases and with heights equal to the height of the curve at the left-hand edge of the rectangle. (These are called *left* rectangles.) Five rectangles, however, do not give a very accurate approximation for the area under the curve: They underestimate the actual area by the small white spaces just above the rectangles.

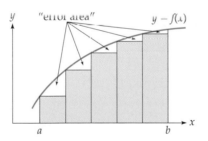

Area under f from a to b
approximated by 5 rectangles

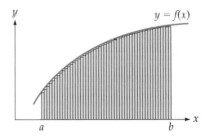

Area under f from a to b
approximated by 50 rectangles

In the second graph, this same area is approximated by fifty rectangles, giving a much better approximation: the white *"error area"* between the curve and the rectangles is so small as to be almost invisible.

These diagrams suggest that more rectangles give a better approximation. In fact, the *exact* area under the curve is defined as the *limit* of the approximations as the number of rectangles *approaches infinity*. The following example shows how to carry out such an approximation for the area under a given curve, and afterward we will find the *exact* area by letting the number of rectangles approach infinity.

EXAMPLE **1**

APPROXIMATING AREA BY RECTANGLES

Approximate the area under the curve $f(x) = x^2$ from 1 to 2 by five rectangles. Use rectangles with equal bases and with heights equal to the height of the curve at the left-hand edge of the rectangles.

Solution

For five rectangles, we divide the distance from $a = 1$ to $b = 2$ into five equal parts, so that each rectangle has width

$$\Delta x = \frac{2 - 1}{5} = \frac{1}{5} = 0.2 \qquad \text{For } n \text{ rectangles,} \quad \Delta x = \frac{b - a}{n}$$

Along the x-axis beginning at $x = 1$ we mark successive points with spacing $\Delta x = 0.2$, giving points 1, 1.2, 1.4, 1.6, 1.8, and 2, as shown below. Above each of the resulting subintervals we draw a rectangle whose height is the height of the curve at the left-hand edge of that rectangle. The curve is $y = x^2$, so these heights are the squares of the corresponding x-values.

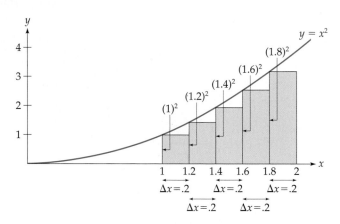

Area from 1 to 2
approximated by 5 rectangles

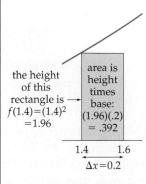

the height
of this
rectangle is
$f(1.4)=(1.4)^2$
$=1.96$

area is
height
times
base:
$(1.96)(.2)$
$= .392$

Middle rectangle

An enlarged view of the middle rectangle is shown on the left. The sum of the areas of the five rectangles, height times base $\Delta x = 0.2$ for each rectangle, is

| 1st rectangle | 2nd rectangle | 3rd rectangle | 4th rectangle | 5th rectangle | Adding height times base for each rectangle |

$$(1)^2 \cdot (0.2) + (1.2)^2 \cdot (0.2) + (1.4)^2 \cdot (0.2) + (1.6)^2 \cdot (0.2) + (1.8)^2 \cdot (0.2)$$

$$= 0.2 + 0.288 + 0.392 + 0.512 + 0.648 = 2.04$$

Multiplying out and summing.

Therefore, the area under the curve is approximately 2.04 square units.

As we saw earlier, using only five rectangles does not give a very accurate approximation for the true area under the curve. For greater accuracy we use more rectangles, calculating the area in the same way. The following table gives the "rectangular approximation" for the area under the curve in Example 1 for larger numbers of rectangles. The calculations were carried out on a graphing calculator using the program on page 384, rounding answers to three decimal places.

Number of Rectangles	Sum of Areas of Rectangles	
5	2.04	← Found in Example 1
10	2.185	
100	2.318	The sum of the areas is approaching
1,000	2.332	
10,000	2.333	$2\frac{1}{3}$

The areas in the right-hand column are approaching 2.333 ... = $2\frac{1}{3}$, which is the *exact* area under the curve (as we will verify later). Therefore:

The area under the curve $f(x) = x^2$ from 1 to 2 is $2\frac{1}{3}$ square units.

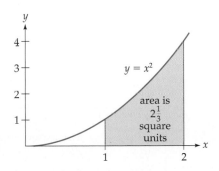

Areas are given in "square units," meaning that if the units on the graph are inches, feet, or some other units, then the area is in *square* inches, *square* feet, or, in general, some other *square* units.

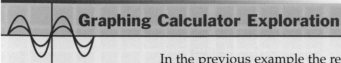

Graphing Calculator Exploration

In the previous example the rectangles touched the curve at the *left-hand* edge of the rectangle (and are called *left* rectangles). Similarly, we may use *midpoint* rectangles (which touch the curve at the *midpoint* of the rectangle), *right* rectangles (which touch the curve at their right-hand edge), or *random* rectangles (which touch the curve at a *random* x-value within the rectangle). The graphing calculator program* RIEMANN draws any of these types of rectangles for any function and calculates the sum of the areas, as shown below. The first screen shows five *midpoint* rectangles for $f(x) = x^2$ on the interval [1, 2]. The second screen shows the sum of the areas for various numbers of midpoint rectangles.

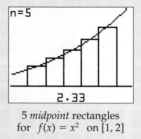

5 *midpoint* rectangles
for $f(x) = x^2$ on [1, 2]

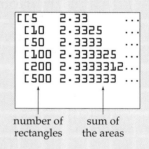

number of   sum of
rectangles   the areas

Notice from the second screen that the areas approach $2\frac{1}{3}$ very quickly, especially when compared with the table on the previous page for *left* rectangles. Do you see why midpoint rectangles are more accurate? [*Hint:* Compare the above graph for *midpoint* rectangles with the following graphs for *left* and *right* rectangles. Which type of rectangles seems to fit the curve most closely?]

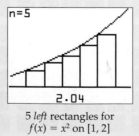

5 *left* rectangles for
$f(x) = x^2$ on [1, 2]

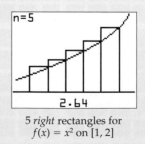

5 *right* rectangles for
$f(x) = x^2$ on [1, 2]

*See the Preface for how to obtain this and other programs.

Definite Integral

Approximating the area under a nonnegative function f by n rectangles means multiplying heights $f(x)$ by widths Δx and adding, obtaining

$$f(x_1) \cdot \Delta x + f(x_2) \cdot \Delta x + f(x_3) \cdot \Delta x + \cdots + f(x_n) \cdot \Delta x$$

The general procedure is as follows:

Area Under f from a to b Approximated by n Left Rectangles

1. Calculate the rectangle width $\Delta x = \dfrac{b - a}{n}$.

2. Find x-values $x_1, x_2, \ldots, x_n$ by successive additions of Δx beginning with $x_1 = a$.

3. Calculate the sum:
$$f(x_1) \cdot \Delta x + f(x_2) \cdot \Delta x + f(x_3) \cdot \Delta x + \cdots + f(x_n) \cdot \Delta x$$

The sum in step 3 is called a *Riemann sum*, after the great German mathematician Georg Bernhard Riemann (1826–1866).* The *limit* of the Riemann sum as the number n of rectangles approaches infinity gives the *area under the curve*, and is called the *definite integral of the function f from a to b*, written $\displaystyle\int_a^b f(x)\, dx$. Formally:

Definite Integral

Let f be a continuous function on an interval $[a, b]$. The definite integral of f from a to b is defined as

$$\int_a^b f(x)\, dx = \lim_{n \to \infty} \left[f(x_1) \cdot \Delta x + f(x_2) \cdot \Delta x + \cdots + f(x_n) \cdot \Delta x \right]$$

where $\Delta x = \dfrac{b - a}{n}$, and $x_1, x_2, \ldots, x_n$ are x-values beginning with $x_1 = a$ and obtained by successive additions of Δx. If f is nonnegative on $[a, b]$, then the definite integral gives the *area under the curve from a to b*.

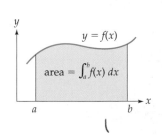

area = $\int_a^b f(x)\, dx$

* Actually, Riemann sums are slightly more general, allowing the subintervals to have different widths, with each width multiplied by the function evaluated at *any* x-value within that subinterval. For a continuous function, any Riemann sum will approach the same limiting value as the rectangles become arbitrarily narrow, so we may restrict our attention to the particular Riemann sums defined above (sometimes called "left Riemann sums").

The numbers a and b are called the *lower* and *upper limits of integration.*

Fundamental Theorem of Integral Calculus

A function followed by a vertical bar $\Big|_a^b$ with numbers a and b means evaluate the function at the *upper* number b and then subtract the evaluation at the *lower* number a.

$$F(x)\,\Big|_a^b \quad = \quad \underbrace{F(b)}_{\substack{\text{Evaluation at} \\ \text{upper number}}} \quad - \quad \underbrace{F(a)}_{\substack{\text{Evaluation at} \\ \text{lower number}}}$$

EXAMPLE 2   **USING THE EVALUATION NOTATION**

$$\underbrace{x^2\,\Big|_3^5}_{\substack{\text{Evaluation} \\ \text{notation}}} = \underbrace{(5)^2}_{\substack{x^2 \\ \text{at } x = 5}} - \underbrace{(3)^2}_{\substack{x^2 \\ \text{at } x = 3}} = \underbrace{25 - 9}_{\text{Simplifying}} = 16$$

Practice Problem 1   Evaluate $\sqrt{x}\,\Big|_4^{25}$ ➤ Solution on page 383

The following *Fundamental Theorem of Integral Calculus* shows how to evaluate definite integrals by using *indefinite* integrals. A geometric and intuitive justification of the theorem is given at the end of this section.

> **Fundamental Theorem of Integral Calculus**
>
> For a continuous function f on an interval $[a, b]$,
>
> $$\int_a^b f(x)\,dx = F(b) - F(a)$$
>
> The right-hand side may be written $F(x)\,\Big|_a^b$
>
> where F is any antiderivative of f.

The theorem is "fundamental" in that it establishes a deep and unexpected connection between definite integrals (limits of Riemann sums) and antiderivatives. It says that definite integrals can be evaluated in two simple steps:

1. Find an *indefinite* integral of the function (omitting the $+ C$).
2. *Evaluate* the result at b and *subtract* the evaluation at a.

| EXAMPLE 3 | **FINDING A DEFINITE INTEGRAL BY THE FUNDAMENTAL THEOREM** |

Find $\displaystyle\int_1^2 x^2\,dx$.

Solution

Because x^2 is continuous on $[1, 2]$, we can use the Fundamental Theorem.

$$\int_1^2 x^2\,dx = \left.\frac{1}{3}x^3\right|_1^2 = \frac{1}{3}2^3 - \frac{1}{3}1^3 = \frac{8}{3} - \frac{1}{3} = \frac{7}{3}$$

$$\underbrace{}_{\substack{\text{Integrating} \\ x^2}} \quad \underbrace{}_{\substack{\text{Evaluating} \\ \frac{1}{3}x^3 \text{ at } x=2}} \quad \underbrace{}_{\substack{\text{And at} \\ x=1}} \quad \underbrace{}_{\text{Subtracting}}$$

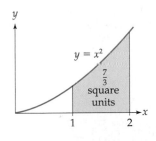

Since definite integrals of nonnegative functions give areas, this result means that the area under $f(x) = x^2$ from $x = 1$ to $x = 2$ is $\frac{7}{3}$ square units.

On pages 370–371 we found this same area by the much more laborious process of calculating Riemann sums and taking the limit as the number of rectangles approached infinity, and our answer here, $\frac{7}{3}$, agrees with the answer there, $2\frac{1}{3}$ square units. Whenever possible, we will calculate definite integrals and find areas in this much simpler way, using the Fundamental Theorem of Integral Calculus.

Graphing Calculator Exploration

a. Graph $y_1 = x^2$ on the graphing window $[0, 2]$ by $[-1, 4]$.

b. Verify the result of Example 3 by having your graphing calculator find the definite integral of x^2 from 1 to 2. [*Hint:* Use a command like FnInt or $\int f(x)\,dx$.]

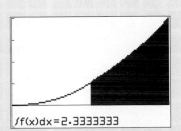

Practice Problem 2   Find the area under $y = x^3$ from 0 to 2 by evaluating $\int_0^2 x^3\, dx.$

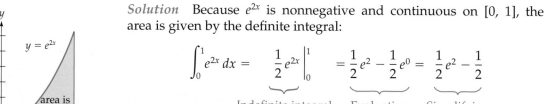

➤ Solution on page 383

EXAMPLE 4   **FINDING THE AREA UNDER A CURVE**

Find the area under $y = e^{2x}$ from $x = 0$ to $x = 1$.

Solution   Because e^{2x} is nonnegative and continuous on $[0, 1]$, the area is given by the definite integral:

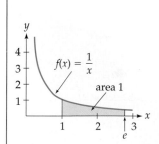

$$\int_0^1 e^{2x}\, dx = \underbrace{\left.\frac{1}{2}e^{2x}\right|_0^1}_{\substack{\text{Indefinite integral} \\ \text{using the formula} \\ \text{for } \int e^{ax}\, dx}} = \underbrace{\frac{1}{2}e^2 - \frac{1}{2}e^0}_{\substack{\text{Evaluating} \\ \text{at } x = 1 \\ \text{and at } x = 0}} = \underbrace{\frac{1}{2}e^2 - \frac{1}{2}}_{\substack{\text{Simplifying} \\ (\text{using } e^0 = 1)}}$$

Therefore, the area is $\frac{1}{2}e^2 - \frac{1}{2}$ square units.

We leave the answer in this "exact" form (in terms of the number e). In an application we would use a calculator to approximate this answer as 3.19 square units.

EXAMPLE 5   **FINDING THE AREA UNDER A CURVE**

Find the area under $f(x) = \dfrac{1}{x}$ from $x = 1$ to $x = e$.

Solution   Because $\dfrac{1}{x}$ is nonnegative and continuous on $[1, e]$, the area is:

$$\int_1^e \frac{1}{x}\, dx = \left.\ln x\right|_1^e = \underbrace{\ln e}_{1} - \underbrace{\ln 1}_{0} = 1$$

Therefore, the area is 1 square unit.

From this example we can give an alternate definition of the number e: e is the number such that the definite integral of $1/x$ from 1 to e is 1.

Graphing Calculator Exploration

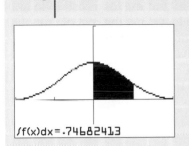

$\int f(x)dx = .74682413$

Use a graphing calculator to evaluate the definite integral

$$\int_0^1 e^{-x^2}\, dx.$$

(This definite integral, which is important in probability and statistics, cannot be evaluated by the Fundamental Theorem because the function e^{-x^2} has no simple antiderivative. Your calculator may find the answer by approximating the area under the curve by modified Riemann sums.)

Definite integrals have many of the properties of indefinite integrals. These properties follow from interpreting definite integrals as limits of Riemann sums.

Properties of Definite Integrals

$$\int_a^b c \cdot f(x)\, dx = c \int_a^b f(x)\, dx$$

A constant may be moved across the integral sign

$$\int_a^b [f(x) \pm g(x)]\, dx = \int_a^b f(x)\, dx \pm \int_a^b g(x)\, dx$$

$\big\lfloor$ Read both upper signs or both lower signs $\big\rfloor$

The integral of a sum is the sum of the integrals (and similarly for differences)

EXAMPLE 6

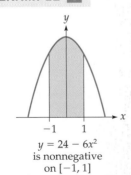

$y = 24 - 6x^2$
is nonnegative
on $[-1, 1]$

USING THE PROPERTIES OF DEFINITE INTEGRALS

Find the area under $y = 24 - 6x^2$ from -1 to 1.

Solution

$$\int_{-1}^1 (24 - 6x^2)\, dx = \left(24x - 6 \cdot \frac{1}{3} x^3\right)\Big|_{-1}^1$$

$$= (24 - 2) - (-24 + 2) = 44$$

Therefore, the area is 44 square units.

Total Cost of a Succession of Units

Given a marginal cost function, to find the *total* cost of producing, say, units 100 to 400, we could proceed as follows: *integrate* the marginal cost to find the total cost, *evaluate* at 400 to find the total cost up to unit 400, and *subtract* the evaluation at 100 to leave just the cost of units 100 to 400. However, these steps of integrating, evaluating, and subtracting are just the steps in evaluating a definite integral by the Fundamental Theorem. Therefore, the cost of a succession of units is equal to the definite integral of the marginal cost function.

Cost of a Succession of Units

For a marginal cost function $MC(x)$:

$$\begin{pmatrix} \text{Total cost of} \\ \text{units } a \text{ to } b \end{pmatrix} = \int_a^b MC(x)\, dx$$

EXAMPLE 7

FINDING THE COST OF A SUCCESSION OF UNITS

A company's marginal cost function is $MC(x) = \dfrac{75}{\sqrt{x}}$, where x is the number of units. Find the total cost of producing units 100 to 400.

Solution

$$\begin{pmatrix} \text{Total cost of} \\ \text{units 100 to 400} \end{pmatrix} = \int_{100}^{400} \frac{75}{\sqrt{x}}\, dx = 75 \int_{100}^{400} x^{-1/2}\, dx \qquad \text{Integrating} \\ \text{marginal cost}$$

$$= (75 \cdot 2 \cdot x^{1/2}) \Big|_{100}^{400}$$

$$= 150 \cdot (400)^{1/2} - 150 \cdot (100)^{1/2}$$

$$= 150 \cdot 20 - 150 \cdot 10 = 1500$$

The cost of producing units 100 to 400 is $1500.

Similarly, integrating *any* rate from a to b gives the *total accumulation* at that rate between a and b.

Total Accumulation at a Given Rate

$$\begin{pmatrix} \text{Total accumulation at} \\ \text{rate } f \text{ from } a \text{ to } b \end{pmatrix} = \int_a^b f(x)\, dx$$

The diagrams below illustrate this idea. In each case, the *curve* represents a *rate*, and the *area under the curve*, given by the definite integral, gives the *total accumulation* at that rate.

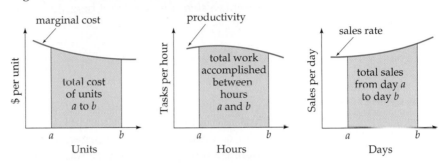

In repetitive tasks, a person's productivity usually increases with practice until it is slowed by monotony.

EXAMPLE 8

FINDING TOTAL PRODUCTIVITY FROM A RATE

A technician can test computer chips at the rate of $-3t^2 + 18t + 15$ chips per hour (for $0 \le t \le 6$), where t is the number of hours after 9:00 A.M. How many chips can be tested between 10:00 A.M. and 1:00 P.M.?

Solution

The total work accomplished is the integral of this rate from $t = 1$ (10 A.M.) to $t = 4$ (1 P.M.):

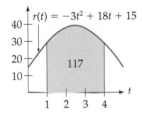

$$\int_1^4 (-3t^2 + 18t + 15) \, dt = (-t^3 + 18 \cdot \frac{1}{2} t^2 + 15t) \Big|_1^4$$

$$= (-64 + 144 + 60) - (-1 + 9 + 15) = 117$$

That is, between 10 A.M. and 1 P.M., 117 chips can be tested.

Integration Notation

The symbol Σ (the Greek letter S) is used in mathematics to indicate a *sum*, and so the Riemann sum can be written $\sum_1^n f(x_k) \, \Delta x$. The fact that the Riemann sum approaches the definite integral can be expressed:

$$\sum_1^n f(x_k) \, \Delta x \longrightarrow \int_a^b f(x) \, dx$$

Σ becomes $\int$
Δ becomes d
as $n \to \infty$

That is, the n approaching infinity changes the Σ (a Greek S) into an integral sign $\int$ (a "stretched out" S), and the Δ (a Greek D) into a d. In other words, the integral notation reminds us that a definite integral represents a *sum of rectangles* of height $f(x)$ and width dx.

Verification of the Fundamental Theorem of Integral Calculus

The following is a geometric and intuitive justification of the Fundamental Theorem of Integral Calculus.

For a continuous and nonnegative function f on an interval $[a, b]$, we define a new function $A(x)$ as the *area under f from a to x*.

$$A(x) = \begin{pmatrix} \text{Area under } f \\ \text{from } a \text{ to } x \end{pmatrix}$$

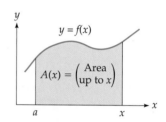

Therefore

$$A(b) = \begin{pmatrix} \text{Area under } f \\ \text{from } a \text{ to } b \end{pmatrix} \qquad \text{Replacing } x \text{ by } b \text{ gives the entire area from } a \text{ to } b$$

and

$$A(a) = 0 \qquad \text{The area "from } a \text{ to } a\text{" must be zero}$$

Subtracting these last two expressions, $A(b) - A(a)$, gives the total area from a to b [since $A(b)$ is the total area, and $A(a)$ is zero]. This same area can be expressed as the definite integral of f from a to b, leading to the equation

$$\int_a^b f(x)\,dx = A(b) - A(a) \qquad \text{Area under } f \text{ from } a \text{ to } b \text{ expressed in two ways}$$

If we can show that $A(x)$ is an antiderivative of $f(x)$, then the equation above will show that the definite integral can be found by an antiderivative evaluated at the upper and lower limits and subtracted, which will verify the Fundamental Theorem. To show that $A(x)$ is an antiderivative of $f(x)$, we show that $A'(x) = f(x)$, differentiating $A(x)$ by the definition of the derivative:

$$A'(x) = \lim_{h \to 0} \frac{A(x + h) - A(x)}{h}$$

In the numerator, $A(x + h)$ is the area under the curve up to $x + h$ and $A(x)$ is the area up to x, so subtracting them, $A(x + h) - A(x)$, leaves just the area from x to $x + h$, as shown below.

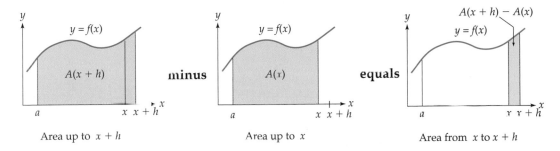

Area up to $x + h$ Area up to x Area from x to $x + h$

When h is small, this last area can be approximated by a rectangle of base h and height $f(x)$, where the approximation becomes exact as h approaches zero. Therefore, in the limit we may replace $A(x + h) - A(x)$ by the area of the rectangle, $h \cdot f(x)$:

$$A'(x) = \lim_{h \to 0} \frac{\overbrace{A(x + h) - A(x)}^{\approx \, h \cdot f(x)}}{h} = \lim_{h \to 0} \frac{h \cdot f(x)}{h} = f(x)$$

This equation says that $A'(x) = f(x)$, showing that $A(x)$ *is* an anti-derivative of $f(x)$. This completes the verification of the Fundamental Theorem of Integral Calculus.

Functions Taking Positive and Negative Values

Riemann sums can be calculated for *any* continuous function, not just nonnegative functions. The following diagram illustrates a Riemann sum for a continuous function that takes positive *and* negative values.

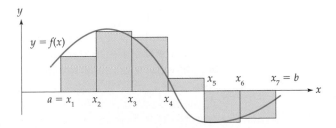

The two rightmost rectangles, where f is *negative*, will make a *negative* contribution to the sum. Taking the limit, the definite integral where the function lies *below* the x-axis will give the *negative* of the area between the curve and the x-axis. Therefore, the definite integral of such a function from a to b will give the *signed area* between the curve and the x-axis: the area *above* the axis minus the area *below* the axis, shown as A_{up} *minus* A_{down} in the diagram below.

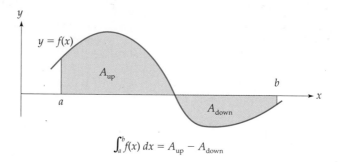

$$\int_a^b f(x)\, dx = A_{up} - A_{down}$$

5.3 Section Summary

We began by approximating the area under a curve from a to b by Riemann sums (sums of rectangles). We then defined the *definite integral* $\int_a^b f(x)\, dx$ as the *limit* of the Riemann sum as the number of rectangles approaches infinity. The Fundamental Theorem of Integral Calculus showed how to evaluate definite integrals much more simply by evaluating *indefinite* integrals:

$$\int_a^b f(x)\, dx = F(x)\,\Big|_a^b \qquad\qquad F \text{ is any antiderivative of } f$$

Distinguish carefully between definite and indefinite integrals: a *definite* integral is a *number,* whereas an *indefinite* integral is a *function* plus an arbitrary constant.

As for the *uses* of definite integrals, the definite integral of a nonnegative function gives the *area under the curve:*

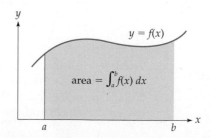

The definite integral of a *rate* gives the *total accumulation* at that rate:

$$\int_a^b f(x)\,dx = \left(\begin{array}{c}\text{Total accumulation at}\\ \text{rate } f \text{ from } a \text{ to } b\end{array}\right)$$

➤ **Solutions to Practice Problems**

1. $\sqrt{x}\,\Big|_4^{25} = \sqrt{25} - \sqrt{4} = 5 - 2 = 3$

2. $\int_0^2 x^3\,dx = \frac{1}{4}x^4\,\Big|_0^2 = \frac{1}{4}\cdot 16 - \frac{1}{4}\cdot 0 = 4$ square units

5.3 **Exercises**

Find the sum of the areas of the shaded rectangles under each graph. Round to two decimal places. [*Hint:* The width of each rectangle is the difference between the x-values at its base. The height of each rectangle is the height of the curve at the left edge of the rectangle.]

1.

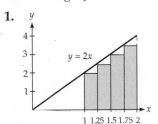

2.

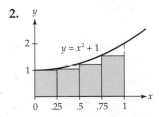

3.

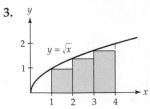

4.

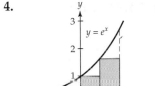

5.

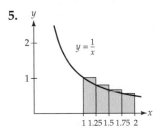

6.

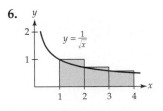

For each function:

i. *Approximate* the area under the curve from a to b by calculating a Riemann sum with the given number of rectangles. Use the method described in Example 1 on pages 370–371, rounding to three decimal places.

ii. Find the *exact* area under the curve from a to b by evaluating an appropriate definite integral using the Fundamental Theorem.

7. $f(x) = 2x$ from $a = 1$ to $b = 2$.
For part (i), use 5 rectangles.

8. $f(x) = x^2 + 1$ from $a = 0$ to $b = 1$.
For part (i), use 5 rectangles.

9. $f(x) = \sqrt{x}$ from $a = 1$ to $b = 4$.
For part (i), use 6 rectangles.

10. $f(x) = e^x$ from $a = -1$ to $b = 1$.
For part (i), use 8 rectangles.

11. $f(x) = \dfrac{1}{x}$ from $a = 1$ to $b = 2$.

For part (i), use 10 rectangles.

12. $f(x) = \dfrac{1}{\sqrt{x}}$ from $a = 1$ to $b = 4$.

For part(i), use 6 rectangles.

Use the graphing calculator program RIEMANN (see page 372), one of the programs given below, or a similar program to find the following Riemann sums.

i. Calculate the Riemann sum for each function for the following values of n: 10, 100, and 1000. Use left, right, or midpoint rectangles, making a table of the answers, rounded to three decimal places.

ii. Find the *exact* value of the area under the curve by evaluating an appropriate definite integral using the Fundamental Theorem. The values of the Riemann sums from part (i) should approach this number.

13. $f(x) = 2x$ from $a = 1$ to $b = 2$

14. $f(x) = x^2 + 1$ from $a = 0$ to $b = 1$

15. $f(x) = \sqrt{x}$ from $a = 1$ to $b = 4$

16. $f(x) = e^x$ from $a = -1$ to $b = 1$

17. $f(x) = \dfrac{1}{x}$ from $a = 1$ to $b = 2$

18. $f(x) = \dfrac{1}{\sqrt{x}}$ from $a = 1$ to $b = 4$

BASIC Program for (Left) Riemann Sums

(Lines 2, 3, 4, and 8 to be completed as indicated in small type)

Riemsum = 0
a =   fill in beginning x-value
b =   fill in ending x-value
n =   fill in number of rectangles
delta = (b − a)/n
x = a
FOR i = 1 to n

Riemsum = Riemsum + $\left(\begin{array}{c}\text{fill in the}\\\text{function}\end{array}\right)$ *delta

x = x + delta
NEXT i
PRINT "RIEMANN SUM IS", Riemsum
END

TI-83 Program for (Left) Riemann Sums

(Enter the function in y_1 before executing)

0 → S
Disp "USES FUNCTION Y1"
Prompt A, B, N
(B-A)/N → D
A → X
For(I, 1, N)
S + Y1*D → S
X + D → X
End
Disp "RIEMANN SUM IS", S

Find the area under each curve between the given *x*-values. For Exercises 19–24, also make a sketch of the curve showing the region.

19. $f(x) = x^2$ from $x = 0$ to $x = 3$

20. $f(x) = x$ from $x = 0$ to $x = 4$

21. $f(x) = 4 - x$ from $x = 0$ to $x = 4$

22. $f(x) = 1 - x^2$ from $x = -1$ to $x = 1$

23. $f(x) = \dfrac{1}{x}$ from $x = 1$ to $x = 2$

24. $f(x) = e^x$ from $x = 0$ to $x = 1$

25. $f(x) = 8x^3$ from $x = 1$ to $x = 3$

26. $f(x) = 6x^2$ from $x = 2$ to $x = 3$

27. $f(x) = 6x^2 + 4x - 1$ from $x = 1$ to $x = 2$

28. $f(x) = 27 - 3x^2$ from $x = 1$ to $x = 3$

29. $f(x) = \dfrac{1}{\sqrt{x}}$ from $x = 4$ to $x = 9$

30. $f(x) = \dfrac{1}{x^2}$ from $x = 1$ to $x = 3$

31. $f(x) = 8 - 4\sqrt[3]{x}$ from $x = 0$ to $x = 8$

32. $f(x) = 9 - 3\sqrt{x}$ from $x = 0$ to $x = 9$

33. $f(x) = \dfrac{1}{x}$ from $x = 1$ to $x = 5$

34. $f(x) = \dfrac{1}{x}$ from $x = e$ to $x = e^3$

35. $f(x) = x^{-1} + x^2$ from $x = 1$ to $x = 2$

36. $f(x) = 6e^{2x}$ from $x = 0$ to $x = 2$

37. $f(x) = 2e^x$ from $x = 0$ to $x = \ln 3$

38. $f(x) = e^{-x}$ from $x = 0$ to $x = 1$

39. $f(x) = e^{x/2}$ from $x = 0$ to $x = 2$

40. $f(x) = e^{x/3}$ from $x = 0$ to $x = 3$

For each function:
a. Integrate ("by hand") to find the area under the curve between the given *x*-values.
b. Verify your answer to part (a) by having your calculator graph the function and find the area (using a command like FnInt or $\int f(x)\, dx$).

41. $f(x) = 12 - 3x^2$ from $x = 1$ to $x = 2$

42. $f(x) = 9x^2 - 6x + 1$ from $x = 1$ to $x = 2$

43. $f(x) = \dfrac{1}{x^3}$ from $x = 1$ to $x = 4$

44. $f(x) = \dfrac{1}{\sqrt[3]{x}}$ from $x = 8$ to $x = 27$

45. $f(x) = 2x + 1 + x^{-1}$ from $x = 1$ to $x = 2$

46. $f(x) = e^x$ from $x = 0$ to $x = 3$

Evaluate each definite integral.

47. $\displaystyle\int_0^1 (x^{99} + x^9 + 1)\, dx$

48. $\displaystyle\int_2^1 (1 + x^{-2})\, dx$

49. $\displaystyle\int_1^2 (6t^2 - 2t^{-2})\, dt$

50. $\displaystyle\int_{-2}^2 (3w^2 - 2w)\, dw$

51. $\displaystyle\int_1^4 \dfrac{1}{y^2}\, dy$

52. $\displaystyle\int_1^4 \dfrac{1}{\sqrt{z}}\, dz$

53. $\displaystyle\int_1^e \dfrac{dx}{x}$

54. $\displaystyle\int_1^{e^2} \dfrac{3}{x}\, dx$

55. $\displaystyle\int_1^3 (9x^2 + x^{-1})\, dx$

56. $\displaystyle\int_1^2 (x^{-1} - 4x^2)\, dx$

57. $\displaystyle\int_{-2}^{-1} 3x^{-1}\, dx$

58. $\displaystyle\int_{-3}^{-1} (1 + x^{-1})\, dx$

59. $\displaystyle\int_0^1 12e^{3x}\, dx$

60. $\displaystyle\int_0^2 3e^{x/2}\, dx$

61. $\displaystyle\int_{-1}^1 5e^{-x}\, dx$

62. $\displaystyle\int_0^1 (6x^2 - 4e^{2x})\, dx$

63. $\displaystyle\int_{\ln 2}^{\ln 3} e^x\, dx$

64. $\displaystyle\int_0^{\ln 5} e^x\, dx$

65. $\displaystyle\int_1^2 \dfrac{(x + 1)^2}{x}\, dx$

66. $\displaystyle\int_1^2 \dfrac{(x + 1)^2}{x^2}\, dx$

Use a graphing calculator to evaluate each definite integral, rounding answers to three decimal places. [*Hint:* Use a command like FnInt or $\int f(x)\, dx$.]

67. $\displaystyle\int_0^2 \dfrac{1}{x^2 + 1}\, dx$

68. $\displaystyle\int_{-1}^1 \sqrt{x^4 + 1}\, dx$

69. $\displaystyle\int_{-1}^1 e^{x^2}\, dx$

70. $\displaystyle\int_{-2}^2 e^{(-1/2)x^2}\, dx$

71. $\displaystyle\int_0^4 \sqrt{x}\, e^x\, dx$

72. $\displaystyle\int_1^4 x^x\, dx$

[handwritten: Integrand is not continuous at x=0]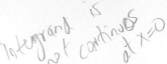

73. OMITTING THE *C* IN DEFINITE INTEGRALS

a. Evaluate the definite integral $\int_0^3 x^2\, dx$.

b. Evaluate the same definite integral by completing the following calculation, in which the antiderivative includes a constant *C*.

$$\int_0^3 x^2\, dx = \left(\frac{1}{3}x^3 + C\right)\Big|_0^3 = \cdots$$

[The constant *C* should cancel out, giving the same answer as in part (a)].

c. Explain why the constant will cancel out of *any* definite integral. (We therefore omit the constant in definite integrals. However, be sure to keep the +*C* in *indefinite* integrals.)

74. Evaluate $\int_1^1 \dfrac{x^{43}e^{\sqrt[3]{x^2}} + x^{-39}}{\ln^{17}\sqrt{x^3 + 11} + x^{199}}\, dx$.

[*Hint:* No work is necessary.]

[handwritten: there is no Area]

75. Show that for any number *a* > 0,

$$\int_1^a \frac{1}{x}\, dx = \ln a$$

This equation is often used as a *definition* of natural logarithms, defining ln *a* as the area under the curve $y = 1/x$ between 1 and *a*.

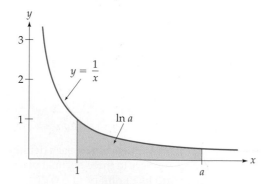

76. a. Try to evaluate the integral $\int_{-1}^1 \dfrac{1}{x^2}\, dx$.

b. Explain why the answer is negative in spite of the fact that the integrand is nonnegative, and so the integral should be positive. [*Hint:* In order to use the Fundamental Theorem, or even to *define* the definite integral, the integrand must be continuous on the interval. Is it?]

77–78. GEOMETRIC INTEGRATION Find

$\int_0^4 f(x)\, dx$ for the function *f(x)* graphed below.

[*Hint:* No calculus is necessary—just as we used integrals to find areas, we can use areas to find integrals. The curves shown are quarter circles.]

77.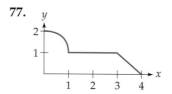

78.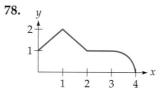

APPLIED EXERCISES

79. GENERAL: Electrical Consumption

On a hot summer afternoon, a city's electricity consumption is $-3t^2 + 18t + 10$ units per hour, where *t* is the number of hours after noon ($0 \le t \le 6$). Find the total consumption of electricity between the hours of 1 and 5 p.m.

80. GENERAL: Weight

An average child of age *x* years gains weight at the rate of $3.9x^{1/2}$ pounds per year (for $0 \le x \le 16$). Find the total weight gain from age 1 to age 9.

81–82. GENERAL: Repetitive Tasks After t hours of work, a bank clerk can process checks at the rate of $r(t)$ checks per hour for the function $r(t)$ given below. How many checks will the clerk process during the first three hours (time 0 to time 3)?

81. $r(t) = -t^2 + 90t + 5$

82. $r(t) = -t^2 + 60t + 9$

83–84. BUSINESS: Cost A company's marginal cost function is $MC(x)$ (given below), where x is the number of units. Find the total cost of the first hundred units ($x = 0$ to $x = 100$).

83. $MC(x) = 6e^{-0.02x}$ **84.** $MC(x) = 8e^{-0.01x}$

85. GENERAL: Price Increase The price of a double-dip ice cream cone is increasing at the rate of $18e^{0.08t}$ cents per year, where t is measured in years and $t = 0$ corresponds to 2000. Find the total change in price between the years 2000 and 2010.

86. BUSINESS: Sales An automobile dealer estimates that the newest model car will sell at the rate of $30/t$ cars per month, where t is measured in months and $t = 1$ corresponds to the beginning of January. Find the number of cars that will be sold from the beginning of January to the beginning of May.

87. BUSINESS: Tin Consumption World consumption of tin is running at the rate of $0.24e^{0.01t}$ million tons per year, where t is measured in years and $t = 0$ corresponds to the beginning of 2000. Find the total consumption of tin from the beginning of 2000 to the beginning of 2010.

88. SOCIOLOGY: Marriages There are approximately $2.2e^{0.01t}$ million marriages per year in the United States, where t is the number of years since 2000. Assuming that this rate continues, find the number of marriages from the year 2000 to the year 2010.

89. BEHAVIORAL SCIENCE: Learning A student can memorize words at the rate of $6e^{-t/5}$ words per minute after t minutes. Find the

total number of words that the student can memorize in the first 10 minutes.

90. BIOMEDICAL: Epidemics An epidemic is spreading at the rate of $12e^{0.2t}$ new cases per day, where t is the number of days since the epidemic began. Find the total number of new cases in the first 10 days of the epidemic.

91. ECONOMICS: Pareto's Law The economist Vilfredo Pareto estimated that the number of people who have an income between A and B dollars ($A < B$) is given by a definite integral of the form

$$N = \int_A^B ax^{-b}\, dx \qquad (b \ne 1)$$

where a and b are constants. Solve this integral.

92. BIOMEDICAL: Poiseuille's Law According to Poiseuille's law, the speed of blood in a blood vessel is given by $V = \dfrac{p}{4Lv}(R^2 - r^2)$, where R is the radius of the blood vessel, r is the distance of the blood from the center of the blood vessel, and p, L, and v are constants determined by the pressure and viscosity of the blood and the length of the vessel. The total blood flow is then given by

$$\left(\begin{array}{c}\text{Total}\\\text{blood flow}\end{array}\right) = \int_0^R 2\pi \frac{p}{4Lv}(R^2 - r^2)r\, dr$$

Find the total blood flow by finding this integral (p, L, v, and R are constants).

93–94. BUSINESS: Capital Value of an Asset The *capital value* of an asset (such as an oil well) that produces a continuous stream of income is the sum of the present value of all future earnings from the asset. Therefore, the capital value of an asset that produces income at the rate of $r(t)$ dollars per year (at a continuous interest rate i) is

$$\left(\begin{array}{c}\text{Capital}\\\text{value}\end{array}\right) = \int_0^T r(t)e^{-it}\, dt$$

where T is the expected life (in years) of the asset.

93. Use the formula in the preceding instructions to find the capital value (at interest rate $i = 0.06$) of an oil well that produces income at the constant rate of $r(t) = 240{,}000$ dollars per year for 10 years.

94. Use the formula in the preceding instructions to find the capital value (at interest rate $i = 0.05$) of a uranium mine that produces income at the rate of $r(t) = 560{,}000t^{1/2}$ dollars per year for 20 years.

95. GENERAL: Area

   a. Use your graphing calculator to find the area between 0 and 1 under the following curves: $y = x$, $y = x^2$, $y = x^3$, and $y = x^4$.

   b. Based on your answers to part (a), conjecture a formula for the area under $y = x^n$ between 0 and 1 for any value of $n > 0$.

   c. Prove your conjecture by evaluating an appropriate definite integral "by hand."

96. GENERAL: Dam Construction Ever since the Johnstown Dam burst in 1889, killing 2200 people, dam construction has become increasingly scientific.

   a. Estimate the amount of concrete needed to build the Snake River Dam by finding the area of the cross section shown in the following diagram and multiplying the result by the 574-foot length of the dam. All dimensions are in feet. [*Hint:* Find the cross-sectional area by integrating and using area formulas.]

   b. If a mixing truck carries 300 cubic feet of concrete, about how many truckloads would be needed to build the dam?

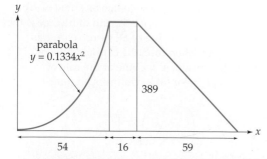

97. BIOMEDICAL: Drug Absorption An oral medication is absorbed into the bloodstream at the rate of $5e^{-0.04t}$ milligrams per minute,

where t is the number of minutes since the medication was taken. Find the total amount of medication absorbed within the first 30 minutes.

98. BIOMEDICAL: Aortic Volume The rate of change of the volume of blood in the aorta t seconds after the beginning of the cardiac cycle is $-kP_0e^{-mt}$ milliliters per second, where k, P_0, and m are constants (depending, respectively, on the elasticity of the aorta, the initial aortic pressure, and various characteristics of the cardiac cycle). Find the total change in volume from time 0 to time T (the end of the cardiac cycle). (Your answer will involve the constants k, P_0, m, and T.)

99. BUSINESS: Sales A dealer predicts that new cars will sell at the rate of $8xe^{-0.1x}$ sales per week in week x. Find the total sales in the first half year (week 0 to week 26).

100. GENERAL: Cigarette Smoking Reread, if necessary, the Application Preview on pages 341–342.

   a. Evaluate the definite integrals

   $$\int_0^1 (300e^{0.025x} - 240e^{0.02x})\, dx$$

   and $\displaystyle\int_7^8 (300e^{0.025x} - 240e^{0.02x})\, dx$

   to verify the answers given there for the amount of tar inhaled from the first and last centimeters of the cigarette.

   b. Evaluate the definite integral

   $$\int_0^8 (300e^{0.025x} - 240e^{0.02x})\, dx$$

   to find the amount of tar inhaled from smoking the entire cigarette.

101. GENERAL: Population A resort community swells at the rate of $100e^{0.4\sqrt{x}}$ new arrivals per day on day x of its "high season." Find the total number of arrivals in the first two weeks (day 0 to day 14).

102. GENERAL: Repetitive Tasks After t hours of work, a medical technician can carry out T-cell counts at the rate of $2x^2e^{-x/4}$ tests per hour. How many tests will the technician process during the first eight hours (time 0 to time 8)?

5.4 **FURTHER APPLICATIONS OF DEFINITE INTEGRALS: AVERAGE VALUE AND AREA BETWEEN CURVES**

Introduction

In this section we will use definite integrals for two important purposes: finding average values of functions and finding areas between curves. Average values are used everywhere. Birth weights of babies are compared with average weights, and retirement benefits are determined by average income. Averages eliminate fluctuations, reducing a collection of numbers to a single "representative" number. Areas between curves are used to find quantities from trade deficits to lives saved by seat belts (see pages 402–403).

Average Value of a Function

The average value of n numbers is found by adding the numbers and dividing by n. For example,

$$\begin{pmatrix} \text{Average of} \\ a, b, \text{ and } c \end{pmatrix} = \frac{1}{3}(a + b + c)$$

How can we find the average value of a *function* on an interval? For example, if a function gives the temperature over a 24-hour period, how can we calculate the *average temperature*? We could, of course, just take the temperature at every hour and then average these 24 values, but this would ignore the temperature at all of the intermediate times. Intuitively, the average should represent a "leveling off" of the curve to a uniform height, the dashed line shown on the left.

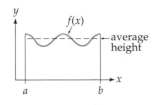

This leveling should use the "hills" to fill in the "valleys," maintaining the same total area under the curve. Therefore, the area under the line (a rectangle with base $(b - a)$ and height up to the line) must equal the area under the curve (the definite integral of the function).

$$\underbrace{(b - a)\begin{pmatrix} \text{Average} \\ \text{height} \end{pmatrix}}_{\text{Area under line}} = \underbrace{\int_a^b f(x)\,dx}_{\text{Area under curve}} \qquad \text{Equating the two areas}$$

Therefore:

$$\begin{pmatrix} \text{Average} \\ \text{height} \end{pmatrix} = \frac{1}{b - a}\int_a^b f(x)\,dx \qquad \text{Dividing by } (b - a)$$

This formula gives the average (or "mean") value of a continuous function on an interval.

Average Value of a Function

$$\left(\begin{array}{c}\text{Average value}\\\text{of } f \text{ on } [a, b]\end{array}\right) = \frac{1}{b - a}\int_a^b f(x)\, dx$$

Average value is the definite integral of the function divided by the length of the interval

---

Finding the average value of a function by integrating and dividing by $b - a$ is analogous to averaging n numbers by adding and dividing by n (since integrals are continuous sums). A derivation of this formula by Riemann sums is given at the end of this section.

EXAMPLE 1

FINDING THE AVERAGE VALUE OF A FUNCTION

Find the average value of $f(x) = \sqrt{x}$ from $x = 0$ to $x = 9$.

Solution

$$\left(\begin{array}{c}\text{Average}\\\text{value}\end{array}\right) = \frac{1}{9 - 0}\int_0^9 \sqrt{x}\, dx = \frac{1}{9}\int_0^9 x^{1/2}\, dx$$

Integral divided by the length of the interval

$$= \frac{1}{9}\frac{2}{3}x^{3/2}\Big|_0^9 = \frac{2}{27}9^{3/2} - \frac{2}{27}0^{3/2}$$

Integrating and evaluating

$$= \frac{2}{27}\left(\sqrt{9}\right)^3 - 0 = \frac{2}{27}27 = 2$$

Simplifying

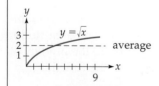

The average value of $f(x) = \sqrt{x}$ over the interval $[0, 9]$ is 2.

EXAMPLE 2

FINDING AVERAGE POPULATION

The population of the United States is predicted to be $P(t) = 281e^{0.012t}$ million people, where t is the number of years since 2000. Predict the average population between the years 2010 and 2020.

Solution We integrate from $t = 10$ (year 2010) to $t = 20$ (year 2020).

$$\left(\begin{matrix}\text{Average} \\ \text{value}\end{matrix}\right) = \frac{1}{20 - 10} \int_{10}^{20} 281 e^{0.012t}\, dt \qquad \begin{matrix}\text{Integral divided} \\ \text{by the length of} \\ \text{the interval}\end{matrix}$$

$$= \frac{281}{10} \int_{10}^{20} e^{0.012t}\, dt$$

$$= 28.1 \frac{1}{0.012} e^{0.012t}\Big|_{10}^{20} \qquad \begin{matrix}\text{Integrating by the} \\ \int e^{ax}\, dx \text{ formula}\end{matrix}$$

$$= 2342 e^{0.24} - 2342 e^{0.12} \approx 337 \qquad \begin{matrix}\text{Evaluating, using} \\ \text{a calculator}\end{matrix}$$

The average population of the United States during the second decade of the twenty-first century will be about 337 million people.

Practice Problem 1

Find the average value of $f(x) = 3x^2$ from $x = 0$ to $x = 2$.

$$\frac{1}{2-0} \int_0^2 3x^2\, dx = \frac{1}{2} x^3 \Big|_0^2 \, dx$$
$$= \frac{1}{2} \left[(2^3) - 0 \right] = 4$$

➤ **Solution on page 398**

Area Between Curves: Integrating "Upper Minus Lower"

We know that definite integrals give areas under curves. To calculate the area *between* two curves, we take the area under the *upper* curve and subtract the area under the *lower* curve.

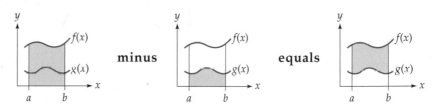

In terms of integrals:

$$\int_a^b f(x)\, dx \qquad - \qquad \int_a^b g(x)\, dx \qquad = \qquad \int_a^b [f(x) - g(x)]\, dx$$

$$\qquad\qquad\qquad\qquad\qquad\qquad\qquad\qquad\qquad\qquad\qquad\qquad \underset{\begin{matrix}\text{Upper} \\ \text{curve}\end{matrix}}{\uparrow} \;\; \underset{\begin{matrix}\text{Lower} \\ \text{curve}\end{matrix}}{\uparrow}$$

Therefore, the area between the curves can be written as a single integral:

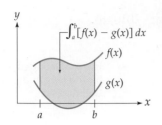

Area Between Curves

The area between two continuous curves $f(x) \geq g(x)$ on $[a, b]$ is

$$\left(\begin{array}{c} \text{Area between} \\ f \text{ and } g \text{ on } [a, b] \end{array}\right) = \int_a^b [f(x) - g(x)]\, dx \qquad \text{Integrate "upper minus lower"}$$

Finding area by integrating "upper minus lower" works regardless of whether one or both curves dip below the x-axis.

EXAMPLE 3

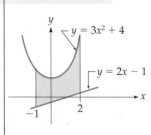

FINDING THE AREA BETWEEN CURVES

Find the area between $y = 3x^2 + 4$ and $y = 2x - 1$ from $x = -1$ to $x = 2$.

Solution The area is shown in the diagram on the left. (You may need to make a similar rough sketch for each problem to see which curve is "upper" and which is "lower.") We integrate "upper minus lower" between the given x-values.

$$\int_{-1}^2 [\underbrace{(3x^2 + 4)}_{\text{Upper}} - \underbrace{(2x - 1)}_{\text{Lower}}]\, dx = \int_{-1}^2 (3x^2 + 4 - 2x + 1)\, dx$$

$$= \int_{-1}^2 (3x^2 - 2x + 5)\, dx \qquad \text{Simplifying}$$

$$= (x^3 - x^2 + 5x)\Big|_{-1}^2 \qquad \text{Integrating}$$

$$= (8 - 4 + 10) - (-1 - 1 - 5) \qquad \text{Evaluating}$$

$$= 21 \text{ square units}$$

Practice Problem 2

Find the area between $y = 2x^2 + 1$ and $y = -x^2 - 1$ from $x = -1$ to $x = 1$.

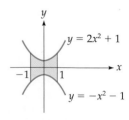

➤ Solution on page 399

$\int_{-1}^{1} [(2x^2+1) - (-x^2-1)] dx = \int_{-1}^{1} (2x^2+1+x^2+1) dx = \int_{-1}^{1} (3x^2+2z) dx$

$= (3(1)^3+2) - (3(-1)^3+2)$

$= 5 - 3+2-1$

$= 4$

If the two curves represent *rates* (one unit per another unit), then the area between the curves gives the *total accumulation* at the upper rate minus the lower rate.

EXAMPLE 4

FINDING SALES FROM EXTRA ADVERTISING

A company marketing high-definition television sets expects to sell them at the rate of $2e^{0.05t}$ thousand sets per month, where t is the number of months since they became available. However, with additional advertising using a sports celebrity, they should sell at the rate of $3e^{0.1t}$ thousand sets per month. How many additional sales would result from the celebrity endorsement during the first year?

Solution

We integrate the difference of the rates from month $t = 0$ (the beginning of the first year) to month $t = 12$ (the end of the year):

$$\int_0^{12} (3e^{0.1t} - 2e^{0.05t})\, dt = \left(3\underbrace{\frac{1}{0.1}}_{30} e^{0.1t} - 2\underbrace{\frac{1}{0.05}}_{40} e^{0.05t} \right)\Bigg|_0^{12}$$

$$= \underbrace{(30e^{1.2} - 40e^{0.6})}_{\text{Evaluating at } t = 12} - \underbrace{(30e^0 - 40e^0)}_{\text{Evaluating at } t = 0}$$

$$\approx \underbrace{26.7 - (-10)}_{\text{Using a calculator}} = 36.7 \qquad \text{In thousands}$$

The celebrity endorsement should result in 36.7 thousand, or 36,700 additional sales during the first year. (The profits from these additional sales must then be compared to the cost of the celebrity endorsement to decide whether it is worthwhile.)

Area Between Curves That Cross

At a point where two curves cross, the upper curve becomes the lower curve and the lower becomes the upper. The area between them must then be calculated by two (or more) integrals, upper minus lower on each interval.

EXAMPLE 5

FINDING THE AREA BETWEEN CURVES THAT CROSS

Find the area between the curves $y = 12 - 3x^2$ and $y = 4x + 5$ from $x = 0$ to $x = 3$.

Solution

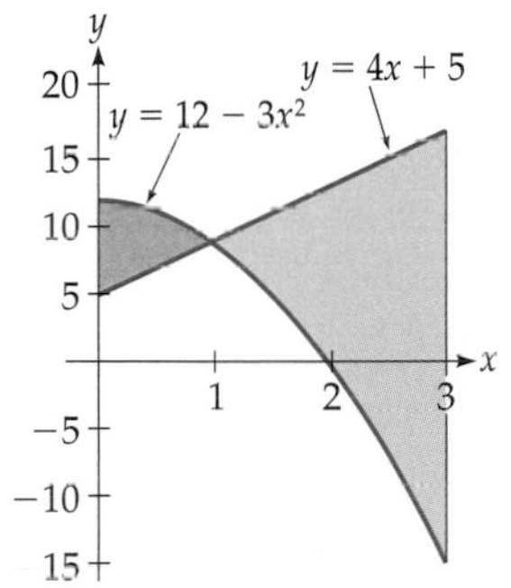

A sketch shows that the curves *do* cross. To find the intersection point, we set the functions equal to each other and solve.

$4x + 5 = 12 - 3x^2$	Equating the two functions
$3x^2 + 4x - 7 = 0$	Combining all terms on the left
$(3x + 7)(x - 1) = 0$	Factoring
$x = 1$	The other solution, $x = -7/3$, is not in the interval $[0, 3]$

The curves cross at $x = 1$, so we must integrate separately over the intervals $[0, 1]$ and $[1, 3]$. The diagram shows which curve is upper and which is lower on each interval.

$$\underbrace{\int_0^1 [\underbrace{(12 - 3x^2)}_{\text{Upper}} - \underbrace{(4x + 5)}_{\text{Lower}}] \, dx} + \int_1^3 [\underbrace{(4x + 5)}_{\text{Upper}} - \underbrace{(12 - 3x^2)}_{\text{Lower}}] \, dx$$

(Integrating upper minus lower on each interval)

$$= \int_0^1 (-3x^2 - 4x + 7) \, dx + \int_1^3 (3x^2 + 4x - 7) \, dx$$

Simplifying the integrands

$$= (-x^3 - 2x^2 + 7x)\Big|_0^1 + (x^3 + 2x^2 - 7x)\Big|_1^3$$

Integrating and simplifying

$$= -1 - 2 + 7 + 27 + 18 - 21 - (1 + 2 - 7) = 32 \quad \text{Evaluating}$$

Therefore, the area between the curves is 32 square units.

Graphing Calculator Exploration

In Example 5 we used *two* integrals, since "upper" and "lower" switched at $x = 1$.

a. To see what happens if you integrate *without* regard to upper and lower, enter $y_1 = 12 - 3x^2$ and $y_2 = 4x + 5$ and graph them on the window [0, 3] by [−20, 20]. Have your calculator find the definite integral of $y_1 - y_2$ on the interval [0, 3]. (Use a command like FnInt or $\int f(x)\, dx$. You should get a negative answer, which cannot be correct for an area.)

b. Explain why the answer was negative. [*Hint:* Look at the graph.]

c. Finally, obtain the correct answer for the area by returning to the calculation in part (a) and integrating the *absolute value* of the difference, $|y_1 - y_2|$ [on some calculators, entered as ABS($y_1 - y_2$)].

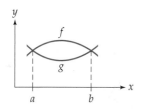

The *x*-values *a* and *b* are where the curves meet.

Areas Bounded by Curves

It is sometimes useful to find the *area bounded by two curves,* without being told the starting and ending *x*-values. In such problems the curves completely enclose an area, and the *x*-values for the upper and lower limits of integration are found by setting the functions equal to each other and solving.

EXAMPLE 6

FINDING AN AREA BOUNDED BY CURVES

Find the area bounded by the curves

$$y = 3x^2 - 12 \quad \text{and} \quad y = 12 - 3x^2$$

Solution The *x*-values for the upper and lower limits of integration are not given, so we find them by setting the functions equal to each other and solving.

$3x^2 - 12 = 12 - 3x^2$	Setting the functions equal to each other
$6x^2 - 24 = 0$	Combining everything on one side
$6(x^2 - 4) = 0$	Factoring
$6(x + 2)(x - 2) = 0$	Factoring further
$x = -2 \text{ and } x = 2$	Solving

The smaller of these, $x = -2$, is the lower limit of integration and the larger, $x = 2$, is the upper limit. To determine which function is "upper" and which is "lower," we choose a "test value" between $x = -2$ and $x = 2$ ($x = 0$ will do), which we substitute into each function to see which is larger. Evaluating each of the original functions at the test point $x = 0$ yields

$$3x^2 - 12 = 3(0)^2 - 12 = -12 \qquad \text{(Smaller)} \qquad \begin{array}{l} 3x^2 - 12 \\ \text{at } x = 0 \end{array}$$

$$12 - 3x^2 = 12 - 3(0)^2 = 12 \qquad \text{(Larger)} \qquad \begin{array}{l} 12 - 3x^2 \\ \text{at } x = 0 \end{array}$$

Therefore, $y = 12 - 3x^2$ is the "upper" function (since it gives a higher y-value) and $y = 3x^2 - 12$ is the "lower" function. We then integrate upper minus lower between the x-values found earlier.

$$\int_{-2}^{2} [\underbrace{12 - 3x^2}_{\text{Upper}} - \underbrace{(3x^2 - 12)}_{\text{Lower}}]\, dx = \int_{-2}^{2} (24 - 6x^2)\, dx \qquad \text{Simplifying}$$

$$= (24x - 2x^3)\,\Big|_{-2}^{2} \qquad \text{Integrating}$$

$$= (48 - 16) - (-48 + 16) = 64 \qquad \text{Evaluating}$$

The area bounded by the two curves is 64 square units.

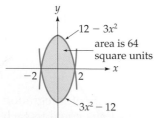

The two curves $y = 12 - 3x^2$ and $y = 3x^2 - 12$ are shown in the graph on the left. Notice that we were able to calculate the area between them without having to graph them.

Practice Problem 3 Find the area bounded by $y = 2x^2 - 1$ and $y = 2 - x^2$.

➤ Solution on page 399

For curves that intersect at *more* than two points, several integrals may be needed, integrating "upper minus lower" on each interval, as in Example 5. Test points in each interval will determine the upper and lower curves on that interval.

5.4 **Section Summary**

The average value of a continuous function over an interval is defined as the definite integral of the function divided by the length of the interval:

$$\left(\begin{array}{c}\text{Average value}\\\text{of } f \text{ on } [a, b]\end{array}\right) - \frac{1}{b-a}\int_a^b f(x)\,dx$$

To find the area between two curves:

1. If the x-values are not given, set the functions equal to each other and solve for the points of intersection.
2. Use a test point within each interval to determine which curve is "upper" and which is "lower."
3. Integrate "upper minus lower" on each interval.

If a curve lies *below* the x-axis, as in the following diagram, then the "upper" curve is the x-axis $(y = 0)$ and the "lower" curve is $y = f(x)$, and so integrating "upper minus lower" results in integrating the *negative* of the function.

$$\int_a^b \underbrace{[0}_{\text{Upper}} - \underbrace{f(x)]}_{\text{Lower}}\,dx = \int_a^b [-f(x)]\,dx = -\int_a^b f(x)\,dx$$

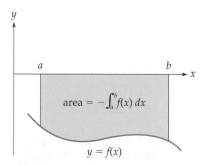

This case need not be remembered separately if you simply remember always to integrate "upper minus lower" over each interval.

Definite integration provides a very powerful method for finding areas, taking us far beyond the few formulas (for rectangles, triangles, and circles) that we knew before studying calculus.

Average Value Formula Derived from Riemann Sums

The formula for the average value of a function (page 390) can be derived using Riemann sums. For a continuous function f we could define a "sample average" by "sampling" the function at n points and averaging the results. From the interval $[a, b]$, we choose n numbers $x_1, x_2, \ldots, x_n$ from successive subintervals of length $\Delta x = \dfrac{b - a}{n}$, sum the resulting values of the function, and divide by n. This gives a "sample average" of the following form:

$$\frac{1}{n}[f(x_1) + f(x_2) + \cdots + f(x_n)] \qquad \text{Sum of } n \text{ values divided by } n$$

$$= \frac{1}{b - a} \cdot \underbrace{\frac{b - a}{n}}_{\Delta x} [f(x_1) + f(x_2) + \cdots + f(x_n)] \qquad \begin{array}{l}\text{Dividing and} \\ \text{multiplying by } b - a\end{array}$$

$$= \frac{1}{b - a} \underbrace{[f(x_1) + f(x_2) + \cdots + f(x_n)] \cdot \Delta x}_{\text{This is a Riemann sum for } \int_a^b f(x)\, dx} \qquad \begin{array}{l}\text{Moving } \Delta x = \dfrac{b - a}{n} \\ \text{to the right}\end{array}$$

To get a more "representative" average, we increase the number n of sample points. Letting n approach infinity makes the above Riemann sum approach the definite integral $\int_a^b f(x)\, dx$, leading to the definition of the average value of a function:

$$\left(\begin{array}{c}\text{Average value} \\ \text{of } f \text{ on } [a, b]\end{array}\right) = \frac{1}{b - a} \int_a^b f(x)\, dx$$

➤ Solutions to Practice Problems

1. $\dfrac{1}{2}\displaystyle\int_0^2 3x^2\, dx = \dfrac{1}{2} \cdot x^3 \Big|_0^2 = \dfrac{1}{2} \cdot 2^3 - \dfrac{1}{2} \cdot 0^3 = \dfrac{1}{2} \cdot 8 = 4$

2. $\displaystyle\int_{-1}^{1} [(2x^2 + 1) - (-x^2 - 1)]\, dx$

$\displaystyle = \int_{-1}^{1} (2x^2 + 1 + x^2 + 1)\, dx$

$\displaystyle = \int_{-1}^{1} (3x^2 + 2)\, dx = (x^3 + 2x)\Big|_{-1}^{1}$

$= (1 + 2) - (-1 - 2) = 6$ square units

3. $2x^2 - 1 = 2 - x^2$

$3x^2 - 3 = 0$

$3(x^2 - 1) = 0$

$3(x + 1)(x - 1) = 0$

$x = 1 \quad$ and $\quad x = -1.$

Test value $x = 0$ shows that $2 - x^2$ is "upper" and $2x^2 - 1$ is "lower."

$\displaystyle\int_{-1}^{1} [(2 - x^2) - (2x^2 - 1)]\, dx$

$\displaystyle = \int_{-1}^{1} (3 - 3x^2)\, dx = (3x - x^3)\Big|_{-1}^{1}$

$= (3 - 1) - (-3 + 1) = 4$ square units

5.4 **Exercises**

Average Value

Find the average value of each function over the given interval.

1. $f(x) = x^2$ on $[0, 3]$ **2.** $f(x) = x^3$ on $[0, 2]$

3. $f(x) = 3\sqrt{x}$ on $[0, 4]$ **4.** $f(x) = \sqrt[3]{x}$ on $[0, 8]$

5. $f(x) = \dfrac{1}{x^2}$ on $[1, 5]$ **6.** $f(x) = \dfrac{1}{x^2}$ on $[1, 3]$

7. $f(x) = 2x + 1$ on $[0, 4]$

8. $f(x) = 4x - 1$ on $[0, 10]$

9. $f(x) = 36 - x^2$ on $[-2, 2]$

10. $f(x) = 9 - x^2$ on $[-3, 3]$

11. $f(x) = 3$ on $[10, 50]$

12. $f(x) = 2$ on $[5, 100]$

13. $f(x) = e^{x/2}$ on $[0, 2]$

14. $f(x) = e^{-2x}$ on $[0, 1]$

15. $f(x) = \dfrac{1}{x}$ on $[1, 2]$

16. $f(x) = \dfrac{1}{x}$ on $[1, 10]$

17. $f(x) = x^n$ on $[0, 1]$, where n is a constant $(n > 0)$

18. $f(x) = e^{kx}$ on $[0, 1]$, where k is a constant $(k \neq 0)$

19. $f(x) = ax + b$ on $[0, 2]$, where a and b are constants

20. $f(x) = \dfrac{1}{x}$ on $[1, c]$, where c is a constant
$(c > 1)$

 21. $f(x) = e^{-x^4}$ on $[-1, 1]$

 22. $f(x) = \sqrt{1 + x^4}$ on $[-2, 2]$

APPLIED EXERCISES ON AVERAGE VALUE

23–24. BUSINESS: Sales A store's sales on day x are given by the function $S(x)$ below. Find the average sales during the first 3 days (day 0 to day 3).

23. $S(x) = 200x + 6x^2$ **24.** $S(x) = 400x + 3x^2$

25. GENERAL: Temperature The temperature at time t hours is $T(t) = -0.3t^2 + 4t + 60$ (for $0 \le t \le 12$). Find the average temperature between time 0 and time 10.

26. BEHAVIORAL SCIENCE: Practice After x practice sessions, a person can accomplish a task in $f(x) = 12x^{-1/2}$ minutes. Find the average time required from the end of session 1 to the end of session 9.

27. ENVIRONMENTAL SCIENCE: Pollution The amount of pollution in a lake x years after the closing of a chemical plant is $P(x) = 100/x$ tons (for $x \ge 1$). Find the average amount of pollution between 1 and 10 years after the closing.

28. GENERAL: Population The population of the United States is predicted to be

$P(t) = 281e^{0.012t}$ million, where t is the number of years after the year 2000. Find the average population between the years 2000 and 2050.

29. BUSINESS: Compound Interest A deposit of $1000 at 5% interest compounded continuously will grow to $V(t) = 1000e^{0.05t}$ dollars after t years. Find the average value during the first 40 years (that is, from time 0 to time 40).

30. BIOMEDICAL: Bacteria A colony of bacteria is of size $S(t) = 300e^{0.1t}$ after t hours. Find the average size during the first 12 hours (that is, from time 0 to time 12).

 31. BUSINESS: Profit The MediGenics Corporation expects its profits to be $1.4e^{0.05x^2}$ million dollars per year x years from now. Find the company's average profit over the next decade (year 0 to year 10).

32. GENERAL: Population The population of a town is predicted to be $5.2e^{\sqrt{x}}$ thousand people x years from now. Find the average population over the next decade (year 0 to year 10).

AREA

33. Find the area between the curve $y = x^2 + 1$ and the line $y = 2x - 1$ (shown below) from $x = 0$ to $x = 3$.

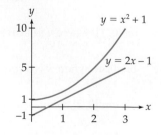

34. Find the area between the curve $y = x^2 + 3$ and the line $y = 2x$ (shown below) from $x = 0$ to $x = 3$.

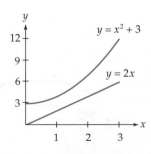

35. Find the area between the curves $y = e^x$ and $y = e^{2x}$ (shown below) from $x = 0$ to $x = 2$. (Leave the answer in its exact form.)

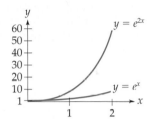

36. Find the area between the curves $y = e^x$ and $y = e^{-x}$ (shown in the following diagram) from $x = 0$ to $x = 1$. (Leave the answer in its exact form.)

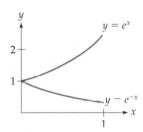

37. a. Sketch the parabola $y = x^2 + 4$ and the line $y = 2x + 1$ on the same graph.
   b. Find the area between them from $x = 0$ to $x = 3$.

38. a. Sketch the parabola $y = x^2 + 5$ and the line $y = 2x + 3$ on the same graph.
   b. Find the area between them from $x = 0$ to $x = 3$.

39. a. Sketch the parabola $y = 3x^2 - 3$ and the line $y = 2x + 5$ on the same graph.
   b. Find the area between them from $x = 0$ to $x = 3$.

40. a. Sketch the parabola $y = 3x^2 - 12$ and the line $y = 2x - 11$ on the same graph.
   b. Find the area between them from $x = 0$ to $x = 3$.

Find the area bounded by the given curves.

41. $y = x^2 - 1$ and $y = 2 - 2x^2$

42. $y = x^2 - 4$ and $y = 8 - 2x^2$

43. $y = 6x^2 - 10x - 8$ and $y = 3x^2 + 8x - 23$

44. $y = 3x^2 - x - 1$ and $y = 5x + 8$

45. $y = x^2$ and $y = x^3$

46. $y = x^3$ and $y = x^4$

47. $y = 7x^3 - 36x$ and $y = 3x^3 + 64x$

48. $y = x^n$ and $y = x^{n-1}$ (for $n > 1$)

 49. $y = e^x$ and $y = x + 3$

[*Hint for Exercises 49–50:* Use INTERSECT to find the intersection points for the upper and lower limits of integration.]

 50. $y = \ln x$ and $y = x - 2$

APPLIED EXERCISES ON AREA

51. GENERAL: Population The birthrate in Africa has increased from $17e^{0.02t}$ million births per year to $22e^{0.02t}$ million births per year, where t is the number of years since 2000. Find the total increase in population that will result from this higher birth rate between 2000 ($t = 0$) and 2020 ($t = 20$).

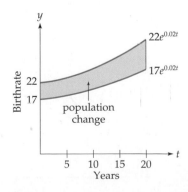

52. BUSINESS: Profit from Expansion A company expects profits of $60e^{0.02t}$ thousand dollars per month, but predicts that if it builds a new and larger factory, its profits will be $80e^{0.04t}$ thousand dollars per month, where t is the number of months from now. Find the extra profits resulting from the new factory during the first two years $(t = 0$ to $t = 24)$. If the new factory will cost $1,000,000, will this cost be paid off during the first two years?

53. BUSINESS: Net Savings A factory installs new machinery that saves $S(x) = 1200 - 20x$ dollars per year, where x is the number of years since installation. However, the cost of maintaining the new machinery is $C(x) = 100x$ dollars per year.

   a. Find the year x at which the maintenance cost $C(x)$ will equal the savings $S(x)$. (At this time, the new machinery should be replaced.)

   b. Find the accumulated net savings [savings $S(x)$ minus cost $C(x)$] during the period from $t = 0$ to the replacement time found in part (a).

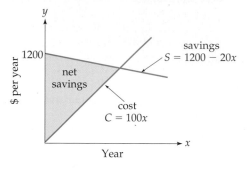

54. GENERAL: Design A graphic design consists of a white square containing the blue shape shown below. Find the area of the blue interior.

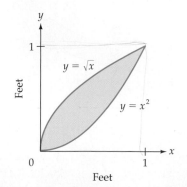

55. ECONOMICS: Balance of Trade A country's annual imports are $I(t) = 30e^{0.2t}$ and its exports are $E(t) = 25e^{0.1t}$, both in billions of dollars, where t is measured in years and $t = 0$ corresponds to the beginning of 2000. Find the country's accumulated trade deficit (imports minus exports) for the 10 years beginning with 2000.

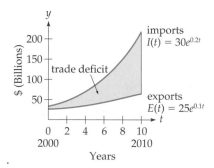

56. ECONOMICS: Cost of Labor Contracts An employer offers to pay workers at the rate of $30,000e^{0.04t}$ dollars per year, while the union demands payment at the rate of $30,000e^{0.08t}$ dollars per year, where $t = 0$ corresponds to the beginning of the contract. Find the accumulated difference in pay between these two rates over the 10-year life of the contract.

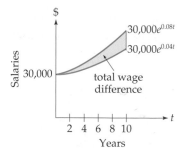

57–58. BUSINESS: Cumulative Profit A company finds that its marginal revenue function is $MR(x) = 700x^{-1}$ and its marginal cost function is $MC(x) = 500x^{-1}$ (both in thousands of dollars), where x is the number of units $(x > 1)$. Find the total profit from

57. $x = 100$ to $x = 200$

58. $x = 200$ to $x = 300$

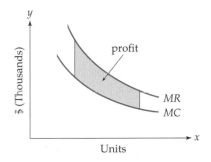

59. GENERAL: Lives Saved by Seat Belts Seat belt use in the United States has now risen to 66%, but nonusers still risk needless expense, injury, and death. The table below gives the number of automobile fatalities per year, and the predicted number of fatalities if everyone wore seat belts. To avoid large numbers, years are listed as years since 1996.

Years Since 1996	Automobile Fatalities	Predicted Fatalities with 100% Seat Belt Use
1996 0	43,649	38,600
1997 1	43,458	38,400
1998 2	41,800	36,900
1999 3	41,300	36,500

Source: National Safety Council

a. Enter the first two columns of numbers into your graphing calculator and make a plot of the resulting points (Years Since 1996 on the x-axis and Fatalities on the y-axis).

b. Have your calculator find the linear regression formula for these data. Then enter the result as y_1, which gives a formula for fatalities each year. Plot the points together with the regression line.

c. Enter the last column of numbers into your calculator and make a plot of the resulting points (Years Since 1996 on the x-axis and Predicted Fatalities on the y-axis), keeping the earlier points too.

d. Have your calculator find the linear regression formula for the new points found in part (c). Then enter the result as y_2, which gives a formula for predicted fatalities each year. Plot both sets of points together with both regression lines.

e. Extend your viewing window to $[0, 10]$ by $[0, 50,000]$ to see what the lines predict for fatalities from 1996 to 2006. The area between these lines represents the lives that could be saved by seat belts.

f. Have your calculator find the area between these lines from 0 to 10, predicting the lives that would be saved by 100% seat-belt use between 1996 and 2006.

REVIEW EXERCISES

These exercises review material that will be helpful in Section 5.6.

Find the derivative of each function.

60. $e^{x^3 + 6x}$

61. $e^{x^2 + 5x}$

62. $\ln(x^3 + 6x)$

63. $\ln(x^2 + 5x)$

5.5   TWO APPLICATIONS TO ECONOMICS: CONSUMERS' SURPLUS AND INCOME DISTRIBUTION

Introduction

In this section we discuss several important economic concepts—consumers' surplus, producers' surplus, and the Gini index of income distribution—each of which is defined as the area between two curves.

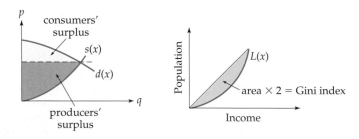

Consumers' Surplus

Imagine that you really liked pizza and were willing to pay $12 for a pizza pie. If, in fact, a pizza costs only $8, then you have, in some sense, "saved" $4, the $12 that you were willing to pay minus the $8 market price. If one were to add up this difference for all pizzas sold in a given period of time (the price that each consumer was willing to pay minus the price actually paid), the total savings would be called the *consumers' surplus* for that product. The consumers' surplus measures the benefit that consumers derive from an economy in which competition keeps prices low.

Demand Functions

Price and quantity are inversely related: if the price of an item rises, the quantity sold generally falls, and vice versa. Through market research, economists can determine the relationship between price and quantity for an item. This relationship can often be expressed as a *demand function* (or demand curve) $d(x)$, so called because it gives the price at which exactly x units will be demanded.

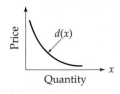

> **Demand Function**
>
> The demand function $d(x)$ for a product gives the price at which exactly x units will be sold.
>
> $$d(x) = \left(\begin{array}{c} \text{Price when} \\ \text{demand is } x \end{array} \right)$$

On page 218 we called $d(x)$ the *price function*.

Mathematical Definition of Consumers' Surplus

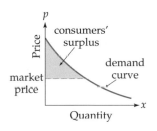

The demand curve gives the price that consumers are *willing* to pay, and the *market price* is what they *do* pay, so the amount by which the demand curve is above the market price measures the benefit or "surplus" to consumers. We add up all of these benefits by integrating, so the area between the demand curve and the market price line gives the *total benefit* that consumers derive from being able to buy at the market price. This total benefit (the shaded area in the diagram on the left) is called the *consumers' surplus*.

Consumers' Surplus

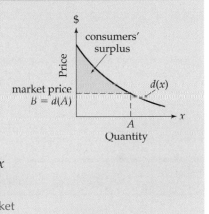

For a demand function $d(x)$ and demand level A, the market price B is the demand function evaluated at $x = A$, so that $B = d(A)$. The consumers' surplus is the area between the demand curve and the market price.

$$\begin{pmatrix} \text{Consumers'} \\ \text{surplus} \end{pmatrix} = \int_0^A [d(x) - B] \, dx$$

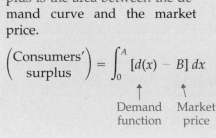

↑ Demand function   ↑ Market price

EXAMPLE 1

FINDING CONSUMERS' SURPLUS FOR ELECTRICITY

If the demand function for electricity is $d(x) = 1100 - 10x$ dollars (where x is in millions of kilowatt-hours, $0 \le x \le 100$), find the consumers' surplus at the demand level $x = 80$.

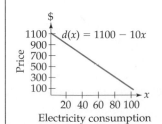

Solution   The market price is the demand function $d(x)$ evaluated at $x = 80$.

$$\begin{pmatrix} \text{Market} \\ \text{price } B \end{pmatrix} = d(80) = 1100 - 10 \cdot 80 = 300 \qquad \begin{matrix} d(x) = 1100 - 10x \\ \text{at } x = 80 \end{matrix}$$

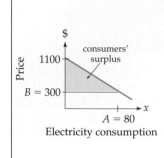

The consumers' surplus is the area between the demand curve and the market price line.

$$\binom{\text{Consumers'}}{\text{surplus}} = \int_0^{80} \underbrace{(1100 - 10x}_{\substack{\text{Demand} \\ \text{price}}} \underbrace{- \ 300)}_{\substack{\text{Market} \\ \text{price}}} dx$$

$$= \int_0^{80} (800 - 10x)\, dx = (800x - 5x^2)\Big|_0^{80}$$

$$= (64{,}000 - 32{,}000) - (0) = 32{,}000$$

Therefore, the consumers' surplus for electricity is $32,000.

How Consumers' Surplus Is Used

In Example 1, at demand level $x = 80$ the consumers' surplus was $32,000. If electricity usage were to increase to $x = 90$, the market price would then drop to $d(90) = 1100 - 10 \cdot 90 = 200$. We could then calculate the consumers' surplus at this higher demand level (and would find that the answer is $40,500). Therefore, a price decrease from $300 to $200 would mean that consumers would benefit by an additional $40{,}500 - \$32{,}000 = \$8500$. This benefit would then be compared to the cost of a new generator to decide whether the expenditure would be worthwhile.

 **Graphing Calculator Exploration**

a. Verify the solution to Example 1 on your graphing calculator by entering $y_1 = 1100 - 10x$ and then using FnInt or $\int f(x)\, dx$ to integrate $y_1(x) - y_1(80)$ from 0 to 80.

b. Find the consumers' surplus at demand level 90 by integrating $y_1(x) - y_1(90)$ from 0 to 90. [*Hint:* Simply return to the calculation of part (a) and replace 80 by 90. Your answer should agree with that in the preceding paragraph.]

Producers' Surplus

Just as the consumers' surplus measures the total benefit to consumers, the *producers' surplus* measures the total benefit that producers derive from being able to sell at the market price. Returning to our pizza example, if a pizza producer might just be willing to remain in business if the price of pizzas dropped to $5, the fact that pizzas can be sold for $8 gives the producer a "benefit" of $3. The sum of all such benefits is the *producers' surplus* for a product.

Supply Functions

Clearly, as the price of an item rises, so does the quantity that producers are willing to supply at that price. The relationship between the price of an item and the quantity that producers are willing to supply at that price can be expressed as a *supply function* (or supply curve) $s(x)$.

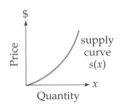

Supply Function

The supply function $s(x)$ for a product gives the price at which exactly x units will be supplied.

$$s(x) = \begin{pmatrix} \text{Price when} \\ \text{supply is } x \end{pmatrix}$$

Mathematical Definition of Producers' Surplus

As before, we integrate to find the total benefit, but now "upper" is the market price and "lower" is the supply curve $s(x)$.

Producers' Surplus

For a supply function $s(x)$ and demand level A, the market price is $B = s(A)$. The producers' surplus is the area between the market price and the supply curve.

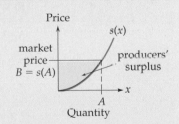

$$\begin{pmatrix} \text{Producers'} \\ \text{surplus} \end{pmatrix} = \int_0^A [B - s(x)]\, dx$$

Market   Supply
price   function

EXAMPLE **2**

FINDING PRODUCERS' SURPLUS

For the supply function $s(x) = 0.09x^2$ dollars and the demand level $x = 200$, find the producers' surplus.

Solution The market price is the supply function $s(x)$ evaluated at $x = 200$.

$$\begin{pmatrix} \text{Market} \\ \text{price } B \end{pmatrix} = s(200) = 0.09(200)^2 = 3600 \qquad \begin{array}{l} s(x) = 0.09x^2 \\ \text{at } x = 200 \end{array}$$

The producers' surplus is the area between the market price line and the supply curve.

$$\begin{pmatrix} \text{Producers'} \\ \text{surplus} \end{pmatrix} = \int_0^{200} (\underbrace{3600}_{\substack{\text{Market} \\ \text{price}}} - \underbrace{0.09x^2}_{\substack{\text{Supply} \\ \text{function}}}) \, dx = (3600x - 0.03x^3)\Big|_0^{200}$$

$$= (720{,}000 - 240{,}000) - (0) = 480{,}000$$

Therefore, the producers' surplus is $480,000.

Consumers' Surplus and Producers' Surplus

The demand x at which the supply and demand curves intersect is called the *market demand*. The consumers' surplus and the producers' surplus can be shown together on the same graph. These two areas together give a numerical measure of the total benefit that consumers and producers derive from competition, showing that both consumers and producers benefit from an open market.

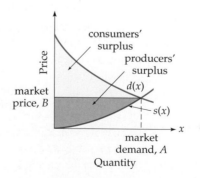

Gini Index of Income Distribution

In any society, some people make more money than others. To measure the "gap" between the rich and the poor, economists calculate the proportion of the total income that is earned by the lowest 20% of the population, and then the proportion that is earned by the lowest 40% of the population, and so on. This information (for the year 2000) is given in the table below (with percentages written as decimals), and is graphed on the right.

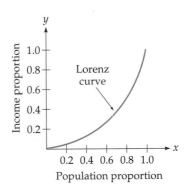

Proportion of Population	Proportion of Income
0.20	0.04
0.40	0.13
0.60	0.27
0.80	0.51
1.00	1.00

Source: U.S. Bureau of the Census

For example, the lowest 20% of the population earns only 4% of the total income, the lowest 40% earns only 13% of the total income, and so on. The curve is known as the *Lorenz curve* (after the American statistician Max Otto Lorenz).

Lorenz Curve

The Lorenz curve $L(x)$ gives the proportion of total income earned by the lowest proportion x of the population.

Gini Index

The Lorenz curve may be compared with two extreme cases of income distribution.

1. *Absolute equality of income* means that everyone earns exactly the same income, and so the lowest 10% of the population earns exactly 10% of the total income, the lowest 20% earns exactly 20% of the income, and so on. This gives the Lorenz curve $y = x$ shown on the following page.

Absolute equality of income

Absolute inequality of income

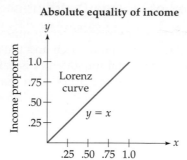

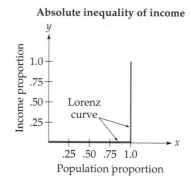

2. *Absolute inequality of income* means that nobody earns any income except one person, who earns all the income. This gives the Lorenz curve shown above on the right.

To measure how the actual distribution differs from absolute equality, we calculate the area between the actual distribution and the line of absolute equality $y = x$. Since this area can be at most $\frac{1}{2}$ (the area of the entire lower triangle), economists multiply the area by 2 to get a number between 0 (absolute equality) and 1 (absolute inequality). This measure is called the *Gini index*. Note that a higher Gini index means greater *in*equality (greater deviation from the line of absolute equality).

Gini Index

For a Lorenz curve $L(x)$, the Gini index is

$$\left(\begin{array}{c}\text{Gini}\\\text{index}\end{array}\right) = 2 \int_0^1 [x - L(x)]\, dx$$

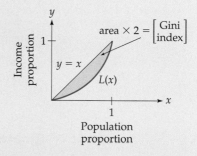

The Gini index varies from 0 (absolute equality) to 1 (absolute inequality).

EXAMPLE 3

FINDING THE GINI INDEX

The Lorenz curve for income distribution in the United States in 2000 was approximately $L(x) = x^{2.4}$. Find the Gini index.

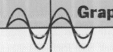

Solution First we calculate the area between the curve of absolute equality $y = x$ and the Lorenz curve $y = x^{2.4}$.

$$\int_0^1 (x - x^{2.4})\, dx = \left(\frac{1}{2} x^2 - \frac{1}{3.4} x^{3.4} \right) \Big|_0^1$$

$$= \frac{1}{2} \cdot 1^2 - \frac{1}{3.4} \cdot 1^{3.4} - 0$$

$$\approx 0.5 - 0.294 = 0.206$$

Multiplying by 2 gives the Gini index.

$$\left(\frac{\text{Gini}}{\text{index}} \right) = 0.41 \qquad \text{Rounding to 2 decimal places.}$$

Practice Problem 1   In 1993, the Gini index for income was 0.39. Since 1993, has income distribution become more equal or less equal?   ➤ Solution on page 412

Graphing Calculator Exploration

In Example 3, how was the Lorenz function of the form x^n found? It was found by a method called "least squares" (discussed in Section 2.6 and more fully in Section 7.4), which minimizes the squared differences between the income proportions (from the table on page 409) and the curve x^n at the x-values 0.2, 0.4, 0.6, and 0.8. This amounts to minimizing the function

$$(0.2^x - 0.04)^2 + (0.4^x - 0.13)^2 + (0.6^x - 0.27)^2 + (0.8^x - 0.51)^2.$$

(Notice that here we are using x for the *exponent*.)

a. Graph this function on your calculator on the window [0, 5] by [−0.2, 1].

b. Find the x that minimizes the function. The resulting value is the exponent. Your answer should agree with the exponent in Example 3.

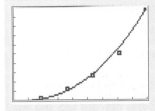

The graph on the left, on the window [0, 1] by [0, 1], shows that the function $x^{2.4}$ fits the points from the table on page 409 reasonably well. (Other types of functions besides x^n could also be used to fit these data.)

The Gini index for *total* wealth can be calculated similarly.

Practice Problem **2**

The Lorenz curve for total wealth in the United States during 2000 was $L(x) = x^{6.9}$. Calculate the Gini index. ➤ **Solution below**

Practice Problem 2 shows that the Gini index for wealth is greater than the Gini index for income. That is, total wealth in the United States is distributed more unequally than total income. One reason for this is that we have an income tax but no "wealth" tax.

➤ **Solutions to Practice Problems**

1. Less equal

2. $\int_0^1 (x - x^{6.9})\, dx = \left(\frac{1}{2}x^2 - \frac{1}{7.9}x^{7.9}\right)\Big|_0^1 = \frac{1}{2} - \frac{1}{7.9} - 0 \approx 0.5 - 0.127 = 0.373$

   Gini index for wealth is 0.75 (multiplying by 2 and rounding)

5.5 **Exercises**

For each demand function $d(x)$ and demand level x, find the consumers' surplus.

1. $d(x) = 4000 - 12x, \quad x = 100$

2. $d(x) = 500 - x, \quad x = 400$

3. $d(x) = 300 - \frac{1}{2}x, \quad x = 200$

4. $d(x) = 200 - \frac{1}{2}x, \quad x = 300$

5. $d(x) = 350 - 0.09x^2, \quad x = 50$

6. $d(x) = 840 - 0.06x^2, \quad x = 100$

7. $d(x) = 200e^{-0.01x}, \quad x = 100$

8. $d(x) = 400e^{-0.02x}, \quad x = 75$

For each supply function $s(x)$ and demand level x, find the producers' surplus.

9. $s(x) = 0.02x, \quad x = 100$

10. $s(x) = 0.4x, \quad x = 200$

11. $s(x) = 0.03x^2, \quad x = 200$

12. $s(x) = 0.06x^2, \quad x = 50$

For each demand function $d(x)$ and supply function $s(x)$:

a. Find the market demand (the positive value of x at which the demand function intersects the supply function).

b. Find the consumers' surplus at the market demand found in part (a).

c. Find the producers' surplus at the market demand found in part (a).

13. $d(x) = 300 - 0.4x, \quad s(x) = 0.2x$

14. $d(x) = 120 - 0.16x, \quad s(x) = 0.08x$

15. $d(x) = 300 - 0.03x^2, \quad s(x) = 0.09x^2$

16. $d(x) = 360 - 0.03x^2, \quad s(x) = 0.006x^2$

17. $d(x) = 300e^{-0.01x}, \quad s(x) = 100 - 100e^{-0.02x}$

18. $d(x) = 400e^{-0.01x}, \quad s(x) = 0.01x^{2.1}$

Find the Gini index for the given Lorenz curve.

19. $L(x) = x^{3.2}$ (the Lorenz curve for U.S. income in 1929)

20. $L(x) = x^3$ (the Lorenz curve for U.S. income in 1935)

21. $L(x) = x^{2.1}$ (the Lorenz curve for income in Sweden in 1990)

22. $L(x) = x^{15.3}$ (the Lorenz curve for wealth in Great Britain in 1990)

23. $L(x) = 0.4x + 0.6x^2$

24. $L(x) = 0.2x + 0.8x^3$

25. $L(x) = x^n$ (for $n > 1$)

26. $L(x) = \frac{1}{2}x + \frac{1}{2}x^n$ (for $n > 1$)

27. $L(x) = \dfrac{e^{x^2} - 1}{e - 1}$

28. $L(x) = 1 - \sqrt{1 - x}$

29. $L(x) = \dfrac{x + x^2 + x^3}{3}$

30. $L(x) = 0.62x^{7.15} + 0.38x^{9.47}$

31–32. The following tables give the distribution of family income in the United States: Exercise 31 is for the year 1977 and Exercise 32 is for 1989. Use the procedure described in the Graphing Calculator Exploration on page 411 to find the Lorenz

function of the form x^n for the data. Then find the Gini index. If you do both problems, did family income become more concentrated or less concentrated from 1977 to 1989? _____

31. _____

Proportion (Lowest) of Families	Proportion of Income (1977)
0.20	0.06
0.40	0.18
0.60	0.34
0.80	0.57

Source: Congressional Budget Office

32. _____

Proportion (Lowest) of Families	Proportion of Income (1989)
0.20	0.04
0.40	0.14
0.60	0.29
0.80	0.51

Source: Congressional Budget Office

REVIEW EXERCISES

These exercises review material that will be helpful in Section 5.6.

Find the derivative of each function.

33. $(x^5 - 3x^3 + x - 1)^4$ **34.** $(x^4 - 2x^2 - x + 1)^5$ **35.** $\ln(x^4 + 1)$ **36.** $\ln(x^3 - 1)$ **37.** e^{x^3} **38.** e^{x^4}

5.6 **INTEGRATION BY SUBSTITUTION**

Introduction

The Chain Rule (page 160) greatly expanded the range of functions that we could differentiate. In this section we will learn a similar technique for integration, called the *substitution method*, which will greatly expand the range of functions that we can integrate. First, however, we must define *differentials*.

Differentials

One of the notations for the derivative of a function $f(x)$ is df/dx. Although written as a fraction, df/dx was not defined as the quotient of two quantities df and dx, but as a single object, the *derivative*. We will now define df and dx separately (they are called *differentials*) so that their quotient $df \div dx$ is equal to the derivative df/dx. We begin with

$$\frac{df}{dx} = f'$$
Since df/dx and f' are both notations for the derivative

$$df = f' \, dx$$
Multiplying each side by dx

This leads to a definition for the differential df.

Differential

For a differentiable function $f(x)$, the differential df is

$$df = f'(x) \, dx$$
df is the derivative times dx

Note that df does *not* mean d times f. The dx is just the notation that appears at the end of integrals, arising from the Δx in the Riemann sum. The reason for finding differentials will be made clear shortly.

EXAMPLE 1

FINDING DIFFERENTIALS

Function $f(x)$	Differential df	
$f(x) = x^2$	$df = 2x \, dx$	
$f(x) = \ln x$	$df = \dfrac{1}{x} \, dx$	Each differential is the derivative of the function times dx
$f(x) = e^{x^2}$	$df = e^{x^2}(2x) \, dx$	
$f(x) = x^4 - 5x + 2$	$df = \underbrace{(4x^3 - 5)}_{f'(x)} \, dx$	

Practice Problem 1 For $f(x) = x^3 - 4x - 2$, find the differential df.

$(3x^2 - 4) \, dx$

➤ Solution on page 424

The differential formula $df = f' \, dx$ is easy to remember because dividing both sides by dx gives

$$\frac{df}{dx} = f'$$

which simply says "the derivative equals the derivative." We may use other letters besides f and x.

EXAMPLE 2

CALCULATING DIFFERENTIALS IN OTHER VARIABLES

Function	Differential	
$u = x^3 + 1$	$du - 3x^2 \, dx$	Differentials end with d followed by the variable
$u = e^{2t} + 1$	$du = 2e^{2t} \, dt$	

Practice Problem 2   For $u = e^{-5t}$, find the differential du.   ► Solution on page 424

Substitution Method

Using differential notation, we can state three very useful integration formulas.

$$\text{(A)} \quad \int u^n \, du = \frac{1}{n+1} u^{n+1} + C \qquad\qquad n \neq -1$$

$$\text{(B)} \quad \int e^u \, du = e^u + C$$

$$\text{(C)} \quad \int \frac{du}{u} = \int \frac{1}{u} \, du = \int u^{-1} \, du = \ln|u| + C$$

These formulas are easy to remember because they are exactly the formulas that we learned earlier (see pages 358 and 365) except that here we use the letter u to stand for a *function*. The du is the differential of the function. Each of these formulas may be justified by differentiating the right-hand side (see Exercises 61–63). A few examples will illustrate their use.

EXAMPLE 3   **INTEGRATING BY SUBSTITUTION**

Find $\displaystyle\int (x^2 + 1)^3 \, 2x \, dx$.

Solution

The integral involves a function to a power:

$$\int (x^2 + 1)^3 \, 2x \, dx \qquad \begin{array}{l} (x^2 + 1)^3 \text{ is a function} \\ \text{to a power} \end{array}$$

$$\downarrow$$

as does formula (A): $\displaystyle\int u^n \, du$ $\begin{array}{l} u^n \text{ is a function to a power in} \\ \displaystyle\int u^n \, du = \frac{1}{n+1} u^{n+1} + C \end{array}$

To make the integral "fit" the formula we take $u = x^2 + 1$ and $n = 3$.

$$\int \underbrace{(x^2 + 1)^3}_{u^3} \, 2x \, dx \qquad\qquad u = x^2 + 1 \quad \text{and} \quad n = 3$$

For $u = x^2 + 1$ the differential is $du = 2x \, dx$, which is exactly the remaining part of the integral. We then write the integral with each x-expression replaced by its equivalent u-expression.

$$\int \underbrace{(x^2 + 1)^3}_{u^3} \underbrace{2x \, dx}_{du} = \int \underbrace{u^3 \, du}_{\substack{\text{Written in} \\ \text{terms of } u}} \qquad \begin{array}{l} \text{Using } u = x^2 + 1 \\ \text{and } du = 2x \, dx \end{array}$$

The last integral we solve by formula (A):

$$\int u^3 \, du = \frac{1}{4} u^4 + C \qquad\qquad \begin{array}{l} \displaystyle\int u^n \, du = \frac{1}{n+1} u^{n+1} + C \\ \text{[formula (A)] with } n = 3 \end{array}$$

Finally, we substitute back to the original variable x, using our relationship $u = x^2 + 1$, to get the answer:

$$\frac{1}{4}(x^2 + 1)^4 + C \qquad\qquad \frac{1}{4} u^4 + C \quad \text{with} \quad u = x^2 + 1$$

The procedure is not as complicated as it might seem. All of these steps may be written together as follows.

$$\int (x^2 + 1)^3 \, 2x \, dx \quad = \quad \int u^3 \, du \quad = \quad \frac{1}{4} u^4 + C \quad = \quad \frac{1}{4}(x^2 + 1)^4 + C$$

u^3 $\quad$ du			
Choosing $u = x^2 + 1$ therefore $du = 2x \, dx$	Substituting $u^3 = (x^2 + 1)^3$ $du = 2x \, dx$	Integrating by formula (A) with $n = 3$	Substituting back to x using $u = x^2 + 1$

We may check this answer by differentiation (using the Generalized Power Rule).

$$\frac{d}{dx}\left[\frac{1}{4}(x^2 + 1)^4 + C \right] = \frac{1}{4} \cdot 4(x^2 + 1)^3 \, 2x = (x^2 + 1)^3 \, 2x$$

Answer $\qquad$ Derivative of the inside $\qquad$ Integrand

Since the result of the differentiation agrees with the original integrand, the integration is correct.

Multiplying Inside and Outside by Constants

If the integral does not exactly match the form $\int u^n \, du$, we may sometimes still solve the integral by multiplying by constants.

EXAMPLE 4

INSERTING CONSTANTS BEFORE SUBSTITUTING

Find $\displaystyle\int (x^2 + 1)^3 x \, dx.$ $\qquad$ Same as Example 3 but without the 2

Solution As before, we use formula (A) with $u = x^2 + 1$, which gives $du = 2x \, dx$. But the integral has only an $x \, dx$, not $\underline{2x} \, dx$, which would allow us to substitute du.

$$\int (x^2 + 1)^3 x \, dx \qquad\qquad u = x^2 + 1, \text{ so } du = 2x \, dx$$

$u^3 \qquad$ not $du = 2x \, dx$ because there is no 2

Therefore, the integral is *not* in the form $\int u^3 \, du$. (The integral must fit the formula exactly: *everything* in the integral must be accounted for either by the u^n or by the du.) However, we may multiply inside the integral by 2 as long as we compensate by also multiplying by $\frac{1}{2}$, and the $\frac{1}{2}$ may be written *outside* the integral (since constants may be moved across the integral sign), leading to the solution:

Multiplying by $\frac{1}{2}$ and 2

$$\frac{1}{2}\int (x^2 + 1)^3\, 2x\, dx = \frac{1}{2}\int u^3\, du = \frac{1}{2}\cdot\frac{1}{4}u^4 + C = \frac{1}{8}(x^2 + 1)^4 + C$$

$u^3 \quad\quad du$ Substituting   Integrating   Substituting back
$u = x^2 + 1$ by formula (A)   to x using
$du = 2x\, dx$ $u = x^2 + 1$

This method of multiplying inside and outside by a constant is very useful, and may be used with the other substitution formulas as well.

EXAMPLE 5   **USING OTHER SUBSTITUTION FORMULAS**

Find $\displaystyle\int e^{x^3}x^2\, dx$.

Solution

The integral involves
e to a function:

$$\int e^{x^3}x^2\, dx$$

         e^{x^3} is e to a function

$\updownarrow$

as does formula (B) on
page 415:

$$\int e^u\, du = e^u + C$$

Matching exponents of e gives $u = x^3$, and the differential of u is $du = 3x^2\, dx$. The differential requires a 3, which is not in the integral, so we multiply inside by 3 and outside by $\frac{1}{3}$.

$$\int e^{x^3}x^2\, dx = \frac{1}{3}\int e^{x^3}\, 3x^2\, dx = \frac{1}{3}\int e^u\, du = \frac{1}{3}e^u + C = \frac{1}{3}e^{x^3} + C$$

                 $e^u \quad du$

$u = x^3$ Substituting   Integrating   Substituting
$du = 3x^2\, dx$ Multiplying   using   back to x
              by $\frac{1}{3}$ and 3   formula (B)   using $u = x^3$

Why did the 1/3 become part of the answer but the 3 disappeared? The 3 became part of the du ($du = 3x^2\, dx$), which was then "used up" in the integration along with the integral sign.

EXAMPLE 6

RECOVERING COST FROM MARGINAL COST

A company's marginal cost function is $MC(x) = \dfrac{x^3}{x^4 + 1}$ and fixed costs are $1000. Find the cost function.

Solution Cost is the integral of marginal cost.

$$C(x) = \int \frac{x^3 \, dx}{x^4 + 1}$$

The differential of the denominator is $4x^3 \, dx$, which except for the 4 is just the numerator. This suggests formula (C), $\displaystyle \int \frac{du}{u} = \ln |u| + C$ with $u = x^4 + 1$. We multiply inside by 4 (to complete the $du = 4x^3 \, dx$ in the numerator) and outside by $\frac{1}{4}$.

$$\int \frac{x^3 \, dx}{x^4 + 1} = \frac{1}{4} \int \frac{\overset{du}{\overbrace{4x^3 \, dx}}}{\underset{u}{\underbrace{x^4 + 1}}} = \frac{1}{4} \int \frac{du}{u} = \frac{1}{4} \ln |u| + C = \frac{1}{4} \ln(x^4 + 1) + C$$

$u = x^4 + 1$	Multiplying		Integrating by	Substituting
$du = 4x^3 \, dx$	by 4 and $\frac{1}{4}$	Substituting	formula (C)	back to x

We dropped the absolute value bars because $x^4 + 1$ is positive. To evaluate the constant C, we set the cost function (evaluated at $x = 0$) equal to the given fixed cost.

$$\frac{1}{4} \underbrace{\ln(1)}_{0} + C = 1000 \qquad \begin{array}{l} \ln(x^4 + 1) + C \ \text{at} \ x = 0 \\ \text{set equal to } 1000 \end{array}$$

This gives $C = 1000$. Therefore, the company's cost function is

$$C(x) = \frac{1}{4} \ln(x^4 + 1) + 1000 \qquad \begin{array}{l} C(x) = \frac{1}{4}\ln(x^4 + 1) + C \\ \text{with } C = 1000 \end{array}$$

Which Formula to Use

The three formulas apply to three different types of integrals.

(A) $\displaystyle \int u^n \, du = \frac{1}{n + 1} u^{n+1} + C \quad (n \neq -1)$
Integrates a function to a constant power (except −1) times the differential of the function

(B) $\displaystyle\int e^u \, du = e^u + C$

Integrates *e to a power* times the differential of the exponent

(C) $\displaystyle\int \frac{du}{u} = \int u^{-1} \, du = \ln|u| + C$

Integrates a *fraction whose top is the differential of the bottom,* or equivalently, *a function to the power* -1 times the differential of the function

To solve an integral by substitution, choose the formula whose left-hand side has the same form as the given integral.

Practice Problem 3

For each of the following integrals, choose the most appropriate formula: (A), (B), or (C). (Do not solve the integral.)

a. $\displaystyle\int e^{5x^2 - 1} x \, dx$ **b.** $\displaystyle\int \frac{x \, dx}{x^2 + 1}$ **c.** $\displaystyle\int (x^4 - 12)^4 x^3 \, dx$

d. $\displaystyle\int (x^4 - 12)^{-1} x^3 \, dx$ ➤ Solutions on page 424

Only Constants Can Be Adjusted

We may multiply inside and outside only by constants, not variables, since only constants can be moved across the integral sign. Therefore, the *du* in a problem must already be "complete" except for adjusting the constant. Otherwise, the problem cannot be solved by a substitution. For example, the following integral cannot be found by a substitution.

$$\int \underbrace{e^{x^3}}_{e^u} \underbrace{x \, dx}_{\text{not } du \text{ (does not have an } x^2)}$$

$$u = x^3$$
$$du = 3x^2 \, dx$$

Practice Problem 4

Which of these integrals can be found by a substitution? [*Hint:* See whether only a constant is needed to complete the *du*.]

a. $\displaystyle\int (x^3 + 1)^3 x^3 \, dx$ **b.** $\displaystyle\int e^{x^2} \, dx$ ➤ Solutions on page 424

EXAMPLE **7**

INTEGRATING BY SUBSTITUTION

Find $\displaystyle\int \sqrt{x^3 - 3x}\,(x^2 - 1)\,dx$.

Solution   Since $\sqrt{x^3 - 3x} = (x^3 - 3x)^{1/2}$ is a *function to a power*, we use the formula for $\int u^n\,du$ with $u = x^3 - 3x$. Comparing the differential $du = (3x^2 - 3)\,dx = 3(x^2 - 1)\,dx$ with the problem shows that we need to multiply by 3.

$$\int (x^3 - 3x)^{1/2}(x^2 - 1)\,dx = \frac{1}{3}\int \underbrace{(x^3 - 3x)^{1/2}}_{u^{1/2}}\underbrace{3(x^2 - 1)\,dx}_{du} \qquad \begin{array}{l}\text{Multiplying}\\ \text{by 3 and } \frac{1}{3}\end{array}$$

$$u = x^3 - 3x$$
$$du = (3x^2 - 3)\,dx$$
$$\quad = 3(x^2 - 1)\,dx$$

$$= \underbrace{\frac{1}{3}\int u^{1/2}\,du}_{\text{Substituting}} = \underbrace{\frac{1}{3}\cdot\frac{2}{3}u^{3/2} + C}_{\substack{\text{Integrating by}\\\text{formula (A)}}} = \underbrace{\frac{2}{9}(x^3 - 3x)^{3/2} + C}_{\substack{\text{Substituting}\\\text{back to } x}}$$

EXAMPLE **8**

INTEGRATING BY SUBSTITUTION

Evaluate $\displaystyle\int e^{\sqrt{x}}\,x^{-1/2}\,dx$.

Solution   The integral involves $e^{\sqrt{x}}$, e to a function, so we use the formula $\int e^u\,du = e^u + C$ with $u = x^{1/2}$.

$$\int e^{x^{1/2}}x^{-1/2}\,dx = 2\int \underbrace{e^{x^{1/2}}}_{e^u}\underbrace{\frac{1}{2}x^{-1/2}\,dx}_{du} \qquad \text{Multiplying by } \tfrac{1}{2} \text{ and 2}$$

$$u = x^{1/2}$$

$$du = \frac{1}{2}x^{-1/2}\,dx$$

$$= \underbrace{2\int e^u\,du}_{\text{Substituting}} = \underbrace{2e^u + C}_{\text{Integrating}} = \underbrace{2e^{x^{1/2}} + C}_{\substack{\text{Substituting}\\\text{back to } x}}$$

Evaluating Definite Integrals by Substitution

Sometimes a definite integral requires a substitution. In such cases changing from x to u also requires changing the limits of integration from x-values to u-values, using the substitution formula for u.

EXAMPLE 9

EVALUATING A DEFINITE INTEGRAL BY SUBSTITUTION

Evaluate $\displaystyle\int_4^5 \frac{dx}{3-x}$.

Solution The differential of the denominator $3-x$ is $-1 \cdot dx$, which except for the -1 is just the numerator. This suggests formula (C), $\displaystyle\int \frac{du}{u} = \ln|u| + C$ with $u = 3 - x$ (from equating the denominators). We multiply inside and outside by -1.

$$u = 3 - 5 = -2$$

$$\int_4^5 \frac{dx}{3-x} = -\int_4^5 \frac{-dx}{3-x} = -\int_{-1}^{-2} \frac{du}{u}$$

$$u = 3 - 4 = -1$$

New upper and lower limits of integration for u are found by evaluating $u = 3 - x$ at the old x limits

$$u = 3 - x$$
$$du = -dx$$

$$= -\ln|u|\ \Big|_{-1}^{-2} = -\ln|-2| - (-\ln|-1|) = -\ln 2 + \ln 1 = -\ln 2$$

Graphing Calculator Exploration

Why is the answer to Example 9 *negative*? Graph $f(x) = \dfrac{1}{3-x}$ on the window $[3, 6]$ by $[-2, 2]$ to see why. Have your calculator find the definite integral from 4 to 5. How would you change the integral so that it gives the *area* between the curve and the x-axis?

Definite integrals are used to find areas, total accumulations, and average values, and any of these uses may require a substitution.

EXAMPLE **10** **FINDING TOTAL POLLUTION FROM A RATE**

Pollution is being discharged into a lake at the rate of $r(t) = 400te^{t^2}$ tons per year, where t is the number of years since measurements began. Find the total amount of pollutant discharged into the lake during the first 2 years.

Solution   The total accumulation is the definite integral from $t = 0$ (the beginning) to $t = 2$ (2 years later). Since the integral involves e to a function, we use the formula for $\int e^u\, du$ with $u = t^2$ (by equating exponents).

$$\int_0^2 400\, t e^{t^2}\, dt \quad = \quad 400 \cdot \frac{1}{2} \int_0^2 2t\, e^{t^2}\, dt = 200 \int_0^4 e^u\, du$$

$u = 2^2 = 4$

Changing the limits to u-values using $u = t^2$

e^u Taking out the constant

$u = t^2$
$du = 2t\, dt$

du
$u = 0^2 = 0$

$$= 200e^u \Big|_0^4 = 200e^4 - 200e^0 \approx 10{,}720$$

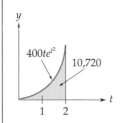

400te^{t^2}

10,720

Therefore, during the first 2 years 10,720 tons of pollutant were discharged into the lake.

Notice that the du does not need to be all together, but can be in several separate pieces, as long as it is all *there*.

EXAMPLE **11** **FINDING AVERAGE WATER DEPTH**

After x months the water level in a newly built reservoir is $L(x) = 40x(x^2 + 9)^{-1/2}$ feet. Find the average depth during the first 4 months.

Solution   The average value is the definite integral from $x = 0$ to $x = 4$ (the end of month 4) divided by the length of the interval.

$$\frac{1}{4} \int_0^4 \underbrace{40x(x^2 + 9)^{-1/2}}_{u^{-1/2}} dx = \frac{1}{4} \cdot 40 \cdot \frac{1}{2} \int_0^4 2x(x^2 + 9)^{-1/2} \, dx = 5 \int_9^{25} u^{-1/2} \, du$$

$$u = x^2 + 9$$
$$du = 2x \, dx$$

Changing the limits to u-values using $u = x^2 + 9$

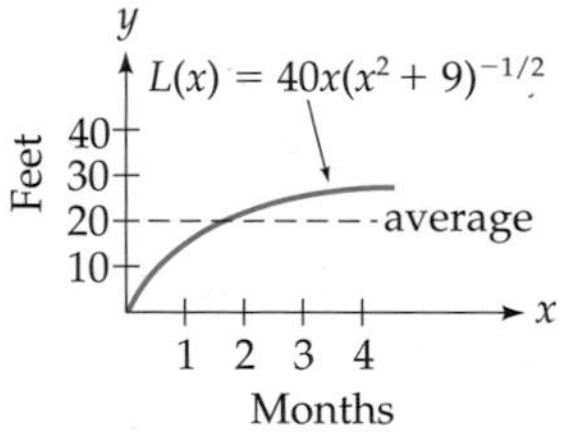

$u = 4^2 + 9 = 25$

$u^{-1/2}$

du

$u = 0^2 + 9 = 9$

$L(x) = 40x(x^2 + 9)^{-1/2}$

$$= 5 \cdot 2u^{1/2} \Big|_9^{25} = 10(25)^{1/2} - 10(9)^{1/2} = 10 \cdot 5 - 10 \cdot 3 = 20$$

That is, the average depth of the reservoir over the last 4 months was 20 feet.

5.6 Section Summary

The three substitution formulas are listed on the inside back cover. Most of the work in using these formulas is making a problem "fit" the left-hand side of one of the formulas (choosing the u and adjusting constants to complete the du). Once a problem fits a left-hand side, the right-hand side immediately gives the answer (except for substituting back to the original variable).

Note that the du now plays a very important role: The du must be correct if the answer is to be correct. For example, the formula $\int e^u \, du = e^u + C$ should not be thought of as the formula for integrating e^u, but as the formula for integrating $e^u \, du$. The du is just as important as the e^u.

▶ **Solutions to Practice Problems**

1. $df = (3x^2 - 4) \, dx$

2. $du = -5e^{-5t} \, dt$

3. a. (B) **b.** (C) **c.** (A) **d.** (C)

4. Neither.

a. Try formula (A) with $u = x^3 + 1$, so $du = 3x^2 \, dx$. The problem has an x^3 for the differential instead of the needed x^2.

b. Try formula (B) with $u = x^2$, so $du = 2x \, dx$. The problem does not have the x that is needed for the differential.

Exercises

Find each indefinite integral. [Integration formulas (A), (B), and (C) are on the inside back cover, numbered 5–7]

1. $\displaystyle\int (x^2 + 1)^9\, 2x\, dx$

[*Hint:* Use $u = x^2 + 1$ and formula 5.]

2. $\displaystyle\int (x^3 + 1)^4\, 3x^2\, dx$

[*Hint:* Use $u = x^3 + 1$ and formula 5.]

3. $\displaystyle\int (x^2 + 1)^9 x\, dx$

[*Hint:* Use $u = x^2 + 1$ and formula 5.]

4. $\displaystyle\int (x^3 + 1)^4 x^2\, dx$

[*Hint:* Use $u = x^3 + 1$ and formula 5.]

5. $\displaystyle\int e^{x^5} x^4\, dx$

[*Hint:* Use $u = x^5$ and formula 7.]

6. $\displaystyle\int e^{x^4} x^3\, dx$

[*Hint:* Use $u = x^4$ and formula 7.]

7. $\displaystyle\int \frac{x^5\, dx}{x^6 + 1}$

[*Hint:* Use $u = x^6 + 1$ and formula 6.]

8. $\displaystyle\int \frac{x^4\, dx}{x^5 + 1}$

[*Hint:* Use $u = x^5 + 1$ and formula 6.]

Show that each integral *cannot* be found by our substitution formulas.

9. $\displaystyle\int \sqrt{x^3 + 1}\, x\, dx$

10. $\displaystyle\int \sqrt{x^5 + 9}\, x^2\, dx$

11. $\displaystyle\int e^{x^4} x^5\, dx$

12. $\displaystyle\int e^{x^3} x^4\, dx$

Find each indefinite integral by the substitution method or state that it cannot be found by our substitution formulas.

13. $\displaystyle\int (x^4 - 16)^5 x^3\, dx$

14. $\displaystyle\int (x^5 - 25)^6 x^4\, dx$

15. $\displaystyle\int e^{-x^2} x\, dx$

16. $\displaystyle\int e^{-x^4} x^3\, dx$

17. $\displaystyle\int e^{3x}\, dx$

18. $\displaystyle\int e^{5x}\, dx$

19. $\displaystyle\int e^{x^2} x^2\, dx$

20. $\displaystyle\int e^{x^3} x\, dx$

21. $\displaystyle\int \frac{dx}{1 + 5x}$

22. $\displaystyle\int \frac{dx}{1 + 3x}$

23. $\displaystyle\int (x^2 + 1)^9 5x\, dx$

24. $\displaystyle\int (x^2 - 4)^6 3x\, dx$

25. $\displaystyle\int \sqrt[4]{z^4 + 16}\, z^3\, dz$

26. $\displaystyle\int \sqrt[3]{z^3 - 8}\, z^2\, dz$

27. $\displaystyle\int \sqrt[4]{x^4 + 16}\, x^2\, dx$

28. $\displaystyle\int \sqrt[3]{x^3 - 8}\, x\, dx$

29. $\displaystyle\int (2y^2 + 4y)^5 (y + 1)\, dy$

30. $\displaystyle\int (3y^2 - 6y)^3 (y - 1)\, dy$

31. $\displaystyle\int e^{x^2 + 2x + 5}(x + 1)\, dx$ **32.** $\displaystyle\int e^{x^3 - 3x + 7}(x^2 - 1)\, dx$

33. $\displaystyle\int \frac{x^3 + x^2}{3x^4 + 4x^3}\, dx$ **34.** $\displaystyle\int \frac{x^2 - x}{2x^3 - 3x^2}\, dx$

35. $\displaystyle\int \frac{x^3 + x^2}{(3x^4 + 4x^3)^2}\, dx$ **36.** $\displaystyle\int \frac{x^2 - x}{(2x^3 - 3x^2)^3}\, dx$

37. $\displaystyle\int \frac{x}{1 - x^2}\, dx$ **38.** $\displaystyle\int \frac{1}{1 - x}\, dx$

39. $\displaystyle\int (2x - 3)^7\, dx$ **40.** $\displaystyle\int (5x + 9)^9\, dx$

41. $\int \dfrac{e^{2x}}{e^{2x} + 1} dx$ **42.** $\int \dfrac{e^{3x}}{e^{3x} - 1} dx$

43. $\int \dfrac{\ln x}{x} dx$ [*Hint:* Let $u = \ln x$.]

44. $\int \dfrac{(\ln x)^2}{x} dx$ [*Hint:* Let $u = \ln x$.]

45. $\int \dfrac{e^{\sqrt{x}}}{\sqrt{x}} dx$ [*Hint:* Let $u = \sqrt{x}$.]

46. $\int \dfrac{e^{1/x}}{x^2} dx$ $\left[\textit{Hint: Let } u = \dfrac{1}{x}. \right]$

Find each integral. [*Hint:* Try some algebra.]

47. $\int (x + 1)x^2 \, dx$ **48.** $\int (x + 4)(x - 2) \, dx$

49. $\int (x + 1)^2 x^3 \, dx$ **50.** $\int (x - 1)^2 \sqrt{x} \, dx$

For each definite integral:

a. Evaluate it "by hand."

b. Check your answer by using a graphing calculator.

51. $\int_0^3 e^{x^2} x \, dx$ **52.** $\int_0^2 e^{x^3} x^2 \, dx$

53. $\int_0^1 \dfrac{x}{x^2 + 1} dx$ **54.** $\int_2^3 \dfrac{x^2}{x^3 - 7} dx$

55. $\int_0^4 \sqrt{x^2 + 9} \, x \, dx$ **56.** $\int_0^3 \sqrt{x^2 + 16} \, x \, dx$

57. $\int_2^3 \dfrac{dx}{1 - x}$ **58.** $\int_3^4 \dfrac{dx}{2 - x}$

59. $\int_1^8 \dfrac{e^{\sqrt[3]{x}}}{\sqrt[3]{x^2}} dx$ **60.** $\int_1^4 \dfrac{e^{\sqrt{x}}}{\sqrt{x}} dx$

61. Prove the integration formula

$$\int u^n \, du = \dfrac{1}{n + 1} u^{n+1} + C \qquad (n \neq -1)$$

as follows.

a. Differentiate the right-hand side of the formula with respect to x (remembering that u is a function of x).

b. Verify that the result of part (a) agrees with the integrand in the formula (after replacing du in the formula by $u' \, dx$).

62. Prove the integration formula $\int e^u \, du = e^u + C$ by following the steps in Exercise 61.

63. Prove the integration formula $\int \dfrac{du}{u} = \ln u + C$ $(u > 0)$ by following the steps in Exercise 61. (Absolute value bars come from applying the same argument to $-u$ for $u < 0$.)

64. Find $\int (x + 1) \, dx$:

a. by using the formula for $\int u^n \, du$ with $n = 1$.

b. by dropping the parentheses and integrating directly.

c. Can you reconcile the two seemingly different answers? [*Hint:* Think of the arbitrary constant.]

APPLIED EXERCISES

65. BUSINESS: Cost A company's marginal cost function is $MC(x) = \dfrac{1}{2x + 1}$ and its fixed costs are 50. Find the cost function.

66. BUSINESS: Cost A company's marginal cost function is $MC(x) = \dfrac{1}{\sqrt{2x + 25}}$ and its fixed costs are 100. Find the cost function.

67. GENERAL: Average Value The population of a city is expected to be $P(x) = x(x^2 + 36)^{-1/2}$ million people after x years. Find the average population between year $x = 0$ and year $x = 8$.

68. GENERAL: Area Find the area between the curve $y = xe^{x^2}$ and the x-axis from $x = 1$ to $x = 3$. (Leave the answer in its exact form.)

69. BUSINESS: Average Sales A company's sales (in millions) during week x are given by $S(x) = \dfrac{1}{x + 1}$. Find the average sales from week $x = 1$ to week $x = 4$.

70. BEHAVIORAL SCIENCE: Repeated Tasks A subject can perform a task at the rate of $\sqrt{2t + 1}$ tasks per minute at time t minutes. Find the total number of tasks performed from time $t = 0$ to time $t = 12$.

71. BIOMEDICAL: Cholesterol An experimental drug lowers a patient's blood serum cholesterol at the rate of $t\sqrt{25 - t^2}$ units per day, where t is the number of days since the drug was administered ($0 \le t \le 5$). Find the total change during the first 3 days.

72. BUSINESS: Total Sales During an automobile sale, cars are selling at the rate of $\dfrac{12}{x + 1}$ cars per day, where x is the number of

days since the sale began. How many cars will be sold during the first 7 days of the sale?

73. BUSINESS: Total Sales A real estate office is selling condominiums at the rate of $100e^{-x/4}$ per week after x weeks. How many condominiums will be sold during the first 8 weeks?

74. BUSINESS: Revenue An aircraft company estimates its marginal revenue function for helicopters to be $MR(x) - \sqrt{x^2 + 80x}(x + 40)$ thousand dollars, where x is the number of helicopters sold. Find the total revenue from the sale of the first 10 helicopters.

75–76. ENVIRONMENTAL SCIENCE: Pollution A factory is discharging pollution into a lake at the rate of $r(t)$ tons per year given below, where t is the number of years that the factory has been in operation. Find the total amount of pollution discharged during the first 3 years of operation.

75. $r(t) = \dfrac{t}{t^2 + 1}$ **76.** $r(t) = t\sqrt{t^2 + 16}$

Chapter Summary with Hints and Suggestions

Reading the text and doing the exercises in this chapter have helped you to master the following skills, which are listed by section (in case you need to review them) and are keyed to particular Review Exercises. Answers for all Review Exercises are given at the back of the book, and full solutions can be found in the Student Solutions Manual.

5.1 Antiderivatives and Indefinite Integrals

- Find an indefinite integral using the Power Rule. *(Review Exercises 1–8.)*

$$\int x^n \, dx = \frac{1}{n + 1} x^{n+1} + C \quad (n \neq -1)$$

- Solve an applied problem involving integration. *(Review Exercises 9–10.)*

5.2 Integration Using Logarithmic and Exponential Functions

- Find an indefinite integral involving e^x or $\dfrac{1}{x}$. *(Review Exercises 11–18.)*

$$\int e^{ax} \, dx = \frac{1}{a} e^{ax} + C$$

$$\int \frac{1}{x} \, dx = \int x^{-1} \, dx = \ln |x| + C$$

■ Solve an applied problem involving integration. *(Review Exercises 19–22.)*

5.3 Definite Integrals and Areas

■ Evaluate a definite integral.
(Review Exercises 23–30.)

$$\int_a^b f(x)\, dx = F(b) - F(a)$$

■ Find the area under a curve.
(Review Exercises 31–36.)

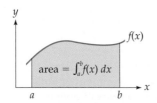

■ Graph a function and find the area under it.
(Review Exercises 37–38.)

■ Solve an applied problem using definite integration. *(Review Exercises 39–44.)*

■ Approximate the area under a curve by rectangles (Riemann sum).
(Review Exercises 45–46.)

$$\int_a^b f(x)\, dx = \lim_{n \to \infty} [f(x_1) \cdot \Delta x + \cdots + f(x_n) \cdot \Delta x]$$

■ Use a Riemann sum program to find Riemann sums. *(Review Exercises 47–48.)*

5.4 Further Applications of Definite Integrals: Average Value and Area Between Curves

■ Find the area bounded by two curves.
(Review Exercises 49–56.)

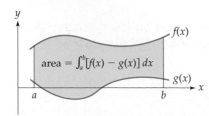

■ Find the average value of a function on an interval. *(Review Exercises 57–60.)*

$$\left(\begin{array}{c} \text{Average of } f \\ \text{from } a \text{ to } b \end{array} \right) = \frac{1}{b - a} \int_a^b f(x)\, dx$$

■ Solve an applied problem involving average value. *(Review Exercises 61–64.)*

■ Solve an applied problem involving area between curves. *(Review Exercises 65–68.)*

5.5 Two Applications to Economics: Consumers' Surplus and Income Distribution

■ Find the consumers' surplus for a product.
(Review Exercises 69–72.)

$$\left(\begin{array}{c} \text{Consumers'} \\ \text{surplus} \end{array} \right) = \int_0^A [d(x) - B]\, dx$$

■ Find the Gini index of income distribution.
(Review Exercises 73–76.)

$$\left(\begin{array}{c} \text{Gini} \\ \text{index} \end{array} \right) = 2 \cdot \int_0^1 [x - L(x)]\, dx$$

5.6 Integration by Substitution

■ Use a substitution to find an integral.
(Review Exercises 77–92.)

$$\int u^n\, du = \frac{1}{n + 1} u^{n+1} + C \qquad (n \neq -1)$$

$$\int e^u\, du = e^u + C \qquad \int \frac{1}{u}\, du = \ln |u| + C$$

■ Use a substitution to find a definite integral.
(Review Exercises 93–100.)

■ Use a substitution to find the area under a curve. *(Review Exercises 101–102.)*

■ Use a substitution to find the average value of a function. *(Review Exercises 103–104.)*

■ Use a substitution to solve an applied problem. *(Review Exercises 105–106.)*

Hints and Suggestions

■ An *indefinite* integral is a function plus a constant, whereas a *definite* integral is a *number* (the signed area between the curve and the x-axis). The Fundamental Theorem of Integral Calculus shows how to evaluate the second in terms of the first.

■ To integrate any power of x *except* x^{-1}, use the Power Rule; for x^{-1}, use the ln rule.

■ *Indefinite* integrals have a $+C$. *Definite* integrals do not.

■ Differentiation *breaks things down into parts*—for example, turning cost into marginal cost (cost per unit). Integration *combines back into a whole*—for example, combining all of the per-unit costs back into a total cost. In fact, the word "integrate" means "make whole."

■ To find the area between two curves, integrate *upper* minus *lower*.

■ For average values, don't forget the $\dfrac{1}{b-a}$.

■ The substitution method can only "fix up" the *constant*; the variable part must already be correct (or else the function cannot be integrated by this technique).

■ A graphing calculator helps by graphing functions, evaluating definite integrals, and finding areas under curves and total accumulations. With an appropriate program it can calculate Riemann sums with many rectangles.

■ **Practice for test:** Review Exercises 1, 5, 9, 15, 21, 23, 27, 31, 37, 39, 43, 49, 55, 57, 61, 69, 73, 81, 83, 93.

Review Exercises for Chapter 5

Practice test exercise numbers are in green.

5.1 Antiderivatives and Indefinite Integrals

Find each indefinite integral.

1. $\displaystyle\int (24x^2 - 8x + 1)\, dx$

2. $\displaystyle\int (12x^3 + 6x - 3)\, dx$

3. $\displaystyle\int \left(6\sqrt{x} - 5\right) dx$

4. $\displaystyle\int \left(8\sqrt[3]{x} - 2\right) dx$

5. $\displaystyle\int \left(10\sqrt[3]{x^2} - 4x\right) dx$

6. $\displaystyle\int \left(5\sqrt{x^3} - 6x\right) dx$

7. $\displaystyle\int (x + 4)(x - 4)\, dx$

8. $\displaystyle\int \frac{3x^3 + 2x^2 + 4x}{x}\, dx$

9. **BUSINESS: Cost** A company's marginal cost function is $MC(x) = x^{-1/2} + 4$, where x is the number of units. If fixed costs are 20,000, find the company's cost function.

10. **GENERAL: Population** The population of a town is now 40,000 and t years from now will be growing at the rate of $300\sqrt{t}$ people per year.

    a. Find a formula for the population of the town t years from now.

    b. Use your formula to find the population of the town 16 years from now.

5.2 Integration Using Logarithmic and Exponential Functions

Find each indefinite integral.

11. $\displaystyle\int e^{x/2}\, dx$ 12. $\displaystyle\int e^{-2x}\, dx$

13. $\int 4x^{-1}\,dx$

14. $\int \dfrac{2}{x}\,dx$

15. $\int \left(6e^{3x} - \dfrac{6}{x}\right)dx$

16. $\int (x - x^{-1})\,dx$

17. $\int (9x^2 + 2x^{-1} + 6e^{3x})\,dx$

18. $\int \left(\dfrac{1}{x^2} + \dfrac{1}{x} + e^{-x}\right)dx$

19. GENERAL: Consumption of Natural Resources World consumption of aluminum is running at the rate of $50e^{0.05t}$ million tons per year, where t is the number of years since 2000.

  a. Find a formula for the total amount of aluminum consumed within t years of 2000.

  b. If consumption continues at this rate, when will the known resources of 7900 million tons of aluminum be exhausted?

20. GENERAL: Total Savings A homeowner installs a solar heating system, which is expected to generate savings at the rate of $200e^{0.1t}$ dollars per year, where t is the number of years since the system was installed.

  a. Find a formula for the total savings in the first t years.

  b. If the system originally cost $1500, when will it "pay for itself"?

21. GENERAL: Consumption of Natural Resources World consumption of zinc is running at the rate of $9e^{0.02t}$ million metric tons per year, where t is the number of years since 2000.

  a. Find a formula for the total amount of zinc consumed within t years of 2000.

  b. If consumption continues at this rate, when will the known resources of 430 million metric tons of zinc be exhausted? (Zinc is used to make protective coatings for iron and steel.)

22. BUSINESS: Profit A company's profit is growing at the rate of $200x^{-1}$ thousand dollars per month after x months, for $x \geq 1$.

  a. Find a formula for the total growth in the profit from month 1 to month x.

  b. When will the total growth in profit reach 600 thousand dollars?

5.3 Definite Integrals and Areas

Evaluate each definite integral.

23. $\displaystyle\int_1^9 \left(x - \dfrac{1}{\sqrt{x}}\right)dx$

24. $\displaystyle\int_2^5 (3x^2 - 4x + 5)\,dx$

25. $\displaystyle\int_1^{e^4} \dfrac{dx}{x}$

26. $\displaystyle\int_1^5 \dfrac{dx}{x}$

27. $\displaystyle\int_0^2 e^{-x}\,dx$

28. $\displaystyle\int_0^2 e^{x/2}\,dx$

29. $\displaystyle\int_0^{100} (e^{0.05x} - e^{0.01x})\,dx$

30. $\displaystyle\int_0^{10} (e^{0.04x} - e^{0.02x})\,dx$

For each function:

  a. Find the area under the curve between the given x-values.

  b. Verify your answer to part (a) by using a graphing calculator to find the area.

31. $f(x) = 6x^2 - 1$, $x = 1$ to $x = 2$

32. $f(x) = 9 - x^2$, $x = -3$ to $x = 3$

33. $f(x) = 12e^{2x}$, $x = 0$ to $x = 3$

34. $f(x) = e^{x/2}$, $x = 0$ to $x = 4$

35. $f(x) = \dfrac{1}{x}$, $x = 1$ to $x = 100$

36. $f(x) = x^{-1}$, $x = 1$ to $x = 1000$

Use a calculator to graph each function and find the area under it between the given x-values.

37. $f(x) = \dfrac{10}{x^4 + 1}$ from $x = -2$ to $x = 2$

38. $f(x) = e^{x^4}$ from $x = -1$ to $x = 1$

39. GENERAL: Weight Gain An average child of age t years gains weight at the rate of $1.7t^{1/2}$ kilograms per year. Find the total weight gain from age 1 to age 9.

40. BEHAVIORAL SCIENCE: Learning A student can memorize foreign vocabulary words at the rate of $2/\sqrt[3]{t}$ words per minute, where t is the number of minutes since the studying began. Find the number of words that can be memorized from time $t = 1$ to time $t = 8$.

41. **ENVIRONMENTAL SCIENCE: Global Warming** The temperature of the earth is rising, as a result of the "greenhouse effect," in which carbon dioxide prevents the escape of heat from the atmosphere. If the temperature is rising at the rate of $0.15e^{0.1t}$ degrees per year, find the total rise in temperature over the next 10 years.

42. **BUSINESS: Cost** A company's marginal cost function is $MC(x) = x^{-1/2} + 4$ dollars, where x is the number of units. Find the total cost of the first 400 units (units $x = 0$ to $x = 400$).

43. **BUSINESS: Cost** A company's marginal cost function is $MC(x) = 22e^{-\sqrt{x}/5}$ dollars, where x is the number of units. Find the total cost of the first hundred units (units $x = 0$ to $x = 100$).

44. **BEHAVIORAL SCIENCE: Repetitive Tasks** A proofreader can read $15xe^{-0.25x}$ pages per hour, where x is the number of hours worked. Find the total number of pages that can be proofread in 8 hours.

Exercises on Riemann Sums

45. **a.** Approximate the area under the curve $f(x) = x^2$ from 0 to 2 using ten left rectangles with equal bases.
    b. Find the *exact* area under the curve between the given x-values by evaluating an appropriate definite integral using the Fundamental Theorem.

46. **a.** Approximate the area under the curve $f(x) = \sqrt{x}$ from 0 to 4 using ten left rectangles with equal bases. (Round calculations to three decimal places.)
    b. Find the *exact* area under the curve between the given x-values by evaluating an appropriate definite integral using the Fundamental Theorem.

47–48. Use the graphing calculator program RIEMANN (see page 372), one of the programs on page 384, or a similar program to find the following Riemann sums.

a. Calculate the Riemann sum for each function below for the following values of n: 10, 100, 1000. Use left, right, or midpoint rectangles, making a table of the answers, keeping three decimal places of accuracy.

b. Find the *exact* value of the area under the curve in each exercise below by evaluating an appropriate definite integral using the Fundamental Theorem. Your answers from part (a) should approach the number found from this integral.

47. $f(x) = e^x$ from $a = -2$ to $b = 2$

48. $f(x) = \dfrac{1}{x}$ from $a = 1$ to $b = 4$

5.4 Further Applications of Definite Integrals: Average Value and Area Between Curves

Find the area bounded by each pair of curves.

49. $y = x^2 + 3x$ and $y = 3x + 1$

50. $y = 12x - 3x^2$ and $y = 6x - 24$

51. $y = x^3$ and $y = x$

52. $y = x^4$ and $y = x$

53. $y = 4x^3$ and $y = 12x^2 - 8x$

54. $y = x^3 + x^2$ and $y = x^2 + x$

55. $y = e^x$ and $y = x + 5$

56. $y = \ln x$ and $y = \dfrac{x^2}{10}$

Find the average value of the function on the given interval.

57. $f(x) = \dfrac{1}{x}$ on $[1, 4]$

58. $f(x) = 6\sqrt{x}$ on $[1, 4]$

59. $f(x) = \sqrt{x^3 + 1}$ on $[0, 5]$

60. $f(x) = \ln(e^{x^2} + 10)$ on $[-2, 2]$

61. **GENERAL: Average Population** The population of the world, now more than 6 billion, is predicted to be $P(t) = 6e^{0.008t}$ billion, where t is the number of years after the year 2000. Find the average population between the years 2000 and 2100.

62. GENERAL: Compound Interest A deposit of $3000 in a bank paying 6% interest compounded continuously will grow to $V(t) = 3000e^{0.06t}$ dollars after t years. Find the average value during the first 20 years ($t = 0$ to $t = 20$).

63. GENERAL: Real Estate The value of a suburban plot of land being considered for rezoning is assessed at $4.3e^{0.01x^2}$ hundred thousand dollars x years from now. Find the average value over the next 10 years (year 0 to year 10).

64. GENERAL: Stock Price The price of a share of stock is expected to be $28e^{0.01x^{1.2}}$ dollars where x is the number of weeks from now. Find the average price over the next year (week 0 to week 52).

65. GENERAL: Art An artist wants to paint the interior of the shape shown below on the side of a building. How much area (in square meters) will the artist need to paint?

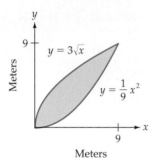

66. ECONOMICS: Balance of Trade A country's annual exports will be $E(t) = 40e^{0.2t}$ and its imports will be $I(t) = 20e^{0.1t}$ (both in billions of dollars per year), where t is the number of years from now. Find the accumulated trade surplus (exports minus imports) over the next 10 years. (See the graph below.)

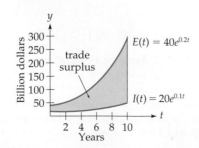

67. BUSINESS: Profit A company's annual revenue and annual costs are expected to be $R(t) = 50e^{0.08t}$ and $C(t) = 20e^{0.04t}$ million dollars per year, where t is the number of years from now. Find the cumulative profit over the next 8 years (year 0 to year 8).

68. GENERAL: Population China, with 21% of the world's population on only 7% of the world's arable land, has taken drastic measures to reduce its population, and has succeeded in reducing its birthrate (births per 1000 population) from 23.3 in 1987 to 16.2 in 2000. China's population is predicted to be $P(x) = 1260e^{0.009x}$ million people x years after 2000.

a. Multiply $P(x)$ by $\frac{23.3}{1000}$ to find the number of births per year at the old birthrate, and enter the result in your calculator as y_1. Then multiply $P(x)$ by $\frac{16.2}{1000}$ to find the number of births per year at the new birthrate, and enter the result as y_2.

b. Graph both y_1 and y_2 on the window $[0, 10]$ by $[0, 35]$, showing the difference in the number of births at the different birthrates during 2000–2010.

c. Find the area between the curves, thereby finding the decrease in the number of births during this period resulting from the lower birthrate.

5.5 Two Applications to Economics: Consumers' Surplus and Income Distribution

69–72. ECONOMICS: Consumers' Surplus For each demand function $d(x)$ and demand level x, find the consumers' surplus.

69. $d(x) = 8000 - 24x$, $x = 200$

70. $d(x) = 1800 - 0.03x^2$, $x = 200$

71. $d(x) = 300e^{-0.2\sqrt{x}}$, $x = 120$

72. $d(x) = \dfrac{100}{1 + \sqrt{x}}$, $x = 100$

73–76. ECONOMICS: Gini Index For each Lorenz curve, find the Gini index.

73. $L(x) = x^{3.5}$

74. $L(x) = x^{2.5}$

75. $L(x) = \dfrac{x}{2 - x}$

76. $L(x) = \dfrac{e^{x^4} - 1}{e - 1}$

5.6 Integration by Substitution

Find each integral or state that it cannot be evaluated by our substitution formulas.

77. $\int x^2 \sqrt[3]{x^3 - 1}\, dx$ **78.** $\int x^3 \sqrt{x^4 - 1}\, dx$

79. $\int x \sqrt[3]{x^3 - 1}\, dx$ **80.** $\int x^2 \sqrt{x^4 - 1}\, dx$

81. $\int \dfrac{dx}{9 - 3x}$ **82.** $\int \dfrac{dx}{1 - 2x}$

83. $\int \dfrac{dx}{(9 - 3x)^2}$ **84.** $\int \dfrac{dx}{(1 - 2x)^2}$

85. $\int \dfrac{x^2}{\sqrt[3]{8 + x^3}}\, dx$ **86.** $\int \dfrac{x}{\sqrt{9 + x^2}}\, dx$

87. $\int \dfrac{w + 3}{(w^2 + 6w - 1)^2}\, dw$

88. $\int \dfrac{t - 2}{(t^2 - 4t + 1)^2}\, dt$

89. $\int \dfrac{(1 + \sqrt{x})^2}{\sqrt{x}}\, dx$ **90.** $\int \dfrac{(1 + \sqrt[3]{x})^2}{\sqrt[3]{x^2}}\, dx$

91. $\int \dfrac{e^r}{e^x - 1}\, dx$ **92.** $\int \dfrac{1}{x \ln x}\, dx$

For each definite integral:

    a. Evaluate it ("by hand") or state that it cannot be evaluated by our substitution formulas.

    b. Verify your answer to part (a) by using a graphing calculator.

93. $\int_0^3 x \sqrt{x^2 + 16}\, dx$ **94.** $\int_0^4 \dfrac{dz}{\sqrt{2z + 1}}$

95. $\int_0^4 \dfrac{w}{\sqrt{25 - w^2}}\, dw$

96. $\int_1^2 \dfrac{x + 1}{(x^2 + 2x - 2)^2}\, dx$

97. $\int_3^9 \dfrac{dx}{x - 2}$ **98.** $\int_4^5 \dfrac{dx}{x - 6}$

99. $\int_0^1 x^3 e^{x^4}\, dx$ **100.** $\int_0^1 x^4 e^{x^5}\, dx$

Find the area under the given curve between the given x-values.

101. $y = \dfrac{x^2 + 6x}{\sqrt[3]{x^3 + 9x^2 + 17}}$ from $x = 1$ to $x = 3$

102. $y = \dfrac{x + 6}{\sqrt{x^2 + 12x + 4}}$ from $x = 0$ to $x = 3$

Find the average value of the function on the given interval.

103. $f(x) = xe^{-x^2}$ on $[0, 2]$

104. $f(x) = \dfrac{x}{x^2 - 3}$ on $[2, 4]$

105. BUSINESS: Cost A company's marginal cost function is $MC(x) = \dfrac{1}{\sqrt{2x + 9}}$ and fixed costs are 100. Find the cost function.

106. BIOMEDICAL: Temperature An experimental drug changes a patient's temperature at the rate of $\dfrac{3x^2}{x^3 + 1}$ degrees per milligram of the drug, where x is the amount of the drug administered. Find the total change in temperature resulting from the first 3 milligrams of the drug. [*Note:* The rate of change of temperature with respect to dosage is called the "drug sensitivity."]

6 Integration Techniques

Application Preview

Improper Integrals and Eternal Recognition

Suppose that after you become rich and famous, you decide to commission a statue of yourself for your hometown. Your town, however, will accept this selfless gesture only if you pay for the perpetual upkeep of the statue by establishing a fund that will generate $2000 annually for every year in the future. Before deciding whether or not to accept this condition, you of course want to know how much it will cost. (Interestingly, we will see that a fund that will generate income forever does *not* require an infinite amount of money.) On page 279 we found that to realize a yield of $2000 t years from now requires only its *present value* deposited now in a bank. At an interest rate of, say, 5% compounded continuously, $2000 in t years requires a deposit now of

$$\left(\begin{array}{l}\text{Present value of \$2000 at 5\%}\\ \text{compounded continuously}\end{array}\right) = 2000e^{-0.05t} \qquad \text{Amount times } e^{-0.05t}$$

Therefore, the size of the fund needed to generate $2000 annually *forever* is found by summing (integrating) this present value over the infinite time interval from zero to infinity (∞).

$$\left(\begin{array}{l}\text{Size of fund to yield an}\\ \text{annual \$2000 forever}\end{array}\right) = \int_0^\infty 2000e^{-0.05t}\,dt \qquad \begin{array}{l}\text{Integrating out}\\ \text{to infinity}\end{array}$$

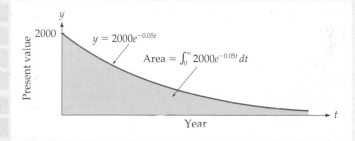

On page 466 we will learn how to evaluate such "improper integrals," finding that the value of this integral is $40,000, which shows that $40,000 deposited in a bank at 5% compounded continuously will generate $2000 every year *forever*. That is, $40,000 would pay for the perpetual upkeep of your statue, buying you (or at least your likeness) a kind of immortality.

Incidentally, to generate $2000 annually for only the first *hundred* years would require $\displaystyle\int_0^{100} 2000e^{-0.05t}\,dt \approx \$39{,}730$ (integrating from 0 to 100). This amount is only $270 less than the amount needed to generate the same sum *forever*. This small additional cost shows that the short term is expensive, but eternity is cheap.

INTEGRATION BY PARTS

Introduction

In this chapter we introduce further techniques for finding integrals: integration by parts, integration by tables, and numerical integration. We also discuss improper integrals (integrals over infinite intervals).

You may have felt that integration is "harder" than differentiation. One reason is that integration is an *inverse* process, carried out by *reversing* differentiation, and inverse processes are generally more difficult. Another reason is that, while we have the product and quotient rules for differentiating complicated expressions, there are no product and quotient rules for integration. The method of integration by parts, explained in this section, is in some sense a "product rule for integration" in that it comes from interpreting the product rule as an integration formula.

Integration by Parts

For two differentiable functions $u(x)$ and $v(x)$, hereafter denoted simply u and v, the Product Rule is

$$(uv)' = u'v + uv'$$

The derivative of a product is the derivative of the first times the second, plus the first times the derivative of the second

If we integrate both sides of this equation, integrating the left side "undoes" the differentiation.

$$uv = \int u'v\, dx + \int uv'\, dx$$

$du \qquad dv$

Using differential notation, $du = u'\, dx, \quad dv = v'\, dx$

$$uv = \int v\, du + \int u\, dv$$

Formula above in differential notation

Solving this equation for the second integral $\int u\, dv$ gives

$$\int u\, dv = uv - \int v\, du$$

This formula is the basis for a technique called "integration by parts."

Integration by Parts

For differentiable functions u and v,

$$\int u\, dv = uv - \int v\, du$$

We use this formula to solve integrals by a "double substitution," substituting u for part of the given integral and dv for the rest, and then expressing the integral in the form $uv - \int v\,du$. The point is to choose the u and the dv so that the resulting integral $\int v\,du$ is *simpler* than the original integral $\int u\,dv$. A few examples will make the method clear.

EXAMPLE 1

INTEGRATING BY PARTS

Use integration by parts to find $\displaystyle\int xe^x\,dx$.

Solution

$$\int \underset{u\ \ dv}{x\,e^x\,dx}$$

Original integral

We choose $u = x$
and $dv = e^x\,dx$

$$\begin{bmatrix} u = x & dv = e^x\,dx \\ \downarrow & \downarrow \\ du = 1\,dx = dx & v = \int e^x\,dx = e^x \end{bmatrix}$$

Differentiating u to find du and integrating dv to find v (omitting the C) (the arrows show which part leads to which other part)

$$= x\,e^x - \int e^x\,dx$$
$$\quad\ \underset{u\ v}{}\quad \underset{v\ du}{}$$

Replacing the original integral $\int u\,dv$ by the right-hand side of the formula, $u\cdot v - \int v\,du$, with $u = x$, $v = e^x$, and $du = dx$

$$= xe^x - e^x + C$$
$$\qquad\ \ \underset{\text{From } \int e^x\,dx}{\uparrow}$$

Finding the new integral $\int e^x\,dx = e^x$ to give the final answer (with $+C$)

The procedure is not as complicated as it might seem. All the steps may be written together as follows:

$$\int \underset{u\ \ dv}{xe^x\,dx} = \underset{uv}{xe^x} - \int \underset{v\ du}{e^x\,dx} = xe^x - e^x + C$$
$$\qquad\qquad\qquad\qquad\qquad\qquad \underset{\text{From } \int e^x\,dx}{}$$

$$\begin{bmatrix} u = x & dv = e^x\,dx \\ du = dx & v = \int e^x\,dx = e^x \end{bmatrix}$$

We can check this answer by differentiation.

Cancel

$$\frac{d}{dx}(xe^x - e^x + C) = e^x + xe^x - e^x = xe^x$$

Differentiating xe^x
by the Product Rule

Agrees with the original integrand, so the integration is correct

Remarks on the Integration by Parts Procedure

i. The differentials du and dv include the dx.

ii. We omit the constant C when we integrate dv to get v because one $+ C$ at the end is enough.

iii. The integration by parts formula does not give a "final answer," but rather expresses the given integral as $uv - \int v \, du$, a product $u \cdot v$ (already integrated) and a new integral $\int v \, du$. That is, integration by parts "exchanges" the original integral $\int u \, dv$ for another integral $\int v \, du$. The hope is that the second integral will be simpler than the first. In our example we "exchanged" $\int xe^x \, dx$ for the simpler $\int e^x \, dx$, which could be integrated immediately by formula 3 (inside back cover).

iv. Integration by parts should be used only if the simpler formulas 1 through 7 (inside back cover) fail to solve the integral.

How to Choose the *u* and the *dv*

In Example 1, the choice of $\begin{cases} u = x \\ dv = e^x \, dx \end{cases}$ "exchanged" the original integral $\int xe^x \, dx$ for the simpler integral $\int e^x \, dx$. If we had instead chosen $\begin{cases} u = e^x \\ dv = x \, dx \end{cases}$ we would have "exchanged" the original integral for $\int x^2 e^x \, dx$ (as you may check), which is *more* difficult than the original (because of the x^2). Therefore, the first choice was the "right" choice in that it led to a solution. Generally, one choice for u and dv will be "best," and finding it may involve some trial and error. While there is no foolproof rule for finding the best u and dv, the following guidelines often help.

Guidelines for Choosing *u* and *dv*

1. Choose dv to be the most complicated part of the integral that can be integrated easily.

2. Choose u so that u' is simpler than u.

| EXAMPLE 2 | **INTEGRATING BY PARTS** |

Find $\displaystyle\int x^2 \ln x \, dx$.

Solution None of the easier formulas (1 through 7 on the inside back cover) will solve the integral, as you may easily check. Therefore, we try integration by parts. The integrand is a product, x^2 times $\ln x$. The guidelines say to choose dv to be the most complicated part that can be easily integrated. We can integrate x^2 but not $\ln x$ (we know how to *differentiate* $\ln x$, but not how to *integrate* it), so we choose $dv = x^2 \, dx$, and therefore $u = \ln x$.

$$\int x^2 \ln x \, dx = (\ln x)\frac{1}{3}x^3 - \int \frac{1}{3}x^3\frac{1}{x}\, dx = \frac{1}{3}x^3 \ln x - \frac{1}{3}\int x^2 \, dx$$

$$\underbrace{}_{\substack{u \\ dv}} \qquad \underbrace{}_{u}\;\underbrace{}_{v} \qquad \underbrace{}_{v}\;\underbrace{}_{du} \qquad \text{Moving the } \tfrac{1}{3}$$

$$\left[\begin{array}{ll} u = \ln x & dv = x^2 \, dx \\[2mm] du = \dfrac{1}{x}\, dx & v = \int x^2 \, dx = \tfrac{1}{3}x^3 \end{array} \right] \qquad \checkmark$$

Moving the $\tfrac{1}{3}$ outside and simplifying $x^3\tfrac{1}{x}$ to x^2

$$= \frac{1}{3}x^3 \ln x - \frac{1}{3}\frac{1}{3}x^3 + C = \frac{1}{3}x^3 \ln x - \frac{1}{9}x^3 + C$$

We may check this answer by differentiation.

$$\frac{d}{dx}\left(\frac{1}{3}x^3 \ln x - \frac{1}{9}x^3 + C\right) = x^2 \ln x + \frac{1}{3}x^3\frac{1}{x} - \frac{1}{3}x^2$$

$$\underbrace{}$$

Differentiating $\tfrac{1}{3}x^3 \ln x$ by the Product Rule

$$= x^2 \ln x + \frac{1}{3}x^2 - \frac{1}{3}x^2 = x^2 \ln x$$

Agrees with the original integrand, so the integration is correct

From $x^3\dfrac{1}{x}$ Cancel

Practice Problem 1   Use integration by parts to find $\displaystyle\int x^3 \ln x \, dx$.   ➤ Solution on page 444

Integration by parts is also useful because it simplifies integrating products of powers of linear functions.

INTEGRATING BY PARTS

Use integration by parts to find $\displaystyle\int (x - 2)(x + 4)^8 \, dx.$

Solution The guidelines recommend that dv be the most complicated part that can be integrated. Both $x - 2$ and $(x + 4)^8$ can be integrated. For example,

$$\int (x + 4)^8 \, dx = \frac{1}{9}(x + 4)^9 + C \qquad \text{By the substitution method with } u = x + 4 \text{ (omitting the details)}$$

Since $(x + 4)^8$ is more complicated than $(x - 2)$, we take $dv = (x + 4)^8 \, dx.$

$$\underbrace{\int \underbrace{(x - 2)}_{u}\underbrace{(x + 4)^8 \, dx}_{dv}}_{} = \underbrace{(x - 2)}_{u}\underbrace{\frac{1}{9}(x + 4)^9}_{v} - \int \underbrace{\frac{1}{9}(x + 4)^9}_{v}\underbrace{dx}_{du}$$

$$\begin{bmatrix} u = x - 2 & dv = (x + 4)^8 \, dx \\ du = dx & v = \int (x + 4)^8 \, dx \\ & = \frac{1}{9}(x + 4)^9 \end{bmatrix}$$

$$= \frac{1}{9}(x - 2)(x + 4)^9 - \frac{1}{9}\int (x + 4)^9 \, dx$$

$$\uparrow$$
$$\text{Taking out the } \tfrac{1}{9}$$

$$= \frac{1}{9}(x - 2)(x + 4)^9 - \underbrace{\frac{1}{9}\frac{1}{10}(x + 4)^{10} + C}_{\text{Integrating by the substitution method}}$$

$$= \frac{1}{9}(x - 2)(x + 4)^9 - \frac{1}{90}(x + 4)^{10} + C$$

Again we could check this answer by differentiation. (Do you see how the Product Rule, applied to the first part of this answer, will give a piece that will cancel with the derivative of the second part?)

Practice Problem 2   Use integration by parts to find $\displaystyle\int (x + 1)(x - 1)^3 \, dx$.

➤ Solution on page 445

Present Value of a Continuous Stream of Income

If a business or some other asset generates income continuously at the rate $C(t)$ dollars per year, where t is the number of years from now, then $C(t)$ is called a *continuous stream of income*.* On page 279 we saw that to find the *present value* of a sum (the amount now that will later yield the stated sum) under continuous compounding we multiply by e^{-rt}, where r is the interest rate and t is the number of years. (We will refer to a rate with continuous compounding as a "continuous interest rate.") Therefore, the present value of the continuous stream $C(t)$ is found by multiplying by e^{-rt} and summing (integrating) over the time period.

> **Present Value of a Continuous Stream of Income**
>
> The present value of the continuous stream of income $C(t)$ dollars per year, where t is the number of years from now, for T years at continuous interest rate r is
>
> $$\begin{pmatrix}\text{Present} \\ \text{value}\end{pmatrix} = \int_0^T C(t)e^{-rt} \, dt$$

EXAMPLE 4

FINDING THE PRESENT VALUE OF A CONTINUOUS STREAM OF INCOME

A business generates income at the rate of $2t$ million dollars per year, where t is the number of years from now. Find the present value of this continuous stream for the next 5 years at the continuous interest rate of 10%.

---

* $C(t)$ must be continuous, meaning that the income is being paid continuously rather than in "lump-sum" payments. However, even lump-sum payments can be approximated by a continuous stream if the payments are frequent enough or last long enough.

Solution

$$\begin{pmatrix} \text{Present} \\ \text{value} \end{pmatrix} = \int_0^5 2te^{-0.1t}\,dt$$

Multiplying $C(t) = 2t$ by $e^{-0.1t}$ (since 10% = 0.1) and integrating from 0 to 5 years

This is a *definite* integral, but we will ignore the limits of integration until after we have found the *indefinite* integral. None of the formulas 1 through 7 on the inside back cover will find this integral, so we try integration by parts with $u = 2t$ and $dv = e^{-0.1t}\,dt$. (Do you see why the guidelines on page 438 suggest this choice?)

$$\int \underbrace{2t}_{u}\,\underbrace{e^{-0.1t}\,dt}_{dv} = \underbrace{(2t)}_{u}\underbrace{(-10e^{-0.1t})}_{v} - \int \underbrace{(-10e^{-0.1t})}_{v}\,\underbrace{2\,dt}_{du}$$

$$\begin{bmatrix} u = 2t & dv = e^{-0.1t}\,dt \\ du = 2\,dt & v = \int e^{-0.1t}\,dt \\ & \quad = -10e^{-0.1t} \end{bmatrix}$$

$$= -20te^{-0.1t} + 20\int e^{-0.1t}\,dt$$

$$= -20te^{-0.1t} + 20(-10)e^{-0.1t} + C$$

$$= -20te^{-0.1t} - 200e^{-0.1t} + C$$

For the *definite* integral, we evaluate this from 0 to 5:

$$(-20te^{-0.1t} - 200e^{-0.1t})\Big|_0^5 = (-20 \cdot 5e^{-0.5} - 200e^{-0.5}) - (\overbrace{-20 \cdot 0e^0}^{0} - \overbrace{200e^0}^{200})$$

$$\underbrace{\qquad\qquad}_{\text{Evaluation at } t = 5} \qquad \underbrace{\qquad\qquad}_{\text{Evaluation at } t = 0}$$

$$= -300e^{-0.5} + 200 \approx 18$$

In millions of dollars (using a calculator)

The present value of the stream of income over 5 years is approximately $18 million.

This answer means that $18 million at 10% interest compounded continuously would generate the continuous stream $C(t) = 2t$ million dollars for 5 years. This method is often used to determine the fair value of a business or some other asset, since it gives the present value of future income.

Graphing Calculator Exploration

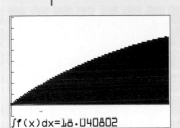

$\int f(x)\,dx=18.040802$

on $[0, 5]$ by $[-2, 8]$

a. Verify the answer to Example 4 by graphing $2xe^{-0.1x}$ and finding the area under the curve from 0 to 5.

b. Can you explain why the curve increases less steeply farther to the right? [*Hint:* Think of the present value of money to be paid in the more distant future.]

Why is it necessary to learn integration by parts if integrals like the one in Example 4 can be evaluated on graphing calculators? One answer is that you should have a *variety* of ways to approach a problem—sometimes geometrically, sometimes analytically, and sometimes numerically (using a calculator). These various approaches mutually support one another; for example, you can solve a problem one way and check it another way. Furthermore, not all applications involve *definite* integrals. Example 4 might have asked for a *formula* for the present value up to any time t, and such formulas may be found by integrating "by hand" but not from most graphing calculators.

Remember that you should use integration by parts *only* if the "easier" formulas (1 through 7 on the inside back cover) fail to solve the integral.

Practice Problem 3

Which of the following integrals require integration by parts, and which can be found by the substitution formula $\int e^u\,du = e^u + C$? (Do not solve the integrals.)

a. $\displaystyle\int xe^x\,dx$ **b.** $\displaystyle\int xe^{x^2}\,dx$

➤ Solution on page 445

Section Summary

The integration by parts formula

$$\int u\, dv = uv - \int v\, du$$

is simply the integration version of the Product Rule. The guidelines on page 438 lead to the following suggestions for choosing u and dv in some commonly occurring integrals.

For Integrals of the Form:	Choose:	
$\int x^n e^{ax}\, dx$	$u = x^n$	$dv = e^{ax}\, dx$
$\int x^n \ln x\, dx$	$u = \ln x$	$dv = x^n\, dx$
$\int (x + a)(x + b)^n\, dx$	$u = x + a$	$dv = (x + b)^n\, dx$

Integration by parts can be useful in any situation involving integrals, such as recovering total cost from marginal cost, calculating areas, average values (the definite integral divided by the length of the interval), continuous accumulations, or present values of continuous income streams.

➤ **Solutions to Practice Problems**

1. $\displaystyle \int x^3 \ln x\, dx = (\ln x)\left(\frac{1}{4}x^4\right) - \int \frac{1}{4}x^4 \frac{1}{x}\, dx$

$$\left[\begin{array}{ll} u = \ln x & dv = x^3\, dx \\ du = \dfrac{1}{x}\, dx & v = \displaystyle\int x^3\, dx = \dfrac{1}{4}x^4 \end{array} \right]$$

$$= (\ln x)\left(\frac{1}{4}x^4\right) - \frac{1}{4}\int x^3\, dx$$

$$= \frac{1}{4}x^4 \ln x - \frac{1}{16}x^4 + C$$

2. $\int (x + 1)(x - 1)^3 \, dx$

$$\left[\begin{array}{ll} u = x + 1 & dv = (x - 1)^3 \, dx \\ du = dx & v = \int (x - 1)^3 \, dx = \frac{1}{4}(x - 1)^4 \end{array} \right]$$

$$= (x + 1)\frac{1}{4}(x - 1)^4 - \int \frac{1}{4}(x - 1)^4 \, dx$$

$$= \frac{1}{4}(x + 1)(x - 1)^4 - \frac{1}{4} \cdot \frac{1}{5}(x - 1)^5 + C$$

$$- \frac{1}{4}(x + 1)(x - 1)^4 - \frac{1}{20}(x - 1)^5 + C$$

3. a. Requires integration by parts

   b. Can be solved by the substitution $u = x^2$

6.1 Exercises

Integration by parts often involves finding integrals like the following when integrating dv to find v. Find the following integrals *without* using integration by parts (using formulas 1 through 7 on the inside back cover). Be ready to find similar integrals during the integration by parts procedure.

1. $\int e^{2x} \, dx$

2. $\int x^5 \, dx$

3. $\int (x + 2) \, dx$

4. $\int (x - 1) \, dx$

5. $\int \sqrt{x} \, dx$

6. $\int e^{-0.5t} \, dt$

7. $\int (x + 3)^4 \, dx$

8. $\int (x - 5)^6 \, dx$

Use integration by parts to find each integral.

9. $\int xe^{2x} \, dx$

10. $\int xe^{3x} \, dx$

11. $\int x^5 \ln x \, dx$

12. $\int x^4 \ln x \, dx$

13. $\int (x + 2)e^x \, dx$ [*Hint:* Take $u = x + 2$.]

14. $\int (x - 1)e^x \, dx$ [*Hint:* Take $u = x - 1$.]

15. $\int \sqrt{x} \ln x \, dx$

16. $\int \sqrt[3]{x} \ln x \, dx$

17. $\int (x - 3)(x + 4)^5 \, dx$

18. $\int (x + 2)(x - 5)^5 \, dx$

19. $\int te^{-0.5t} \, dt$

20. $\int te^{-0.2t} \, dt$

21. $\int \frac{\ln t}{t^2} \, dt$

22. $\int \frac{\ln t}{\sqrt{t}} \, dt$

23. $\int s(2s + 1)^4 \, ds$

24. $\int \frac{x + 1}{e^{3x}} \, dx$

25. $\int \frac{x}{e^{2x}} \, dx$

26. $\int \frac{\ln(x + 1)}{\sqrt{x + 1}} \, dx$

27. $\int \frac{x}{\sqrt{x + 1}} \, dx$

28. $\int x\sqrt{x + 1} \, dx$

29. $\int xe^{ax} \, dx$ $(a \neq 0)$

30. $\int (x + b)e^{ax} dx \qquad (a \neq 0)$

31. $\int x^n \ln ax \, dx \qquad (a \neq 0, \ n \neq -1)$

32. $\int (x + a)^n \ln(x + a) \, dx \qquad (n \neq -1)$

33. $\int \ln x \, dx \qquad$ [*Hint:* Take $u = \ln x, \quad dv = dx$.]

34. $\int \ln x^2 \, dx \qquad$ [*Hint:* Take $u = \ln x^2, \quad dv = dx$.]

35. $\int x^3 e^{x^2} dx \qquad$ [*Hint:* Take $u = x^2, \quad dv = xe^{x^2},$
and use a substitution to find v from dv.]

36. $\int x^3 (x^2 - 1)^6 dx \qquad$ [*Hint:* Take $u = x^2,$

$dv = x(x^2 - 1)^6 \, dx,$ and use a substitution to find v from dv.]

Find each integral by integration by parts or a substitution, as appropriate.

37. **a.** $\int xe^{x^2} dx$ **b.** $\int \dfrac{(\ln x)^3}{x} dx$

    c. $\int x^2 \ln 2x \, dx$ **d.** $\int \dfrac{e^x}{e^x + 4} dx$

38. **a.** $\int \sqrt{\ln x} \, \dfrac{1}{x} dx$ **b.** $\int x^2 e^{x^3} dx$

    c. $\int x^7 \ln 3x \, dx$ **d.** $\int xe^{4x} dx$

Evaluate each definite integral using integration by parts. (Leave answers in exact form.)

39. $\displaystyle\int_0^2 xe^x \, dx$ 40. $\displaystyle\int_0^3 xe^x \, dx$

41. $\displaystyle\int_1^3 x^2 \ln x \, dx$ 42. $\displaystyle\int_1^2 x \ln x \, dx$

43. $\displaystyle\int_0^2 z(z - 2)^4 \, dz$ 44. $\displaystyle\int_0^4 z(z - 4)^6 \, dz$

45. $\displaystyle\int_0^{\ln 4} te^t \, dt$ 46. $\displaystyle\int_1^e \ln x \, dx$

Find in two different ways.

47. $\int x(x - 2)^5 \, dx$

   a. Use integration by parts.
   b. Use the substitution $u = x - 2$ (so x is replaced by $u + 2$) and then multiply out the integrand.

48. $\int x(x + 4)^6 \, dx$

   a. Use integration by parts.
   b. Use the substitution $u = x + 4$ (so x is replaced by $u - 4$) and then multiply out the integrand.

Derive each formula by using integration by parts on the left-hand side. (Assume $n > 0$.)

49. $\int x^n e^x \, dx = x^n e^x - n \int x^{n-1} e^x \, dx$

50. $\int (\ln x)^n \, dx = x(\ln x)^n - n \int (\ln x)^{n-1} \, dx$

51. Use the formula in Exercise 49 to find the integral $\int x^2 e^x \, dx$. [*Hint:* Apply the formula twice.]

52. Use the formula in Exercise 50 to find the integral $\int (\ln x)^2 \, dx$. [*Hint:* Apply the formula twice.]

53. **a.** Find the integral $\int x^{-1} \, dx$ by integration by parts (using $u = x^{-1}$ and $dv = dx$), obtaining

$$\int x^{-1} \, dx = x^{-1}x - \int (-x^{-2})x \, dx$$

which gives

$$\int x^{-1} \, dx = 1 + \int x^{-1} \, dx$$

   b. Subtract the integral from both sides of this last equation, obtaining $0 = 1$. Explain this apparent contradiction.

54. We omit the constant of integration when we integrate dv to get v. Including the constant C

in this step simply replaces v by $v + C$, giving the formula

$$\int u \, dv = u(v + C) - \int (v + C) \, du$$

Multiplying out the parentheses and expanding the last integral into two gives

$$\int u \, dv = uv + Cu - \int v \, du - C \int du$$

Show that the second and fourth terms on the right cancel, giving the "old" integration by parts formula $\int u \, dv = uv - \int v \, du$. This shows that including the constant in the dv to v step gives the same formula. *One constant of integration at the end is enough.*

APPLIED EXERCISES

55. BUSINESS: Revenue If a company's marginal revenue function is $MR(x) = xe^{x/4}$, find the revenue function. [*Hint:* Evaluate the constant C so that revenue is 0 at $x = 0$.]

56. BUSINESS: Cost A company's marginal cost function is $MC(x) = xe^{-x/2}$ and fixed costs are 200. Find the cost function. [*Hint:* Evaluate the constant C so that the cost is 200 at $x = 0$.]

57. BUSINESS: Present Value of a Continuous Stream of Income An electronics company generates a continuous stream of income of $4t$ million dollars per year, where t is the number of years that the company has been in operation. Find the present value of this stream of income over the first 10 years at a continuous interest rate of 10%.

For Exercises 58–59:

a. Solve *without* using a graphing calculator.
b. Verify your answer to part (a) using a graphing calculator.

58. BUSINESS: Present Value of a Continuous Stream of Income An oil well generates a continuous stream of income of $60t$ thousand dollars per year, where t is the number of years that the rig has been in operation. Find the present value of this stream of income over the first 20 years at a continuous interest rate of 5%.

59. BIOMEDICAL: Drug Dosage A drug taken orally is absorbed into the bloodstream at the rate of $te^{-0.5t}$ milligrams per hour, where t is the number of hours since the drug was taken. Find the total amount of the drug absorbed during the first 5 hours.

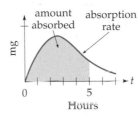

60. ENVIRONMENTAL SCIENCE: Pollution Contamination is leaking from an underground waste-disposal tank at the rate of $t \ln t$ thousand gallons per month, where t is the number of months since the leak began. Find the total leakage from the end of month 1 to the end of month 4.

61. GENERAL: Area Find the area under the curve $y = x \ln x$ and above the x-axis from $x = 1$ to $x = 2$.

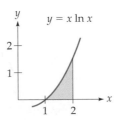

62. POLITICAL SCIENCE: Fund Raising A politician can raise campaign funds at the rate

(*continues*)

of $50te^{-0.1t}$ thousand dollars per week during the first t weeks of a campaign. Find the average amount raised during the first 5 weeks.

63. **BUSINESS: Product Recognition** A company begins advertising a new product and finds that after t weeks the product is gaining customer recognition at the rate of $t^2 \ln t$ thousand customers per week (for $t \geq 1$). Find the total gain in recognition from the end of week 1 to the end of week 6.

64. **GENERAL: Population** The population of a town is increasing at the rate of $400te^{0.02t}$ people per year, where t is the number of years from now. Find the total gain in population during the next 5 years.

Repeated Integration by Parts

Sometimes an integral requires two or more integrations by parts. As an example, we apply integration by parts to the integral $\int x^2 e^x \, dx$.

$$\int x^2 e^x \, dx = x^2 e^x - \int e^x \, 2x \, dx = x^2 e^x - 2 \int x e^x \, dx$$

$$\begin{array}{cccc} u \; dv & u \; v & v \; du \end{array}$$

$$\left[\begin{array}{ll} u = x^2 & dv = e^x \, dx \\ du = 2x \, dx & v = \int e^x \, dx = e^x \end{array} \right]$$

The new integral $\int x e^x \, dx$ is solved by a second integration by parts. Continuing with the previous solution, we choose new u and du:

$$= x^2 e^x - 2\left(\int x e^x \, dx \right) \qquad \left[\begin{array}{ll} u = x & dv = e^x \, dx \\ du = dx & v = e^x \end{array} \right]$$

$$\qquad\qquad u \; dv$$

$$= x^2 e^x - 2\left(x e^x - \int e^x \, dx \right)$$

$$\qquad\qquad uv \qquad v \; du$$

$$= x^2 e^x - 2(x e^x - e^x) + C$$

$$= x^2 e^x - 2x e^x + 2e^x + C$$

After reading the preceding explanation, find each integral by repeated integration by parts.

65. $\displaystyle\int x^2 e^{-x} dx$

66. $\displaystyle\int x^2 e^{2x} dx$

67. $\displaystyle\int (x+1)^2 e^x \, dx$

68. $\displaystyle\int (\ln x)^2 \, dx$

69. $\displaystyle\int x^2 (\ln x)^2 \, dx$

70. $\displaystyle\int x^3 e^x \, dx$

71–72. For each definite integral:

a. Evaluate it by integration by parts. (Give answer in its *exact* form.)

b. Verify your answer to part (a) using a graphing calculator.

71. $\displaystyle\int_0^2 x^2 e^x \, dx$

72. $\displaystyle\int_1^5 (\ln x)^2 \, dx$

Repeated Integration by Parts Using a Table

The solution to a repeated integration by parts problem can be organized in a table. As an example, we solve $\int x^2 e^{3x} \, dx$. We begin by choosing

$$u = x^2 \qquad dv = v' \, dx = e^{3x} \, dx$$

We then make a table consisting of the following three columns:

Alternating Signs	$u = x^2$ and Its Derivatives	$v' = e^{3x}$ and Its Antiderivatives
$+$	x^2	e^{3x}
$-$	$2x$	$\frac{1}{3} e^{3x}$ Using
$+$	2	$\frac{1}{9} e^{3x}$ the formula for
$-$	0	$\frac{1}{27} e^{3x}$ $\int e^{ax} dx$

Stop when you get to 0

Finally, the solution is found by adding the *signed* products of the diagonals shown in the table:

$$\int x^2 e^{3x} \, dx = \frac{1}{3} x^2 e^{3x} - \frac{2}{9} x e^{3x} + \frac{2}{27} e^{3x} + C$$

After reading the preceding example, find each integral by repeated integration by parts using a table.

73. $\displaystyle\int x^2 e^{-x} dx$

74. $\displaystyle\int x^2 e^{2x} \, dx$

75. $\displaystyle\int x^3 e^{2x}\, dx$ **76.** $\displaystyle\int x^3 e^{-x}\, dx$ **77.** $\displaystyle\int (x-1)^3 e^{3x}\, dx$ **78.** $\displaystyle\int (x+1)^2(x+2)^5\, dx$

6.2 INTEGRATION USING TABLES

Introduction

There are many techniques of integration, and only some of the most useful ones will be discussed in this book. Many of the advanced techniques lead to integration formulas, which can then be collected into a "table of integrals." In this section we will see how to find integrals by choosing an appropriate formula from such a table.*

On the inside back cover is a short table of integrals that we shall use. The formulas are grouped according to the type of integrand (for example, "Forms Involving $x^2 - a^2$"). Look at the table now (formulas 9 through 23) to see how it is organized.

Using Integral Tables

Given a particular integral, we first look for a formula that fits it exactly.

EXAMPLE 1

INTEGRATING USING AN INTEGRAL TABLE

Find $\displaystyle\int \frac{1}{x^2 - 4}\, dx$.

Solution The denominator $x^2 - 4$ is of the form $x^2 - a^2$ (with $a = 2$), so we look in the table of integrals under "Forms Involving $x^2 - a^2$." Formula 15,

$$\int \frac{1}{x^2 - a^2}\, dx = \frac{1}{2a} \ln \left| \frac{x-a}{x+a} \right| + C \qquad \text{Formula 15}$$

with $a = 2$ becomes our answer:

$$\int \frac{1}{x^2 - 4}\, dx = \frac{1}{4} \ln \left| \frac{x-2}{x+2} \right| + C \qquad \begin{array}{l}\text{Formula 15 with} \quad a = 2 \\ \text{substituted on both sides}\end{array}$$

---

* Another way is to use a more advanced graphing calculator (see pages 455–456) or a computer software package such as Maple, Mathcad, or Mathematica.

Note that the expression $x^2 - a^2$ does not require that the last number be a "perfect square." For example, $x^2 - 3$ can be written $x^2 - a^2$ with $a = \sqrt{3}$.

Integral tables are useful in many applications, such as integrating a rate to find the total accumulation.

| EXAMPLE 2 | **FINDING TOTAL SALES FROM THE SALES RATE** |

A company's sales rate is $\dfrac{x}{\sqrt{x + 9}}$ sales per week after x weeks. Find a formula for the total sales after x weeks.

Solution To find the *total* sales $S(x)$ we integrate the *rate* of sales.

$$S(x) = \int \frac{x}{\sqrt{x + 9}} \, dx$$

In the table on the inside back cover under "Forms Involving $\sqrt{ax + b}$" we find

$$\int \frac{x}{\sqrt{ax + b}} \, dx = \frac{2ax - 4b}{3a^2} \sqrt{ax + b} + C \qquad \text{Formula 13}$$

This formula with $a = 1$ and $b = 9$ gives the integral

$$S(x) = \int \frac{x}{\sqrt{x + 9}} \, dx = \frac{2x - 36}{3} \sqrt{x + 9} + C \qquad \begin{array}{l}\text{Formula 13 with} \\ a = 1 \text{ and } b = 9\end{array}$$

$$= \left(\frac{2}{3} x - 12\right) \sqrt{x + 9} + C \qquad \text{Simplifying}$$

To evaluate the constant C we use the fact that total sales at time $x = 0$ must be zero: $S(0) = 0$.

$$(-12) \sqrt{9} + C = 0 \qquad \begin{array}{l}(\frac{2}{3} x - 12) \sqrt{x + 9} + C \\ \text{at } x = 0 \text{ set equal to zero}\end{array}$$

$$-36 + C = 0 \qquad \text{Simplifying}$$

Therefore, $C = 36$. Substituting this into $S(x)$ gives the formula for the total sales in the first x weeks.

$$S(x) = \left(\frac{2}{3} x - 12\right) \sqrt{x + 9} + 36 \qquad \begin{array}{l}S(x) = (\frac{2}{3} x - 12) \sqrt{x + 9} + C \\ \text{with } C = 36\end{array}$$

Modern methods of biotechnology are being used to develop many new products, including powerful antibiotics, disease-resistant crops, and bacteria that literally "eat" oil spills. These "gene splicing" techniques require evaluating definite integrals such as the following.

EXAMPLE 3

GENETIC ENGINEERING

Under certain circumstances, the number of generations of bacteria needed to increase the frequency of a gene from 0.2 to 0.5 is

$$n = 2.5 \int_{0.2}^{0.5} \frac{1}{q^2(1 - q)} \, dq$$

Find n (rounded to the nearest integer).

Solution Formula 12 (inside back cover) integrates a similar-looking fraction.

$$\int \frac{1}{x^2(ax + b)} \, dx = -\frac{1}{b}\left(\frac{1}{x} + \frac{a}{b} \ln\left|\frac{x}{ax + b}\right|\right) + C \qquad \text{Formula 12}$$

To make $(ax + b)$ into $(1 - x)$, we take $a = -1$ and $b = 1$, so the left-hand side of the formula becomes

$$\int \frac{1}{x^2(-x + 1)} \, dx \qquad \text{or} \qquad \int \frac{1}{x^2(1 - x)} \, dx \qquad \begin{array}{l}\text{From formula 12}\\\text{with } a = -1, \ b = 1\end{array}$$

Except for replacing x by q, this is the same as our integral. Therefore, the indefinite integral is found by formula 12 with $a = -1$ and $b = 1$ (which we express in the variable q).

$$\int \frac{1}{q^2(1 - q)} \, dq = -\left(\frac{1}{q} - \ln\left|\frac{q}{1 - q}\right|\right) + C \qquad \begin{array}{l}\text{Formula 12 with}\\a = -1 \text{ and } b = 1\end{array}$$

For the *definite* integral from 0.2 to 0.5, we evaluate and subtract.

$$\underbrace{-\left(\frac{1}{0.5} - \ln\left|\frac{0.5}{1 - 0.5}\right|\right)}_{\text{Evaluation at } q = 0.5} - \underbrace{\left[-\left(\frac{1}{0.2} - \ln\left|\frac{0.2}{1 - 0.2}\right|\right)\right]}_{\text{Evaluation at } q = 0.2}$$

$$= -\underbrace{(2 - \ln 1)}_{0} + \left(5 - \underbrace{\ln\frac{0.2}{0.8}}_{\ln 0.25}\right) = -2 + 5 - \ln 0.25 \approx \underbrace{4.39}_{\substack{\text{Using a}\\\text{calculator}}}$$

We multiply this by the 2.5 in front of the original integral.

$$(2.5)(4.39) \approx 10.98$$

Therefore, 11 generations are needed to raise the gene frequency from 0.2 to 0.5.

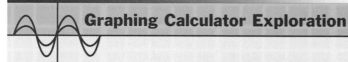

Graphing Calculator Exploration

on [0, 1] by [0, 100]
(shaded from 0.2 to 0.5)

a. Verify the answer to Example 3 by finding the definite integral of
$$y = \frac{2.5}{x^2(1-x)} \quad \text{from 0.2 to 0.5.}$$

b. From the graph of the function shown on the left, which would require more generations: increasing the gene frequency from 0.1 to 0.2 or from 0.5 to 0.6?

c. Can you think of a reason for this? [*Hint:* Genes reproduce from other similar genes.]

Sometimes a substitution is needed to transform a formula to fit a given integral. In such cases both the x and the dx must be transformed. A few examples will make the method clear.

EXAMPLE 4

USING A TABLE WITH A SUBSTITUTION

Find $\displaystyle\int \frac{x}{\sqrt{x^4 + 1}} \, dx$.

Solution   The table of integrals on the inside back cover has no formula involving x^4. However, $x^4 = (x^2)^2$, so a formula involving x^2, along with a substitution, might work. Formula 18 looks promising:

$$\int \frac{1}{\sqrt{x^2 \pm a^2}} \, dx = \ln \left| x + \sqrt{x^2 \pm a^2} \right| + C$$

The $\pm$ means: use either the upper sign or the lower sign on both sides

$$\int \frac{1}{\sqrt{x^2 + 1}} \, dx = \ln \left| x + \sqrt{x^2 + 1} \right| + C$$

Formula 18 with $a = 1$ and the upper sign

With the substitution

$$x = z^2$$
$$dx = 2z\,dz \qquad \text{Differential of } x = z^2$$

this becomes

$$\int \frac{1}{\sqrt{z^4 + 1}}\, 2z\,dz = \ln\left| z^2 + \sqrt{z^4 + 1} \right| + C \qquad \begin{array}{l}\text{Above formula with} \\ x = z^2 \text{ and } dx = 2z\,dz\end{array}$$

Dividing by 2 and replacing z by x gives the integral that we wanted:

$$\int \frac{x}{\sqrt{x^4 + 1}}\, dx = \frac{1}{2} \ln\left(x^2 + \sqrt{x^4 + 1}\right) + C \qquad \begin{array}{l}\text{Dropping the absolute} \\ \text{value bars since} \\ x^2 + \sqrt{x^4 + 1} \text{ is positive}\end{array}$$

Given a particular integral, how do we choose a formula?

How To Choose a Formula

Find a formula that matches the *most complicated part* of the integral, making appropriate substitutions to change the formula into the given integral.

For instance, in Example 4 we matched the $\sqrt{x^4 + 1}$ in the given integral with the $\sqrt{x^2 \pm a^2}$ in the formula, and the rest of the integral followed from the differential.

Practice Problem

Find $\displaystyle\int \frac{t}{9t^4 - 1}\, dt.$ [*Hint:* Use formula 15 with $x = 3t^2$.]

➤ Solution on page 456

EXAMPLE 5

USING A TABLE WITH A SUBSTITUTION

Find $\displaystyle\int \frac{e^{-2t}}{e^{-t} + 1}\, dt.$

Solution Looking in the table of integrals under "Forms Involving e^{ax} and ln x," none of the formulas looks anything like this integral. However, replacing e^{-t} by x would make the denominator of our integral into $x + 1$, so formula 9 might help. This formula with $a = 1$ and $b = 1$ is

$$\int \frac{x}{x + 1}\, dx = x - \ln|x + 1| + C \qquad \text{Formula 9 with } a = 1 \text{ and } b = 1$$

With the substitution

$$x = e^{-t}$$
$$dx = -e^{-t}\, dt \qquad \text{Differential of } x = e^{-t}$$

formula 9 becomes

$$\int \frac{e^{-t}}{e^{-t} + 1} (-e^{-t})\, dt = e^{-t} - \ln|e^{-t} + 1| + C$$

or

$$-\int \frac{e^{-2t}}{e^{-t} + 1}\, dt = e^{-t} - \ln(e^{-t} + 1) + C$$

Except for the negative sign, this is the given integral. Multiplying through by -1 gives the final answer.

$$\int \frac{e^{-2t}}{e^{-t} + 1}\, dt = -e^{-t} + \ln(e^{-t} + 1) + C$$

Reduction Formulas

Sometimes we must apply a formula several times to simplify an integral in stages.

EXAMPLE 6

USING A REDUCTION FORMULA

Find $\displaystyle\int x^3 e^{-x}\, dx.$

Solution In the integral table, we find formula 21.

$$\int x^n e^{ax}\, dx = \frac{1}{a} x^n e^{ax} - \frac{n}{a} \int x^{n-1} e^{ax}\, dx \qquad \text{Formula 21. With } n = 3 \text{ and } a = -1, \text{ the left side fits our integral}$$

The right-hand side of this formula involves a new integral, but with a *lower* power of x. We will apply formula 21 several times, each time reducing the power of x until we eliminate it completely. Applying formula 21 with $n = 3$ and $a = -1$,

$$\int x^3 e^{-x}\, dx = -x^3 e^{-x} + 3 \int x^2 e^{-x}\, dx \qquad \text{After one application}$$

↑ The power has been reduced

$$= -x^3 e^{-x} + 3\left(-x^2 e^{-x} + 2 \int x^1 e^{-x}\, dx\right) \qquad \begin{array}{l}\text{Applying formula 21}\\ \text{again (now with } n = 2)\\ \text{to the last integral above}\end{array}$$

$$= -x^3 e^{-x} - 3x^2 e^{-x} + 6 \int x^1 e^{-x}\, dx \qquad \text{Multiplying out}$$

$$= -x^3 e^{-x} - 3x^2 e^{-x} + 6\left(-xe^{-x} + \int x^0 e^{-x}\, dx\right) \qquad \begin{array}{l}\text{Using formula 21 a third time}\\ \text{(now with } n = 1)\end{array}$$

$$1$$

$$= -x^3 e^{-x} - 3x^2 e^{-x} - 6xe^{-x} + 6 \int e^{-x}\, dx \qquad \begin{array}{l}\text{Now solve this last integral by the}\\ \text{formula } \int e^{ax}\, dx = \dfrac{1}{a} e^{ax} + C\end{array}$$

$$= -x^3 e^{-x} - 3x^2 e^{-x} - 6xe^{-x} - 6e^{-x} + C \qquad \begin{array}{l}\text{The solution, after three applica-}\\ \text{tions of formula 21}\end{array}$$

$$= -e^{-x}(x^3 + 3x^2 + 6x + 6) + C \qquad \text{Factoring}$$

We used formula 21 three times, reducing the x^3 in steps, first down to x^2, then to x^1, and finally to $x^0 = 1$, at which point we could solve the integral easily. If the power of x in the integral had been higher, more applications of formula 21 would have been necessary. Formulas such as 21 and 22 are called *reduction formulas*, since they express an integral in terms of a similar integral but with a smaller power of x.

Graphing Calculator Exploration

Some advanced graphing calculators can find *indefinite* integrals. For example, the Texas Instruments *TI-89* graphing calculator finds the integrals in Examples 5 and 6 as follows.

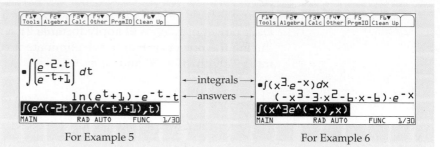

For Example 5   For Example 6

With some algebra, you can verify that these answers agree with those found in Examples 5 and 6, even though they do not look the same. (Notice that this calculator omits the arbitrary constant of integration.)

6.2 Section Summary

To find a formula that "fits" a given integral, we look for the formula whose left-hand side most closely matches the most complicated part of the integral. Then we choose constants (and possibly a substitution) to make the formula fit exactly. Although we have been using a very brief table, the technique is the same with a more extensive table. Many integral tables have been published, some of which are book-length, containing several thousand formulas.*

➤ Solution to Practice Problem

Formula 15 with the substitution $x = 3t^2$, $dx = 6t\, dt$, and $a = 1$ becomes

$$\int \frac{1}{9t^4 - 1}\, 6t\, dt = \frac{1}{2} \ln \left| \frac{3t^2 - 1}{3t^2 + 1} \right| + C$$

Dividing each side by 6 gives the answer:

$$\int \frac{1}{9t^4 - 1}\, dt = \frac{1}{12} \ln \left| \frac{3t^2 - 1}{3t^2 + 1} \right| + C$$

* A useful table of integrals containing more than 400 formulas is found in *CRC Standard Mathematical Tables*, CRC Press, Boca Raton, Florida.

6.2   **Exercises**

For each integral, state the number of the integration formula (from the inside back cover) and the values of the constants a and b so that the formula fits the integral. (Do not evaluate the integral.)

1. $\displaystyle\int \frac{1}{x^2(5x-1)}\,dx$ **2.** $\displaystyle\int \frac{x}{2x-3}\,dx$

3. $\displaystyle\int \frac{1}{x\sqrt{-x+7}}\,dx$ **4.** $\displaystyle\int \frac{x}{\sqrt{2x+1}}\,dx$

5. $\displaystyle\int \frac{x}{1-x}\,dx$ **6.** $\displaystyle\int \frac{1}{x\sqrt{1-4x}}\,dx$

Find each integral by using the integral table on the inside back cover.

7. $\displaystyle\int \frac{1}{9-x^2}\,dx$
[*Hint:* Use formula 16 with $a=3$.]

8. $\displaystyle\int \frac{1}{x^2-25}\,dx$
[*Hint:* Use formula 15 with $a=5$.]

9. $\displaystyle\int \frac{1}{x^2(2x+1)}\,dx$
[*Hint:* Use formula 12 with $a=2,\ b=1$.]

10. $\displaystyle\int \frac{x}{x+2}\,dx$
[*Hint:* Use formula 9 with $a=1,\ b=2$.]

11. $\displaystyle\int \frac{x}{1-x}\,dx$ [*Hint:* Use formula 9.]

12. $\displaystyle\int \frac{x}{\sqrt{1-x}}\,dx$ [*Hint:* Use formula 13.]

13. $\displaystyle\int \frac{1}{(2x+1)(x+1)}\,dx$

14. $\displaystyle\int \frac{x}{(x+1)(x+2)}\,dx$

15. $\displaystyle\int \sqrt{x^2-4}\,dx$ **16.** $\displaystyle\int \frac{1}{\sqrt{x^2-1}}\,dx$

17. $\displaystyle\int \frac{1}{z\sqrt{1-z^2}}\,dz$ **18.** $\displaystyle\int \frac{\sqrt{4+z^2}}{z}\,dz$

19. $\displaystyle\int x^3\,e^{2x}\,dx$ **20.** $\displaystyle\int x^{99}\,\ln x\,dx$

21. $\displaystyle\int x^{-101}\ln x\,dx$ **22.** $\displaystyle\int (\ln x)^2\,dx$

23. $\displaystyle\int \frac{1}{x(x+3)}\,dx$ **24.** $\displaystyle\int \frac{1}{x(x-3)}\,dx$

25. $\displaystyle\int \frac{z}{z^4-4}\,dz$ **26.** $\displaystyle\int \frac{z}{9-z^4}\,dz$

27. $\displaystyle\int \sqrt{9x^2+16}\,dx$ **28.** $\displaystyle\int \frac{1}{\sqrt{16x^2-9}}\,dx$

29. $\displaystyle\int \frac{1}{\sqrt{4-e^{2t}}}\,dt$ **30.** $\displaystyle\int \frac{e^t}{9-e^{2t}}\,dt$

31. $\displaystyle\int \frac{e^t}{e^{2t}-1}\,dt$ **32.** $\displaystyle\int \frac{e^{2t}}{1-e^t}\,dt$

33. $\displaystyle\int \frac{x^3}{\sqrt{x^8-1}}\,dx$ **34.** $\displaystyle\int x^2\sqrt{x^6+1}\,dx$

35. $\displaystyle\int \frac{1}{x\sqrt{x^3+1}}\,dx$ **36.** $\displaystyle\int \frac{\sqrt{1-x^6}}{x}\,dx$

37. $\displaystyle\int \frac{e^t}{(e^t-1)(e^t+1)}\,dt$ **38.** $\displaystyle\int \frac{e^{2t}}{(e^t-1)(e^t+1)}\,dt$

39. $\displaystyle\int x\,e^{x/2}\,dx$ **40.** $\displaystyle\int \frac{x}{e^x}\,dx$

41. $\displaystyle\int \frac{1}{e^{-x}+4}\,dx$ **42.** $\displaystyle\int \frac{1}{\sqrt{e^{-x}+4}}\,dx$

For each definite integral:
a. Evaluate it using the table of integrals on the inside back cover. (Leave answers in *exact* form.)
b. Use a graphing calculator to verify your answer to part (a).

43. $\displaystyle\int_4^5 \sqrt{x^2-16}\,dx$ **44.** $\displaystyle\int_0^4 \frac{1}{\sqrt{x^2+9}}\,dx$

45. $\displaystyle\int_2^3 \frac{1}{x^2-1}\,dx$ **46.** $\displaystyle\int_2^4 \frac{1}{1-x^2}\,dx$

47. $\displaystyle\int_3^5 \frac{\sqrt{25-x^2}}{x}\,dx$ **48.** $\displaystyle\int_3^4 \frac{1}{x\sqrt{25-x^2}}\,dx$

Find each integral by whatever means are necessary (either substitution or tables).

49. $\displaystyle\int \frac{1}{2x + 6}\, dx$

50. $\displaystyle\int \frac{x}{x^2 - 4}\, dx$

51. $\displaystyle\int \frac{x}{2x + 6}\, dx$

52. $\displaystyle\int \frac{1}{4 - x^2}\, dx$

53. $\displaystyle\int x\sqrt{1 - x^2}\, dx$

54. $\displaystyle\int \frac{x}{\sqrt{1 - x^2}}\, dx$

55. $\displaystyle\int \frac{\sqrt{1 - x^2}}{x}\, dx$

56. $\displaystyle\int \frac{1}{\sqrt{x^2 - 1}}\, dx$

Find each integral. [*Hint:* Separate each integral into two integrals, using the fact that the numerator is a sum or difference, and find the two integrals by two different formulas.]

57. $\displaystyle\int \frac{x - 1}{(3x + 1)(x + 1)}\, dx$

58. $\displaystyle\int \frac{x - 1}{x^2(x + 1)}\, dx$

59. $\displaystyle\int \frac{x + 1}{x\sqrt{1 + x^2}}\, dx$

60. $\displaystyle\int \frac{x - 1}{x\sqrt{x^2 + 4}}\, dx$

61. $\displaystyle\int \frac{x + 1}{x - 1}\, dx$ [*Hint:* After separating into two integrals, find one by a formula and the other by a substitution.]

62. $\displaystyle\int \frac{x + 1}{\sqrt{x^2 + 1}}\, dx$ [*Hint:* After separating into two integrals, find one by a formula and the other by a substitution.]

APPLIED EXERCISES

63. BUSINESS: Total Sales A company's sales rate is $x^2 e^{-x}$ million sales per month after x months. Find a formula for the total sales in the first x months. [*Hint:* Integrate the sales rate to find the total sales and determine the constant C so that total sales are zero at time $x = 0$.]

64. GENERAL: Population The population of a city is expected to grow at the rate of $x/\sqrt{x + 9}$ thousand people per year after x years. Find the total change in population from year 0 to year 27.

65. BIOMEDICAL: Gene Frequency Under certain circumstances, the number of generations necessary to increase the frequency of a gene from 0.1 to 0.3 is

$$n = 3 \int_{0.1}^{0.3} \frac{1}{q^2(1 - q)}\, dq$$

Find n (rounded to the nearest integer).

66. BEHAVIORAL SCIENCE A subject in a psychology experiment gives responses at the rate of $t/\sqrt{t + 1}$ correct answers per minute after t minutes.

  a. Find the total number of correct responses from time $t = 0$ to time $t = 15$.

  b. Verify your answer to part (a) using a graphing calculator.

67. BUSINESS: Cost The marginal cost function for a computer chip manufacturer is $MC(x) = 1/\sqrt{x^2 + 1}$, and fixed costs are $2000. Find the cost function.

68. SOCIAL SCIENCE: Employment An urban job placement center estimates that the number of residents seeking employment t years from now will be $t/(2t + 4)$ million people.

  a. Find the average number of job seekers during the period $t = 0$ to $t = 10$.

  b. Verify your answer to part (a) using a graphing calculator.

6.3 IMPROPER INTEGRALS

Introduction

In this section we define integrals over intervals that are infinite in length. Such "improper" integrals have many applications, such as in the Application Preview on page 435 at the beginning of this chapter.

Limits as x Approaches $\pm\infty$

The notation $x \to \infty$ ("x approaches infinity") means that x takes on arbitrarily large values.

Limits Approaching $\pm\infty$

$x \to \infty$	means:	x takes values arbitrarily far to the *right* on the number line.
$x \to -\infty$	means:	x takes values arbitrarily far to the *left* on the number line.

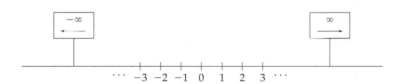

$$\cdots \quad -3 \;\; -2 \;\; -1 \;\; 0 \;\; 1 \;\; 2 \;\; 3 \quad \cdots$$

Evaluating limits as x approaches positive or negative infinity is simply a matter of thinking about large and small numbers. The reciprocal of a large number is a small number. For example,

$$\frac{1}{1,000,000} = 0.000001$$

One over a million is one one-millionth

Similarly:

$$\lim_{x \to \infty} \frac{1}{x^2} = 0$$

$$\lim_{x \to \infty} \frac{1}{e^x} = \lim_{x \to \infty} e^{-x} = 0$$

As the denominator approaches infinity (with the numerator constant), the value approaches zero

These examples illustrate the following general rules.

$$\lim_{x \to \infty} \frac{1}{x^n} = 0 \qquad (n > 0)$$ As x approaches infinity, 1 over x to a *positive* power approaches zero

$$\lim_{x \to \infty} e^{-ax} = 0 \qquad (a > 0)$$ As x approaches infinity, e to a *negative* number times x approaches zero

EXAMPLE 1

EVALUATING LIMITS

a. $\lim_{b \to \infty} \dfrac{1}{b^2} = 0$ Using the first rule in the box above

b. $\lim_{b \to \infty} \left(3 - \dfrac{1}{b}\right) = 3$ Because the $\dfrac{1}{b}$ approaches zero

c. $\lim_{b \to \infty} (e^{-2b} - 5) = -5$ Because the e^{-2b} approaches zero

Similar rules hold for x approaching *negative* infinity.

$$\lim_{x \to -\infty} \frac{1}{x^n} = 0 \qquad \text{(for integer } n > 0)$$ As x approaches *negative* infinity, 1 over x to a positive integer approaches zero

$$\lim_{x \to -\infty} e^{ax} = 0 \qquad (a > 0)$$ As x approaches negative infinity, e to a positive number times x approaches zero (because the exponent is approaching $-\infty$)

 Graphing Calculator Exploration

a. Define $y_1 = 1/x^2$ and $y_2 = e^{-x}$ and use the TABLE feature of your calculator to evaluate these functions at x-values like 1, 2, 3, 5, 10, 100, and 1000. Which of the limit rules above do the results verify? [*Note:* An answer such as 3E⁻5 means $3 \cdot 10^{-5} = 0.00003$.]

b. Change y_2 to be $y_2 = e^x$ and use the TABLE to evaluate y_1 and y_2 at negative x-values like -1, -2, -3, -5, -10, and -100. Which limit rules do the results verify?

Some quantities become arbitrarily large, and so do not have limits. (For a limit to exist, it must be *finite*.)

The following limits do not exist:

$$\lim_{x \to \infty} x^n \qquad (n > 0)$$
As x approaches infinity, x to a positive power has no limit

$$\lim_{x \to \infty} e^{ax} \qquad (a > 0)$$
As x approaches infinity, e to a positive number times x has no limit

$$\lim_{x \to \infty} \ln x$$
As x approaches infinity, the natural logarithm of x has no limit

EXAMPLE 2

FINDING WHETHER A LIMIT EXISTS

a. $\lim_{b \to \infty} b^3$ does not exist.

Because b^3 becomes arbitrarily large as b approaches infinity

b. $\lim_{b \to \infty} (\sqrt{b} - 1)$ does not exist.

Because $\sqrt{b}$ becomes arbitrarily large as b approaches infinity

Practice Problem 1

Evaluate the following limits (if they exist).

a. $\lim_{b \to \infty} \left(1 - \dfrac{1}{b}\right)$ **b.** $\lim_{b \to \infty}\left(\sqrt[3]{b} + 3\right)$

➤ Solutions on page 469

Improper Integrals

A definite integral over an interval of infinite length is an *improper integral*. As a first example, we evaluate the improper integral $\displaystyle\int_{1}^{\infty} \dfrac{1}{x^2}\, dx$, which gives the area under the curve $y = \dfrac{1}{x^2}$ from $x = 1$ extending arbitrarily far to the right.

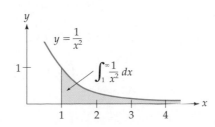

EXAMPLE 3

EVALUATING AN IMPROPER INTEGRAL

Evaluate $\displaystyle\int_1^\infty \frac{1}{x^2}\, dx$.

Solution   To integrate to infinity, we first integrate over a *finite* interval, from 1 to some number b (think of b as some very large number), and then take the limit as b approaches ∞. First integrate from 1 to b.

$$\int_1^b \frac{1}{x^2}\, dx = \int_1^b x^{-2}\, dx = \left(-x^{-1}\right)\Big|_1^b = \left(-\frac{1}{x}\right)\Big|_1^b \qquad \text{Using the Power Rule}$$

$$= -\frac{1}{b} - \left(-\frac{1}{1}\right) = -\frac{1}{b} + 1 \qquad \begin{array}{l}\text{Evaluating and}\\ \text{simplifying}\end{array}$$

$$\underbrace{\phantom{-\frac{1}{b}}}_{\text{at } x = b}\ \underbrace{\phantom{-\left(-\frac{1}{1}\right)}}_{\text{at } x = 1}$$

Then take the limit of this answer as $b \to \infty$.

$$\lim_{b \to \infty}\left(-\frac{1}{b} + 1\right) = 1 \qquad \begin{array}{l}\text{Limit as } b \to \infty \text{ (the } \frac{1}{b}\\ \text{approaches zero)}\end{array}$$

This gives the answer:

$$\int_1^\infty \frac{1}{x^2}\, dx = 1 \qquad \text{Integral from 1 to } \infty \text{ equals 1}$$

Since the limit exists, we say that the improper integral is *convergent*. Geometrically, this procedure amounts to finding the area under the curve from 1 to some number b, shown on the left below, and then letting $b \to \infty$ to find the area arbitrarily far to the right.

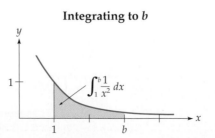

Integrating to b

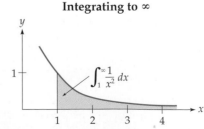

Integrating to ∞

> ### Improper Integrals—Integrating to ∞
>
> If f is continuous and nonnegative for $x \geq a$, we define
>
> $$\int_a^\infty f(x)\, dx = \lim_{b \to \infty} \int_a^b f(x)\, dx$$
>
>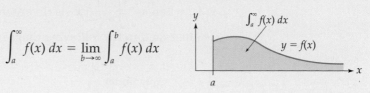
>
> provided that the limit exists. The improper integral is said to be *convergent* if the limit exists, and *divergent* if the limit does not exist.

It is possible to define improper integrals for functions that take negative values, and even for discontinuous functions, but we shall not do so in this book, since most applications involve functions that are positive and continuous.

Practice Problem 2 Evaluate $\displaystyle\int_2^\infty \frac{1}{x^2}\, dx.$

➤ Solution on page 469

EXAMPLE 4

FINDING WHETHER AN INTEGRAL DIVERGES

Evaluate $\displaystyle\int_1^\infty \frac{1}{\sqrt{x}}\, dx.$

Solution Integrating up to b:

$$\int_1^b \frac{1}{\sqrt{x}}\, dx = \int_1^b x^{-1/2}\, dx \;=\; \underbrace{2 \cdot x^{1/2}\Big|_1^b}_{\substack{\text{Integrating by} \\ \text{the Power Rule}}} \;=\; \underbrace{2\sqrt{b} - 2\sqrt{1}}_{\text{Evaluating}} = 2\sqrt{b} - 2$$

Letting b approach infinity:

$$\lim_{b \to \infty}\left(2\sqrt{b} - 2\right) \quad \text{does not exist}$$

Because $\sqrt{b}$ becomes infinite as $b \to \infty$

Therefore,

$$\int_1^\infty \frac{1}{\sqrt{x}}\, dx \text{ is } divergent$$

The integral cannot be evaluated

Notice from Examples 3 and 4 that $\displaystyle\int_1^\infty \frac{1}{x^2}\,dx$ is convergent (its value is 1), whereas $\displaystyle\int_1^\infty \frac{1}{\sqrt{x}}\,dx$ is divergent (it does not have a finite value), as illustrated in the following diagrams.

Area under $\dfrac{1}{x^2}$

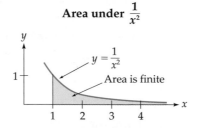

Area under $\dfrac{1}{\sqrt{x}}$

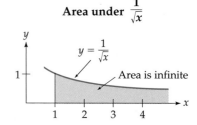

Intuitively, the areas differ because the curve $\dfrac{1}{x^2}$ lies much closer to the x-axis than does the curve $\dfrac{1}{\sqrt{x}}$ for large values of x, as shown above, and so has a smaller area under it.

Spreadsheet Exploration

The following spreadsheet* allows us to numerically "see" that one of these integrals converges while the other diverges. From Example 3 on page 462 we know that $\displaystyle\int_1^\infty \frac{1}{x^2}\,dx = \lim_{b \to \infty}\left(1 - \frac{1}{b}\right)$, while from Example 4 on page 463 we have $\displaystyle\int_1^\infty \frac{1}{\sqrt{x}}\,dx = \lim_{b \to \infty}\left(2\sqrt{b} - 2\right)$. The increasingly large values of b shown in column **A** are used to evaluate $1 - \dfrac{1}{b}$ in column **B** and $2\sqrt{b} - 2$ in column **C**.

* See the Preface for how to obtain this and other Excel spreadsheets.

	C7		▼		=	=2*$A7^(1/2)-2

	A	B	C
1	b	Integral of x ^(-2)	Integral of x ^(-1/2)
2	1	0	0.000
3	5	0.8	2.472
4	10	0.9	4.325
5	50	0.98	12.142
6	100	0.99	18.000
7	500	0.998	42.721
8	1000	0.999	61.246
9	5000	0.9998	139.421
10	10000	0.9999	198.000
11	50000	0.99998	445.214
12	100000	0.99999	630.456
13	500000	0.999998	1412.214

Is it clear which of these integrals (areas) is converging and which is diverging?

We may combine the two steps of integrating up to b and letting $b \to \infty$ into a single line. We show how to do this by evaluating the integral from Example 3 again, but more briefly.

$$\int_1^\infty x^{-2}\, dx = \lim_{b\to\infty} \int_1^b x^{-2}\, dx = \lim_{b\to\infty} \left. (-x^{-1}) \right|_1^b = \lim_{b\to\infty}\left[-\frac{1}{b} - \left(-\frac{1}{1}\right)\right] = 1$$

Integrating;
now use
$x^{-1} = \dfrac{1}{x}$

Approaches
0

Permanent Endowments

Funds that generate steady income forever are called *permanent endowments.*

EXAMPLE **5**

FINDING THE SIZE OF A PERMANENT ENDOWMENT

In the Application Preview on page 435 we found that the size of the fund necessary to generate $2000 annually forever (at 5% interest compounded continuously) is $\int_0^\infty 2000e^{-0.05t}\, dt$. Find the size of this permanent endowment by evaluating the integral.

Solution

$$\int_0^\infty \underbrace{2000}_{\substack{\text{Annual}\\ \text{income}}} e^{-0.05t}\, dt \;=\; \lim_{b\to\infty}\left(2000 \int_0^b e^{\underset{\substack{\text{Continuous}\\ \text{interest rate}}}{-0.05t}}\, dt\right)$$

Annual income, Continuous interest rate, Moved outside

$$= \lim_{b\to\infty}\left[2000(-20)e^{-0.05t}\,\Big|_0^b\right] \qquad \begin{array}{l}\text{Integrating by}\\ \int e^{ax}\, dx = \dfrac{1}{a}e^{ax}\end{array}$$

$$= \lim_{b\to\infty}\,(\underbrace{-40{,}000e^{-0.05b}}_{\substack{\text{Approaches}\\ 0}} + 40{,}000\underbrace{e^{0}}_{1}) = 40{,}000$$

Therefore, the size of the permanent endowment that will pay the $2000 annual maintenance forever is $40,000.

Permanent endowments are used to estimate the ultimate cost of anything that requires continuous long-term funding, from buildings to government agencies to toxic waste sites.

Finding Improper Integrals Using Substitutions

Solving an improper integral may require a substitution. In such cases we apply the substitution not only to the integrand but also to the differential and the upper and lower limits of integration.

EXAMPLE **6**

FINDING AN IMPROPER INTEGRAL USING A SUBSTITUTION

Evaluate $\displaystyle\int_2^\infty \frac{x}{(x^2+1)^2}\, dx$.

Solution We use the substitution $u = x^2 + 1$, so $du = 2x\,dx$, requiring multiplication by 2 and by $\frac{1}{2}$. Notice how the substitution changes the limits.

$$\text{as } x \to \infty, u = x^2 + 1 \to \infty$$

$$\int_2^\infty \frac{x}{(x^2+1)^2}\,dx = \frac{1}{2}\int_2^\infty \frac{2x}{(x^2+1)^2}\,dx = \frac{1}{2}\int_5^\infty \frac{du}{u^2} = \frac{1}{2}\lim_{b\to\infty}\int_5^b u^{-2}\,du$$

$$\left[\begin{array}{c} u = x^2 + 1 \\ du = 2x\,dx \end{array}\right] \qquad u^2 \qquad u - 2^2 + 1 = 5$$

$$= \frac{1}{2}\lim_{b\to\infty}\left[-u^{-1}\right]\Big|_5^b = \frac{1}{2}\lim_{b\to\infty}\left[-\frac{1}{b} - \left(-\frac{1}{5}\right)\right] = \frac{1}{2}\cdot\frac{1}{5} = \frac{1}{10}$$

Integrating   Approaches 0

To integrate over an interval that extends arbitrarily far to the *left*, we again integrate over a finite interval and then take the limit.

> ### Improper Integrals—Integrating to $-\infty$
>
> If f is continuous and nonnegative for $x \le b$, we define
>
> $$\int_{-\infty}^b f(x)\,dx = \lim_{a\to-\infty}\int_a^b f(x)\,dx$$
>
>
> provided that the limit exists. The improper integral is *convergent* if the limit exists, and *divergent* if the limit does not exist.

To integrate over the *entire* x-axis, from $-\infty$ to ∞, we use two integrals, one from $-\infty$ to 0, and the other from 0 to ∞, and then add the results.

Improper Integrals—Integrating from $-\infty$ to ∞

If f is continuous and nonnegative for *all* values of x, we define

$$\int_{-\infty}^{\infty} f(x)\, dx = \lim_{a \to -\infty} \int_{a}^{0} f(x)\, dx + \lim_{b \to \infty} \int_{0}^{b} f(x)\, dx$$

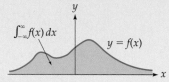

The improper integral is *convergent* if both limits exist, and *divergent* if either limit does not exist.

EXAMPLE **7**

INTEGRATING TO $-\infty$

Evaluate $\displaystyle\int_{-\infty}^{3} 4e^{2x}\, dx$.

Solution

$$\int_{-\infty}^{3} 4e^{2x}\, dx = \lim_{a \to -\infty} \int_{a}^{3} 4e^{2x}\, dx$$

$$= \lim_{a \to -\infty} \left[4 \cdot \frac{1}{2} \cdot e^{2x} \Big|_{a}^{3} \right] \qquad \text{Integrating}$$

$$= \lim_{a \to -\infty} (2e^{6} - 2e^{2a}) = 2e^{6}$$

$\underbrace{\qquad\qquad}$
Approaches 0
as $a \to -\infty$

Practice Problem **3**   Evaluate the improper integral $\displaystyle\int_{-\infty}^{1} 12e^{3x}\, dx$.   ➤ Solution on page 469

6.3 **Section Summary**

A definite integral in which one or both limits of integration are infinite is called an "improper" integral. The improper integral of a continuous nonnegative function is defined as the *limit* of the integral over a finite interval. The integral is *convergent* if the limit exists, and *divergent* otherwise. This idea of dealing with the infinite by "dropping back to the finite and then taking the limit" is a standard technique in mathematics.

Several particular limits are helpful in evaluating improper integrals.

Approaching Infinity	*Approaching Negative Infinity*
$\lim\limits_{x \to \infty} \dfrac{1}{x^n} = 0$ $(n > 0)$	$\lim\limits_{x \to -\infty} \dfrac{1}{x^n} = 0$ (for integer $n > 0$)
$\lim\limits_{x \to \infty} e^{-ax} = 0$ $(a > 0)$	$\lim\limits_{x \to -\infty} e^{ax} = 0$ $(a > 0)$

Improper integrals give continuous sums over infinite intervals. For example, the total future output of an oil well can be found by integrating the production rate out to infinity, and the value of an asset that lasts indefinitely (such as land) can be found by integrating the present value of the future income out to infinity. Even when infinite duration is unrealistic, ∞ is used to represent "long-term behavior."

➤ **Solutions to Practice Problems**

1. a. $\lim\limits_{b \to \infty} \left(1 - \dfrac{1}{b}\right) = 1$ because $\dfrac{1}{b}$ approaches 0

   b. $\lim\limits_{b \to \infty} \left(\sqrt[3]{b} + 3\right)$ does not exist   because $\sqrt[3]{b}$ gets arbitrarily large

2. $\displaystyle\int_2^b x^{-2}\, dx = (-x^{-1}) \Big|_2^b = -\dfrac{1}{b} - \left(-\dfrac{1}{2}\right) = \dfrac{1}{2} - \dfrac{1}{b}$

   $\lim\limits_{b \to \infty} \left(\dfrac{1}{2} - \dfrac{1}{b}\right) = \dfrac{1}{2}$

   Therefore, $\displaystyle\int_2^\infty \dfrac{1}{x^2}\, dx = \dfrac{1}{2}$

3. $\displaystyle\int_{-\infty}^1 12e^{3x}\, dx = \lim\limits_{a \to -\infty} \int_a^1 12e^{3x}\, dx = \lim\limits_{a \to -\infty} \left[12 \cdot \dfrac{1}{3} e^{3x} \Big|_a^1\right] = \lim\limits_{a \to -\infty} (4e^3 - 4e^{3a}) = 4e^3$

6.3 Exercises

Evaluate each limit (or state that it does not exist).

1. $\lim\limits_{x \to \infty} \dfrac{1}{x^2}$

2. $\lim\limits_{b \to \infty} \left(\dfrac{1}{\sqrt{b}} - 8 \right)$

3. $\lim\limits_{b \to \infty} (1 - 2e^{-5b})$

4. $\lim\limits_{b \to \infty} (3e^{3b} - 4)$

5. $\lim\limits_{x \to \infty} (2 - e^{x/2})$

6. $\lim\limits_{x \to \infty} (1 - e^{-x/3})$

7. $\lim\limits_{b \to \infty} (3 + \ln b)$

8. $\lim\limits_{b \to \infty} (2 - \ln b^2)$

Evaluate each improper integral or state that it is divergent.

9. $\displaystyle\int_1^\infty \dfrac{1}{x^3}\, dx$

10. $\displaystyle\int_1^\infty \dfrac{1}{\sqrt[3]{x^4}}\, dx$

11. $\displaystyle\int_2^\infty 3x^{-4}\, dx$

12. $\displaystyle\int_0^\infty e^{-t}\, dt$

13. $\displaystyle\int_2^\infty \dfrac{1}{x}\, dx$

14. $\displaystyle\int_1^\infty \dfrac{1}{x^{0.99}}\, dx$

15. $\displaystyle\int_1^\infty \dfrac{1}{x^{1.01}}\, dx$

16. $\displaystyle\int_{10}^\infty e^{-x/5}\, dx$

17. $\displaystyle\int_0^\infty e^{-0.05t}\, dt$

18. $\displaystyle\int_0^\infty e^{0.01t}\, dt$

19. $\displaystyle\int_5^\infty \dfrac{1}{(x-4)^3}\, dx$

20. $\displaystyle\int_0^\infty \dfrac{x}{(x^2+1)^2}\, dx$

21. $\displaystyle\int_0^\infty \dfrac{x}{x^2+1}\, dx$

22. $\displaystyle\int_0^\infty \dfrac{x^2}{x^3+1}\, dx$

23. $\displaystyle\int_0^\infty x^2 e^{-x^3}\, dx$

24. $\displaystyle\int_e^\infty (\ln x)^{-2}\dfrac{1}{x}\, dx$

25. $\displaystyle\int_{-\infty}^0 e^{3x}\, dx$

26. $\displaystyle\int_{-\infty}^0 \dfrac{x^4}{(x^5-1)^2}\, dx$

27. $\displaystyle\int_{-\infty}^1 \dfrac{1}{2-x}\, dx$

28. $\displaystyle\int_{-\infty}^0 \dfrac{1}{1-x}\, dx$

29. $\displaystyle\int_{-\infty}^\infty \dfrac{e^x}{(1+e^x)^2}\, dx$

30. $\displaystyle\int_{-\infty}^\infty \dfrac{e^{-x}}{(1+e^{-x})^3}\, dx$

31. $\displaystyle\int_{-\infty}^\infty \dfrac{e^x}{1+e^x}\, dx$

32. $\displaystyle\int_{-\infty}^\infty \dfrac{e^{-x}}{1+e^{-x}}\, dx$

33. Use a graphing calculator to estimate the improper integrals $\displaystyle\int_0^\infty e^{\sqrt{x}}\, dx$ and $\displaystyle\int_0^\infty e^{-x^2}\, dx$ (if they converge) as follows:

a. Define y_1 to be the definite integral (using FnInt) of $e^{\sqrt{x}}$ from 0 to x.

b. Define y_2 to be the definite integral of e^{-x^2} from 0 to x.

c. y_1 and y_2 then give the *areas* under these curves out to any number x. Make a TABLE of values of y_1 and y_2 for x-values such as 1, 10, 100, and 500. Which integral converges (and to what number, approximated to five decimal places) and which diverges?

34. Use a graphing calculator to estimate the improper integrals $\displaystyle\int_0^\infty \dfrac{1}{x^2+1}\, dx$ and

$\displaystyle\int_0^\infty \dfrac{1}{\sqrt{x}+1}\, dx$ (if they converge) as follows:

a. Define y_1 to be the definite integral (using FnInt) of $\dfrac{1}{x^2+1}$ from 0 to x.

b. Define y_2 to be the definite integral of $\dfrac{1}{\sqrt{x}+1}$ from 0 to x.

c. y_1 and y_2 then give the *areas* under these curves out to any number x. Make a TABLE of values of y_1 and y_2 for x-values such as 1, 10, 100, 500, and 10,000. Which integral converges (and to what number, approximated to five decimal places) and which diverges?

APPLIED EXERCISES

35. GENERAL: Permanent Endowments Find the size of the permanent endowment needed to generate an annual $12,000 forever at a continuous interest rate of 6%.

36. GENERAL: Permanent Endowments Show that the size of the permanent endowment needed to generate an annual C dollars forever at interest rate r compounded continuously is C/r dollars.

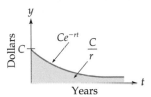

37. GENERAL: Permanent Endowments
   a. Find the size of the permanent endowment needed to generate an annual $1000 forever at a continuous interest rate of 10%.
   b. At this same interest rate, the size of the fund needed to generate an annual $1000 for precisely 100 years is $\int_0^{100} 1000e^{-0.1t}\,dt$. Evaluate this integral (it is not an improper integral), approximating your answer using a calculator.
   c. Notice that the cost for the first 100 years is almost the same as the cost forever. This illustrates again the principle that in endowments, the short term is expensive, but eternity is cheap.

38–40. BUSINESS: Capital Value of an Asset The capital value of an asset is defined as the present value of all future earnings. For an asset that may last indefinitely (such as real estate or a corporation), the capital value is

$$\begin{pmatrix}\text{Capital}\\\text{value}\end{pmatrix} = \int_0^{\infty} C(t)e^{-rt}\,dt$$

where $C(t)$ is the income per year and r is the continuous interest rate. Find the capital value of a piece of property that will generate an annual income of $C(t)$, for the function $C(t)$ given below, at a continuous interest rate of 5%.

38. $C(t) = 8000$ dollars

39. $C(t) = 50\sqrt{t}$ thousand dollars

40. $C(t) = 59\,t^{0.1}$ thousand dollars

41. BUSINESS: Oil Well Output An oil well is expected to produce oil at the rate of $50e^{-0.05t}$ thousand barrels per month indefinitely, where t is the number of months that the well has been in operation. Find the total output over the lifetime of the well by integrating this rate from 0 to ∞. [*Note:* The owner will shut down the well when production falls too low, but it is convenient to estimate the total output as if production continued forever.]

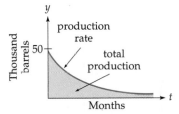

42. GENERAL: Duration of Telephone Calls Studies have shown that the proportion of telephone calls that last longer than t minutes is approximately $\displaystyle\int_t^{\infty} 0.3e^{-0.3s}\,ds.$ Use this formula to find the proportion of telephone calls that last longer than 4 minutes.

43. AREA Find the area between the curve $y = 1/x^{3/2}$ and the x-axis from $x = 1$ to ∞.

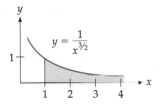

44. AREA Find the area between the curve $y = e^{-4x}$ and the x-axis from $x = 0$ to ∞.

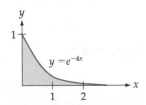

45. AREA Find the area between the curve $y = e^{-ax}$ (for $a > 0$) and the x-axis from $x = 0$ to ∞.

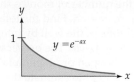

46. AREA Find the area between the curve $y = 1/x^n$ (for $n > 1$) and the x-axis from $x = 1$ to ∞.

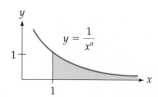

47. BEHAVIORAL SCIENCE: Mazes In a psychology experiment, rats were placed in a T-maze, and the proportion of rats who required more than t seconds to reach the end was $\int_t^\infty 0.05e^{-0.05s} \, ds$. Use this formula to find the proportion of rats who required more than 10 seconds.

48. SOCIOLOGY: Prison Terms If the proportion of prison terms that are longer than t years is given by the improper integral $\int_t^\infty 0.2e^{-0.2s} \, ds$, find the proportion of prison terms that are longer than 5 years.

49. BUSINESS: Product Reliability The proportion of light bulbs that last longer than t hours is predicted to be $\int_t^\infty 0.001e^{-0.001s} \, ds$. Use this formula to find the proportion of light bulbs that will last longer than 1200 hours.

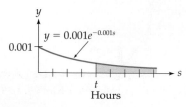

50. BUSINESS: Warranties When a company sells a product with a lifetime guarantee, the number of items returned for repair under the guarantee usually decreases with time. A company estimates that the annual rate of returns after t years will be $800e^{-0.2t}$. Find the total number of returns by summing (integrating) this rate from 0 to ∞.

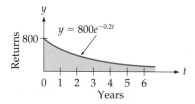

51. BUSINESS: Sales A publisher estimates that a book will sell at the rate of $16{,}000e^{-0.8t}$ books per year t years from now. Find the total number of books that will be sold by summing (integrating) this rate from 0 to ∞.

52. BIOMEDICAL: Drug Absorption To determine how much of a drug is absorbed into the body, researchers measure the difference between the dosage D and the amount of the drug excreted from the body. The total amount excreted is found by integrating the excretion rate $r(t)$ from 0 to ∞. Therefore, the amount of the drug absorbed by the body is

$$D - \int_0^\infty r(t) \, dt.$$

If the initial dose is $D = 200$ milligrams (mg), and the excretion rate is $r(t) = 40e^{-0.5t}$ mg per hour, find the amount of the drug absorbed by the body.

53–54. GENERAL: Permanent Endowments
The formula for integrating the exponential function a^{bx} is $\int a^{bx} \, dx = \dfrac{1}{b \ln a} a^{bx} + C$ for constants $a > 0$ and b, as may be verified by using the differentiation formulas on page 320.

53. Use the formula above to find the size of the permanent endowment needed to generate an annual $2000 forever at 5% interest compounded annually.

[*Hint:* Find $\int_0^\infty 2000 \cdot 1.05^{-x} \, dx$.] Compare your answer with that found in Example 5 (page 466) for the same interest rate but compounded continuously.

54. Use the formula above to find the size of the permanent endowment needed to generate an annual $12,000 forever at 6% interest compounded annually.

[*Hint:* Find $\displaystyle\int_0^\infty 12{,}000 \cdot 1.06^{-x}\, dx$.]   Compare your answer with that found in Exercise 35 (page 470) for the same interest rate but compounded continuously.

55. **BUSINESS: Preferred Stock** Since preferred stock can remain outstanding indefinitely, the present value per share is the limit of the present value of an annuity* paying that share's dividend D at interest rate r:

$$\begin{pmatrix}\text{Present} \\ \text{value}\end{pmatrix} = \lim_{t \to \infty} D\left(\frac{1 - (1 + r)^{-t}}{r}\right)$$

Find this limit in terms of D and r.

56. **BIOMEDICAL: Population Growth** The *Gompertz growth curve* models the size $N(t)$ of a population at time $t \geq 0$ as

$$N(t) = Ke^{-ae^{-bt}}$$

where K and b are positive constants. Find $\displaystyle\lim_{t \to \infty} N(t)$.

* See Section 1.4 of *Calculus with Finite Mathematics* by the same authors for a derivation of this formula.

6.4 NUMERICAL INTEGRATION

Introduction

In spite of the many techniques of integration, there are still integrals that cannot be found by *any* method (as finite combinations of elementary functions). One example is the integral $\int e^{-x^2}\, dx$, which is closely related to the famous "bell-shaped curve" of probability and statistics. For *definite* integrals, however, it is always possible to *approximate* the actual value by interpreting it as the area under a curve, a process called *numerical integration*. We will discuss two of the most useful methods of numerical integration, based on approximating areas by *trapezoids* and by *parabolas* (known as "Simpson's Rule"). Both methods can be programmed on a calculator or computer. In fact, when you evaluate a definite integral on a graphing calculator using a command like FnInt, the calculator is using a numerical integration procedure similar to those described here. This section explains the mathematics behind such operations.

Rectangular Approximation and Riemann Sums

In Section 5.3 we approximated the area under a curve by rectangles, and we called the sum of the areas of the rectangles a *Riemann sum*.

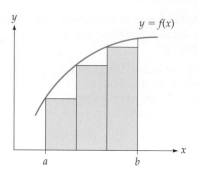

Area under $y = f(x)$ from a to b
approximated by three rectangles

However, these rectangles underestimate the true area under the curve by the white "error area" just above the rectangles. For greater accuracy we could increase the number of rectangles (as we did in Section 5.3), but this would involve more calculation and consequently more roundoff errors. Instead, we will replace the rectangles by shapes that fit the curve more closely.

Trapezoidal Approximation

We modify the approximating rectangles by allowing their tops to slant with the curve, as shown below, giving a much better "fit" to the curve.

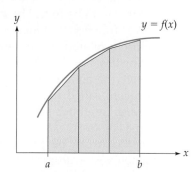

Area under $y = f(x)$ from a to b
approximated by three trapezoids

Such shapes, in which two sides are parallel, are called *trapezoids*. Clearly, approximating the area under the curve by trapezoids is more accurate than approximating it by rectangles: the white "error area" is much smaller.

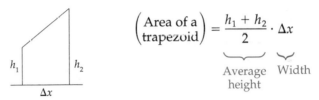

Area under a curve approximated by a
trapezoid (left) and by a rectangle (right).

The area of a trapezoid is the average of the two heights times the width.

$$\begin{pmatrix}\text{Area of a}\\\text{trapezoid}\end{pmatrix} = \underbrace{\frac{h_1 + h_2}{2}}_{\text{Average height}} \cdot \underbrace{\Delta x}_{\text{Width}}$$

If we use trapezoids that all have the same width Δx, we may add up all of the average heights first and then multiply by Δx. Furthermore, averaging the heights means dividing each height by 2 since $\frac{h_1 + h_2}{2} = \frac{h_1}{2} + \frac{h_2}{2}$. However, a side between two rectangles will be counted twice, once for the trapezoid on either side, thereby canceling the division by 2. Therefore, in adding up the heights, only the two outside heights, at a and b, should be divided by 2. This leads to the following procedure for trapezoidal approximation.

Trapezoidal Approximation

To approximate $\displaystyle\int_a^b f(x)\,dx$ by using n trapezoids:

1. Calculate $\Delta x = \dfrac{b - a}{n}.$ Trapezoid width

2. Find numbers $x_1, x_2, \ldots, x_{n+1}$ starting with $x_1 = a$ and successively adding Δx, ending with $x_{n+1} = b$.

3. The approximation for the integral is

$$\int_a^b f(x)\,dx \approx \left[\frac{1}{2}f(x_1) + f(x_2) + \cdots + f(x_n) + \frac{1}{2}f(x_{n+1})\right]\Delta x$$

This last formula calculates the total area of the n trapezoids shown below.

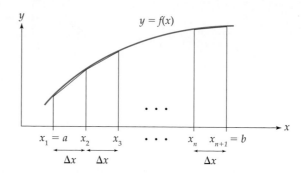

It is easiest to carry out the calculation in a table.

EXAMPLE 1

APPROXIMATING AN INTEGRAL USING TRAPEZOIDS

Approximate $\displaystyle\int_1^2 x^2\, dx$ using four trapezoids.

Solution   The limits of integration are $a = 1$ and $b = 2$, and we are using $n = 4$ trapezoids. The method consists of the following six steps.

1. Calculate the trapezoid width $\Delta x = \dfrac{b - a}{n} = \dfrac{2 - 1}{4} = 0.25.$

2. List the x-values a through b with spacing Δx.

3. Apply $f(x)$ to each x-value.

x	$f(x) = x^2$
Initial point a 1	$(1)^2$ $= \cancel{1}$ 0.5
add Δx 1.25	$(1.25)^2 \approx 1.56$
add Δx 1.5	$(1.5)^2 \; = 2.25$
add Δx 1.75	$(1.75)^2 \approx 3.06$
add Δx Final point b 2.0	$(2)^2 \; = \cancel{4}$ 2

4. Take half of first and last entries.

5. Sum last column.

6. Multiply by Δx.

$$9.37 \cdot (0.25) \approx 2.34$$

Final answer

Therefore, the estimate using four trapezoids is $\displaystyle\int_1^2 x^2\, dx \approx 2.34.$

Earlier (see page 375) we evaluated this integral exactly, obtaining $\int_1^2 x^2 \, dx = \frac{7}{3} \approx 2.33$, and we can use this result to assess the accuracy of the trapezoidal method: our approximation of 2.34 is very accurate in spite of the fact that we used only four trapezoids. The relative error (the actual error, 0.01, divided by the actual value, 7/3) is $\frac{0.01}{7/3} \approx 0.004$. Expressed as a percentage, this is 0.4% (four tenths of one percent), which is also remarkably small. Notice also that the trapezoidal approximation of 2.34 using four trapezoids is far more accurate than the Riemann sum of 2.04 that we found on pages 370–371 using five (left) rectangles.

Error in Trapezoidal Approximation

The maximum error in trapezoidal approximation obeys the following formula.

Trapezoidal Error

For the trapezoidal approximation of $\displaystyle\int_a^b f(x) \, dx$ with n trapezoids,

$$(\text{Error}) \leq \frac{(b-a)^3}{12n^2} \max_{a \leq x \leq b} |f''(x)|$$

This formula is very difficult to use, because it involves maximizing the absolute value of the second derivative. We will not make further use of it except to observe that the n^2 in the denominator means that doubling the number of trapezoids reduces the maximum error by a factor of *four*.

Trapezoidal approximation is most easily carried out on a calculator or a computer, as shown in the following Graphing Calculator Exploration.

Graphing Calculator Exploration

The graphing calculator program* TRAPEZOD calculates the trapezoidal approximation of a definite integral and, for small values of n, draws the approximating trapezoids. The first screen

* See the Preface for how to obtain this and other programs.

below shows the approximation of $\int_1^2 x^2\,dx$ using four trapezoids, giving a value 2.34375 that agrees with the 2.34 found in the previous example. Even for only four trapezoids, the curve $y = x^2$ is almost indistinguishable from the tops of the trapezoids, the difference appearing only as a slight thickening of the line where they diverge slightly.

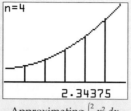

N	TRAP APPRX	
[[5	2.34	...
[10	2.335	...
[50	2.3334	...
[100	2.33335	...
[200	2.3333375	...
[500	2.333334	...

Approximating $\int_1^2 x^2\,dx$ by four trapezoids

For 500 trapezoids, the approximation is about $2\frac{1}{3}$

The screen on the right above shows the trapezoidal approximations as the number n of trapezoids increases from 5 to 500. The approximations do seem to approach the *exact* value $2\frac{1}{3}$ found in Example 1 by evaluating the definite integral.

Try using a command like FnInt to find the area under $y = x^2$ from $x = 1$ to $x = 2$. How do the answers compare? (FnInt uses a method similar to those described in this section, but is faster since it is designed into the calculator.)

The error formula given on the previous page is of limited usefulness on a computer since maximizing the second derivative is itself subject to error. What is done instead is to calculate the approximations for larger and larger values of n until successive approximations agree to the desired degree of accuracy. For example, the last two results in the above table give answers agreeing to five decimal places, indicating an error of less than 0.00001. While this procedure does not *guarantee* this accuracy, it is often accepted in practice.

Simpson's Rule (Parabolic Approximation)

For even greater accuracy, we could increase n (resulting in more calculations and more roundoff errors) or we could change the trapezoids, replacing the tops by *curves* chosen to fit the given curve more closely. Replacing the tops of the trapezoids by *parabolas* that pass through three points of the given curve leads to an even more accurate method of approximation, called Simpson's Rule.* The following diagram shows such a curve and its approximation by two parabolas.

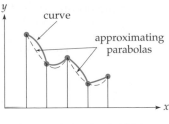

A curve approximated by two
parabolas

Since each parabola spans two intervals, the number of intervals must be even. The area under the approximating parabolas is easily found by integration. The procedure is described as follows and is illustrated in Example 2. A justification of Simpson's Rule is given in Exercise 37.

Simpson's Rule (Parabolic Approximation)

To approximate $\int_{a}^{b} f(x)\,dx$ by Simpson's Rule using n intervals:

1. Calculate $\Delta x = \dfrac{b - a}{n}$. n must be even

2. Find numbers $x_1, x_2, x_3, \ldots, x_{n+1}$ starting with $x_1 = a$ and successively adding Δx, ending with $x_{n+1} = b$.

3. The approximation for the integral is

$$\int_{a}^{b} f(x)\,dx \approx [\,f(x_1) + \underbrace{4f(x_2) + 2f(x_3) + \cdots + 4f(x_n)}_{\text{Alternating 4's and 2's}} + f(x_{n+1})\,]\,\frac{\Delta x}{3}$$

The function values are multiplied by "weights," which, written out by themselves, are

$$\underset{\substack{\uparrow \\ \text{Initial 1}}}{1} \quad \underbrace{4 \quad 2 \quad 4 \quad 2 \quad 4 \quad \cdots \quad 4 \quad 2 \quad 4}_{\substack{\text{Alternating 4's and 2's,} \\ \text{beginning and ending with 4}}} \quad \underset{\substack{\uparrow \\ \text{Final 1}}}{1}$$

* Named after Thomas Simpson (1701–1761), an early user, but not the discoverer, of the formula.

EXAMPLE 2

APPROXIMATING AN INTEGRAL USING SIMPSON'S RULE

Approximate $\displaystyle\int_{3}^{5} \frac{1}{x}\,dx$ by Simpson's Rule with $n = 4$. $n = 4$ means 2 parabolas

Solution   The method consists of the following six steps.

1. Calculate $\Delta x = \dfrac{b - a}{n} = \dfrac{5 - 3}{4} = 0.5$ n must be even

2. List x-values a through b with spacing Δx.
3. Apply $f(x)$ to each x-value.
4. Multiply by the weights to get

x	$f(x) = \dfrac{1}{x}$	weights	$f(x) \cdot$ **weight**
a 3	0.33333	1 ← Initial 1	0.33333
add Δx 3.5	0.28571	4	1.14284
add Δx 4	0.25	2	0.50000
add Δx 4.5	0.22222	4	0.88888
add Δx *b* 5	0.2	1 ← Final 1	0.20000

Alternating 4's and 2's

5. Sum last column.
6. Multiply by $\dfrac{\Delta x}{3}$.

$$3.06505 \cdot \left(\frac{0.5}{3}\right) \approx 0.51084$$

Final answer

Therefore, $\displaystyle\int_{3}^{5} \frac{1}{x}\,dx \approx 0.51084.$

This integral, too, can be found exactly. The answer is $\ln 5 - \ln 3 \approx$ 0.51083,   for an error of only 0.00001. The *relative* error (actual error divided by actual value) is

$$\frac{0.00001}{0.51083} \approx 0.00002 = 0.002\%$$

which is also extremely small.

IQ Distribution

Although it is increasingly clear that human intelligence cannot be measured by a single number, IQ tests are still widely used. (IQ stands for Intelligence Quotient, and is defined as mental age divided by chronological age, multiplied by 100.) The average American IQ is 100, and the distribution of IQs follows the famous "bell-shaped curve" so often used in statistics.*

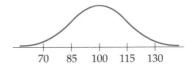

70   85   100   115   130

The proportion of Americans with IQs between two numbers A and B (with $A < B$) is given by the following integral:

$$\left(\begin{array}{l}\text{Proportion of Americans} \\ \text{with IQs between } A \text{ and } B\end{array}\right) \approx \frac{1}{\sqrt{2\pi}} \int_{(A-100)/15}^{(B-100)/15} e^{-x^2/2}\,dx$$

For example, the proportion of Americans who have IQs between 115 and 145 is found by substituting $A = 115$ and $B = 145$ into the lower and upper limits in the formula above:

$$\left(\begin{array}{l}\text{Proportion of IQs} \\ \text{between 115 and 145}\end{array}\right) = \frac{1}{\sqrt{2\pi}} \int_{1}^{3} e^{-x^2/2}\,dx$$

3 from $\dfrac{B-100}{15} = \dfrac{145-100}{15} = \dfrac{45}{15} = 3$

1 from $\dfrac{A-100}{15} = \dfrac{115-100}{15} = \dfrac{15}{15} = 1$

This integral cannot be found "by hand" (as a finite combination of elementary functions). It can be approximated by Simpson's Rule, just as in Example 2, but it is easier to carry out the calculation on a graphing calculator.

*For those familiar with statistics, IQ scores are normally distributed with mean 100 and standard deviation 15.

Graphing Calculator Exploration

The graphing calculator program[†] SIMPSON uses Simpson's Rule to approximate definite integrals and, for small values of n, draws the approximating parabolas. The first screen below shows

† See the Preface for how to obtain this and other programs.

the approximation of $\dfrac{1}{\sqrt{2\pi}}\displaystyle\int_1^3 e^{-x^2/2}\,dx$ using $n = 4$ intervals (2 parabolas), giving approximately 0.157. Simpson's Rule is so accurate that the curve below is almost indistinguishable from the approximating parabolas, the difference again appearing only as a slight thickening of the line where they diverge slightly.

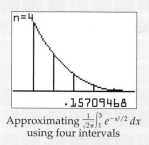

.15709468

Approximating $\frac{1}{\sqrt{2\pi}}\int_1^3 e^{-x^2/2}\,dx$
using four intervals

N	SMPSN APPRX
[[4	.15709468...
[10	.15730028...
[20	.15730504...
[50	.15730534...
[100	.15730535...
[200	.15730535...

For 100 and 200 intervals the approximations agree to 8 decimal places

The screen on the right shows that Simpson's Rule converges extremely rapidly.

Try using a command like FnInt to find the area under $y = 1/\sqrt{2\pi}\,e^{-x^2/2}$ from $x = 1$ to $x = 3$. How do the answers compare?

Converting the answer 0.157 to a percentage, about 16% of all Americans have IQs in the range 115 to 145.

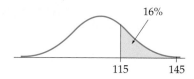

16%

115   145

The Error in Simpson's Rule

The maximum error in Simpson's Rule obeys the following formula.

Error in Simpson's Rule

In approximating $\displaystyle\int_a^b f(x)\,dx$ by Simpson's Rule with n intervals,

$$(\text{Error}) \le \frac{(b-a)^5}{180n^4}\max_{a\le x\le b}\left|f^{(4)}(x)\right|$$

This formula is difficult to use because it involves maximizing the absolute value of the fourth derivative. It does, however, show that Simpson's Rule is *exact* for cubics (third-degree polynomials), since cubics have zero fourth derivative, and that doubling the value of n reduces the maximum error by a factor of 16 (because of the n^4).

6.4 Section Summary

Some definite integrals cannot be evaluated exactly because it is impossible (or very difficult) to find an antiderivative. However, any definite integral can be *approximated* by numerical integration, and two of the most useful methods are trapezoidal approximation and Simpson's Rule (parabolic approximation). Each method involves choosing a number n of intervals and calculating a "weighted average" of function values at the endpoints of these intervals. Higher values of n generally give greater accuracy, but also involve more calculation (and therefore more roundoff errors). In practice, Simpson's Rule is generally the method of choice, since it gives greater accuracy for only slightly more effort. Trapezoidal approximation, however, has the advantage of having a simpler formula for its error. It is particularly appropriate to seek an approximation if the "exact" answer involves logarithms or exponentials, since these functions will probably be approximated anyway in the evaluation step.

It is curious that in Chapter 5 we used integrals to evaluate areas, and here we are using areas to evaluate integrals. This is typical of mathematics, in which any equivalence (such as definite integrals and areas) is exploited in both directions.

Programs for Trapezoidal Approximation and Simpson's Rule

On the following page are two programs for trapezoidal approximation and two for Simpson's Rule for the *TI-83* calculator and in BASIC. They may be adapted for other calculators or computers. The small print is explanation, not part of the program. These programs give only single numerical answers, whereas TRAPEZOD and SIMPSON draw graphs and make tables showing the approximations for several values of n.

TI-83 Program for Trapezoidal Approximation

(Enter the function in y_1 before executing.)

Disp "USES FUNCTION Y_1"
Prompt A, B, N
$(B - A)/N \rightarrow D$
$(Y_1(A) + Y_1(B))/2 \rightarrow S$
$A + D \rightarrow X$
For(I,1,N − 1)
$S + Y_1 \rightarrow S$
$X + D \rightarrow X$
End
$S*D \rightarrow S$
Disp "TRAP APPROX IS ", S

BASIC Program for Trapezoidal Approximation

(Lines 2, 3, 4, and 8 to be completed as indicated in small type)

approx = 0
a =   fill in beginning x-value
b =   fill in ending x-value
n =   fill in number of trapezoids
delta = (b − a)/n
x = a
FOR i = 1 TO n + 1

$f = \left(\begin{smallmatrix}\text{fill in the}\\\text{function}\end{smallmatrix}\right)$

IF i = 1 OR i = n + 1 THEN f = f/2
approx = approx + f
x = x + delta
NEXT i
approx = approx*delta
PRINT "TRAP APPROX IS", approx
END

TI-83 Program for Simpson's Rule

(Enter the function in y_1 before executing.)

Disp "USES FUNCTION Y_1"
Disp "N MUST BE EVEN"
Prompt A, B, N
$(B - A)/N \rightarrow D$
$Y_1(A)–Y_1(B) \rightarrow S$
$A \rightarrow X$
For(I,1,N/2)
$S + 4*Y_1(X + D) + 2*Y_1(X + 2D) \rightarrow S$
$X + 2D \rightarrow X$
End
$S*D/3 \rightarrow S$
Disp "SIMP APPROX IS " , S

BASIC Program for Simpson's Rule

(Lines 1, 2, 3, and 9 to be completed as indicated in small type)

a =   fill in beginning x-value
b =   fill in ending x-value
n =   fill in number of intervals (even)
delta = (b − a)/n
approx = 0
x = a
FOR i = 1 TO n + 1
IF i = 1 OR i = n + 1 THEN weight = 1

approx = approx + $\left(\begin{smallmatrix}\text{fill in the}\\\text{function}\end{smallmatrix}\right)$ * weight

x = x + delta
IF weight = 4 THEN weight = 2 ELSE weight = 4
NEXT i
approx = approx*delta/3
PRINT "SIMP APPROX IS", approx
END

6.4 **Exercises**

EXERCISES ON TRAPEZOIDAL APPROXIMATION

For each definite integral:

a. Approximate it "by hand," using trapezoidal approximation with $n = 4$ trapezoids. Round calculations to three decimal places.

b. Evaluate the integral exactly using antiderivatives, rounding to three decimal places.

c. Find the actual error (the difference between the actual value and the approximation).

d. Find the relative error (the actual error divided by the actual value, expressed as a percent).

1. $\int_1^3 x^2 \, dx$ **2.** $\int_1^2 x^3 \, dx$

3. $\int_2^4 \frac{1}{x} \, dx$ **4.** $\int_1^3 \frac{1}{x} \, dx$

Approximate each integral using trapezoidal approximation "by hand" with the given value of n. Round all calculations to three decimal places.

5. $\int_0^1 \sqrt{1 + x^2} \, dx$, $n = 3$

6. $\int_0^1 \sqrt{1 + x^3} \, dx$, $n = 3$

7. $\int_0^1 e^{-x^2} \, dx$, $n = 4$ **8.** $\int_0^1 e^{x^2} \, dx$, $n = 4$

Approximate each integral using the graphing calculator program TRAPEZOD (see pages 477–478) or one of the trapezoidal approximation programs on the previous page or a similar program. Use the following values for the numbers of intervals: 10, 50, 100, 200, 500. Then give an estimate for the value of the definite integral, keeping as many decimal places as the last two approximations agree (when rounded).

9. $\int_1^2 \sqrt{\ln x} \, dx$ **10.** $\int_0^1 \ln(x^2 + 1) \, dx$

11. $\int_{-1}^1 \sqrt{16 + 9x^2} \, dx$ **12.** $\int_{-1}^1 \sqrt{25 - 9x^2} \, dx$

13. $\int_{-1}^1 e^{x^2} \, dx$ **14.** $\int_0^4 \sqrt{1 + x^4} \, dx$

APPLIED EXERCISES ON TRAPEZOIDAL APPROXIMATION

15–16. GENERAL: IQs Use the formula on page 481 and TRAPEZOD or another trapezoidal approximation program (see the previous page) to find the proportion of Americans with IQs between the following two numbers. Use successively higher values of n until answers agree to four decimal places.

15. 100 and 130   **16.** 130 and 145

17–18. BUSINESS: Investment Growth An investment grows at a rate of $3.2 \, e^{\sqrt{t}}$ thousand dol-

lars per year, where t is the number of years since the beginning of the investment. Use TRAPEZOD or another trapezoidal approximation program to estimate the total growth of the investment during the period stated below. Use successively higher values of n until answers agree to two decimal places.

17. In the first 2 years (year 0 to year 2)

18. In the first 3 years (year 0 to year 3)

EXERCISES ON SIMPSON'S RULE

Estimate each definite integral "by hand," using Simpson's Rule with $n = 4$. Round all calculations to three decimal places. Exercises 19–26 correspond to Exercises 1–8, in which the same

integrals were estimated using trapezoids. If you did the corresponding exercise, compare your Simpson's Rule answer with your trapezoidal answer.

(See instructions on the previous page.)

19. $\int_1^3 x^2\,dx$

20. $\int_1^2 x^3\,dx$

21. $\int_2^4 \frac{1}{x}\,dx$

22. $\int_1^3 \frac{1}{x}\,dx$

23. $\int_0^1 \sqrt{1+x^2}\,dx$

24. $\int_0^1 \sqrt{1+x^3}\,dx$

25. $\int_0^1 e^{-x^2}\,dx$

26. $\int_0^1 e^{x^2}\,dx$

Approximate each integral using the graphing calculator program SIMPSON (see pages 481–482) or one of the Simpson's Rule approximation programs on page 484 or a similar program. Use the following values for the numbers of intervals: 10, 20, 50, 100, 200. Then give an estimate for the value of the definite integral, keeping as many decimal places as the last two approximations agree to (when rounded). Exercises 27–32 correspond to Exercises 9–14 in which the same integrals were estimated using trapezoids. If you did the corre-

sponding exercise, compare your Simpson's Rule answer with your trapezoidal answer.

27. $\int_1^2 \sqrt{\ln x}\,dx$

28. $\int_0^1 \ln(x^2+1)\,dx$

29. $\int_{-1}^1 \sqrt{16+9x^2}\,dx$

30. $\int_{-1}^1 \sqrt{25-9x^2}\,dx$

31. $\int_{-1}^1 e^{x^2}\,dx$

32. $\int_0^4 \sqrt{1+x^4}\,dx$

33–34. APPROXIMATION OF IMPROPER INTEGRALS For each improper integral:

a. Make it a "proper" integral by using the substitution $x = \frac{1}{t}$ and simplifying.

b. Approximate the proper integral using Simpson's Rule (either "by hand" or using a program) with $n = 4$ intervals, rounding your answer to three decimal places.

33. $\int_1^\infty \frac{1}{x^3+1}\,dx$

34. $\int_1^\infty \frac{x}{x^3+1}\,dx$

APPLIED EXERCISES ON SIMPSON'S RULE

35. GENERAL: Suspension Bridges The cable of a suspension bridge hangs in a parabolic curve. The equation of the cable shown below is $y = \frac{x^2}{2000}$. Its length in feet is given by the integral

$$\int_{-400}^{400} \sqrt{1+\left(\frac{x}{1000}\right)^2}\,dx$$

Approximate this integral using Simpson's Rule, using successively higher values of n until answers agree to the nearest whole number.

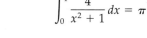

36. APPROXIMATION OF π The number π is the ratio of the circumference of a circle to its diameter (since $C = \pi D$). It can be shown that the following definite integral is equal to π.

$$\int_0^1 \frac{4}{x^2+1}\,dx = \pi$$

Find π by approximating this integral using a Simpson's Rule program, using successively higher values of n until answers agree to four decimal places.

JUSTIFICATION OF SIMPSON'S RULE

37. JUSTIFICATION OF SIMPSON'S RULE Justify Simpson's Rule by carrying out the following steps, which lead to the formula for

Simpson's Rule (page 479) in a simple case.

i. Observe that if the three points shown in the following diagram lie on the parabola

$f(x) = ax^2 + bx + c$, then the following three equations hold:

$$a(-d)^2 + b(-d) + c = y_1 \quad \text{Since} \quad f(-d) = y_1$$
$$a(0)^2 + b(0) + c = y_2 \quad \text{Since} \quad f(0) = y_2$$
$$a(d)^2 + b(d) + c = y_3 \quad \text{Since} \quad f(d) = y_3$$

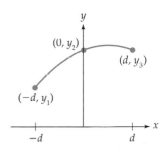

ii. Simplify these three equations to obtain

$$ad^2 - bd + c = y_1$$
$$c = y_2$$
$$ad^2 + bd + c = y_3$$

iii. Add the first and last equation plus four times the middle equation to obtain

$$2ad^2 + 6c = y_1 + 4y_2 + y_3$$

You will use this equation in step v.

iv. Evaluate the integral $\displaystyle\int_{-d}^{d} (ax^2 + bx + c)\, dx$ and simplify to show that the area under the parabola from $-d$ to d is

$$\text{Area} = \frac{2}{3}ad^3 + 2cd = \left(2ad^2 + 6c\right)\frac{d}{3}$$

v. Use the equation found in step iii to write this area as

$$\text{Area} = (y_1 + 4y_2 + y_3)\frac{d}{3}$$

vi. Use $y_1 = f(-d)$, $y_2 = f(0)$, $y_3 = f(d)$ and the fact that the spacing Δx is equal to d to rewrite this last equation as

$$\text{Area} = [f(-d) + 4f(0) + f(d)]\frac{\Delta x}{3}$$

This is exactly the formula for Simpson's Rule using one parabola ($n = 2$). For several parabolas placed next to each other, the function values *between* two neighboring parabolas are added twice (once for each side), and so should have weight 2. Therefore, successive function values are multiplied by the weights given on page 479:

$$1 \quad 4 \quad 2 \quad 4 \quad 2 \quad 4 \quad \cdots \quad 4 \quad 1$$

Chapter Summary with Hints and Suggestions

Reading the text and doing the exercises in this chapter have helped you to master the following skills, which are listed by section (in case you need to review them) and are keyed to particular Review Exercises. Answers for all Review Exercises are given at the back of the book, and full solutions can be found in the Student Solutions Manual.

6.1 Integration by Parts

■ Find an integral using integration by parts. *(Review Exercises 1–14.)*

$$\int u\, dv = uv - \int v\, du$$

■ Find an integral by whatever technique is necessary. *(Review Exercises 15–22.)*

■ Solve an applied problem using integration by parts. *(Review Exercises 23–24.)*

$$\left(\begin{array}{c}\text{Present}\\\text{value}\end{array}\right) = \int_0^T C(t)e^{-rt}\, dt$$

$$\left(\begin{array}{c}\text{Total}\\\text{accumulation}\end{array}\right) = \int_0^T r(t)\, dt$$

6.2 Integration Using Tables

■ Find an integral using a table of integrals. (*Review Exercises 25–36.*)

■ Solve an applied problem using an integral table. (*Review Exercises 37–38.*)

6.3 Improper Integrals

■ Evaluate an improper integral (if it is convergent). (*Review Exercises 39–56.*)

$$\int_a^\infty f(x)\,dx = \lim_{b\to\infty} \int_a^b f(x)\,dx$$

■ Solve an applied problem involving an improper integral. (*Review Exercises 57–60.*)

■ Predict whether an improper integral converges, and then check by evaluating it. (*Review Exercises 61–62.*)

6.4 Numerical Integration

■ Approximate an integral using trapezoidal approximation "by hand." (*Review Exercises 63–68.*)

$$\int_a^b f(x)\,dx \approx \left[\frac{1}{2}f(x_1) + f(x_2) + \right.$$
$$\left. \cdots + f(x_n) + \frac{1}{2}f(x_{n+1}) \right] \cdot \Delta x$$

■ Use a program to approximate an integral by trapezoidal approximation. (*Review Exercises 69–74.*)

■ Approximate an integral using Simpson's Rule "by hand." (*Review Exercises 75–80.*)

$$\int_a^b f(x)\,dx \approx [f(x_1) + 4f(x_2) + 2f(x_3) +$$
$$\cdots + 4f(x_n) + f(x_{n+1})] \cdot \frac{\Delta x}{3}$$

■ Use a program to approximate an integral by Simpson's Rule. (*Review Exercises 81–86.*)

■ Approximate an improper integral using trapezoidal approximation. (*Review Exercises 87–88.*)

Hints and Suggestions

■ The unifying idea of this chapter is extensions of the concept of integration.

■ Integration by parts takes one integral and gives another integral that, it is hoped, is simpler than the original integral. The formula is simply an integration version of the Product Rule. When using it, try to choose dv (including the dx) to be the most complicated part of the integrand that you can integrate, and, if possible, choose the u to be something that simplifies when differentiated.

■ There are tables of integrals that are much longer than the one on the inside back cover. Longer tables, however, require much more time to search for the "right" formula. Other techniques (such as a substitution, integration by parts, or use of a formula more than once) may be used with an integral table.

■ To find an integral, try the following methods. First try the "basic" formulas 1 through 4 on the inside back cover. Then try a substitution (formulas 5 through 7). If these methods fail, try integration by parts or an integral table. Remember that some integrals *cannot* be integrated (in terms of elementary functions). A *definite* integral can always be approximated by numerical methods.

■ Before "evaluating" an improper integral, be sure that the integrand is defined over the interval, and that the integral is convergent. If the integral diverges, then it has no value and we simply state that it is divergent.

■ Numerical integration involves approximating the area under a curve using geometric figures such as trapezoids or parabolas (Simpson's Rule). In practice, the calculations are usually carried out on a calculator or computer, but doing some "by hand" helps to make the method clear.

■ A graphing calculator is very helpful for approximating definite integrals by trapezoidal approximation or Simpson's Rule for large values of n. Graphing calculators also have their own built-in numerical procedures for approximating integrals when you use FnInt.

■ **Practice for Test:** Review Exercises 3, 5, 13, 25, 27, 33, 39, 41, 57, 63, 71, 79, 85.

Review Exercises for Chapter 6
Practice test exercise numbers in green.

6.1 Integration by Parts

Find each integral using integration by parts.

1. $\int xe^{2x}\, dx$

2. $\int xe^{-x}\, dx$

3. $\int x^8 \ln x\, dx$

4. $\int \sqrt[4]{x} \ln x\, dx$

5. $\int (x-2)(x+1)^5\, dx$

6. $\int (x+3)(x-1)^4\, dx$

7. $\int \dfrac{\ln t}{\sqrt{t}}\, dt$

8. $\int x^7 e^{x^4}\, dx$

9. $\int x^2 e^x\, dx$

10. $\int (\ln x)^2\, dx$

11. $\int x(x+a)^n\, dx$ (for constants a and $n > 0$)

12. $\int x(1-x)^n\, dx$ (for constant $n > 0$)

13. $\int_0^5 xe^x\, dx$

14. $\int_1^e x \ln x\, dx$

Find each integral by a substitution or by integration by parts, as appropriate.

15. $\int \dfrac{dx}{1-x}$

16. $\int xe^{-x^2}\, dx$

17. $\int x^3 \ln 2x\, dx$

18. $\int \dfrac{dx}{(1-x)^2}$

19. $\int \dfrac{\ln x}{x}\, dx$

20. $\int \dfrac{e^{2x}}{e^{2x}+1}\, dx$

21. $\int \dfrac{e^{\sqrt{x}}}{\sqrt{x}}\, dx$

22. $\int (e^{2x}+1)^3 e^{2x}\, dx$

23. BUSINESS: Present Value of a Continuous Stream of Income A company generates a continuous stream of income of $25t$ million dollars per year, where t is the number of years that the company has been in operation.

a. Find the present value of this stream for the first 10 years at 5% interest compounded continuously. (Do not use a graphing calculator.)

b. Verify your answer to part (a) using FnInt on a graphing calculator.

24. ENVIRONMENTAL SCIENCE: Pollution Radioactive waste is leaking out of cement storage vessels at the rate of $te^{0.2t}$ hundred gallons per month, where t is the number of months since the leak began.

a. Find the total leakage during the first 3 months. (Do not use a graphing calculator.)

b. Verify your answer to part (a) using FnInt on a graphing calculator

6.2 Integration Using Tables

Use the integral table on the inside back cover to find each integral.

25. $\int \dfrac{1}{25-x^2}\, dx$

26. $\int \dfrac{1}{x^2-4}\, dx$

27. $\int \dfrac{x}{(x-1)(x-2)}\, dx$

28. $\int \dfrac{1}{(x-1)(x-2)}\, dx$

29. $\int \dfrac{1}{x\sqrt{x+1}}\, dx$

30. $\int \dfrac{x}{\sqrt{x+1}}\, dx$

31. $\int \dfrac{1}{\sqrt{x^2+9}}\, dx$

32. $\int \dfrac{1}{\sqrt{x^2+16}}\, dx$

33. $\int \dfrac{z^3}{\sqrt{z^2+1}}\, dz$

34. $\int \dfrac{e^{2t}}{e^t+2}\, dt$

35. $\int x^2 e^{2x}\, dx$

36. $\int (\ln x)^4\, dx$

37. BUSINESS: Cost A company's marginal cost function is $MC(x) = \dfrac{1}{(2x+1)(x+1)}$ and fixed costs are 1000 (all in dollars). Find the company's cost function.

38. GENERAL: Population The population of a town is growing at the rate of $\sqrt{t^2 + 1600}$ people per year, where t is the number of years from now.

a. Find the total increase in population during the first 30 years. (Do not use a graphing calculator.)

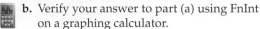 b. Verify your answer to part (a) using FnInt on a graphing calculator.

6.3 Improper Integrals

Find the value of each improper integral or state that it is divergent.

39. $\int_1^\infty \frac{1}{x^5}\,dx$ **40.** $\int_1^\infty \frac{1}{x^6}\,dx$ **41.** $\int_1^\infty \frac{1}{\sqrt[5]{x}}\,dx$

42. $\int_1^\infty \frac{1}{\sqrt[6]{x}}\,dx$ **43.** $\int_0^\infty e^{-2x}\,dx$

44. $\int_4^\infty e^{-0.5x}\,dx$ **45.** $\int_0^\infty e^{2x}\,dx$

46. $\int_4^\infty e^{0.5x}\,dx$ **47.** $\int_0^\infty e^{-t/5}\,dt$

48. $\int_{100}^\infty e^{-t/10}\,dt$ **49.** $\int_0^\infty \frac{x^3}{(x^4 + 1)^2}\,dx$

50. $\int_0^\infty \frac{x^4}{(x^5 + 1)^2}\,dx$ **51.** $\int_{-\infty}^0 e^{2t}\,dt$

52. $\int_{-\infty}^0 e^{4t}\,dt$ **53.** $\int_{-\infty}^4 \frac{1}{(5 - x)^2}\,dx$

54. $\int_{-\infty}^8 \frac{1}{(9 - x)^2}\,dx$ **55.** $\int_{-\infty}^\infty \frac{e^{-x}}{(1 + e^{-x})^4}\,dx$

56. $\int_{-\infty}^\infty \frac{e^{-x}}{(1 + e^{-x})^3}\,dx$

57. GENERAL: Permanent Endowments Find the size of the permanent endowment needed to generate an annual $6000 forever at an interest rate of 10% compounded continuously.

58. GENERAL: Automobile Age Insurance records indicate that the proportion of cars on the road that are more than x years old is approximated by the integral $\int_x^\infty 0.21e^{-0.21t}\,dt$. Find the proportion of cars that are more than 5 years old.

59. BUSINESS: Book Sales A publisher estimates that the demand for a certain book will be $12e^{-0.05t}$ thousand copies per year, where t is the number of years since the book's publication. Find the total number of books that will be sold from the publication date onward.

60. GENERAL: Resource Consumption If the rate of consumption of a certain mineral is $300e^{-0.04t}$ million tons per year (where t is the number of years from now), find the total amount of the mineral that will be consumed from now on.

For each improper integral, use a graphing calculator to evaluate it (or to show that it diverges) as follows:

a. Define y_1 to be the definite integral (using FnInt) of the integrand from 1 to x.

b. Make a TABLE of values of y_1 for x-values such as 1, 10, 100, and 1000. Does the integral converge (and if so, to what number) or does it diverge?

c. Verify your answers to part (b) by evaluating the improper integral "by hand."

61. $\int_1^\infty \frac{1}{x^3}\,dx$ **62.** $\int_1^\infty \frac{1}{\sqrt[3]{x}}\,dx$

6.4 Numerical Integration

Estimate each integral using trapezoidal approximation with the given value of n. (Round all calculations to three decimal places.)

63. $\int_0^1 \sqrt{1 + x^4}\,dx,\ \ n = 3$

64. $\int_0^1 \sqrt{1 + x^5}\,dx,\ \ n = 3$

65. $\int_0^1 e^{x^2/2}\,dx,\ \ n = 4$

66. $\int_0^1 e^{-x^2/2}\,dx,\ \ n = 4$

67. $\int_{-1}^1 \ln(1 + x^2)\,dx,\ \ n = 4$

68. $\int_{-1}^1 \ln(x^3 + 2)\,dx,\ \ n = 4$

Use a trapezoidal approximation program such as TRAPEZOD to approximate each integral. Use successively higher values of n until the results agree to three decimal places (rounded).

69. $\int_0^1 \sqrt{1 + x^4}\, dx$ **70.** $\int_0^1 \sqrt{1 + x^5}\, dx$

71. $\int_0^1 e^{x^2/2}\, dx$ **72.** $\int_0^1 e^{-x^2/2}\, dx$

73. $\int_{-1}^1 \ln(1 + x^2)\, dx$ **74.** $\int_{-1}^1 \ln(x^3 + 2)\, dx$

Estimate each integral using Simpson's Rule (parabolic approximation) with the given value of n. (Round all calculations to four decimal places.)

75. $\int_0^1 \sqrt{1 + x^4}\, dx, \quad n = 4$

76. $\int_0^1 \sqrt{1 + x^5}\, dx, \quad n = 4$

77. $\int_0^1 e^{x^2/2}\, dx, \quad n = 4$

78. $\int_0^1 e^{-x^2/2}\, dx, \quad n = 4$

79. $\int_{-1}^1 \ln(1 + x^2)\, dx, \quad n = 4$

80. $\int_{-1}^1 \ln(x^3 + 2)\, dx, \quad n = 4$

 Use a Simpson's Rule approximation program such as SIMPSON to approximate each integral. Use successively higher values of n until the rounded results agree to six decimal places.

81. $\int_0^1 \sqrt{1 + x^4}\, dx$ **82.** $\int_0^1 \sqrt{1 + x^5}\, dx$

83. $\int_0^1 e^{x^2/2}\, dx$ **84.** $\int_0^1 e^{-x^2/2}\, dx$

85. $\int_{-1}^1 \ln(1 + x^2)\, dx$ **86.** $\int_{-1}^1 \ln(x^3 + 2)\, dx$

For each improper integral:

a. Make it a "proper" integral by using the substitution $x = \dfrac{1}{t}$ and simplifying.

b. Approximate the proper integral using trapezoidal approximation with $n = 4$. Keep three decimal places.

87. $\int_1^\infty \dfrac{1}{x^2 + 1}\, dx$ **88.** $\int_1^\infty \dfrac{x^2}{x^4 + 1}\, dx$

7 Calculus of Several Variables

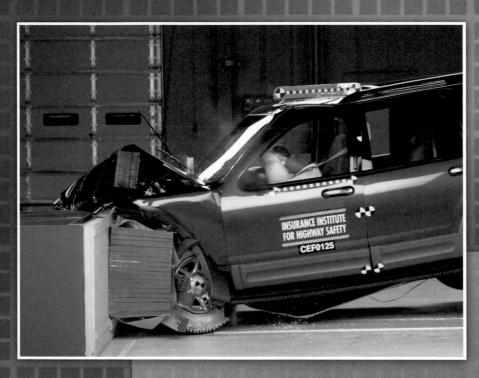

Application Preview

Safe Cars, Unsafe Streets

Since 1970, improvements in automobile design and increased use of seat belts have lowered the number of automobile-related fatalities in the United States. On the other hand, during this same period the greater availability of handguns, together with other socioeconomic causes, has increased the number of gunshot fatalities. The data are shown in the following table and graph.

	Years Since 1970	Automobile Fatalities	Gunshot Fatalities
1970	0	55,800	26,800
1975	5	47,100	33,400
1980	10	53,200	32,900
1985	15	45,800	31,800
1990	20	47,600	36,000
1995	25	43,900	37,900
2000	30	41,300	32,200

Source: National Center for Health Statistics

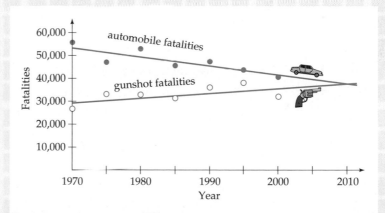

Two lines have been fit to these data points by a technique called "least squares." The lines show that at some time guns will overtake automobiles as the leading cause of accidental death in America. (In Exercise 31 on page 547 you will find the precise year.)

In this chapter we will discuss the least squares technique for fitting lines or curves to data points, as well as other useful applications of functions of several variables.

7.1   FUNCTIONS OF SEVERAL VARIABLES

Introduction

Many quantities depend on *several* variables. For example, the "wind-chill" factor announced by the weather bureau during the winter depends on two variables: temperature and wind speed. The cost of a telephone call depends on *three* variables: distance, duration, and time of day.

In this chapter we define functions of two or more variables and learn how to differentiate and integrate them. We use derivatives for calculating rates of change and optimizing functions, and integrals for finding volumes, continuous sums, and average values.

Graphing calculators will be less useful in this chapter because of their small screens and limited computing power, but computer-drawn pictures of three-dimensional surfaces will be very useful.

Functions of Two Variables

A function f that depends on *two* variables, x and y, is written $f(x, y)$ (read: f of x and y). The *domain* of the function is the set of all ordered pairs (x, y) for which it is defined. The *range* is the set of all resulting values of the function. Formally:

Function of Two Variables

A function f of two variables is a rule such that to each ordered pair (x, y) in the domain of f there corresponds one and only one number $f(x, y)$.

If the domain is not stated, it will always be taken to be the largest set of ordered pairs for which the function is defined (the "natural domain").

EXAMPLE 1

FINDING THE DOMAIN OF A FUNCTION

For $f(x, y) = \dfrac{\sqrt{x}}{y^2}$, find   **a.** the domain   **b.** $f(9, 21)$

Solution

a. $\{(x, y) \mid x \geq 0, y \neq 0\}$

For $\sqrt{x}/y^2$, x cannot be negative (because of the $\sqrt{\ }$), and y cannot be zero

b. $f(9, -1) = \dfrac{\sqrt{9}}{(-1)^2} = \dfrac{3}{1} = 3$

$f(x, y) = \sqrt{x}/y^2$
with $x = 9$ and $y = -1$

EXAMPLE 2

FINDING THE DOMAIN OF A FUNCTION INVOLVING LOGARITHMS AND EXPONENTIALS

For $g(u, v) = e^{uv} - \ln u$, find **a.** the domain **b.** $g(1, 2)$

Solution

a. $\{(u, v) \mid u > 0\}$

u must be positive so that its logarithm is defined

b. $g(1, 2) = e^{1 \cdot 2} - \underbrace{\ln 1}_{0} = e^2 - 0 = e^2$

$g(u, v) = e^{uv} - \ln u$
with $u = 1$ and $v = 2$

Practice Problem 1   For $f(x, y) = \dfrac{\ln x}{e^{\sqrt{y}}}$, find   **a.** the domain   **b.** $f(e, 4)$

➤ Solutions on page 504

Functions of two variables are used in many applications.

EXAMPLE 3

FINDING A COMPANY'S COST FUNCTION

A company manufactures three-speed and ten-speed bicycles. It costs $100 to make each three-speed bicycle, it costs $150 to make each ten-speed bicycle, and fixed costs are $2500. Find the cost function, and use it to find the cost of producing 15 three-speed bicycles and 20 ten-speed bicycles.

Solution   Let:

$$x = \text{the number of three-speed bicycles}$$
$$y = \text{the number of ten-speed bicycles}$$

The cost function is

$$C(x, y) = 100x \quad + \quad 150y \quad + \quad 2500$$

Unit cost | Quan-tity | Unit cost | Quan-tity | Fixed costs

The cost of producing 15 three-speed bicycles and 20 ten-speed bicycles is found by evaluating $C(x, y)$ at $x = 15$ and $y = 20$:

$$C(15, 20) = 100 \cdot 15 + 150 \cdot 20 + 2500$$
$$= 1500 + 3000 + 2500 = 7000$$

Producing 15 three-speed and 20 ten-speed bicycles costs $7000.

The variables x and y in the preceding example stand for numbers of bicycles, and so should take only integer values. Instead, however, we will allow x and y to be "continuous" variables, and round to integers at the end if necessary.

Some other "everyday" examples of functions of two variables are as follows:

$A(l, w) = lw$ — Area of a rectangle of length l and width w

$f(w, v) = kwv^2$ — Length of the skid marks for a car of weight w and velocity v skidding to a stop (k is a constant depending on the road surface)

Cobb–Douglas Production Functions

A function used to model the output of a company or a nation is called a *production function*, and the most famous is the Cobb–Douglas production function*

$$P(L, K) = aL^b K^{1-b} \qquad \text{For constants } a > 0 \text{ and } 0 < b < 1$$

This function expresses the total production P as a function of L, the number of units of labor, and K, the number of units of capital. (Labor

---

* First used by Charles Cobb and Paul Douglas in a landmark study of the American economy published in 1928.

is measured in work-hours, and capital means *invested* capital, including the cost of buildings, equipment, and raw materials.)

EXAMPLE **4**

EVALUATING A COBB–DOUGLAS PRODUCTION FUNCTION

Cobb and Douglas modeled the output of the American economy by the function $P(L, K) = L^{0.75}K^{0.25}$. Find $P(150, 220)$.

Solution

$$P(150, 220) - (150)^{0.75}(220)^{0.25}$$

$P(L, K) = L^{0.75}K^{0.25}$ with $L = 150$ and $K = 220$

$$\approx (42.9)(3.85) \approx 165$$

Using a calculator

That is, 150 units of labor and 220 units of capital should result in approximately 165 units of production.

Graphing Calculator Exploration

The windchill index (revised in 2001), announced by the weather bureau during the winter to measure the combined effect of wind and cold, is calculated from the formula below, where x is wind speed (miles per hour) and y is temperature (degrees Fahrenheit).

$$W(x, y) = 35.74 + 0.6215y - 35.75x^{0.16} + 0.4275yx^{0.16}$$

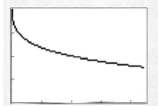

a. Enter this function into your graphing calculator but with y (temperature) replaced by 32 so that it becomes a function of one variable, x (wind speed). Then graph the function on the window [0, 45] by [0, 40]. The graph shows how the perceived temperature decreases as wind speed increases.

b. Find the wind speed that makes it feel like 18 degrees (that is, find x where $y = 18$).

c. Notice that the graph drops more steeply for low wind speeds than for high wind speeds. What does this mean about the effect of an extra 5 miles per hour of wind on a calm day as opposed to a windy day? (Exercises 36 and 37 continue this analysis.)

Functions of Three or More Variables

Functions of three (or more) variables are defined analogously. Some examples are:

$$V(l, w, h) = lwh$$

Volume of a rectangular solid of length l, width w, and height h

$$W(P, r, n) = Pe^{rn}$$

Worth of P dollars invested at a continuous interest rate r for n years

$$f(w, x, y, z) = \frac{w + x + y + z}{4}$$

Average of four numbers

EXAMPLE 5

FINDING THE DOMAIN OF A FUNCTION

For $f(x, y, z) = \dfrac{\sqrt{x}}{y} + \ln\dfrac{1}{z}$, find **a.** the domain **b.** $f(4, -1, 1)$

Solution

a. In $f(x, y, z) = \dfrac{\sqrt{x}}{y} + \ln\dfrac{1}{z}$ we must have $x \geq 0$ (because of the square root), $y \neq 0$ (since it is a denominator), and $z > 0$ (so that $1/z$ has a logarithm). Therefore, the domain is

$$\{(x, y, z) \mid x \geq 0, y \neq 0, z > 0\}$$

b. $f(4, -1, 1) = \dfrac{\sqrt{4}}{-1} + \ln\dfrac{1}{1} = \dfrac{2}{-1} + \underbrace{\ln 1}_{0} = -2$

EXAMPLE 6

FINDING THE VOLUME AND AREA OF A DIVIDED BOX

An open-top box is to have a center divider, as shown in the diagram. Find formulas for the volume V of the box and for the total amount of material M needed to construct the box.

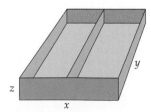

Solution The volume is length times width times height.

$$V = xyz$$

The box consists of a bottom, a front and back, two sides, and a divider, whose areas are shown in the diagram. Therefore, the total amount of material (the area) is

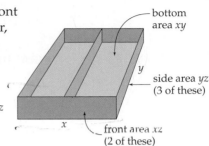

bottom area xy

side area yz (3 of these)

front area xz (2 of these)

$$M = \underset{\substack{\diagup \\ \text{Bottom}}}{xy} + \underset{\substack{\mid \\ \text{Back} \\ \text{and front}}}{2xz} + \underset{\substack{\setminus \\ \text{Sides and} \\ \text{divider}}}{3yz}$$

Practice Problem 2

Find a formula for the total amount of material M needed to construct an open-top box with three parallel dividers. Use the variables shown in the diagram.

➤ **Solution on page 504**

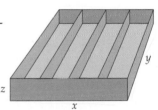

Graph of a Function of Two Variables

Graphing a function of two variables requires a three-dimensional coordinate system. We draw three perpendicular axes as shown on the right.* We will usually draw only the positive half of each axis, although each axis extends infinitely far in the negative direction as well. The plane at the base is called the *x-y plane*.

A point in a three-dimensional coordinate system is specified by three coordinates, giving its distances from the origin in the x, y, and z directions. For example, the point

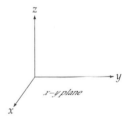

x-y plane

The three-dimensional ("right-handed") coordinate system

$$(2, 3, 4)$$

$\uparrow\ \uparrow\ \uparrow$

z-coordinate
y-coordinate
x-coordinate

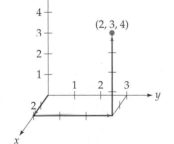

The point $(2, 3, 4)$

is plotted by starting at the origin, moving 2 units in the x direction, 3 units in the y direction, and then 4 units in the (vertical) z direction.

---

*This is called a "right-handed" coordinate system because the x, y, and z axes correspond to the first two fingers and thumb of the right hand.

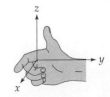

To graph a function of two variables we choose values for x and y, calculate z-values from $z = f(x, y)$, and plot the points (x, y, z).

EXAMPLE **7**

GRAPHING A FUNCTION OF TWO VARIABLES

To graph $f(x, y) = 18 - x^2 - y^2$, we set z equal to the function.

$$z = 18 - x^2 - y^2 \qquad\qquad z \text{ replaces } f(x, y)$$

Then we choose values for x and y. Choosing $x = 1$ and $y = 2$ gives

$$z = 18 - 1^2 - 2^2 = 13 \qquad \begin{array}{l} z = 18 - x^2 - y^2 \text{ with} \\ x = 1 \text{ and } y = 2 \end{array}$$

for the point

$$(1, 2, 13) \qquad \begin{array}{l} \text{The chosen } x = 1, \ y = 2, \\ \text{and the calculated } z \end{array}$$

Choosing $x = 2$ and $y = 3$ gives

$$z = 18 - 2^2 - 3^2 = 5 \qquad \begin{array}{l} z = 18 - x^2 - y^2 \text{ with} \\ x = 2 \text{ and } y = 3 \end{array}$$

for the point

$$(2, 3, 5) \qquad \begin{array}{l} \text{The chosen } x = 2, \ y = 3, \\ \text{and the calculated } z \end{array}$$

These points $(1, 2, 13)$ and $(2, 3, 5)$ are plotted on the graph on the left below. The completed graph of the function is shown on the right.

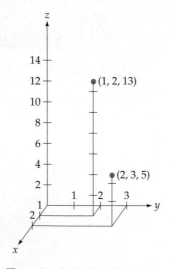

The points $(1, 2, 13)$ and $(2, 3, 5)$
of the function
$f(x, y) = 18 - x^2 - y^2$

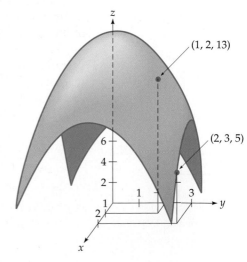

The graph of $f(x, y) = 18 - x^2 - y^2$

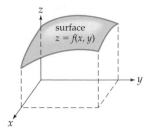

The graph of a function of two variables is a surface whose height above the point (x, y) in the $x\,y$ plane is $z = f(x, y)$.

In general, the graph of a function of *two* variables is a *surface* above or below the *x-y* plane, just as the graph of a function of *one* variable is a *curve* above or below the *x*-axis.

Graphing functions of two variables involves drawing three-dimensional graphs, which is very difficult. Graphing functions of *more* than two variables requires *more* than three dimensions, and is impossible. For this reason we will not graph functions of several variables. We will, however, often speak of a function of two variables as representing a *surface* in three-dimensional space.

Spreadsheet Exploration

The following spreadsheet* graph of $f(x, y) = 18 - x^2 - y^2$ from the previous example is a chart showing the values of the function that were calculated for values of x and y between -5 and 5.

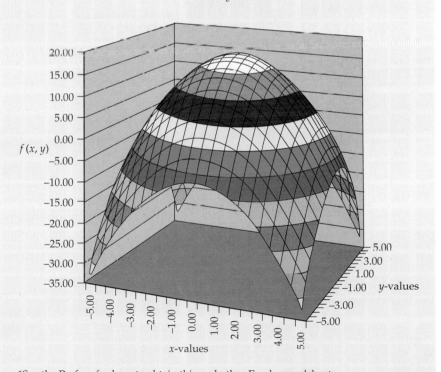

*See the Preface for how to obtain this and other Excel spreadsheets.

Just as with functions of one variable, useful graphs can be constructed only if we know the values of the variables corresponding to points of interest. As we have already seen, these values are easily found using calculus.

Relative Extreme Points and Saddle Points

Certain points on a surface of such a graph are of special importance.

Relative Maximum Point

A point (a, b, c) on a surface $z = f(x, y)$ is a *relative maximum point* if $f(a, b) \geq f(x, y)$ for all (x, y) in some region surrounding (a, b).

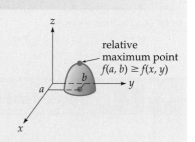

Relative Minimum Point

A point (a, b, c) on a surface $z = f(x, y)$ is a *relative minimum point* if $f(a, b) \leq f(x, y)$ for all (x, y) in some region surrounding (a, b).

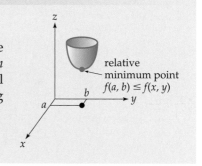

As before, the term *relative extreme point* means a point that is either a relative maximum or a relative minimum point. A surface can have any number of relative extreme points, even none.

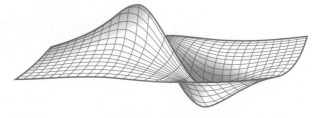

A surface with two relative extreme points: one relative maximum and one relative minimum.

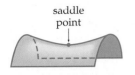

saddle
point

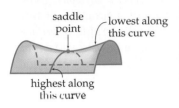

saddle
point

lowest along
this curve

highest along
this curve

The point shown on the left is called a *saddle point* (so named because the diagram resembles a saddle).

A saddle point is a point that is the highest point along one curve of the surface and the lowest point along another curve. A saddle point is *not* a relative extreme point.

If we think of a surface $z = f(x, y)$ as a landscape, then relative maximum and minimum points correspond to "hilltops" and "valley bottoms," and a saddle point corresponds to a "mountain pass" between two peaks.

Gallery of Surfaces

The following are the graphs of a few functions of two variables.

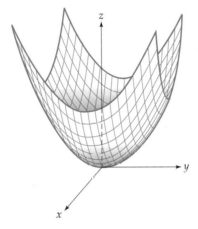

The surface $f(x, y) = x^2 + y^2$ has a relative minimum point at the origin.

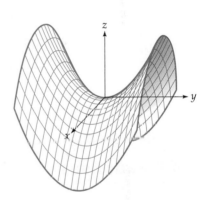

The surface $f(x, y) = y^2 - x^2$ has a saddle point at the origin.

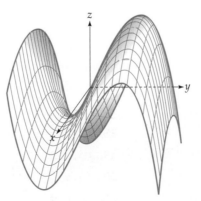

The surface $f(x, y) = 12y + 6x - x^2 - y^3$ has a saddle point and a relative maximum point.

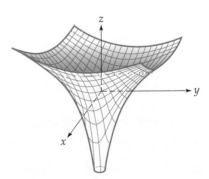

The surface $f(x, y) = \ln (x^2 + y^2)$ has no relative extreme points. It is undefined at $(0, 0)$.

Section Summary

Just as for a function of one variable, a *function of several variables* gives exactly one value for each point in its domain. For a function of *two* variables, denoted $f(x, y)$, the values $z = f(x, y)$ determine a *surface* above or below the *x-y* plane. The surface may have *relative maximum points* and *relative minimum points* ("hilltops" and "valley bottoms") or even *saddle points* (high points along one curve and low points along another), as shown on the previous pages.

➤ **Solutions to Practice Problems**

1. a. $\{(x, y) \mid x > 0, y \geq 0\}$

 b. $f(e, 4) = \dfrac{\ln e}{e^{\sqrt{4}}} = \dfrac{1}{e^2} = e^{-2}$

2. $M = xy + 2xz + 5yz$

Exercises

For each function, find the domain.

1. $f(x, y) = \dfrac{1}{xy}$

2. $f(x, y) = \dfrac{\sqrt{x}}{\sqrt{y}}$

3. $f(x, y) = \dfrac{1}{x - y}$

4. $f(x, y) = \dfrac{\sqrt[3]{x}}{\sqrt[3]{y}}$

5. $f(x, y) = \dfrac{\ln x}{y}$

6. $f(x, y) = \dfrac{x}{\ln y}$

7. $f(x, y, z) = \dfrac{e^{1/y} \ln z}{x}$

8. $f(x, y, z) = \dfrac{\sqrt{x} \ln y}{z}$

For each function, evaluate the given expression.

9. $f(x, y) = \sqrt{99 - x^2 - y^2}$, find $f(3, -9)$

10. $f(x, y) = \sqrt{75 - x^2 - y^2}$, find $f(5, -1)$

11. $g(x, y) = \ln(x^2 + y^4)$, find $g(0, e)$

12. $g(x, y) = \ln(x^3 - y^2)$, find $g(e, 0)$

13. $w(u, v) = \dfrac{1 + 2u + 3v}{uv}$, find $w(-1, 1)$

14. $w(u, v) = \dfrac{2u + 4u}{v - u}$, find $w(1, -1)$

15. $h(x, y) = e^{xy + y^2 - 2}$, find $h(1, -2)$

16. $h(x, y) = e^{x^2 - xy - 4}$, find $h(1, -2)$

17. $f(x, y) = xe^y - ye^x$, find $f(1, -1)$

18. $f(x, y) = xe^y + ye^x$, find $f(-1, 1)$

19. $f(x, y, z) = xe^y + ye^z + ze^x$, find $f(1, -1, 1)$

20. $f(x, y, z) = xe^y + ye^z + ze^x$, find $f(-1, 1, -1)$

21. $f(x, y, z) = z \ln \sqrt{xy}$, find $f(-1, -1, 5)$

22. $f(x, y, z) = z\sqrt{x} \ln y$, find $f(4, e, -1)$

APPLIED EXERCISES

23. BUSINESS: Stock Yield The *yield* of a stock is defined as $Y(d, p) = \dfrac{d}{p}$ where d is the dividend per share and p is the price of a share of stock. Find the yield of a stock that sells for $140 and offers a dividend of $2.20.

24. BUSINESS: Price-Earnings Ratio The price-earnings ratio of a stock is defined as $R(P, E) = \dfrac{P}{E}$ where P is the price of a share of stock and E is its earnings. Find the price-earnings ratio of a stock that is selling for $140 with earnings of $1.70.

25. GENERAL: Scuba Diving The maximum duration of a scuba dive (in minutes) can be estimated from the formula

$$T(v, d) = \frac{33v}{d + 33}$$

where v is the volume of air (at sea-level pressure) in the tank and d is the depth of the dive. Find $T(90, 33)$.

26. SOCIAL SCIENCE: Cephalic Index Anthropologists define the *cephalic index* to distinguish the head shapes of different people. For a head of width W and length L (measured from above), the cephalic index is

$$C(W, L) = 100\frac{W}{L}$$

Calculate the cephalic index for a head of width 8 inches and length 10 inches.

27. ECONOMICS: Cobb–Douglas Functions A company's production is estimated to be $P(L, K) = 2L^{0.6}K^{0.4}$. Find $P(320, 150)$.

28. BIOMEDICAL: Body Area The surface area (in square feet) of a person of weight w pounds and height h feet is approximated by the function $A(w, h) = 0.55 \, w^{0.425}h^{0.725}$. Use this function to estimate the surface area of a person who weighs 160 pounds and who is 6 feet tall. (Such estimates are important in certain medical procedures.)

29. ECONOMICS: Cobb–Douglas Functions Show that the Cobb–Douglas production function $P(L, K) = aL^{b}K^{1-b}$ satisfies the equation $P(2L, 2K) = 2 \cdot P(L, K)$. This shows that doubling the amounts of labor and capital doubles production, a property called *returns to scale.*

30. ECONOMICS: Cobb–Douglas Functions Show that the Cobb–Douglas function $P(L, K) = aL^{b}K^{1-b}$ with $0 < b < 1$ satisfies

$$P(2L, K) < 2P(L, K) \quad \text{and} \quad P(L, 2K) < 2P(L, K)$$

This shows that doubling the amounts of either labor or capital alone results in *less* than double production, a property called *diminishing returns.*

31. GENERAL: Telephone Calls For two cities with populations x and y that are d miles apart, the number of telephone calls per hour between them can be estimated by the function of three variables

$$f(x, y, d) = \frac{3xy}{d^{2.4}}$$

(This is called the *gravity model.*) Use the gravity model to estimate the number of calls between two cities of populations 40,000 and 60,000 that are 600 miles apart.

32. ENVIRONMENTAL SCIENCE: Tag and Recapture Estimates Ecologists estimate the size of animal populations by capturing and tagging a few animals, and then releasing them. After the first group has mixed with the population, a second group of animals is captured, and the number of tagged animals in this group is counted. If originally T animals were tagged, and the second group is of size S and contains t tagged animals, then the population is estimated by the function of three variables

$$P(T, S, t) = \frac{TS}{t}$$

Estimate the size of a deer population if 100 deer were tagged, and then a second group of 250 contained 20 tagged deer.

33. BUSINESS: Cost Function It costs an appliance company $210 to manufacture each washer and $180 to manufacture each dryer, and fixed costs are $4000. Find the company's cost function $C(x, y)$, using x and y for the numbers of washers and dryers, respectively.

34–35. GENERAL: Box Design For each open-top box shown below, find formulas for:

a. the volume.
b. the total amount of material (the area).

34.

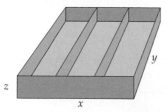

35.

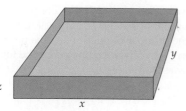

36. GENERAL: Windchill Enter the formula for windchill from the Graphing Calculator Exploration on page 497, but with y (temperature) replaced by 32 so that it becomes a function of one variable, x (wind speed).

a. Graph the function on the window $[0, 45]$ by $[0, 40]$.
b. Use NDERIV or dy/dx to find the slope of this curve at $x = 5$. Interpret the answer.

(continues)

c. Repeat part (b) but for $x = 20$, interpreting the answer.
d. What do the answers indicate about the windchill effect of additional wind on a calm day as opposed to on an already windy day?

37. GENERAL: Windchill Enter the formula for windchill index from the Graphing Calculator Exploration on page 497, but with y (temperature) replaced by *several* temperatures: 20, 30, 40, and 50 degrees (using y_1, y_2, y_3, and y_4).

a. Graph the functions on the window $[0, 45]$ by $[-10, 60]$.
b. Notice that the lower temperature curves slope downward more steeply than the others. What does this mean about the effect of wind on a colder day?
c. Use NDERIV or dy/dx to find the slope of the lowest and the highest curves at $x = 10$. Interpret the answers.
d. Do your answers to part (c) support your conclusion in part (b)?

38. BIOMEDICAL: Oxygen Consumption The oxygen consumption of a well-insulated non-sweating mammal can be estimated from the formula $f(t_b, t_a, w) = 2.5(t_b - t_a)w^{-0.67}$, where t_b is the animal's body temperature, t_a is the air temperature (both in degrees Celsius), and w is the animal's weight (in kilograms).* Find the oxygen consumption of a 40-kilogram animal whose body temperature is 35 degrees when the air temperature is 5 degrees.

*See Duane J. Clow and N. Scott Urquhart, *Mathematics in Biology*, W.W. Norton, 1974, p. 412.

7.2 PARTIAL DERIVATIVES

Introduction

Functions of several variables have several derivatives, one for each variable. In this section you will learn how to calculate and interpret these derivatives.

Partial Derivatives

Before defining partial derivatives, we review the rules governing derivatives and constants.

$$\frac{d}{dx} c = 0$$

For a constant *standing alone,* the derivative is zero.

$$\frac{d}{dx}(cx^3) = c \cdot 3x^2$$

For a constant *multiplying* a function, the constant is carried along.

Carry along the constant $\quad$ Derivative of x^3

These rules will be very useful in this section.

A function $f(x, y)$ has two derivatives, called *partial derivatives,* one with respect to x and the other with respect to y.

Partial Derivatives

$$\frac{\partial}{\partial x} f(x, y) = \lim_{h \to 0} \frac{f(x + h, y) - f(x, y)}{h}$$

Partial derivative of f with respect to x (x is changed by h, y is held constant)

$$\frac{\partial}{\partial y} f(x, y) = \lim_{h \to 0} \frac{f(x, y + h) - f(x, y)}{h}$$

Partial derivative of f with respect to y (y is changed by h, x is held constant)

(provided that the limits exist).

Partial derivatives are written with a "curly" ∂, $\partial/\partial x$ instead of d/dx, and are often called "partials." (The Greek letter ∂ is a lowercase delta, and we have already used the capital delta, Δ, to denote change.)

Rewriting the first formula, but without the y's, gives

$$\lim_{h \to 0} \frac{f(x + h) - f(x)}{h}$$

$$\lim_{h \to 0} \frac{f(x + h, y) - f(x, y)}{h}$$
but omitting the y's

which is just the "ordinary derivative" from page 102. Therefore, the partial derivative with respect to x is just the ordinary derivative with respect to x with y held constant. Similarly, the partial with respect to y is just the ordinary derivative with respect to y, but now with x held constant.

$$\frac{\partial}{\partial x} f(x, y) = \left(\begin{array}{l} \text{Derivative of } f \text{ with respect} \\ \text{to } x \text{ with } y \text{ held constant} \end{array} \right)$$

$$\frac{\partial}{\partial y} f(x, y) = \left(\begin{array}{l} \text{Derivative of } f \text{ with respect} \\ \text{to } y \text{ with } x \text{ held constant} \end{array} \right)$$

EXAMPLE 1

FINDING A PARTIAL DERIVATIVE WITH RESPECT TO x

Find $\dfrac{\partial}{\partial x} x^3 y^4$.

Solution The $\partial/\partial x$ means differentiate with respect to x, holding y (and therefore y^4) constant. We therefore differentiate the x^3 and carry along the "constant" y^4:

Derivative of x^3

$$\frac{\partial}{\partial x} x^3 y^4 = 3x^2 y^4$$

Carry along the "constant" y^4

$\partial/\partial x$ means differentiate with respect to x, treating y (and therefore y^4) as a constant

EXAMPLE 2

FINDING A PARTIAL WITH RESPECT TO y

Find $\dfrac{\partial}{\partial y} x^3 y^4$.

Solution

$$\frac{\partial}{\partial y} x^3 y^4 = x^3 4y^3 \qquad = \qquad 4x^3 y^3$$

Carry along the "constant" x^3 — Derivative of y^4 — Writing the 4 first

$\partial/\partial y$ means differentiate with respect to y, treating x (and therefore x^3) as a constant

Practice Problem 1 Find **a.** $\dfrac{\partial}{\partial x} x^4 y^2$ **b.** $\dfrac{\partial}{\partial y} x^4 y^2$

➤ Solutions on page 518

EXAMPLE 3

FINDING A PARTIAL DERIVATIVE

Find $\dfrac{\partial}{\partial x} y^4$.

Solution

$$\dfrac{\partial}{\partial x} y^4 = 0$$

The derivative of a constant is zero
(since $\partial / \partial x$ means hold y constant)

Partial with respect to x — Function of y alone

Practice Problem 2   Find $\dfrac{\partial}{\partial y} x^2$.

➤ Solution on page 519

EXAMPLE 4

FINDING A PARTIAL OF A POLYNOMIAL IN TWO VARIABLES

Find $\dfrac{\partial}{\partial x} (2x^4 - 3x^3y^3 - y^2 + 4x + 1)$.

Solution

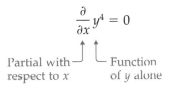

$$\dfrac{\partial}{\partial x}(2x^4 - 3x^3y^3 - y^2 + 4x + 1) = 8x^3 - 9x^2y^3 + 4$$

Differentiating
with respect to
x, so each y is
held constant

$$\dfrac{\partial}{\partial x} y^2 = 0$$

Practice Problem 3   Find $\dfrac{\partial}{\partial y} (2x^4 - 3x^3y^3 - y^2 + 4x + 1)$.

➤ Solution on page 519

Subscript Notation for Partial Derivatives

Partial derivatives are often denoted by subscripts: a subscript x means the partial with respect to x, and a subscript y means the partial with respect to y.*

---

* Sometimes subscripts 1 and 2 are used to indicate partial derivatives with respect to the first and second variables: $f_1(x, y)$ means $f_x(x, y)$ and $f_2(x, y)$ means $f_y(x, y)$. We will not use this notation in this book.

$$f_x(x, y) = \frac{\partial}{\partial x} f(x, y)$$ f_x means the partial of f with respect to x

$$f_y(x, y) = \frac{\partial}{\partial y} f(x, y)$$ f_y means the partial of f with respect to y

---

EXAMPLE 5

USING SUBSCRIPT NOTATION

Find $f_x(x, y)$ if $f(x, y) = 5x^4 - 2x^2y^3 - 4y^2$.

Solution

$$f_x(x, y) = 20x^3 - 4xy^3$$ Differentiating with respect to x, holding y constant

---

EXAMPLE 6

FINDING A PARTIAL INVOLVING LOGS AND EXPONENTIALS

Find both partials of $f = e^x \ln y$.

Solution

$$f_x = e^x \ln y$$ The derivative of e^x is e^x (times the "constant" $\ln y$)

$$f_y = e^x \frac{1}{y}$$ The derivative of $\ln y$ is $\frac{1}{y}$ (times the "constant" e^x)

EXAMPLE 7

FINDING A PARTIAL OF A FUNCTION TO A POWER

Find f_y if $f = (xy^2 + 1)^4$.

Solution

Using the Generalized Power Rule (the derivative of f^n is $nf^{n-1}f'$, but with f' meaning a *partial*)

$$f_y = 4(xy^2 + 1)^3(x2y)$$

Partial of the inside with respect to y

$$= 8xy(xy^2 + 1)^3$$ Simplifying

EXAMPLE 8

FINDING A PARTIAL OF A QUOTIENT

Find $\dfrac{\partial g}{\partial x}$ if $g = \dfrac{xy}{x^2 + y^2}$.

Solution

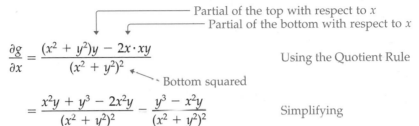

$$\frac{\partial g}{\partial x} = \frac{(x^2 + y^2)y - 2x \cdot xy}{(x^2 + y^2)^2} \qquad \text{Using the Quotient Rule}$$

Bottom squared

$$= \frac{x^2y + y^3 - 2x^2y}{(x^2 + y^2)^2} = \frac{y^3 - x^2y}{(x^2 + y^2)^2} \qquad \text{Simplifying}$$

EXAMPLE 9

FINDING A PARTIAL OF THE LOGARITHM OF A FUNCTION

Find $f_x(x, y)$ if $f(x, y) = \ln(x^2 + y^2)$.

Solution

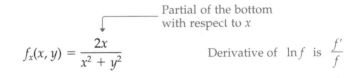

$$f_x(x, y) = \frac{2x}{x^2 + y^2} \qquad \text{Derivative of } \ln f \text{ is } \frac{f'}{f}$$

An expression such as $f_x(2, 5)$, which involves both differentiation and evaluation, means *first differentiate and then evaluate.**

EXAMPLE 10

EVALUATING A PARTIAL DERIVATIVE

Find $f_y(1, 3)$ if $f(x, y) = e^{x^2+y^2}$.

---

* $f_x(2, 5)$ may be written $\dfrac{\partial f}{\partial x}(2, 5)$ or $\dfrac{\partial f}{\partial x}\Big|_{(2, 5)}$, again meaning first differentiate, then evaluate.

Solution

$$f_y(x, y) = e^{x^2+y^2}(2y)$$

Derivative of e^f is $e^f \cdot f'$

Partial of the exponent with respect to y

$$f_y(1, 3) = e^{1^2+3^2}(2 \cdot 3)$$

Evaluating at $x = 1$ and $y = 3$

$$= 6e^{10}$$

Simplifying

Practice Problem 4   Find $f_y(1, 2)$ if $f(x, y) = e^{x^3+y^3}$.   ➤ Solution on page 519

Partial Derivatives in Three or More Variables

Partial derivatives in three or more variables are defined similarly. That is, the partial derivative of $f(x, y, z)$ with respect to any one variable is the "ordinary" derivative with respect to that variable, holding all other variables constant.

EXAMPLE 11   **FINDING A PARTIAL OF A FUNCTION OF THREE VARIABLES**

$$\frac{\partial}{\partial x}(x^3y^4z^5) = 3x^2y^4z^5$$

$\partial/\partial x$ means differentiate with respect to x, holding y and z constant

Hold constant   Derivative of x^3

Practice Problem 5   Find $\dfrac{\partial}{\partial y}(x^3y^4z^5)$.   ➤ Solution on page 519

EXAMPLE 12   **EVALUATING A PARTIAL IN THREE VARIABLES**

Find $f_z(1, 1, 1)$ if $f(x, y, z) = e^{x^2+y^2+z^2}$.

Solution

$$f_z(x, y, z) = e^{x^2+y^2+z^2}(2z)$$

Partial with respect to z

$$= 2ze^{x^2+y^2+z^2}$$

Writing the $2z$ first

$$f_z(1, 1, 1) = 2 \cdot 1 \cdot e^{1^2+1^2+1^2} = 2e^3$$

Evaluating

Interpreting Partial Derivatives as Rates of Change

Since partials are just "ordinary" derivatives with the other variable(s) held constant, they give *instantaneous rates of change* with respect to one variable at a time.

Partials as Rates of Change

$$f_x(x, y) = \begin{pmatrix} \text{Instantaneous rate of change of } f \text{ with} \\ \text{respect to } x \text{ when } y \text{ is held constant} \end{pmatrix}$$

$$f_y(x, y) = \begin{pmatrix} \text{Instantaneous rate of change of } f \text{ with} \\ \text{respect to } y \text{ when } x \text{ is held constant} \end{pmatrix}$$

This is why they are called *partial* derivatives: not all the variables are changed at once, only a "partial" change is made.

Cobb–Douglas Production Functions

Recall that a Cobb–Douglas production function $P(L, K) = aL^bK^{1-b}$ expresses production P as a function of L (units of labor) and K (units of capital). Each partial therefore gives the rate of increase of production with respect to one of these variables while the other is held constant.

INTERPRETING THE PARTIALS OF A COBB–DOUGLAS PRODUCTION FUNCTION

Find and interpret $P_L(120, 200)$ and $P_K(120, 200)$ for the Cobb–Douglas function $P(L, K) = 20L^{0.6}K^{0.4}$.

Solution

$$P_L = 12L^{-0.4}K^{0.4}$$

Partial with respect to L (the 12 is 20 times 0.6)

$$P_L(120, 200) = 12(120)^{-0.4}(200)^{0.4} \approx 14.7$$

Substituting $L = 120$ and $K = 200$, and evaluating using a calculator

Interpretation: $P_L = 14.7$ means that production increases by about 14.7 units for each additional unit of labor (when $L = 120$ and $K = 200$). This is called the *marginal productivity of labor.*

$$P_K = 8L^{0.6}K^{-0.6}$$ Partial with respect to K (the 8 is 20 times 0.4)

$$P_K(120, 200) = 8(120)^{0.6}(200)^{-0.6} \approx 5.9$$ Substituting $L = 120$ and $K = 200$, and evaluating using a calculator

Interpretation: $P_K = 5.9$ means that production increases by about 5.9 units for each additional unit of capital (when $L = 120$ and $K = 200$). This is called the *marginal productivity of capital.*

These numbers show that to increase production, additional units of labor are more than twice as effective as additional units of capital (at the levels $L = 120$ and $K = 200$).

Partial derivatives give the marginals for one product at a time.

Interpreting Partials as Marginals

Let $C(x, y)$ be the (total) cost function for x units of product A and y units of product B. Then

$$C_x(x, y) = \left(\begin{matrix} \text{Marginal cost function for product A when} \\ \text{production of product B is held constant} \end{matrix} \right)$$

$$C_y(x, y) = \left(\begin{matrix} \text{Marginal cost function for product B when} \\ \text{production of product A is held constant} \end{matrix} \right)$$

Similar statements hold, of course, for revenue and profit functions: the partials give the marginals for one variable at a time when the other variables are held constant.

EXAMPLE 14 **INTERPRETING PARTIALS OF A PROFIT FUNCTION**

A company's profit from producing x radios and y televisions per day is $P(x, y) = 4x^{3/2} + 6y^{3/2} + xy$. Find the marginal profit functions. Then find and interpret $P_y(25, 36)$.

Solution

$$P_x(x, y) = 6x^{1/2} + y$$

Marginal profit for radios when television production is held constant

$$P_y(x, y) = 9y^{1/2} + x$$

Marginal profit for televisions when radio production is held constant

$$P_y(25, 36) = 9\underbrace{(36)^{1/2}}_{6} + 25 = 79$$

Evaluating P_y at $x = 25$ and $y = 36$

Interpretation: Profit increases by about $79 per additional television (when producing 25 radios and 36 televisions per day).

Interpreting Partials Geometrically

A function $f(x, y)$ represents a surface in three-dimensional space, and the partial derivatives are the slopes along the surface in different directions: $\partial f / \partial x$ gives the slope of the surface "in the x-direction," and $\partial f / \partial y$ gives the slope of the surface "in the y-direction" at the point (x, y).

To put this colloquially, if you were on the surface $z = f(x, y)$, then $\partial f / \partial x$ would be the steepness of the surface *in the x-direction*, and $\partial f / \partial y$ would be the steepness of the surface *in the y-direction* from the point (x, y).

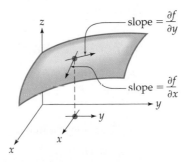

Partial derivatives are slopes.

For example, the following graph shows gridlines in the x direction (roughly up and down the page, with the positive x direction being down) and in the y direction (roughly across the page, with the positive y direction being to the right).

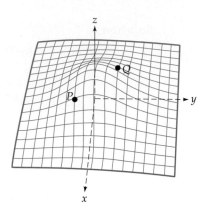

Therefore, at the point P on the graph the partials would have the following signs:

$$\frac{\partial f}{\partial x} < 0$$

Walking from P in the positive x direction would mean walking *downhill*

$$\frac{\partial f}{\partial y} > 0$$

Walking from P in the positive y direction would mean walking *uphill*

Practice Problem **6**   On the preceding graph at the point Q,

a. Is $\dfrac{\partial f}{\partial x}$ positive or negative?

b. Is $\dfrac{\partial f}{\partial y}$ positive or negative?

[*Hint:* Leaving Q in those directions, would you be walking uphill or downhill?]   ➤ Solutions on page 519

Higher-Order Partial Derivatives

We can differentiate a function more than once to obtain *higher-order* partials.

Second-Order Partials		
Subscript Notation	*∂ Notation*	*In Words*
f_{xx}	$\dfrac{\partial^2}{\partial x^2} f$	Differentiate twice with respect to x
f_{yy}	$\dfrac{\partial^2}{\partial y^2} f$	Differentiate twice with respect to y

Subscript Notation	∂ Notation	In Words
f_{xy}	$\dfrac{\partial^2}{\partial y \, \partial x} f$	Differentiate first with respect to x, then with respect to y
f_{yx}	$\dfrac{\partial^2}{\partial x \, \partial y} f$	Differentiate first with respect to y, then with respect to x

Each notation means differentiate first with respect to the variable *closest* to f.

Calculating a "second partial" such as f_{xy} is a two-step process: first calculate f_x, and then differentiate the *result* with respect to y.

EXAMPLE 15

FINDING SECOND-ORDER PARTIALS

Find all four second-order partials of $f(x, y) = x^4 + 2x^2y^2 + x^3y + y^4$.

Solution   First we calculate

$$f_x = 4x^3 + 4xy^2 + 3x^2y \qquad \text{Partial of } f \text{ with respect to } x$$

Then from this we find f_{xx} and f_{xy}:

$$f_{xx} = 12x^2 + 4y^2 + 6xy \qquad \begin{array}{l}\text{Differentiating } f_x = 4x^3 + 4xy^2 + 3x^2y \\ \text{with respect to } x\end{array}$$

$$f_{xy} = 8xy + 3x^2 \qquad \begin{array}{l}\text{Differentiating } f_x = 4x^3 + 4xy^2 + 3x^2y \\ \text{with respect to } y\end{array}$$

Now, returning to the original function $f = x^4 + 2x^2y^2 + x^3y + y^4$, we calculate

$$f_y = 4x^2y + x^3 + 4y^3 \qquad \text{Partial of } f \text{ with respect to } y$$

Then, from this,

$$f_{yy} = 4x^2 + 12y^2 \qquad \begin{array}{l}\text{Differentiating } f_y = 4x^2y + x^3 + 4y^3 \\ \text{with respect to } y\end{array}$$

$$f_{yx} = 8xy + 3x^2 \qquad \begin{array}{l}\text{Differentiating } f_y = 4x^2y + x^3 + 4y^3 \\ \text{with respect to } x\end{array}$$

Notice that these "mixed partials" are equal:

$$\left.\begin{array}{l}f_{xy} = 8xy + 3x^2 \\ f_{yx} = 8xy + 3x^2\end{array}\right] \text{Equal}$$

That is, $f_{xy} = f_{yx}$, so reversing the order of differentiation (first x, then y, or first y, then x) makes no difference. This is not true for all functions, but it is true for all the functions that we will encounter in this book, and it is also true for all functions that you are likely to encounter in applications.*

Practice Problem 7   For the function $f(x, y) = x^3 - 3x^2y^4 + y^3$,   find

a. f_x **b.** f_{xy} **c.** f_y **d.** f_{yx} ➤ Solutions on page 519

7.2 Section Summary

For a function $f(x, y)$ the partial derivatives are:

$$\frac{\partial}{\partial x} f(x, y) = f_x(x, y) = \lim_{h \to 0} \frac{f(x + h, y) - f(x, y)}{h} \quad \text{Partial derivative of } f \text{ with respect to } x$$

$$\frac{\partial}{\partial y} f(x, y) = f_y(x, y) = \lim_{h \to 0} \frac{f(x, y + h) - f(x, y)}{h} \quad \text{Partial derivative of } f \text{ with respect to } y$$

provided that the limits exist. The partial derivative with respect to either variable is found by applying the usual differentiation rules to that variable and treating the other variable as a constant. Partial derivatives (or "partials") can be interpreted as instantaneous rates of change:

$$f_x(x, y) = \begin{pmatrix} \text{Instantaneous rate of change of } f \\ \text{with respect to } x \text{ when } y \text{ is held constant} \end{pmatrix}$$

$$f_y(x, y) = \begin{pmatrix} \text{Instantaneous rate of change of } f \\ \text{with respect to } y \text{ when } x \text{ is held constant} \end{pmatrix}$$

Partials can also be interpreted as marginals for one product while production of the other is held constant (see page 514). Geometrically, a function $z = f(x, y)$ of two variables represents a *surface* in three-dimensional space, and the partial $\partial f / \partial x$ gives the *slope* or *steepness* along the surface *in the x direction*, and the partial $\partial f / \partial y$ gives the slope or steepness *in the y-direction*, as shown in the diagram on page 515.

➤ **Solutions to Practice Problems**

1. a. $\dfrac{\partial}{\partial x} x^4 y^2 = 4x^3 y^2$ **b.** $\dfrac{\partial}{\partial y} x^4 y^2 = x^4 2y = 2x^4 y$

---

* $f_{xy} = f_{yx}$ if these partials are continuous. A more detailed statement can be found in an advanced calculus book.

2. $\frac{\partial}{\partial y} x^2 = 0$

3. $\frac{\partial}{\partial y}(2x^4 - 3x^3y^3 - y^2 + 4x + 1)$

$= 0 - 3x^3 3y^2 - 2y + 0 = -9x^3y^2 - 2y$

4. $f_y(x, y) = e^{x^3+y^3}(3y^2) = 3y^2 e^{x^3+y^3}$

$f_y(1, 2) = (3 \cdot 4)e^{1^3+2^3} = 12e^9$

5. $\frac{\partial}{\partial y}(x^3y^4z^5) = x^3 4y^3 z^5 = 4x^3y^3z^5$

6. a. Positive **b.** Negative

7. a. $f_x = 3x^2 - 6xy^4$

b. $f_{xy} = -6x4y^3 = -24xy^3$

c. $f_y = -3x^2 4y^3 + 3y^2 = -12x^2y^3 + 3y^2$

d. $f_{yx} = -24xy^3$

7.2 Exercises

For each function, find the partials **a.** $f_x(x, y)$ and **b.** $f_y(x, y)$.

1. $f(x, y) = x^3 + 3x^2y^2 - 2y^3 - x + y$

2. $f(x, y) = 2x^4 - 7x^3y^2 - xy + 1$

3. $f(x, y) = 12x^{1/2}y^{1/3} + 8$

4. $f(x, y) = x^{-1}y + xy^{-2}$

5. $f(x, y) = 100x^{0.05}y^{0.02}$ **6.** $f(x, y) = \dfrac{x}{y}$

7. $f(x, y) = (x + y)^{-1}$

8. $f(x, y) = (x^2 + xy + 1)^4$

9. $f(x, y) = \ln(x^3 + y^3)$ **10.** $f(x, y) = x^2e^y$

11. $f(x, y) = 2x^3e^{-5y}$ **12.** $f(x, y) = e^{x+y}$

13. $f(x, y) = e^{xy}$ **14.** $f(x, y) = \ln(xy^3)$

15. $f(x, y) = \ln\sqrt{x^2 + y^2}$ **16.** $f(x, y) = \dfrac{xy}{x + y}$

For each function, find **a.** $\dfrac{\partial w}{\partial u}$ and **b.** $\dfrac{\partial w}{\partial v}$.

17. $w = (uv - 1)^3$ **18.** $w = (u - v)^3$

19. $w = e^{(u^2-v^2)/2}$ **20.** $w = \ln(u^2 + v^2)$

For each function, evaluate the stated partials.

21. $f(x, y) = 4x^3 - 3x^2y^2 - 2y^2$, find $f_x(-1, 1)$ and $f_y(-1, 1)$

22. $f(x, y) = 2x^4 - 5x^2y^3 - 4y$, find $f_x(1, -1)$ and $f_y(1, -1)$

23. $f(x, y) = e^{x^2+y^2}$, find $f_x(0, 1)$ and $f_y(0, 1)$

24. $g(x, y) = (xy - 1)^5$, find $g_x(1, 0)$ and $g_y(1, 0)$

25. $h(x, y) = x^2y - \ln(x + y)$, find $h_x(1, 1)$

26. $f(x, y) = \sqrt{x^2 + y^2}$, find $f_y(8, -6)$

For each function, find the second-order partials **a.** f_{xx}, **b.** f_{xy}, **c.** f_{yx}, and **d.** f_{yy}.

27. $f(x, y) = 5x^3 - 2x^2y^3 + 3y^4$

28. $f(x, y) = 4x^2 - 3x^3y^2 + 5y^5$

29. $f(x, y) = 9x^{1/3}y^{2/3} - 4xy^3$

30. $f(x, y) = 32x^{1/4}y^{3/4} - 5x^3y$

31. $f(x, y) = ye^x - x \ln y$

32. $f(x, y) = y \ln x + xe^y$

For each function, calculate the third-order partials
a. f_{xxy}, **b.** f_{xyx}, and **c.** f_{yxx}.

33. $f(x, y) = x^4y^3 - e^{2x}$ **34.** $f(x, y) = x^3y^4 - e^{2y}$

For each function of three variables, find the partials **a.** f_x, **b.** f_y, and **c.** f_z.

35. $f = xy^2z^3$ **36.** $f = x^2y^3z^4$

37. $f = (x^2 + y^2 + z^2)^4$ **38.** $f = (xyz + 1)^3$

39. $f = e^{x^2+y^2+z^2}$ **40.** $f = \ln(x^2 - y^3 + z^4)$

For each function, evaluate the stated partial.

41. $f = 3x^2y - 2xz^2$, find $f_x(2, -1, 1)$

42. $f = 2yz^3 - 3x^2z$, find $f_z(2, -1, 1)$

43. $f = e^{x^2+2y^2+3z^2}$, find $f_y(-1, 1, -1)$

44. $f = e^{2x^3+3y^3+4z^3}$, find $f_y(1, -1, 1)$

APPLIED EXERCISES

45–46. BUSINESS: Marginal Profit An electronics company's profit $P(x, y)$ from making x tape decks and y CD players per day is given below.

a. Find the marginal profit function for tape decks.
b. Evaluate your answer to part (a) at $x = 200$ and $y = 300$ and interpret the result.
c. Find the marginal profit function for CD players.
d. Evaluate your answer to part (c) at $x = 200$ and $y = 100$ and interpret the result.

45. $P(x, y) = 2x^2 - 3xy + 3y^2 + 150x + 75y + 200$

46. $P(x, y) = 3x^2 - 4xy + 4y^2 + 80x + 100y + 200$

47–48. BUSINESS: Cobb–Douglas Production Functions A company's production is given by the Cobb–Douglas function $P(L, K)$ below, where L is the number of units of labor and K is the number of units of capital.

a. Find $P_L(27, 125)$ and interpret this number.
b. Find $P_K(27, 125)$ and interpret this number.
c. From your answers to parts (a) and (b), which will increase production more: an additional unit of labor or an additional unit of capital?

47. $P(L, K) = 270L^{1/3}K^{2/3}$

48. $P(L, K) = 225L^{2/3}K^{1/3}$

49. BUSINESS: Sales A store's TV sales depend on x, the price of the televisions, and y, the amount spent on advertising, according to the function $S(x, y) = 200 - 0.1x + 0.2y^2$. Find and interpret the marginals S_x and S_y.

50. ECONOMICS: Value of an MBA A 1973 study found that a businessperson with a master's degree in business administration (MBA) earned an average salary of $S(x, y) = 10,990 + 1120x + 873y$ dollars, where x is the number of years of work experience before the MBA, and y is the number of years of work experience after the MBA. Find and interpret the marginals S_x and S_y.

51. SOCIOLOGY: Status A study found that a person's status in a community depends on the person's income and education according to the function $S(x, y) = 7x^{1/3}y^{1/2}$, where x is income (in thousands of dollars) and y is years of education beyond high school.

a. Find $S_x(27, 4)$ and interpret this number.
b. Find $S_y(27, 4)$ and interpret this number.

52. BIOMEDICAL: Blood Flow The resistance of blood flowing through an artery of radius r and length L (both in centimeters) is $R(r, L) = 0.08Lr^{-4}$.

a. Find $R_r(0.5, 4)$ and interpret this number.
b. Find $R_L(0.5, 4)$ and interpret this number.

53. GENERAL: Highway Safety The length in feet of the skid marks from a truck of weight w (tons) traveling at velocity v (miles per hour) skidding to a stop on a dry road is $S(w, v) = 0.027wv^2$.

a. Find $S_w(4, 60)$ and interpret this number.
b. Find $S_v(4, 60)$ and interpret this number.

54. GENERAL: Windchill Temperature The windchill temperature (revised in 2001) announced by the weather bureau during the cold weather measures how cold it "feels" for a given temperature and wind speed. The formula is $C(t, w) = 35.74 + 0.6215t - 35.75w^{0.16} + 0.4275tw^{0.16}$ where t is the temperature (in degrees Fahrenheit) and w is the wind speed (in miles per hour). Find and interpret $C_w(30, 20)$.

COMPETITIVE AND COMPLEMENTARY COMMODITIES

55. ECONOMICS: Competitive Commodities
Certain commodities (such as butter and margarine) are called "competitive" or "substitute" commodities because one can substitute for the other. If $B(b, m)$ gives the daily sales of butter as a function of b, the price of butter, and m, the price of margarine:

a. Give an interpretation of $B_b(b, m)$.
b. Would you expect $B_b(b, m)$ to be positive or negative? Explain.
c. Give an interpretation of $B_m(b, m)$.
d. Would you expect $B_m(b, m)$ to be positive or negative? Explain.

56. ECONOMICS: Complementary Commodities Certain commodities (such as washing machines and clothes dryers) are called "complementary" commodities because they are often used together. If $D(d, w)$ gives the monthly sales of dryers as a function of d, the price of dryers, and w, the price of washers:

a. Give an interpretation of $D_d(d, w)$.
b. Would you expect $D_d(d, w)$ to be positive or negative? Explain.
c. Give an interpretation of $D_w(d, w)$.
d. Would you expect $D_w(d, w)$ to be positive or negative? Explain.

7.3   OPTIMIZING FUNCTIONS OF SEVERAL VARIABLES

Introduction

The graph of a function of two variables is a *surface*, with relative maximum and minimum points ("hilltops" and "valley bottoms") and saddle points ("mountain passes"). In this section we will see how to maximize and minimize such functions by finding critical points and using a two-variable version of the second-derivative test.

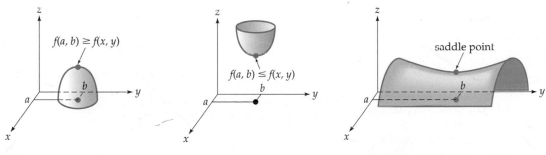

f has a relative *maximum* value at (a, b).

f has a relative *minimum* value at (a, b).

f has a saddle point at (a, b) (neither a maximum nor a minimum).

For simplicity, we will consider only functions whose first and second partials are defined everywhere. Such optimization techniques have many applications, such as maximizing profit for a company that makes several products.

Critical Points

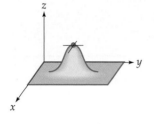

A critical point
(both partials are zero)

At the very top of a smooth hill, the slope or steepness in any direction must be zero. That is, a straight stick would balance horizontally at the top. Since the partials f_x and f_y are the slopes in the x and y directions, these partials must both be zero at a relative maximum point, and similarly for a relative minimum point or a saddle point. A point (a, b) at which both partials are zero is called a *critical point* of the function.*

Critical Point

(a, b) is a critical point of $f(x, y)$ if

$$f_x(a, b) = 0 \quad \text{and} \quad f_y(a, b) = 0 \qquad \text{Both first partials are zero}$$

Relative maximum and minimum values can occur only at critical points.

EXAMPLE **1**

FINDING CRITICAL POINTS

Find all critical points of

$$f(x, y) = 3x^2 + y^2 + 3xy + 3x + y + 6$$

Solution   We want all points at which both partials are zero.

$$f_x = 6x + 3y + 3 \quad \text{and} \quad f_y = 2y + 3x + 1 \qquad \text{Partials}$$

$$\left. \begin{aligned} 6x + 3y + 3 &= 0 \\ 3x + 2y + 1 &= 0 \end{aligned} \right\} \begin{aligned} &\text{Partials} \\ &\text{set equal to zero} \end{aligned}$$

└ Reordered so the x- and y-terms line up

To solve these equations simultaneously, we multiply the second by -2 so that the x-terms drop out when we add.

---

* We use the term "critical point" for the *pair* of values (a, b) at which the two partials are zero. The corresponding point on the graph has *three* coordinates $(a, b, f(a,b))$.

$$6x + 3y + 3 = 0 \qquad \text{First equation}$$
$$\underline{-6x - 4y - 2 = 0} \qquad \text{Second equation times } -2$$
$$-y + 1 = 0 \qquad \text{Adding } (x \text{ drops out})$$
$$y = 1 \qquad \text{From solving } -y + 1 = 0$$

Substituting $y = 1$ into either equation gives x:

$$6x + 3 + 3 = 0 \qquad \begin{array}{l}\text{Substituting } y = 1 \text{ into} \\ 6x + 3y + 3 - 0\end{array}$$
$$6x = -6 \qquad \text{Simplifying}$$
$$x = -1 \qquad \text{Solving}$$

These x- and y-values give one critical point.

$$\text{CP: } (-1, 1) \qquad \text{From } x = -1, \ y = 1$$

Second-Derivative Test for Functions $f(x, y)$: The D-Test

To determine whether $f(x, y)$ has a relative maximum, a relative minimum, or a saddle point at a critical point, we use the following D-test, which is a generalization of the second-derivative test on page 208.

D-Test

If (a,b) is a critical point of the function $f(x,y)$, then for D defined by

$$D = f_{xx}(a, b) \cdot f_{yy}(a, b) - [f_{xy}(a, b)]^2$$

More briefly,
$$D = f_{xx}f_{yy} - (f_{xy})^2$$

i. f has a relative *maximum* at (a, b) if $D > 0$ and $f_{xx}(a, b) < 0$.

ii. f has a relative *minimum* at (a, b) if $D > 0$ and $f_{xx}(a, b) > 0$.

$\left. \right\}$ Different signs for $f_{rr}(a, b)$

iii. f has a *saddle point* at (a, b) if $D < 0$.

The following observations may help in understanding the D-test.

1. The D-test is used only *after* finding the critical points. The test is then applied to each critical point, one at a time.

2. $D > 0$ appears only in parts i and ii, and so guarantees that the function has either a relative maximum or a relative minimum. Then all that remains to be done is to use the "old"

second-derivative test (checking the sign of f_{xx}) to determine which one (maximum or minimum) occurs.

3. $D < 0$ means a saddle point, regardless of the sign of f_{xx}. (A saddle point is neither a maximum nor a minimum.)

4. $D = 0$ means that the D-test is inconclusive—the function may have a maximum, a minimum, or a saddle point at the critical point.

EXAMPLE 2

FINDING RELATIVE EXTREME VALUES OF A POLYNOMIAL

Find the relative extreme values of

$$f(x, y) = 3x^2 + y^2 + 3xy + 3x + y + 6$$

Solution We find critical points by setting the two partials equal to zero and solving. But we did this for the same function in Example 1, finding one critical point, $(-1, 1)$. For the D-test, we calculate the second partials.

$$f_{xx} = 6 \qquad\qquad \text{From } f_x = 6x + 3y + 3$$
$$f_{yy} = 2 \qquad\qquad \text{From } f_y = 2y + 3x + 1$$
$$f_{xy} = 3 \qquad\qquad \text{From } f_x = 6x + 3y + 3$$

Calculating D:

$$D = 6 \cdot 2 - (3)^2 = 12 - 9 = 3 \qquad D = f_{xx}f_{yy} - (f_{xy})^2$$

$$\underset{f_{xx} \quad f_{yy} \quad f_{xy}}{\diagup \;\; \diagdown \quad \diagdown} \qquad\qquad \text{Positive}$$

D is positive and f_{xx} is positive (since $f_{xx} = 6$), so f has a *relative minimum* (part ii of the D-test) at the critical point $(-1, 1)$. (If f_{xx} had been negative, there would have been a relative maximum.) The relative minimum *value* is found by evaluating $f(x, y)$ at $(-1, 1)$.

$$f(-1, 1) = 3 + 1 - 3 - 3 + 1 + 6 \qquad \begin{array}{l} f = 3x^2 + y^2 + 3xy + 3x + y + 6 \\ \text{evaluated at } x = -1, \ y = 1 \end{array}$$

$$= 5$$

Relative minimum value: $f = 5$ at $x = -1$, $y = 1$.

EXAMPLE 3

FINDING RELATIVE EXTREME VALUES OF AN EXPONENTIAL FUNCTION

Find the relative extreme values of $f(x, y) = e^{x^2 - y^2}$.

Solution

$$f_x = e^{x^2-y^2}(2x) \atop f_y = e^{x^2-y^2}(-2y)\Bigg\}\ \text{Partials}$$

$$e^{x^2-y^2}(2x) = 0 \atop e^{x^2-y^2}(-2y) = 0\Bigg\}\ \text{Partials set equal to zero}$$

$$\text{CP: } (0,0)\ \ \text{Since } \begin{cases} e^{x^2-y^2}(2x) = 0 \\ e^{x^2-y^2}(-2y) = 0 \end{cases} \text{only at } x = 0,\ y = 0$$

For the *D*-test we calculate the second partials:

$$f_{xx} = e^{x^2-y^2}(2x)(2x) + e^{x^2-y^2}(2)$$
From $f_x = e^{x^2-y^2}(2x)$ using the Product Rule

$$= 4x^2 e^{x^2-y^2} + 2e^{x^2-y^2}$$
Simplifying

$$f_{yy} = e^{x^2-y^2}(-2y)(-2y) + e^{x^2-y^2}(-2)$$
From $f_y = e^{x^2-y^2}(-2y)$ using the Product Rule

$$= 4y^2 e^{x^2-y^2} - 2e^{x^2-y^2}$$
Simplifying

$$f_{xy} = e^{x^2-y^2}(2x)(-2y) = -4xye^{x^2-y^2}$$
From $f_x = e^{x^2-y^2}(2x)$ treating x as a constant

Evaluating at the critical point (0, 0):

$$f_{xx}(0,0) = 0e^0 + 2e^0 - 0 + 2 = 2 \qquad f_{xx} = 4x^2 e^{x^2-y^2} + 2e^{x^2-y^2}\ \text{at } (0,0)$$
$$f_{yy}(0,0) = 0e^0 - 2e^0 = 0 - 2 = -2 \qquad f_{yy} = 4y^2 e^{x^2-y^2} - 2e^{x^2-y^2}\ \text{at } (0,0)$$
$$f_{xy}(0,0) = 0e^0 = 0 \qquad f_{xy} = -4xye^{x^2-y^2}\ \text{at } (0,0)$$

Therefore *D* is

$$D = (2)(-2) - (0)^2 = -4 - 0 = -4 \qquad\qquad D = f_{xx}f_{yy} - (f_{xy})^2$$
$$\underset{\substack{\diagup\quad\diagdown\quad\diagdown \\ f_{xx}\quad f_{yy}\quad f_{xy}}}{}$$
Negative

Since *D* is negative, the function has a *saddle point* (part iii of the *D*-test) at the critical point (0, 0).

> *f* has no relative extreme values
> (it has a saddle point at $x = 0$, $y = 0$).

Maximizing Profit

If a company makes too few of its products, the resulting lost sales will lower the company's profits. On the other hand, making too many will "flood the market" and depress prices, again resulting in lower profits. Therefore, any realistic profit function must have a maximum at some "intermediate" point (x, y), which therefore must be a *relative* maximum point. Many applied problems are solved in this way: by knowing that the *absolute* extreme values (the highest and lowest values on the entire domain) must exist, and finding them as *relative* extreme points.

Suppose that a company produces two products, A and B, and that the two price functions are

$$p(x) = \begin{pmatrix} \text{Price at which exactly } x \text{ units} \\ \text{of product A will be sold} \end{pmatrix}$$

and

$$q(y) = \begin{pmatrix} \text{Price at which exactly } y \text{ units} \\ \text{of product B will be sold} \end{pmatrix}$$

If $C(x, y)$ is the (total) cost function, then the company's profit will be

$$\underbrace{P(x, y)}_{\text{Profit}} = \underbrace{p(x) \cdot x}_{\substack{\text{Price times} \\ \text{quantity for} \\ \text{product A}}} + \underbrace{q(y) \cdot y}_{\substack{\text{Price times} \\ \text{quantity for} \\ \text{product B}}} - \underbrace{C(x, y)}_{\text{Cost}}$$

Revenue for each product (price times quantity) minus the cost function

---

EXAMPLE 4

MAXIMIZING PROFIT FOR A COMPANY

Universal Motors makes compact and midsized cars. The price function for compacts is $p = 17 - 2x$ (for $0 \le x \le 8$), and the price function for midsized cars is $q = 20 - y$ (for $0 \le y \le 20$), both in thousands of dollars, where x and y are, respectively, the numbers of compact and midsized cars produced per hour. If the company's cost function is

$$C(x, y) = 15x + 16y - 2xy + 5$$

thousand dollars, find how many of each car should be produced and the prices that should be charged in order to maximize profit. Also find the maximum profit.

Solution   The profit function is

$$P(x, y) = (17 - 2x)x + (20 - y)y - (15x + 16y - 2xy + 5)$$

$$\underbrace{\overset{\text{Price} \quad \text{Quantity}}{}}_{\text{For compacts}} \quad \underbrace{\overset{\text{Price} \quad \text{Quantity}}{}}_{\text{For midsized}} \quad \overset{\text{Cost}}{\underbrace{}}$$

$$= 17x - 2x^2 + 20y - y^2 - 15x - 16y + 2xy - 5 \quad \text{Multiplying out}$$
$$= -2x^2 - y^2 + 2xy + 2x + 4y - 5 \quad\quad\quad \text{Simplifying}$$

We maximize $P(x, y)$ in the usual way:

$$P_x = -4x + 2y + 2$$
$$P_y = -2y + 2x + 4 \quad \Big\} \text{ Partials}$$

$-4x + 2y + 2 = 0$	Partials set equal to zero
$\underline{2x - 2y + 4 = 0}$	Rearranged to line up x's and y's
$-2x \qquad + 6 = 0$	Adding (the y's cancel)
$x = 3$	From solving $-2x + 6 = 0$
$y = 5$	From substituting $x = 3$ into either equation (omitting the details)

These two values give one critical point.

$$\text{CP: } (3, 5)$$

For the D-test we calculate the second partials:

$$P_{xx} = -4 \qquad P_{xy} = 2 \qquad P_{yy} = -2 \qquad \begin{array}{l}\text{From } P_x = -4x + 2y + 2 \\ \text{and } P_y = 2y + 2x + 4\end{array}$$

$$D = (-4)(-2) - (2)^2 = 4 \qquad\qquad D = P_{xx}P_{yy} - (P_{xy})^2$$

D is positive and $P_{xx} = -4$ is negative, so profit is indeed *maximized* at $x = 3$ and $y = 5$. To find the prices, we evaluate the price functions:

$$p - 17 - 2 \cdot 3 = 11 \qquad \text{(thousand dollars)} \qquad \begin{array}{l}p = 17 - 2x \\ \text{evaluated at } x = 3\end{array}$$

$$q = 20 - 5 = 15 \qquad \text{(thousand dollars)} \qquad \begin{array}{l}q = 20 - y \\ \text{evaluated at } y = 5\end{array}$$

The profit comes from the profit function:

$$P(3, 5) = \underbrace{-2 \cdot 3^2 - 5^2 + 2 \cdot 3 \cdot 5 + 2 \cdot 3 + 4 \cdot 5 - 5}_{\begin{array}{c}P = -2x^2 - y^2 + 2xy + 2x + 4y - 3 \\ \text{evaluated at } x = 3, \ y = 5\end{array}}$$

$$= 8 \text{ (thousand dollars)}$$

Profit is maximized when the company produces 3 compacts per hour, selling them for $11,000 each, and 5 midsized cars per hour, selling them for $15,000 each. The maximum profit will be $8000 per hour.

The D-test ensures that you have *maximized* profit rather than minimized it.*

Some functions have *more* than one critical point.

---

* How are the price and cost functions in such problems found? Price functions may be constructed by the methods used on pages 229–231 or by more sophisticated techniques of market research. Cost functions may be found simply as the sum of the unit cost times the number of units for each product, or by regression techniques based on the least squares method described in the following section for constructing functions from data.

| EXAMPLE 5 | **FINDING RELATIVE EXTREME VALUES** |

Find the relative extreme values of

$$f(x, y) = x^2 + y^3 - 6x - 12y$$

Solution

$$2x - 6 = 0 \qquad\qquad f_x = 0$$
$$3y^2 - 12 = 0 \qquad\qquad f_y = 0$$

The first gives

$$x = 3 \qquad\qquad \text{Solving } 2x - 6 = 0$$

and the second gives

$$3y^2 = 12 \qquad\qquad \begin{array}{l}\text{Adding 12 to each}\\\text{side of } 3y^2 - 12 = 0\end{array}$$

$$y^2 = 4 \qquad\qquad \text{Dividing by 3}$$

$$y = \pm 2 \qquad\qquad \text{Taking square roots}$$

From $x = 3$ and $y = \pm 2$ we get *two* critical points:

$$\text{CP: } (3, 2) \quad \text{and} \quad (3, -2)$$

Calculating the second partials and substituting them into D:

$$D = (2)(6y) - (0)^2 = 12y \qquad \begin{array}{l}D = (f_{xx})(f_{yy}) - (f_{xy})^2 \text{ with}\\ f_{xx} = 2,\ f_{yy} = 6y,\ f_{xy} = 0\end{array}$$

We apply the D-test to the critical points one at a time.

At $(3, 2)$: $D = 12 \cdot 2 > 0$ $\qquad\qquad \begin{array}{l}D = 12y \text{ evaluated}\\\text{at } (3, 2)\end{array}$

$\qquad\qquad f_{xx} = 2 > 0$

$\qquad\qquad$ *relative minimum* at $x = 3,\ y = 2$ $\qquad \begin{array}{l}\text{Since } D \text{ and } f_{xx} \text{ are}\\\text{both positive}\end{array}$

At $(3, -2)$: $D = 12 \cdot (-2) < 0$ $\qquad\qquad \begin{array}{l}D = 12y \text{ evaluated}\\\text{at } (3, -2)\end{array}$

$\qquad\qquad$ *saddle point* at $x = 3,\ y = -2$ $\qquad$ Since D is negative

Answer:

Relative minimum value: $f = -25$ at $\begin{cases} x = 3 \\ y = 2 \end{cases}$ $\begin{array}{l}f = -25 \text{ from}\\ f = x^2 + y^3 - 6x - 12y\\\text{at } (3, 2)\end{array}$

(saddle point at $x = 3,\ y = -2$)

Competition and Collusion

In 1938, the French economist Antoine Cournot published the following comparison of a monopoly (a market with only one supplier) and a duopoly (a market with two suppliers).

Monopoly. Imagine that you are selling spring water from your own spring (or any product whose cost of production is negligible). Since you are the only supplier in town, you have a "monopoly." Suppose that your price function is $p = 6 - 0.01x$, where p is the price in dollars at which you will sell precisely x gallons ($0 \leq x \leq 600$). Your revenue is then

$$R(x) = (6 - 0.01x)x = 6x - 0.01x^2$$

Price $(6 - 0.01x)$ times quantity x

You maximize revenue by setting its derivative equal to zero:

$$R'(x) = 6 - 0.02x = 0$$

$$x = \frac{6}{0.02} = 300$$

Solving $6 - 0.02x = 0$

Therefore, you should sell 300 gallons per day. (The second-derivative test will verify that revenue is maximized.) The price will be

$$p = 6 - 0.01 \cdot 300 = 6 - 3 = 3$$

$p = 6 - 0.01x$
evaluated at $x = 300$

or $3 dollars per gallon. Your maximum revenue will be

$$R(300) = 6 \cdot 300 - 0.01(300)^2 = \$900$$

$R(x) = 6x - 0.01x^2$
evaluated at $x = 300$

Duopoly. Suppose now that your neighbor opens a competing spring water business. (A market such as this with two suppliers is called a "duopoly.") Now both of you must share the same market. If your neighbor sells y gallons per day (and you sell x), you must both sell at price

$$p = 6 - 0.01(x + y) = 6 - 0.01x - 0.01y$$

Price function $p = 6 - 0.01x$ with x replaced by the combined quantity $x + y$

Each of you calculates revenue as price times quantity:

$$\left(\begin{array}{c}\text{Your}\\\text{revenue}\end{array}\right) = p \cdot x = (6 - 0.01x - 0.01y)x = 6x - 0.01x^2 - 0.01xy$$

$$\left(\begin{array}{c}\text{Neighbor's}\\\text{revenue}\end{array}\right) = p \cdot y = (6 - 0.01x - 0.01y)y = 6y - 0.01xy - 0.01y^2$$

You each want to maximize revenue, so you set the partials equal to zero:

$$6 - 0.02x - 0.01y = 0$$

Partial of $6x - 0.01x^2 - 0.01xy$ with respect to x, set equal to zero

$$6 - 0.01x - 0.02y = 0$$

Partial of $6y - 0.01xy - 0.01y^2$ with respect to y, set equal to zero

These are easily solved by multiplying one of them by 2 and subtracting the other. The solution (omitting the details) is

$$x = 200 \qquad y = 200$$

so each of you sells 200 gallons per day. The selling price for both will be

$$p = 6 - 0.01(200 + 200) = 6 - 4 = 2 \qquad \begin{array}{l} p = 6 - 0.01\,(x + y) \\ \text{evaluated at} \quad x = 200, \ y = 200 \end{array}$$

or $2 per gallon. Revenue is price ($2) times quantity (200), resulting in $400 for each of you.

Comparison of the Monopoly and the Duopoly

The two systems may be compared as follows:

	Monopoly	Duopoly
Quantity:	300 gallons	200 gallons each, 400 total
Price:	$3 per gallon	$2 per gallon
Revenue:	$900	$400 each, $800 total

Notice that the duopoly produces *more* than the monopoly (400 gallons versus only 300 in the monopoly), and does so at a *lower price* ($2 versus $3 in the monopoly). Cournot therefore concluded that consumers benefit more from a duopoly than from a monopoly.

However, the smart duopolist will notice that the revenue of $400 is less than half of the monopoly revenue of $900. Therefore, it benefits each duopolist to cooperate and share the market as a single monopoly. This is called *collusion*. It is for this reason that markets with only a few suppliers tend toward collusion rather than competition.

Optimizing Functions of Three or More Variables

Functions of *more* than two variables are optimized in the same way: by finding critical points (where all of the first partials are zero). For example, to optimize a function $f(x, y, z)$ we would set the three partials equal to zero and solve:

$$f_x = 0$$
$$f_y = 0$$
$$f_z = 0$$

The second-derivative test for functions of three or more variables is very complicated, and we shall not discuss it.

Section Summary

To optimize a function $f(x, y)$ of two variables, first find all critical points, the points (a, b) where both first partials are zero.

$$f_x(a, b) = 0$$
$$f_y(a, b) = 0$$

Then apply the D-test (page 523) to each critical point to determine whether the function has a relative maximum, a relative minimum, or a saddle point at that critical point.

Exercises

Find the relative extreme values of each function.

1. $f(x, y) = x^2 + 2y^2 + 2xy + 2x + 4y + 7$

2. $f(x, y) = 2x^2 + y^2 + 2xy + 4x + 2y + 5$

3. $f(x, y) = 2x^2 + 3y^2 + 2xy + 4x - 8y$

4. $f(x, y) = 3x^2 + 2y^2 + 2xy + 8x - 4y$

5. $f(x, y) = 3xy - 2x^2 - 2y^2 + 14x - 7y - 5$

6. $f(x, y) = 2xy - 2x^2 - 3y^2 + 4x - 12y + 5$

7. $f(x, y) = xy + 4x - 2y + 1$

8. $f(x, y) = 5xy - 2x^2 - 3y^2 + 5x - 7y + 10$

9. $f(x, y) = 3x - 2y - 6$

10. $f(x, y) = 5x - 4y + 5$

11. $f(x, y) = e^{(x^2 + y^2)/2}$ **12.** $f(x, y) = e^{5(x^2 + y^2)}$

13. $f(x, y) = \ln(x^2 + y^2 + 1)$

14. $f(x, y) = \ln(2x^2 + 3y^2 + 1)$

15. $f(x, y) = -x^3 - y^2 + 3x - 2y$

16. $f(x, y) = x^3 - y^2 - 3x + 6y$

17. $f(x, y) = y^3 - x^2 - 2x - 12y$

18. $f(x, y) = -x^2 - y^3 - 6x + 3y + 4$

19. $f(x, y) = x^3 - 2xy + 4y$

20. $f(x, y) = y^3 - 2xy - 4x$

APPLIED EXERCISES

21. BUSINESS: Maximum Profit A company manufactures two products. The price function for product A is $p = 12 - \frac{1}{2}x$ (for $0 \le x \le 24$), and for product B is $q = 20 - y$ (for $0 \le y \le 20$), both in thousands of dollars, where x and y are the amounts of products A and B, respectively. If the cost function is

$$C(x, y) = 9x + 16y - xy + 7$$

thousands of dollars, find the quantities and the prices of the two products that maximize profit. Also find the maximum profit.

22. BUSINESS: Maximum Profit A company manufactures two products. The price function for product A is $p = 16 - x$ (for $0 \le x \le 16$), and for product B is $q = 19 - \frac{1}{2}y$ (for $0 \le y \le 38$), both in thousands of dollars, where x and y are the amounts of product A and B, respectively. If the cost function is

$$C(x, y) = 10x + 12y - xy + 6$$

thousand dollars, find the quantities and the prices of the two products that maximize profit. Also find the maximum profit.

23. **BUSINESS: Price Discrimination** An automobile manufacturer sells cars in America and Europe, charging different prices in the two markets. The price function for cars sold in America is $p = 20 - 0.2x$ thousand dollars (for $0 \le x \le 100$), and the price function for cars sold in Europe is $q = 16 - 0.1y$ thousand dollars (for $0 \le y \le 160$), where x and y are the numbers of cars sold per day in America and Europe, respectively. The company's cost function is

$$C = 20 + 4(x + y) \qquad \text{thousand dollars}$$

a. Find the company's profit function. [*Hint:* Profit is revenue from America plus revenue from Europe minus costs, where each revenue is price times quantity.]
b. Find how many cars should be sold in each market to maximize profit. Also find the price for each market.

24. **BIOMEDICAL: Drug Dosage** In a laboratory test the combined antibiotic effect of x milligrams of medicine A and y milligrams of medicine B is given by the function

$$f(x, y) = xy - 2x^2 - y^2 + 110x + 60y$$

(for $0 \le x \le 55, 0 \le y \le 60$). Find the amounts of the two medicines that maximize the antibiotic effect.

25. **PSYCHOLOGY: Practice and Rest** A subject in a psychology experiment who practices a skill for x hours and then rests for y hours achieves a test score of $f(x, y) = xy - x^2 - y^2 + 11x - 4y + 120$ (for $0 \le x \le 10$, $0 \le y \le 4$). Find the numbers of hours of practice and rest that maximize the subject's score.

26. **SOCIOLOGY: Absenteeism** The number of office workers near a beach resort who call in "sick" on a warm summer day is

$$f(x, y) = xy - x^2 - y^2 + 110x + 50y - 5200$$

where x is the air temperature ($70 \le x \le 100$) and y is the water temperature ($60 \le y \le 80$). Find the air and water temperatures that maximize the number of absentees.

27–28. ECONOMICS: Competition and Collusion Compare the outputs of a monopoly and a duopoly by repeating the analysis on pages 528–530 for the following price function.

That is:

a. For a monopoly, calculate the quantity x that maximizes your revenue. Also calculate the price p and the revenue R.
b. For the duopoly, calculate the quantities x and y that maximize revenue for each duopolist. Calculate the price p and the two revenues.
c. Are more goods produced under a monopoly or a duopoly?
d. Is the price lower under a monopoly or a duopoly?

27. $p = 12 - 0.005x$ dollars $\qquad (0 \le x \le 2400)$

28. $p = a - bx$ dollars (for positive numbers a and b with $0 \le x \le a/b$)

29. **BUSINESS: Maximum Profit** An automobile dealer can sell 8 sedans per day at a price of $20,000 and 4 SUVs (sport utility vehicles) per day at a price of $25,000. She estimates that for each $400 decrease in price of the sedans she can sell two more per day, and for each $600 decrease in price for the SUVs she can sell one more. If each sedan costs her $16,800 and each SUV costs her $19,000, and fixed costs are $1000 per day, what price should she charge for the sedans and the SUVs to maximize profit? How many of each type will she sell at these prices? [*Hint:* Let x be the number of $400 price decreases for sedans and y be the number of $600 price decreases for SUVs, and use the method of Examples 1 and 2 on pages 229–231 for each type of car.]

30. **BUSINESS: Maximum Revenue** An airline flying to a Midwest destination can sell 20 coach-class tickets per day at a price of $250 and six business-class tickets per day at a price of $750. It finds that for each $10 decrease in the price of the coach ticket, it will sell four more per day, and for each $50 decrease in the business-class price, it will sell two more per day. What prices should the airline charge for the coach- and business-class tickets to maximize revenue? How many of each type will be sold at these prices? [*Hint:* Let x be the number of $10 price decreases for coach tickets and y be the number of $50 price decreases for business-class tickets, and use the method of Examples 1 and 2 on pages 229–231 for each type of ticket.]

OPTIMIZING FUNCTIONS OF THREE VARIABLES

31. BUSINESS: Price Discrimination An automobile manufacturer sells cars in America,
Europe, and Asia, charging a different price in
each of the three markets. The price function
for cars sold in America is $p = 20 - 0.2x$ (for
$0 \leq x \leq 100$), the price function for cars sold
in Europe is $q = 16 - 0.1y$ (for $0 \leq y \leq 160$),
and the price function for cars sold in Asia is
$r = 12 - 0.1z$ (for $0 \leq z \leq 120$), all in thousands of dollars, where x, y, and z are the numbers of cars sold in America, Europe, and Asia,
respectively. The company's cost function is
$C = 22 + 4(x + y + z)$ thousand dollars.

  a. Find the company's profit function
$P(x, y, z)$. [*Hint:* The profit will be revenue
from America plus revenue from Europe
plus revenue from Asia minus costs, where
each revenue is price times quantity.]

  b. Find how many cars should be sold in each
market to maximize profit. [*Hint:* Set the
three partials P_x, P_y, and P_z equal to zero
and solve. Assuming that the maximum
exists, it must occur at this point.]

32. ECONOMICS: Competition and Collusion
Suppose that in the discussion of competition
and collusion (pages 528–530), *two* of your
neighbors began selling spring water. Use the
price function $p = 36 - 0.01x$ (for
$0 \leq x \leq 3600$) and repeat the analysis, but
now comparing a monopoly with competition among *three* suppliers (a "triopoly").
That is:

  a. For a monopoly, calculate the quantity x
that maximizes your revenue. Also calculate the price p and the revenue R.

  b. For a triopoly, find the quantities x, y, and z
for the three suppliers that maximize rev-
enue for each. Also calculate the price p and
the three revenues. [*Hint:* Find the three
revenue functions, one for each supplier,
and maxmimize each with respect to that
supplier's variable.]

  c. Are more goods produced under a monopoly or under a triopoly?

  d. Is the price lower under a monopoly or under a triopoly?

EXERCISES WITH MORE THAN ONE CRITICAL POINT

Find the relative extreme values of each function.

33. $f(x, y) = x^3 + y^3 - 3xy$

34. $f(x, y) = x^5 + y^5 - 5xy$

35. $f(x, y) - 12xy - x^3 - 6y^2$

36. $f(x, y) = 6xy - x^3 - 3y^2$

37. $f(x, y) = 2x^4 + y^2 - 12xy$

38. $f(x, y) = 16xy - x^4 - 2y^2$

7.4   LEAST SQUARES

Introduction

You may have wondered how the mathematical models in this book
were developed. For example, how are the constants a and b in the
Cobb–Douglas production function $P = aL^bK^{1-b}$ determined for a
particular company or nation? The problem of finding the function
that fits a collection of data can be viewed geometrically as the problem of fitting a curve to a collection of points. The simplest case of this

problem is fitting a straight line to a collection of points, and the most widely used method for doing this is called *least squares*. Least squares is the method that graphing calculators use to find regression lines and curves. Even if you continue to use your calculator to carry out the calculations, this section is important in that it explains what you are actually doing.

Least squares lines are used extensively in forecasting and for detecting underlying trends in data.

A First Example

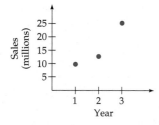

The graph on the left shows a company's annual sales (in millions) over a 3-year period. How can we fit a straight line to these three points? Clearly, these points do not lie exactly on a line, and so rather than an "exact" fit, we want the line $y = ax + b$ that fits these three points most closely.

Let d_1, d_2, and d_3 stand for the vertical distances between the three points and the line $y = ax + b$. The line that minimizes the sum of the squares of these vertical deviations is called the *least squares line* or the *regression line*.* Squaring the deviations ensures that none are negative, so that a deviation below the line does not "cancel" one above the line.

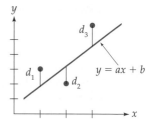

EXAMPLE 1

MINIMIZING THE SQUARED DEVIATIONS

The table below gives a company's sales (in millions) in year x. Find the least squares line for the data. Then use the line to predict sales in year 4.

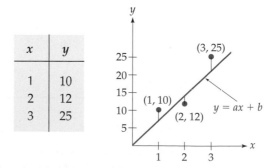

x	y
1	10
2	12
3	25

---

* The word "regression" comes from an early use of this technique to determine whether unusually tall parents have unusually tall children. It seems that tall parents do have tall offspring, but not quite as tall, with successive generations exhibiting a "regression" toward the average height of the population.

Solution The vertical deviations are found by calculating the heights of the line $y = ax + b$ at each x-value minus the y-values from the table. These differences are then squared and summed.

$$S = (a \cdot 1 + b - 10)^2 + (a \cdot 2 + b - 12)^2 + (a \cdot 3 + b - 25)^2$$

| Height of the line $y = ax + b$ at $x = 1$ | y-value of the point at $x = 1$ | Height of the line $y = ax + b$ at $x = 2$ | y-value of the point at $x = 2$ | Height of the line $y = ax + b$ at $x = 3$ | y-value of the point at $x = 3$ |

This sum S depends on a and b, the numbers that determine the line $y = ax + b$. To minimize S, we set its partials with respect to a and b equal to zero:

$$\frac{\partial S}{\partial a} = 2(a + b - 10) + 2(2a + b - 12) \cdot 2 + 2(3a + b - 25) \cdot 3$$
Differentiating each part of S by the Generalized Power Rule

$$= 2a + 2b - 20 + 8a + 4b - 48 + 18a + 6b - 150$$
Multiplying out

$$= 28a + 12b - 218$$
Combining terms

$$\frac{\partial S}{\partial b} = 2(a + b - 10) + 2(2a + b - 12) + 2(3a + b - 25)$$
Differentiating each part of S by the Generalized Power Rule

$$= 2a + 2b - 20 + 4a + 2b - 24 + 6a + 2b - 50$$
Multiplying out

$$= 12a + 6b - 94$$
Combining terms

We set the two partials equal to zero and solve:

$$28a + 12b - 218 = 0$$
$$12a + 6b - 94 = 0$$
Partials set equal to zero

$$28a + 12b - 218 = 0$$
First equation

$$-24a - 12b + 188 = 0$$
Second multiplied by -2

$$4a \qquad - 30 = 0$$
Adding (the b's drop out)

$$a = \frac{30}{4} = 7.5$$
Solving $4a - 30 = 0$

$$b = \frac{4}{6} \approx 0.67$$
From substituting $a = 7.5$ into $12a + 6b - 94 = 0$ and solving for b

These values for a and b give the least squares line. (The D-test would show that S has indeed been minimized.)

$$y = 7.5x + 0.67$$

$\uparrow$

$y = ax + b$ with
$a = 7.5$ and $b = 0.67$

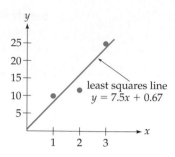

least squares line
$y = 7.5x + 0.67$

To predict the sales in year 4, we evaluate the least squares line at $x = 4$:

$$y = 7.5(4) + 0.67 = 30.67$$

$y = 7.5x + 0.67$
evaluated at $x = 4$

Prediction: 30.67 million sales in year 4.

The slope of the line is 7.5, meaning that the linear trend in the company's sales is a growth of 7.5 million sales per year.

Graphing Calculator Exploration

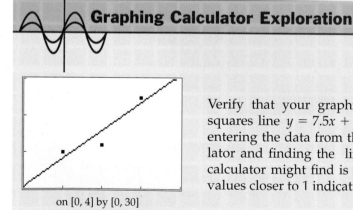

on [0, 4] by [0, 30]

Verify that your graphing calculator gives the same least squares line $y = 7.5x + 0.67$ that we found in Example 1 by entering the data from the table on page 534 into your calculator and finding the linear regression line. The r that your calculator might find is a measure of "goodness of fit," with values closer to 1 indicating a better fit.

Least Squares Line for *n* Points

Example 1 used only three points, which is too few for most realistic applications. If we carry out the same steps for n points, we would obtain the following formulas (in which Σ stands for sum).

Least Squares Line

x	y
x_1	y_1
x_2	y_2
.	.
.	.
.	.
x_n	y_n

For data ... calculate

$$a = \frac{n\sum xy - (\sum x)(\sum y)}{n\sum x^2 - (\sum x)^2}$$

$$b = \frac{1}{n}\left(\sum y - a\sum x\right)$$

n = number of data points
$\sum x$ = sum of x's
$\sum y$ = sum of y's
$\sum xy$ = sum of products $x \cdot y$
$\sum x^2$ = sum of squares of x's

The least squares line is then $y = ax + b$

From now on we will find least squares lines by using these formulas, a derivation of which is given at the end of this section.

EXAMPLE 2

FINDING THE LEAST SQUARES LINE

A 1955 study compared cigarette smoking with the mortality rate for lung cancer in several countries. Find the least squares line that fits these data. Then use the line to predict lung cancer deaths if per capita cigarette consumption is 600 cigarettes per month (a pack a day).

	Cigarette Consumption (per capita)	Lung Cancer Deaths (per million)
Norway	250	90
Sweden	300	120
Denmark	350	170
Canada	500	150

Solution   The procedure for calculating a and b consists of six steps, beginning with the following table.

1. List the x- and y-values   2. Multiply $x \cdot y$ 3. Square each x

x	y	xy	x^2
250	90	22,500	62,500
300	120	36,000	90,000
350	170	59,500	122,500
500	150	75,000	250,000
1400	530	193,000	525,000
‖	‖	‖	‖
$\sum x$	$\sum y$	$\sum xy$	$\sum x^2$

← 4. Sum each column

5. Calculate a and b using the formulas on the preceding page.

$$a = \frac{(4)(193{,}000) - (1400)(530)}{(4)(525{,}000) - 1400^2}$$

$$= \frac{30{,}000}{140{,}000} \approx 0.21$$

n = number of points

$$a = \frac{n \sum xy - (\sum x)(\sum y)}{n \sum x^2 - (\sum x)^2}$$

$$b = \frac{1}{4}[530 - 0.21(1400)] = 59$$

$$b = \frac{1}{n}(\sum y - a \sum x)$$

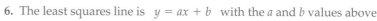

6. The least squares line is $y = ax + b$ with the a and b values above

$$y = 0.21x + 59 \qquad y = ax + b \quad \text{with} \quad a = 0.21 \quad \text{and} \quad b = 59$$

This line is graphed on the left, and shows that lung cancer mortality increases with cigarette smoking.

To predict the lung cancer deaths if per capita cigarette consumption reaches 600, we evaluate the least squares line at $x = 600$.

$$y = 0.21 \cdot 600 + 59 = 185 \qquad y = 0.21x + 59 \quad \text{with} \quad x = 600$$

Predicted annual mortality: 185 deaths per million.

Criticism of Least Squares

Least squares is the most widely used method for fitting lines to points, but it does have one weakness: the vertical deviations from the line are squared, so one large deviation, when squared, can have an unexpectedly large influence on the line. For example, the graph on the right shows four points and their least squares line. The line fits the points quite closely.

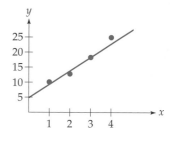

The second graph adds a fifth point, one quite out of line with the others, and shows the least squares line for the five points. The added point has an enormous effect on the line, causing it to slope downward even though all of the other points suggest an upward slope. In actual applications, one should inspect the points for such "outliers" before calculating the least squares line.

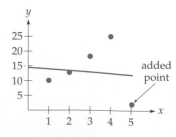

Fitting Exponential Curves by Least Squares

It is not always appropriate to fit a straight line to a set of points. Sometimes a collection of points will suggest a *curve* rather than a line, such as one of the following exponential curves.

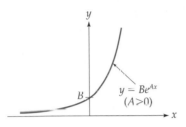

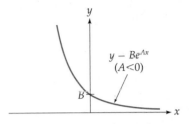

Least squares can be used to fit an exponential curve of the form $y = Be^{Ax}$ to a collection of points as follows. Taking natural logs of both sides of $y = Be^{Ax}$ gives

$$\ln y = \ln(Be^{Ax}) = \ln B + \ln e^{Ax} = \ln B + Ax \qquad \text{Using the properties of natural logarithms}$$

If we introduce a new variable $Y = \ln y$, and also let $b = \ln B$, then $\ln y = \ln B + Ax$ becomes

$$Y = b + Ax \qquad\qquad \text{Or, equivalently,} \quad Y = Ax + b$$

Therefore, we fit a straight line to the *logarithms* of the y-values. (Now we are minimizing not the squared deviations, but the squared deviations of the logarithms.) The procedure consists of the eight steps shown in the following example.

EXAMPLE 3

FITTING AN EXPONENTIAL CURVE BY LEAST SQUARES

The world population* since 1850 is shown in the table below. Fit an exponential curve to these data and predict the world population in the year 2050.

Year	Population (billions)
1850	1.26
1900	1.65
1950	2.52
2000	6.08

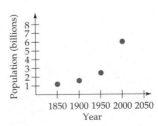

* *Source:* U.S. Census Bureau

Solution We number the years 1 through 4 (since they are evenly spaced).

1. List the x- and y-values   **2.** Take ln of the y-values (call them *capital Y*)   **3.** Multiply $x \cdot Y$ (capital Y ⟶)   **4.** Square each x

x	y	$Y = \ln y$	xY	x^2
1	1.26	0.23	0.23	1
2	1.65	0.50	1.00	4
3	2.52	0.92	2.76	9
4	6.08	1.81	7.24	16
10		3.46	11.23	30
$\parallel$		$\parallel$	$\parallel$	$\parallel$
Σx		ΣY	ΣxY	Σx^2

5. Add each column (except y)

6. Calculate A and b (n = number of points)

$$A = \frac{4(11.23) - 10(3.46)}{4(30) - 10^2} = \frac{10.32}{20} = 0.516 \qquad A = \frac{n\Sigma xY - (\Sigma x)(\Sigma Y)}{n\Sigma x^2 - (\Sigma x)^2}$$

$$b = \frac{1}{4}(3.46 - 0.516 \cdot 10) = -0.425 \qquad b = \frac{1}{n}(\Sigma Y - A \Sigma x)$$

$$B = e^b = e^{-0.425} \approx 0.65 \qquad \longleftarrow \text{7. Calculate } B = e^b \text{ (since } b = \ln B)$$

The exponential curve is

$$y = 0.65e^{0.516x} \qquad \longleftarrow \text{8. The curve is } y = Be^{Ax} \text{ with the } A \text{ and } B \text{ found above}$$

The population in 2050 is this function evaluated at $x = 5$:

$$y = 0.65e^{0.516(5)} \approx 8.6$$

Therefore, world population in the year 2050 is predicted to be 8.6 billion.

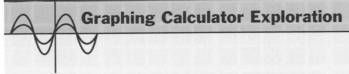

Graphing Calculator Exploration

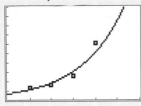

Verify that your graphing calculator gives the same curve $y = 0.65e^{0.516x}$ that we found in Example 3 by entering the x- and y-values from the previous table into your calculator and finding the exponential regression curve. [You may find $y = 0.65 \cdot 1.675^x$, which is the same curve, since $e^{0.516} \approx 1.675$]

7.4 Section Summary

The technique of least squares can be used to determine the equation of a line that "best fits" a collection of points in the sense that it minimizes the sum of the squared vertical deviations. The method is described on pages 537–538. Least squares can also be used to fit a *curve* to a collection of points. The procedure for fitting an exponential curve $y = Be^{Ax}$ is described on pages 539–540.

To determine whether to fit a line or a curve to a collection of points, you should first graph the points. For example, the points in the left-hand graph below suggest a line, while those in the right-hand graph suggest an exponential curve.

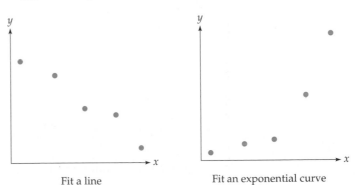

Fit a line   Fit an exponential curve

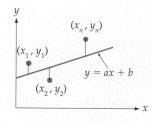

Derivation of the Formula for the Least Squares Line

The formulas for a and b in the least squares line come from minimizing the squared vertical deviations, just as in Example 1, but now for the n points $(x_1, y_1), (x_2, y_2), \cdots, (x_n, y_n)$.

$$S = (ax_1 + b - y_1)^2 + (ax_2 + b - y_2)^2 + \cdots + (ax_n + b - y_n)^2$$

The partials are

$$\frac{\partial S}{\partial a} = 2(ax_1 + b - y_1)x_1 + 2(ax_2 + b - y_2)x_2 + \cdots + 2(ax_n + b - y_n)x_n$$

$$= 2ax_1^2 + 2bx_1 - 2x_1y_1 + 2ax_2^2 + 2bx_2 - 2x_2y_2 + \cdots \qquad \text{Multiplying out}$$
$$\quad + 2ax_n^2 + 2bx_n - 2x_ny_n$$

$$= 2a(x_1^2 + x_2^2 + \cdots + x_n^2) + 2b(x_1 + x_2 + \cdots + x_n) \qquad \text{Regrouping}$$
$$\quad - 2(x_1y_1 + x_2y_2 + \cdots + x_ny_n)$$

$$= 2a \sum x^2 + 2b \sum x - 2 \sum xy \qquad \text{Using } \Sigma \text{ for sum}$$

$$\frac{\partial S}{\partial b} = 2(ax_1 + b - y_1) + 2(ax_2 + b - y_2) + \cdots + 2(ax_n + b - y_n)$$

$$= 2ax_1 + 2b - 2y_1 + 2ax_2 + 2b - 2y_2 + \cdots + 2ax_n + 2b - 2y_n \qquad \text{Multiplying out}$$

$$= 2a(x_1 + x_2 + \cdots + x_n) + 2b(1 + 1 + \cdots + 1) \qquad \text{Regrouping}$$
$$\quad - 2(y_1 + y_2 + \cdots + y_n)$$

$$= 2a \sum x + 2bn - 2 \sum y \qquad \text{Using } \Sigma \text{ for sum}$$

We set the partials equal to zero:

$$a \sum x^2 + b \sum x - \sum xy = 0$$
$$a \sum x + bn - \sum y = 0 \qquad \text{Dividing each by 2}$$

To solve for a we multiply the first of these equations by n, the second by $\sum x$, and subtract:

$$an \sum x^2 + bn \sum x - n \sum xy = 0$$

$$a \left(\sum x \right)^2 + bn \sum x - \left(\sum x \right) \left(\sum y \right) = 0 \qquad \textit{Be careful! } \Sigma\, xy \textit{ is not the same as } (\Sigma\, x)(\Sigma\, y)$$

$$a \left(n \sum x^2 - \left(\sum x \right)^2 \right) - \left(n \sum xy - \left(\sum x \right) \left(\sum y \right) \right) = 0 \qquad \text{Subtracting the last two equations}$$

Solving this last equation for a gives:

$$a = \frac{n \sum xy - \left(\sum x \right) \left(\sum y \right)}{n \sum x^2 - \left(\sum x \right)^2} \qquad \text{Formula for } a$$

For b we obtain:

$$b = \frac{1}{n}\left(\sum y - a \sum x\right) \qquad a \sum x + nb - \sum y = 0 \;\; \text{solved for } b$$

These are the formulas in the box on page 537.

7.4 Exercises

EXERCISES ON LEAST SQUARES LINES

Find the least squares line for each table of points.

1.

x	y
1	2
2	5
3	9

2.

x	y
1	2
2	4
3	7

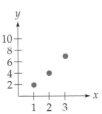

3.

x	y
1	6
3	4
6	2

4.

x	y
1	9
4	6
5	1

5.

x	y
0	7
1	10
2	10
3	15

6.

x	y
0	5
1	8
2	8
3	12

7.

x	y
−1	10
0	8
1	5
3	0
5	2

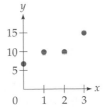

8.

x	y
−2	12
0	10
2	6
4	0
5	−3

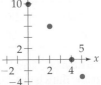

APPLIED EXERCISES ON LEAST SQUARES LINES

9. BUSINESS: Sales A company's annual sales are shown in the following table. Find the least squares line. Use your line to predict the sales in the next year ($x = 5$).

Year	1	2	3	4
Sales (millions)	7	10	11	14

10. GENERAL: Automobile Costs The following table gives the cost per mile of operating a compact car, depending upon the number of miles driven per year. Find the least squares line for these data. Use your answer to predict the cost per mile for a car driven 25,000 miles annually ($x = 5$).

Annual Mileage	x	Cost per Mile (cents)
5,000	1	50
10,000	2	35
15,000	3	27
20,000	4	25

11. SOCIOLOGY: Crime A sociologist finds the following data for the number of felony arrests per year in a town. Find the least squares line. Then use it to predict the number of felony arrests in the next year.

Year	1	2	3	4
Arrests	120	110	90	100

12. GENERAL: Farming A farmer's wheat yield (bushels per acre) depends on the amount of fertilizer (hundreds of pounds per acre) according to the following table. Find the least squares line. Then use the line to predict the yield using 3 hundred pounds of fertilizer per acre.

Fertilizer	1.0	1.5	2.0	2.5
Yield	30	35	38	40

13. BASEBALL: The Disappearance of the .400 Hitter Between 1901 and 1930, baseball boasted several .400 hitters (Lajoie, Cobb, Jackson, Sisler, Heilmann, Hornsby, and Terry), but only one since then (Ted Williams in 1941). The decline of the "heavy hitter" is evidenced by the following data, showing the highest batting average for each 20-year period (National or American League). Find the least squares line for these data and use it to predict the highest batting average for the period 2001–2020 ($x = 6$).

	x	Highest Average
1901–1920	1	.422
1921–1940	2	.424
1941–1960	3	.406
1961–1980	4	.390
1981–2000	5	.394

14. ECONOMICS: Consumer Price Index The consumer price index (CPI) is shown in the following table. Fit a least squares line to the data. Then use the line to predict the CPI in the year 2010 ($x = 7$).

	x	CPI
1980	1	82.4
1985	2	107.6
1990	3	130.7
1995	4	152.4
2000	5	173.2

15–16. GENERAL: Percentage of Smokers The following tables show the percentage of smokers among the adult population in the United States for every five years since 1980 (the first table is males, the second females). Find the least squares lines for these data, and use your answer to predict the percentage of that sex who will smoke in the year 2010 ($x = 7$).

15.

	x	Percent Males
1980	1	38
1985	2	32
1990	3	28
1995	4	27
2000	5	28

16.

x		Percent Females
1980	1	30
1985	2	28
1990	3	23
1995	4	22
2000	5	23

Source: National Center for Health Statistics

17. GENERAL: Smoking and Longevity The following data show the life expectancy of a 25-year-old male based on the number of cigarettes smoked daily. Find the least squares line for these data. The slope of the line estimates the years lost per extra cigarette per day.

Cigarettes Smoked Daily	Life Expectancy
0	73.6
5	69.0
15	68.1
30	67.4
40	65.3

18. GENERAL: Pollution and Absenteeism The following table shows the relationship between the sulfur dust content of the air (in micrograms per cubic meter) and the number of female absentees in industry. (Only absences of at least 7 days were counted.) Find the least squares line for these data. Use your answer to predict absences in a city with a sulfur dust content of 25.

	Sulfur	Absences per 1000 Employees
Cincinnati	7	19
Indianapolis	13	44
Woodbridge	14	53
Camden	17	61
Harrison	20	88

EXERCISES ON FITTING EXPONENTIAL CURVES

Use least squares to find the exponential curve $y = Be^{Ax}$ for the following tables of points.

19.

x	y
1	2
2	4
3	7

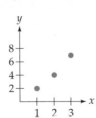

20.

x	y
1	3
2	6
3	11

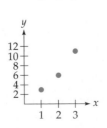

21.

x	y
1	10
3	5
6	1

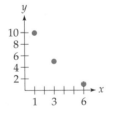

22.

x	y
1	12
4	3
5	2

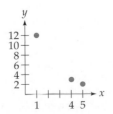

23.

x	y
0	1
1	2
2	5
3	10

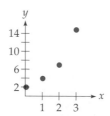

24.

x	y
0	2
1	4
2	7
3	15

25.

x	y
−1	20
0	18
1	15
3	4
5	1

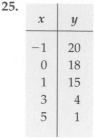

26.

x	y
−2	20
0	12
2	9
4	6
5	5

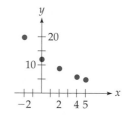

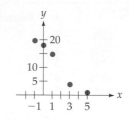

APPLIED EXERCISES ON FITTING EXPONENTIAL CURVES

27. POLITICAL SCIENCE: Cost of a Congressional Victory The following table shows average amounts spent by winners of seats in the House of Representatives in presidential election years. Fit an exponential curve to these data and use it to predict the cost of a House seat in the years 2004 and 2008.

	x	Cost ($1000)
1992	1	544
1996	2	764
2000	3	833

Source: Center for Responsive Politics

28. GENERAL: Drunk Driving The following table shows how a driver's blood-alcohol level (% grams per dekaliter) affects the probability of being in a collision. A collision factor of 3 means that the probability of a collision is 3 times as large as normal. Fit an exponential curve to the data. Then use your curve to estimate the collision factor for a blood-alcohol level of 15.

Blood-Alcohol Level	Collision Factor
0	1
6	1.1
8	3
10	6

29. BUSINESS: Super Bowl Advertising The following table shows the cost of television advertising during the Super Bowl in *thousands of dollars per second.* Fit an exponential curve to the data and then use it to predict the cost in the year 2004.

	x	Cost ($1000 per Second)
1997	1	40
1998	2	40.3
1999	3	53.3
2000	4	73.3
2001	5	80

Source: Nielsen Media Research

30. GENERAL: Stamp Prices The following table shows the cost of a first-class postage stamp in past years. Fit an exponential curve to these data and use your answer to predict the cost of a stamp in the year 2020.

	x	Postage (cents)
1940	1	3
1960	2	4
1980	3	15
2000	4	34

31. GENERAL: Safe Cars, Unsafe Streets Reread the Application Preview on page 493 at the beginning of this chapter and carry out the following steps to find when gunshot fatalities will overtake automobile fatalities in the United States.

a. Enter into your calculator the data from the table on page 493, *Years since 1970* into list L_1, *Automobile Fatalities* into L_2, and *Gunshot Fatalities* into L_3.

b. Have your calculator find the linear regression formula for L_1 and L_2 and enter it into function y_1, giving an approximate linear formula for the number of gunshot fatalities x years after 1970. *(continues)*

c. Have your calculator find the linear regression formula for L_1 and L_3 and enter it into function y_2, giving an approximate linear formula for the number of automobile fatalities x years after 1970. Graph all of the data and both functions on the window $[0, 50]$ by $[0, 60000]$.

d. Have your calculator find where the curves intersect, and interpret the x-value (years after 1970) as the year when gunshot fatalities will overtake automobile fatalities.

32. GENERAL: Cost of College Education The following table shows the cost of one year of tuition at a private four-year college. Fit an exponential curve to the data and then use it to predict the cost for the academic year 2005–2006.

	x	Cost ($1000)
1975–76	1	2.3
1980–81	2	3.6
1985–86	3	5.4
1990–91	4	9.4
1995–96	5	12.4
2000–01	6	16.3

Source: The College Board

7.5 LAGRANGE MULTIPLIERS AND CONSTRAINED OPTIMIZATION

Introduction

In Section 7.3 we optimized functions of several variables. Some problems, however, involve maximizing or minimizing a function subject to a *constraint*. For example, a company might want to maximize production subject to the constraint of staying within its budget, or a soft drink distributor might want to design the least expensive aluminum can subject to the constraint that it hold exactly 12 ounces of soda. In this section we solve such "constrained optimization" problems by the method of Lagrange multipliers, invented by the French mathematician Joseph Louis Lagrange (1736–1813).

Constraints

If a constraint can be written as an equation, such as

$$x^2 + y^2 = 100$$

then by moving all the terms to the left-hand side, it can be written with zero on the right-hand side:

$$\underbrace{x^2 + y^2 - 100}_{g(x,\, y)} = 0$$

In general, any equation can be written with all terms moved to the left-hand side, and we will write all constraints in the form

$$g(x, y) = 0$$

Practice Problem 1   Write the constraint $2y^3 = 3x - 1$ in the form $g(x, y) = 0$.

➤ Solution on page 559

First Example of Lagrange Multipliers

We illustrate the method of Lagrange multipliers by an example. The method requires a new variable, and it is customary to use λ ("lambda," the Greek letter l, in honor of Lagrange).

EXAMPLE 1

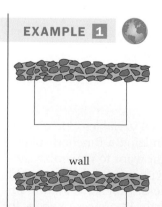

MAXIMIZING THE AREA OF AN ENCLOSURE

A farmer wants to build a rectangular enclosure along an existing stone wall. If the side along the wall needs no fence, find the dimensions of the largest enclosure that can be made using only 400 feet of fence.

Solution

We want the enclosure of largest *area*. Let

$$x = \text{width (perpendicular to the wall)}$$
$$y = \text{length (parallel to the wall)}$$

Two widths and one length must be made from the 400 feet of fence, so the constraint is

$$2x + y = 400$$

Therefore, the problem becomes:

maximize $A = xy$ \hfill Area is length times width

subject to $2x + y - 400 = 0$ \hfill Constraint $2x + y = 400$ written with zero on the right

$\underbrace{}$

$\quad\quad g(x,y)$

We write a new function $F(x, y, \lambda)$, called the *Lagrange function*, which consists of the function to be maximized plus λ times the constraint function:

$$F(x, y, \lambda) = xy + \lambda(2x + y - 400)$$

$F(x, y, \lambda) = \begin{pmatrix} \text{Function to} \\ \text{be optimized} \end{pmatrix}$

$\quad\quad + \lambda\begin{pmatrix} \text{Constraint} \\ \text{function} \end{pmatrix}$

$$= xy + \lambda 2x + \lambda y - \lambda 400$$ \hfill Multiplied out

The Lagrange function $F(x, y, \lambda)$ is a function of three variables, x, y, and λ, and we begin as usual by setting its partials with respect to each variable equal to zero.

$F_x = y + 2\lambda \quad\quad = 0$ \hfill Partial of $xy + \lambda 2x + \lambda y - \lambda 400$ with respect to x

$F_y = x + \lambda \quad\quad = 0$ \hfill Partial of $xy + \lambda 2x + \lambda y - \lambda 400$ with respect to y

$F_\lambda = \underbrace{2x + y - 400}_{} = 0$ \hfill Partial of $xy + \lambda 2x + \lambda y - \lambda 400$ with respect to λ

$\quad\quad\quad$ Constraint $g = 0$

We solve the first two of these equations for λ:

$$\lambda = -\frac{1}{2}y$$ \hfill Solving $y + 2\lambda = 0$ for λ

$$\lambda = -x$$ \hfill Solving $x + \lambda = 0$ for λ

Then we set these two expressions for λ equal to each other:

$$-\frac{1}{2}y = -x$$ \hfill Equating $\lambda = -\frac{1}{2}y$ and $\lambda = 2x$

$$y = 2x$$ \hfill Multiplying by -2

We use this relationship $y = 2x$ to eliminate the y in the equation $2x + y - 400 = 0$ (which came from the third partial $F_\lambda = 0$):

$2x + 2x - 400 = 0$ \hfill $2x + y - 400 = 0$ with y replaced by $2x$

$4x = 400$ \hfill Simplifying

$x = 100$ \hfill Dividing by 4

$y = 200$ \hfill From $y = 2x$ with $x = 100$

The largest possible enclosure has width 100 feet (perpendicular to the wall) and length 200 feet (parallel to the wall).

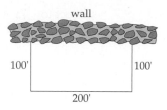

Method of Lagrange Multipliers

In general, the function to be maximized or minimized is called the *objective function,* because the "objective" of the whole procedure is to optimize it. (In Example 1 the objective function was the area function $A = xy$ which was to be maximized.) The variable λ is called the *Lagrange multiplier.* The entire method may be summarized as follows. A justification of Lagrange's method is given at the end of this section.

Lagrange Multipliers

To optimize $f(x, y)$ subject to $g(x, y) = 0$:
Objective function and constraint

1. Write $F(x, y, \lambda) = f(x, y) + \lambda g(x, y)$.
Objective function plus λ times the constraint function

2. Set the partials of F equal to zero:

$$F_x = 0 \qquad F_y = 0 \qquad F_\lambda = 0$$

and solve for the critical points.

3. The solution to the original problem (if it exists) will occur at one of these critical points.

It is important to realize that Lagrange's method only finds *critical points*—it does not tell whether the function is maximized, minimized, or neither at the critical point. (The *D*-test, which involves calculating $D = f_{xx}f_{yy} - (f_{xy})^2$, is for *un*constrained optimization, and cannot be used with constraints.) In each problem we must *know* that the maximum or minimum (whichever is requested) *does* exist, and it then follows that it must occur at a critical point (found by Lagrange multipliers).

How do we know that the maximum or minimum does exist? Most reasonable applied problems *do* have solutions. In Example 1, for instance, making the width or length too small would make the area (length times width) small, so the area must be largest for some "intermediate" length and width. Therefore, the problem *does* have a solution, which is what Lagrange multipliers find.

We could have solved Example 1 *without* Lagrange multipliers by solving the constraint equation $2x + y = 400$ for y and using it to eliminate the y in the objective function, as we did in Sections 3.3 and 3.4. However, Lagrange's method has two advantages: it eliminates the need to solve the constraint equation (which can sometimes be difficult), and it is symmetric in the variables, thereby allowing you to solve for whichever variable is easier to find.

Hints for Solving the Partial Equations

Solving the partial derivative equations $F_x = 0$, $F_y = 0$, and $F_\lambda = 0$ can sometimes be difficult. The following strategy (used in each example in this section) may be helpful.

1. Solve each of $F_x = 0$ and $F_y = 0$ for λ.
2. Set the two expressions for λ equal to each other.
3. Solve the equation resulting from step 2 together with $F_\lambda = 0$ to find x and y.

EXAMPLE 2

DESIGNING THE MOST EFFICIENT CONTAINER

A container company wants to design an aluminum can that requires the least amount of aluminum but that contains exactly 12 fluid ounces (21.3 cubic inches). Find the radius and height of the can.

Solution Minimizing the amount of aluminum means minimizing the surface area (top plus bottom plus sides) of the cylindrical can. If we let r and h stand for the radius and height (in inches) of the can, then the diagram on the next page shows that the area is:

$$A = \underbrace{2\pi r^2}_{\substack{\text{Top and} \\ \text{bottom area}}} + \underbrace{2\pi rh}_{\substack{\text{Side} \\ \text{area}}}$$

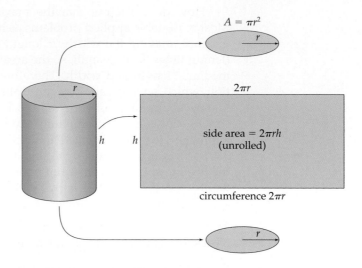

The volume is

$$V = \pi r^2 h$$

Therefore, the problem becomes

minimize $\quad A = 2\pi r^2 + 2\pi rh$ Area = (top and bottom) + side

subject to $\quad \pi r^2 h = 21.3$ Volume must equal 21.3 in^3

The Lagrange function is

$$F = 2\pi r^2 + 2\pi rh + \lambda(\pi r^2 h - 21.3)$$

Objective function A plus λ times the constraint

$$F_r = 4\pi r + 2\pi h + \lambda 2\pi rh = 0$$
$$F_h = 2\pi r + \lambda \pi r^2 \qquad\quad = 0$$
$$F_\lambda = \pi r^2 h - 21.3 \qquad\quad = 0$$

Partials set equal to zero

Solving, we have

$$\lambda = -\frac{4\pi r + 2\pi h}{2\pi rh} = -\frac{2r + h}{rh}$$

$4\pi r + 2\pi h + \lambda 2\pi rh = 0$
solved for λ

$$\lambda = -\frac{2\pi r}{\pi r^2} = -\frac{2}{r}$$

$2\pi r + \lambda \pi rh^2 = 0$
solved for λ

Equating λ's yields

$$\frac{2r + h}{rh} = \frac{2}{r}$$

Equating the two expressions for λ and multiplying each side by -1

$$2r + h = 2h$$

Multiplying each side by rh

$$2r = h$$

Subtracting h from each side

Therefore,

$$\pi r^2(2r) = 21.3$$

$\pi r^2 h - 21.3 = 0$ (the third partial equation) with $h = 2r$

$$2\pi r^3 = 21.3$$

Simplifying

$$r^3 = \frac{21.3}{2\pi} \approx 3.39$$

Dividing by 2π (using a calculator)

$$r \approx \sqrt[3]{3.39} \approx 1.5$$

Taking cube roots

$$h = 3$$

From $h = 2r$ with $r = 1.5$

The most economical 12-fluid-ounce can has radius $r = 1.5$ inches and height $h = 3$ inches.

Notice that the height (3 inches) is twice the radius (1.5 inches), *so the height equals the diameter.* This shows that the most efficient can (least area for given volume) has a "squarish" shape.

It is interesting to consider why so few cans are shaped like this. The most common 12-ounce can for soft drinks is about twice as tall as it is across, requiring about 67% more aluminum than the most efficient can. This results in an enormous waste for the millions of cans manufactured each year. It seems that beverage companies prefer taller cans because they have more area for advertising, and they are easier to hold. Some products, however, are sold in "efficient" cans, with the height equal to the diameter.

Practice Problem 2 Which of the cans pictured below is most "efficiently" proportioned?

cola   paint   motor oil   tuna

➤ Solution on page 559

With Lagrange multipliers we can maximize a company's output subject to a budget constraint.

EXAMPLE 3

MAXIMIZING PRODUCTION

A company's output is given by the Cobb–Douglas production function $P = 600L^{2/3}K^{1/3}$, where L and K are the numbers of units of labor and capital. Each unit of labor costs the company $40, and each unit of capital costs $100. If the company has a total of $3000 for labor and capital, how much of each should it use to maximize production?

Solution   The budget constraint is

$$40L + 100K = 3000$$

Cost of   Cost of   Total
labor   capital   budget

Unit cost times the number of units for labor and capital

Therefore, the problem becomes:

maximize $P = 600L^{2/3}K^{1/3}$

subject to $40L + 100K - 3000 = 0$ Constraint with zero on the right

$F(L, K, \lambda) = 600L^{2/3}K^{1/3} + \lambda(40L + 100K - 3000)$ Lagrange function

$$\left.\begin{array}{l} F_L = 400L^{-1/3}K^{1/3} + 40\lambda = 0 \\ F_K = 200L^{2/3}K^{-2/3} + 100\lambda = 0 \\ F_\lambda = 40L + 100K - 3000 = 0 \end{array}\right\}\ \begin{array}{l}\text{Partials set equal}\\ \text{to zero}\end{array}$$

$\lambda = -10L^{-1/3}K^{1/3}$ Solving the first equation for λ

$\lambda = -2L^{2/3}K^{-2/3}$ Solving the second equation for λ

$-10L^{-1/3}K^{1/3} = -2L^{2/3}K^{-2/3}$ Equating the two expressions for λ

$5L^{-1/3}K^{1/3} = L^{2/3}K^{-2/3}$ Dividing each side by -2

$5K = L$ Multiplying by $L^{1/3}K^{2/3}$ on each side

$40(5K) + 100K - 3000 = 0$ Substituting $L = 5K$ into $40L + 100K - 300 = 0$

$300K = 3000$ Simplifying

$K = 10$ Solving for K

$L = 50$ From $L = 5K$ with $K = 10$

The company should use 50 units of labor and 10 units of capital.

Meaning of the Lagrange Multiplier

The Lagrange multiplier λ has a useful interpretation. If we call the units of the objective function "objective units" and the units of the constraint function "constraint units," then λ (or, more precisely, its absolute value) has the following interpretation, which will be justified in the next section (see pages 569–570).

> **Interpretation of λ**
>
> $$|\lambda| = \left(\begin{array}{c} \text{Number of additional objective units} \\ \text{for each additional constraint unit} \end{array} \right)$$

As an illustration, in the preceding example we maximized production (units) subject to a budget constraint (dollars), so $|\lambda|$ gives *the number of additional units produced per additional dollar*, or, in the jargon of economics, *the marginal productivity of money*. We can calculate the value of λ from either of the expressions for it.

$$\underbrace{\lambda = -10L^{-1/3}K^{1/3}}_{\text{From Example 3}} = \underbrace{-10(50)^{-1/3}(10)^{1/3}}_{\substack{\text{Substituting} \\ L - 50, \ K = 10}} \approx \underbrace{-5.8}_{\substack{\text{Using a} \\ \text{calculator}}} \qquad \begin{array}{l} \text{From the} \\ \text{previous page} \end{array}$$

Therefore* $|\lambda| \approx 5.8$, meaning that production increases by about 5.8 units for each additional dollar in the budget. For example, an additional \$100 would result in about 580 extra units of production.

Practice Problem 3

In Example 1 we maximized the area of an enclosure subject to the constraint of having only 400 feet of fence.

a. Interpret the meaning of $|\lambda|$ in Example 1. [*Hint:* The objective function is area, and the constraint units are feet of fence.]

b. Calculate $|\lambda|$ in Example 1. [*Hint:* Use either expression for λ on page 549.]

c. Use this number to approximate the additional area that could be enclosed by an additional 5 feet of fence. ➤ **Solutions on page 559**

Geometry of Constrained Optimization

A constrained optimization problem can be visualized as a surface (the objective function), with a curve along it determined by the constraint. The *constrained* maximum is the highest point along the *curve*, whereas the *uncon-*strained maximum is the highest point on the *entire surface*.

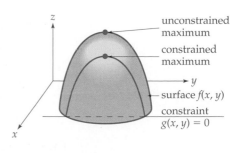

unconstrained maximum

constrained maximum

surface $f(x, y)$

constraint $g(x, y) = 0$

---

* The negative sign in λ comes from our defining the Lagrange function as $F = f + \lambda g$. We could equally well have defined it as $F = f - \lambda g$, in which case λ would have been positive.

If there are several critical points, we evaluate the objective function at each of them. The maximum and minimum values (which we assume to exist) are the largest and smallest of the resulting values of the function. In this way we can find both the maximum and minimum values in a constrained optimization problem.

EXAMPLE 4

FINDING BOTH EXTREME VALUES

Maximize *and* minimize $f(x, y) = 4xy$
subject to the constraint $x^2 + y^2 = 50$

Solution

$$F(x, y, \lambda) = 4xy + \lambda(x^2 + y^2 - 50)$$ Lagrange function

$$\left.\begin{array}{l} F_x = 4y + \lambda 2x \quad = 0 \\ F_y = 4x + \lambda 2y \quad = 0 \\ F_\lambda = x^2 + y^2 - 50 = 0 \end{array}\right\} \text{Partials set equal to zero}$$

$$\lambda = -\frac{4y}{2x} = -\frac{2y}{x} \qquad \begin{array}{l}\text{Solving } 4y + \lambda 2x = 0 \text{ for } \lambda \\ \text{(and simplifying)}\end{array}$$

$$\lambda = -\frac{4x}{2y} = -\frac{2x}{y} \qquad \begin{array}{l}\text{Solving } 4x + \lambda 2y = 0 \text{ for } \lambda \\ \text{(and simplifying)}\end{array}$$

$$\frac{2y}{x} = \frac{2x}{y} \qquad \begin{array}{l}\text{Equating the two } \lambda\text{'s} \\ \text{(multplying by } -1)\end{array}$$

$$2y^2 = 2x^2 \qquad \begin{array}{l}\text{Multiplying both sides by } xy \\ \text{(or "cross-multiplying")}\end{array}$$

$$y^2 = x^2 \qquad \text{Dividing by 2}$$

$$y = \pm x \qquad \begin{array}{l}\text{Taking square roots} \\ (+ \text{ and } -)\end{array}$$

$$x^2 + x^2 - 50 = 0 \qquad \begin{array}{l}x^2 + y^2 - 50 = 0 \text{ (the third} \\ \text{partial equation) with } y = \pm x\end{array}$$

$$2x^2 - 50 = 0 \qquad \text{Simplifying}$$

$$2(x^2 - 25) = 0 \qquad \text{Factoring}$$

$$\underbrace{\qquad\qquad}$$
$$(x + 5)(x - 5)$$

$$x = \pm 5 \qquad \text{Solving}$$
$$y = \pm 5 \qquad \text{From } y = \pm x$$

Thus there are *four* critical points (all possible combinations of $x = \pm 5$ and $y = \pm 5$). We evaluate the objective function $f(x, y) = 4xy$ at each of them.

Critical point	$f(x, y) = 4xy$	
(5, 5)	100	} 100 is the highest value of $f = 4xy$,
(−5, −5)	100	occurring at (5, 5) and at (−5, −5)
(5, −5)	−100	} −100 is the lowest value of $f = 4xy$,
(−5, 5)	−100	occurring at (−5, 5) and at (5, −5)

Maximum value of f is 100, occurring at $\begin{cases} x = 5 \\ y = 5 \end{cases}$ and $\begin{cases} x = -5 \\ y = -5 \end{cases}$

Minimum value of f is −100, occurring at $\begin{cases} x = -5 \\ y = 5 \end{cases}$ and $\begin{cases} x = 5 \\ y = -5 \end{cases}$

Constrained Optimization of Functions of Three or More Variables

To optimize a function f of any number of variables subject to a constraint $g = 0$, we proceed just as before, writing the Lagrange function $F = f + \lambda g$ and setting the partial with respect to each variable equal to zero. For example, to maximize $f(x, y, z)$ subject to $g(x, y, z) = 0$, we would write

$$F(x, y, z, \lambda) = f(x, y, z) + \lambda g(x, y, z) \qquad \text{Lagrange function}$$

and solve the partial equations

$$F_x = 0$$
$$F_y = 0$$
$$F_z = 0$$
$$F_\lambda = 0$$

Justification of Lagrange's Method

The following is a geometric justification of Lagrange's method for maximizing an objective function $f(x, y)$ subject to a constraint $g(x, y) = 0$. In the graph on the next page, each of the parallel curves represents a curve along which the objective function f takes a constant value, with successively higher curves corresponding to higher values of the constant. Maximizing f subject to $g = 0$ means finding the highest of these "objective" curves that still meets the constraint curve.

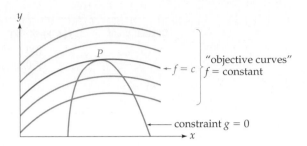

Observe that at the point where the highest objective curve meets the constraint curve (the point P), *the two curves have the same slope.* In Exercise 37 on page 572 we will show that the slope of the curve $f = c$ is $-f_x/f_y$, and the slope of the curve $g = 0$ is $-g_x/g_y$. Equating these slopes gives

$$-\frac{f_x}{f_y} = -\frac{g_x}{g_y} \quad \text{or, equivalently,} \qquad -\frac{f_x}{g_x} = -\frac{f_y}{g_y} \qquad \begin{array}{l}\text{Multiplying by } f_y \\ \text{and dividing by } g_x\end{array}$$

The two sides of this last equation being equal is equivalent to equating each side to a number λ:

$$\lambda = -\frac{f_x}{g_x} \quad \text{and} \quad \lambda = -\frac{f_y}{g_y}$$

With a little algebra (clearing fractions, and moving everything to one side), these equations become

$$f_x + \lambda g_x = 0 \quad \text{and} \quad f_y + \lambda g_y = 0$$

However, these two equations, together with the constraint $g = 0$, are precisely the equations obtained from Lagrange's method by setting the partials of $F = f + \lambda g$ equal to zero. This completes the justification of Lagrange's method, and also exhibits its simplicity and power by showing all that is accomplished by the familiar steps of setting the partials equal to zero.

7.5 Section Summary

To optimize (maximize or minimize) an *objective* function $f(x, y)$ subject to a *constraint* $g(x, y) = 0$ we form the *Lagrange function* $F(x, y, \lambda) = f(x, y) + \lambda g(x, y)$ and set its partials equal to zero:

$$f_x(x, y) + \lambda g_x(x, y) = 0 \qquad\qquad F_x = 0$$
$$f_y(x, y) + \lambda g_y(x, y) = 0 \qquad\qquad F_y = 0$$
$$g(x, y) = 0 \qquad\qquad F_\lambda = 0$$

We then find all solutions (x, y) of these equations. The maximum or minimum (if they exist) of $f(x, y)$ subject to $g(x, y) = 0$ will occur at one of these solutions.

The absolute value of the *Lagrange multiplier* λ found from solving these equations gives the approximate change in the *objective* function that would result from changing the *constraint* function by one unit.

➤ **Solutions to Practice Problems**

1. $2y^3 - 3x + 1 = 0$ Moving all terms to the left-hand side

2. The paint can

3. **a.** The approximate additional area for each additional foot of fence

    b. $|\lambda| = 100$ (using $\lambda = -x$ with $x = 100$)

    c. Approximately 500 more square feet of area (from $5 \cdot 100$)

7.5 Exercises

Use Lagrange multipliers to maximize each function $f(x, y)$ subject to the constraint. (The maximum values *do* exist.)

1. $f(x, y) = 3xy, \quad x + 3y = 12$

2. $f(x, y) = 2xy, \quad 2x + y = 20$

3. $f(x, y) = 6xy, \quad 2x + 3y = 24$

4. $f(x, y) = 3xy, \quad 3x + 2y = 60$

5. $f(x, y) = xy - 2x^2 - y^2, \quad x + y = 8$

6. $f(x, y) = 12xy - 3y^2 - x^2, \quad x + y = 16$

7. $f(x, y) = x^2 - y^2 + 3, \quad 2x + y = 3$

8. $f(x, y) = y^2 - x^2 - 5, \quad x + 2y = 9$

9. $f(x, y) = \ln(xy), \quad x + y = 2e$

10. $f(x, y) = e^{xy}, \quad x + 2y = 8$

Use Lagrange multipliers to minimize each function $f(x, y)$ subject to the constraint. (The minimum values *do* exist.)

11. $f(x, y) = x^2 + y^2, \quad 2x + y = 15$

12. $f(x, y) = x^2 + y^2, \quad x + 2y = 30$

13. $f(x, y) = xy, \quad y = x + 8$

14. $f(x, y) = xy, \quad y = x + 6$

15. $f(x, y) = x^2 + y^2, \quad 2x + 3y = 26$

16. $f(x, y) = 5x^2 + 6y^2 - xy, \quad x + 2y = 24$

17. $f(x, y) = \ln(x^2 + y^2), \quad 2x + y = 25$

18. $f(x, y) = \sqrt{x^2 + y^2 + 5}, \quad 2x + y = 10$

19. $f(x, y) = e^{x^2 + y^2}, \quad x + 2y = 10$

20. $f(x, y) = 2x + y, \quad 2\ln x + \ln y = 12$

Use Lagrange multipliers to maximize *and* minimize each function subject to the constraint. (The maximum and minimum values *do* exist.)

21. $f(x, y) = 2xy, \quad x^2 + y^2 = 8$

22. $f(x, y) = 2xy, \quad x^2 + y^2 = 18$

23. $f(x, y) = x + 2y, \quad 2x^2 + y^2 = 72$

24. $f(x, y) = 12x + 30y, \quad x^2 + 5y^2 = 81$

APPLIED EXERCISES

Solve each using Lagrange multipliers. (The stated extreme values *do* exist.)

25. **GENERAL: Fences** A parking lot, divided into two equal parts, is to be constructed against a building, as shown in the diagram. Only 6000 feet of fence are to be used, and the side along the building needs no fence.

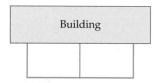

Building

a. What are the dimensions of the largest area that can be so enclosed?
b. Evaluate and give an interpretation for $|\lambda|$.

26. **GENERAL: Fences** Three adjacent rectangular lots are to be fenced in, as shown in the diagram using 12,000 feet of fence. What is the largest total area that can be so enclosed?

27–28. GENERAL: Container Design A cylindrical tank without a top is to be constructed with the least amount of material (bottom plus side area). Find the dimensions if the volume is to be:

open top

h

27. 160 cubic feet

28. 120 cubic feet

29. **GENERAL: Postal Regulations** The U.S. Postal Service will accept a package if its length plus its girth is not more than 84 inches. Find the dimensions and volume of the largest package with a square end that can be mailed. (*See the diagram in the next column.*)

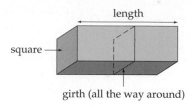

length

square

girth (all the way around)

30. **GENERAL: Postal Regulations** Solve Exercise 29, but now for a package with a round end, so that the package is a cylinder rather than a rectangular solid. Compare the volume with that of Exercise 29.

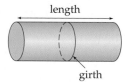

length

girth

31. **BUSINESS: Maximum Production** A company's output is given by the Cobb–Douglas production function $P = 200L^{3/4}K^{1/4}$, where L and K are the numbers of units of labor and capital. Each unit of labor costs $50 and each unit of capital costs $100, and $8000 is available to pay for labor and capital.

a. How many units of labor and of capital should be used to maximize production?
b. Evaluate and give an interpretation for $|\lambda|$.

32. **BUSINESS: Production Possibilities** A company manufactures two products, in quantities x and y. Because of limited materals and capital, the quantities produced must satisfy the equation $2x^2 + 5y^2 = 32,500$. (This curve is called a *production possibilities curve*.) If the company's profit function is $P = 4x + 5y$ dollars, how many of each product should be made to maximize profit? Also find the maximum profit.

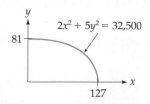

33. GENERAL: Package Design A metal box with a square base is to have a volume of 45 cubic inches. If the top and bottom cost 50 cents per square inch and the sides cost 30 cents per square inch, find the dimensions that minimize the cost. [*Hint:* The cost of the box is the area of each part (top, bottom, and sides) times the cost per square inch for that part. Minimize this subject to the volume constraint.]

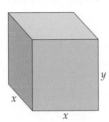

34. GENERAL: Building Design A one-story building is to have 8000 square feet of floor space. The front of the building is to be made of brick, which costs $120 per linear foot, and the back and sides are to be made of cinderblock, which costs only $80 per linear foot.

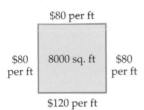

$80 per ft

$80 per ft   8000 sq. ft $80 per ft

$120 per ft

a. Find the length and width that minimize the cost of the building. [*Hint:* The cost of the building is the length of the front, back, and sides, each times the cost per foot for that part. Minimize this subject to the area constraint.]

b. Evaluate and give an interpretation for $|\lambda|$.

Functions of Three Variables

(The stated extreme values *do* exist.)

35. Minimize $f(x, y, z) = x^2 + y^2 + z^2$
subject to $2x + y - z = 12$.

36. Minimize $f(x, y, z) = x^2 + y^2 + z^2$
subject to $x - y + 2z = 6$.

37. Maximize $f(x, y, z) = x + y + z$
subject to $x^2 + y^2 + z^2 = 12$.

38. Maximize $f(x, y, z) = xyz$
subject to $x^2 + y^2 + z^2 = 12$.

39. GENERAL: Building Design A one-story storage building is to have a volume of 250,000 cubic feet. The roof costs $32 per square foot, the walls $10 per square foot, and the floor $8 per square foot. Find the dimensions that minimize the cost of the building.

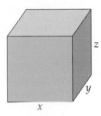

40. GENERAL: Container Design An open-top box with two parallel partitions, as in the diagram, is to have volume 64 cubic inches. Find the dimensions that require the least amount of material.

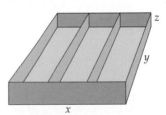

<hr/>

7.6 **TOTAL DIFFERENTIALS AND APPROXIMATE CHANGES**

Introduction

In this section we define the *total differential* of a function of several variables, and use it to approximate the change in the function resulting from changes in the independent variables. In addition to giving several applications, we will justify the interpretation of the Lagrange variable λ introduced on page 555.

Total Differential of a Function of Two Variables

For a function $f(x)$ of one variable, we defined on page 414 the differential to be the *derivative multiplied by dx*, $dy = f'(x) \cdot dx$. For a function $f(x, y)$ of *two* variables, we define the total differential analogously as the partial derivatives multiplied by dx and dy and added.*

Total Differential of $f(x, y)$

For a function $f(x, y)$, the total differential df is

$$df = f_x(x, y) \cdot dx + f_y(x, y) \cdot dy$$

More briefly,
$df = f_x \cdot dx + f_y \cdot dy$

The total differential can also be written with the partials in the ∂ notation:

$$df = \frac{\partial f}{\partial x} \cdot dx + \frac{\partial f}{\partial y} \cdot dy$$

Total differential in ∂ notation

EXAMPLE 1

FINDING A TOTAL DIFFERENTIAL

Find the total differential df of $\quad f(x, y) = 5x^3 - 4xy^{-1} + 3y^4$.

Solution   The partials are

$$f_x = 15x^2 - 4y^{-1}$$

Partial of $5x^3 - 4xy^{-1} + 3y^4$ with respect to x

$$f_y = 4xy^{-2} + 12y^3$$

Partial of $5x^3 - 4xy^{-1} + 3y^4$ with respect to y

Then df is

$$df = \underbrace{(15x^2 - 4y^{-1})}_{f_x(x, y)} \cdot dx + \underbrace{(4xy^{-2} + 12y^3)}_{f_y(x, y)} \cdot dy$$

$df = f_x \, dx + f_y \, dy$ with the above partials

---

* Technically, the partials f_x and f_y must be continuous. However, since we have not defined continuity for functions of two variables, we will not discuss this requirement further, except to say that it is satisfied by all functions in this section and by most functions encountered in applications.

If dx and dy are considered as new variables, then the total differential df is a function of the *four* variables x, y, dx, and dy. The total differential of $z = f(x, y)$ is denoted dz.

EXAMPLE 2

FINDING THE TOTAL DIFFERENTIAL OF A LOGARITHMIC FUNCTION

Find the total differential of $z = \ln(x^2 + y^3)$.

Solution The partials are

Partials of denominator

$$z_x = \frac{2x}{x^2 + y^3} \qquad z_y = \frac{3y^2}{x^2 + y^3} \qquad \text{Derivative of } \ln f \text{ is } \frac{f'}{f}$$

Therefore, dz is

$$dz = \frac{2x}{x^2 + y^3} \cdot dx + \frac{3y^2}{x^2 + y^3} \cdot dy \qquad \text{Partials times } dx \text{ and } dy$$

Practice Problem

Find the total differential of $g(x, y) = x^2 e^{5y}$. ➤ Solution on page 571

Approximating Changes by Total Differentials

For a function $f(x, y)$, changing the value of x by Δx and y by Δy generally changes the value of the function. The change Δf in the function is found by evaluating f at the "changed" values and subtracting f evaluated at the original values.

Change in $f(x, y)$

$$\Delta f = \underbrace{f(x + \Delta x, y + \Delta y)}_{\substack{f \text{ at the} \\ \text{"new" values}}} - \underbrace{f(x, y)}_{\substack{f \text{ at the} \\ \text{"old" values}}}$$

For *independent* variables (such as x and y) we use "Δ" and "d" interchangeably to denote changes. That is,

$$\Delta x = dx \qquad \text{and} \qquad \Delta y = dy \qquad \begin{array}{l} \text{For independent variables,} \\ \text{``}\Delta\text{''} = \text{``}d\text{''} \end{array}$$

However, for *dependent* variables, "Δ" and "*d*" have different meanings: Δ indicates the *actual* change, and *d* indicates the total differential.

For some functions, calculating the actual change Δ*f* can be complicated. The total differential provides a simple *linear approximation* for the actual change. The partials f_x and f_y give the rate of change of *f* per unit change in *x* and *y*, respectively. Changing *x* by Δ*x* units changes *f* by approximately $f_x \cdot \Delta x$ units, and changing *y* by Δ*y* units changes *f* by approximately $f_y \cdot \Delta y$ units. Therefore, changing *both x* and *y* should change *f* by approximately the *sum* of these changes. In symbols:

Differential Approximation Formula

$$\underbrace{f(x + \Delta x, y + \Delta y) - f(x, y)}_{\text{Change in } f} \quad \approx \quad \underbrace{f_x \cdot \Delta x + f_y \cdot \Delta y}_{\substack{\text{Total differential of } f \\ (\text{since } \Delta x = dx \text{ and } \Delta y = dy)}}$$

Written more compactly:

$$\Delta f \approx df$$

Since $\Delta f = f(x + \Delta x, y + \Delta y) - f(x, y)$
and $df = f_x(x, y) \cdot dx + f_y(x, y) \cdot dy$

The partials in the first formula are evaluated at (x, y), and the approximation improves as Δ*x* and Δ*y* approach zero. (See pages 567–568 for a geometric explanation of the approximation $\Delta f \approx df$.)

EXAMPLE 3

APPROXIMATING AN ACTUAL CHANGE BY A DIFFERENTIAL

For $f(x, y) = x^2 + 4xy + y^3$ and values $x = 3$, $y = 2$, $\Delta x = 0.2$, and $\Delta y = -0.1$, find: **a.** Δ*f*   **b.** *df*

Solution

a. From the given values,

$$x + \Delta x = 3 + 0.2 = 3.2 \quad \text{and} \quad y + \Delta y = 2 - 0.1 = 1.9$$

The change Δ*f* is

$$\Delta f = f(3.2, 1.9) - f(3, 2) \qquad \qquad \Delta f = f(x + \Delta x, y + \Delta y) - f(x, y)$$

$$= \underbrace{3.2^2 + 4 \cdot (3.2) \cdot (1.9) + 1.9^3}_{f(3.2, 1.9)} - \underbrace{(3^2 + 4 \cdot 3 \cdot 2 + 2^3)}_{f(3, 2)} \qquad \text{Using } f(x, y) = x^2 + 4xy + y^3$$

$$= 10.24 + 24.32 + 6.859 - (9 + 24 + 8) \qquad \text{Evaluating}$$

$$= 41.419 - 41 = 0.419 \qquad \qquad \text{Change in } f \text{ is } \Delta f = 0.419$$

b. The total differential df is

$$df = \overbrace{(2x + 4y)}^{f_x(x, y)} \cdot dx + \overbrace{(4x + 3y^2)}^{f_y(x, y)} \cdot dy$$

Partials of $x^2 + 4xy + y^3$ times dx and dy

$$= (2 \cdot 3 + 4 \cdot 2) \cdot (0.2) + (4 \cdot 3 + 3 \cdot 2^2) \cdot (-0.1)$$

Evaluating at $x = 3$, $y = 2$, $dx = 0.2$, and $dy = -0.1$

$$= 2.8 - 2.4 = 0.4$$

Total differential is $df = 0.4$

We found $\begin{cases} \Delta f = 0.419 \\ df = 0.4 \end{cases}$. The total differential $df = 0.4$ provides a reasonably accurate approximation for the actual change $\Delta f = 0.419$. The approximation would be even more accurate for smaller values of Δx and Δy.

Why should we bother calculating an *approximation df* when with a calculator we can easily find the *exact* change Δf? The total differential df has the advantage of *linearity:* If we were to *double* the changes Δx and Δy in the independent variables, then the differential would also double, from 0.4 to 0.8; if we were to *halve* the changes Δx and Δy, then the differential would also be halved, from 0.4 to 0.2. The *actual* change Δf admits no such simple modification—whenever Δx and Δy change it has to be recalculated "from scratch" using the formula $\Delta f = f(x + \Delta x, y + \Delta y) - f(x, y)$. The linearity of the total differential df is an advantage when you are trying to understand how various changes in x and y would affect the value of f.

EXAMPLE **4**

ESTIMATING ADDITIONAL PROFIT

The American Farm Machinery Company finds that if it manufactures x economy tractors and y heavy-duty tractors per month, then its profit (in thousands of dollars) will be $P(x, y) = 3x^{4/3} + 0.05xy + 4y$. If the company now manufactures 125 economy tractors and 100 heavy-duty tractors per month, find an approximation for the additional profit from manufacturing 3 more economy tractors and 2 more heavy-duty tractors per month.

Solution   We want an estimate for the change in the profit $P(x, y)$ when production increases above the levels $x = 125$ and $y = 100$ by amounts $\Delta x = 3$ and $\Delta y = 2$. The partials are

$$P_x = 4x^{1/3} + 0.05y \qquad P_y = 0.05x + 4$$

Partials of $P = 3x^{4/3} + 0.05xy + 4y$ with respect to x and y

The total differential is

$$dP = (4x^{1/3} + 0.05y) \cdot dx + (0.05x + 4) \cdot dy$$

$dP = P_x \cdot dx + P_y \cdot dy$

$$= \underbrace{(4 \cdot 125^{1/3}}_{\sqrt[3]{125} \,=\, 5} + 0.05 \cdot 100) \cdot 3 + (0.05 \cdot 125 + 4) \cdot 2$$

Substituting
$x = 125$, $y = 100$,
$\Delta x = 3$, and $\Delta y = 2$

$$= (20 + 5) \cdot 3 + (6.25 + 4) \cdot 2 = 75 + 20.5 = 95.5$$

In thousands
of dollars

Therefore, producing 3 more economy tractors and 2 more heavy-duty tractors will generate about $95,500 in additional profit.

The actual change in the profit function, found by applying the Δf formula on page 563, is $\Delta P = P(125 + 3, 100 + 2) - P(125, 100) \approx 96.04$, so the approximation of 95.5 is indeed quite accurate.

The *linearity* of the total differential means that if sometime later the company wanted to estimate the additional profit from *doubling* these changes—making 6 more economy tractors and 4 more heavy-duty tractors—they would need only to double the estimate of $95,500 to immediately obtain $191,000.

Estimating Errors

No physical measurement can ever be made with perfect accuracy. If you can estimate the maximum error in a measurement, then you can use differentials to estimate the resulting error in a calculation. The measurement errors may be expressed as *percentage* errors or *actual* numbers (as in Example 6, to be discussed shortly). The following example estimates the percentage error in calculating the volume of a cylinder. Such calculations have applications from predicting variations in soft-drink cans to ensuring safety margins for artificial arteries.

EXAMPLE 5

ESTIMATING THE ERROR IN CALCULATING VOLUME

A cylinder is measured to have radius r and height h, but these measurements may be in error by up to 1%. Estimate the resulting percentage error in calculating the volume of the cylinder.

Solution   The height and radius being in error by 1% means that

$$\Delta r = 0.01r$$
$$\Delta h = 0.01h$$

For each, the change is 1% of the value

The volume of a cylinder is $V = \pi r^2 h$,   and the total differential is

$$dV = 2\pi rh \cdot dr + \pi r^2 \cdot dh$$

Partials of $V = \pi r^2 h$ times dr and dh

$$= 2\pi rh \cdot 0.01r + \pi r^2 \cdot 0.01h$$

Substituting $dr = \Delta r = 0.01r$ and $dh = \Delta r = 0.01h$

$$= \pi r^2 h \underbrace{(2 \cdot 0.01 + 0.01)}_{0.03}$$

Factoring out $\pi r^2 h$

$$= \underbrace{0.03 \pi r^2 h}_{} = 0.03 \cdot V$$

Change is 3% of volume

Volume V of the cylinder

This result, $dV = 0.03 \cdot V$,   means that the volume may be in error by as much as 3% if the radius and height are "off" by 1%.

Geometric Visualization of *df* and Δ*f*

A function $f(x, y)$ represents a *surface* in three-dimensional space, and the change Δf represents the change in height when moving from one point to another along the surface, as shown below.

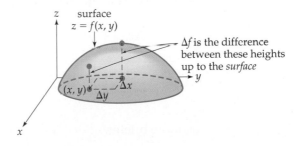

The total differential *df* represents the change in height when moving from one point to another along the *plane* that best fits the surface at the first point, as shown on the following page. This plane is called the *tangent plane* to the surface at the point.

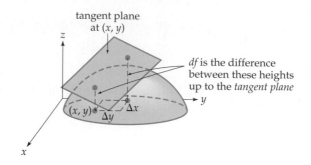

Since the tangent plane fits the surface closely near the point (x, y), changes df in height along the *tangent plane* should be very close to changes Δf in height along the *surface,* which is why the total differential df closely approximates the actual change Δf for small changes in x and y.

Total Differential of a Function of Three Variables

The total differential may be generalized to apply to functions of three (or more) variables, multiplying each partial by "d" of that variable and adding.

Total Differential of $f(x, y, z)$

For a function $f(x, y, z)$, the total differential df is

$$df = f_x(x, y, z) \cdot dx + f_y(x, y, z) \cdot dy + f_z(x, y, z) \cdot dz \qquad \text{Partials times } dx, dy, \text{ and } dz$$

The total differential of a function $f(x, y, z)$ gives an estimate for the change Δf when the variables are changed by amounts Δx, Δy, and Δz:

$$\underbrace{f(x + \Delta x, y + \Delta y, z + \Delta z) - f(x, y, z)}_{\text{Change in } f} \approx \underbrace{\frac{\partial f}{\partial x} \cdot \Delta x + \frac{\partial f}{\partial y} \cdot \Delta y + \frac{\partial f}{\partial z} \cdot \Delta z}_{\substack{\text{Total differential of } f \\ (\text{since } \Delta x = dx, \\ \Delta y = dy,\ \Delta z = dz)}}$$

or, more briefly:

$$\Delta f \approx df$$

The partials in the formula above are evaluated at (x, y, z), and the approximation is increasingly accurate for smaller values of Δx, Δy, and Δz.

EXAMPLE

ESTIMATING THE ERROR IN CALCULATING VOLUME

A rectangular box is measured to be 30 inches long, 24 inches wide, and 10 inches high. If the maximum errors in measuring the length, width, and height of the box are, respectively, 0.3, 0.2, and 0.1 inch, estimate the maximum error in calculating its volume.

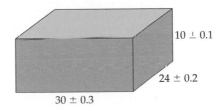

Solution The volume of the box is length times width times height, $V = x \cdot y \cdot z$. We want to estimate the change in volume resulting from changing the dimensions $x = 30$, $y = 24$, and $z = 10$ by the amounts $\Delta x = 0.3$, $\Delta y = 0.2$, and $\Delta z = 0.1$. The total differential is

$$df = \overbrace{y \cdot z}^{V_x} \cdot dx + \overbrace{x \cdot z}^{V_y} \cdot dy + \overbrace{x \cdot y}^{V_z} \cdot dz$$

Partials of $V = x \cdot y \cdot z$ times $dx, dy,$ and dz

$$= 24 \cdot 10 \cdot 0.3 + 30 \cdot 10 \cdot 0.2 + 30 \cdot 24 \cdot 0.1$$

Substituting $x = 30$, $y = 24$, $z = 10$, $dx = 0.3$, $dy = 0.2$, and $dz = 0.1$

$$= 72 + 60 + 72 = 204$$

Multiplying out and adding

That is, the maximum error in calculating the volume is approximately 204 cubic inches.

While an error of 204 cubic inches may seem large, the *relative* error (the error divided by the volume, expressed as a percentage) is only

$$\frac{204}{30 \cdot 24 \cdot 10} = \frac{204}{7200} \approx 0.028 = 2.8\%$$

Relative error is 2.8%

Justification of the Interpretation of the Lagrange Multiplier λ

In the preceding section we used Lagrange multipliers to optimize an objective function $f(x, y)$ subject to a constraint $g(x, y) = 0$. We did so by setting the partials of $F(x, y, \lambda) = f(x, y) + \lambda g(x, y)$ equal to zero, and we interpreted the absolute value of the "Lagrange multiplier" $|\lambda|$ as *the number of additional objective units per additional constraint unit*. This interpretation may be justified as follows. Setting the partials of F with respect to x and y equal to zero gives

$$\begin{cases} f_x + \lambda g_x = 0 \\ f_y + \lambda g_y = 0 \end{cases} \quad \text{or, equivalently,} \quad \begin{cases} f_x = -\lambda g_x \\ f_y = -\lambda g_y \end{cases}$$

If we increase x and y by amounts Δx and Δy, the resulting change in the objective function $f(x, y)$ can be approximated by the total differential:

$$\underbrace{f(x + \Delta x, y + \Delta y) - f(x, y)}_{\Delta f} \approx f_x \, \Delta x + f_y \, \Delta y \qquad \Delta f \approx df$$

$$= -\lambda g_x \, \Delta x - \lambda g_y \, \Delta y \qquad \begin{array}{l} \text{Substituting} \\ f_x = -\lambda g_x \text{ and} \\ f_y = -\lambda g_y \end{array}$$

$$= -\lambda \underbrace{(g_x \, \Delta x + g_y \, \Delta y)}_{dg} \qquad \begin{array}{l} \text{Factoring out } -\lambda \\ \text{leaves the total} \\ \text{differential of } g \end{array}$$

$$\approx -\lambda \cdot \Delta g \qquad \begin{array}{l} \text{Replacing } dg \text{ by } \Delta g \\ (\text{since } dg \approx \Delta g) \end{array}$$

These equations, read from beginning to end, say that $\Delta f \approx -\lambda \cdot \Delta g$, or

$$\frac{\Delta f}{\Delta g} \approx -\lambda \qquad \text{Dividing by } \Delta g$$

The left-hand side of this is the ratio of the change in the objective function f to the change in the constraint function g. Taking absolute values and letting Δx and Δy approach zero (to make the approximation exact) shows that $|\lambda|$ is *the number of objective units per additional constraint unit*, as stated on page 555.

7.6 Section Summary

For a function $f(x, y)$, the total differential df is defined as the partials multiplied by dx and dy and added:

$$df = f_x(x, y) \cdot dx + f_y(x, y) \cdot dy \qquad df = \frac{\partial f}{\partial x} dx + \frac{\partial f}{\partial y} dy$$

The actual change in the function when x and y change by amounts Δx and Δy is

$$\Delta f = f(x + \Delta x, y + \Delta y) - f(x, y)$$

For small values of $\Delta x = dx$ and $\Delta y = dy$, the actual change Δf can be approximated by the total differential df:

$$f(x + \Delta x, y + \Delta y) - f(x, y) \approx f_x(x, y) \cdot dx + f_y(x, y) \cdot dy \qquad \Delta f \approx df$$

The actual change Δf may depend on Δx and Δy in very complicated ways, but the total differential df is *linear* in $\Delta x = dx$ and $\Delta y = dy$, and is therefore easier to calculate.

➤ **Solution to Practice Problem**

$g_x = 2xe^{5y} \qquad g_y = x^2 5 e^{5y}$ Partials

$dg = 2xe^{5y} \cdot dx + 5x^2 e^{5y} \cdot dy$ Total differential

7.6 Exercises

Find the total differential of each function.

1. $f(x, y) = x^2 y^3$

2. $f(x, y) = x^4 y^{-1}$

3. $f(x, y) = 6x^{1/2} y^{1/3} + 8$

4. $f(x, y) = 100 x^{0.05} y^{0.02} - 7$

5. $g(x, y) = \dfrac{x}{y}$

6. $g(x, y) = \dfrac{x}{x + y}$

7. $g(x, y) = (x - y)^{-1}$

8. $g(x, y) = \sqrt{x^2 + y^2}$

9. $z = \ln(x^3 - y^2)$

10. $z = x^2 \ln y$

11. $z = xe^{2y}$

12. $z = e^{3x - 2y}$

13. $w = 2x^3 + xy + y^2$

14. $w = 3x^2 - xy^{-1} + y^3$

15. $f(x, y, z) = 2x^2 y^3 z^4$

16. $f(x, y, z) = xy + yz + xz$

17. $f(x, y, z) = \ln(xyz)$

18. $f(x, y, z) = \ln(x^2 + y^2 + z^2)$

19. $f(x, y, z) = e^{xyz}$

20. $f(x, y, z) = e^{x^2 + y^2 + z^2}$

For the given function and values, find:

a. Δf **b.** df

21. $f(x, y) = x^2 + xy + y^3$, $x = 4$, $\Delta x = dx = 0.2$, $y = 2$, $\Delta y = dy = -0.1$

22. $f(x, y) = x^3 + xy + y^3$, $x = 5$, $\Delta x = dx = 0.01$, $y = 3$, $\Delta y = dy = -0.01$

23. $f(x, y) = e^x + xy + \ln y$, $x = 0$, $\Delta x = dx = 0.05$, $y = 1$, $\Delta y = dy = 0.01$

24. $f(x, y) = \ln(x^2 + y^2)$, $x = 6$, $\Delta x = dx = 0.1$, $y = 8$, $\Delta y = dy = 0.2$

25. $f(x, y, z) = xy + z^2$, $x = 3$, $\Delta x = dx = 0.03$, $y = 2$, $\Delta y = dy = 0.02$, $z = 1$, $\Delta z = dz = 0.01$

26. $f(x, y, z) = x^2 + y^2 + z^2$, $x = 3$, $\Delta x = dx = 0.1$, $y = 4$, $\Delta y = dy = 0.1$, $z = 5$, $\Delta z = dz = 0.1$

APPLIED EXERCISES

Use total differentials to solve the following exercises.

27. GENERAL: Measurement Errors A rectangle is measured to be 150 feet by 100 feet, but each measurement may be "off" by half a foot. Estimate the error in calculating the area. Then estimate the error in calculating the area if each measurement is "off" by one foot.

28. GENERAL: Telephone Calls For two cities with populations x and y (in thousands) that are 500 miles apart, the number of telephone calls per day between them can be modeled by the function $12xy$. For two cities with populations 40 thousand and 60 thousand, estimate the number of additional telephone calls if each city grows by 1 thousand people. Then estimate the number of additional calls if instead each city were to grow by only 500 people.

29–30. BUSINESS: Profit An electronics company's profit in dollars from making x tape decks and y CD players per day is given by the following profit function $P(x, y)$. If the company currently produces 200 tape decks and 300 CD players, estimate the extra profit that would result from producing five more tape decks and four more CD players.

29. $P(x, y) = 2x^2 - 3xy + 3y^2$

30. $P(x, y) = 3x^2 - 4xy + 4y^2$

31. GENERAL: Highway Safety The emergency stopping distance in feet for a truck of weight w tons traveling at v miles per hour on a dry road is $S = 0.027wv^2$. For a truck that weighs 4 tons and is usually driven at 60 miles per hour, estimate the extra stopping distance if it has an extra half ton of load and is traveling 5 miles per hour faster than usual.

32. GENERAL: Scuba Diving The maximum duration (in minutes) of a scuba dive can be estimated by the formula $T = \dfrac{33v}{d + 33}$, where v is the volume of air in the tank (in cubic feet at sea-level pressure) and d is the depth (in feet) of the dive. For values $v = 100$ and $d = 67$, estimate the change in duration if an extra 20 cubic feet of air is added and the dive is 10 feet deeper.

33. GENERAL: Relative Error in Calculating Area A rectangle is measured to have length x and width y, but each measurement may be in error by 1%. Estimate the percentage error in calculating the area.

34. GENERAL: Relative Error in Calculating Volume A rectangular solid is measured to have length x, width y, and height z, but each measurement may be in error by 1%. Estimate the percentage error in calculating the volume.

35. BIOMEDICAL: Cardiac Output Medical researchers calculate the quantity of blood pumped through the lungs (in liters per minute) by the formula $C = \dfrac{x}{y - z}$, where x is the amount of oxygen absorbed by the lungs (in milliliters per minute), and y and z are, respectively, the concentrations of oxygen in the blood just after and just before passing through the lungs (in milliliters of oxygen per liter of blood). Typical measurements are $x = 250$, $y = 160$, and $z = 150$. Estimate the error in calculating the cardiac output C if each measurement may be "off" by 5 units.

36. GENERAL: Windchill The windchill index (revised in 2001) announced during the winter by the weather bureau measures how cold it "feels" for a given temperature t (in degrees Fahrenheit) and wind speed w (in miles per hour). It is calculated by the formula $C(t, w) = 35.74 + 0.6215t - 35.75w^{0.16} + 0.4275tw^{0.16}$. If the temperature is 30 degrees and the wind speed is 10 miles per hour, estimate the change in the windchill temperature if the wind speed increases by 4 miles per hour and the temperature drops by 5 degrees.

37. THE SLOPE OF $f(x, y) = c$ On page 558 we used the fact that the slope in the x-y plane of the curve defined by $f(x, y) = c$ (for constant c) is given by the formula $-\dfrac{f_x}{f_y}$.

Verify this formula by justifying the following five steps.

a. If $f(x, y) = c$ can be solved explicitly for a function $y = F(x)$, then we may write $f(x, F(x)) = c$.

Justify: $f(x + \Delta x, F(x + \Delta x)) - f(x, F(x)) = 0$.

b. Justify: $f(x + \Delta x, F(x + \Delta x) - F(x) + F(x)) - f(x, F(x)) = 0$.

c. Defining ΔF by $\Delta F = F(x + \Delta x) - F(x)$, we may write the equation above as

$$f(x + \Delta x, \Delta F + F(x)) - f(x, F(x)) = 0$$

Then, writing F for $F(x)$, this becomes

$$f(x + \Delta x, F + \Delta F) - f(x, F) = 0$$

Justify: $f_x \Delta x + f_y \Delta F \approx 0$.

d. Justify: $\dfrac{\Delta F}{\Delta x} \approx -\dfrac{f_x}{f_y}$.

e. Justify: $\dfrac{dF}{dx} = -\dfrac{f_x}{f_y}$.

This shows that the slope of $F(x)$, and therefore the slope of $f(x, y) = c$, is $-\dfrac{f_x}{f_y}$.

38. BIOMEDICAL: Blood Vessels A section of an artery is measured to have length 12 centimeters and diameter 0.8 centimeter. If each measurement may be off by 0.1 centimeter, find the volume with an estimate of the error.

7.7 MULTIPLE INTEGRALS

Introduction

This section discusses *integration* of functions of several variables. We define the *double integral* of a function by considering the volume under a surface $z = f(x, y)$. We then evaluate double integrals by "iterated" (repeated) single integrations. Finally, we use double integrals to calculate volumes, average values, and total accumulations. We restrict our attention to continuous functions (surfaces that have no holes or breaks), since most functions encountered in applications satisfy this restriction.

Rectangular Regions, Volumes, and Double Integrals

The points (x, y) in the plane with x taking values between numbers a and b and y taking values between numbers c and d determine a *rectangular region R*.

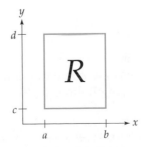

Rectangular region $R = \{(x,y) \mid a \le x \le b, c \le y \le d\}$

The graph below shows a nonnegative function $f(x, y)$ defined on a rectangular region R. We want to find the *volume* under the surface f and above the region R.

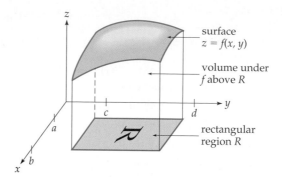

Volume under the surface $z = f(x, y)$
lying above a rectangular region R

We begin by *approximating* the volume by rectangular solids ("boxes") extending from R up to the surface. We divide R into small rectangles by drawing lines parallel to the x- and y-axes with spacing Δx and Δy, as shown below.* On each of these small rectangles we erect a rectangular solid with height $f(x_i, y_j)$, the height of the surface at some point (x_i, y_j) in the base rectangle.

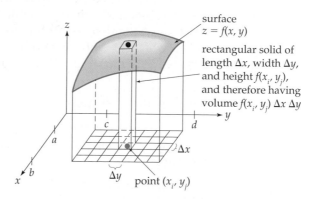

Volume under f over R showing one of the rectangular solids

The volume of the rectangular solid is $f(x_i, y_j) \cdot \Delta x \cdot \Delta y$ (height times length times width), and the sum of the volumes of all such rectangular

---

* Technically, the parallel lines need not have equal spacing. However, we will be letting the spacings approach zero, and the final results are the same for equal or unequal spacing.

solids approximates the volume under f above R:

$$\begin{pmatrix}\text{Volume under}\\ f \text{ above } R\end{pmatrix} \approx \sum f(x_i, y_j) \cdot \Delta x \cdot \Delta y$$

Σ means sum over
all base rectangles

The *limit* of this sum as both Δx and Δy approach zero (so that the base rectangles become smaller and more numerous) gives the *exact* volume, and is called the *double integral of f over R*, denoted $\iint\limits_R f(x, y)\, dx\, dy$.

Double Integrals

The double integral of a continuous function $f(x, y)$ on a rectangular region R is

$$\iint\limits_R f(x, y)\, dx\, dy = \lim_{\Delta x, \Delta y \to 0} \sum f(x_i, y_j) \cdot \Delta x \cdot \Delta y$$

The sum is over
all rectangles in R,
each containing
one (x_i, y_j)

If $f(x, y)$ is nonnegative on R, then the double integral gives the *volume under f over R*.

Iterated Integrals

Evaluating double integrals from the definition is difficult. Fortunately, double integrals can be evaluated by two separate "single" integrations, integrating with respect to one variable at a time while holding the other variable constant. (This is analogous to partial differentiation with respect to one variable, holding the other variable constant.) Such repeated integrals are called *iterated* integrals ("iterated" means "repeated"). A proof that double integrals can be evaluated as iterated integrals can be found in a more theoretical calculus book.

EXAMPLE 1

EVALUATING AN ITERATED INTEGRAL

Evaluate the iterated integral $\displaystyle\int_0^1 \int_0^2 (3x^2 + 6xy^2)\, dx\, dy$.

Solution The two separate integrations will be clearer if we use parentheses:

$$\int_0^1 \left(\int_0^2 (3x^2 + 6xy^2)\, dx \right) dy$$

An inner x-integral and
an outer y-integral

The inner integral gives

$$\int_0^2 (3x^2 + 6xy^2)\, dx \;=\; \left(x^3 + 6\cdot\frac{1}{2}x^2y^2\right)\Bigg|_{x=0}^{x=2}$$

dx means integrate with respect to x holding y constant

Integral of $3x^2$

Integral of x

Held constant

$$= 2^3 + 3\cdot 2^2y^2 \;-\; 0 \;=\; 8 + 12y^2$$

Evaluated at $x = 2$

And at $x = 0$

Simplified

We now apply the outer y-integral to this result:

$$\int_0^1 (8 + 12y^2)\, dy = (8y + 4y^3)\Bigg|_{y=0}^{y=1} = 8 + 4 \;-\; 0 \;=\; 12$$

Result of the inner integral

dy means integrate with respect to y

$12\cdot\frac{1}{3}$

Evaluated at $y = 1$

And at $y = 0$

Final answer

Therefore:

$$\int_0^1 \int_0^2 (3x^2 + 6xy^2)\, dx\, dy = 12$$

The iterated integral equals 12

Always solve an iterated integral "from the inside out."

$$\int_0^1 \left(\int_0^2 (3x^2 + 6xy^2)\, dx\right) dy$$

Limits for y

Limits for x

First integrate with respect to x

Then with respect to y

In Example 1 we integrated first with respect to x and then with respect to y. The next example shows that switching the order of integration gives the same answer, provided that we also switch the x and y limits of integration. That is,

$$\int_0^1 \int_0^2 (3x^2 + 6xy^2)\, dx\, dy$$

is equal to

$$\int_0^2 \int_0^1 (3x^2 + 6xy^2)\, dy\, dx$$

EXAMPLE 2

REVERSING THE ORDER OF INTEGRATION

Evaluate $\displaystyle\int_0^2 \int_0^1 (3x^2 + 6xy^2)\, dy\, dx.$ Same as Example 1, but with the order of integration reversed

Solution First we evaluate the inner y-integral:

$$\int_0^1 (3x^2 + 6xy^2)\, dy \;=\; (3x^2 y \;+\; 2xy^3)\,\Big|_{y=0}^{y=1} \;=\; 3x^2 + 2x \;-\; 0$$

Integrate with respect to y x held constant $\frac{1}{3}\cdot 6$ Evaluated at $y-1$ And at $y=0$

$$= 3x^2 + 2x$$

Then we apply the outer x-integral to this expression:

$$\int_0^2 (3x^2 + 2x)\, dx = (x^3 + x^2)\,\Big|_{x=0}^{x=2} \;=\; 8 + 4 \;-\; 0 \;=\; 12$$

From inner integration   Evaluated at $x=2$ And at $x=0$ Final answer

Therefore:

$$\int_0^2 \int_0^1 (3x^2 + 6xy^2)\, dy\, dx = 12$$

Notice that Examples 1 and 2 (in which the order of integration was reversed) gave the same answer, 12. Reversing the order of integration *always* gives the same answer, provided that the function is continuous.

Reversing the Order of Integration

For a continuous $f(x, y)$

$$\int_c^d \int_a^b f(x, y)\, dx\, dy$$

is equal to

$$\int_a^b \int_c^d f(x, y)\, dy\, dx$$

Double Integrals and Volumes

Earlier, we defined double integrals over rectangular regions. Double integrals can be evaluated by *either* of two iterated integrals (integrating in either order). The limits of integration are taken directly from the region R.

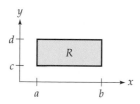

Evaluating Double Integrals

The double integral $\displaystyle\iint\limits_{R} f(x, y)\, dx\, dy$

over the region $R = \{(x, y) \mid a \le x \le b, c \le y \le d\}$

can be evaluated by finding *either* of the iterated integrals

$$\int_c^d \int_a^b f(x, y)\, dx\, dy \qquad \text{or} \qquad \int_a^b \int_c^d f(x, y)\, dy\, dx$$

The limits of integration come from the definition of the rectangle R.

EXAMPLE 3

EVALUATING A DOUBLE INTEGRAL

Evaluate $\displaystyle\iint\limits_{R} y^2 e^{-x}\, dx\, dy$ where $R = \{(x, y) \mid 0 \le x \le 2, -1 \le y \le 1\}$.

Solution This double integral can be evaluated by finding either of the iterated integrals

$$\int_{-1}^{1} \int_0^2 y^2 e^{-x}\, dx\, dy \qquad \text{or} \qquad \int_0^2 \int_{-1}^1 y^2 e^{-x}\, dy\, dx \qquad \text{Limits of integration come from } R$$

We find the second one, beginning with the inner integral.

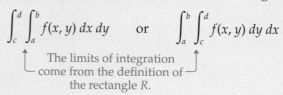

$$\int_{-1}^1 y^2 e^{-x}\, dy = \left(\frac{1}{3}y^3 e^{-x}\right)\Big|_{y=-1}^{y=1} = \frac{1}{3}e^{-x} - \left(\frac{1}{3}(-1)e^{-x}\right)$$

Integrated ⎯ Held constant ⎯ Evaluated at $y = 1$ ⎯ And at $y = -1$

$$= \frac{1}{3}e^{-x} + \frac{1}{3}e^{-x} = \frac{2}{3}e^{-x}$$

Then we integrate this with respect to x:

$$\int_0^2 \frac{2}{3} e^{-x}\, dx = -\frac{2}{3} e^{-x} \Big|_{x=0}^{x=2} = -\frac{2}{3} e^{-2} - \left(-\frac{2}{3} e^0\right) = -\frac{2}{3} e^{-2} + \frac{2}{3}$$

Practice Problem

Show that evaluating this same double integral by the iterated integral in the *other* order gives the same answer. That is, evaluate the iterated integral

$$\int_{-1}^1 \int_0^2 y^2 e^{-x}\, dx\, dy$$

➤ Solution on page 584

The volume under a surface can be found by evaluating a double integral (since this is how double integrals were defined).

Volume by Double Integrals

For a nonnegative continuous function $f(x, y)$, the volume under the surface $z = f(x, y)$ and above a rectangular region R in the x-y plane is

$$\binom{\text{Volume under}}{f \text{ above } R} = \iint\limits_R f(x, y)\, dx\, dy$$

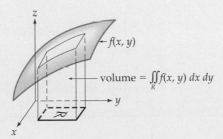

If the surface lies *below* the x-y plane, this integral gives the *negative* of the volume.

EXAMPLE 4

FINDING THE VOLUME UNDER A SURFACE

A modernistic tent with closed sides is constructed according to the function shown below. To design ventilation and heating systems, it is necessary to know the volume under the tent. Find the volume under the tent on the indicated rectangle.

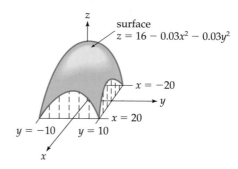

Solution

The volume is the integral of the function over the rectangle *R*:

$$\int_{-10}^{10} \int_{-20}^{20} (16 - 0.03x^2 - 0.03y^2) \, dx \, dy$$

Limits of integration come from *R*

The inner integral is

$$\int_{-20}^{20} (16 - 0.03x^2 - 0.03y^2) \, dx = (16x - 0.01x^3 - 0.03y^2x) \Big|_{x=-20}^{x=20}$$

$$= \underbrace{320 - 80 - 0.6y^2}_{\text{Evaluated at } x = 20} - \underbrace{(-320 + 80 + 0.6y^2)}_{\text{Evaluated at } x = -20} = 480 - 1.2y^2$$

Integrating this with respect to *y*:

From integrating $1.2y^2$

$$\int_{-10}^{10} (480 - 1.2y^2) \, dy = (480y - 0.4y^3) \Big|_{-10}^{10}$$

$$= 4800 - 400 - (-4800 + 400) = 8800$$

Therefore, the volume under the tent is 8800 cubic feet.

Average Value

On page 390 the average value of a function of *one* variable over an interval was defined as the definite integral of the function divided by the length of the interval. For similar reasons, the average value of a function $f(x, y)$ of *two* variables over a region is defined as the *double integral divided by the area* of the region.

Average Value

$$\left(\begin{array}{c}\text{Average value}\\\text{of } f \text{ over } R\end{array}\right) = \frac{1}{\text{Area of } R} \iint\limits_{R} f(x, y) \, dx \, dy$$

Double integral over the region divided by the area of the region

---

For a rectangular region R, the area of R is simply length times width.

EXAMPLE 5   **FINDING THE AVERAGE TEMPERATURE OVER A REGION**

The temperature x miles east and y miles north of a weather station is $T(x, y) = 60 + 2x - 4y$ degrees. Find the average temperature over the rectangular region R extending 2 miles north and south from the station and 5 miles east (as shown in the following diagram).

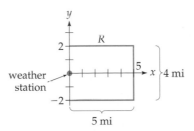

Solution The area of the region R is $4 \cdot 5 = 20$ square miles (length times width). The average temperature is the double integral divided by 20:

$$\text{Average} = \frac{1}{20} \int_{-2}^{2} \int_{0}^{5} (60 + 2x - 4y) \, dx \, dy$$

The inner integral is

$$\int_{0}^{5} (60 + 2x - 4y) \, dx = (60x + x^2 - 4yx)\Big|_{x=0}^{x=5}$$

$$= \underbrace{300 + 25 - 20y}_{\text{Evaluated at } x = 5} - 0 = 325 - 20y$$

The (outer) integral of this expression is

$$\int_{-2}^{2} (325 - 20y) \, dy = (325y - 10y^2) \Big|_{y=-2}^{y=2} = 650 - 40 - (-650 - 40)$$

$$= 610 + 690 = 1300$$

Finally, for the average we divide by 20 (the area of the region):

$$\frac{1300}{20} = 65$$

The average temperature over the region is 65 degrees.

Integrating over More General Regions

We can also integrate over regions R that are bounded by curves, provided that the curves as well as the function being integrated are continuous.

Double Integrals over Regions Between Curves

Let R be the region bounded by a lower curve $y = g(x)$ and an upper curve $y = h(x)$ from $x = a$ to $x = b$, as shown below. Then the double integral of $f(x, y)$ over R is

$$\iint\limits_{R} f(x, y) \, dx \, dy = \int_{a}^{b} \int_{g(x)}^{h(x)} f(x, y) \, dy \, dx$$

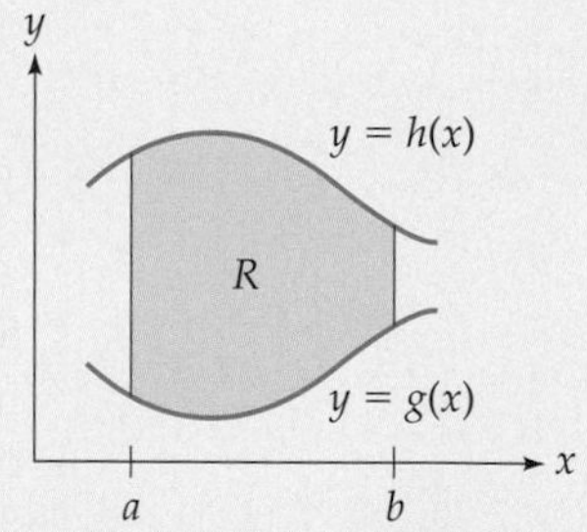

If f is nonnegative, this double integral gives the volume under the surface $f(x, y)$ above R.

EXAMPLE 6

FINDING THE VOLUME UNDER A SURFACE

Find the volume under the surface $f(x, y) = 12xy$ and above the region R shown on the following page.

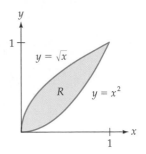

Solution The region R is bounded by the upper curve $h(x) = \sqrt{x}$ and the lower curve $g(x) = x^2$ from $x = 0$ to $x = 1$. From the boxed definition on the previous page, the volume is given by the following integral:

$$\text{Volume} = \int_0^1 \int_{x^2}^{\sqrt{x}} 12xy \, dy \, dx$$

We solve the inner integral first:

$$\int_{x^2}^{\sqrt{x}} 12xy \, dy = 12x\frac{1}{2}y^2 \Big|_{y=x^2}^{y=\sqrt{x}} = 6xy^2 \Big|_{y=x^2}^{y=\sqrt{y}}$$

$$= \underbrace{6x(\sqrt{x})^2}_{\substack{\text{Evaluating} \\ \text{at } y = \sqrt{x}}} - \underbrace{6x(x^2)^2}_{\substack{\text{Evaluating} \\ \text{at } y = x^2}} = \underbrace{6x^2 - 6x^5}_{\text{Simplified}}$$

Now we integrate the result with respect to x:

$$\int_0^1 (6x^2 - 6x^5) \, dx = \left(6 \cdot \frac{1}{3}x^3 - x^6\right) \Big|_0^1$$

$$= (2x^3 - x^6) \Big|_0^1 = 2 - 1 - (0) = 1$$

Therefore, the volume under the surface and above the region R is 1 cubic unit.

Section Summary

The double integral of a nonnegative function $f(x, y)$ over a region R gives the *volume* under the surface above R. (This is analogous to defining the definite integral of a function $f(x)$ of *one* variable as the *area* under the curve.)

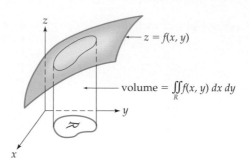

We evaluate a double integral by evaluating an iterated (repeated) integral:

$$\iint\limits_{R} f(x, y)\, dx\, dy \qquad \text{over} \qquad R = \{(x, y) \mid a \le x \le b, c \le y \le d\}$$

is found by evaluating either of the two iterated integrals

$$\int_{c}^{d} \int_{a}^{b} f(x, y)\, dx\, dy \qquad \text{or} \qquad \int_{a}^{b} \int_{c}^{d} f(x, y)\, dy\, dx \qquad \text{\small Integrating in either order}$$

Note the distinction: *double* integrals are written with R (which must be specified) under the double integral, and *iterated* integrals are written with upper and lower *limits* on each integral sign.

Double integrals do more than just find volumes; they give *continuous sums* (as illustrated in Exercises 43 and 44), and when divided by the area of the region they give the *average value of the function over the region*.

Exercises 47–50 discuss "triple" integrals of functions $f(x, y, z)$ of three variables. Triple integrals are evaluated by *iterated* integrals (but *three* of them), integrating successively with respect to one variable at a time, holding all others constant.

➤ **Solution to Practice Problem**

$$\int_{0}^{2} y^2 e^{-x}\, dx = -y^2 e^{-x} \Big|_{x=0}^{x=2} = -y^2 e^{-2} + y^2 e^{0}$$

$$= -y^2 e^{-2} + y^2$$

$$\int_{-1}^{1} (-y^2 e^{-2} + y^2)\, dy = \left(-\frac{1}{3} y^3 e^{-2} + \frac{1}{3} y^3 \right)\Big|_{-1}^{1}$$

$$= -\frac{1}{3} e^{-2} + \frac{1}{3} - \left(\frac{1}{3} e^{-2} - \frac{1}{3} \right) = -\frac{2}{3} e^{-2} + \frac{2}{3} \qquad \text{(same as before)}$$

7.7   Exercises

Evaluate each (single) integral.

1. $\displaystyle\int_1^{x^2} 8xy^3\, dy$

2. $\displaystyle\int_1^{y^2} 10x^4\, dx$

3. $\displaystyle\int_{-y}^{y} 9x^2y\, dx$

4. $\displaystyle\int_{-x}^{x} 6xy^2\, dy$

5. $\displaystyle\int_0^x (6y - x)\, dy$

6. $\displaystyle\int_0^y (4x - y)\, dx$

Evaluate each iterated integral.

7. $\displaystyle\int_0^2 \int_0^1 4xy\, dx\, dy$

8. $\displaystyle\int_0^2 \int_0^1 8xy\, dy\, dx$

9. $\displaystyle\int_0^2 \int_0^1 x\, dy\, dx$

10. $\displaystyle\int_0^4 \int_0^3 y\, dx\, dy$

11. $\displaystyle\int_0^1 \int_0^2 x^3y^7\, dx\, dy$

12. $\displaystyle\int_0^1 \int_0^3 x^8y^2\, dy\, dx$

13. $\displaystyle\int_1^3 \int_0^2 (x + y)\, dy\, dx$

14. $\displaystyle\int_1^2 \int_0^4 (x - y)\, dx\, dy$

15. $\displaystyle\int_{-1}^1 \int_0^3 (x^2 - 2y^2)\, dx\, dy$

16. $\displaystyle\int_{-1}^1 \int_0^3 (2x^2 + y^2)\, dy\, dx$

17. $\displaystyle\int_{-3}^3 \int_0^3 y^2e^{-x}\, dy\, dx$

18. $\displaystyle\int_{-2}^2 \int_0^2 xe^{-y}\, dx\, dy$

19. $\displaystyle\int_{-2}^2 \int_{-1}^1 ye^{xy}\, dx\, dy$

20. $\displaystyle\int_{-1}^1 \int_{-1}^1 xe^{xy}\, dy\, dx$

21. $\displaystyle\int_0^2 \int_x^1 12xy\, dy\, dx$

22. $\displaystyle\int_0^1 \int_y^1 4xy\, dx\, dy$

23. $\displaystyle\int_3^5 \int_0^y (2x - y)\, dx\, dy$

24. $\displaystyle\int_2^4 \int_0^x (x - 2y)\, dy\, dx$

25. $\displaystyle\int_{-3}^3 \int_0^{4x} (y - x)\, dy\, dx$

26. $\displaystyle\int_{-1}^1 \int_0^{2y} (x + y)\, dx\, dy$

27. $\displaystyle\int_0^1 \int_{-y}^y (x + y^2)\, dx\, dy$

28. $\displaystyle\int_0^2 \int_{-x}^x (x^2 - y)\, dy\, dx$

For each double integral:

a. Write the *two* iterated integrals that are equal to it.

b. Evaluate *both* iterated integrals (the answers should agree).

29. $\displaystyle\iint_R 3xy^2\, dx\, dy$

with $R = \{\, (x, y) \mid 0 \le x \le 2, 1 \le y \le 3 \,\}$

30. $\displaystyle\iint_R 6x^2y\, dx\, dy$

with $R = \{\, (x, y) \mid 0 \le x \le 1, 1 \le y \le 2 \,\}$

31. $\displaystyle\iint_R ye^x\, dx\, dy$

with $R = \{\, (x, y) \mid -1 \le x \le 1, 0 \le y \le 2 \,\}$

32. $\displaystyle\iint_R xe^y\, dx\, dy$

with $R = \{\, (x, y) \mid 0 \le x \le 1, -2 \le y \le 2 \,\}$

Use integration to find the volume under each surface $f(x, y)$ above the region R.

33. $f(x, y) = x + y$
$R = \{\, (x, y) \mid 0 \le x \le 2, 0 \le y \le 2 \,\}$

34. $f(x, y) = 8 - x - y$
$R = \{\, (x, y) \mid 0 \le x \le 4, 0 \le y \le 4 \,\}$

35. $f(x, y) = 2 - x^2 - y^2$
$R = \{\, (x, y) \mid 0 \le x \le 1, 0 \le y \le 1 \,\}$

36. $f(x, y) = x^2 + y^2$
$R = \{\, (x, y) \mid 0 \le x \le 2, 0 \le y \le 2 \,\}$

Use integration to find the volume under each surface $f(x, y)$ above the region R.

37. $f(x, y) = 2xy$
over the region
shown in the
graph.

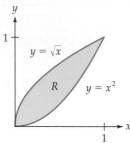

38. $f(x, y) = 3xy^2$
over the region
shown in the
graph.

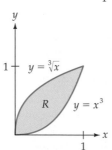

39. $f(x, y) = e^y$
over the region
shown in the
graph.

40. $f(x, y) = e^y$
over the region
shown in the
graph.

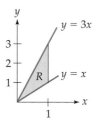

APPLIED EXERCISES

41. GENERAL: Average Temperature The temperature x miles east and y miles north of a weather station is given by the function $f(x, y) = 48 + 4x - 2y$. Find the average temperature over the region R shown below.

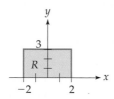

42. ENVIRONMENTAL SCIENCE: Average Air Pollution The air pollution near a chemical refinery is $f(x, y) = 20 + 6x^2y$ parts per million (ppm), where x and y are the numbers of miles east and north of the refinery. Find the average pollution level for the region R shown below.

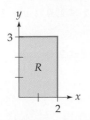

43. GENERAL: Total Population of a Region The population density (people per square mile) x miles east and y miles north of the center of a city is $P(x, y) = 12{,}000e^{x-y}$. Find the total population of the region R shown in the following diagram. [*Hint:* Integrate the population density over the region R. This is an example of a double integral as a *sum*, giving a *total* population over a region.]

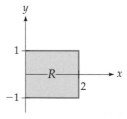

44. BUSINESS: Value of Mineral Deposit The value of an offshore mineral deposit x miles east and y miles north of a certain point is $f(x, y) = 4x + 6y^2$ million dollars per square mile. Find the total value of the tract shown on next page. [*Hint:* Integrate the function over the region R. This is an example of a double integral as a *sum*, giving a *total* value over a region.]

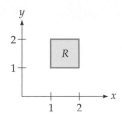

45–46. GENERAL: Volume of a Building To estimate heating and air conditioning costs, it is necessary to know the volume of a building.

45. A conference center has a curved roof whose height is $f(x, y) = 40 - 0.006x^2 + 0.003y^2$. The building sits on a rectangle extending from $x = -50$ to $x = 50$ and $y = -100$ to $y = 100$. Use integration to find the volume of the building. (All dimensions are in feet.)

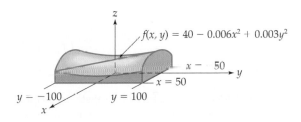

46. An airplane hangar has a curved roof whose height is $f(x, y) = 40 - 0.03x^2$. The building sits on a rectangle extending from $x = -20$ to $x = 20$ and $y = -100$ to $y = 100$. Use integration to find the volume of the building. (All dimensions are in feet.)

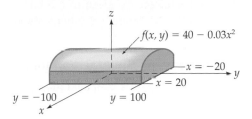

Triple Integrals

Evaluate each triple iterated integral. [*Hint:* Integrate with respect to one variable at a time, treating the other variables as constants, working from the inside out.]

47. $\displaystyle\int_1^2 \int_0^3 \int_0^1 (2x + 4y - z^2)\, dx\, dy\, dz$

48. $\displaystyle\int_1^2 \int_0^3 \int_0^2 (6x - 2y + z^2)\, dx\, dy\, dz$

49. $\displaystyle\int_1^2 \int_0^2 \int_0^1 2xy^2z^3\, dx\, dy\, dz$

50. $\displaystyle\int_1^3 \int_0^1 \int_0^2 12x^3y^2z\, dx\, dy\, dz$

Chapter Summary with Hints and Suggestions

Reading the text and doing the exercises in this chapter have helped you to master the following skills, which are listed by section (in case you need to review them) and are keyed to particular Review Exercises. Answers for all Review Exercises are given at the back of the book, and full solutions can be found in the Student Solutions Manual.

7.1 Functions of Several Variables

- Find the domain of a function of two variables. (*Review Exercises 1–4.*)

7.2 Partial Derivatives

- Find the first and second partials of a function of two variables. (*Review Exercises 5–12.*)

$$\frac{\partial}{\partial x} f(x, y) = f_x(x, y) = \lim_{h \to 0} \frac{f(x + h, y) - f(x, y)}{h}$$

$$\frac{\partial}{\partial y} f(x, y) = f_y(x, y) = \lim_{h \to 0} \frac{f(x, y + h) - f(x, y)}{h}$$

- Evaluate the first partials of a function of two variables. (*Review Exercises 13–16.*)

- Solve an applied problem involving partials, and interpret the answer. (*Review Exercises 17–18.*)

$$\frac{\partial}{\partial x} f(x, y) = \begin{pmatrix} \text{Rate of change of } f \text{ with respect} \\ \text{to } x \text{ when } y \text{ is held constant} \end{pmatrix}$$

$$\frac{\partial}{\partial y} f(x, y) = \begin{pmatrix} \text{Rate of change of } f \text{ with respect} \\ \text{to } y \text{ when } x \text{ is held constant} \end{pmatrix}$$

7.3 Optimizing Functions of Several Variables

- Find the relative extreme values of a function. (*Review Exercises 19–30.*)

$$\begin{cases} f_x = 0 \\ f_y = 0 \end{cases} \quad D = f_{xx}f_{yy} - (f_{xy})^2$$

- Solve an applied problem by optimizing a function of two variables. (*Review Exercises 31–32.*)

7.4 Least Squares

- Find a least squares line "by hand." (*Review Exercises 33–34.*)

- Find the least squares line for actual data, and use it to make a prediction. (*Review Exercises 35–36.*)

7.5 Lagrange Multipliers and Constrained Optimization

- Solve a constrained maximum or minimum problem using Lagrange multipliers. (*Review Exercises 37–42.*)

$$F(x, y, \lambda) = f(x, y) + \lambda g(x, y) \quad \begin{cases} F_x = 0 \\ F_y = 0 \\ F_\lambda = 0 \end{cases}$$

- Find *both* extreme values in a constrained optimization problem using Lagrange multipliers. (*Review Exercises 43–44.*)

- Solve an applied problem using Lagrange multipliers. (*Review Exercises 45–48.*)

7.6 Total Differentials and Approximate Changes

- Find the total differential of a function. (*Review Exercises 49–54.*)

$$df = \frac{\partial f}{\partial x} dx + \frac{\partial f}{\partial y} dy$$

- Solve an applied problem by using the total differential to estimate an actual change. (*Review Exercise 55.*)

$$\Delta f = f(x + \Delta x, y + \Delta y) - f(x, y) \qquad \Delta f \approx df$$

■ Estimate the relative error in an area calculation. (*Review Exercise 56.*)

7.7 Multiple Integrals

■ Evaluate an iterated integral.
(*Review Exercises 57–60.*)

$$\int_c^d \int_a^b f(x, y)\, dx\, dy = \int_a^b \int_c^d f(x, y)\, dy\, dx$$

■ Find the volume under a surface above a region using a double integral.
(*Review Exercises 61–64.*)

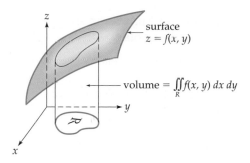

■ Find the average population over a region using a double integral. (*Review Exercise 65.*)

■ Find the total value of a region using a double integral. (*Review Exercise 66.*)

Hints and Suggestions

■ The graph of a function of two variables is represented by a *surface* above (or below) the *x-y* plane. Such three-dimensional graphs are difficult to draw "by hand" but can be shown on some graphing calculators and computer screens. Such surfaces have maximum and minimum points (just as with functions of one variable), but also a new phenomenon, *saddle points* (see page 521).

■ The graph of a function of *three* or more variables would require *four* or more dimensions, and so cannot be drawn.

■ When finding partial derivatives, remember which is the variable of differentiation and then treat the other variable as a constant. Other than this, the differentiation formulas are the same.

■ Partials give instantaneous rates of change with respect to one variable while the other is held constant. They also give *marginals* with respect to one product while production of the other is held constant.

■ The *D*-test (page 523) applies only to critical points, where the first partials are zero. These critical points must be found first, and then the *D*-test is applied to each, one at a time.

■ Least squares is carried out automatically by a graphing calculator or computer when it finds the linear regression line.

■ Solving a constrained optimization problem by Lagrange multipliers is often easier than eliminating one of the variables, as was done in Sections 3.3 and 3.4.

■ Multiple integrals give volume under a function if the function is nonnegative.

■ When finding the average value of a function of two variables, don't forget to divide by the area of the region over which you are integrating.

■ **Practice for Test:** Review Exercises 3, 7, 15, 17, 19, 33, 41, 45, 49, 51, 57, 65.

Review Exercises for Chapter 7

Practice test exercise numbers are in green.

7.1 Functions of Several Variables

For each function, state the domain.

1. $f(x, y) = \dfrac{\sqrt{x}}{\sqrt[3]{y}}$ **2.** $f(x, y) = \dfrac{\sqrt[3]{x}}{\sqrt{y}}$

3. $f(x, y) = e^{1/x} \ln y$ **4.** $f(x, y) = \dfrac{\ln y}{x}$

7.2 Partial Derivatives

For each function f, calculate
a. f_x, **b.** f_y, **c.** f_{xy}, and **d.** f_{yx}.

5. $f(x, y) = 2x^5 - 3x^2 y^3 + y^4 - 3x + 2y + 7$

6. $f(x, y) = 3x^4 + 5x^3 y^2 - y^6 - 6x + y - 9$

7. $f(x, y) = 18x^{2/3} y^{1/3}$

8. $f(x, y) = \ln(x^2 + y^3)$

9. $f(x, y) = e^{x^3 - 2y^3}$ **10.** $f(x, y) = 3x^2 e^{-5y}$

11. $f(x, y) = ye^{-x} - x \ln y$

12. $f(x, y) = x^2 e^y + y \ln x$

For each function, calculate
a. $f_x(1, -1)$ and **b.** $f_y(1, -1)$.

13. $f(x, y) = \dfrac{x + y}{x - y}$ **14.** $f(x, y) = \dfrac{x}{x^2 + y^2}$

15. $f(x, y) = (x^3 + y^2)^3$ **16.** $f(x, y) = (2xy - 1)^4$

17. BUSINESS: Marginal Productivity A company's production is given by the Cobb–Douglas function $P(L, K) = 160L^{3/4} K^{1/4}$, where L is the number of units of labor and K is the number of units of capital.

  a. Find $P_L(81, 16)$ and interpret this number.
  b. Find $P_K(81, 16)$ and interpret this number.
  c. From your answers to parts (a) and (b), which will increase production more, an additional unit of labor or an additional unit of capital?

18. BUSINESS: Advertising A clothing designer's sales S depend on x, the amount spent on television advertising, and y, the amount spent on print advertising (both in thousands of dollars), according to the function

$$S(x, y) = 60x^2 + 90y^2 - 6xy + 200$$

Find $S_x(2, 3)$ and $S_y(2, 3)$, and interpret these numbers.

7.3 Optimizing Functions of Several Variables

For each function, find all relative extreme values.

19. $f(x, y) = 2x^2 - 2xy + y^2 - 4x + 6y - 3$

20. $f(x, y) = x^2 - 2xy + 2y^2 - 6x + 4y + 2$

21. $f(x, y) = 2xy - x^2 - 5y^2 + 2x - 10y + 3$

22. $f(x, y) = 2xy - 5x^2 - y^2 + 10x - 2y + 1$

23. $f(x, y) = 2xy + 6x - y + 1$

24. $f(x, y) = 4xy - 4x + 2y - 4$

25. $f(x, y) = e^{-(x^2 + y^2)}$ **26.** $f(x, y) = e^{2(x^2 + y^2)}$

27. $f(x, y) = \ln(5x^2 + 2y^2 + 1)$

28. $f(x, y) = \ln(4x^2 + 3y^2 + 10)$

29. $f(x, y) = x^3 - y^2 - 12x - 6y$

30. $f(x, y) = y^2 - x^3 + 12x - 4y$

31. BUSINESS: Maximum Profit A boatyard builds 18-foot and 22-foot sailboats. Each 18-foot boat costs $3000 to build, each 22-foot boat costs $5000 to build, and the company's fixed costs are $6000. The price function for the 18-foot boats is $p = 7000 - 20x$, and that for the 22-foot boats is $q = 8000 - 30y$ (both in dollars), where x and y are the numbers of 18-foot and 22-foot boats, respectively.

  a. Find the company's cost function $C(x, y)$.
  b. Find the company's revenue function $R(x, y)$.
  c. Find the company's profit function $P(x, y)$.
  d. Find the quantities and prices that maximize profit. Also find the maximum profit.

32. BUSINESS: Price Discrimination A company sells farm equipment in America and Europe, charging different prices in the two markets. The price function for harvesters sold in America is $p = 80 - 0.2x$, and the price function for harvesters sold in Europe is $q = 64 - 0.1y$ (both in thousands of dollars), where x and y are the numbers sold per day in America and Europe, respectively. The company's cost function is $C = 100 + 12(x + y)$ thousand dollars.

a. Find the company's profit function.
b. Find how many harvesters should be sold in each market to maximize profit. Also find the price for each market.

7.4 Least Squares

For each exercise:

a. Find the least squares line "by hand."
b. Check your answer using a graphing calculator.

33.

x	y
1	−1
3	6
4	6
5	10

34.

x	y
1	7
2	4
4	2
5	−1

35. GENERAL: The Aging of America The population of Americans who are over 65 years old is growing faster than the population at large. Find the least squares line for the following data for the over-65 population, and use it to predict the size of the over-65 population in the year 2010 ($x = 6$).

	x	y = Population (millions)
1960	1	16.7
1970	2	20.1
1980	3	25.5
1990	4	31.2
2000	5	34.7

36. ECONOMICS: Unemployment The unemployment rate in the United States for several years is given in the following table. Find the least squares line for these data and use it to predict the average unemployment rate for the year 2010 ($x = 7$).

	x	y = Percent Unemployed
1980	1	7.1
1985	2	7.2
1990	3	5.6
1995	4	5.6
2000	5	4.1

7.5 Lagrange Multipliers and Constrained Optimization

Use Lagrange multipliers to optimize each function subject to the given constraint. (The stated extreme values *do* exist.)

37. Maximize $f(x, y) = 6x^2 - y^2 + 4$
subject to $3x + y = 12$.

38. Maximize $f(x, y) = 4xy - x^2 - y^2$
subject to $x + 2y = 26$.

39. Minimize $f(x, y) = 2x^2 + 3y^2 - 2xy$
subject to $2x + y = 18$.

40. Minimize $f(x, y) = 12xy - 1$
subject to $y - x = 6$.

41. Minimize $f(x, y) = e^{x^2+y^2}$
subject to $x + 2y = 15$.

42. Maximize $f(x, y) = e^{-x^2-y^2}$
subject to $2x + y = 5$.

Use Lagrange multipliers to find the maximum *and* minimum values of each function subject to the given constraint. (Both extreme values *do* exist.)

43. $f(x, y) = 6x - 18y$
subject to $x^2 + y^2 = 40$

44. $f(x, y) = 4xy$
subject to $x^2 + y^2 = 32$

45. BUSINESS: Maximum Profit A company's profit is $P = 300x^{2/3}y^{1/3}$, where x and y are, respectively, the amounts spent on production and advertising. The company has a total of $60,000 to spend.

   a. Use Lagrange multipliers to find the amounts for production and advertising that maximize profit.

   b. Evaluate and give an interpretation for $|\lambda|$.

46. BIOMEDICAL: Nutrition A nursing home uses two vitamin supplements, and the nutritional value of x ounces of the first together with y ounces of the second is $4x + 2xy + 8y$. The first costs $2 per ounce, the second costs $1 per ounce, and the nursing home can spend only $8 per patient per day.

   a. Use Lagrange multipliers to find how much of each supplement should be used to maximize the nutritional value subject to the budget constraint.

   b. Evaluate and give an interpretation for $|\lambda|$.

47. ECONOMICS: Least Cost Rule A company's production is given by the Cobb–Douglas function $P = 60L^{2/3}K^{1/3}$, where L and K are the numbers of units of labor and capital. Each unit of labor costs $25 and each unit of capital costs $100. The company wants to produce exactly 1920 units.

   a. Find the numbers of units of labor and capital that meet the production requirements at the lowest cost.

   b. Find the marginal productivity of labor and the marginal productivity of capital. [*Hint:* This means the partials of P with respect to L and K.]

   c. Show that at the values found in part (a), the following relationship holds:

$$\frac{\text{Marginal productivity of labor}}{\text{Marginal productivity of capital}} = \frac{\text{Price of labor}}{\text{Price of capital}}$$

   This is called the "least cost rule."

48. GENERAL: Container Design An open-top box with a square base and two perpendicular dividers, as shown in the diagram, is to have a volume of 576 cubic inches. Use Lagrange multipliers to find the dimensions that require the least amount of material.

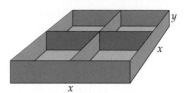

7.6 Total Differentials and Approximate Changes

Find the total differential of each function.

49. $f(x, y) = 3x^2 + 2xy + y^2$

50. $f(x, y) = x^2 + xy - 3y^2$

51. $g(x, y) = \ln(xy)$

52. $g(x, y) = \ln(x^3 + y^3)$

53. $z = e^{x-y}$ **54.** $z = e^{xy}$

55. BUSINESS: Sales A clothing designer's sales S depend on x, the amount spent on television advertising, and y, the amount spent on print advertising (all in thousands of dollars) according to the formula $S(x, y) = 60x^2 - 6xy + 90y^2 + 200$. If the company now spends 2 thousand dollars on television advertising and 3 thousand dollars on print advertising, use the total differential to estimate the change in sales if television advertising is increased by $500 and print advertising is decreased by $500. [*Hint:* Δx and Δy must be in thousands of dollars, just as x and y are.] Then estimate the change in sales if each of the changes in advertising were halved.

56. GENERAL: Relative Error in Calculating Area A triangular piece of real estate is measured to have length x feet and altitude y feet, but each measurement may be in error by 1%. Estimate the percentage error in calculating the area by using a total differential.

7.7 Multiple Integrals

Evaluate each iterated integral.

57. $\displaystyle\int_0^4 \int_{-1}^1 2xe^{2y}\,dy\,dx$

58. $\displaystyle\int_{-1}^1 \int_0^3 (x^2 - 4y^2)\,dx\,dy$

59. $\displaystyle\int_{-1}^1 \int_{-y}^y (x + y)\,dx\,dy$

60. $\displaystyle\int_{-2}^2 \int_x^x (x + y)\,dy\,dx$

Find the volume under the surface $f(x, y)$ above the region R.

61. $f(x, y) = 8 - x - y$
$R = \{(x, y) \mid 0 \le x \le 2, 0 \le y \le 4\}$

62. $f(x, y) = 6 - x - y$
$R = \{(x, y) \mid 0 \le x \le 4, 0 \le y \le 2\}$

63. $f(x, y) = 12xy^3$
over the region
shown in the
graph.

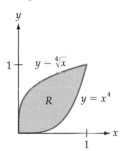

64. $f(x, y) = 15xy^4$
over the region
shown in the
graph.

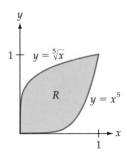

65. GENERAL: Average Population of a Region The population per square mile x miles east and y miles north of the center of a city is $P(x, y) = 12{,}000 + 100x - 200y$. Find the *average* population over the region shown below.

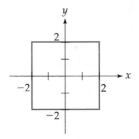

66. GENERAL: Total Value of a Region
The value of land x blocks east and y blocks north of the center of a town is $V(x, y) = 40 - 4x - 2y$ hundred thousand dollars per block. Find the total value of the parcel of land shown in the graph above.

Cumulative Review for Chapters 1–7

The following exercises review some of the basic techniques that you learned in Chapters 1–7. Answers to all of these cumulative review exercises are given in the answer section at the back of the book. A graphing calculator is suggested but not required.

1. Draw the graph of the function
$$f(x) = \begin{cases} 2x - 1 & \text{if } x \le 3 \\ 7 - x & \text{if } x > 3 \end{cases}$$

2. Simplify: $\left(\dfrac{1}{8}\right)^{-2/3}$

3. Use the definition of the derivative
$$f'(x) = \lim_{h \to 0} \frac{f(x + h) - f(x)}{h}$$
to find the derivative of $f(x) = \dfrac{1}{x}$.

4. If $f(x) = 12\sqrt[3]{x^2} - 4$, find $f'(8)$.

5. A camera store finds that if it sells disposable cameras at a price of p dollars each, it will sell

$$S(p) = \frac{800}{p + 8} \text{ of them per week. Find } S'(12)$$

and interpret the answer.

6. Find $\dfrac{d}{dx}[x^2 + (2x + 1)^4]^3$.

7. Make sign diagrams for the first and second derivatives and graph the function $f(x) = x^3 + 9x^2 - 48x - 148$, showing all relative extreme points and inflection points.

8. Make a sign diagram for the first derivative

and graph the function $f(x) = \dfrac{1}{x^2 - 4x}$,

showing all asymptotes and relative extreme points.

9. A homeowner wants to use 80 feet of fence to make a rectangular enclosure along an existing stone wall. If the side along the existing wall needs no fence, what are the dimensions of the enclosure that has the largest possible area?

10. An open-top box with a square base is to have a volume of 108 cubic feet. Find the dimensions of the box of this type that can be made with the least amount of material.

11. A spherical balloon is being inflated at the constant rate of 128 cubic feet per minute. Find how fast the radius is increasing at the moment when the radius is 4 feet.

12. A sum of $1000 is deposited in a bank account paying 8% interest. Find the value of the account after 3 years if the interest is compounded:

   a. quarterly   **b.** continuously

13. In t years the population of a county is predicted to be $P(t) = 12,000e^{0.02t}$, and the population of a neighboring county is predicted to be $Q(t) = 9000e^{0.04t}$. In how many years will the second county overtake the first in population?

14. A sum is deposited in a bank account paying 6% interest compounded monthly. How many

years will it take for the value to increase by 50%?

15. Make sign diagrams for the first and second derivatives and graph the function $f(x) = e^{-x^2/2}$, showing all relative extreme points and inflection points.

16. Find $\displaystyle\int (12x^2 - 4x + 1)\,dx$.

17. Pollution is being discharged into a lake at the rate of $18e^{0.02t}$ million gallons per year, where t is the number of years from now. Find a formula for the total amount of pollution that will be discharged into the lake during the next t years.

18. Find the area bounded by $y = 20 - x^2$ and $y = 8 - 4x$.

19. Find the average value of $f(x) = 12\sqrt{x}$ over the interval $[0, 4]$.

20. Find: **a.** $\displaystyle\int \frac{x^2\,dx}{x^3 + 1}$ **b.** $\displaystyle\int \frac{e^{\sqrt{x}}\,dx}{\sqrt{x}}$

21. Find by integration by parts: $\displaystyle\int xe^{4x}\,dx$.

22. Use the integral table on the inside back cover

to find $\displaystyle\int \frac{\sqrt{4 - x^2}}{x}\,dx$.   Check your answer by

differentiating and then simplifying.

23. Evaluate $\displaystyle\int_1^{\infty} \frac{1}{x^3}\,dx$.

24. Use trapezoidal approximation with $n = 4$ trapezoids to approximate $\int_0^1 \sqrt{x^2 + 1}\,dx$. (If you use a graphing calculator, compare your answer with the value obtained by using FnInt.)

25. Use Simpson's Rule with $n = 4$ to approximate $\int_0^1 \sqrt{x^2 + 1}\,dx$. (If you use a graphing calculator, compare your answer with the value obtained by using FnInt.)

26. For the function $f(x, y) = \dfrac{3\sqrt{x} + \ln y}{x - y}$, find:

   a. the domain
   b. $f(4, 1)$

27. Find the first partial derivatives of the function $f(x, y) = x \ln y + ye^{2x}$.

28. Find the relative extreme values of
$f(x, y) = 2x^2 - 2xy + y^2 + 4x - 6y + 12$.

29. Find the least squares line for the following points:

x	y
1	-3
3	1
5	3
7	8

30. Use Lagrange multipliers to find the minimum value of $f(x, y) = 3x^2 + 2y^2 - 2xy$ subject to the constraint $x + 2y = 18$.

31. Find the total differential of $f(x, y) = 2x^2 + xy - 3y^2 + 4$.

32. Find the volume under the surface $f(x, y) = 12 - x - 2y$ over the rectangle
$$R = \{ (x, y) \mid 0 \le x \le 2, 0 \le y \le 3 \}.$$

8 Trigonometric Functions

Application Preview

Tones, Temperature, and Trigonometry

This chapter introduces the sine and cosine functions, "wavy" curves that repeat themselves at regular intervals. Such functions have many applications. For example, musical sounds are caused by vibrations in the air, which can be picked up by a microphone and displayed on an oscilloscope. The following graph shows the sound wave from a saxophone playing a G#. This curve can be expressed as the sum of modified sine and cosine curves.*

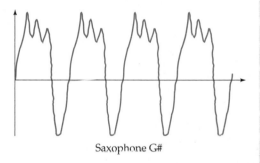

Saxophone G#

The next graph shows the daily mean temperatures in New York City for the years 2000 and 2001 (the jagged line), and the normal daily high and low temperatures (the lighter blue band), which may be closely approximated by sine and cosine curves.

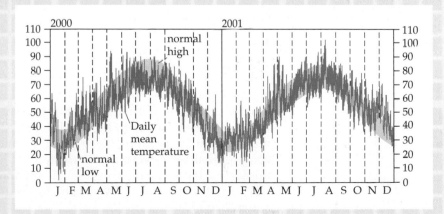

* See Dorothea Bone, "Music and the Circular Functions," UMAP unit 588, COMAP, Lexington, Mass., and John R. Pierce, *The Science of Musical Sound*, Scientific American Books, W. H. Freeman & Co., New York, 1983.

Tones, Temperature, and Trigonometry
(continued)

The graph below shows the rising level of carbon dioxide in the atmosphere as measured at the Mauna Loa Observatory in Hawaii. Carbon dioxide contributes to the "greenhouse effect," which causes global warming. This curve is a sine curve (the seasonal fluctuations) added to an exponential function (the upward trend).

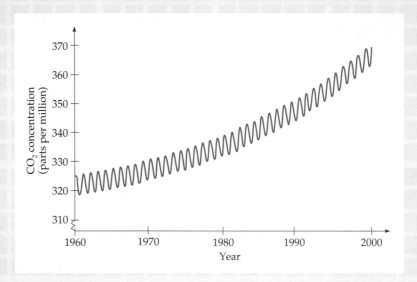

8.1 TRIANGLES, ANGLES, AND RADIAN MEASURE

Introduction

The graph on the previous page shows the annual cycle of temperatures, varying "periodically" from winter cold to summer heat and back to winter cold. In this chapter we will study trigonometric functions and use them to model this and other periodic behavior, from musical sounds to business cycles. It is assumed that you have studied trigonometry at some time in the past. The first two sections review trigonometry from the beginning, but selectively rather than comprehensively, covering only the topics that will be needed for applications. Later sections discuss differentiation and integration of trigonometric functions. We begin by discussing triangles, angles, and radian measure.

Right Triangles

A triangle with a right (90°) angle is called a *right triangle*. The lengths of the sides of a right triangle are related by the Pythagorean Theorem.*

Pythagorean Theorem

In a right triangle, the square of the hypotenuse equals the sum of the squares of the other two sides.

$$a^2 + b^2 = c^2$$

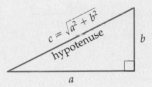

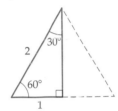

A proof of the Pythagorean Theorem is outlined in Exercise 27. We first use the theorem as follows: given an equilateral triangle of side 2, cutting it in half, as shown on the left, leaves a right triangle with hypotenuse 2 and side 1. The third side, b, can then be found from the Pythagorean Theorem as follows.

$$1^2 + b^2 = 2^2 \qquad \text{$a^2 + b^2 = c^2$ with } a = 1 \text{ and } c = 2$$

$$1 + b^2 = 4 \qquad \text{Simplifying}$$

$$b^2 = 3 \qquad \text{Subtracting 1}$$

Therefore, the third side has length $b = \sqrt{3}$, leading to the following result.

30°–60°–90° Triangle

A triangle with angles 30°, 60°, 90° has sides in proportion to 1, 2, $\sqrt{3}$.

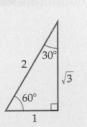

---

* This theorem was first attributed to the school of the Greek philosopher Pythagoras (sixth century B.C.), although the Babylonian tablet "Plimpton 322" shows that it was known several centuries earlier.

We say "in proportion to" because the numbers 1, 2, $\sqrt{3}$ can be doubled, tripled, halved, or multiplied by *any* positive number, to obtain other 30°–60°–90° triangles. Note that the smallest side (1) goes opposite the smallest angle (30°), and the largest side (2) goes opposite the largest angle (90°).

We now consider a triangle with angles 45°, 45°, and 90°, as shown on the left. The two 45° angles make the triangle isosceles, so it has two equal sides, which we take to be of length 1. We find the length of the hypotenuse from the Pythagorean Theorem, as follows.

$$1^2 + 1^2 = c^2 \qquad \qquad a^2 + b^2 = c^2 \text{ with } a = b = 1$$

$$2 = c^2 \qquad \qquad \text{Simplifying}$$

Therefore, the hypotenuse has length $c = \sqrt{2}$, giving the following result.

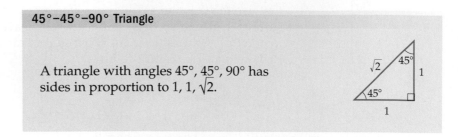

45°–45°–90° Triangle

A triangle with angles 45°, 45°, 90° has sides in proportion to 1, 1, $\sqrt{2}$.

Again, the largest side $\left(\sqrt{2}\right)$ goes opposite the largest angle (90°). These two particular right triangles will be used extensively in the next section.

Angles and Radian Measure

Trigonometry requires a precise definition of the word "angle." An *angle* is formed when a line segment is rotated around one of its endpoints (the *vertex*), from an *initial side* to a *terminal side*. We use an arrow to indicate the rotation.

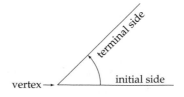

An angle in the *x-y*-plane is in *standard position* if its vertex is at the origin and its initial side lies along the positive *x*-axis. An angle is *positive* if its rotation is *counterclockwise*, and *negative* if its rotation is *clockwise*. The first three angles in the following diagram are positive angles, and the last is a negative angle.

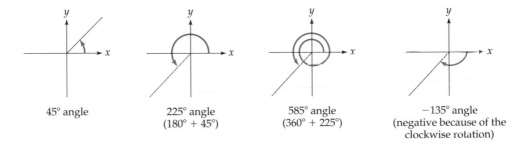

| 45° angle | 225° angle (180° + 45°) | 585° angle (360° + 225°) | −135° angle (negative because of the clockwise rotation) |

It is not enough to specify the initial and terminal sides of an angle—the rotation must also be specified. For example, the second, third, and fourth angles in the above diagrams have the same initial and terminal sides but are different because they have different rotations. Angles that have the same initial and terminal sides (even if formed by different rotations) are called *coterminal* angles. Any two coterminal angles differ by a multiple of 360°.

The choice of 360 for the number of degrees in a full circle dates back to the ancient Babylonians, whose numbering system was based on 60, the "sexagesimal" system. (We still use their system for time: 60 seconds in a minute and 60 minutes in an hour.) In calculus we use another method for measuring angles, called *radian measure*. We use radian measure because it leads to simpler formulas for differentiation and integration in Section 8.3.

The radian measure of an angle is found by placing the angle at the center of a circle of radius 1 and measuring the length of the arc of the circle cut off by the angle: the length of the arc gives the measure of the angle.

Radian Measure

An angle of *x* radians is an angle that cuts off an arc of length *x* on a circle of radius 1 centered at the vertex.

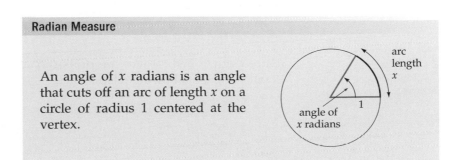

As usual, a positive angle means counterclockwise rotation. The following diagrams show angles of $\frac{1}{2}$ radian, 1 radian, and 2 radians.

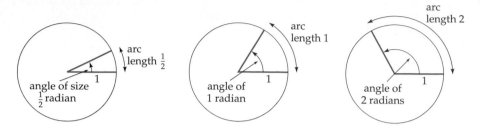

Radian measure of angle equals the arc length cut off on a unit circle.

The circumference formula $C = 2\pi r$ with $r = 1$ shows that the circumference of the unit circle is 2π. Therefore, a full circle of 360° is equivalent to 2π radians.

$$360° = 2\pi \text{ radians}$$

Dividing each side by 2 shows that $180° = \pi$ radians, and dividing again shows that $90° = \pi/2$ radians. The degree and radian measures of the following *quadrant angles* should be memorized.

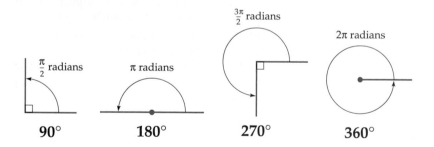

Converting Between Degrees and Radians

Since degrees and radians are proportional, we can convert between them using the following relationship.

$$\frac{\text{Degrees}}{180} = \frac{\text{Radians}}{\pi}$$

This equation may be solved for either degrees or radians, resulting in the two formulas

$$\text{Radians} = \frac{\pi}{180} \cdot \text{degrees}$$

To change from degrees to radians, multiply by $\frac{\pi}{180}$

and

$$\text{Degrees} = \frac{180}{\pi} \cdot \text{radians}$$

To change from radians to degrees, multiply by $\frac{180}{\pi}$

EXAMPLE 1

CONVERTING DEGREES TO RADIANS

Find the radian measure of each angle: **a.** 45°   **b.** 150°   **c.** −60°

Solution

To convert to radians, multiply degrees by $\frac{\pi}{180}$.

a. $45° = \overset{1}{\cancel{45}} \cdot \frac{\pi}{\underset{4}{\cancel{180}}} = \frac{\pi}{4}$ radians

Multiplying by $\frac{\pi}{180}$

b. $150° = \overset{5}{\cancel{150}} \cdot \frac{\pi}{\underset{6}{\cancel{180}}} = \frac{5\pi}{6}$ radians

c. $-60° = \overset{-1}{\cancel{-60}} \cdot \frac{\pi}{\underset{3}{\cancel{180}}} = -\frac{\pi}{3}$ radians

---

We leave π in our answers to keep them exact. For an approximation, we could replace π by 3.14.

Graphing Calculator Exploration

a. Try using a graphing calculator to convert from degrees to radians and check your answers to Example 1. (On some calculators, use MODE to select RADIANS and then enter degrees with the ° symbol and press ENTER.) Since your calculator approximates π, you should get answers like 0.785 (for 45°), 2.618 (for 150°), and −1.047 (for −60°).

b. Then try converting your answers back to degrees. [On some calculators, use MODE to select DEGREES and enter, for example, $(\pi \div 4)^r$ to obtain 45.]

EXAMPLE 2

CONVERTING RADIANS TO DEGREES

Find the degree measure of each angle:

a. $\dfrac{\pi}{6}$ radians   **b.** $\dfrac{3\pi}{4}$ radians   **c.** $-\dfrac{\pi}{2}$ radians   **d.** 1 radian

Solution

To convert to degrees, multiply radians by $\dfrac{180}{\pi}$.

a. $\dfrac{\pi}{6}$ radians $= \dfrac{\pi}{6} \cdot \dfrac{\overset{30}{\cancel{180}}}{\cancel{\pi}} = 30°$ Multiplying by $\dfrac{180}{\pi}$

b. $\dfrac{3\pi}{4}$ radians $= \dfrac{3\pi}{4} \cdot \dfrac{\overset{45}{\cancel{180}}}{\cancel{\pi}} = 3 \cdot 45° = 135°$

c. $-\dfrac{\pi}{2}$ radians $= -\dfrac{\pi}{2} \cdot \dfrac{\overset{90}{\cancel{180}}}{\cancel{\pi}} = -90°$

d. 1 radian $= 1 \cdot \dfrac{180}{\pi} \approx 57°$

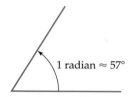

1 radian $\approx 57°$

Practice Problem

a. Find the radian measure of an angle of 60°.

b. Find the degree measure of an angle of $\dfrac{2\pi}{3}$ radians.

➤ Solutions on page 609

The angles 30°, 45°, and 60° occur so frequently that their radian measures should be memorized.

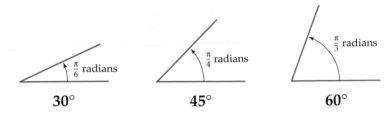

30°   **45°**   **60°**

These angles may be added to the quadrant angles to find corresponding angles in other quadrants. For example, the largest angle shown in the diagram on the right has measure

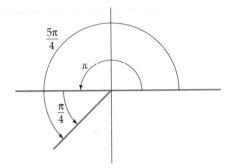

$$\pi + \frac{\pi}{4} = \frac{5\pi}{4}$$

Other angles resulting from such additions are shown in the following diagram.

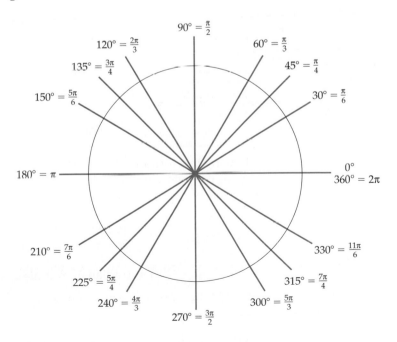

Alternative Definition of Radian Measure

The circumference formula $C = 2\pi r$ shows that the circumference C of a circle is proportional to the radius r. Therefore, we may define radian measure using a circle of *any* radius r, as long as arc length is measured in "radius" units, which means dividing the actual arc length by the radius r.

Alternative Definition of Radian Measure

The angle that cuts off an arc of length s on a circle of radius r has radian measure

$$\begin{pmatrix} \text{Radian} \\ \text{measure} \end{pmatrix} = \frac{s}{r} \quad \begin{matrix} \leftarrow \text{arc length} \\ \leftarrow \text{radius} \end{matrix}$$

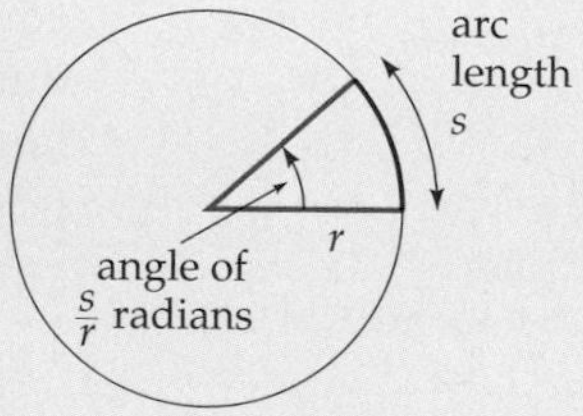

arc length s

r

angle of $\frac{s}{r}$ radians

Since radian measure is a *quotient*, arc length over radius, any units (inches, feet, etc.) will cancel out. For this reason, radians are called *dimensionless* units.

Arc Length Along a Circle

Radian measure calculates angles according to the formula

$$\text{Radians} = \frac{\text{arc length}}{\text{radius}}$$

Solving this formula for arc length (multiplying each side by "radius") gives the following formula.

Arc Length

$$\begin{pmatrix} \text{Arc} \\ \text{length} \end{pmatrix} = (\text{radians}) \cdot (\text{radius})$$

Arc length is radians times radius

This formula gives the arc length along any circle.

Speed of Rotation of the Earth

How fast are you moving if you "stand still" on the equator? The earth makes one full rotation (2π radians) every 24 hours, so during 1 hour it rotates $\dfrac{2\pi}{24} = \dfrac{\pi}{12}$ radian. Since the earth's radius is approximately 4000 miles, during 1 hour a point on the equator will move a distance of

$$\begin{pmatrix} \text{Arc} \\ \text{length} \end{pmatrix} = \frac{\pi}{12} \cdot 4000 \approx 1047 \text{ miles} \qquad \text{Radians times radius}$$

That is, standing on the equator you are actually rotating around the center of the earth at a speed of over 1000 miles per hour. You don't feel it because everything around you is moving with you.

Graphing Calculator Exploration

(For those familiar with the cosine function)

What is your speed if you live somewhere other than the equator? It depends on your latitude. If you are at latitude x, then your speed is $1047 \cos x$.

a. Set your calculator for DEGREES and graph $y_1 = 1047 \cos x$ on the window [0, 90] by [0, 1047]. Observe how the speed varies from 1047 miles per hour at the equator (0° latitude) to 0 miles per hour at the North Pole (90° latitude).

b. Find the latitude where you are and evaluate $y_1 = 1047 \cos x$ at that number to find your speed of rotation.

EXAMPLE 3

INTERCITY DISTANCES

The cities of Washington, DC and Moscow determine a central angle of 1.22 radians (at the center of the earth). Find the distance between these cities measured along the surface of the earth. (Assume that the earth is a sphere of radius 4000 miles.)

Solution

Since arc length is radians times radius,

$$\begin{pmatrix} \text{Distance from} \\ \text{Washington, DC} \\ \text{to Moscow} \end{pmatrix} = 1.22 \cdot 4000 = 4880$$

$$\underbrace{}_{\text{radians}} \ \underbrace{}_{\text{radius}}$$

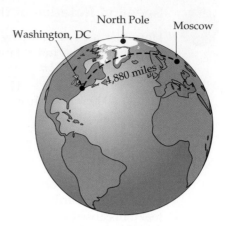

Therefore, an airplane flying between Washington, DC and Moscow must fly approximately 4880 miles (since the altitude of the airplane is negligible compared to the radius of the earth).

8.1 Section Summary

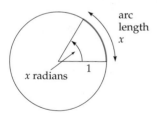

We began by finding the sides of two particular triangles: a $30°–60°–90°$ triangle (sides 1, 2, $\sqrt{3}$) and a $45°–45°–90°$ triangle (sides 1, 1, $\sqrt{2}$). We then defined the *radian measure* of an angle as the arc length cut off on a unit circle centered at the vertex of the angle.

Alternatively, using a circle of *any* radius, the radian measure is the arc length divided by the radius. To put this colloquially, if we were walking along the circumference of a circle, then walking a distance of x radii determines an angle of x radians. The diagram on page 605 shows the measures of many common angles in both degrees and radians.

To convert between degrees and radians, we use the relationship

$$\frac{\text{Degrees}}{180} = \frac{\text{Radians}}{\pi}$$

The arc length of a circle intercepted by a central angle measured in radians can be calculated from the formula

$$\begin{pmatrix} \text{Arc} \\ \text{length} \end{pmatrix} = (\text{radians}) \cdot (\text{radius})$$

> **Solution to Practice Problem**

a. $60° = \overset{1}{\cancel{60}} \cdot \dfrac{\pi}{\underset{3}{\cancel{180}}} = \dfrac{\pi}{3}$ radians

b. $\dfrac{2\pi}{3}$ radians $= \dfrac{2\pi}{\underset{1}{3}} \cdot \dfrac{\overset{60}{\cancel{180}}}{\cancel{\pi}} = 120°$

8.1 Exercises

Find the degree and radian measure of each angle shown below.

1.
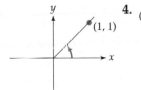

2.
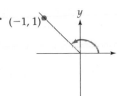

3.

4. $(-1, 1)$
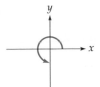

5.
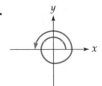

6. $(2, 2)$
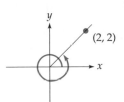

7.
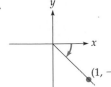
$(1, -1)$

8.
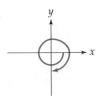

For each angle below:

i. Find the radian measure (without using a calculator).

ii. Check your answers using a graphing calculator.

9. a. $30°$ **b.** $225°$ **c.** $-180°$

10. a. $15°$ **b.** $210°$ **c.** $-90°$

11. a. $60°$ **b.** $315°$ **c.** $-120°$

12. a. $45°$ **b.** $240°$ **c.** $-150°$

13. a. $540°$ **b.** $135°$ **c.** $1°$

14. a. $750°$ **b.** $150°$ **c.** $10°$

For each angle below:

i. Find the degree measure (without using a calculator).

ii. Check your answers using a graphing calculator.

15. a. $\dfrac{\pi}{4}$ **b.** $\dfrac{2\pi}{3}$ **c.** $-\dfrac{5\pi}{6}$

16. a. $\dfrac{\pi}{6}$ **b.** $\dfrac{3\pi}{2}$ **c.** $-\dfrac{5\pi}{4}$

17. a. $\dfrac{\pi}{3}$ **b.** $\dfrac{5\pi}{4}$ **c.** $-\dfrac{7\pi}{6}$

18. a. $\dfrac{3\pi}{4}$ **b.** $\dfrac{4\pi}{3}$ **c.** $-\dfrac{7\pi}{4}$

19. a. $\dfrac{9\pi}{4}$ **b.** -3π **c.** 5

20. a. $\dfrac{7\pi}{3}$ **b.** -5π **c.** 3

APPLIED EXERCISES

21. GENERAL: Arc Length A 27-inch-long pendulum swings through an angle of $\frac{\pi}{9}$ radians. Find the length of arc through which it swings. (Give your answer in terms of π, and then approximate it using $\pi \approx 3.14$.)

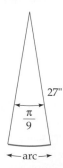

27"

$\frac{\pi}{9}$

←arc→

22. GENERAL: Building Swaying Modern sky-scrapers are designed to bend in high winds. As a general rule, a 100 mile per hour wind can cause a tall building to bend by 0.002 radian. Find the length of arc through which the 1450 foot tall Sears Tower in Chicago would bend in such a wind. [*Note:* Because of back-and-forth motion, the swaying might be twice this great.]

23. GENERAL: Airplanes An airliner is flying in a circular holding pattern with a radius of 12 miles. Find the central angle (in radians) that the airliner will have flown through by flying 18 miles along this circular path.

24. GENERAL: Distance Find the distance along the surface of the earth between Los Angeles and Honolulu if the two cities determine a central angle of 0.64 radian. (Assume that the earth is a sphere of radius 4000 miles.)

25. GENERAL: Distance Find the distance along the surface of the earth between New York City and Paris if the two cities determine a central angle of 0.9 radian. (Assume that the earth is a sphere of radius 4000 miles.)

26. GENERAL: Moon Diameter Estimate the diameter of the moon if it is 240,000 miles from the earth and it forms an angle of 0.009 radian when viewed from the earth. [*Hint:* For very small angles, the diameter can be approximated by arc length.]

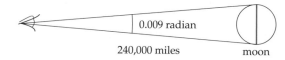

0.009 radian

240,000 miles   moon

27. PROOF OF THE PYTHAGOREAN THEOREM Use the diagram below to prove the Pythagorean Theorem for the right triangle with sides a and b and hypotenuse c as follows:

  i. Calculate the area of the outer square by multiplying length times width. [*Hint:* Length and width are each $a + b$.]

  ii. Calculate the area of the outer square again, but now as the inner square plus four right triangles, each of area $\frac{1}{2} ab$.

  iii. Set the two expressions for the area equal to each other and simplify, obtaining $a^2 + b^2 = c^2$.

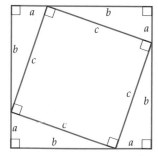

SINE AND COSINE FUNCTIONS

Introduction

In this section we define the sine and cosine functions, and use them to model quantities that fluctuate periodically.

Sine and Cosine Functions

We use the Greek letter θ ("theta") to represent an angle (always measured in *radians* unless otherwise indicated). For an angle θ in standard position, we choose a point (x, y) on its terminal side and define the sine and cosine functions as follows.

---

sin θ and cos θ

Let (x, y) be a point on the terminal side of an angle of θ radians in standard position, and let r be its distance from the origin. We define sin θ and cos θ as follows:

$$\sin \theta = \frac{y}{r}$$

$$\cos \theta = \frac{x}{r}$$

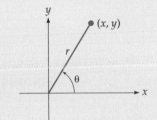

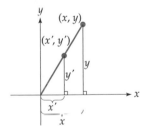

It does not matter which point on the terminal side (other than the origin) we choose; any two points (x, y) and (x', y') determine similar triangles, so the ratios of the sides will be the same.

If we choose the point (x, y) to be a distance $r = 1$ from the origin, then the definitions of sin θ and cos θ simplify to

$$\sin \theta = y \qquad\qquad \sin \theta = \frac{y}{r} \text{ with } r = 1$$

$$\cos \theta = x \qquad\qquad \cos \theta = \frac{x}{r} \text{ with } r = 1$$

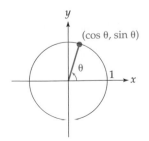

To put this another way, a point on the circle of radius 1 centered at the origin can be represented as $(\cos \theta, \sin \theta)$, where θ is the angle at the origin, as shown in the diagram on the left.

EXAMPLE 1

FINDING VALUES OF SINE AND COSINE

Evaluate $\sin \theta$ and $\cos \theta$ for θ equal to 0, $\dfrac{\pi}{2}$, π, and $\dfrac{3\pi}{2}$.

Solution The angles are shown in the following graphs, along with the point (x, y) on the terminal side a distance $r = 1$ from the origin. The sine and cosine are then calculated below, using $\sin \theta = y$ and $\cos \theta = x$ (since $r = 1$).

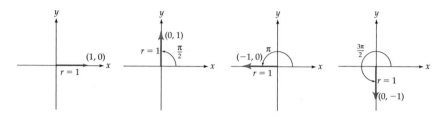

$\sin 0 = 0$	$\sin \dfrac{\pi}{2} = 1$	$\sin \pi = 0$	$\sin \dfrac{3\pi}{2} = -1$ from the y-coordinates
$\cos 0 = 1$	$\cos \dfrac{\pi}{2} = 0$	$\cos \pi = -1$	$\cos \dfrac{3\pi}{2} = 0$ from the x-coordinates

EXAMPLE 2

FINDING VALUES OF SINE AND COSINE

Find $\sin \dfrac{\pi}{6}$ and $\cos \dfrac{\pi}{6}$.

Solution

Draw an angle of $\frac{\pi}{6}$ radians (30°) in standard position.

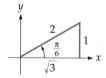

Complete the sides of the 30°-60°-90° triangle with sides 1, 2, $\sqrt{3}$.

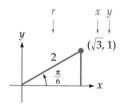

Write the coordinates of a point based on the triangle sides.

From these values of x, y, and r we find the sine and cosine:

$$\sin \frac{\pi}{6} = \frac{1}{2} \qquad\qquad \sin \theta = \frac{y}{r}$$

$$\cos \frac{\pi}{6} = \frac{\sqrt{3}}{2} \qquad\qquad \cos \theta = \frac{x}{r}$$

Practice Problem Find $\sin \dfrac{\pi}{4}$ and $\cos \dfrac{\pi}{4}$.

➤ Solution on page 622

Similar calculations lead to sines and cosines of the following angles. You should either memorize this table or be able to calculate these values by remembering the sides of 30°–60°–90° and 45°–45°–90° triangles.

θ	0	$\dfrac{\pi}{6}$	$\dfrac{\pi}{4}$	$\dfrac{\pi}{3}$	$\dfrac{\pi}{2}$
$\sin \theta$	0	$\dfrac{1}{2}$	$\dfrac{\sqrt{2}}{2}$	$\dfrac{\sqrt{3}}{2}$	1
$\cos \theta$	1	$\dfrac{\sqrt{3}}{2}$	$\dfrac{\sqrt{2}}{2}$	$\dfrac{1}{2}$	0

Notice the patterns:

$$\dfrac{\sqrt{0}}{2} \quad \dfrac{\sqrt{1}}{2} \quad \dfrac{\sqrt{2}}{2} \quad \dfrac{\sqrt{3}}{2} \quad \dfrac{\sqrt{4}}{2}$$

$$\dfrac{\sqrt{4}}{2} \quad \dfrac{\sqrt{3}}{2} \quad \dfrac{\sqrt{2}}{2} \quad \dfrac{\sqrt{1}}{2} \quad \dfrac{\sqrt{0}}{2}$$

EXAMPLE 3

FINDING SINE AND COSINE IN OTHER QUADRANTS

Find $\sin \dfrac{2\pi}{3}$ and $\cos \dfrac{2\pi}{3}$.

Solution

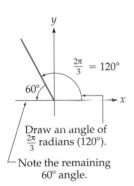

Draw an angle of $\dfrac{2\pi}{3}$ radians (120°).
└ Note the remaining 60° angle.

Draw a perpendicular to an axis to make a 30°-60°-90° triangle with sides 1, 2, $\sqrt{3}$

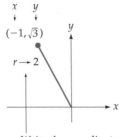

Write the coordinates of a point based on the triangle sides (*with the correct signs*).

From these values of x, y, and r:

$$\sin \dfrac{2\pi}{3} = \dfrac{\sqrt{3}}{2}$$

$$\cos \dfrac{2\pi}{3} = \dfrac{-1}{2} = -\dfrac{1}{2}$$

$$\sin \theta = \dfrac{y}{r}$$

$$\cos \theta = \dfrac{x}{r}$$

Graphing Calculator Exploration

```
sin(2π/3)
        .8660254038
√(3)/2
        .8660254038
cos(2π/3)
               -.5
```

Your calculator should find a value of $\cos\dfrac{2\pi}{3}$ that agrees with the result found above, and a value for $\sin\dfrac{2\pi}{3}$ that agrees with $\dfrac{\sqrt{3}}{2}$ to as many decimal places as the calculator shows.

EXAMPLE 4

FINDING SINE AND COSINE OF A NEGATIVE NUMBER

Find $\sin\left(-\dfrac{\pi}{4}\right)$ and $\cos\left(-\dfrac{\pi}{4}\right)$.

Solution Using only one graph:

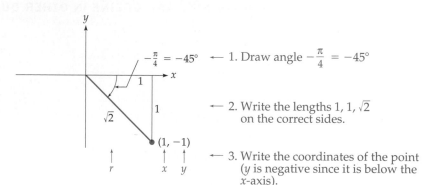

← 1. Draw angle $-\dfrac{\pi}{4} = -45°$

← 2. Write the lengths $1, 1, \sqrt{2}$ on the correct sides.

← 3. Write the coordinates of the point (y is negative since it is below the x-axis).

From these values of x, y, and r:

$$\sin\left(-\dfrac{\pi}{4}\right) = \dfrac{-1}{\sqrt{2}} = -\dfrac{\sqrt{2}}{2} \qquad\qquad \sin\theta = \dfrac{y}{r}$$

Multiplying numerators and denominators by $\sqrt{2}$

$$\cos\left(-\dfrac{\pi}{4}\right) = \dfrac{1}{\sqrt{2}} = \dfrac{\sqrt{2}}{2} \qquad\qquad \cos\theta = \dfrac{x}{r}$$

Sine and Cosine of an Acute Angle

An *acute* angle is an angle between 0 and $\pi/2$ radians. For an acute angle θ, $\sin \theta$ and $\cos \theta$ can be expressed as ratios of sides of a right triangle (called hyp, opp, and adj for the *hyp*otenuse, *opp*osite side, and *adj*acent side).

Right Triangle Formulas for sin θ and cos θ

$$\sin \theta = \frac{\text{opp}}{\text{hyp}}$$

$$\cos \theta = \frac{\text{adj}}{\text{hyp}}$$

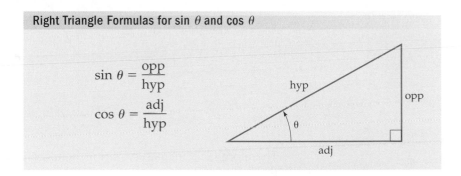

EXAMPLE 5

CALCULATING DISTANCES

How long a ladder is needed to reach a roof 20 feet above the ground if the ladder makes a 60° angle with the ground?

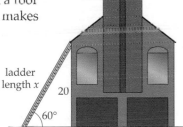

Solution　The ladder and building form the hypotenuse and opposite side of the right triangle shown above, suggesting that we use $\sin \theta = \dfrac{\text{opp}}{\text{hyp}}$ with $\theta = \dfrac{\pi}{3}$.

$$\sin \frac{\pi}{3} = \frac{20}{x} \qquad\qquad \sin \theta = \frac{\text{opp}}{\text{hyp}}$$

$$\frac{\sqrt{3}}{2} = \frac{20}{x} \qquad\qquad \text{Using } \sin \frac{\pi}{3} = \frac{\sqrt{3}}{2}$$

$$\sqrt{3}\,x = 40 \qquad\qquad \text{Cross-multiplying}$$

$$x = \frac{40}{\sqrt{3}} \approx 23.09 \qquad \text{Solving for } x, \text{ then using a calculator}$$

Therefore, the ladder must be about 24 feet long.

Graphs of the Sine and Cosine Functions

On page 611, we saw that the coordinates of a point on a circle of radius 1 centered at the origin can be represented as $(\cos\theta, \sin\theta)$, where θ is the angle at the origin.

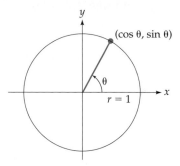

Therefore, as θ increases, $\sin\theta$ behaves just like the y-coordinate of the point (x, y) as it moves around the circle—first increasing from 0 to 1, then decreasing to 0 to -1, then increasing back to 0. Plotting θ on the horizontal axis, we obtain the following graph.

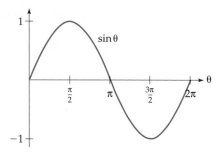

This pattern of values repeats itself each time the angle completes another full circle of 2π radians, giving the following graph of $\sin t$ (where we have replaced θ by t).

Graph of $\sin t$

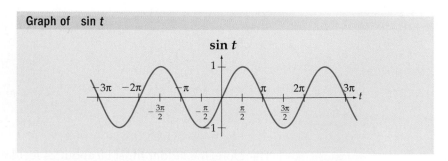

Similarly, $\cos \theta$ behaves just like the x-coordinate of the point (x, y) as it moves around the circle—first decreasing from 1 to 0 to -1, then increasing back to 1.

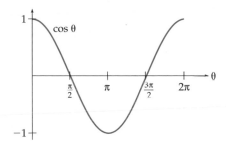

Again, this cycle repeats itself each time the angle completes another full circle of 2π radians, giving the following graph of $\cos t$.

Graph of $\cos t$

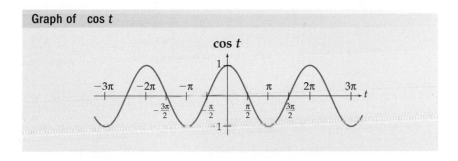

(These graphs could also have been found by the more tedious process of plotting points.) Sin t and $\cos t$ are said to be *periodic with period 2π*, meaning that they repeat their values every 2π units. We will now think of $\sin t$ and $\cos t$ as *ordinary functions*, unrelated to angles, functions that oscillate between $+1$ and -1 in a "wavy" pattern. We will use $\sin t$ and $\cos t$ to model quantities that undergo regular fluctuations, such as seasonal sales.

Modified Sine and Cosine Curves—Changing Amplitude and Period

The *amplitude* of a sine or cosine curve is half the distance between its highest and lowest values (sometimes called its *half-height*). Therefore, $\sin t$ and $\cos t$ have amplitude 1. By multiplying by constants, we may create similar "wavy" functions with any amplitude and period. For positive numbers a and b, the function $a \sin bt$ oscillates between the values a and $-a$, and so has amplitude a. The period of

$a \sin bt$ is found by setting $bt = 2\pi$, giving period $t = \dfrac{2\pi}{b}$. Similar statements hold for the function $a \cos bt$.

Modified Sine and Cosine Functions

$\left.\begin{array}{l} y = a \sin bt \\ y = a \cos bt \end{array}\right\}$ have amplitude a and period $\dfrac{2\pi}{b}$.

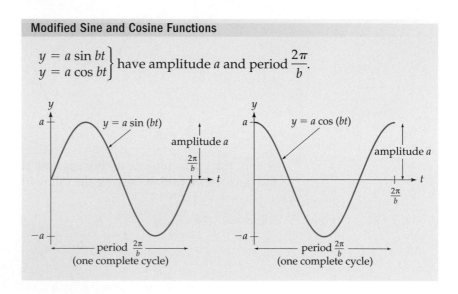

EXAMPLE 6

GRAPHING A MODIFIED SINE CURVE

Sketch the graph of $f(t) = 3 \sin 2t$.

Solution   The amplitude is 3 (since the sine is multiplied by 3), and the period is π (the sine's period 2π divided by the 2). Therefore, the graph is a sine curve, but between heights 3 and -3, with a complete cycle occurring between 0 and π, as shown below.

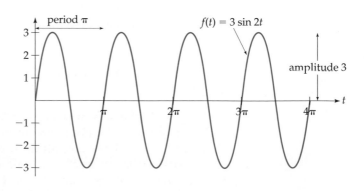

Graphing Calculator Exploration

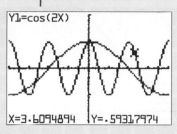

Y1=cos(2X)

X=3.6094894   Y=.59317974

Graph $y_1 = \cos 2x$ and $y_2 = \cos 0.5x$ on the window $[-2\pi, 2\pi]$ by $[-2, 2]$. (Be sure that your calculator is set for radians.) Notice how the number in front of the x changes the period: a larger number makes the curve "wiggle" faster.

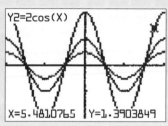

Y2=2cos(X)

X=5.4810765   Y=1.3903849

Graph $y_1 = \cos x$, $y_2 = 2 \cos x$, and $y_3 = 0.5 \cos x$ on the same window. Notice how the number in front of the *function* changes the amplitude: a larger number makes the amplitude larger.

Predict what the curve $y = 2 \cos 2x$ would look like. Then check your prediction by graphing the curve. What about $y = -2 \cos x$? What about $y = -2 \sin 2x$?

Seasonal Sales

The sine and cosine functions with modified amplitude and period are often used to model quantities that fluctuate in a regular pattern, such as seasonal sales.

EXAMPLE **7** **PREDICTING SEASONAL REVENUE**

A travel agency specializing in Caribbean vacations models its daily revenue (in dollars) on day n of the year by

$$R(n) = 600 + 500 \cos\left(\frac{2\pi n}{365}\right)$$

Find the company's revenue on March 1 (day $n = 60$).

Solution

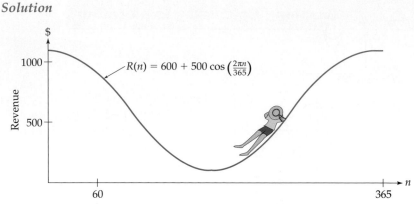

$$R(60) = 600 + 500 \cos\left(\underbrace{\frac{2\pi \cdot 60}{365}}_{\text{0.512 (using a calculator set for radians)}}\right) \approx 856 \qquad \begin{array}{l} R(n) = 600 + 500 \cos\left(\dfrac{2\pi n}{365}\right) \\ \text{with} \quad n = 60 \end{array}$$

Therefore, the travel agency's daily revenue on March 1 will be about $856.

Notice that the revenue in the above graph never reaches zero because of the vertical shift by 600 (see page 59).

Trigonometric Identities

A formula that holds for *all* values of the variable is called an *identity*. There are many trigonometric identities, and we will mention only a few. The first is

$$\sin^2 t + \cos^2 t = 1 \qquad \begin{array}{l} \sin^2 t \ \text{ means } \ (\sin t)^2 \\ \cos^2 t \ \text{ means } \ (\cos t)^2 \end{array}$$

To prove this identity, recall that $\sin t = \dfrac{y}{r}$ and $\cos t = \dfrac{x}{r}$, so

$$\sin^2 t + \cos^2 t = \left(\frac{y}{r}\right)^2 + \left(\frac{x}{r}\right)^2 = \frac{x^2 + y^2}{r^2} = \frac{r^2}{r^2} = 1 \quad \text{Since} \ \ x^2 + y^2 = r^2$$

This and other useful identities are listed below, and the remaining proofs are developed in Exercises 42–44.

Trigonometric Identities

$$\sin^2 t + \cos^2 t = 1 \qquad\qquad \text{Proved above}$$

$$\sin(t + 2\pi) = \sin t$$
$$\cos(t + 2\pi) = \cos t \qquad\qquad \begin{array}{l}\text{Periodicity}\\\text{identities}\end{array}$$

$$\sin(-t) = -\sin t$$
$$\cos(-t) = \cos t \qquad\qquad \begin{array}{l}\text{Negative angle}\\\text{identities}\end{array}$$

$$\sin t = \cos\left(\frac{\pi}{2} - t\right)$$
$$\cos t = \sin\left(\frac{\pi}{2} - t\right) \qquad\qquad \begin{array}{l}\text{Complementary}\\\text{angle identities}\end{array}$$

$$\sin(s \pm t) = (\sin s)\cdot(\cos t) \pm (\cos s)\cdot(\sin t)$$
$$\cos(s \pm t) = (\cos s)\cdot(\cos t) \mp (\sin s)\cdot(\sin t) \qquad\qquad \begin{array}{l}\text{Sum and difference}\\\text{identities}\end{array}$$

In the last two identities, use the upper signs throughout the equation, or the lower signs throughout.

8.2 Section Summary

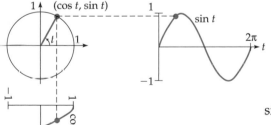

The sine and cosine functions are defined in terms of the coordinates (x, y) of a point on a unit circle forming an angle of θ radians:
$$\sin\theta = y \quad \text{and} \quad \cos\theta = x.$$
For acute angles, these functions can also be interpreted in terms of the sides of a right triangle: $\sin\theta = \dfrac{\text{opp}}{\text{hyp}}$ and $\cos\theta = \dfrac{\text{adj}}{\text{hyp}}$. For a few particular angles (such as 30°, 45°, and 60°) we can evaluate these functions *exactly*, whereas for other angles we use a calculator. We then graphed these functions and used them to model quantities that undergo regular fluctuations. The trigonometric identities will be used in the exercises and in the following section.

➤ **Solution to Practice Problem**

$$\sin\frac{\pi}{4} = \frac{1}{\sqrt{2}} = \frac{\sqrt{2}}{2}$$

$$\cos\frac{\pi}{4} = \frac{1}{\sqrt{2}} = \frac{\sqrt{2}}{2}$$

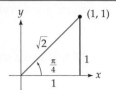

8.2 Exercises

Find the sine and cosine of each angle θ. When r is not given, calculate it from the Pythagorean Theorem.

1.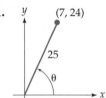

2.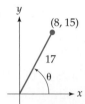

3.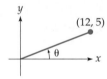

4.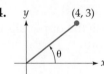

5.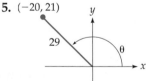

6.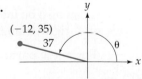

7.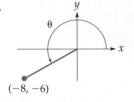

8.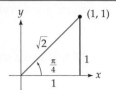

9.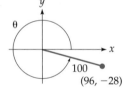

10.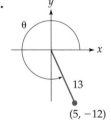

Evaluate *without* using a calculator, leaving answers in exact form.

11. **a.** $\sin\frac{\pi}{3}$ **b.** $\cos\frac{\pi}{3}$

   c. $\sin\frac{\pi}{4}$ **d.** $\cos\frac{\pi}{4}$

12. **a.** $\sin\frac{3\pi}{4}$ **b.** $\cos\frac{3\pi}{4}$

   c. $\sin\frac{5\pi}{3}$ **d.** $\cos\frac{5\pi}{3}$

13. **a.** $\sin \dfrac{5\pi}{4}$ **b.** $\cos \dfrac{5\pi}{4}$

   c. $\sin \dfrac{5\pi}{6}$ **d.** $\cos \dfrac{5\pi}{6}$

14. **a.** $\sin \dfrac{11\pi}{6}$ **b.** $\cos \dfrac{11\pi}{6}$

   c. $\sin \dfrac{7\pi}{4}$ **d.** $\cos \dfrac{7\pi}{4}$

15. **a.** $\sin\left(-\dfrac{\pi}{3}\right)$ **b.** $\cos\left(-\dfrac{\pi}{3}\right)$

   c. $\sin\left(-\dfrac{\pi}{4}\right)$ **d.** $\cos\left(-\dfrac{\pi}{4}\right)$

16. **a.** $\sin \dfrac{5\pi}{2}$ **b.** $\cos \dfrac{5\pi}{2}$

   c. $\sin 3\pi$ **d.** $\cos 3\pi$

Approximate using a calculator (set for *radians*). Round answers to two decimal places.

17. **a.** $\sin \dfrac{\pi}{10}$ **b.** $\cos \dfrac{\pi}{10}$

   c. $\sin 1$ **d.** $\cos 1$

18. **a.** $\sin \dfrac{\pi}{5}$ **b.** $\cos \dfrac{\pi}{5}$

   c. $\sin 4$ **d.** $\cos 4$

Sketch the graph of each function showing the amplitude and period.

19. $y = 4 \sin t$ 20. $y = 6 \cos t$

21. $y = \sin 4t$ 22. $y = \cos 6t$

23. $y = 3 \cos 4t$ 24. $y = 5 \sin 4t$

25. $y = -\cos 2t$ 26. $y = -\sin 2t$

Use the sum and difference identities on page 621 to find the following sines and cosines. Do *not* use a calculator.

27. **a.** $\sin \dfrac{5\pi}{12}$ **b.** $\cos \dfrac{5\pi}{12}$

   [*Hint:* Take $s = \dfrac{\pi}{4}$ and $t = \dfrac{\pi}{6}$ with upper signs.]

28. **a.** $\sin \dfrac{7\pi}{12}$ **b.** $\cos \dfrac{7\pi}{12}$

   [*Hint:* Take $s = \dfrac{\pi}{4}$ and $t = \dfrac{\pi}{3}$ with upper signs.]

APPLIED EXERCISES

29. **GENERAL: Ladder Height** A 30-foot ladder leaning against a wall makes a 70° angle with the ground. How high up the wall does the ladder reach? (See the diagram below).

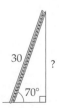

30. **GENERAL: Kite Flying** A straight kite string makes an angle of 40° with the gound, as shown in the diagram. If 250 feet of string has been let out, how high is the kite?

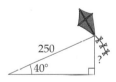

31. **GENERAL: Distance** A boat sailing due west toward a point 200 feet directly north of you makes an angle of 48°, as shown in the diagram. Find the distance between you and the boat.

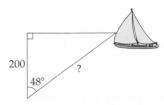

32. GENERAL: Wheelchair Ramp According to the American National Standards Institute, a wheelchair ramp should make an angle of 4.75° with the horizontal. Find the length of such a ramp that rises 2 feet.

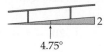

33. BUSINESS: Seasonal Sales An ice cream parlor estimates that the number of ice cream cones that it will sell on day n of the year is

$$S(n) = 700 - 600 \cos \frac{2\pi n}{365}$$

Find the number of ice cream cones that will be sold on August 1 (day 213).

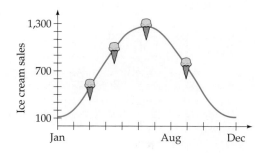

34. BUSINESS: Seasonal Sales A store estimates that the number of ski parkas that it will sell on day n of the year is

$$S(n) = 27 + 25 \cos \frac{2\pi n}{365}$$

Find the number of ski parkas that will be sold on February 1 (day 32).

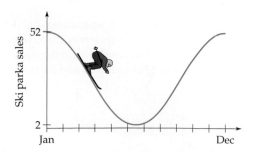

35. BIOMEDICAL: Breathing When you breathe normally, you inhale and exhale about 0.8 liter of air approximately every 4 seconds, with about 4.5 liters remaining in your lungs. Under these conditions, the total amount of air in your lungs t seconds after you last inhaled is

$$L(t) = 4.9 + 0.4 \cos \frac{\pi t}{2} \qquad \text{liters}$$

Find (*without* using a calculator) the amount of air in your lungs at time

a. $t = 0$ (just after inhaling)
b. $t = 2$ (just after exhaling)

36. COSINE AND SINE ON THE UNIT CIRCLE

a. Set your calculator for PARAMETRIC equations (on some calculators, use MODE), and enter $x_1 = \cos T$, $y_1 = \sin T$. (Be sure your calculator is set for *radians*.)

b. GRAPH the curve on the window with T in $[0, 2\pi]$, x in $[-2, 2]$ and y in $[-1.5, 1.5]$.

c. Use TRACE to move a point around the circle, with X and Y showing the cosine and sine of the angle. Observe how the sine and cosine change as the point moves around the circle.

d. Use EVALUATE to check the sine and cosine at T-values such as $\frac{\pi}{6}, \frac{\pi}{4}$, and $\frac{\pi}{3}$.

(Reset your calculator to FUNCTION mode after this exercise.)

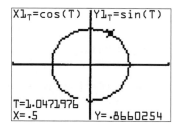

37. MODIFIED SINE CURVES

a. Graph the curves $y_1 = \sin x$, $y_2 = \sin 2x$, and $y_3 = \sin 0.5x$ on the window $[-2\pi, 2\pi]$ by $[-2, 2]$. (Be sure your calculator is set for *radians*.) Notice how a number in front of x changes the *period*: a larger number makes the curve "wiggle faster."

b. Now graph $y_1 = \sin x$, $y_2 = 2 \sin x$, and $y_3 = 0.5 \sin x$ on the same window. Notice how a number in front of the function changes the *amplitude*: a larger number makes the amplitude larger.

(*continues*)

c. Predict what the curve $y = 2 \sin 2x$ would look like. Then check your answer by graphing it. What about $y = -2 \sin x$? What about $y = -2 \sin 2x$?

38. THREE SIMILAR CURVES Graph

$$y_1 = 1 - x^2, \quad y_2 = \sqrt{1 - x^2}, \quad \text{and} \quad y_3 = \cos \frac{\pi x}{2}$$

on the window $[-1, 1]$ by $[0, 1.3]$. Which curve is a parabola, which is a semicircle, and which is a trigonometric function? (This shows that trigonometric curves should be drawn slightly more "pointed" than circles.)

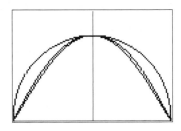

39. GENERAL: Temperature The mean temperature in Fairbanks, Alaska, on day x of the year has been estimated to be

$$T(x) = 25 + 37 \sin \frac{2\pi(x - 101)}{365}$$

degrees Fahrenheit.

a. Graph this function on the window $[0, 365]$ by $[-20, 70]$. (Be sure your calculator is set for *radians*.)
b. Find the temperature on January 10 (day 10).
c. Find the temperature on July 11 (day 192).

40. VERIFICATION OF $\sin^2 x + \cos^2 x = 1$
Graph $y_1 = (\sin x)^2$, $y_2 = (\cos x)^2$, and $y_3 = y_1 + y_2$ on the window $[-2\pi, 2\pi]$ by $[-1, 2]$. Why does the fact that y_3 is a horizontal line at height 1 verify the identity $\sin^2 x + \cos^2 x = 1$?

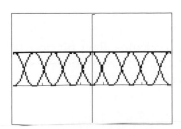

41. GENERAL: Greenhouse Gases The "greenhouse effect," which may be increasing global temperatures and, through the melting of polar ice, raising sea levels, is caused by carbon dioxide and other atmospheric gases. The carbon dioxide curve graphed on page 598 is given by $C(x) = 315 + 6.8e^{0.054x} + 3.5 \sin 2\pi x$ ppm (parts per million), where x is the number of years since January 1, 1960.

a. Graph this function on the window $[40, 50]$ by $[360, 400]$.

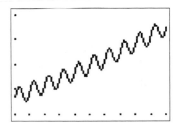

b. Zoom in on a point somewhere in the middle of the graph. Notice how the sine curve (seasonal variation) follows an upward trend.
c. Use this formula to predict the carbon dioxide levels on January 1, 2010.

42. PROOFS OF NEGATIVE ANGLE IDENTITIES As illustrated in the following diagram, changing the sign of an angle (from t to $-t$) changes the sign of the y-coordinate of a point on the terminal side, but not the sign of the x-coordinate. Use this fact together with the definitions of sin t and cos t on page 611 to prove the negative angle identities: $\sin(-t) = -\sin t$ and $\cos(-t) = \cos t$.

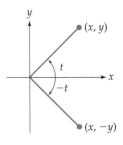

43. PROOFS OF COMPLEMENTARY ANGLE IDENTITIES Prove the identities

$$\sin t = \cos\left(\frac{\pi}{2} - t\right) \quad \text{and} \quad \cos t = \sin\left(\frac{\pi}{2} - t\right)$$

for an acute angle t as follows.

a. Show that if one acute angle of a right triangle is of size t radians, then the other acute angle is of size $\frac{\pi}{2} - t$ radians.

b. Observe that the "opposite" side for angle t is the "adjacent" side for angle $\frac{\pi}{2} - t$, and vice versa. Use these facts with the expressions for $\sin t$ and $\cos t$ on page 615 to prove the identities above.

44. PROOFS OF THE SUM AND DIFFERENCE IDENTITIES Prove the identity

$$\sin(s + t) = (\sin s)(\cos t) + (\cos s)(\sin t)$$

for acute angles whose sum is also acute by using the diagram on the right as follows (the notation AB means the length of the line segment between points A and B, and similarly for any other points).

a. Show that $AB = \sin(s + t)$.

b. Show that $AC = \sin t$.

c. Show that $CD = \cos t$.

d. Show that $CE = \dfrac{\dfrac{DE}{CD}}{\dfrac{CD}{DE}} \cdot CD = \dfrac{\sin s}{\cos s} \cdot \cos t$.

e. Show that

$$AB = (\cos s)(AC + CE)$$
$$= (\cos s)\left(\sin t + \frac{\sin s}{\cos s}\cos t\right)$$

f. From (a) and (e), conclude that

$$\sin(s + t) = (\sin s)(\cos t) + (\cos s)(\sin t)$$

[*Note:* The identity

$$\cos(s + t) = (\cos s)(\cos t) - (\sin s)(\sin t)$$

will be proved in Exercise 71 on page 638. The identities for $\sin(s - t)$ and $\cos(s - t)$ are derived by changing t to $-t$ in the identities for $\sin(s + t)$ and $\cos(s + t)$ and then using the negative angle identities.]

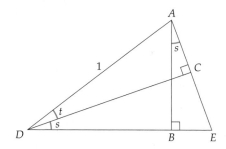

DERIVATIVES OF SINE AND COSINE FUNCTIONS

Introduction

The last two sections contained no calculus, their purpose being to introduce the sine and cosine functions. In this section we will differentiate these functions, and use these derivatives to optimize functions and calculate rates of change.

Derivatives of the Sine and Cosine Functions

The derivatives of the sine and cosine functions are given by the following formulas.

Derivatives of $\sin t$ and $\cos t$	
$\dfrac{d}{dt} \sin t = \cos t$	The derivative of sine is cosine (for t in radians)
$\dfrac{d}{dt} \cos t = -\sin t$	The derivative of cosine is negative sine

To see graphically that the derivative (the slope) of $\sin t$ is $\cos t$, observe in the diagram below that, at any particular t-value, the *slope* of the *sine* function is the same as the *value* (height) of the *cosine* function.

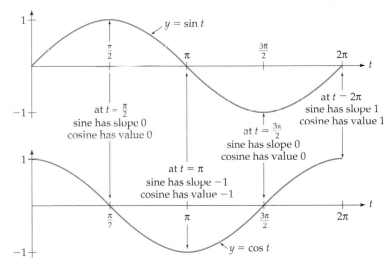

Similarly, to see that the derivative of $\cos t$ is $-\sin t$, observe in the diagram above that the *slope* of the *cosine* curve is the negative of the *value* of the *sine* curve. Justifications of these two formulas are given at the end of this section. The following examples show how to use these formulas.

EXAMPLE 1

DIFFERENTIATING A QUOTIENT INVOLVING SINE

Differentiate $f(t) = \dfrac{\sin t}{t}$.

Solution

Derivative of $\sin t$

Derivative of t

$$\frac{d}{dt}\frac{\sin t}{t} = \frac{t \cdot \cos t - 1 \cdot \sin t}{t^2} = \frac{t \cos t - \sin t}{t^2} \qquad \text{Using the Quotient Rule}$$

EXAMPLE 2

DIFFERENTIATING A PRODUCT INVOLVING COSINE

Differentiate $g(t) = t^3 \cos t$.

Solution

$$\frac{d}{dt}(t^3 \cdot \cos t) = 3t^2 \cdot \cos t + t^3 \cdot (-\sin t) = 3t^2 \cos t - t^3 \sin t \qquad \begin{array}{l}\text{Using the}\\ \text{Product}\\ \text{Rule}\end{array}$$

Derivative   Second   First   Derivative
of the first   left alone   left alone   of $\cos t$

Practice Problem 1

Differentiate $f(t) = t \sin t$.   ➤ Solution on page 636

These rules for differentiating sine and cosine, combined with the Chain Rule, show how to differentiate the sine and cosine of a *function*. For a differentiable function $f(t)$:

Derivative of $\sin f(t)$ and $\cos f(t)$

$$\frac{d}{dt}\sin f(t) = \cos f(t) \cdot \frac{df}{dt}$$

The derivative of the sine of a function is the cosine of the function times the derivative of the function

$$\frac{d}{dt}\cos f(t) = -\sin f(t) \cdot \frac{df}{dt}$$

The derivative of the cosine of a function is negative the sine of the function times the derivative of the function

---

EXAMPLE 3

DIFFERENTIATING THE COSINE OF A FUNCTION

Differentiate $g(t) = \cos(t^2 + 1)$.

Solution

$$\frac{d}{dt}\underbrace{\cos(t^2 + 1)}_{f(t)} = \underbrace{-\sin(t^2 + 1)}_{-\sin f(t)} \cdot \underbrace{2t}_{\frac{df}{dt}} \qquad \text{Using } \frac{d}{dt}\cos f = -\sin f \cdot \frac{df}{dt}$$

$$= -2t \sin(t^2 + 1)$$

EXAMPLE 4

FINDING A SECOND DERIVATIVE

Find the *second* derivative of $g(t) = \sin 4t$.

Solution

$$g'(t) = \frac{d}{dt}(\sin 4t) = \cos 4t \cdot (4) = 4 \cos 4t \qquad \text{Using } \frac{d}{dt}\sin f = \cos f \cdot \frac{df}{dt}$$

Derivative of $4t$

$$g''(t) = \frac{d}{dt}(4 \cos 4t) = 4[-\sin 4t \cdot (4)] \qquad \frac{d}{dt}\cos f = -\sin f \cdot \frac{df}{dt}$$

$$= -16 \sin 4t$$

Practice Problem 2   Differentiate $f(t) = \sin\dfrac{t}{2}$. $\left[\textit{Hint: Write } \dfrac{t}{2} \textit{ as } \dfrac{1}{2}t.\right]$

➤ Solution on page 636

EXAMPLE 5

EVALUATING A DERIVATIVE

For $f(x) = \dfrac{\sin x}{\cos x}$, find $f'\left(\dfrac{\pi}{3}\right)$.

Solution   Differentiating:

Derivative of $\sin x$
Derivative of $\cos x$

$$f'(x) = \frac{\cos x \cdot \cos x - (-\sin x) \cdot \sin x}{\cos^2 x} \qquad \text{Using the Quotient Rule}$$

$$= \frac{\cos^2 x + \sin^2 x}{\cos^2 x} = \frac{1}{\cos^2 x} \qquad \text{Using the identity } \sin^2 x + \cos^2 x = 1$$

Evaluating at the given $\frac{\pi}{3}$:

$$f'\left(\frac{\pi}{3}\right) = \frac{1}{\cos^2 \frac{\pi}{3}} = \frac{1}{\left(\cos \frac{\pi}{3}\right)^2} = \frac{1}{\left(\frac{1}{2}\right)^2} = \frac{1}{\left(\frac{1}{4}\right)} = 4 \qquad \text{Using} \quad \cos \frac{\pi}{3} = \frac{1}{2}$$

EXAMPLE 6

DIFFERENTIATING AN EXPONENTIAL TRIGONOMETRIC FUNCTION

If $f(z) = e^{\sin z}$, find $f'(\pi)$.

Solution

$$f'(z) = e^{\sin z} \cos z \qquad\qquad \text{Using} \quad \frac{d}{dt} e^f = e^f \cdot \frac{df}{dt}$$

$$f'(\pi) = \underbrace{e^{\sin \pi}}_{0} \underbrace{\cos \pi}_{-1} = e^0 \underbrace{(-1)}_{1} = -1 \qquad \text{Evaluating at } \pi$$

Practice Problem 3

Differentiate $f(x) = \ln(1 + \cos x)$.

➤ Solution on page 636

EXAMPLE 7

DIFFERENTIATING A TRIGONOMETRIC FUNCTION

Differentiate $f(t) = \sin^2 t^3$.

Solution

We write $\sin^2 t^3$ as $(\sin t^3)^2$ and then differentiate using the Generalized Power Rule.

$$\frac{d}{dt}(\sin t^3)^2 = 2(\sin t^3)^1 \cdot \underbrace{\cos t^3 \cdot 3t^2}_{\substack{\text{Derivative of} \\ \sin t^3}} = 6t^2 \sin t^3 \cos t^3$$

Growth of Sales

In the preceding section we used sine and cosine functions to model the fluctuations in sales of seasonal goods. The *derivative* of such a sales function then gives the *rate of change of sales*, showing how rapidly sales are increasing or decreasing.

EXAMPLE **8**

FINDING GROWTH OF SEASONAL SALES

A department store predicts that the number of bathing suits that it will sell during week t of the year (with $t = 1$ meaning the first week of January) is given by the formula

$$S(t) = 400 - 400 \cos \frac{2\pi t}{52}$$

Find the rate of change of the sales during the first week of May (week $t = 18$).

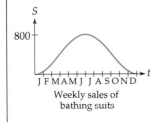

S

800

J F M A M J J A S O N D

Weekly sales of
bathing suits

Solution   The rate of change means the derivative, and the differentiation is easier if we write $\dfrac{2\pi t}{52}$ as $\dfrac{2\pi}{52} t$, a constant times the variable.

$$S'(t) = \frac{d}{dt}\left(400 - 400 \cos \frac{2\pi}{52} t\right)$$

$$= 400 \sin \frac{2\pi}{52} t \cdot \left(\frac{2\pi}{52}\right) \qquad \text{Using } \frac{d}{dt}\cos f = -\sin f \cdot \frac{df}{dt}$$

$$= 400 \frac{2\pi}{52} \sin \frac{2\pi}{52} t \qquad\qquad \text{Rearranging}$$

Evaluating at week $t = 18$,

$$S'(18) = 400 \underbrace{\frac{2\pi}{52}}_{48.3} \sin \underbrace{\left(\frac{2\pi}{52} 18\right)}_{0.823} \approx 39.8 \qquad \begin{array}{l}\text{Using a calculator} \\ \text{set for radians}\end{array}$$

Interpretation: Sales will be increasing by about 40 per week.

Graphing Calculator Exploration

dy/dx=39.776616

Verify the answer to the previous example as follows:

a. Graph the function $y_1 = 400 - 400 \cos (2\pi x/52)$ on the window $[0, 52]$ by $[-200, 1000]$.

b. Find the derivative at $x = 18$ (using dy/dx or NDERIV). Your answer should agree with that found in the preceding example.

As before, we optimize a function by setting its first derivative equal to zero and solving, then using the second-derivative test to determine whether we have maximized or minimized.

EXAMPLE 9

MAXIMIZING VOLUME

A water trough consists of two boards, each 10 feet long and 1 foot wide, attached along one edge at an angle of θ radians, and closed at both ends, as shown below. Find the angle θ that maximizes the volume of the trough.

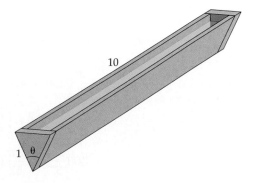

Solution

The volume will be maximized if we maximize the cross-sectional area (shown at the right).

    Turning this triangle on its side gives a triangle of base 1 and height $\sin \theta$

$$\left(\text{since} \quad \sin \theta = \frac{\text{opp}}{\text{hyp}} = \frac{h}{1} = h\right).$$

The area of the triangle is

$$A = \frac{1}{2} \cdot 1 \cdot \sin \theta = \frac{1}{2} \sin \theta \qquad \text{One-half base times height}$$

Differentiating,

$$A' = \frac{1}{2} \cos \theta = 0 \qquad \begin{array}{l}\text{To maximize, set the} \\ \text{derivative equal to zero}\end{array}$$

The only angle between 0 and π whose cosine is zero (think of the graph of $\cos x$) is

$$\theta = \frac{\pi}{2} \qquad \text{Solution to} \quad \cos \theta = 0$$

The second-derivative test gives

$$A'' = -\frac{1}{2}\sin\theta \qquad\qquad \text{Differentiating } A' = \tfrac{1}{2}\cos\theta$$

$$= -\frac{1}{2}\sin\frac{\pi}{2} = -\frac{1}{2} \qquad\qquad \text{Evaluating at } \frac{\pi}{2}$$

Negative

Since A'' is negative, the area is *maximized* at $\theta = \dfrac{\pi}{2}$, or 90°.

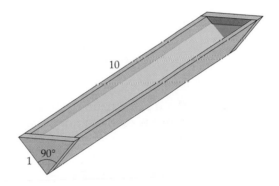

Therefore, joining the boards at an angle of 90° will maximize the volume of the trough.

Why We Use Radians Instead of Degrees

We can now explain why we use radians instead of degrees in calculus. On page 603 we saw that multiplying by $\dfrac{\pi}{180}$ converts degrees into *radians*. Therefore, the sine and cosine functions, written as functions of t *degrees*, are $\sin\left(\dfrac{\pi}{180}t\right)$ and $\cos\left(\dfrac{\pi}{180}t\right)$. The derivatives of these "degree" sine and cosine functions are then

$$\frac{d}{dt}\sin\left(\frac{\pi}{180}t\right) = \cos\left(\frac{\pi}{180}t\right)\cdot\frac{\pi}{180} \qquad \text{Using } \frac{d}{dt}\sin f = \cos f\cdot\frac{df}{dt}$$

and

$$\frac{d}{dt}\cos\left(\frac{\pi}{180}t\right) = -\sin\left(\frac{\pi}{180}t\right)\cdot\frac{\pi}{180} \qquad \text{Using } \frac{d}{dt}\cos f = -\sin f\cdot\frac{df}{dt}$$

Using the notations $\sin t°$ and $\cos t°$ for the "degree" sine and cosine functions, these derivative formulas become:

$$\frac{d}{dt}\sin t° = \frac{\pi}{180}\cos t°$$

and

$$\frac{d}{dt}\cos t° = -\frac{\pi}{180}\sin t°$$

It is the extra factor $\pi/180$ in these formulas that makes them more complicated than the simpler *radian* formulas on page 627. This is why we use radians: they simplify the formulas for differentiation and (as we shall see on page 640) integration.

Justification of the Differentiation Formulas for $\sin t$ and $\cos t$

The differentiation formulas for the sine and cosine functions come from the definition of the derivative. Before we can show how this is done, we need to establish the following two limits.

$$\lim_{h\to 0}\frac{\sin h}{h} = 1 \qquad \frac{\sin h}{h} \text{ approaches 1 as } h \text{ approaches 0}$$

and

$$\lim_{h\to 0}\frac{\cos h - 1}{h} = 0 \qquad \frac{\cos h - 1}{h} \text{ approaches 0 as } h \text{ approaches 0}$$

Instead of presenting formal proofs of these limits, we justify them by using a graphing calculator to calculate values of the functions as the variable takes values approaching zero.

Graphing Calculator Exploration

Find the two preceding limits as follows, using the TABLE feature of a graphing calculator (using x instead of h for ease of entry).

a. Enter $y_1 = (\sin x)/x$ and $y_2 = (\cos x - 1)/x$.

b. Set the TABLE feature to ASK you for the x-values.

c. Enter x-values approaching zero, such as 1, .1, .01, .001, .0001, ...
or their negatives. You should get a table such as the one shown
below. Observe that the values of $y_1 = (\sin x)/x$ approach 1
and the values of $y_2 = (\cos x - 1)/x$ approach zero.

X	Y₁	Y₂	
1	.84147	-.4597	
.1	.99833	-.05	
.01	.99998	-.005	
.001	1	-5E-4	← $-5 \times 10^{-4} = -.0005$
1E-4	1	-5E-5	← $-5 \times 10^{-5} = -.00005$

X =

1 E 4 means
$1 \times 10^{-4} = .0001$ →

The derivative of $f(t) = \sin t$ is then found as follows.

$$f'(t) = \lim_{h \to 0} \frac{f(t + h) - f(t)}{h} = \lim_{h \to 0} \frac{\sin(t + h) - \sin t}{h}$$

Definition of $f'(t)$,
with $f(t) = \sin t$

$$= \lim_{h \to 0} \frac{(\sin t)(\cos h) + (\cos t)(\sin h) - \sin t}{h}$$

Using the identity
$\sin (s + t) =$
$(\sin s)(\cos t) +$
$(\cos s)(\sin t)$

$$= \lim_{h \to 0} \left(\frac{(\sin t)(\cos h) - \sin t}{h} + \frac{(\cos t)(\sin h)}{h} \right)$$

Rearranging the
numerator, and
then separating
into two fractions

$$= \lim_{h \to 0} \left(\sin t \frac{\cos h - 1}{h} + \cos t \frac{\sin h}{h} \right)$$

Fractoring $\sin t$ from
the first numerator,
then moving $\sin t$
and $\cos t$ out front

$$= (\sin t) \left(\underbrace{\lim_{h \to 0} \frac{\cos h - 1}{h}}_{0} \right) + (\cos t) \left(\underbrace{\lim_{h \to 0} \frac{\sin h}{h}}_{1} \right)$$

Separating the limit
into two limits,
then taking $\sin t$
and $\cos t$ outside of
the h limits

$$= \underbrace{(\sin t) \cdot (0)}_{0} + \underbrace{(\cos t) \cdot (1)}_{\cos t} = \cos t$$

Using the two
limits from the
previous page

This is the result that we wanted, showing that the derivative of
$f(t) = \sin t$ is $f'(t) = \cos t$. While the derivative formula for the

cosine can be derived similarly (see Exercise 68), it can be found even more easily by using the identities $\cos t = \sin\left(\dfrac{\pi}{2} - t\right)$ and $\sin t = \cos\left(\dfrac{\pi}{2} - t\right)$ from page 621:

$$\frac{d}{dt}\cos t = \frac{d}{dt}\sin\left(\frac{\pi}{2} - t\right) = \cos\left(\frac{\pi}{2} - t\right)\cdot(-1) = -\sin t$$

Using	Using	Using
$\cos t = \sin\left(\dfrac{\pi}{2} - t\right)$	$\dfrac{d}{dt}\sin f = \cos f \cdot f'$	$\sin t = \cos\left(\dfrac{\pi}{2} - t\right)$

8.3 Section Summary

The derivatives of the sine and cosine functions are given by the following formulas.

$$\frac{d}{dt}\sin t = \cos t \qquad\qquad \frac{d}{dt}\sin f(t) = \cos f(t)\cdot\frac{df}{dt}$$

$$\frac{d}{dt}\cos t = -\sin t \qquad\qquad \frac{d}{dt}\cos f(t) = -\sin f(t)\cdot\frac{df}{dt}$$

In the formulas on the right, $f(t)$ is any differentiable function.

We use derivatives to find the rate of change of daily sales (showing how rapidly sales are changing), and to optimize functions (maximizing volume) by setting the derivative equal to zero.

➤ **Solutions to Practice Problems**

1. $\dfrac{d}{dt}(t\cdot\sin t) = 1\cdot\sin t + t\cdot\cos t = \sin t + t\cos t$

2. $\dfrac{d}{dt}\sin\dfrac{1}{2}t = \left(\cos\dfrac{1}{2}t\right)\left(\dfrac{1}{2}\right) = \dfrac{1}{2}\cos\dfrac{t}{2}$

3. $\dfrac{d}{dx}\ln(1 + \cos x) = \dfrac{-\sin x}{1 + \cos x}$

8.3 Exercises

Differentiate each function.

1. $f(t) = t^3 \sin t$

2. $f(t) = t^2 \cos t$

3. $f(t) = \dfrac{\cos t}{t}$

4. $f(t) = (1 + \sin t)^4$

5. $f(t) = \sin(t^3 + 1)$

6. $f(t) = \cos(t^3 + t + 1)$

7. $f(t) = \sin 5t$

8. $f(t) = \cos 4t$

9. $f(t) = 2 \cos 3t$

10. $f(t) = 6 \sin 2t$

11. $f(t) = \cos\left(\dfrac{\pi}{180} t\right)$

12. $f(t) = \sin\left(\dfrac{\pi}{200} t\right)$

13. a. $f(t) = t^2 + \sin \pi t$
b. Find $f'(0)$

14. a. $f(t) = t \cos \pi t$
b. Find $f'(0)$

15. a. $f(t) = 4 \cos \dfrac{t}{2}$
b. Find $f'\left(\dfrac{\pi}{3}\right)$

16. a. $f(t) = 48 \sin \dfrac{t}{12}$
b. Find $f'(4\pi)$

17. a. $f(t) = \sin(\pi - t)$
b. Find $f'\left(\dfrac{\pi}{2}\right)$

18. a. $f(t) = \cos(2\pi - t)$
b. Find $f'(\pi)$

19. $f(t) = \cos(t + \pi)^3$

20. $f(t) = \sin^3(\pi + t)$

21. $f(t) = \cos^3(t + \pi)$

22. $f(t) = \sin(\pi + t)^3$

23. $f(t) = \sin(\pi - t)^2$

24. $f(t) = \sin^2(\pi - t)$

25. $f(t) = \sin^2 t^3$

26. $f(t) = \cos^4 t^2$

27. $f(x) = \cos \sqrt{x + 1}$

28. $f(x) = \sin \sqrt[3]{x}$

29. $f(x) = 4 \sin x + \sin 4x + \sin^4 x$

30. $f(x) = 4 \cos x + \cos 4x + \cos^4 x$

31. $f(x) = \sin e^x$

32. $f(x) = e^{\cos x}$

33. $f(z) = e^{z + \cos z}$

34. $f(z) = \cos(z + e^z)$

35. $f(z) = (1 - \cos z)^4$

36. $f(z) = (\sin z + \cos z)^2$

37. a. $f(z) = \ln(\sin z)$
b. Find $f'\left(\dfrac{\pi}{2}\right)$

38. a. $f(z) = \cos(\ln z)$
b. Find $f'(1)$

39. $f(z) = \sin(z + \ln z)$

40. $f(z) = \ln(z + \sin z)$

41. $f(x) = 3 \sin^2 x - 4 \cos^2 x$

42. $f(x) = 4 \cos^2 x + 5 \sin^2 x$

43. $f(x) = \dfrac{1}{x} \cos x^2$

44. $f(x) = x^2 \sin \dfrac{1}{x}$

45. $f(x) = e^{-x} \sin e^x$

46. $f(x) = e^x \cos e^{-x}$

47. a. $f(x) = \dfrac{\cos x}{\sin x}$
b. Find $f'\left(\dfrac{\pi}{3}\right)$

48. a. $f(x) = \dfrac{\sin 2x}{\cos 2x}$
b. Find $f'\left(\dfrac{\pi}{8}\right)$

49. $f(t) = \sin^2 t - \cos^2 t$

50. $f(t) = \cos^2 t + \sin^2 t$

51. a. $f(t) = \dfrac{1 + \sin t}{\cos t}$
b. Find $f'(\pi)$

52. a. $f(t) = \dfrac{1 - \cos t}{\sin t}$
b. Find $f'\left(\dfrac{\pi}{2}\right)$

53. $f(x) = \sin(\cos x)$

54. $f(x) = \cos(\sin x)$

55. $f(x) = \left(\cos \dfrac{x}{2} - 1\right)^2$

56. $f(x) = 2x \cos \dfrac{x}{2}$

57. $f(x) = x^2 \sin x + 2x \cos x - 2 \sin x$

58. $f(x) = x^2 \cos x - 2x \sin x - 2 \cos x$

59. $f(t) = \sin^2(t^2 + 1)$

60. $f(t) = \cos^2 \sqrt{t}$

Find the *second* derivative of each function.

61. $f(t) = t \sin t$

62. $f(t) = \sin at + \cos bt$ (a and b are constants)

63. $f(x) = e^{\sin x}$

64. $f(x) = e^{\cos x}$

65. $f(z) = z \sin z + \cos z$

66. $f(z) = \sin z - z \cos z$

67. DERIVATIVE FORMULA FOR COS x
Derive the formula for the derivative of $\cos x$ (assuming that the derivative exists) from the formula for the derivative of $\sin x$ as follows.

a. Differentiate both sides of the identity
$$\sin^2 x + \cos^2 x = 1, \text{ obtaining}$$
$$2 \sin x \cos x + 2 \cos x \dfrac{d}{dx} \cos x = 0$$
(continues)

b. Solve this equation for $\dfrac{d}{dx}\cos x$ to obtain

$$\frac{d}{dx}\cos x = -\sin x.$$

68. Use the definition of the derivative to show that the derivative of $f(t) = \cos t$ is $f'(t) = -\sin t$. [*Hint:* The steps are similar to those on page 635, but using the identity $\cos(t + h) = (\cos t)(\cos h) - (\sin t)(\sin h)$.]

69. Show that the formulas on the right side of the box on page 636 reduce to the formulas on the left side in the case that the function f is simply $f(t) = t$.

70. Why do the functions $\sin^2 x$ and $-\cos^2 x$ have the same derivative?

71. PROOF OF cos($s + t$) IDENTITY Prove the identity

$$\cos(s + t) = (\cos s)(\cos t) - (\sin s)(\sin t)$$

by differentiating the identity from page 621,

$$\sin(s + t) = (\sin s)(\cos t) + (\cos s)(\sin t)$$

with respect to s, holding t constant.

72. DIFFERENTIATION FORMULAS FOR sin x AND cos x

a. Verify the formula $\dfrac{d}{dx}\sin x = \cos x$ by defining $y_1 = \sin x$ and y_2 as its derivative (using NDERIV). Then graph y_1 and y_2 on the window $[-2\pi, 2\pi]$ by $[-2, 2]$, observing that y_2 appears to be the cosine curve.

b. Then change y_1 to $\cos x$ to verify that $\dfrac{d}{dx}\cos x = -\sin x$ by observing that y_2 appears to be the negative of the sine curve.

Note that NDERIV finds derivatives *numerically*, using the difference quotient with a small value of h, so this is an independent verification of the formulas.

APPLIED EXERCISES

73–74. BUSINESS: Seasonal Sales A manufacturer estimates that the number of air conditioners sold during week t of the year will be

$$S(t) = 500 - 500\cos\left(\frac{\pi}{26}t\right)$$

Find the rate of change of sales at

73. $t = 19$ (mid May)

74. $t = 37$ (mid September)

75. GENERAL: Maximizing Volume of a Tent A tent made by hanging a square sheet over a ridge pole at angle θ as shown will enclose a volume of

$$V(\theta) = 1000\sin\theta$$

cubic feet. Find the angle θ that maximizes the volume $V(\theta)$.

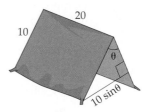

76. BUSINESS: Maximum Temperature The temperature in a chemical refining tower after t hours is $T(t) = 100 + 60(\sin t + \cos t)$ degrees Fahrenheit (for $0 \le t \le \pi$). Find when the temperature is the greatest.

77. GENERAL: Honeycombs A honeycomb consists of cells like the one shown, where θ is the angle between the upper faces and the axis of the cell. Minimizing the surface of the area is equivalent to minimizing the function

$$S(\theta) = \frac{\sqrt{3} - \cos\theta}{\sin\theta}$$

Minimize this function by setting its derivative equal to zero and solving, deriving the relationship $\cos\theta = \dfrac{1}{\sqrt{3}}$. The acute angle that solves this equation is approximately 54.7°, which is the actual angle used by honeybees.*

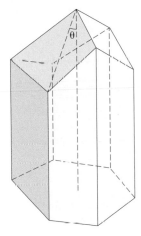

78. Maximizing Area

   a. Find a formula for the area of a right triangle (shown below) with hypotenuse 1 and one acute angle of x radians ($0 < x < \pi/2$).

   b Find the angle x that maximizes the area. [*Hint:* Either graph the area function and use MAXIMUM, or use NDERIV to find where its derivative equals zero.]

   c. Convert x into degrees by multiplying by $\dfrac{180}{\pi}$. Did you expect to obtain this answer?

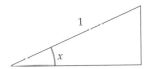

79. GENERAL: Maximizing Distance of a Thrown Ball A ball thrown with an initial velocity of 60 miles per hour at an angle of inclination with the ground of x radians, as shown

below, will travel a horizontal distance of
$$D(x) = 242(\sin x)(\cos x)$$
feet (neglecting air resistance).

   a. Find the angle x that maximizes this distance [*Hint:* Either graph the distance function and use MAXIMUM, or use NDERIV to find where its derivative equals zero.]

   b. Convert x into degrees by multiplying by $\dfrac{180}{\pi}$. [*Note:* The angle that you find holds for all velocities.*]

80. GENERAL: Temperature The mean temperature in Fairbanks, Alaska on day x of the year is approximately
$$T(x) = 25 + 37\sin\!\left(\frac{2\pi(x-101)}{365}\right)$$
degrees Fahrenheit.[†]

   a. Graph this function and its derivative (using NDERIV) on the window [0, 365] by [−20, 70].

   b. Find the rate of change of temperature on April 15 (day 105) and on October 15 (day 288) and interpret these answers.

   c. Find the days of the highest and lowest temperatures.

81. BIOMEDICAL: Blood Pressure A person's blood pressure varies periodically according to the formula $p(t) = 90 + 15\sin(2.5\pi t)$, where t is the number of seconds since the beginning of a cardiac cycle. *(continues)*

---

* The result of this exercise applies to a *thrown ball*. For a ball hit by a bat, the best angle is 35°. This is because a batted ball can be given a backspin of 2000 rpm or more, giving it lift through what is called the Magnus effect. See Robert K. Adair, *The Physics of Baseball*, Harper & Row, 1990.

† Barbara and Clifton Lando, "Is the Graph of Temperature Variation a Sine Curve?" *The Mathematics Teacher* 70:534–537, September 1977. Further information from this article is used elsewhere in this chapter.

* See D'Arcy Wentworth Thompson, *On Growth and Form,* Columbia University Press, 1917; Dover Publications, 1992. See also E. Batschelet, *Introduction to Mathematics for Life Scientists,* Springer-Verlag, 1971.

a. Graph the function on the window [0, 1.6] by [0, 120].
b. When is blood pressure the *greatest* for $0 \le t \le 1.6$, and what is the maximum blood pressure?
c. When is blood pressure the *lowest* for $0 \le t \le 1.6$, and what is the minimum blood pressure?

82. ATHLETICS: Basketball To make a basket, a basketball player must launch the ball with just the right angle and initial velocity for it to pass through the hoop. It can be shown from Newton's laws of motion that the initial velocity v and angle θ must satisfy the equation

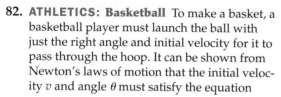

$$v = \sqrt{\dfrac{16d}{\cos^2 \theta \left(\dfrac{\sin\theta}{\cos\theta} - \dfrac{h}{d} \right)}}$$

where the constants d and h are as shown in the following diagram. To minimize the chances of the ball bouncing off the rim, coaches recommend shooting with the smallest possible velocity, called the "soft shot."

Assume that for a free throw the constants are $d = 14$ and $h = 3$, and graph the resulting function on the window $[0, \pi/2]$ by [0, 100]. Then find the "soft shot" angle θ that minimizes the function.*

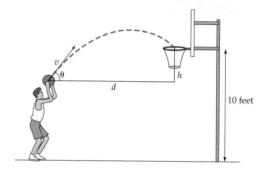

10 feet

*See Peter J. Brancazio, "Physics of Basketball," *American Journal of Physics* 49(4):356–365, 1981. In general, the angle for the soft shot is $\theta = 45° + \dfrac{1}{2} \left(\begin{array}{c} \text{Angle of elevation} \\ \text{from ball to rim} \end{array} \right)$, found by setting the derivative of v equal to zero and solving for θ.

8.4 **INTEGRALS OF SINE AND COSINE FUNCTIONS**

Introduction

In this section we will integrate sine and cosine functions and use their integrals to find *areas under curves, total accumulations,* and *average values.*

Integrals of Sine and Cosine

The integrals of the sine and cosine functions are given by the following formulas, which are proved by differentiating their right-hand sides.

Integrals of $\sin t$ and $\cos t$	
$$\int \sin t \, dt = -\cos t + C$$	The integral of sine is negative cosine (plus a constant)
$$\int \cos t \, dt = \sin t + C$$	The integral of cosine is sine (plus a constant)

These formulas are easily remembered. The derivative of sine is cosine, so the integral of cosine is sine, which is just the second formula. The first formula is similar, but with a negative sign (from the differentiation formula for cosine).

EXAMPLE 1

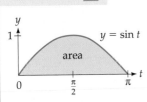

FINDING AREA BY INTEGRATION

Find the area under one arch of the sine curve $y = \sin t$.

Solution For the area under one arch of $\sin t$, we integrate from 0 to π:

$$\text{Area} = \int_0^{\pi} \sin t \, dt = \underbrace{-\cos t \Big|_0^{\pi}}_{\substack{\text{Using} \\ \int \sin t \, dt = -\cos t + C}} = \underbrace{-\cos \pi}_{\substack{\text{Evaluating} \\ \text{at } \pi}} - \underbrace{(-\cos 0)}_{\substack{\text{And} \\ \text{at } 0}}$$

$$= -(-1) - (-1) = 1 + 1 = 2$$

The area is 2 square units.

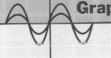 **Graphing Calculator Exploration**

Use a graphing calculator to verify the answer to the example above as follows:

a. Graph $y_1 = \sin x$ on the window $[0, \pi]$ by $[-1, 2]$.

b. Integrate this function from 0 to π (using FnInt, $\int f(x) \, dx$, or a similar command). Your answer should agree with that found in the preceding example.

The substitution method (introduced on page 415) greatly expands the class of functions that we can integrate. As before, we write the integration formulas with x replaced by u, where u stands for a differentiable function.

$$\int \sin u \, du = -\cos u + C$$

$$\int \cos u \, du = \sin u + C$$

To evaluate a particular integral, we represent part of the integrand by u, calculate its differential du, and multiply inside and outside by constants (if necessary) to complete the du. The right-hand side of the formula then gives the solution, after substituting back to the original variable.

EXAMPLE 2

USING THE SUBSTITUTION METHOD

Find $\int x^2 \sin x^3 \, dx$.

Solution The integrand involves $\sin x^3$, the sine of a function, suggesting the $\int \sin u \, du$ formula with $u = x^3$. The differential is $du = 3x^2 \, dx$ (the derivative times dx).

$$\int x^2 \sin x^3 \, dx \;=\; \frac{1}{3} \int \sin x^3 \, 3x^2 \, dx \;=\; \frac{1}{3} \int \sin u \, du$$

$$\underbrace{}_{\substack{\sin u \\ u = x^3 \\ du = 3x^2 \, dx}} \qquad \underbrace{}_{\sin u} \; \underbrace{}_{du} \qquad \underbrace{}_{\text{Substituting}}$$

Multiplying by 3 and $\frac{1}{3}$

$$=\; \frac{1}{3}(-\cos u) + C \;=\; -\frac{1}{3}\cos x^3 + C$$

Integrating by $\int \sin u \, du = -\cos u + C$ Substituting back to x using $u = x^3$

We can check this answer by differentiation:

$$\frac{d}{dx}\left(-\frac{1}{3}\cos x^3 + C\right) \;=\; -\frac{1}{3}(-\sin x^3)(3x^2) \;=\; x^2 \sin x^3$$

It checks: the derivative of the answer is the integrand

Answer   Using $\dfrac{d}{dx}\cos f = \sin f \cdot \dfrac{df}{dt}$ Simplifying

Practice Problem 1 Find $\int \cos \dfrac{2t}{3} \, dt$. [*Hint:* Write $\dfrac{2t}{3}$ as $\dfrac{2}{3}t$, a constant times the variable.]

➤ Solution on page 648

Other Substitution Formulas

An integral may require one of the earlier integration formulas (from page 415), but now with u being a trigonometric function.

(A) $\displaystyle\int u^n \, du = \frac{1}{n+1} u^{n+1} + C$ $(n \neq -1)$

(B) $\displaystyle\int e^u \, du = e^u + C$

(C) $\displaystyle\int u^{-1} \, du = \int \frac{du}{u} = \ln |u| + C$

EXAMPLE 3

USING ANOTHER SUBSTITUTION FORMULA

Find $\displaystyle\int \sin^5 2t \cos 2t \, dt$.

Solution   The integrand includes $\sin^5 2t = (\sin 2t)^5$, a function to a power, suggesting the $\int u^n \, du$ formula with $u = \sin 2t$.

$$\int \underbrace{(\sin 2t)^5}_{u^5} \cos 2t \, dt = \frac{1}{2} \int \underbrace{(\sin 2t)^5}_{u^5} \underbrace{2 \cos 2t \, dt}_{du} = \frac{1}{2} \int \underbrace{u^5 \, du}_{\text{Substituting}}$$

$u = \sin 2t$ Multiplying by 2 and $\frac{1}{2}$
$du = \cos 2t \cdot 2dt$

$$= \underbrace{\frac{1}{2}\frac{1}{6} u^6 + C}_{\substack{\text{Integrating} \\ \text{by formula (A)}}} = \underbrace{\frac{1}{12} \sin^6 2t + C}_{\substack{\text{Substituting back} \\ \text{using } u = \sin 2t}}$$

Practice Problem 2   Find $\displaystyle\int \cos^2 x \sin x \, dx$.   ➤ Solution on page 648

EXAMPLE 4

EVALUATING A DEFINITE INTEGRAL

Evaluate the definite integral $\displaystyle\int_0^{\pi/2} e^{\sin z} \cos z \, dz$.

Solution   The integrand involves $e^{\sin z}$, e to a function, suggesting the $\int e^u \, du$ formula with $u = \sin z$. With definite integrals, we must also change the upper and lower limits of integration to their corresponding u-values.

$$u = \sin \frac{\pi}{2} = 1$$

$$\int_0^{\pi/2} \underbrace{e^{\sin z}}_{\substack{e^u \\ u = \sin z \\ du = \cos z\, dz \\ u = \sin 0 = 0}} \underbrace{\cos z\, dz}_{du} = \underbrace{\int_0^1 e^u\, du}_{\text{Substituting}} = \underbrace{e^u \Big|_0^1}_{\substack{\text{Integrating} \\ \text{by formula (B)}}} = \underbrace{e^1 - e^0}_{\text{Evaluating}} = \underbrace{e - 1}_{\substack{\text{Final} \\ \text{answer}}}$$

We leave the answer in its *exact* form, $e - 1$, although we could use $e \approx 2.718$ to obtain the approximate answer 1.718.

EXAMPLE 5

INTEGRATING A QUOTIENT OF TRIGONOMETRIC FUNCTIONS

Find $\displaystyle\int \frac{1 + \cos t}{t + \sin t}\, dt.$

Solution The derivative of the denominator is the numerator: $\dfrac{d}{dt}(t + \sin t) = 1 + \cos t,$ suggesting the $\displaystyle\int \frac{du}{u}$ formula with $u = t + \sin t.$

$$\int \underbrace{\frac{1 + \cos t}{t + \sin t}}_{\substack{u = t + \sin t \\ du = (1 + \cos t)\, dt}}\, dt = \underbrace{\int \frac{du}{u}}_{\text{Substituting}} = \underbrace{\ln|u| + C}_{\substack{\text{Integrating} \\ \text{by formula (C)}}} = \underbrace{\ln|t + \sin t| + C}_{\substack{\text{Substituting back} \\ \text{using } u = t + \sin t}}$$

Practice Problem 3   Evaluate $\displaystyle\int \frac{\sin t}{\cos t}\, dt.$ ➤ Solution on page 648

Many applications involve the *modified* trigonometric functions sin *at* and cos *at*, with a *scale constant a* that changes the period (see page 618). These functions are easily integrated by the substitution method, and the integrals occur so often that we list them as separate formulas.

$$\int \sin at \, dt = -\frac{1}{a} \cos at + C$$

For constant $a \neq 0$

$$\int \cos at \, dt = \frac{1}{a} \sin at + C$$

For example, Practice Problem 1 on page 642 is evaluated by the second formula in one step:

$$\int \cos \frac{2}{3} t \, dt = \frac{3}{2} \sin \frac{2}{3} t + C \qquad \text{Using the second formula above}$$

$$\underset{a}{\underbrace{\quad}} \quad \underset{\frac{1}{a}}{\underbrace{\quad}}$$

Recall that definite integration represents *continuous summation*, summing a rate to give the *total accumulation* at that rate (see page 378). For example, integrating a sales rate (such as sales per week) from *a* to *b* gives the *total sales* from week *a* to week *b*.

EXAMPLE 6

SUMMARIZING TOTAL SALES OF A SEASONAL PRODUCT

A clothing store estimates that the number of overcoats sold per week will be

$$S(t) = 30 + 30 \cos \frac{\pi t}{26} \qquad \text{Coats per week}$$

where *t* is the number of weeks into the winter. Find the total number of overcoats that will be sold during the first 8 weeks of winter.

Solution   To find the total for the first 8 weeks, we integrate the rate from 0 to 8:

$$\int_0^8 \left(30 + 30 \cos \frac{\pi}{26} t\right) dt = \left(30t + 30 \frac{26}{\pi} \sin \frac{\pi}{26} t\right)\Big|_0^8$$

Writing $\frac{\pi t}{26}$

as $\frac{\pi}{26} t$

Using

$\int \cos at\, dt = \frac{1}{a} \sin at + C$

$$= \left(30 \cdot 8 + 30 \frac{26}{\pi} \sin \frac{8\pi}{26}\right) - 0 \approx 444$$

240   ≈ 204

Evaluating   And at
at $t = 8$ $t = 0$

Approximately 444 overcoats will be sold during the first 8 weeks.

This answer is shown graphically below as the area under the sales curve from 0 to 8.

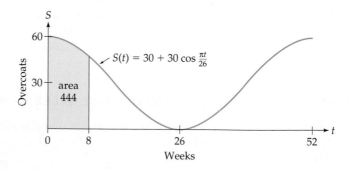

Integrals also give average values when multiplied by $\dfrac{1}{b - a}$ (see page 390).

EXAMPLE **7**   **FINDING AVERAGE TEMPERATURE**

The mean daily temperature in Fairbanks, Alaska, on day x of the year is

$$T(x) = 25 + 37 \sin \frac{2\pi(x - 101)}{365}$$ Degrees Fahrenheit

Find the average summer temperature, June 21 to September 23 (day 172 to day 266).

Solution   The average value formula (page 390) gives

$$\frac{1}{266 - 172} \int_{172}^{266} \left(25 + 37 \sin \frac{2\pi(x - 101)}{365} \right) dx$$ The average is the integral from a to b divided by (b a).

We could calculate this "by hand," but instead we use a graphing calculator, graphing the function $T(x)$ on the window $[0, 365]$ by $[-20, 70]$ and using the $\int f(x)dx$ command to integrate from 172 to 266.

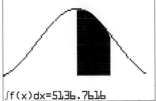

$\int f(x)dx=5136.7616$

Dividing the result by $(266 - 172)$ gives the average summer temperature in Fairbanks:

$$\frac{5137}{266 - 172} \approx 54.6$$ About 55 degrees!

8.4 **Section Summary**

The differentiation formulas for the sine and cosine functions (page 627) led immediately to the integration formulas

$$\int \sin t \, dt = -\cos t + C$$

$$\int \cos t \, dt = \sin t + C$$

These two formulas extend immediately to their "u forms" by the substitution method (see page 641) and to a "scale constant" a (see page 645). We use integration for *areas*, for *continuous summations*, for *total accumulations*, and for *average values*.

➤ **Solutions to Practice Problems**

1. $\int \underbrace{\cos\frac{2}{3}t}_{\cos u}\, dt = \frac{3}{2} \int \underbrace{\cos\frac{2}{3}t}_{\cos u}\, \underbrace{\frac{2}{3}dt}_{du}$

$u = \frac{2}{3}t$

$du = \frac{2}{3}dt$

$= \frac{3}{2}\int \cos u\, du = \frac{3}{2}\sin u + C = \frac{3}{2}\sin\frac{2t}{3} + C$

2. $\int \underbrace{\cos^2 x}_{u^2} \sin x\, dx = -\int \underbrace{\cos^2 x}_{u^2}\, \underbrace{(-\sin x)\, dx}_{du}$

$u = \cos x$

$du = -\sin x\, dx$

$= -\int u^2\, du = -\frac{1}{3}u^3 + C = -\frac{1}{3}\cos^3 x + C$

3. $\int \frac{\sin t}{\cos t}\, dt = -\int \frac{-\sin t}{\cos t}\, dt$

$u = \cos t$

$du = -\sin t\, dt$

$= -\int \frac{du}{u} = -\ln|u| + C$

$= -\ln|\cos t| + C$

8.4 Exercises

Find each integral.

1. $\int (\cos t - \sin t)\, dt$ **2.** $\int (\sin t + \cos t)\, dt$

3. $\int (1 + \sin x)\, dx$ **4.** $\int (\cos x - 1)\, dx$

5. $\int (t + 1)\cos(t^2 + 2t)\, dt$

6. $\int t^2 \sin(t^3 + 1)\, dt$

7. $\int \sin \pi t\, dt$ **8.** $\int \cos 2t\, dt$

9. $\int \cos \frac{\pi t}{2}\, dt$ **10.** $\int \sin \frac{2t}{5}\, dt$

11. $\int \sin(\pi - t)\, dt$ **12.** $\int \cos(1 - t)\, dt$

13. $\int \cos \frac{2\pi(t + 20)}{365}\, dt$ **14.** $\int \sin \frac{\pi(t + 3)}{26}\, dt$

15. $\int \sin^2 t \cos t\, dt$ **16.** $\int \cos^3 t \sin t\, dt$

17. $\int \frac{\sin t}{\cos^3 t}\, dt$ **18.** $\int \sqrt{\sin t} \cos t\, dt$

19. $\int e^{1 + \sin x} \cos x\, dx$ **20.** $\int e^{2x} \cos e^{2x}\, dx$

21. $\int \dfrac{\sin t}{\cos t}\,dt$ **22.** $\int \dfrac{\cos t}{1 + \sin t}\,dt$

23. $\int \dfrac{\sin w}{\sqrt{1 - \cos w}}\,dw$

24. $\int \sqrt[3]{1 + \sin w}\,\cos w\,dw$

25. $\int \dfrac{\sin y \cos y}{\sqrt{1 + \sin^2 y}}\,dy$ **26.** $\int \dfrac{\cos \sqrt{z}}{\sqrt{z}}\,dz$

27. $\int \left(x^2 - \dfrac{1}{x} + \cos 2x \right) dx$

28. $\int (e^{2x} + \sqrt{x} - \sin 3x)\,dx$

For each definite integral:

a. Evaluate it "by hand," leaving the answer in *exact form*.

b. Check your answer to part (a) using a graphing calculator.

29. $\displaystyle\int_0^{1/2} \sin \pi t\,dt$ **30.** $\displaystyle\int_0^{\pi/4} \cos 2t\,dt$

31. $\displaystyle\int_0^{\pi} (t + \cos t)\,dt$ **32.** $\displaystyle\int_0^{\pi} (1 - \sin t)\,dt$

33. $\displaystyle\int_0^{1} (t + \sin \pi t)\,dt$ **34.** $\displaystyle\int_0^{1} (t^2 - \cos \pi t)\,dt$

35. $\displaystyle\int_{\pi/4}^{3\pi/4} \cos\left(t + \dfrac{\pi}{4}\right) dt$ **36.** $\displaystyle\int_{\pi/2}^{\pi} e^{\sin t} \cos t\,dt$

37. Find the area under the curve $y = \cos \dfrac{x}{2}$ and above the *x*-axis from $x = 0$ to $x = \pi$.

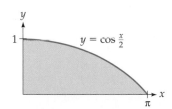

38. Find the area under the curve $y = x + \sin x$ and above the *x*-axis from $x = 0$ to $x = \pi$. Leave your answer in its *exact* form. If you are using a graphing calculator, verify your answer numerically using FnInt or $\int f(x)dx$. (See the following diagram.)

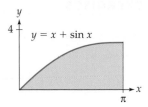

39. Find the area under $y = \sin x + \cos x$ and above the *x*-axis from $x = 0$ to $x = \pi/2$. Leave your answer in its *exact* form. If you are using a graphing calculator, verify your answer numerically using FnInt or $\int f(x)dx$.

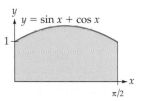

40. Find the area under the curve $y = x \sin x^2$ and above the *x*-axis from $x = 0$ to $x = \sqrt{\pi}$. Leave your answer in its *exact* form. If you are using a graphing calculator, verify your answer numerically using FnInt or $\int f(x)dx$.

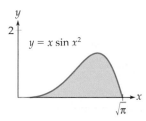

41. Prove the integration formulas in the box on page 645 by differentiating the right-hand sides to obtain the integrands.

42. Find the integral $\int \sin t \cos t\,dt$ in two ways:

  a. Using the substitution method with $u = \sin t$.

  b. Using the substitution method with $u = \cos t$.

  c. Can you reconcile the two seemingly different answers? [*Hint:* Try using a trigonometric identity.]

APPLIED EXERCISES

43–44. BUSINESS: Seasonal Sales A ski shop predicts sales of

$$S(t) = 10 + 9 \cos \frac{\pi t}{26}$$

thousand dollars during week t of the year, with $t = 0$ corresponding to the beginning of January.

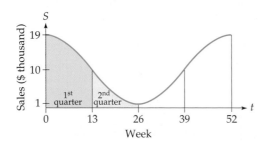

Predict the company's total sales during:

43. The first quarter of the year $(t = 0$ to $t = 13)$.

44. The second quarter of the year $(t = 13$ to $t = 26)$.

45. GENERAL: Pollution A chemical refinery discharges $P(t) = 5 + 3 \sin \frac{\pi t}{6}$ tons of pollution each month, where t is the number of months that the plant has been in operation. For the first 18 months $(t = 0$ to $t = 18)$, find:

a. The *total* amount of pollutant discharged.
b. The *average* amount of pollutant discharged.

46. BUSINESS: Summing an Income Stream An investment generates a stream of income at the rate of $S(t) = 50 + 10 \sin \frac{\pi t}{6}$ thousand dollars per month, where t is the number of months since the investment was made. For the first 6 months $(t = 0$ to $t = 6)$, find:

a. The *total* income from the income stream.
b. The *average* income from the income stream.

 47. ENVIRONMENTAL: Bird Counts The number of migratory birds flying north past a checkpoint is $300 - 300 \cos\left(\frac{\pi}{6} t\right)$ per month,

where t is the number of months since the beginning of the year. Find the total number for the year.

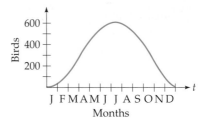

48. BIOMEDICAL: Blood Flow Blood flows through an artery at the rate of $30 - 30 \cos(2.4\pi t)$ cubic centimeters per second, where t is the number of seconds after the beginning of the cardiac cycle. Find the total amount of blood flowing through the artery in the first 5/6 second (one complete cardiac cycle).

49. GENERAL: Cost of Heating a Home Suppose that the temperature outside of a house on day t of the year is $T(t) = 60 - 30 \cos \frac{2\pi t}{365}$ degrees Fahrenheit. If it costs \$1 to raise the temperature of the house by 1° for 1 day, then setting the thermostat at 75° will result in an annual heating bill equal to the area of the shaded region shown below.

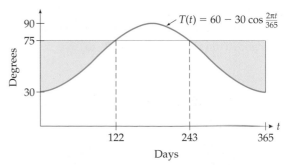

This area is given by the integral below (with the two equal areas calculated as twice the first). Use a graphing calculator to evaluate this integral to find the annual heating cost.

$$\left(\begin{array}{c}\text{Heating} \\ \text{costs at 75°}\end{array}\right) = 2 \int_0^{122} \left[75 - \left(60 - 30 \cos \frac{2\pi t}{365} \right) \right] dt$$

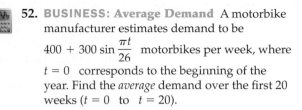

50. GENERAL: Cost of Heating a Home (*continuation*) If the thermostat is lowered to 68°, the annual heating cost is equal to the area shown below.

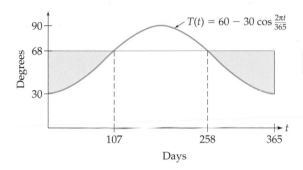

The annual heating cost is given by:

$$\left(\begin{array}{c}\text{Heating}\\\text{costs at }68°\end{array}\right) = 2\int_0^{107}\left[68 - \left(60 - 30\cos\frac{2\pi t}{365}\right)\right]dt$$

a. Use a graphing calculator to evaluate this integral to find the new heating cost.

b. Compare your answers from part (a) with your answer from Exercise 49 to find how much is saved by lowering the thermostat.

51. BUSINESS: Hotel Occupancy A Caribbean hotel estimates that it will have an occupancy of $200 + 100\cos\dfrac{\pi t}{26}$ guests during week t of the year. Find the *average* weekly occupancy during the first 10 weeks of the year ($t = 0$ to $t = 10$).

52. BUSINESS: Average Demand A motorbike manufacturer estimates demand to be $400 + 300\sin\dfrac{\pi t}{26}$ motorbikes per week, where $t = 0$ corresponds to the beginning of the year. Find the *average* demand over the first 20 weeks ($t = 0$ to $t = 20$).

53. GENERAL: Temperature in New York The mean temperature in New York City on day x of the year is

$$54 + 24\sin\frac{2\pi(x - 101)}{365}$$

degrees Fahrenheit. Use a graphing calculator to find:

a. The average temperature during January and February (days 0 to 59).

b. The average temperature during July and August (days 181 to 243).

54. GENERAL: Temperature in Minneapolis/ St. Paul The mean temperature in Minneapolis/ St. Paul on day x of the year is

$$45 + 32\sin\frac{2\pi(x - 101)}{365}$$

degrees Fahrenheit. Use a graphing calculator to find:

a. The average temperature during January and February (days 0 to 59).

b. The average temperature during July and August (days 181 to 243).

8.5 OTHER TRIGONOMETRIC FUNCTIONS

Introduction

In this section we define the tangent, cotangent, secant, and cosecant functions, differentiate and integrate them, and discuss some of their applications.

The Tangent, Cotangent, Secant, and Cosecant Functions

The functions $\tan\theta$, $\cot\theta$, $\sec\theta$, and $\csc\theta$ are defined as follows:

tan θ, cot θ, sec θ, and csc θ

Let (x, y) be any point on the terminal side of an angle of θ radians in standard position, and let r be its distance from the origin. We define:

$$\tan \theta = \frac{\sin \theta}{\cos \theta} = \frac{y}{x}$$

$$\cot \theta = \frac{1}{\tan \theta} = \frac{x}{y}$$

$$\sec \theta = \frac{1}{\cos \theta} = \frac{r}{x}$$

$$\csc \theta = \frac{1}{\sin \theta} = \frac{r}{y}$$

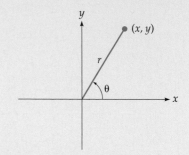

The domain of each function is the set of numbers such that the denominator is not zero. Unlike the sine and cosine functions, these four functions can take very large (positive) values and very small (negative) values.

EXAMPLE **1**

EVALUATING TRIGONOMETRIC FUNCTIONS AT A QUADRANT ANGLE

Find tan π, cot π, sec π, and csc π.

Solution

An angle of π radians is shown below, together with a point (x, y) on its terminal side.

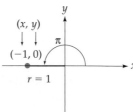

From these x, y, and r values, we calculate the values of the trigonometric functions:

$$\tan \pi = \frac{0}{-1} = 0 \qquad\qquad\qquad \tan \theta = \frac{y}{x}$$

$$\cot \pi = \frac{-1}{0} \quad \text{Undefined!} \qquad\qquad \cot \theta = \frac{x}{y}$$

$$\sec \pi = \frac{1}{-1} = -1 \qquad\qquad\quad \sec \theta = \frac{r}{x}$$

$$\csc \pi = \frac{1}{0} \quad \text{Undefined!} \qquad\qquad \csc \theta = \frac{r}{y}$$

EXAMPLE 2

EVALUATING TRIGONOMETRIC FUNCTIONS FROM A GRAPH

Find $\tan \theta$, $\cot \theta$, $\sec \theta$, and $\csc \theta$ for the angle θ shown below.

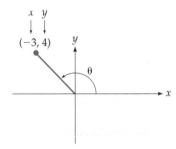

Solution

The distance from the point $(-3, 4)$ to the origin is found from the Pythagorean Theorem:

$$r = \sqrt{(-3)^2 + 4^2} = \sqrt{9 + 16} = \sqrt{25} = 5$$

From these x, y, and r values we calculate

$$\tan \theta = \frac{4}{-3} = -\frac{4}{3} \qquad\qquad \tan \theta = \frac{y}{x}$$

$$\cot \theta = \frac{-3}{4} = -\frac{3}{4} \qquad\qquad \cot \theta = \frac{x}{y}$$

$$\sec \theta = \frac{5}{-3} = -\frac{5}{3} \qquad\qquad \sec \theta = \frac{r}{x}$$

$$\csc \theta = \frac{5}{4} \qquad\qquad\qquad \csc \theta = \frac{r}{y}$$

EXAMPLE 3

EVALUATING TRIGONOMETRIC FUNCTIONS

Find $\tan\dfrac{\pi}{3}$, $\cot\dfrac{\pi}{3}$, $\sec\dfrac{\pi}{3}$, and $\csc\dfrac{\pi}{3}$.

Solution

Recall that $\sin\dfrac{\pi}{3} = \dfrac{\sqrt{3}}{2}$ and $\cos\dfrac{\pi}{3} = \dfrac{1}{2}$.   See page 613

Therefore:

$$\tan\frac{\pi}{3} = \frac{\sin\dfrac{\pi}{3}}{\cos\dfrac{\pi}{3}} = \frac{\dfrac{\sqrt{3}}{2}}{\dfrac{1}{2}} = \sqrt{3}$$

$$\cot\frac{\pi}{3} = \frac{1}{\tan\dfrac{\pi}{3}} = \frac{1}{\sqrt{3}} = \frac{\sqrt{3}}{3}$$ Multiplying numerator and denominator by $\sqrt{3}$

$$\sec\frac{\pi}{3} = \frac{1}{\cos\dfrac{\pi}{3}} = \frac{1}{\dfrac{1}{2}} = 2$$

$$\csc\frac{\pi}{3} = \frac{1}{\sin\dfrac{\pi}{3}} = \frac{1}{\dfrac{\sqrt{3}}{2}} = \frac{2}{\sqrt{3}} = \frac{2\sqrt{3}}{3}$$ Multiplying numerator and denominator by $\sqrt{3}$

Using the $\boxed{\sin}$, $\boxed{\cos}$, and $\boxed{\tan}$ keys on your calculator, together with the $\boxed{1/x}$ or $\boxed{x^{-1}}$ key for reciprocals, you can evaluate trigonometric functions at *any* values.

EXAMPLE 4

APPROXIMATING USING A CALCULATOR

Use a calculator to approximate $\cot\dfrac{\pi}{9}$.

Solution

$$\cot\frac{\pi}{9} = \frac{1}{\tan\dfrac{\pi}{9}} \approx 2.747$$ Using a calculator (set for radians)

Practice Problem **1** Use a calculator to find $\csc\dfrac{\pi}{7}$. ➤ Solution on page 663

For *acute* angles θ, these trigonometric functions can be expressed as ratios of sides of a right triangle. We simply take the earlier definitions and replace x, y, and r by adj, opp, and hyp.

Right Triangle Formulas for $\tan\theta$, $\cot\theta$, $\sec\theta$, **and** $\csc\theta$

$$\tan\theta = \frac{\text{opp}}{\text{adj}}$$

$$\cot\theta = \frac{\text{adj}}{\text{opp}}$$

$$\sec\theta = \frac{\text{hyp}}{\text{adj}}$$

$$\csc\theta = \frac{\text{hyp}}{\text{opp}}$$

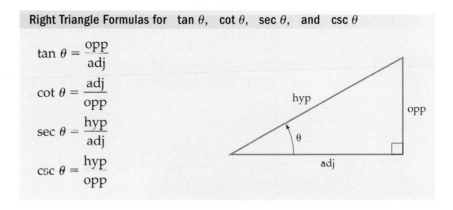

EXAMPLE **5**

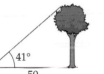

ESTIMATING HEIGHT

Estimate the height of a tree if its top makes an angle of 41° with the ground when seen from 50 feet away.

Solution

Since the height and the known distance (50 feet) are the opposite and adjacent sides of a right triangle, we use the tangent (opp/adj) of the angle. Letting h = height of tree:

$$\tan 41° = \frac{h}{50} \qquad\qquad \tan\theta = \frac{\text{opp}}{\text{adj}}$$

Solving for h:

$$h = 50\tan 41° \approx 50\cdot 0.869 \approx 43.5 \qquad \begin{array}{l}\tan 41° \approx 0.869 \text{ (from a}\\ \text{calculator set for } \textit{degrees)}\end{array}$$

The tree is about 43.5 feet high.

If an object is thrown with initial velocity v (feet per second) and with an initial angle θ to the horizontal, then its height y when it has gone a horizontal distance x from its starting point is given by the following formula:*

* Neglecting air resistance and backspin. See Philippe de Mezieres, *The Mathematics of Projectiles in Sports*, Australian Academy of Sciences, 1990.

$$y = x \tan \theta - \left(\frac{4x}{v} \sec \theta \right)^2$$

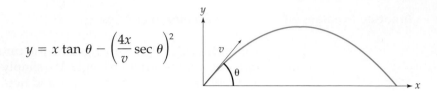

EXAMPLE 6

HOME RUNS

A baseball is hit at an angle of 35° with the horizontal with an initial velocity of 120 feet per second. Will it make a home run by clearing the outfield fence 400 feet away if the fence is 10 feet higher than the initial height of the ball?

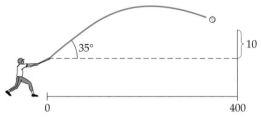

Solution

The formula with the given data becomes

$$y = 400 \tan 35° - \left(\frac{4 \cdot 400}{120} \sec 35° \right)^2 \quad \begin{array}{l} \text{Previous formula with distance} \\ x = 400, \ \text{angle} \ \ \theta = 35°, \ \text{and} \\ \text{initial velocity} \ \ v = 120 \end{array}$$

$$\underbrace{}_{0.700} \qquad \underbrace{}_{1.22} \qquad \text{Using a calculator set for } degrees$$

$$\approx 15.1$$

The ball will be 15 feet high, easily clearing the 10-foot fence and making a home run.

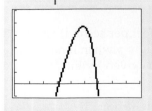

Graphing Calculator Exploration

What is the best angle to get the baseball as high as possible over the outfield fence (as before, neglecting air resistance and backspin)? Set your calculator for *degrees* and enter the expression for y from the preceding example into your graphing calculator as y_1 but with 35° replaced by x. Then graph y_1 on the window $[0, 90]$ by $[-10, 60]$ and use MAXIMUM to find the best angle x. (Afterward, reset your calculator for *radians*.)

Graphs of Tangent, Cotangent, Secant, and Cosecant Functions

The graphs of $\tan t$, $\cot t$, $\sec t$, and $\csc t$ (for t in *radians*) are as follows:

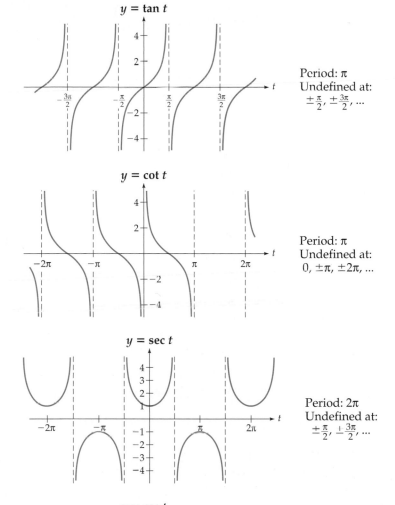

$y = \tan t$

Period: π
Undefined at:
$\pm\frac{\pi}{2}, \pm\frac{3\pi}{2}, \ldots$

$y = \cot t$

Period: π
Undefined at:
$0, \pm\pi, \pm2\pi, \ldots$

$y = \sec t$

Period: 2π
Undefined at:
$\pm\frac{\pi}{2}, \pm\frac{3\pi}{2}, \ldots$

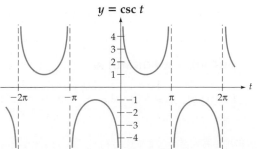

$y = \csc t$

Period: 2π
Undefined at:
$0, \pm\pi, \pm2\pi, \ldots$

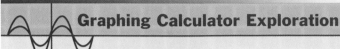

Graphing Calculator Exploration

Verify the graphs above by setting the window of your graphing calculator to $[-2\pi, 2\pi]$ by $[-4, 4]$, and graphing successively:

$$\tan x, \quad \cot x \left(\text{as } \frac{1}{\tan x}\right), \quad \sec x \left(\text{as } \frac{1}{\cos x}\right), \quad \text{and} \quad \csc x \left(\text{as } \frac{1}{\sin x}\right)$$

Derivatives of Trigonometric Functions

Derivatives of the tangent, cotangent, secant, and cosecant functions (always in *radians*) are given by the formulas in the first column in the following box. Combining these formulas with the Chain Rule gives the formulas in the second column, in which $f(t)$ stands for a differentiable function.

Derivatives of $\tan t$, $\cot t$, $\sec t$, and $\csc t$

$$\frac{d}{dt}\tan t = \sec^2 t \qquad\qquad \frac{d}{dt}\tan f(t) = \sec^2 f(t) \cdot \frac{df}{dt}$$

$$\frac{d}{dt}\cot t = -\csc^2 t \qquad\qquad \frac{d}{dt}\cot f(t) = -\csc^2 f(t) \cdot \frac{df}{dt}$$

$$\frac{d}{dt}\sec t = \sec t \cdot \tan t \qquad\qquad \frac{d}{dt}\sec f(t) = \sec f(t) \cdot \tan f(t) \cdot \frac{df}{dt}$$

$$\frac{d}{dt}\csc t = -\csc t \cdot \cot t \qquad\qquad \frac{d}{dt}\csc f(t) = -\csc f(t) \cdot \cot f(t) \cdot \frac{df}{dt}$$

Note the pattern: adding "co" to each function and attaching a negative sign gives a new formula. For example, the derivative of the tangent is the secant squared, so the derivative of the *co*tangent is the *negative* of the *co*secant squared.

EXAMPLE **7**

PROVING A FORMULA

Prove the first formula: $\dfrac{d}{dt}\tan t = \sec^2 t$.

Solution

$$\frac{d}{dt}\tan t = \frac{d}{dt}\left(\frac{\sin t}{\cos t}\right) \qquad \text{Since} \quad \tan t = \frac{\sin t}{\cos t}$$

Derivative of $\sin t$

Derivative of $\cos t$

$$= \frac{\cos t \cos t - (-\sin t)\sin t}{\cos^2 t} \qquad \text{Using the Quotient Rule}$$

$$= \frac{\cos^2 t + \sin^2 t}{\cos^2 t} = \frac{1}{\cos^2 t} \qquad \text{Simplifying and using } \sin^2 x + \cos^2 x = 1$$

$$= \sec^2 t \qquad \text{Since} \quad \frac{1}{\cos t} = \sec t$$

This proves the first formula. The other three formulas are proved similarly (see Exercises 38–40).

EXAMPLE 8

DIFFERENTIATING THE TANGENT OF A FUNCTION

Find $\dfrac{d}{dt}\tan(t^3 + 1)$

Solution

$$\frac{d}{dt}\tan(t^3 + 1) = \sec^2(t^3 + 1)\cdot(3t^2) \qquad \text{Using} \quad \frac{d}{dt}\tan f = \sec^2 f \cdot \frac{df}{dt}$$

Derivative of $t^3 + 1$

$$= 3t^2 \sec^2(t^3 + 1)$$

Practice Problem 2   Find $\dfrac{d}{dt}\sec \pi t$.   ➤ Solution on page 663

Integrals of Trigonometric Functions

The differentiation formulas in the preceding box, when read in reverse, become the following integration formulas.

Integrals of Trigonometric Functions

$$\int \sec^2 t \, dt = \tan t + C \qquad\qquad \text{Since} \quad \frac{d}{dt} \tan t = \sec^2 t$$

$$\int \csc^2 t \, dt = -\cot t + C \qquad\qquad \text{Since} \quad \frac{d}{dt} (-\cot t) = \csc^2 t$$

$$\int \sec t \tan t \, dt = \sec t + C \qquad\qquad \text{Since} \quad \frac{d}{dt} \sec t = \sec t \tan t$$

$$\int \csc t \cot t \, dt = -\csc t + C \qquad\qquad \text{Since} \quad \frac{d}{dt} (-\csc t) = \csc t \cot t$$

Replacing the variable t by u gives analogous formulas for the substitution method.

EXAMPLE 9

INTEGRATING USING A SUBSTITUTION

Find $\displaystyle\int \sec^2 5t \, dt$.

Solution We use the first formula in the box above, but in its *"u"* form.

$$\int \sec^2 5t \, dt = \frac{1}{5} \int \sec^2 5t \, 5 \, dt = \frac{1}{5} \int \sec^2 u \, du$$

$$\qquad\qquad u \qquad\qquad\qquad u \quad du \qquad \text{Substituting}$$

$$u = 5t \qquad\qquad \text{Multiplying}$$
$$du = 5 \, dt \qquad\qquad \text{and}$$
$$\text{dividing by 5}$$

$$= \quad \frac{1}{5} \tan u + C \quad = \quad \frac{1}{5} \tan 5t + C$$

$$\qquad\qquad \text{Integrating by} \qquad \text{Substituting back}$$
$$\qquad \int \sec^2 u \, du = \tan u + C \qquad \text{using} \quad u = 5t$$

EXAMPLE 10

INTEGRATING USING A SUBSTITUTION

Find $\displaystyle\int \tan^3 t \sec^2 t \, dt$.

Solution   The derivative of $\tan t$ is $\sec^2 t$, so the integral is of the form $\int u^3 \, du$.

$$\int \tan^3 t \, \sec^2 t \, dt \; = \; \int u^3 \, du \; = \; \frac{1}{4} u^4 + C \; = \; \frac{1}{4} \tan^4 t + C$$

$\underbrace{}_{\substack{u^3 \\ u \,=\, \tan t \\ du \,=\, \sec^2 t \, dt}} \underbrace{}_{du}$ $\underbrace{}_{\substack{\text{Substituting}}}$ $\underbrace{}_{\substack{\text{Integrating by}\\ \text{the Power Rule}}}$ $\underbrace{}_{\substack{\text{Substituting back}\\ \text{using } u = \tan t}}$

Exercises 59–61 show how to derive the following integration formulas.

Integrals of $\tan t$, $\cot t$, $\sec t$ **and** $\csc t$

$$\int \tan t \, dt = -\ln|\cos t| + C$$

$$\int \cot t \, dt = \ln|\sin t| + C$$

$$\int \sec t \, dt = \ln|\sec t + \tan t| + C$$

$$\int \csc t \, dt = \ln|\csc t - \cot t| + C$$

Graphing Calculator Exploration

One of the difficulties of trigonometry is that the trigonometric identities allow results to be written in many different forms. Some advanced graphing calculators can find trigonometric integrals. For example, the Texas Instruments *TI-89* graphing calculator finds the integral $\int \sec t \, dt$ as follows.

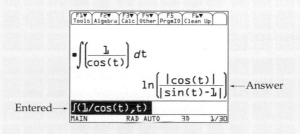

Show that this answer agrees with the third formula in the box on the previous page. [*Hint:* Begin with the answer in the screen, multiply numerator and denominator by $(\sin t + 1)$, and simplify.]

For a quick reference, all of the integration formulas used in this book are summarized on the inside back cover.

EXAMPLE 11

FINDING AREA

Find the area under $y = \tan x$ from $x = 0$ to $x = \dfrac{\pi}{4}$.

Solution Since $\tan x$ is nonnegative between $x = 0$ and $x = \dfrac{\pi}{4}$, we integrate from 0 to $\dfrac{\pi}{4}$:

$$\int_0^{\pi/4} \tan t\, dt = \underbrace{-\ln|\cos t|\,\Big|_0^{\pi/4}}_{\substack{\text{Integrating by the}\\\text{first boxed formula}}} = -\ln\underbrace{\left|\cos\frac{\pi}{4}\right|}_{\frac{\sqrt{2}}{2}} + \ln\underbrace{|\cos 0|}_{1}$$

$$= -\ln\frac{\sqrt{2}}{2} + \underbrace{\ln 1}_{0} = -\ln\frac{\sqrt{2}}{2}$$

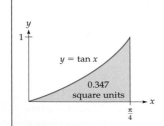

The area is $-\ln\dfrac{\sqrt{2}}{2} \approx 0.347$ square unit.

8.5 Section Summary

In this section we defined the remaining four trigonometric functions: tangent, cotangent, secant, and cosecant. If (x, y) is any point on the terminal side of an angle of θ radians in standard position, and r is the distance from the origin, then:

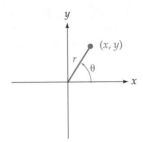

$$\tan \theta = \frac{\sin \theta}{\cos \theta} = \frac{y}{x} \qquad \sec \theta = \frac{1}{\cos \theta} = \frac{r}{x}$$

$$\cot \theta = \frac{\cos \theta}{\sin \theta} = \frac{x}{y} \qquad \csc \theta = \frac{1}{\sin \theta} = \frac{r}{y}$$

Tangent and cotangent are particularly useful in applications since they give the ratio of the sides of a right triangle with acute angle θ. The graphs of these functions are shown on page 657. We often regard these as functions of a variable t, without regard to angles. The derivative and corresponding integral formulas are given below.

$$\frac{d}{dt} \tan t = \sec^2 t \qquad \int \sec^2 t \, dt = \tan t + C$$

$$\frac{d}{dt} \cot t = -\csc^2 t \qquad \int \csc^2 t \, dt = -\cot t + C$$

$$\frac{d}{dt} \sec t = \sec t \cdot \tan t \qquad \int \sec t \cdot \tan t \, dt = \sec t + C$$

$$\frac{d}{dt} \csc t = -\csc t \cdot \cot t \qquad \int \csc t \cdot \cot t \, dt = -\csc t + C$$

Some other integral formulas are given on page 661.

> **Solutions to Practice Problems**

1. $\csc \dfrac{\pi}{7} \approx 2.305$

2. $\dfrac{d}{dt} \sec \pi t = (\sec \pi t)(\tan \pi t) \pi = \pi \sec \pi t \tan \pi t$

8.5 **Exercises**

For each angle θ, find:

a. $\tan \theta$ **b.** $\cot \theta$ **c.** $\sec \theta$ **d.** $\csc \theta$.

1.

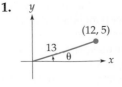

2.

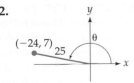

3.

4.

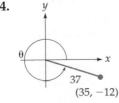

Evaluate without using a calculator.

5. a. $\tan \dfrac{\pi}{6}$ **b.** $\csc \dfrac{\pi}{6}$

6. a. $\cot \dfrac{\pi}{6}$ **b.** $\sec \dfrac{\pi}{6}$

7. a. $\tan \dfrac{3\pi}{2}$ **b.** $\csc \dfrac{3\pi}{2}$

8. a. $\cot \dfrac{3\pi}{2}$ **b.** $\sec \dfrac{3\pi}{2}$

9. a. $\tan \dfrac{\pi}{4}$ **b.** $\cot \dfrac{\pi}{4}$

10. a. $\tan \dfrac{3\pi}{4}$ **b.** $\sec \dfrac{3\pi}{4}$

11. a. $\tan \dfrac{7\pi}{6}$ **b.** $\csc \dfrac{7\pi}{6}$

12. a. $\tan \dfrac{5\pi}{3}$ **b.** $\cot \dfrac{5\pi}{3}$

Use a calculator to approximate each value.

13. $\cot \dfrac{\pi}{5}$ **14.** $\sec \dfrac{\pi}{7}$ **15.** $\sec \dfrac{5\pi}{12}$ **16.** $\csc \dfrac{9\pi}{52}$

17. GENERAL: Flagpole Height Estimate the height of a flagpole if its top makes an angle of 62° with the ground when seen from 30 feet away.

18. GENERAL: River Width Estimate the width of a river if two points on opposite shores make a 35° angle for someone standing 40 feet down the shore.

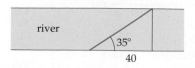

19. GENERAL: Slope and Angle of Inclination Show that the slope of a line is the *tangent* of the "angle of inclination" between the line and the x-axis.

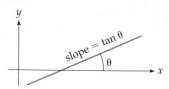

20. ATHLETICS: Baseball A baseball is hit so that it leaves the bat with an initial velocity of 115 feet per second and at an angle of 50° with the horizontal. Will it make a home run by clearing the outfield fence 400 feet away if the fence is 8 feet higher than the initial height of the ball? [*Hint:* Use the formula on page 656.]

21. ATHLETICS: Football A football is kicked from the ground with an initial velocity of 60 feet per second and at an angle of 40° with the horizontal. Will it make a field goal by clearing the 10-foot-high goal post that is 90 feet away? [*Hint:* Use the formula on page 656.]

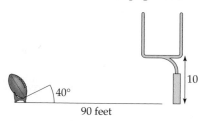

22. ATHLETICS: Baseball (*continuation of 20*) At what angle should the ball be hit (with initial velocity 115 feet per second) so as to maximize its height above the outfield fence? Set your calculator for degrees and enter the height function that you used in Exercise 20 as y_1, but with 50° replaced by x. Then graph y_1 on an appropriate window and use MAXIMUM to maximize it and find the angle. (Afterward, reset your calculator for *radians*.)

23. ATHLETICS: Football (*continuation of 21*) At what angle should the football be kicked (with initial velocity 60 feet per second) so as to maximize its height over the goal post? Set your calculator for degrees and enter the height function that you used in Exercise 21 as y_1, but with 40° replaced by x. Then graph y_1

on an appropriate window and use MAXI-MUM to maximize it and find the angle. (Afterward, reset your calculator for *radians*.)

24. Graph $y_1 = \csc^2 x$ on the window $[-2\pi, 2\pi]$ by $[-4, 4]$. What is its period? Where is it undefined? (This graph will be used in Exercise 38.)

25. Graph $y_1 = \sec x \tan x$ on the window $[-2\pi, 2\pi]$ by $[-4, 4]$. What is its period? Where is it undefined? (This graph will be used in Exercise 39.)

26. Graph $y_1 = \csc x \cot x$ on the window $[-2\pi, 2\pi]$ by $[-4, 4]$. What is its period? Where is it undefined? (This graph will be used in Exercise 40.)

27. Derive the identity $\tan^2 t + 1 = \sec^2 t$. [*Hint:* Divide the identity $\sin^2 t + \cos^2 t = 1$ by $\cos^2 t$.]

28. Derive the identity $1 + \cot^2 t = \csc^2 t$. [*Hint:* Divide the identity $\sin^2 t + \cos^2 t = 1$ by $\sin^2 t$.]

Differentiate each function.

29. $f(t) = t \cot t$

30. $f(t) = t^2 \sec t$

31. $f(t) = \tan(t^3 + 1)$

32. $f(t) = \cot(2t - 1)$

33. $f(x) = \sec(\pi x + 1)$

34. $f(x) = \tan\left(\dfrac{x}{2} - \pi\right)$

35. $f(z) = \cot^2 z^3$

36. $f(z) = \csc^4 \sqrt{z}$

37. Why do $\tan^2 t$ and $\sec^2 t$ have the same derivative?

38. a. Derive the formula $\dfrac{d}{dt}\cot t = -\csc^2 t$ by writing $\cot t = \dfrac{\cos t}{\sin t}$ and differentiating by the Quotient Rule.
  b. Verify this formula on a graphing calculator by entering $y_1 = \cot x$ [entered as $(\tan x)^{-1}$], graphing its derivative (using NDERIV), and observing that the result is the negative of the graph of $\csc^2 x$ found in Exercise 24.

39. a. Derive the formula $\dfrac{d}{dt}\sec t = \sec t \tan t$ by writing $\sec t = (\cos t)^{-1}$ and differentiating by the Generalized Power Rule.

b. Verify this formula on a graphing calculator by entering $y_1 = \sec x$ [entered as $(\cos x)^{-1}$], graphing its derivative (using NDERIV), and observing that the result is the same as the graph of $\sec x \tan x$ found in Exercise 25.

40. a. Derive the formula $\dfrac{d}{dt}\csc t = -\csc t \cot t$ by writing $\csc t = (\sin t)^{-1}$ and differentiating by the Generalized Power Rule.
  b. Verify this formula on a graphing calculator by entering $y_1 = \csc x$ [entered as $(\sin x)^{-1}$], graphing its derivative (using NDERIV), and observing that the result is the negative of the graph of $\csc x \cot x$ found in Exercise 26.

41. GENERAL: Length of Shadow When the sun is x degrees above the horizon, a 6-foot-tall person will cast a shadow of length $S(x) = 6 \cot \dfrac{\pi x}{180}$ feet. Calculate $S'(10)$ and interpret the resulting number.

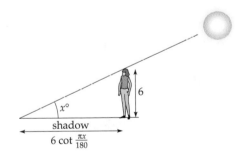

42. GENERAL: Ladder Reach A sufficiently long extension ladder whose base is 4 feet from a wall and that makes an angle of x degrees with the ground will reach to a height of $H(x) = 4 \tan \dfrac{\pi x}{180}$ feet. Find $H'(75)$ and interpret the resulting number.

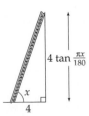

Find each integral.

43. $\int t^2 \sec^2(t^3 + 1)\, dt$ **44.** $\int \csc^2 \pi t\, dt$

45. $\int \csc \pi t \cot \pi t\, dt$

46. $\int t \sec(t^2 + 1) \tan(t^2 + 1)\, dt$

47. $\int \tan(1 - t)\, dt$ **48.** $\int \cot \dfrac{t}{2}\, dt$

49. $\int x^4 \csc x^5\, dx$ **50.** $\int x^{-1/2} \sec x^{1/2}\, dx$

51. $\int \cot^4 x \csc^2 x\, dx$ **52.** $\int \sqrt{\tan x}\, \sec^2 x\, dx$

Evaluate each definite integral.

53. $\displaystyle\int_{\pi/4}^{\pi/2} \cot t\, dt$ **54.** $\displaystyle\int_{0}^{\pi/4} \sec^2 t\, dt$

55. a. Find (without using a graphing calculator) the area under the curve $y = \sec x$ and above the x-axis from $x = 0$ to $x = \dfrac{\pi}{4}$.

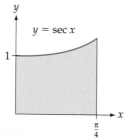

$y = \sec x$

b. Use a graphing calculator to verify your answer (using FnInt, $\int f(x)\, dx$, or a similar command).

56. a. Find (without using a graphing calculator) the area under the curve $y = \sec x \tan x$ and above the x-axis between $x = 0$ and $x = \pi/4$.

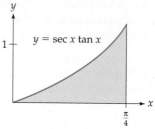

$y = \sec x \tan x$

b. Use a graphing calculator to verify your answer (using FnInt, $\int f(x)\, dx$, or a similar command).

57. Find the integral $\int \tan t \sec^2 t\, dt$ in two ways:

a. Using the substitution method with $u = \tan t$.

b. Using the substitution method with $u = \sec t$.

c. Can you reconcile the two seemingly different answers?

58. Find the integral $\int \cot t \csc^2 t\, dt$ in two ways:

a. Using the substitution method with $u = \cot t$.

b. Using the substitution method with $u = \csc t$.

c. Can you reconcile the two seemingly different answers?

59. Find $\int \tan t\, dt$ and $\int \cot t\, dt$.

[*Hint:* Write $\tan t = \dfrac{\sin t}{\cos t}$ and $\cot t = \dfrac{\cos t}{\sin t}$ and integrate by the substitution method.]

60. Find $\int \sec t\, dt$.

[*Hint:* Multiply the integrand by $\dfrac{\sec t + \tan t}{\sec t + \tan t}$ and integrate by the substitution method.]

61. Find $\int \csc t\, dt$.

[*Hint:* Multiply the integrand by $\dfrac{\csc t - \cot t}{\csc t - \cot t}$ and integrate by the substitution method.]

Chapter Summary with Hints and Suggestions

Reading the text and doing the exercises in this chapter have helped you to master the following skills, which are listed by section (in case you need to review them) and are keyed to particular Review Exercises. Answers for all Review Exercises are given at the back of the book, and full solutions can be found in the Student Solutions Manual.

8.1 Triangles, Angles, and Radian Measure

- Convert angle measures between degrees and radians. *(Review Exercises 1–2.)*

$$\frac{\text{Degrees}}{180} = \frac{\text{Radians}}{\pi}$$

- Calculate the length of a section of a circle. *(Review Exercises 3–4.)*

$$\left(\begin{matrix}\text{Arc}\\\text{length}\end{matrix}\right) = (\text{radians})(\text{radius})$$

8.2 Sine and Cosine Functions

- Find the exact value of a sine or cosine function. *(Review Exercises 5–6.)*

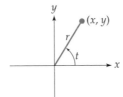

$$\sin t = \frac{y}{r}$$

$$\cos t = \frac{x}{r}$$

- Sketch the graph of a modified sine or cosine function. *(Review Exercises 7–8.)*

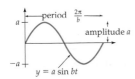
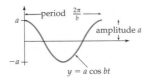

- Solve an applied problem involving a trigonometric function. *(Review Exercises 9–12.)*

8.3 Derivatives of Sine and Cosine Functions

- Find the derivative of a sine or cosine function. *(Review Exercises 13–20.)*

$$\frac{d}{dt}\sin t = \cos t \qquad \frac{d}{dt}\cos t = -\sin t$$

- Solve an applied problem by differentiating a sine or cosine function. *(Review Exercises 21–26.)*

8.4 Integrals of Sine and Cosine Functions

- Find the indefinite integral of a sine or cosine function. *(Review Exercises 27–36.)*

$$\int \sin t \, dt = -\cos t + C$$

$$\int \cos t \, dt = \sin t + C$$

- Find the definite integral of a sine or cosine function. *(Review Exercises 37–38.)*

- Find the average value of a sine or cosine function on an interval. *(Review Exercises 39–40.)*

- Find an area involving a sine or cosine function. *(Review Exercises 41–42.)*

- Solve an applied problem by integrating a sine or cosine function. *(Review Exercises 43–44.)*

8.5 Other Trigonometric Functions

- Find the exact value of one of the other trigonometric functions. *(Review Exercises 45–46.)*

$$\tan t = \frac{\sin t}{\cos t} = \frac{y}{x}$$

(continues)

$$\cot t = \frac{1}{\tan t} = \frac{x}{y}$$

$$\sec t = \frac{1}{\cos t} = \frac{r}{x}$$

$$\csc t = \frac{1}{\sin t} = \frac{r}{y}$$

■ Solve an applied problem involving one of the other trigonometric functions. *(Review Exercises 47–48.)*

■ Find the derivative of one of the other functions. *(Review Exercises 49–54.)*

$$\frac{d}{dt}\tan t = \sec^2 t$$

$$\frac{d}{dt}\cot t = -\csc^2 t$$

$$\frac{d}{dt}\sec t = \sec t \tan t$$

$$\frac{d}{dt}\csc t = -\csc t \cot t$$

■ Find the indefinite integral of one of the other trigonometric functions. *(Review Exercises 55–60.)*

■ Find the definite integral of one of the other trigonometric functions. *(Review Exercises 61–62.)*

■ Find the area under a trigonometric curve. *(Review Exercises 63–64.)*

Sections 8.3, 8.4, and 8.5

■ Solve a more involved graphing, differentiation, integration, or numerical integration problem involving a trigonometric function. *(Review Exercises 65–84.)*

Hints and Suggestions

■ We measure angles in *radians* because it simplifies the differentiation and integration formulas. All equivalent measures can be found by using the equation $180° = \pi$ radians.

■ Think of π as a natural unit for radian measure, just as it occurs naturally in the formulas for the circumference and area of a circle.

■ The "triangle" definitions such as $\sin t = \dfrac{\text{opp}}{\text{hyp}}$ hold only for *acute* angles, whereas the "coordinate" definitions such as $\sin t = \dfrac{y}{r}$ hold for *all* angles.

■ We can find $\sin t$ and $\cos t$ *exactly* for some angles, such as $\dfrac{\pi}{6}, \dfrac{\pi}{4},$ and $\dfrac{\pi}{3},$ using $30°-60°-90°$ and $45°-45°-90°$ triangles. For other angles we approximate the value using a calculator.

■ Try not to think of $\sin t$ and $\cos t$ as functions of *angles,* but as ordinary "numerical" functions, so that you can use them to model any quantity that varies periodically, such as a business cycle. Each has period $2\pi,$ but $\sin t$ begins at the origin and $\cos t$ begins one unit up.

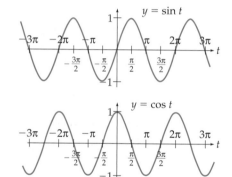

■ The uses of differentiation remain the same (slopes, rates of change, marginals, and velocities) and the uses of integration remain the same (areas, total accumulations, and average values), but we can now apply these ideas to *periodic* behavior.

■ The derivative formulas for the other trigonometric functions can be derived from those for $\sin t$ and $\cos t$ by using the Quotient Rule.

■ A graphing calculator helps by graphing trigonometric functions and finding derivatives (for optimization) and definite integrals (for areas, total accumulations, and average values).

■ **Practice for Test:** Review Exercises 1, 5, 7, 9, 15, 19, 21, 25, 27, 33, 39, 41, 43, 45, 49, 55, 63.

Review Exercises for Chapter 8

Practice test exercise numbers are in green.

8.1 Triangles, Angles, and Radian Measure

1. **a.** Convert 150° into radians.
   b. Convert $\dfrac{\pi}{9}$ radians into degrees.

2. **a.** Convert 120° into radians.
   b. Convert $\dfrac{\pi}{5}$ radians into degrees.

3–4. ARC LENGTH: Clock Hands The largest single-faced clock in the world is the Colgate clock on the Hudson River in Jersey City, New Jersey, with a minute hand that is 26 feet long. Find the arc length through which the tip of the minute hand travels between the following times. [*Hint:* Find the angle in radians. Give an approximate answer using $\pi \approx 3.14$.]

3. Between noon and 12:40 P.M.

4. Between 3 P.M. and 3:45 P.M.

8.2 Sine and Cosine Functions

Evaluate *without* using a calculator (leaving answers in *exact* form).

5. **a.** $\sin\dfrac{2\pi}{3}$ **b.** $\cos\dfrac{2\pi}{3}$

   c. $\sin\dfrac{5\pi}{4}$ **d.** $\cos\dfrac{5\pi}{4}$

6. **a.** $\sin\dfrac{7\pi}{6}$ **b.** $\cos\dfrac{7\pi}{6}$

   c. $\sin\dfrac{3\pi}{4}$ **d.** $\cos\dfrac{3\pi}{4}$

Sketch the graph of each function. (If you use a graphing calculator, make a hand-drawn sketch showing the amplitude and period.)

7. $y = 7\cos 2x$ 8. $y = 6\sin 2x$

9. **GENERAL: Building Construction** A roof is to make an angle of 30° with the horizontal, and rise 12 feet in the middle. Find the length of the rafters.

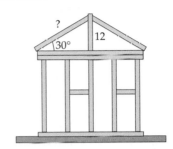

10. **GENERAL: Water Depth** A 40-foot anchor line makes an angle of 30° with the surface. How deep is the water?

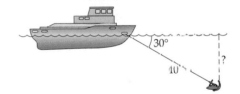

 11. **GENERAL: Daylight** The number of hours of daylight in Seattle, Washington, on day x of the year is approximated by the function

$$D(x) = 12.25 + 3.75\sin\frac{2\pi(x-81)}{365}$$

   a. Graph this function on the window [0, 365] by [0, 20]. (Be sure your calculator is set for *radians.*)
   b. Find the number of hours of daylight on May 1 (day 121).
   c. Find the number of hours of daylight on December 1 (day 335).

 12. **GENERAL: Daylight** (*continuation*) For the "daylight" function in Exercise 11, find the days with the largest and smallest amounts of daylight. [*Hint:* Use MAXIMUM and MINIMUM.]

8.3 Derivatives of Sine and Cosine Functions

Find the derivative of each function.

13. $f(t) = 2\cos(4t - 1)$ **14.** $f(t) = \sin(2 - 3t)$

15. $f(x) = x^3\sin x + 3x^2\cos x - 6x\sin x - 6\cos x$

16. $f(t) = \dfrac{1 + \sin t}{1 + \cos t}$ **17.** $f(t) = \sin^3 t^2 - \cos\dfrac{\pi}{3}$

18. $f(x) = \sin e^{x^2}$ **19.** $f(t) = \sqrt{\sin t}$

20. $f(t) = \sin\sqrt{t}$

21. BUSINESS: Seasonal Sales A manufacturer estimates that the number of overcoats sold during week t of the year is

$$S(t) = 250 + 250\cos\frac{\pi t}{26}$$

a. Graph this function on the window $[0, 52]$ by $[-100, 600]$.
b. Find the rate of change of sales at week $t = 37$ (mid September).

22. BUSINESS: Seasonal Sales (*continuation*) For the sales function in Exercise 21, find when the rate of increase in sales is the greatest. [*Hint:* Use NDERIV and MAXIMUM.)

23. GENERAL: Temperature The normal mean temperature in Eureka, California on day x of the year is approximately

$$T(x) = 55 - 24\cos\frac{2\pi(x - 20)}{365}$$

degrees Fahrenheit.

a. Graph this function on the window $[0, 365]$ by $[0, 80]$.
b. Find the rate of change of temperature on October 15 (day 288).

24. GENERAL: Temperature (*continuation*) For the temperature function in Exercise 23, find when the rate of increase is the greatest. [*Hint:* Use NDERIV and MAXIMUM.]

25. GENERAL: Maximizing Volume An enclosure made by leaning a sheet of plywood against a wall at an angle θ with the ground, will have a volume of $V(\theta) = 128\sin\theta\cos\theta$

cubic feet. Find the angle θ that maximizes the volume $V(\theta)$.

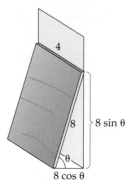

26. BIOMEDICAL: Vascular Branching A surgeon wants to attach a new artery to an existing one at the angle θ that minimizes resistance of blood flow.

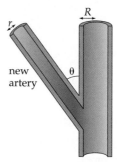

Under certain conditions, minimizing the resistance is equivalent to minimizing the function

$$R(\theta) = \frac{r^{-4} - R^{-4}\cos\theta}{\sin\theta}$$

Derive the relationship $\cos\theta = \dfrac{r^4}{R^4}$ by setting the derivative R' equal to zero and solving.

8.4 Integrals of Sine and Cosine Functions

Find each integral.

27. $\displaystyle\int\cos\frac{\pi t}{5}\,dt$ **28.** $\displaystyle\int t\sin(t^2 + 1)\,dt$

29. $\displaystyle\int \sin(\pi - t)\, dt$

30. $\displaystyle\int e^{-x} \sin e^{-x}\, dx$

31. $\displaystyle\int e^{\cos x} \sin x\, dx$

32. $\displaystyle\int \frac{\cos t}{\sqrt{\sin t}}\, dt$

33. $\displaystyle\int \frac{\cos 3t}{1 + \sin 3t}\, dt$

34. $\displaystyle\int (x^3 - e^{-x} - \cos \pi x)\, dx$

35. $\displaystyle\int \frac{\cos(\ln y)}{y}\, dy$

36. $\displaystyle\int \frac{\sin \frac{1}{z}}{z^2}\, dz$

For each definite integral:

   a. Evaluate it "by hand."

   b. Check your answer to part (a) by evaluating using a graphing calculator.

37. $\displaystyle\int_{\pi/6}^{\pi/2} \sin 2t\, dt$

38. $\displaystyle\int_{\pi/6}^{\pi} \sin^2 t \cos t\, dt$

39. Find the average value of $f(t) = \sin \pi t$ on the interval $[0, 3]$. (Give an exact and also an approximate answer.)

40. Find the average value of $f(t) = 200 + 100 \cos \pi t$ on the interval $[0, 3]$.

41. Find the area under the curve $y = \sin^2 x \cos x$ and above the x-axis from $x = 0$ to $x = \dfrac{\pi}{2}$.

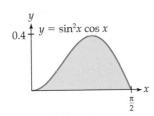

42. Find the area between the curves $y = \cos x$ and $y = \sin x$ from $x = 0$ to $x = \dfrac{\pi}{4}$, as shown in the next column. (Give the answer in exact form and also approximate it.)

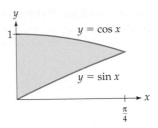

43. BUSINESS: Total Sales A boatyard estimates that sales during week x of the year will be $S(x) = 25 - 20 \cos \dfrac{\pi x}{26}$ thousand dollars, where $x = 0$ corresponds to the beginning of the year.

   a. Graph the sales function on $[0, 52]$ by $[-10, 50]$.

   b. Find the total sales during the first half of the year ($x = 0$ to $x = 26$).

44. GENERAL: Average Temperature The temperature at time x hours in a warehouse is given by the formula

$$T(x) = 60 - 10 \sin \frac{\pi x}{12}$$

degrees Fahrenheit.

   a. Graph the function on $[0, 24]$ by $[0, 70]$.

   b. Find the average temperature during the first 6 hours.

8.5   Other Trigonometric Functions

Evaluate *without* using a calculator (leaving answers in *exact* form).

45. a. $\tan \dfrac{\pi}{3}$ **b.** $\cot \dfrac{\pi}{3}$

   c. $\sec \dfrac{\pi}{3}$ **d.** $\csc \dfrac{\pi}{3}$

46. a. $\tan \dfrac{3\pi}{4}$ **b.** $\cot \dfrac{3\pi}{4}$

   c. $\sec \dfrac{3\pi}{4}$ **d.** $\csc \dfrac{3\pi}{4}$

47. GENERAL: Calculating Height Estimate the height of a monument if its top, from 600 feet away, makes an angle of 40° with the ground.

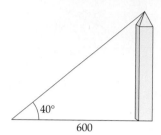

40°

600

48. GENERAL: Calculating Distance Estimate the distance from a boat to an 80-foot-tall lighthouse if the top of the lighthouse makes an angle of 20° with the water.

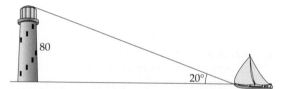

80

20°

Differentiate each function.

49. $f(x) = x \tan x^2$ **50.** $f(x) = x^2 \cot x$

51. $f(t) = \csc(\pi - t)$ **52.** $f(t) = \sec \sqrt{t}$

53. $f(t) = e^{\tan \pi t}$

54. Why do $\cot^2 t$ and $\csc^2 t$ have the same derivative?

Find each integral.

55. $\int \sec^2 \dfrac{\pi t}{2} \, dt$ **56.** $\int x \tan x^2 \, dx$

57. $\int t^2 \sec t^3 \tan t^3 \, dt$

58. $\int \csc(\pi - t) \cot(\pi - t) \, dt$

59. $\int \sqrt{\cot x} \, \csc^2 x \, dx$ **60.** $\int \dfrac{\sec^2 x}{\sqrt{\tan x}} \, dx$

For each definite integral:

a. Evaluate it "by hand."

 b. Check your answer to part (a) by evaluating using a graphing calculator.

61. $\displaystyle\int_0^{\pi/4} \sec t \, dt$ **62.** $\displaystyle\int_{\pi/4}^{\pi/2} \csc^2 t \, dt$

63. a. AREA UNDER CURVE Find "by hand" the area under the curve $y = \cot x$ and above the x-axis from $x = \dfrac{\pi}{4}$ to $x = \dfrac{\pi}{2}$.

 b. Verify your answer to part (a) using a graphing calculator.

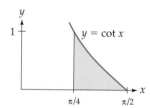

64. a. AREA UNDER CURVE Find "by hand" the area under the curve $y = \csc x$ and above the x-axis from $x = \dfrac{\pi}{4}$ to $x = \dfrac{\pi}{2}$.

 b. Verify your answer to part (a) using a graphing calculator.

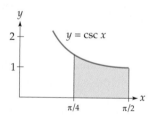

Additional Topics

65–66. GRAPHING TRIGONOMETRIC FUNC-TIONS Graph each function for $0 \le x \le 2\pi$ (using an appropriate range of y-values). Make a hand-drawn sketch, showing all relative extreme points.

65. $f(x) = \sin^2 x$ **66.** $f(x) = \sin x + \cos x$

67–68. IMPLICIT DIFFERENTIATION For each equation:

a. Find dy/dx by implicit differentiation.
b. Evaluate dy/dx at the given point.

67. $\sin 2x + \cos y = 0$

Evaluate dy/dx at $\left(\dfrac{\pi}{2}, \dfrac{\pi}{2}\right)$.

68. $\sin y - \cos 4x = 0$

Evaluate dy/dx at $\left(\dfrac{\pi}{8}, 0\right)$.

69. RELATED RATES: Radar Tracking A helicopter, rising vertically at the rate of 10 feet per second, is tracked by a radar station 200 feet away, as shown in the diagram. How fast is the angle θ of the radar beam changing at the moment when the angle is $\dfrac{\pi}{3}$?

$\left[\text{Hint: } \tan \theta = \dfrac{y}{200}.\right]$

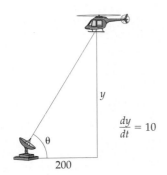

$\dfrac{dy}{dt} = 10$

y

θ

200

70. RELATED RATES: Radar Tracking A plane is flying at a constant altitude of 2 miles toward a radar station at the rate of 120 miles per hour, as shown in the diagram. How fast is the angle θ of the radar beam changing at the moment when the angle is $\dfrac{\pi}{6}$?

$\left[\text{Hint: } \cot \theta = \dfrac{x}{2}.\right]$

2

$\dfrac{dx}{dt} = -120$

θ

x

71–74. FUNCTIONS OF SEVERAL VARIABLES
For each function, find the partials: **a.** $f_x(x, y)$ and **b.** $f_y(x, y)$.

71. $f(x, y) = \sin 2x - y \cos x$

72. $f(x, y) = \ln \cos(x - y)$

73. $f(x, y) = (\sin x)e^{\cos y}$ **74.** $f(x, y) = \sin(x^3 y)$

75–80. INTEGRATION BY PARTS Evaluate each integral by integration by parts.

75. $\displaystyle\int t \sin t \, dt$ **76.** $\displaystyle\int t \cos t \, dt$

77. $\displaystyle\int x \sec^2 x \, dx$ **78.** $\displaystyle\int x \sec x \tan x \, dx$

79. $\displaystyle\int e^t \sin t \, dt$ **80.** $\displaystyle\int e^t \cos t \, dt$

[*Hint for Exercises* 79 *and* 80: Integrate by parts twice and solve the resulting equation for the original integral.]

81–82. NUMERICAL INTEGRATION Approximate each integral using trapezoidal approximation with $n = 10$, $n = 100$, and $n = 500$ trapezoids. Give the most accurate approximation that your answers will support.

81. $\displaystyle\int_0^\pi e^{\sin x} \, dx$ **82.** $\displaystyle\int_0^{\pi/2} \sqrt{\sin x} \, dx$

83–84. NUMERICAL INTEGRATION Approximate each integral using Simpson's Rule with $n = 10$, $n = 100$, and $n = 500$. Give the most accurate approximation that your answers will support.

83. $\displaystyle\int_0^\pi e^{\sin x} \, dx$ **84.** $\displaystyle\int_0^{\pi/2} \sqrt{\sin x} \, dx$

9 Differential Equations

Application Preview

Personal Wealth and Differential Equations

Like most people, you probably have (or soon will have) income that comes from two sources: salary and personal investments (such as interest on savings). From this income, you pay "necessary" expenses (housing and food), spend some of the remainder on "luxuries" (vacations and entertainment), and save the rest (increasing your investments). Given some simplifying assumptions, we can construct a mathematical model for your wealth at any time t. Suppose that after paying for necessities, you spend a fixed percentage (typically 75%) of the remainder on luxuries, and that your investments generate income at a fixed rate. Let

$$s = \text{your salary}$$
$$W(t) = \text{your wealth (savings), which is a function of time } t$$
$$r = \text{rate of interest on your wealth (savings)}$$
$$n = \text{amount spent on necessities}$$
$$p = \text{proportion of your income after necessities that you spend on luxuries}$$

A model of your wealth is then given by the following "differential equation," which states that income must equal outflow:

Income   Outflow

$$s + r \cdot W = n + p(s + r \cdot W - n) + \frac{dW}{dt}$$

Salary   Interest   Necessities   Luxuries (a fixed   Change
on wealth   proportion of income   in wealth
(savings)   after necessities)

With a little algebra, this "differential equation" can be written

$$\frac{dW}{dt} = (1 - p)(s - n + r \cdot W)$$

The *solution* to this differential equation (assuming zero initial wealth) is the "wealth" function

$$W(t) = \frac{s - n}{r}(e^{(1-p)rt} - 1)$$

**Personal Wealth and
Differential Equations**
(continued)

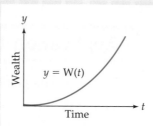

Given numbers for salary, necessities, interest rate, and proportion spent on luxuries, this solution could be used to predict when your savings will reach a particular level (for example, enough to buy a house). It could also be used to show how your wealth would change if you lowered your "luxury proportion" from, say, 75% to 70% (see Exercise 73 on page 693).

This is but one illustration of the wide variety of applications of differential equations, which are discussed in this and the following sections.

(see Exercise 73 on page 693)

9.1 SEPARATION OF VARIABLES

Introduction

A differential equation is simply an equation involving derivatives. Practically any relationship involving rates of change can be expressed using a differential equation, and in this chapter we will use differential equations to model the growth of populations, the spread of epidemics, and the reduction of air pollution, to mention only a few applications. We will solve differential equations by the techniques of separation of variables and integrating factors, and for differential equations not solvable by either approach we will use Euler's method to obtain *approximate* solutions.

We will take y to be a function of x, sometimes writing it as $y(x)$. We will write the derivative of y as either y' or $\dfrac{dy}{dx}$.

Differential Equation $y' = f(x)$

We have actually been solving differential equations since the beginning of Chapter 5. For example, the differential equation

$$y' = 2x$$

saying that the derivative of a function is $2x$, is solved simply by integrating:

$$y = \int 2x\, dx = x^2 + C \qquad\qquad y = x^2 + C \ \text{ satisfies } \ y' = 2x$$

In general, a differential equation of the form

$$y' = f(x)$$

is solved by integrating:

$$y = \int f(x)\, dx$$

Therefore, whenever we integrate a marginal cost function to find cost, or integrate a rate to find the total accumulation, we are solving a differential equation of the form $y' = f(x)$.

General and Particular Solutions

The solution of the differential equation $y' = 2x$ is $y = x^2 + C$, with an arbitrary constant C. We call $y = x^2 + C$ the *general solution* because taking all possible values of the constant C gives *all* solutions of the differential equation. If we take C to be a particular number, we get a *particular solution*. Some particular solutions of the differential equation $y' = 2x$ are

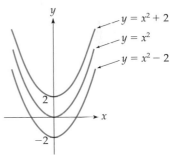

$$y = x^2 + 2 \qquad \text{(taking } C = 2\text{)}$$
$$y = x^2 \qquad \text{(taking } C = 0\text{)}$$
$$y = x^2 - 2 \qquad \text{(taking } C = -2\text{)}$$

Three particular solutions from
the general solution $y = x^2 + C$.

The different values of the arbitrary constant C give a "family" of curves, and the general solution $y = x^2 + C$ may be thought of as the entire family.

The solution of a differential equation is a *function*. The *general* solution contains an arbitrary constant, and a *particular* solution has this constant replaced by a particular number.

Verifying Solutions

Verifying that a function is a solution of a differential equation is simply a matter of calculating the necessary derivatives, substituting them into the differential equation, and checking that the two sides of the equation are equal.

EXAMPLE 1

VERIFYING A SOLUTION OF A DIFFERENTIAL EQUATION

Verify that $y = e^{2x} + e^{-x} - 1$ is a solution of the differential equation

$$y'' - y' - 2y = 2$$

Solution The differential equation involves y, y', and y'', so we calculate

$y = e^{2x} + e^{-x} - 1$	Given function
$y' = 2e^{2x} - e^{-x}$	Derivative
$y'' = 4e^{2x} + e^{-x}$	Second derivative

Then we substitute these into the differential equation.

$$y'' - y' - 2y = 2$$

$$(4e^{2x} + e^{-x}) - (2e^{2x} - e^{-x}) - 2(e^{2x} + e^{-x} - 1) \stackrel{?}{=} 2$$

$\stackrel{?}{=}$ means the equation may not be true

$$4e^{2x} + e^{-x} - 2e^{2x} + e^{-x} - 2e^{2x} - 2e^{-x} + 2 \stackrel{?}{=} 2 \qquad \text{Expanding}$$

$$4e^{2x} + e^{-x} - 2e^{2x} + e^{-x} - 2e^{2x} - 2e^{-x} + 2 \stackrel{?}{=} 2 \qquad \text{Canceling}$$

$$2 = 2 \qquad \text{It checks!}$$

Since the equation is satisfied, the given function is indeed a solution of the differential equation.

If the two sides had not turned out to be exactly the same, the given function y would *not* have been a solution of the differential equation.

Practice Problem 1

Verify that $y = e^{-x} + e^{3x}$ is a solution of the differential equation

$$y'' - 2y' - 3y = 0$$

➤ Solution on page 689

The differential equations in Example 1 and Practice Problem 1 are called *second-order* differential equations because they involve second derivatives but no higher-order derivatives. From here on we will restrict our attention to *first-order* differential equations—that is, differential equations involving only the *first* derivative. Many first-order differential equations can be solved by a method called *separation of variables*.

Separation of Variables

A differential equation is said to be "separable" if the variables can be "separated" by moving every x and dx to one side of the equation and every y and dy to the other side. We may then solve the differential equation by integrating both sides. Several examples will make the method clear.

EXAMPLE 2

FINDING A GENERAL SOLUTION

Find the general solution of the differential equation $\dfrac{dy}{dx} = 2xy^2$.

Solution

$$dy = 2xy^2\,dx$$

Multiplying both sides of the differential equation by dx

$$\frac{dy}{y^2} = 2x\,dx$$

Dividing each side by y^2 $(y \neq 0)$; the variables are now separated: every y on one side, every x on the other

$$y^{-2}\,dy = 2x\,dx$$

In power form

$$\int y^{-2}\,dy = \int 2x\,dx$$

Integrating both sides

$$-y^{-1} = x^2 + C$$

Using the Power Rule (writing one C for both integrations)

$$\frac{1}{y} = -x^2 - C$$

Writing y^{-1} as $\dfrac{1}{y}$ and multiplying by -1

$$y = \frac{1}{-x^2 - C}$$

Taking reciprocals of both sides

This is the general solution of the differential equation. The solution may be left in this form, but if we replace the arbitrary constant C by $-c$, another constant but with the opposite sign, this solution may be written

$$y = \frac{1}{-x^2 + c}$$

or, slightly shorter,

$$y = \frac{1}{c - x^2}$$

Reversing the order, giving the general solution of the differential equation

The differential equation $y' = 2xy^2$ has another solution: the function that is identically zero, $y(x) \equiv 0$. This is known as a *singular solution* and cannot be obtained from the general solution. We will not consider singular solutions further in this book except to say that $y = 0$ will generally be a solution whenever separating the variables results in dy being divided by a positive power of y, as in the previous example.

Separable Differential Equations

A first-order differential equation is *separable* if it can be written in the following form for some functions $f(x)$ and $g(y)$:

$$\frac{dy}{dx} = \frac{f(x)}{g(y)} \qquad \text{for } g(y) \neq 0$$

It is solved by separating variables and integrating:

$$\int g(y)\, dy = \int f(x)\, dx \qquad \text{Multiplying by } dx \text{ and } g(y) \text{ and integrating}$$

The preceding example asked for the *general* solution of a differential equation. Sometimes we will be given a differential equation together with some additional information that selects a *particular* solution from the general solution, information that determines the value of the arbitrary constant in the general solution. This additional information is called an *initial condition.*

EXAMPLE 3

FINDING A PARTICULAR SOLUTION

Solve the differential equation $y' = \dfrac{6x}{y^2}$ with the initial condition $y(1) = 2$.

Solution   First we find the general solution by separating the variables.

$$\frac{dy}{dx} = \frac{6x}{y^2} \qquad \text{Replacing } y' \text{ by } \frac{dy}{dx}$$

$$y^2\, dy = 6x\, dx \qquad \text{Multiplying both sides by } dx \text{ and } y^2 \text{ (the variables are separated)}$$

$$\int y^2\, dy = \int 6x\, dx \qquad \text{Integrating both sides}$$

$$\frac{1}{3}y^3 = 3x^2 + C \qquad \text{Using the Power Rule}$$

$$y^3 = 9x^2 + \underbrace{3C}_{c} \qquad \text{Multiplying by 3}$$

$$\text{(3 times a constant is just another constant)}$$

$$y^3 = 9x^2 + c \qquad \text{Replacing } 3C \text{ by } c$$

$$y = \sqrt[3]{9x^2 + c} \qquad \text{Taking cube roots}$$

This is the general solution, with arbitrary constant c. The initial condition $y(1) = 2$ says that $y = 2$ when $x = 1$. We substitute these values into the general solution $y = \sqrt[3]{9x^2 + c}$ and solve for c.

$$2 = \sqrt[3]{9 + c} \qquad\qquad y = \sqrt[3]{9x^2 + c} \text{ with } x = 1 \text{ and } y = 2$$
$$8 = 9 + c \qquad\qquad\qquad \text{Cubing each side}$$
$$-1 = c \qquad\qquad\qquad \text{Solving for } c \text{ gives } c = -1$$

Therefore, we replace c by -1 in the general solution to obtain the particular solution:

$$y = \sqrt[3]{9x^2 - 1} \qquad\qquad y = \sqrt[3]{9x^2 + c} \text{ with } c = -1$$

This solution $y = \sqrt[3]{9x^2 - 1}$ with $x = 1$ gives $y = \sqrt[3]{9 - 1} = \sqrt[3]{8} = 2$,

so the initial condition $y(1) = 2$ is indeed satisfied. We could also verify that the differential equation is satisfied by substituting this solution into it.

In general, solving a differential equation with an initial condition just means finding the general solution and then using the initial condition to evaluate the constant. Several solutions of the differential equation $y' = 6x/y^2$ are shown below, with the particular solution $y = \sqrt[3]{9x^2 - 1}$ shown in color.

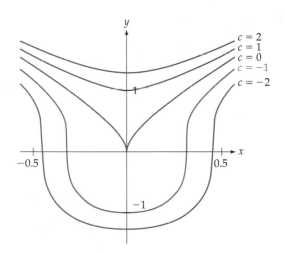

The solution $y = \sqrt[3]{9x^2 + c}$
for various values of c.

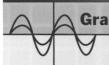

Graphing Calculator Exploration

A graphing calculator can help you to *see* what a differential equation is saying. The differential equation in the preceding example,

$$\frac{dy}{dx} = \frac{6x}{y^2} \qquad \text{Slope equals } \frac{6x}{y^2}$$

says that the slope at any point (x, y) in the plane is given by the formula $6x/y^2$. The graphing calculator program* SLOPEFLD draws a "slope field" consisting of many small line segments with the slopes specified by the differential equation. SLOPEFLD can also graph a particular solution. The slope field of the above differential equation is shown on the left below, and next to it is the same slope field with the particular solution found in the preceding example.

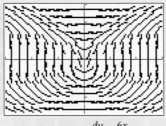

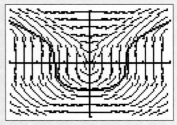

Slopefield for $\dfrac{dy}{dx} = \dfrac{6x}{y^2}$ on $[-0.6, 0.6]$ by $[-1.5, 1.5]$

The same slopefield with the solution $y = \sqrt[3]{9x^2 - 1}$

Look back at the graph of the five solutions on the previous page and see how each of them follows the prescribed slopes in the above slope field. In fact, starting at *any* point in the plane you could draw a curve following the indicated slopes, thereby geometrically constructing a *solution curve*.

*See the Preface for how to obtain this and other programs. SLOPEFLD can draw the slope field of any differential equation of the form $y' = f(x, y)$.

Practice Problem 2

Solve the differential equation and initial condition

$$\begin{cases} y' = \dfrac{6x^2}{y^4} \\ y(0) = 2 \end{cases}$$ ➤ Solution on page 689

Recall that to solve for y in the logarithmic equation

$$\ln y = f(x)$$

we exponentiate both sides and simplify

$$y = e^{f(x)}$$

Using $e^{\ln y} = y$ on the left side

This idea will be useful in the next example.

EXAMPLE 4

FINDING A PARTICULAR SOLUTION

Solve the differential equation and initial condition $\begin{cases} \dfrac{dy}{dx} = xy \\ y(0) = 2 \end{cases}$

Solution

$$dy = xy\, dx$$ Multiplying by dx

$$\frac{dy}{y} = x\, dx$$ Dividing by y ($y \neq 0$). The variables are separated

$$\int \frac{dy}{y} = \int x\, dx$$ Integrating

$$\ln y = \frac{1}{2}x^2 + C$$ Evaluating the integrals (assuming $y > 0$)

$$y = e^{\frac{1}{2}x^2 + C}$$ Solving for y by exponentiating

$$y = e^{\frac{1}{2}x^2}\underbrace{e^C}_{c}$$ e to a sum can be expressed as a product

e^C is a positive constant c

$$y = ce^{x^2/2}$$ Replacing e^C by c and $e^{\frac{1}{2}x^2}$ by $e^{x^2/2}$

This is the general solution. To satisfy the initial condition $y(0) = 2$, we substitute $y = 2$ and $x = 0$ and solve for c.

$$2 = ce^0$$ $y = ce^{x^2/2}$ with $y = 2$ and $x = 0$

$$2 = c$$ Since $e^0 = 1$

Substituting $c = 2$ into the general solution gives the particular solution

$$y = 2e^{x^2/2}$$ $y = ce^{x^2/2}$ with $c = 2$

We wrote the solution of the integral $\displaystyle\int \frac{dy}{y}$ as $\ln y$, without absolute value bars, thereby assuming that $y > 0$. *Keeping* the absolute values would have given $|y| = e^{\frac{1}{2}x^2 + C} = ce^{\frac{1}{2}x^2}$ so $y = \pm ce^{\frac{1}{2}x^2}$ where c is a positive constant, showing that the general solution is $y = ce^{x^2/2}$, just as before, but now where c is *any* (positive or negative) constant. We will make similar simplifying assumptions in similar situations.

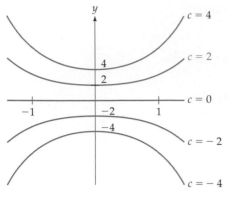

Solutions $y = ce^{x^2/2}$
for various values of c.

Note that the constant that was *added* in the integration step became a constant *multiplier* in the solution $y = ce^{x^2/2}$. In general, the constant may appear *anywhere* in the solution. The diagram on the left shows the solutions $y = ce^{x^2/2}$ for various values of c, with the particular solution $y = 2e^{x^2/2}$ shown in color.

Graphing Calculator Exploration

The slope field for the differential equation in the preceding example, found by entering $y_1 = xy$, setting the window to $[-2, 2]$ by $[-6, 6]$ and running SLOPEFLD, is shown on the left below. Notice how the slopes are determined by the function $x \cdot y$: in the first quadrant they increase when either x or y increases, and they are positive in the first and third quadrants (where x and y have the same sign) and negative in the other quadrants.

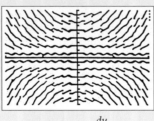

Slopefield for $\dfrac{dy}{dx} = xy$

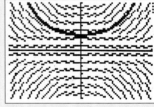

With the solution $y = 2e^{x^2/2}$

The screen on the right shows the same slope field with the particular solution found in the example. Do you see how the other solutions graphed above this Graphing Calculator Exploration follow these slopes?

Practice Problem 3

Solve the differential equation and initial condition

$$\begin{cases} \dfrac{dy}{dx} = x^2 y \\ y(0) = 5 \end{cases}$$

➤ Solution on page 689

EXAMPLE **5**

FINDING A GENERAL SOLUTION

Find the general solution of the differential equation

$$yy' - x = 0$$

Solution　We are asked for the general solution rather than a particular solution.

$$y \frac{dy}{dx} - x$$ 　　　Replacing y' by dy/dx and moving the x to the other side

$$y \, dy = x \, dx$$ 　　　Multiplying by dx

$$\int y \, dy = \int x \, dx$$ 　　　Integrating

$$\frac{1}{2}y^2 = \frac{1}{2}x^2 + C$$ 　　　Using the Power Rule

$$y^2 = x^2 + \underbrace{2C}_{c}$$ 　　　Multiplying by 2

To solve for y we take the square root of each side. However, there are *two* square roots, one positive and one negative.

$$y = \sqrt{x^2 + c} \qquad \text{and} \qquad y = -\sqrt{x^2 + c}$$

These two solutions together are the general solution:

$$y = \pm \sqrt{x^2 + c}$$

Graphing Calculator Exploration

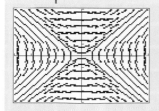

Use SLOPEFLD or a similar program to have your calculator draw the slope field for the differential equation above, $y' = x/y$, using the window $[-4, 4]$ by $[-4, 4]$. Try to draw a curve through the point $(0, 1)$ following the slopes. This would be the particular solution satisfying $y(0) = 1$.

For some differential equations, the integration step requires a substitution.

| EXAMPLE 6 | **FINDING A PARTICULAR SOLUTION USING A SUBSTITUTION** |

Solve the differential equation and initial condition $\begin{cases} y' = xy - x \\ y(0) = 4 \end{cases}$

Solution

$$\frac{dy}{dx} = xy - x \qquad\qquad\qquad \text{Replacing } y' \text{ by } \frac{dy}{dx}$$

$$\frac{dy}{dx} = x(y - 1) \qquad\qquad \text{Factoring (to separate variables)}$$

$$\frac{dy}{y-1} = x\,dx \qquad\qquad \text{Dividing by } y - 1 \text{ and multiplying by } dx$$

$$\int \frac{dy}{y - 1} = \int x\,dx \qquad\qquad \text{Integrating}$$

$$\left[\begin{array}{l} u = y - 1 \\ du = dy \end{array} \right] \qquad\qquad \text{Using a substitution}$$

$$\int \frac{du}{u} = \int x\,dx \qquad\qquad \text{Substituting}$$

$$\ln u = \frac{1}{2}x^2 + C \qquad\qquad \text{Integrating (assuming } u > 0)$$

$$\ln(y - 1) = \frac{1}{2}x^2 + C \qquad\qquad \text{Substituting back to } y \text{ using } u = y - 1$$

$$y - 1 = e^{\frac{1}{2}x^2 + C} \qquad\qquad \text{Solving for } y - 1 \text{ by exponentiating}$$

$$y - 1 = e^{\frac{1}{2}x^2} \cdot \underbrace{e^C}_{c} = c \cdot e^{x^2/2} \qquad \text{Replacing } e^C \text{ by } c \text{ and } e^{\frac{1}{2}x^2} \text{ by } e^{x^2/2}$$

$$y = c \cdot e^{x^2/2} + 1 \qquad\qquad \text{Adding 1 to each side}$$

This is the general solution. To satisfy the initial condition $y(0) = 4$ we substitute $y = 4$ and $x = 0$.

$$4 = ce^0 + 1 \qquad\qquad \begin{array}{l} y = ce^{x^2/2} + 1 \text{ with} \\ y = 4 \text{ and } x = 0 \end{array}$$

$$4 = c + 1 \qquad\qquad \text{Since } e^0 = 1$$

This gives $c = 3$. Therefore, the particular solution is

$$y = 3e^{x^2/2} + 1 \qquad\qquad \begin{array}{l} y = ce^{x^2/2} + 1 \\ \text{with } c = 3 \end{array}$$

Accumulation of Wealth

The examples so far have *given* us a differential equation to solve. In this application we will first *derive* a differential equation to represent a situation and then solve it.

EXAMPLE **7**

PREDICTING WEALTH

Suppose that you have saved $5000, and that you expect to save an additional $3000 during each year. If you deposit these savings in a bank account paying 5% interest compounded continuously, find a formula for your bank balance after t years.

Solution Let $y(t)$ stand for your bank balance (in thousands of dollars) after t years. Each year $y(t)$ grows by 3 (thousand dollars) plus 5% interest. This growth can be modeled by a differential equation:

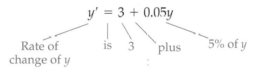

$$y' = 3 + 0.05y$$

Rate of change of y is   3   plus   5% of y

y increases by 3 plus 5% of itself

Before continuing, be sure that you understand how this differential equation models the changes due to savings and interest.

We solve it by separating variables.

$$\frac{dy}{dt} = 3 + 0.05y$$
    Replacing y' by $\frac{dy}{dt}$

$$\int \frac{dy}{3 + 0.05y} = \int dt$$
    Dividing by $3 + 0.05y$, multiplying by dt, and then integrating

$$\begin{bmatrix} u = 3 + 0.05y \\ du = 0.05dy \end{bmatrix}$$
    Using a substitution

$$20 \int \frac{du}{u} = \int dt$$
    Substituting (the 20 comes from $\frac{1}{0.05} = 20$)

$$20 \ln u = t + C$$
    Integrating (assuming $u > 0$)

$$\ln(3 + 0.05y) = 0.05t + 0.05C$$
    Substituting $u = 3 + 0.05y$ and dividing by 20

$$\underbrace{\qquad}_{c}$$

$$\ln(3 + 0.05y) = 0.05t + c$$
    Replacing $0.05C$ by c

$$3 + 0.05y = e^{0.05t+c} = e^{0.05t} \underbrace{e^c}_{k} = ke^{0.05t}$$
    Exponentiating and then simplifying constants

$$0.05y = ke^{0.05t} - 3 \qquad \text{Subtracting 3}$$

$$y = 20ke^{0.05t} - 60 \qquad \text{Dividing by 0.05}$$

$$\underbrace{\quad}_{b} \qquad \text{(Always simplify arbitrary constants)}$$

$$y = be^{0.05t} - 60 \qquad \text{With arbitrary constant } b$$

You began at time $t = 0$ with 5 thousand dollars, which gives the initial condition $y(0) = 5$. We substitute $y = 5$ and $t = 0$.

$$5 = be^0 - 60 \qquad \begin{array}{l} y = be^{0.05t} - 60 \text{ with} \\ y = 5 \text{ and } t = 0 \end{array}$$

$$5 = b - 60 \qquad \text{Since } e^0 = 1$$

Therefore, $b = 65$, which we substitute into the general solution, obtaining a formula for your accumulated wealth after t years.

$$y = 65e^{0.05t} - 60 \qquad \begin{array}{l} y = be^{0.05t} - 60 \text{ with} \\ b = 65 \end{array}$$

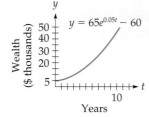

For example, to find your wealth after 10 years, we evaluate the solution of the differential equation at $t = 10$.

$$y = 65e^{0.5} - 60 \approx 47.167 \qquad y = 65e^{0.05t} - 60 \text{ with } t = 10$$

Therefore, after 10 years, you will have $47,167 in the bank.

9.1 ## Section Summary

We solved separable differential equations by separating the variables (moving every y to one side and every x to the other, with dx and dy in the numerators) and integrating both sides. The *general* solution of a differential equation includes an arbitrary constant, while a *particular* solution results from evaluating the constant, usually by applying an initial condition. The *slope field* of a differential equation allows you to construct geometrically a solution from any point by following the slopes away from the point.

We also saw how to *derive* a differential equation from information about how a quantity changes. For example, we can "read" the differential equation as follows.

$$\underset{\substack{\text{Rate of} \\ \text{change}}}{\underbrace{y'}} \quad \underset{\substack{\text{is}}}{\underbrace{=}} \quad \underset{\substack{\text{a constant times} \\ \text{the amount present}}}{\underbrace{ay}} + \underset{\substack{\text{plus a constant} \\ \text{amount added}}}{\underbrace{b}}$$

The applied exercises show how differential equations lead to important formulas in a wide variety of fields.

➤ **Solutions to Practice Problems**

1. $y = e^{-x} + e^{3x}$

    $y' = -e^{-x} + 3e^{3x}$

    $y'' = e^{-x} + 9e^{3x}$

    Substituting these expressions into the differential equation:

    $(e^{-x} + 9e^{3x}) - 2(-e^{-x} + 3e^{3x}) - 3(e^{-x} + e^{3x}) \stackrel{?}{=} 0$

    $e^{-x} + 9e^{3x} + 2e^{-x} - 6e^{3x} - 3e^{-x} - 3e^{3x} \stackrel{?}{=} 0$

    $\qquad\qquad\qquad\qquad\qquad 0 = 0 \qquad\qquad\qquad$ it checks

2. $\dfrac{dy}{dx} = \dfrac{6x^2}{y^4}$

    $y^4 \, dy = 6x^2 \, dx$

    $\displaystyle\int y^4 \, dy = \int 6x^2 \, dx$

    $\tfrac{1}{5}y^5 = 2x^3 + C$

    $y^5 = 10x^3 + 5C = 10x^3 + c$

    $y = \sqrt[5]{10x^3 + c}$

    The initial condition gives

    $2 = \sqrt[5]{0 + c}$

    $2 = \sqrt[5]{c}$

    $32 = c \qquad\qquad$ (raising each side to the fifth power)

    Solution: $y = \sqrt[5]{10x^3 + 32}$

3. $\dfrac{dy}{dx} = x^2 y$

    $\displaystyle\int \dfrac{dy}{y} = \int x^2 \, dx$

    $\ln y = \tfrac{1}{3}x^3 + C$

    $y = e^{\frac{1}{3}x^3 + C} = e^{\frac{1}{3}x^3} e^C = ce^{x^3/3}$

    The initial condition gives

    $5 = ce^0 = c$

    Solution: $y = 5e^{x^3/3}$

9.1 Exercises

Verify that the function y satisfies the given differential equation.

1. $y = e^{2x} - 3e^x + 2$
$y'' - 3y' + 2y = 4$

2. $y = e^{5x} - 4e^x + 1$
$y'' - 6y' + 5y = 5$

3. $y = ke^{ax} - \dfrac{b}{a}$ (for constants a, b, and k)
$y' = ay + b$

4. $y = ax^2 + bx$ (for constants a and b)
$y' = \dfrac{y}{x} + ax$

Find the general solution of each differential equation or state that the differential equation is not separable. If the exercise says "and check," verify that your answer is a solution.

5. $y^2 y' = 4x$ **6.** $y^4 y' = 8x$

7. $y' = x + y$ **8.** $y' = xy - 1$

9. $y' = 6x^2 y$ and check

10. $y' = 12x^3 y$ and check

11. $y' = \dfrac{y}{x}$ and check

12. $y' = \dfrac{y^2}{x^2}$ and check

13. $yy' = 4x$ **14.** $yy' = 6x^2$

15. $y' = e^{xy}$ **16.** $y' = e^x + y$

17. $y' = 9x^2$ **18.** $y' = 6e^{-2x}$

19. $y' = \dfrac{x}{x^2 + 1}$ **20.** $y' = xy^2$

21. $y' = x^2 y$ **22.** $y' = \dfrac{x}{y}$

23. $y' = x^m y^n$ (for $m > 0$, $n \neq 1$)

24. $y' = x^m y$ (for $m > 0$)

25. $y' = 2\sqrt{y}$ **26.** $y' = 5 + y$

27. $xy' = x^2 + y^2$ **28.** $y' = \sqrt{x + y}$

29. $y' = xy + x$ **30.** $y' = x - 2xy$

31. $y' = ye^x - e^x$ **32.** $y' = ye^x - y$

33. $y' = ay^2$ (for constant $a > 0$)

34. $y' = axy$ (for constant a)

35. $y' = ay + b$ (for constants a and b)

36. $y' = (ay + b)^2$ (for constants $a \neq 0$ and b)

Solve each differential equation and initial condition and verify that your answer satisfies both the differential equation and the initial condition.

37. $\begin{cases} y^2 y' = 2x \\ y(0) = 2 \end{cases}$ **38.** $\begin{cases} y^4 y' = 3x^2 \\ y(0) = 1 \end{cases}$

39. $\begin{cases} y' = xy \\ y(0) = -1 \end{cases}$ **40.** $\begin{cases} y' = y^2 \\ y(2) = -1 \end{cases}$

41. $\begin{cases} y' = 2xy^2 \\ y(0) = 1 \end{cases}$ **42.** $\begin{cases} y' = 2xy^4 \\ y(0) = 1 \end{cases}$

43. $\begin{cases} y' = \dfrac{y}{x} \\ y(1) = 3 \end{cases}$ **44.** $\begin{cases} y' = \dfrac{2y}{x} \\ y(1) = 2 \end{cases}$

45. $\begin{cases} y' = 2\sqrt{y} \\ y(1) = 4 \end{cases}$ **46.** $\begin{cases} y' = \sqrt{y}\,e^x - \sqrt{y} \\ y(0) = 1 \end{cases}$

47. $\begin{cases} y' = y^2 e^x + y^2 \\ y(0) = 1 \end{cases}$ **48.** $\begin{cases} y' = xy - 5x \\ y(0) = 4 \end{cases}$

49. $\begin{cases} y' = ax^2 y \\ y(0) = 2 \end{cases}$ (for constant $a > 0$)

50. $\begin{cases} y' = axy \\ y(0) = 4 \end{cases}$ (for constant $a > 0$)

APPLIED EXERCISES

51. BUSINESS: Elasticity For a demand function $D(p)$, the elasticity of demand (see page 326) is defined as $E = \dfrac{-pD'}{D}$. Find demand functions $D(p)$ that have constant elasticity by solving the differential equation $\dfrac{-pD'}{D} = k$, where k is a constant.

52. BIOMEDICAL: Cell Growth A cell receives nutrients through its surface, and its surface area is proportional to the two-thirds power of its weight. Therefore, if $w(t)$ is the cell's weight at time t, then $w(t)$ satisfies $w' = aw^{2/3}$, where a is a positive constant. Solve this differential equation with the initial condition $w(0) = 1$ (initial weight 1 unit).

53–54. BUSINESS: Continuous Annuities An annuity is a fund into which one makes equal payments at regular intervals. If the fund earns interest at rate r compounded continuously, and deposits are made continuously at the rate of d dollars per year (a "continuous annuity"), then the value $y(t)$ of the fund after t years satisfies the differential equation $y' = d + ry$. (Do you see why?)

53. Solve the differential equation above for the continuous annuity $y(t)$ with deposit rate $d = \$1000$ and continuous interest rate $r = 0.05$, subject to the initial condition $y(0) = 0$ (zero initial value).

54. Solve the differential equation above for the continuous annuity $y(t)$, where d and r are unknown constants, subject to the initial condition $y(0) = 0$ (zero initial value).

55. GENERAL: Crime A medical examiner called to the scene of a murder will usually take the temperature of the body. A corpse cools at a rate proportional to the difference between its temperature and the temperature of the room. If $y(t)$ is the temperature (in degrees Fahrenheit) of the body t hours after the murder, and if the room temperature is 70°, then y satisfies

$$y' = -0.32(y - 70)$$

$y(0) = 98.6$ (body temperature initially 98.6°)

a. Solve this differential equation and initial condition.

b. Use your answer to part (a) to estimate how long ago the murder took place if the temperature of the body when it was discovered was 80°. [*Hint:* Find the value of t that makes your solution equal 80°.]

56. BIOMEDICAL: Glucose Levels Hospital patients are often given glucose (blood sugar) through a tube connected to a bottle suspended over their beds. Suppose that this "drip" supplies glucose at the rate of 25 mg

per minute, and each minute 10% of the accumulated glucose is consumed by the body. Then the amount $y(t)$ of glucose (in excess of the normal level) in the body after t minutes satisfies

$$y' = 25 - 0.1y \quad \text{(Do you see why?)}$$

$y(0) = 0 \quad$ (zero excess glucose at $t = 0$)

Solve this differential equation and initial condition.

57. GENERAL: Friendships Suppose that you meet 30 new people each year, but each year you forget 20% of all of the people that you know. If $y(t)$ is the total number of people who you remember after t years, then y satisfies the differential equation $y' - 30 - 0.2y$. (Do you see why?) Solve this differential equation subject to the condition $y(0) = 0$ (you knew no one at birth).

58. ENVIRONMENTAL SCIENCE: Pollution For more than 75 years the Flexfast Rubber Company in Massachusetts discharged toxic toluene solvents into the ground at a rate of 5 tons per year. Each year approximately 10% of the accumulated pollutants evaporated into the air. If $y(t)$ is the total accumulation of pollution in the ground after t years, then y satisfies

$$y' = 5 - 0.1y \quad \text{(Do you see why?)}$$

$y(0) = 0 \quad$ (initial accumulation zero)

Solve this differential equation and initial condition to find a formula for the accumulated pollutant after t years.

59. GENERAL: Accumulation of Wealth Suppose that you now have \$6000, you expect to save an additional \$3000 during each year, and all of this is deposited in a bank paying 10% interest compounded continuously. Let $y(t)$ be your bank balance (in thousands of dollars) t years from now.

a. Write a differential equation that expresses the fact that your balance will grow by 3 (thousand dollars) and also by 10% of itself. [*Hint:* See Example 7.]

b. Write an initial condition to say that at time zero the balance is 6 (thousand dollars).

(continues)

c. Solve your differential equation and initial condition.

d. Use your solution to find your bank balance $t = 25$ years from now.

60. GENERAL: Accumulation of Wealth
Suppose that you now have $2000, you expect to save an additional $6000 during each year, and all of this is deposited in a bank paying 4% interest compounded continuously. Let $y(t)$ be your bank balance (in thousands of dollars) t years from now.

a. Write a differential equation that expresses the fact that your balance will grow by 6 (thousand dollars) and also by 4% of itself. [*Hint:* See Example 7.]

b. Write an initial condition to say that at time zero the balance is 2 (thousand dollars).

c. Solve your differential equation and initial condition.

d. Use your solution to find your bank balance $t = 20$ years from now.

61. BUSINESS: Sales Your company has developed a new product, and your marketing department has predicted how it will sell. Let $y(t)$ be the (monthly) sales of the product after t months.

a. Write a differential equation that says that the rate of growth of the sales will be four times the one-half power of the sales.

b. Write an initial condition that says that at time $t = 0$ sales were 10,000.

c. Solve this differential equation and initial value.

d. Use your solution to predict the sales at time $t = 12$ months.

62. BUSINESS: Sales Your company has developed a new product, and your marketing department has predicted how it will sell. Let $y(t)$ be the (monthly) sales of the product after t months.

a. Write a differential equation that says that the rate of growth of the sales will be six times the two-thirds power of the sales.

b. Write an initial condition that says that at time $t = 0$ sales were 1000.

c. Solve this differential equation and initial value.

d. Use your solution to predict the sales at time $t = 12$ months.

63. BIOMEDICAL: Bacterial Colony Let $y(t)$ be the size of a colony of bacteria after t hours.

a. Write a differential equation that says that the rate of growth of the colony is equal to eight times the three-fourths power of its present size.

b. Write an initial condition that says that at time zero the colony is of size 10,000.

c. Solve the differential equation and initial condition.

d. Use your solution to find the size of the colony at time $t = 6$ hours.

64. GENERAL: Value of a Building Let $y(t)$ be the value of a commercial building (in millions of dollars) after t years.

a. Write a differential equation that says that the rate of growth of the value of the building is equal to two times the one-half power of its present value.

b. Write an initial condition that says that at time zero the value of the building is 9 million dollars.

c. Solve the differential equation and initial condition.

d. Use your solution to find the value of the building at time $t = 5$ years.

65. BIOMEDICAL: Heart Function In the *reservoir model*, the heart is viewed as a balloon that swells as it fills with blood (during a period called the *systole*), and then at time t_0 it shuts a valve and contracts to force the blood out (the *diastole*). Let $p(t)$ represent the pressure in the heart at time t.

a. During the diastole, which lasts from t_0 to time T, $p(t)$ satisfies the differential equation

$$\frac{dp}{dt} = -\frac{K}{R}p$$

Find the general solution $p(t)$ of this differential equation. (K and R are positive constants determined, respectively, by the strength of the heart and the resistance of the arteries. The differential equation states that as the heart contracts, the pressure decreases (dp/dt is negative) in proportion to itself.)

b. Find the particular solution that satisfies the condition $p(t_0) = p_0$. (p_0 is a constant representing the pressure at the transition time t_0.)

c. During the systole, which lasts from time 0 to time t_0, the pressure $p(t)$ satisfies the differential equation

$$\frac{dp}{dt} = KI_0 - \frac{K}{R}p$$

Find the general solution of this differential equation. (I_0 is a positive constant representing the constant rate of blood flow into the heart while it is expanding.) [*Hint:* Use the same u-substitution technique that was used in Example 7.]

d. Find the particular solution that satisfies the condition $p(t_0) = p_0$.

e. In parts (b) and (d) you found the formulas for the pressure $p(t)$ during the diastole $(t_0 \leq t \leq T)$ and the systole $(0 \leq t \leq t_0)$. Since the heart behaves in a cyclic fashion, these functions must satisfy $p(T) = p(0)$. Equate the solutions at these times (use the correct formula for each time) to derive the important relationship

$$R = \frac{p_0}{I_0} \frac{1 - e^{-KT/R}}{1 - e^{-KI_0/R}}$$

66. **BIOMEDICAL: Fick's Law** Fick's Law governs the diffusion of a solute across a cell membrane. According to Fick's Law, the concentration $y(t)$ of the solute inside the cell at time t satisfies $\dfrac{dy}{dt} = \dfrac{kA}{V}(C_0 - y)$, where k is the diffusion constant, A is the area of the cell membrane, V is the volume of the cell, and C_0 is the concentration outside the cell.

a. Find the general solution of this differential equation. (Your solution will involve the constants k, A, V, and C_0.)

b. Find the particular solution that satisfies the initial condition $y(0) = y_0$, where y_0 is the initial concentration inside the cell.

 The following exercises require the use of a slope field program.

For each differential equation and initial condition:

a. Use SLOPEFLD or a similar program to graph the slope field for the differential equation on the window $[-5, 5]$ by $[-5, 5]$.

b. Sketch the slope field on a piece of paper and draw a solution curve that follows the slopes and that passes through the point $(0, 2)$.

c. Solve the differential equation and initial condition.

d. Use SLOPEFLD or a similar program to graph the slope field and the solution that you found in part (c). How good was the sketch that you made in part (b) compared with the solution graphed in part (d)?

67. $\begin{cases} \dfrac{dy}{dx} = \dfrac{6x^2}{y^4} \\ y(0) = 2 \end{cases}$
68. $\begin{cases} \dfrac{dy}{dx} = \dfrac{x^2}{y^2} \\ y(0) = 2 \end{cases}$
69. $\begin{cases} \dfrac{dy}{dx} = \dfrac{4x}{y^3} \\ y(0) = 2 \end{cases}$

 The following exercises require the use of a slope field program.

For each differential equation:

a. Use SLOPEFLD or a similar program to graph the slope field for the differential equation on the window $[-5, 5]$ by $[-5, 5]$.

b. Sketch the slope field on a piece of paper and draw a solution curve that follows the slopes and that passes through the given point.

70. $\dfrac{dy}{dx} = x - y^2$
  point: $(0, 1)$

71. $\dfrac{dy}{dx} = \dfrac{x}{y^2 + 1}$
  point: $(0, -1)$

72. $\dfrac{dy}{dx} = x \ln(y^2 + 1)$
  point: $(0, -2)$

73. **PERSONAL FINANCE: Wealth** Reread the Application Preview on pages 675–676.

a. Make up some reasonable numbers of s (annual salary) and n (annual cost of necessities) and substitute them together with $r = 0.06$ and $p = 0.75$ into the formula for $W(t)$. Then evaluate $W(20)$ to predict your wealth in 20 years.

b. Change p to 0.70 to see how your wealth $W(20)$ would change if you lived more frugally.

c. Give yourself a raise by increasing s by 20% to see how that affects $W(20)$.

d. Suppose that you take out a mortgage to buy a house, thereby increasing n by 15%. How does $W(20)$ change?

FURTHER APPLICATIONS OF DIFFERENTIAL EQUATIONS: THREE MODELS OF GROWTH

Introduction

This section continues our study of differential equations, but with a different approach. Instead of solving individual differential equations, we will solve three important classes of differential equations (for *unlimited*, *limited*, and *logistic* growth) and remember their solutions. This will enable us to solve many problems by identifying the appropriate differential equation and then immediately writing the solution. In this section we begin *to think in terms of differential equations*.

The circumference of a circle is proportional to its diameter.

Proportion

We say that one quantity is *proportional* to another quantity if the first quantity is a *constant multiple* of the second. That is, y is proportional to x if $y = ax$ for some "proportionality constant" a. For example, the formula $C = \pi D$ for the circumference of a circle shows that the circumference C is proportional to the diameter D.

Unlimited Growth

In many situations the growth of a quantity is proportional to its present size. For example, a population of cells will grow in proportion to its present size, and a bank account earns interest in proportion to its current value. If a quantity y grows so that its rate of growth y' is proportional to its present size y, then y satisfies the differential equation

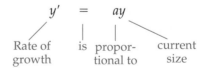

$$y' \quad = \quad ay$$

Rate of growth   is propor-   current
                   tional to   size

We solve this differential equation by separating variables.

$$\frac{dy}{dt} = ay \qquad \text{Replacing } y' \text{ by } \frac{dy}{dt}$$

$$\int \frac{dy}{y} = \int a\,dt \qquad \text{Dividing by } y \ (y \neq 0), \text{ multiplying by } dt, \text{ and integrating}$$

$$\ln y = at + C \qquad \text{Integrating } (y > 0)$$

$$y = e^{at+C} = e^{at}e^{C} = ce^{at} \qquad \text{Solving for } y \text{ by exponentiating and then replacing } e^{C} \text{ by } c$$

$$\underset{c}{\rule{0pt}{0pt}}$$

At time $t = 0$ this becomes

$$y(0) = ce^0 = c \qquad\qquad y(t) = ce^{at} \text{ with } t = 0$$

Summarizing:

Unlimited Growth

The differential equation $\qquad y' = ay$

with initial condition $\qquad\quad y(0) = c$

is solved by $\qquad\qquad\qquad y = ce^{at}$

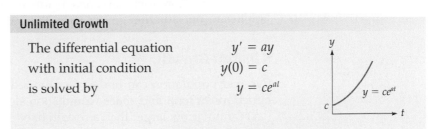

Such growth, where the rate of growth is proportional to the present size, is called *unlimited growth* because the solution y grows arbitrarily large. Given this result, whenever we encounter a differential equation of the form $y' = ay$, we can immediately write the solution $y = ce^{at}$, where c is the initial value.

EXAMPLE 1

PREDICTING ART APPRECIATION (UNLIMITED GROWTH)

An art collection, initially worth $25,000, continuously grows in value at the rate of 5% a year. Express this growth as a differential equation and find a formula for the value of the collection after t years. Then estimate the value of the art collection after 10 years.

Solution   Growing continuously at the rate of 5% means that the value $y(t)$ grows by 5% of itself

$$y' \quad = \quad 0.05y$$

Rate of   is   5% of the
growth   current value

This differential equation is of the form $y' = ay$ (unlimited growth) with $a = 0.05$ and initial value 25,000, so we may immediately write its solution.

$$y(t) = 25{,}000e^{0.05t} \qquad\quad \begin{array}{l} y = ce^{at} \text{ with } a = 0.05 \\ \text{and } c = 25{,}000 \end{array}$$

This formula gives the value of the art collection after t years. To find the value after 10 years, we evaluate at $t = 10$:

$$y(10) = 25{,}000e^{0.05 \cdot 10} \qquad\qquad y(t) = 25{,}000e^{0.05t} \text{ with } t = 10$$
$$= 25{,}000e^{0.5} \approx 41{,}218 \qquad\quad \text{Using a calculator}$$

In ten years the art collection will be worth $41,218.

The solution ce^{at} is the same as the continuous compounding formula Pe^{rn} (except for different letters). On page 282 we derived the formula Pe^{rn} rather laboriously, using the discrete interest formula $P(1 + r)^n$, replacing r by r/n, and taking the limit as $n \to \infty$. The present derivation, using differential equations, is much simpler and shows that ce^{at} applies to *any* situation governed by the differential equation $y' = ay$.

Limited Growth

No real population can undergo unlimited growth for very long. Restrictions of food and space would soon slow its growth. If a quantity $y(t)$ cannot grow larger than a certain fixed maximum size M, and if its growth rate y' is proportional to how far it is from its upper limit, then y satisfies

$$y' = a(M - y) \qquad\qquad a > 0$$

Rate of growth   is proportional to   distance below upper bound M

We solve this by separating variables.

$\dfrac{dy}{dt} = a(M - y)$	Replacing y' by $\dfrac{dy}{dt}$
$\displaystyle\int \dfrac{dy}{M - y} = \int a\,dt$	Dividing by $M - y$, multiplying by dt, and integrating
$\begin{bmatrix} u = M - y \\ du = -dy \end{bmatrix}$	Using a substitution
$-\displaystyle\int \dfrac{du}{u} = \int a\,dt$	Substituting
$-\ln u = at + C$	Integrating ($u > 0$)
$\ln(M - y) = -at - C$	Multiplying by -1 and replacing u by $(M - y)$
$M - y = e^{-at-C} = e^{-at}e^{-C} = ce^{-at}$	Solving for $M - y$ by exponentiating and simplifying
$y = M - ce^{-at}$	Subtracting M and multiplying by -1

We impose the initial condition $y(0) = 0$ (size zero at time $t = 0$).

$$0 = M - ce^0 = M - c \qquad y = M - ce^{-at} \text{ with } y = 0 \text{ and } t = 0$$

Therefore, $c = M$, which gives the solution

$$y = M - Me^{-at} = M(1 - e^{-at}) \qquad y = M - ce^{-at} \text{ with } c = M$$

Summarizing:

Limited Growth

The differential equation

$$y' = a(M - y)$$

with initial condition

$$y(0) = 0$$

is solved by

$$y = M(1 - e^{-at})$$

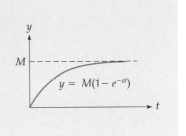

This type of growth, in which the rate of growth is proportional to the distance below an upper limit M, is called *limited growth* because the solution $y = M(1 - e^{-at})$ approaches the limit M as $t \rightarrow \infty$. Given this result, whenever we encounter a differential equation of the form

$$y' = a(M - y)$$

with initial value zero, we may immediately write the solution

$$y = M(1 - e^{-at})$$

Diffusion of Information by Mass Media

If a news bulletin is repeatedly broadcast over radio and television, the news spreads quickly at first, but later more slowly when most people have already heard it. Sociologists often assume that the rate at which news spreads is proportional to the number who have not yet heard the news. Let M be the population of a city, and let $y(t)$ be the number of people who have heard the news within t time units. Then y satisfies the differential equation

$$y' = a(M - y)$$

Rate of growth | is | propor-tional to | number who have not heard the news

We recognize this as the differential equation for limited growth, whose solution is $y = M(1 - e^{-at})$. It remains only to determine the values of the constants M and a.

EXAMPLE **2**

PREDICTING SPREAD OF INFORMATION (LIMITED GROWTH)

An important news bulletin is broadcast to a town of 50,000 people, and after 2 hours 30,000 people have heard the news. Find a formula for the number of people who have heard the bulletin within t hours. Then find how many people will have heard the news within 6 hours.

Solution If $y(t)$ is the number of people who have heard the news within t hours, then y satisfies

$$y' = a(\underbrace{50{,}000 - y})$$
$\qquad\qquad y' = a(M - y) \quad \text{with} \quad M = 50{,}000$

Number who have not heard the news

This is the differential equation for limited growth with $M = 50{,}000$, so the solution (from the preceding box) is

$$y = 50{,}000(1 - e^{-at}) \qquad y = M(1 - e^{-at}) \quad \text{with} \quad M = 50{,}000$$

To find the value of the constant a, we use the given information that 30,000 people have heard the news within 2 hours.

$30{,}000 = 50{,}000(1 - e^{-a \cdot 2})$ $\quad y = 50{,}000(1 - e^{-at}) \quad \text{with} \quad y = 30{,}000$ and $t = 2$

$0.6 = 1 - e^{-2a}$ $\qquad$ Dividing each side by 50,000

$0.4 = e^{-2a}$ $\qquad$ Subtracting 1 and then multiplying by -1

$-0.916 \approx -2a$ $\qquad$ Taking natural logs (using $\ln e^x = x$ on the right)

$a \approx 0.46$ $\qquad$ Dividing by -2

Therefore, the number of people who have heard the news within t hours is

$$y(t) = 50{,}000(1 - e^{-0.46t}) \qquad y = 50{,}000(1 - e^{-at}) \quad \text{with} \quad a = 0.46$$

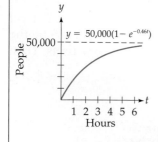

To find the number who have heard the news within 6 hours, we evaluate this solution at $t = 6$.

$y(6) = 50{,}000(1 - e^{-(0.46)(6)}) = 50{,}000(1 - e^{-2.76}) \approx 46{,}835$ $\quad$ Using a calculator

Within 6 hours about 46,800 people have heard the news.

We found the value of the constant a by substituting the given data into the solution, simplifying, and taking logs. Similar steps will be required in many other problems, and in the future we shall omit the details.

Learning Theory

Psychologists have found that there seems to be an upper limit to the number of meaningless words that a person can memorize, and that memorizing becomes increasingly difficult approaching that bound. If M is this upper limit, and if $y(t)$ is the number of words that can be memorized in t minutes, then the situation is modeled by the differential equation for limited growth.

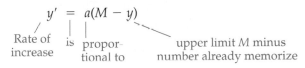

$$y' \;=\; a(M - y)$$

Rate of increase   is   proportional to   upper limit M minus number already memorize

This can be interpreted as saying that the rate at which new words can be memorized is proportional to the "unused memory capacity."

EXAMPLE 3

PREDICTING MEMORIZATION (LIMITED GROWTH)

Suppose that a person can memorize at most 100 meaningless words, and that after 15 minutes 10 words have been memorized. How long will it take to memorize 50 words?

Solution   The number $y(t)$ of words that can be memorized in t minutes satisfies

$$y' = a(100 - y)$$

     $y' = a(M - y)$
     with $M = 100$

This is the differential equation for limited growth, so the solution is

$$y = 100(1 - e^{-at})$$

     $y = M(1 - e^{-at})$
     with $M = 100$

To evaluate the constant a, we use the given information that 10 words have been memorized in 15 minutes.

$$10 = 100(1 - e^{-a \cdot 15})$$

     $y = 100(1 - e^{-at})$ with
     $y = 10$ and $t = 15$

Solving this equation for the constant a (omitting the details, which are the same as in Example 2) gives $a = 0.007$.

$$y(t) = 100(1 - e^{\,0.007t})$$

     $y = 100(1 - e^{-at})$
     with $a = 0.007$

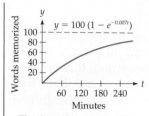

$y = 100 (1 - e^{-0.007t})$

The slope (the rate at which additional words can be memorized) decreases near the upper limit.

This solution gives the number of words that can be memorized in t minutes. To find how long it takes to memorize 50 words, we set this solution equal to 50 and solve for t.

$$50 = 100(1 - e^{-0.007t})$$

$y = 100(1 - e^{-0.007t})$
with $y = 50$

Solving for t (again the details are similar to those in Example 2) gives $t = 99$ minutes. Therefore, 50 words can be memorized in about 1 hour and 39 minutes.

Logistic Growth

Some quantities grow in proportion to both their present size *and* their distance from an upper limit M.

$$y' = ay(M - y)$$

Rate of   is   propor-   present   upper limit M
growth   tional to   size   minus present size

This differential equation can be solved by separation of variables (see Exercise 55) to give the solution

$$y = \frac{M}{1 + ce^{-aMt}}$$

where c and a are positive constants. This function is called the *logistic function*, governing *logistic growth*.

Logistic Growth

The differential equation

$$y' = ay(M - y)$$

with initial condition

$$y(0) = \frac{M}{1 + c}$$

is solved by

$$y = \frac{M}{1 + ce^{-aMt}}$$

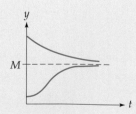

(The upper and lower curves in the box represent solutions whose initial values are, respectively, greater than or less than M.) As before, this result enables us to solve a differential equation "at sight," leaving only the evaluation of constants.

The lower curve in the box that rises to M is called a *sigmoidal* or *S-shaped curve*, and is used to model growth that begins slowly, then becomes more rapid, and finally slows again near the upper limit. Many different quantities grow according to sigmoidal curves.

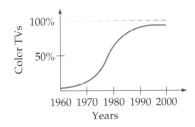

Percent of households owning a color television. (*Source*: Census Bureau)

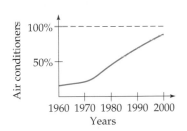

Percent of households with air conditioning. (*Source*: WorldWatch)

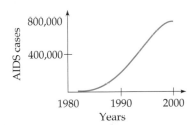

Total U.S. AIDS cases by year of diagnosis (*Source*: Centers for Disease Control)

Environmental Science

For an animal environment (such as a lake or a forest), the population that it can support will have an upper limit, called the *carrying capacity of the environment*. Ecologists often assume that an animal population grows in proportion to both its present size and its distance below the carrying capacity of the environment.

$$y' = ay(M - y)$$

Rate of growth   is   propor- tional to   current size   carrying capacity minus present size

Since this is the logistic differential equation, we know that the solution is the logistic function

$$y = \frac{M}{1 + ce^{-aMt}}$$

EXAMPLE **4**

PREDICTING AN ANIMAL POPULATION (LOGISTIC GROWTH)

Ecologists estimate that an artificial lake can support a maximum of 2500 fish. The lake is initially stocked with 500 fish, and after 6 months the fish population is estimated to be 1500. Find a formula for the number of fish in the lake after t months, and estimate the fish population at the end of the first year.

Solution   Letting $y(t)$ stand for the number of fish in the lake after t months, the situation is modeled by the logistic differential equation.

$$y' = ay(2500 - y)$$

$\qquad y' = ay(M - y)$
$\qquad$ with $M = 2500$

Rate of is propor- current carrying capacity
growth tional to size minus present size

The solution is the logistic function with $M = 2500$.

$$y = \frac{2500}{1 + ce^{-a2500t}} \qquad\qquad y = \frac{M}{1 + ce^{-aMt}} \text{ with } M = 2500$$

$$= \frac{2500}{1 + ce^{-bt}} \qquad\qquad \text{Replacing } a \cdot 2500 \text{ by another constant } b$$

To evaluate the constants c and b, we use the fact that the lake was originally stocked with 500 fish.

$$500 = \frac{2500}{1 + ce^0} \qquad\qquad y = \frac{2500}{1 + ce^{-bt}} \text{ with } y = 500, \ t = 0$$

$$500 = \frac{2500}{1 + c} \qquad\qquad \text{Simplifying}$$

Solving this for c (omitting the details—the first step is to multiply both sides by $1 + c$) gives $c = 4$, so the logistic function becomes

$$y = \frac{2500}{1 + 4e^{-bt}} \qquad\qquad y = \frac{2500}{1 + ce^{-bt}} \text{ with } c = 4$$

To evaluate b, we substitute the information that the population is $y = 1500$ at $t = 6$.

$$1500 = \frac{2500}{1 + 4e^{-b \cdot 6}} \qquad\qquad y = \frac{2500}{1 + 4e^{-bt}} \text{ with } \begin{array}{l} y = 1500, \\ t = 6 \end{array}$$

Solving for b (again omitting the details—the first step is to multiply both sides by $1 + 4e^{-b \cdot 6}$) gives $b = 0.30$, so the logistic function becomes

$$y(t) = \frac{2500}{1 + 4e^{-0.3t}} \qquad\qquad y = \frac{2500}{1 + 4e^{-bt}} \text{ with } b = 0.3$$

This is the formula for the population after t months. To find the population after a year, we evaluate at $t = 12$.

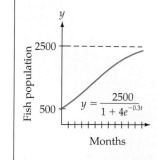

$$y(12) = \frac{2500}{1 + 4e^{-0.3(12)}} = \frac{2500}{1 + 4e^{-3.6}} \approx 2254 \qquad \text{Using a calculator}$$

Therefore, the population at the end of the first year is 2254, which is about 90% of the carrying capacity of the lake.

Epidemics

Many epidemics spread at a rate proportional to both the number of people already infected (the "carriers") and also the number who have yet to catch the disease (the "susceptibles"). If $y(t)$ is the number of infected people at time t from a population of size M, then $y(t)$ satisfies the logistic differential equation

$$y' = ay(M - y)$$

Rate of is propor- number number
growth tional to infected susceptible

Therefore, the size of the infected population is given by the logistic function

$$y = \frac{M}{1 + ce^{-aMt}}$$

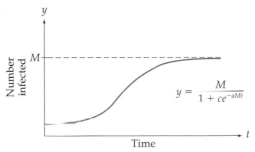

The constants are evaluated just as in Example 4, using the (initial) number of cases reported at time $t = 0$, and also the number of cases at some later time.

Spread of Rumors

Sociologists have found that rumors spread at a rate proportional to the number who have heard the rumor (the "informed") and the number who have not heard the rumor (the "uninformed"). Therefore, in a population of size M, the number $y(t)$ who have heard the rumor within t time units satisfies the logistic differential equation

$$y' = ay(M - y)$$

Rate of is propor- number number
growth tional to informed uninformed

The solution $y(t)$ is then the logistic function

$$y = \frac{M}{1 + ce^{-aMt}}$$

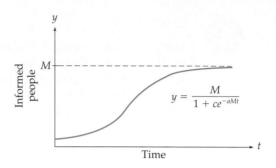

It remains only to evaluate the constants. The spread of a rumor is analogous to the spread of a disease, with an "informed" person being one who has been "infected."

Limited and Logistic Growth of Sales

The sales of a product whose total sales will approach an upper limit (market saturation) can be modeled by either the limited or the logistic equation. Which do you use when? For a product advertised over mass media, sales will at first grow rapidly, indicating a *limited* model. For a product becoming known only through "word of mouth," sales will at first grow slowly, indicating a *logistic* model.

Practice Problem

The graphs below show the total sales through day t for two different products, A and B. Which of these products was advertised and which became known only by "word of mouth"? State an appropriate differential equation for each curve.

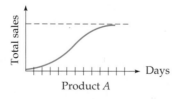

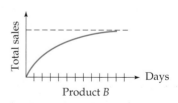

➤ Solution on page 706

9.2 Section Summary

The unlimited, limited, and logistic growth models are summarized in the following table. If one of these differential equations governs a particular situation, we can write the solution immediately, evaluating the constants from the given data.

Three Models of Growth

Type	Differential Equation	Solution	Graph	Examples
Unlimited Growth is proportional to present size	$y' = ay$	$y = ce^{at}$		Investments Bank accounts Unlimited populations
Limited Growth (starting at 0) is proportional to maximum size M minus present size.	$y' = a(M - y)$	$y = M(1 - e^{-at})$		Information spread by mass media Memorizing random information Total sales (advertised)
Logistic Growth is proportional to present size and to maximum size M minus present size.	$y' = ay(M - y)$	$y = \dfrac{M}{1 + ce^{-aMt}}$		Confined populations Epidemics Rumors Total sales (unadvertised)

To decide which (if any) of the three models applies in a given situation, think of whether the growth is proportional to *size*, to *unused capacity*, or to *both* (as shown in the chart). Notice that the differential equation gives much more insight into how the growth occurs than does the solution. This is what we meant at the beginning of the section by "thinking in terms of differential equations."

Graphing Calculator Exploration

Use the program* SLOPEFLD or a similar program to graph the slope fields for the following differential equations (one at a time) on the window [0, 5] by [0, 5]:

* See the Preface for how to obtain this and other programs.

a. $y' = y$
(unlimited)

b. $y' = 3 - y$
(limited)

c. $y' = y(3 - y)$
(logistic)

Do you see how the slope field gives a "picture" of the differential equation? Each graph is on the window [0, 5] by [0, 5].

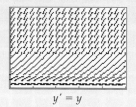

$y' = y$

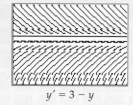

$y' = 3 - y$

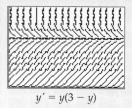

$y' = y(3 - y)$

▶ **Solution to Practice Problem**

Product *A*, whose growth begins slowly, was not advertised, and the differential equation is logistic: $y' = ay(M - y)$. Product *B*, whose growth begins rapidly, was advertised, and the differential equation is limited: $y' = a(M - y)$.

9.2 Exercises

1. Verify that $y(t) = ce^{at}$ solves the differential equation for unlimited growth, $y' = ay$, with initial condition $y(0) = c$.

2. Verify that $y(t) = M(1 - e^{-at})$ solves the differential equation for limited growth, $y' = a(M - y)$, with initial condition $y(0) = 0$.

Determine the type of each differential equation: *unlimited* growth, *limited* growth, *logistic* growth, or *none* of these. (Do not solve, just identify the type.)

3. $y' = 0.02y$

4. $y' = 5(100 - y)$

5. $y' = 30(0.5 - y)$

6. $y' = 0.4y(0.01 - y)$

7. $y' = 2y^2(0.5 - y)$

8. $y' = 6y$

9. $y' = y(6 - y)$

10. $y' = 0.01(100 - y^2)$

11. $y' = 4y(0.04 - y)$

12. $y' = 4500(1 - y)$

Find the solution $y(t)$ by recognizing each differential equation as determining unlimited, limited, or logistic growth, and then finding the constants.

13. $y' = 6y$
    $y(0) = 1.5$

14. $y' = 0.25y$
    $y(0) = 4$

15. $y' = -y$
    $y(0) = 100$

16. $y' = \frac{y}{2}$
    $y(0) = 8$

17. $y' = -0.45y$
    $y(0) = -1$

18. $y' = 0$
    $y(0) = 5$

19. $y' = 2(100 - y)$
    $y(0) = 0$

20. $y' = 48(2 - y)$
    $y(0) = 0$

21. $y' = 0.05(0.25 - y)$
    $y(0) = 0$

22. $y' = \frac{2}{3}(1 - y)$
    $y(0) = 0$

23. $y' = 80 - 2y$
    $y(0) = 0$

24. $y' = 27 - 3y$
    $y(0) = 0$

[*Hint for Exercises 23–26:* Use factoring to write the differential equation in the form $y' = a(M - y)$.]

25. $y' = 2 - 0.01y$
    $y(0) = 0$

26. $y' = 6 - 8y$
    $y(0) = 0$

27. $y' = 5y(100 - y)$
    $y(0) = 10$

28. $y' = y(1 - y)$
    $y(0) = \frac{1}{2}$

29. $y' = 0.25y(0.5 - y)$
    $y(0) = 0.1$

30. $y' = \frac{1}{3}y(\frac{1}{2} - y)$
    $y(0) = \frac{1}{6}$

31. $y' = 3y(10 - y)$
    $y(0) = 20$

32. $y' = y(2 - y)$
    $y(0) = 4$

33. $y' = 6y - 2y^2$
    $y(0) = 1$

34. $y' = 3y - 6y^2$
    $y(0) = \frac{1}{6}$

[*Hint for Exercises 33–34:* Use factoring to write the differential equation in the form $y' = ay(M - y)$.]

APPLIED EXERCISES

Write the differential equation (unlimited, limited, or logistic) that applies to the situation described. Then use its solution to solve the problem.

35. GENERAL: Appreciation The value of a stamp collection, initially worth $1500, grows continuously at the rate of 8% per year. Find a formula for its value after t years.

36. GENERAL: Appreciation The value of a home, originally worth $25,000, grows continuously at the rate of 6% per year. Find a formula for its value after t years.

37. BUSINESS: Total Sales A manufacturer estimates that he can sell a maximum of 100,000 digital tape recorders in a city. His total sales grow at a rate proportional to the distance below this upper limit. If after 5 months total sales are 10,000, find a formula for the total sales after t months. Then use your answer to estimate the total sales at the end of the first year.

38. BUSINESS: Product Recognition Let $p(t)$ be the number of people in a city who have heard of a new product after t weeks of advertising. The city is of size 1,000,000, and $p(t)$ grows at a rate proportional to the number of people in the city who have *not* heard of the product. If after 8 weeks 250,000 people have heard of the product, find a formula for $p(t)$. Use your formula to estimate the number of people who will have heard of the product after 20 weeks of advertising.

39. GENERAL: Fund-Raising In a drive to raise $5000, fund-raisers estimate that the rate of contributions is proportional to the distance from the goal. If $1000 was raised in 1 week, find a formula for the amount raised in t weeks. How many weeks will it take to raise $4000?

40. GENERAL: Learning A person can memorize at most 40 two-digit numbers. If that person can memorize 15 numbers in the first 20 minutes, find a formula for the number that can be memorized in t minutes. Use your answer to estimate how long the person will take to memorize 30 numbers.

41. BUSINESS: Sales A telephone company estimates the maximum market for car phones in a city to be 10,000. Total sales are proportional to both the number already sold and the size of the remaining market. If 100 phones have been sold at time $t = 0$ and after 6 months 2000 have been sold, find a formula for the total sales after t months. Use your answer to estimate the total sales at the end of the first year.

42. BIOMEDICAL: Epidemics During a flu epidemic in a city of 1,000,000, a flu vaccine sells in proportion to both the number of people already inoculated and the number not yet inoculated. If 100 doses have been sold at time $t = 0$ and after 4 weeks 2000 doses have been sold, find a formula for the total number of doses sold within t weeks. Use your formula to predict the sales after 10 weeks.

43. SOCIOLOGY: Rumors One person at an airport starts a rumor that a plane has been hijacked, and within 10 minutes 200 people have heard the rumor. If there are 800 people in the airport, find a formula for the number who have heard the rumor within t minutes. Use your answer to estimate how many will have heard the rumor within 15 minutes.

44. GENERAL: Epidemics A flu epidemic on a college campus of 4000 students begins with 12 cases, and after 1 week has grown to 100 cases. Find a formula for the size of the epidemic after t weeks. Use your answer to estimate the size of the epidemic after 2 weeks.

45. ENVIRONMENTAL SCIENCE: Deer Population A wildlife refuge is initially stocked with 100 deer, and can hold at most 800 deer. If 2 years later the deer population is 160, find a formula for the deer population after t years. Use your answer to estimate when the deer population will reach 400.

46. POLITICAL SCIENCE: Voting Suppose that a bill in the U.S. Senate gains votes in proportion to the number of votes that it already has and to the number of votes that it does not have. If it begins with one vote (from its sponsor) and after 3 days it has 30 votes, find a formula for the number of votes that it will have after t days. (*Note:* The number of votes in the Senate is 100.) When will the bill have "majority support" of 51 votes?

Solve Exercises 47–50 by recognizing that the differential equation is one of the three types whose solutions we know.

47. BIOMEDICAL: Drug Absorption A drug injected into a vein is absorbed by the body at a rate proportional to the amount remaining in the blood. For a certain drug, the amount $y(t)$ remaining in the blood after t hours satisfies $y' = -0.15y$ with $y(0) = 5$. Find $y(t)$ and use your answer to estimate the amount present after 2 hours.

48. BUSINESS: Stock Value One model for the growth of the value of stock in a corporation assumes that the stock has a limiting "market value" L, and that the value $v(t)$ of the stock

on day t satisfies the differential equation $v' = a(L - v)$ for some constant a. Find a formula for the value $v(t)$ of a stock whose market value is $L = 40$ if on day $t = 10$ it was selling for $v = 30$.

49. GENERAL: Dam Sediment A hydroelectric dam generates electricity by forcing water through turbines. Sediment accumulating behind the dam, however, will reduce the flow and eventually require dredging. Let $y(t)$ be the amount of sediment (in thousands of tons) accumulated in t years. If sediment flows in from the river at the constant rate of 20 thousand tons annually, but each year 10% of the accumulated sediment passes through the turbines, then the amount of sediment remaining satisfies the differential equation $y' = 20 - 0.1y$.

   a. By factoring the right-hand side, write this differential equation in the form $y' = a(M - y)$. Note the value of M, the maximum amount of sediment that will accumulate.

   b. Solve this (factored) differential equation together with the initial condition $y(0) = 0$ (no sediment until the dam was built).

   c. Use your solution to find when the accumulated sediment will reach 95% of the value of M found in step (a). This is when dredging is required.

50. BIOMEDICAL: Glucose Levels Solve Exercise 56 on page 691 by factoring the right-hand side of the differential equation to write it in the form $y' = a(M - y)$.

OTHER GROWTH MODELS

Solve each differential equation by separation of variables.

51. GENERAL: Population If $y(t)$ is the size of a population at time t, then $\dfrac{y'}{y}$ is the population growth rate divided by the size of the population, and is called the *individual birthrate*. Suppose that the individual birthrate is proportional to the size of the population,

$\dfrac{y'}{y} = ay$ for some constant a. Find a formula for the size of the population after t years.

52. GENERAL: Gompertz Curve Another differential equation that is used to model the growth of a population $y(t)$ is $y' = bye^{-at}$, where a and b are constants. Solve this differential equation.

53. GENERAL: Allometry Solve the differential equation of allometric growth: $y' = \dfrac{ay}{x}$ (where a is a constant). This differential equation governs the relative growth rates of different parts of the same animal.

54. GENERAL: Population Suppose that a population $y(t)$ in a certain environment grows in proportion to the square of the difference between the carrying capacity M and the present population, that is, $y' = a(M - y)^2$, where a is a constant. Solve this differential equation.

LOGISTIC GROWTH FUNCTION

55. Solve the logistic differential equation $y' = ay(M - y)$ as follows:

a. Separate variables to obtain

$$\frac{dy}{y(M - y)} = a\,dt$$

b. Integrate, using on the left-hand side the integration formula

$$\int \frac{dy}{y(M - y)} = \frac{1}{M}\ln\left(\frac{y}{M - y}\right)$$

(which may be checked by differentiation).

c. Exponentiate to solve for $\dfrac{y}{(M - y)}$ and then solve for y.

d. Show that the solution can be expressed as

$$y = \frac{M}{1 + ce^{-aMt}}$$

56. Find the inflection point of the logistic curve

$$f(x) = \frac{M}{1 + ce^{-aMx}}$$

and show that it occurs at midheight between $y = 0$ and the upper limit $y = M$. [*Hint: Do you already know $f'(x)$?*]

57. GENERAL: Raindrops (*Requires Slope Field Program*) Why do larger-sized raindrops fall faster than smaller ones? It depends on the resistance they encounter as they fall through the air. For large raindrops, the resistance to gravity's acceleration is proportional to the *square* of the velocity, whereas for small droplets, the resistance is proportional to the *first power* of the velocity. More precisely, their velocities obey the following differential equa-

tions, with each differential equation leading to a different *terminal velocity* for the raindrop:

i. $\dfrac{dv}{dt} = 32.2 - 0.1115v^2$ Downpour droplets, about 0.05 inch in diameter

ii. $\dfrac{dv}{dt} = 32.2 - 52.6v$ Drizzle droplets, about 0.003 inch in diameter

iii. $\dfrac{dv}{dt} = 32.2 - 5260v$ Fog droplets, about 0.0003 inch in diameter

(The 32.2 represents the force of gravity, and the other constant is determined experimentally.*)

a. Use a slope field program to graph the slope field of differential equation (i) on the window [0, 3] by [0, 20] (using x and y instead of t and v). From the slope field, must the solution curves rising from the bottom level off at a particular y-value? Estimate the value. This number is the terminal velocity (in feet per second) for a downpour droplet.

b. Do the same for differential equation (ii), but on the window [0, 0.1] by [0, 1]. What is the terminal velocity for a drizzle droplet?

c. Do the same for differential equation (iii), but on the window [0, 0.001] by [0, 0.01]. What is the terminal velocity for a fog droplet?

d. At this speed [from part (c)], how long would it take a fog droplet to fall 1 foot? This shows why fog clears so slowly.

---

* R. Gunn and Gilbert D. Kinzer, "The Terminal Velocity of Fall for Water Droplets in Stagnant Air," *Journal of Meteorology* 6:243, 1949.

9.3 FIRST-ORDER LINEAR DIFFERENTIAL EQUATIONS

Introduction

Not all differential equations can be solved by separation of variables. For example, the differential equation

$$y' = x + y$$

cannot be solved by separating variables (as you should check for yourself). But it can be solved by the method of *integrating factors*, which is the subject of the section. The method involves a clever use of the Product Rule, which we now review.

Using the Product Rule Backwards

When we apply the Product Rule to an expression containing an unknown function y, we must write the derivative of the unknown function y as y'. For example:

$$\frac{d}{dx}(x^2y) \quad = \quad 2xy \quad + \quad x^2y' \qquad\qquad \text{Using the Product Rule} \quad \frac{d}{dx}(f\cdot g) = f'\cdot g + f\cdot g'$$

Both are functions of x — Derivative of the first — Second left alone — First left alone — Derivative of the second

Now we will reverse this process: beginning with the two terms on the right (even if written in a different order), we will express them in the form of the derivative on the left.

EXAMPLE 1

USING THE PRODUCT RULE BACKWARDS

Express $x^5y' + 5x^4y$ as the derivative of a product.

Solution

$$x^5y' + 5x^4y \quad = \quad \frac{d}{dx}(x^5y) \qquad\qquad \text{Recognizing} \quad \frac{d}{dx}(x^5y) = 5x^4y + x^5y' \quad \text{in the reverse order}$$

Original expression — Written as the derivative of a product

Practice Problem 1

Express $e^{2x}y' + 2e^{2x}y$ as the derivative of a product.

➤ Solution on page 719

First-Order Linear Differential Equations and Integrating Factors

This technique will be useful for solving a type of differential equation called *first-order linear,* which is defined as follows:

First-Order Linear Differential Equations

A differential equation is *first-order linear* if it can be written in the form

$$y' + p(x) \cdot y = q(x)$$

where $p(x)$ and $q(x)$ are functions of x alone.

The previous equation is called the *standard form* of a first-order linear differential equation.* We solve it by multiplying both sides by an *integrating factor,* enabling us to write the left-hand side as the derivative of a product. A few examples will demonstrate the method.

EXAMPLE 2

SOLVING A FIRST-ORDER LINEAR DIFFERENTIAL EQUATION

Solve the differential equation $y' + \dfrac{2}{x}y = x$.

Solution

This *is* in standard form $y' + p(x) \cdot y = q(x)$ with $\begin{cases} p(x) = \dfrac{2}{x} \\ q(x) = x \end{cases}$.

We multiply through by x^2. (The choice of x^2 will be explained shortly.)

$$\underbrace{x^2y' + 2xy}_{\dfrac{d}{dx}(x^2y)} = x^3 \qquad\qquad y' + \dfrac{2}{x}y = x \ \text{ times } \ x^2$$

The left-hand side is now the derivative of x^2y

---

*Technically, "first-order" means that only the *first* derivative of y appears, and "linear" means that the equation is linear in y' and y. If $p(x)$ and $q(x)$ are continuous on a closed interval, then the differential equation is guaranteed to have a solution on that interval.

Therefore

$$\frac{d}{dx}(x^2 y) = x^3$$

We integrate both sides, on the left side simply dropping the d/dx, and on the right side using the Power Rule:

$$x^2 y = \frac{1}{4}x^4 + C$$

⌣ From $\int x^3\, dx = \frac{1}{4}x^4 + C$

Finally, we solve for the unknown function y by dividing both sides by x^2, obtaining the general solution of the differential equation:

$$y = \frac{1}{4}x^2 + \frac{C}{x^2} \qquad\qquad x^2 y = \frac{1}{4}x^4 + C \quad \text{divided by } x^2$$

How to Choose the Integrating Factor

In Example 2, the x^2 that we multiplied by is called an "integrating factor" because it enabled us to write the left-hand side as a derivative, which could then be integrated. How do we choose the integrating factor? There is a simple formula.

Integrating Factor

For the first-order linear differential equation

$$y' + p(x) \cdot y = q(x)$$

the integrating factor is $I(x) = e^{\int p(x)\, dx}$.

We omit the constant of integration in calculating the integrating factor, since it has no effect on the solution (as is shown in Exercise 33).

EXAMPLE 3

FINDING AN INTEGRATING FACTOR AND SOLVING

Solve the differential equation $y' + y = e^{3x}$.

Solution The equation is already in standard form $y' + p(x) \cdot y = q(x)$

with $\begin{cases} p(x) = 1 \\ q(x) = e^{3x} \end{cases}$ The integrating factor is

$$I(x) = e^{\int 1\,dx} = e^x \qquad\qquad \text{Integrating factor } I(x) = e^{\int p(x)\,dx}$$

Multiplying by $I(x) = e^x$,

$$\underbrace{e^x y' + e^x y}_{\frac{d}{dx}(e^x y)} = e^{4x} \qquad\qquad \begin{matrix} y' + y = e^{3x} \\ \text{multiplied by } e^x \end{matrix}$$

From $e^x \cdot e^{3x}$

$$\frac{d}{dx}(e^x y) = e^{4x} \qquad\qquad \begin{matrix}\text{Writing the left-hand}\\ \text{side as the derivative}\\ \text{of a product}\end{matrix}$$

$$e^x y = \frac{1}{4} e^{4x} + C \qquad\qquad \begin{matrix}\text{Integrating both sides}\\ \left(\text{dropping the } \dfrac{d}{dx}\right)\end{matrix}$$

From $\int e^{4x}\,dx = \frac{1}{4} e^{4x} + C$

$$y = \frac{1}{4} e^{3x} + C e^{-x} \qquad\qquad \begin{matrix}\text{Solving for } y \text{ by dividing}\\ \text{both sides by } e^x. \text{ This is}\\ \text{the general solution.}\end{matrix}$$

Note that the constant of integration C (from integrating the right-hand side) does not stand "by itself," but is multiplied by e^{-x}. In general, the constant C may appear *anywhere* in the solution.

Finding integrating factors may involve using the properties of logarithms. For example, the integrating factor for the differential equation $y' + \frac{2}{x} y = x$ in Example 2 is found as follows:

$$I(t) = e^{\int p(x)\,dx} = e^{\int(2/x)\,dx} = e^{2\int(1/x)\,dx} = e^{2\ln x}$$

$p(x)$ Taking out the constant Integrating $1/x$ gives $\ln x$

$$= e^{\ln x^2} = x^2$$

Using $2\ln x = \ln x^2$ Using $e^{\ln z} = z$

You should study the steps in this calculation carefully, since they occur frequently with integrating factors. The last two steps used natural logarithm properties 8 and 4 (see the inside back cover). We omitted the absolute value bars when we integrated $1/x$ to get $\ln x$, since in most applications the independent variable is positive, often representing time. Furthermore, in this particular case, including the absolute value bars leads to the same integrating factor as *without* the absolute value bars, since the x is squared.

Practice Problem 2   Find the integrating factor for $y' + \dfrac{4}{x}y = x$.   ➤ **Solution on page 719**

As before, an initial condition is used to evaluate the arbitrary constant.

EXAMPLE 4

FINDING A PARTICULAR SOLUTION

Solve the differential equation and initial condition $\begin{cases} xy' - 3y = x^2 \\ y(1) = 5 \end{cases}$

Solution

$$y' - \frac{3}{x}y = x$$

Dividing by x to write the equation in standard form with $p(x) = -3/x$

→ Try to skip to here →

$$I(x) = e^{\int \frac{3}{x}\,dx} = e^{-3\int \frac{1}{x}\,dx} = e^{-3\ln x} = e^{\ln x^{-3}} = x^{-3}$$

Integrating factor is x^{-3}

$$x^{-3}y' - 3x^{-4}y = x^{-2}$$

$y' - \dfrac{3}{x}y = x$ multiplied by x^{-3}

$$\frac{d}{dx}(x^{-3}y) = x^{-2}$$

Expressing the left side as the derivative of a product

$$x^{-3}y = \underbrace{-x^{-1} + C}$$

Integrating both sides

From $\int x^{-2}\,dx = -x^{-1} + C$

$$y = -x^2 + Cx^3$$

Solving for y by multiplying by x^3

$$5 = -1 + C$$

Substituting $y = 5$ and $x = 1$ [from the given $y(1) = 5$]

Therefore $C = 6$, giving

$$y = -x^2 + 6x^3$$

$y = -x^2 + Cx^3$ with $C = 6$ (the particular solution)

The general procedure for solving first-order linear differential equations is as follows:

Solving First-Order Linear Differential Equations

1. Write the differential equation in standard form
   $$y' + p(x) \cdot y = q(x).$$

2. Find the integrating factor $I(x) = e^{\int p(x)\,dx}$ (omitting the constant of integration).

3. Multiply the equation through by the integrating factor $I(x)$.

4. Express the left-hand side as the derivative of a product,
   $$\frac{d}{dx}[I(x) \cdot y].$$

5. Integrate both sides (dropping the prime on the left side, and writing the constant of integration on the right side).

6. Solve the resulting equation for y. If there is an initial condition, use it to evaluate the arbitrary constant of integration.

This general procedure works because multiplying the integrating factor changes the left-hand side of the differential equation into

$$e^{\int p(x)\,dx}y' + e^{\int p(x)\,dx}p(x) \cdot y \qquad \text{which equals} \qquad \frac{d}{dx}\left(e^{\int p(x)\,dx}y\right)$$

This last expression can then be integrated and the equation solved for y. The next example shows that integrating the right-hand side may require a substitution or some other integration technique.

EXAMPLE 5

USING AN INTEGRATING FACTOR AND A SUBSTITUTION

Solve the differential equation and initial condition $\begin{cases} y' = 6x - 2xy \\ y(0) = 2 \end{cases}$

Solution

$$y' + 2xy = 6x \qquad\qquad \text{Standard form, with } p(x) = 2x$$

$$I(x) = e^{\int 2x\,dx} = e^{x^2} \qquad\qquad \text{Integrating factor}$$

$$e^{x^2}y' + e^{x^2}2xy = e^{x^2}6x \qquad\qquad y' + 2xy = 6x \text{ times } e^{x^2}$$

$$\frac{d}{dx}(e^{x^2}y) = e^{x^2}6x \qquad\qquad \text{Expressing the left side as the derivative of a product}$$

$$e^{x^2}y = \int e^{x^2}6x \, dx = 3 \int e^{x^2} 2x \, dx = 3e^{x^2} + C$$

Integrating both sides (using the substitution method on $\int e^{x^2} 6x \, dx$)

$\underbrace{\quad}_{e^u} \quad \underbrace{\quad}_{du}$ Using
$\int e^u \, du = e^u + C$

$$y = 3 + Ce^{-x^2}$$

Solving for y by multiplying $e^{x^2}y = 3e^{x^2} + C$ by e^{-x^2}

$$2 = 3 + C$$

Substituting $y = 2$ and $x = 0$ (from the given $y(0) = 2$)

Therefore, $C = -1$, giving

$$y = 3 - e^{-x^2}$$

$y = 3 + Ce^{-x^2}$ with $C = -1$ (the particular solution)

Radon Pollution

Radon is a colorless, odorless gas formed by the natural decay of uranium in the earth. When dispersed in the atmosphere it is harmless, but if allowed to concentrate in a room or basement, it becomes a serious health hazard. Indoor radon pollution is second only to smoking as a certified cause of lung cancer, killing an estimated 7000 to 30,000 Americans each year.*

EXAMPLE 6

VENTING RADON POLLUTION

A 1000-cubic-foot room has a radon level of 500 pCi (picocuries) per cubic foot. A ventilation system is installed, which each hour brings in 100 cubic feet of outside air (containing 4 pCi per cubic foot), while an equal volume of air leaves the room. Assuming that the air in the room mixes thoroughly, find how soon the radon level will fall to 112 pCi per cubic foot, the safety level set by the Environmental Protection Agency (EPA).

Solution

Let

$$y(t) = \begin{pmatrix} \text{Amount of radon in} \\ \text{room after } t \text{ hours} \end{pmatrix}$$

---

* See Bernard Cohen, *Radon, a Homeowner's Guide to Detection and Control* (Consumer Reports Books, 1987). His book measures radon per *liter*, whereas we measure it per *cubic foot*. To change our numbers to "per liter," multiply them by 0.0353.

so

$$y(0) = 500{,}000$$

Initial amount (500 per cubic foot times 1000 cubic feet)

The concentration of radon *in the incoming air* is given as 4, and the concentration *in the room* at any time is $y/1000$ (the amount (y) of radon in the room divided by the volume (1000) of the room). Multiplying these concentrations by 100, the volume of air flowing in and out, gives the rates at which radon enters and leaves the room each hour. We show these rates in an inflow-outflow diagram for a 1-hour time period.

$$\text{IN} \quad \begin{array}{c} 4 \text{ per cu ft} \\ \times\ 100 \text{ cu ft} \\ = 400 \end{array} \longrightarrow \quad y(t) = \left\{ \begin{array}{c} \text{Total radon} \\ \text{in room} \end{array} \right\} \quad \begin{array}{c} \dfrac{y}{1000} \text{ per cu ft} \\ \times\ 100 \text{ cu ft} \\ = \dfrac{y}{10} = 0.1y \end{array} \longrightarrow \quad \text{OUT}$$

(originally 500,000)

volume: 1000 cu ft

This leads to the following differential equation:

$$y' = 400 - 0.1y \qquad \left(\begin{array}{c} \text{Rate of} \\ \text{change} \end{array} \right) = \left(\begin{array}{c} \text{Amount} \\ \text{added} \end{array} \right) - \left(\begin{array}{c} \text{Amount} \\ \text{removed} \end{array} \right)$$

Rate of change of y \quad is \quad 400 added \quad 0.1y removed

$$y' + 0.1y = 400 \qquad\qquad \text{In standard form}$$

$$I(t) = e^{\int 0.1\,dt} = e^{0.1t} \qquad\qquad \text{Integrating factor}$$

$$e^{0.1t}y' + 0.1e^{0.1t}y = 400e^{0.1t} \qquad\qquad y' + 0.1y = 400 \quad \text{times } e^{0.1t}$$

$$\frac{d}{dt}\left(e^{0.1t}y\right) = 400e^{0.1t} \qquad\qquad \begin{array}{l} \text{Left-hand side as the} \\ \text{derivative of a product} \end{array}$$

$$e^{0.1t}y = 400 \cdot 10e^{0.1t} + C \qquad\qquad \text{Integrating}$$

From $\int e^{0.1t}\,dt = \dfrac{1}{0.1}e^{0.1t} + C$

$$y = 4000 + Ce^{-0.1t} \qquad\qquad \begin{array}{l} \text{Solving for } y \text{ by} \\ \text{multiplying by } e^{-0.1t} \end{array}$$

$$500{,}000 = 4000 + C \qquad\qquad \begin{array}{l} \text{Substituting } y = 500{,}000 \\ \text{and } x = 0 \\ (\text{from } y(0) = 500{,}000) \end{array}$$

Therefore, $C = 496{,}000$, giving

$$y = 4000 + 496{,}000e^{-0.1t} \qquad\qquad \begin{array}{l} y = 4000 + Ce^{-0.1t} \\ \text{with } C = 496{,}000 \end{array}$$

Amount of radon after t hours

To find when the EPA limit of 112 pCi per cubic foot is reached, we solve

$$\underbrace{4000 + 496{,}000e^{-0.1t}}_{\substack{\text{Total amount of radon} \\ \text{(according to formula)}}} = \underbrace{112{,}000}_{\substack{\text{Total radon at EPA safety level} \\ \text{(112 per cubic foot times 1000 cubic feet)}}}$$

We could solve this equation for t as we did in the preceding section, or we could use a graphing calculator (see the following Graphing Calculator Exploration). Either way, we would find that the radon concentration will fall to the EPA level in about 15.2 hours.

The amount of radon remaining in the room after t hours is shown by the following graph.

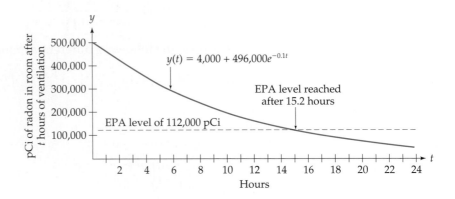

Graphing Calculator Exploration

Graph $y_1 = 4000 + 496{,}000e^{-0.1x}$ and $y_2 = 112{,}000$ on the window $[0, 24]$ by $[0, 500{,}000]$. Use INTERSECT to find where they meet.

The "long-run" amount of radon in the room is found by taking the limit as t approaches infinity:

$$\lim_{t \to \infty} (4000 + \underbrace{496{,}000e^{-0.1t}}_{\substack{\text{Approaches 0} \\ \text{as } t \to \infty}}) = 4000 \qquad \text{4000 pCI}$$

This same number can be found as follows: outside air (with 4 pCi per cubic foot) filling the 1000-cubic-foot room will eventually result in $4 \cdot 1000 = 4000$ pCi of radon. This agreement serves as a check of the solution.

Example 6 is a typical "mixture problem," in which two mixtures (here, outside and inside air), each having a different concentration of a particular substance (radon), are combined and you want to find the concentration at some later time. The same technique works for *any* mixture, whether it is air pollution in a room or chemicals in a container. It is generally easiest to let $y(t)$ stand for the *total amount* of the substance in the room (or container) at any time, so that the *concentration* can be expressed $\dfrac{y}{\text{Total volume}}$. Making an *inflow-outflow* diagram for the gain and loss per time unit leads to a differential equation, which you then solve. (This differential equation could also have been solved by separation of variables.)

9.3 Section Summary

A first-order linear differential equation can be written in the standard form

$$y' + p(x) \cdot y = q(x) \qquad \qquad p \text{ and } q \text{ are functions of } x \text{ alone}$$

We solve first-order linear differential equations by multiplying both sides by the *integrating factor* $I(x) = e^{\int p(x)\,dx}$, expressing the left-hand side as the derivative of a product, integrating both sides, and then solving for y.

➤ Solutions to Practice Problems

1. $e^{2x}y' + 2e^{2x}y = \dfrac{d}{dx}(e^{2x}y)$ $\qquad\qquad$ Using $\dfrac{d}{dx}e^{2x} = 2e^{2x}$

➤ Try to skip to here ➟

2. $I(x) = e^{\int \frac{4}{x}dx} = e^{4\int \frac{1}{x}dx} = e^{4\ln x} = e^{\ln x^4} = x^4$

9.3 Exercises

Solve each first-order linear differential equation.

1. $y' + 2y = 8$ $\qquad$ **2.** $y' - y = 2$

3. $y' - 2y = e^{-2x}$ $\qquad$ **4.** $y' + 3y = e^{3x}$

5. $y' + \dfrac{5}{x}y = 24x^2$ $\qquad$ **6.** $y' + \dfrac{4}{x}y = 12x$

7. $xy' - y = x^2$ $\qquad$ **8.** $xy' - y = x$

9. $y' + 3x^2y = 9x^2$ **10.** $y' - 4x^3y = 8x^3$

11. $y' - 2xy = 0$ **12.** $y' + xy = 0$

13. $(x + 1)y' + y = 2x$ **14.** $y' + \dfrac{2x}{x^2 + 1}y = 3$

15. $y' - \dfrac{2}{x}y = 6x^3 - 9x^2$

16. $xy' - y = x^3e^{x^2}$ **17.** $y' = x + y$

18. $xy' + y = x^2 \ln x$

[*Hint for Exercises 17 and 18:* Use integration by parts or an integral table.]

19. $y' + (\cos x)y = \cos x$

20. $y' - (\sin x)y = \sin x$

Solve each differential equation with the given initial condition.

21. $y' + 3y = 12e^x$ **22.** $y' + 4y = e^{-3x}$
    $y(0) = 5$ $y(0) = 4$

23. $xy' + 2y = 14x^5$ **24.** $xy' + 4y = 10x$
    $y(1) = 1$ $y(1) = 0$

25. $xy' = 2y + x^2$ **26.** $xy' = 3y + x^4$
    $y(1) = 3$ $y(1) = 7$

27. $y' + 2xy = 4x$ **28.** $y' - 3x^2y = 6x^2$
    $y(0) = 0$ $y(0) = 1$

Solve each differential equation in two ways: first by using an integrating factor, and then by separation of variables (as in Section 9.1)

29. $y' = y + 1$ **30.** $y' = xy$

Each exercise below has one differential equation that can be solved by an integrating factor and one that can be solved by separation of variables (as in Section 9.1). Solve each by the appropriate technique.

31. a. $y' + xy^2 = 0$ **32. a.** $yy' = x$
 b. $y' = y + x^2e^x$ **b.** $y' + y = e^{-x}$

33. Show that the constant of integration in the integrating factor can be omitted. [*Hint:* Show that multiplying by the integrating factor $e^{\int p(x)\,dx + C}$ just multiplies the differential equation by a nonzero constant, which does not change the solution.]

34. Derive the formula $\quad y(x) = \dfrac{1}{I(x)} \displaystyle\int I(x)q(x)dx$

for the solution of the first-order linear differential equation $y' + p(x)y = q(x)$, where $I(x) = e^{\int p(x)\,dx}$. (*Note:* Rather than memorizing this formula, it is better to understand the method, which generalizes to an even larger class of differential equations.)

APPLIED EXERCISES

35. BUSINESS: Value of a Company A company grows in value by 10% each year, and also gains 20% of a growing market estimated at $100e^{0.1t}$ million dollars, where t is the number of years that the company has been in business. Therefore, the value $y(t)$ of the company (in millions of dollars) satisfies

$\quad y' = 0.1y + 20e^{0.1t}$

$y(0) = 5$ (Initial value $5 million)

a. Solve this differential equation and initial condition to find a formula for the value of the company after t years.

b. Use your solution to find the value of the company after 2 years.

36. ECONOMICS: Exports A country's cumulative exports $y(t)$ (in millions of dollars) grow in

proportion to its *average* size y/t over the last t years, plus a fixed growth rate (10), and so satisfy

$\quad y' = \dfrac{1}{t}y + 10$ (for $t \geq 1$)

$y(1) = 8$ ($8 million after 1 year)

a. Solve this differential equation and initial condition to find the country's cumulative exports after t years.

b. Use your solution to find the country's cumulative exports after 5 years.

37. BIOMEDICAL: AIDS According to data from the Centers for Disease Control, the total number $y(t)$ of cases of AIDS (acquired immunodeficiency syndrome) diagnosed in a certain region after t years satisfies

$$y' - \frac{3}{t}y = 0 \qquad \text{(for } t \geq 1\text{)}$$
$$y(1) = 125 \qquad \text{(125 cases after 1 year)}$$

a. Solve this differential equation and initial conditon. (Your solution will show that AIDS does not spread logistically, as do most epidemics, but like a power. This means that AIDS will spread more slowly, which seems to result from its being transmitted at different rates within different subpopulations.

b. Use your solution to predict the number of AIDS cases in the region by the year $t = 15$.

38. ENVIRONMENTAL SCIENCE: Global Warming The burning of coal and oil is increasing the amount of carbon dioxide in the atmosphere, which is expected to trap more solar radiation and increase global temperatures (the "greenhouse effect"), possibly raising the sea level by melting polar ice and flooding coastal regions. The current 3200 billion tons of carbon dioxide in the atmosphere is growing by about 50 billion tons per year, while about 1% of the accumulated carbon dioxide is removed by natural processes. Therefore, the amount $y(t)$ of carbon dioxide (in billions of tons) t years from now satisfies

$$y' = -0.01y + 50$$
$$y(0) = 3200 \qquad \text{(Initial amount)}$$

a. Solve this differential equation and initial condition.

b. Graph the solution on a graphing calculator and find when the amount of carbon dioxide will reach 4000 billion tons, at which time global temperatures are expected to have risen by 3 degrees Fahrenheit.

c. Use your solution to find the long-run amount of carbon dioxide in the atmosphere.

39. ENVIRONMENTAL SCIENCE: Algae Bloom An algae bloom in a lake is a sudden growth of algae that consumes nutrients and blocks sunlight, killing other life in the water. Suppose that the number of tons $y(t)$ of algae after t weeks satisfies

$$y' = ty + t$$
$$y(0) = 2 \qquad \text{(Initial tonnage)}$$

a. Solve this differential equation and initial condition.

b. Use your solution to find the amount of algae after 2 weeks.

c. Graph the solution on a graphing calculator and find when the algae bloom will reach 40 tons.

40. BIOMEDICAL: Drug Absorption A patient's ability to absorb a drug sometimes changes with time, and the dosage must therefore be adjusted. Suppose that the number of milligrams $y(t)$ of a drug remaining in the patient's bloodstream after t hours satisfies

$$y' = -\frac{1}{t}y + t \qquad \text{(for } t \geq 1\text{)}$$
$$y(3) = 5 \qquad \text{(5 mg after 3 hours)}$$

Solve this differential equation and initial condition to find the amount remaining in the bloodstream after t hours.

41. ENVIRONMENTAL SCIENCE: Radon Pollution A 10,000-cubic-foot-room has an initial radon level of 800 pCi (picocuries) per cubic foot. A ventilation system is installed that each hour brings in 500 cubic feet of outside air (containing 5 pCi per cubic foot), while an equal volume of air leaves the room. Assume that the air in the room mixes thoroughly.

a. Find a differential equation and initial condition that govern the total amount $y(t)$ of radon in the room after t hours.

b. Solve this differential equation and initial conditon.

c. Graph the solution on a graphing calculator and find how soon the radon level will fall to the EPA safety level of 112 pCi per cubic foot.

42. ENVIRONMENTAL SCIENCE: Smoke Pollution A 5000-cubic-foot room has 400 smoke particles per cubic foot. A ventilation system is turned on that each minute brings in 50 cubic feet of outside air (containing 10 smoke particles per cubic foot), while an equal volume of air leaves the room. Assume that the air in the room mixes thoroughly.

a. Find a differential equation and initial condition that govern the total number $y(t)$ of smoke particles in the room after t minutes.

(continues)

b. Solve this differential equation and initial condition.

c. Graph the solution on a graphing calculator and find how soon the smoke level will fall to 100 smoke particles per cubic foot.

43. **ENVIRONMENTAL SCIENCE: Smoke Pollution** A 12,000-cubic-foot room has 500 smoke particles per cubic foot. A ventilation system is turned on that each minute brings in 600 cubic feet of smoke-free air, while an equal volume of air leaves the room. Also, during each minute, smokers in the room add a total of 10,000 particles of smoke to the room. Assume that the air in the room mixes thoroughly.

a. Find a differential equation and initial condition that govern the total number $y(t)$ of smoke particles in the room after t minutes.

b. Solve this differential equation and initial condition.

c. Find how soon the smoke level will fall to 100 smoke particles per cubic foot.

44. **GENERAL: Mixture** A 500-gallon tank is filled with water containing 0.2 ounce of impurities per gallon. Each hour, 200 gallons of water (containing 0.01 ounce of impurities per gallon) is added and mixed into the tank, while an equal volume of water is removed.

a. Write a differential equation and initial condition that describe the total amount $y(t)$ of impurities in the tank after t hours.

b. Solve this differential equation and initial condition.

c. Use your solution to find when the impurities will reach 0.05 ounce per gallon, at which time the water may be used for drinking.

d. Use your solution to find the "long-run" amount of impurities in the tank.

45. **GENERAL: Mixture** The water in a 100,000-gallon reservoir contains 0.1 gram of pesticide per gallon. Each hour, 2000 gallons of water (containing 0.01 gram of pesticide per gallon) is added and mixed into the reservoir, and an equal volume of water is drained off.

a. Write a differential equation and initial condition that describe the amount $y(t)$ of pesticide in the reservoir after t hours.

b. Solve this differential equation and initial condition.

c. Graph your solution on a graphing calculator and find when the amount of pesticide will reach 0.02 gram per gallon, at which time the water is safe to drink.

d. Use your solution to find the "long-run" amount of pesticide in the reservoir.

46. **GENERAL: Banking** You deposit $8000 into a bank account paying 5% interest compounded continuously, and you withdraw funds continuously at the rate of $1000 per year. Therefore, the amount $y(t)$ in the account after t years satisfies

$$y' = 0.05y - 1000$$
$$y(0) = 8000$$

a. Solve this differential equation and initial value.

b. Graph your solution on a graphing calculator and find how long it will take until the account is empty.

47. **GENERAL: Dieting** A person's weight $w(t)$ after t days of eating c calories per day can be modeled by the following differential equation*

$$w' + 0.005w = \frac{c}{3500}$$

a. If a person initially weighing 170 pounds goes on a diet of 2100 calories per day, find a formula for the person's weight after t days.

b. Use your solution to find when the person will have lost 15 pounds.

c. Find the "limiting weight" that will be approached if the person continues on this diet indefinitely.

48. **GENERAL: Dieting** (*continuation*) Use the differential equation in Exercise 47 to find the number of calories a person weighing 170 pounds should eat each day in order to lose 6 pounds in 21 days.

49–50. **BIOMEDICAL: Drug Dynamics** When you swallow a pill, the medication passes through your stomach lining into your bloodstream, where some is absorbed by the cells of your body and the rest continues to circulate for future absorption. The amount $y(t)$ of medication remaining in the

* The 0.005 represents the proportional weight loss per day when eating nothing, and 3500 is the conversion rate for calories into pounds. See Arthur Segal, "A Linear Diet Model," *College Mathematics Journal* 18:44–45, January 1987.

bloodstream after t hours can be modeled by the differential equation

$$\frac{dy}{dt} = abe^{-bt} - cy$$

for constants a, b, and c (respectively the dosage of the pill, the dissolution constant of the pill, and the absorption constant of the medication). For the given values of the constants:

a. Substitute the constants into the stated differential equation.

b. Solve the differential equation (with the initial condition of having no medicine in the bloodstream at time $t = 0$) to find a formula for the amount of medicine in the bloodstream at any time t (hours). *(continues)*

c. Use your solution to find the amount of medicine in the bloodstream at time $t = 2$ hours.

d. Graph your solution on a graphing calculator and find when the amount of medication in the bloodstream is maximized.

49. $a = 10\,\text{mg}$, $b = 3$, $c = 0.2$

50. $a = 30\,\text{mg}$, $b = 5$, $c = 0.3$

51. BIOMEDICAL: Drug Dynamics Read the explanation in the preceding exercise. Then solve the stated differential equation with the initial condition $y(0) = 0$ (initially no medication in the bloodstream). Show that your solution can be written in the form

$$y(t) = \frac{ab}{b - c}(e^{-ct} - e^{-bt}).$$

9.4 APPROXIMATE SOLUTIONS OF DIFFERENTIAL EQUATIONS: EULER'S METHOD

Introduction

There are many techniques for solving differential equations besides those described in this book. Despite all of the techniques, however, there are many differential equations that cannot be solved by *any* method (as finite combinations of elementary functions). Even if a differential equation *can* be solved, its solution may be too complicated to be useful. In such cases we must settle for an *approximation* to the actual solution, and one way of finding such an approximation is *Euler's method*. Leonhard Euler (1707–1783; his name is pronounced "Oiler") was a prolific Swiss mathematician in whose honor the constant e was named.

Euler's Method

A differential equation together with an initial condition is called an *initial value problem.* For an initial value problem of the form

$$y' = g(x, y)$$

$$y(a) = y_0$$

Initial value $y = y_0$
at $x = a$

Euler's method finds an approximation to the solution on a closed interval beginning at $x = a$.* Euler observed that if the solution $y(x)$ passes through a point (x, y) in the plane, then at that point the

---

* Technically, if g and $\dfrac{\partial g}{\partial y}$ are continuous for all x in an interval $[a, b]$ and all values of y, then the initial value problem has a solution for x-values in some subinterval beginning at a. Less stringent conditions can also be given.

solution has slope $g(x, y)$, since $y' = g(x, y)$. This idea is used to construct a polygonal approximation (consisting of connected line segments) that is tangent to the true solution at the initial point.

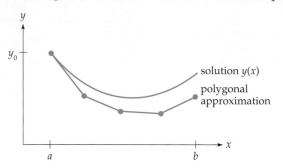

The actual (but generally unknown) solution $y(x)$, and below it a polygonal approximation with $n = 4$ segments.

To construct the Euler approximation on an interval $[a, b]$ using n segments, we divide $[a, b]$ into n subintervals by equally spaced points

$$a = x_0 < x_1 < x_2 < \cdots < x_n = b \quad \text{with spacing (step size)} \quad h = \frac{b - a}{n}.$$

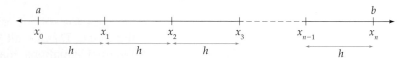

Beginning with $x_0 = a$, successive x-values are found by adding h to the previous x-value until $x_n = b$ is reached. The segments are constructed on these subintervals as follows.

The initial condition $y(a) = y_0$ gives the initial point (x_0, y_0) (since $x_0 = a$), and the differential equation gives the slope at this point as $y' = g(x_0, y_0)$. The first segment is constructed from this point, (x_0, y_0), with this slope, $g(x_0, y_0)$, extending a step size h to the right, as illustrated below.

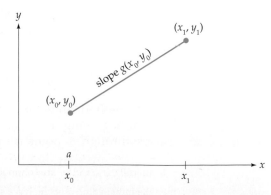

The first segment of the Euler approximation.

The equation of this segment comes from the point-slope form for a line, with the given point and slope.

$$y - y_0 = g(x_0, y_0) \cdot (x - x_0)$$

$y - y_1 = m(x - x_1)$
with point (x_0, y_0) and
slope $m = g(x_0, y_0)$

or

$$y = y_0 + g(x_0, y_0) \cdot (x - x_0)$$

Adding y_0 to each
side (for $x_0 \le x \le x_1$)

The y-coordinate y_1 at the end of the first segment is found by evaluating at $x = x_1$:

$$y_1 = y_0 + g(x_0, y_0) \cdot \underbrace{(x_1 - x_0)}_{h} = y_0 + g(x_0, y_0) \cdot h$$

Formula above
at $x = x_1$

The second segment is constructed similarly, beginning at the previous end point (x_1, y_1) and with slope $g(x_1, y_1)$ [from $y' - g(x, y)$ evaluated at this same point], as illustrated in the following diagram.

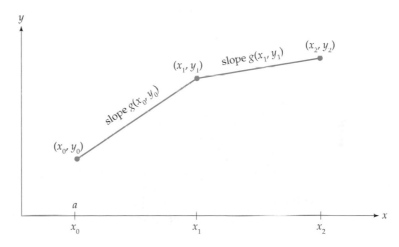

First two segments of the Euler approximation.
The slope of each segment is $g(x, y)$ evaluated
at the left-hand endpoint of the segment.

Just as before, this leads to a formula for the second segment:

$$y = y_1 + g(x_1, y_1) \cdot (x - x_1)$$

For $x_1 \le x \le x_2$

This process is then repeated, with each new segment beginning at the previous endpoint and having slope $y' = g(x, y)$ evaluated at this same point. The process continues until it reaches the final x-value $x_n - b$. An example will illustrate the method.

EXAMPLE **1**

FINDING AN APPROXIMATE SOLUTION

For the initial value problem $\begin{cases} y' = 3x - y \\ y(0) = 4 \end{cases}$, find the Euler approximation to the solution on the interval $[0, 1.5]$ using $n = 3$ segments. State the approximation to the solution at $x = 1.5$.

Solution

The function g is

$$g(x, y) = 3x - y$$

Right-hand side of the differential equation

The step size is

$$h = \frac{1.5 - 0}{3} = \frac{1.5}{3} = 0.5$$

$h = \dfrac{b - a}{n}$ with $a = 0$, $b = 1.5$, and $n = 3$

The initial point is

$$x_0 = 0, \qquad y_0 = 4$$

From the given $y(0) = 4$

Segment 1
Slope $\quad g(x_0, y_0) = g(0, 4) = 3 \cdot 0 - 4 = -4$

$g(x, y) = 3x - y$ with $x = 0$ and $y = 4$

Endpoint $\quad x_1 = 0.5, \quad y_1 = 4 + (-4) \cdot (0.5) = 2$

$x_1 = x_0 + h,$
$y_1 = y_0 + g(x_0, y_0) \cdot h$

Segment 2
Slope $\quad g(x_1, y_1) = g(0.5, 2) = 3 \cdot (0.5) - 2 = -0.5$

$g(x, y) = 3x - y$ with $x = 0.5$ and $y = 2$

Endpoint $\quad x_2 = 1, \quad y_2 = 2 + (-0.5) \cdot (0.5) = 1.75$

$x_2 = x_1 + h,$
$y_2 = y_1 + g(x_1, y_1) \cdot h$

Segment 3
Slope $\quad g(x_2, y_2) = g(1, 1.75) = 3 \cdot 1 - 1.75 = 1.25$

$g(x, y) = 3x - y$ with $x = 1$ and $y = 1.75$

Endpoint $\quad x_3 = 1.5, \quad y_3 = 1.75 + (1.25) \cdot (0.5) = 2.375$

$x_3 = x_2 + h,$
$y_3 = y_2 + g(x_2, y_2) \cdot h$

The points $(0, 4)$, $(0.5, 2)$, $(1, 1.75)$, and $(1.5, 2.375)$ determine the polygonal approximation of the solution. The y-values are the approximations of the solution at the x-values. In particular, the approximation at the endpoint $x = 1.5$ is

$$y(1.5) = y_3 = 2.375$$

The polygonal approximation of the solution to this problem is graphed below.

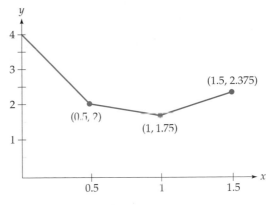

Euler approximation to the solution

of $\begin{cases} y' = 3x - y \\ y(0) = 4 \end{cases}$ with $n = 3$ segments

Graphing Calculator Exploration

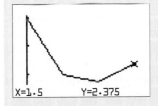

X=1.5   Y=2.375

The graphing calculator program* EULER carries out the Euler approximation for any number of segments, draws the polygonal approximation, and shows the coordinates of the points. The screen on the left shows the approximation for the initial value problem in the preceding example, and shows the coordinates of the final point.

*See the Preface for how to obtain this and other programs.

Accuracy of the Euler Approximation

In Example 1 we used only $n = 3$ segments, so we should not expect great accuracy. Small values of n like 3 are only for demonstration purposes. More realistic values of n are in the hundreds or thousands, with the calculations carried out on a programmable calculator or computer. The following diagram shows the Euler approximation for $n = 3$ that we found, and also the approximations for $n = 6$ and $n = 100$, together with the actual solution (found by using an integrating factor).

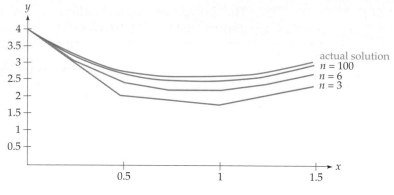

Actual solution to $\begin{cases} y' = 3x - y \\ y(0) = 4 \end{cases}$ and

the Euler approximations using 3, 6, and 100 segments.

The accuracy for $n = 100$ segments may be seen numerically from the following table.

x-Values	Euler Approximation ($n = 100$)	Actual y-Values
0	4	4
0.5	2.73	2.75
1	2.56	2.58
1.5	3.04	3.06

Inaccuracies in Euler's approximation come from two sources. First, the *actual* solution changes slope *continuously,* whereas the approximation changes slope only at a few isolated points. Second, the points at which the new slopes are calculated are not generally on the curve (except for the initial point), and so do not give the true slope of the solution curve. For greater accuracy one may increase n (but this involves more calculation and therefore more roundoff errors), or use a more advanced method (such as the Runge-Kutta method).

Practice Problem

For the initial value problem $\begin{cases} y' = 7x - 2y \\ y(3) = 6 \end{cases}$, find:

a. the initial point (x_0, y_0)

b. the slope of the solution at the initial point   ➤ Solutions on page 730

The following differential equation cannot be solved by either of the "exact" methods of this chapter, so we use EULER. (We could also use one of the programs listed on page 730.)

Graphing Calculator Exploration

Let $y(x)$ be the value (in millions of dollars) of a computer software company x years after it was founded. The company is expected to grow, but its growth will be slowed by inefficiency as its size increases. Suppose that the company's value $y(x)$ satisfies

$$y' = 2.6x - e^{0.3y}$$

$$y(0) = 1 \qquad \text{(Starting value is \$1 million)}$$

Use Euler's method with $n = 50$ to estimate the value of the company after 5 years.

The screen on the left shows the graph from the program EULER on the interval $[0, 5]$ beginning with $y(0) = 1$ and using 50 line segments, along with the coordinates of the final point, which gives the value of the company.

The value of the company will be \$8.35 million.

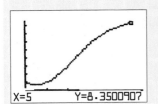

X=5   Y=8.3500907

9.4 Section Summary

For an initial value problem of the form $\begin{cases} y' = g(x, y) \\ y(a) = y_0 \end{cases}$, Euler's method gives an *approximation* of the solution on an interval $[a, b]$. The approximation consists of n connected line segments, each on an interval of length $h = (b - a)/n$, beginning from the initial point at $x_0 = a$, with slope calculated from $y' = g(x, y)$ evaluated at the left-hand endpoint of the segment. The endpoints of these segments, after the initial point (x_0, y_0) with $x_0 = a$, are calculated successively from the formulas

$$\left. \begin{array}{l} x_i = x_{i-1} + h \\ y_i = y_{i-1} + g(x_{i-1}, y_{i-1}) \cdot h \end{array} \right\} \quad i = 1, 2, \ldots, n$$

The accuracy generally increases as the number of segments n increases, but at the cost of more calculation. While the calculations may be carried out "by hand" (as in Example 1), they are easily done on a graphing calculator or computer.

Calculator and Computer Programs for Euler's Method

The following are brief programs for Euler's method, one for the TI-83 Graphing Calculator and one in BASIC. (The small print in blue is explanation, not part of the program.)

TI-83 Program for Euler's Method		BASIC Program for Euler's Method	
Disp "USES FUNCTION Y1"		a=	fill in beginning x-value
Prompt A, B	beginning, ending x-values	b=	fill in ending x-value
Prompt Y	beginning y-value	y=	fill in beginning y-value
Prompt N	number of segments	n=	fill in number of segments
(B-A)/N→H		h=(b-a)/n	
A→X		x=a	
For (I, 1, N)		FOR i=1 TO n	
Y+Y1*H→Y		$y=y+\left(\begin{array}{c}\text{fill in the}\\\text{function}\end{array}\right)*h$	
X+H→X		x=x+h	
Disp "X=", X		PRINT x, y	
Disp "Y=", Y		NEXT i	
Disp "PRESS ENTER"		END	
PAUSE			
END			

➤ **Solutions to Practice Problem**

   a. $(3, 6)$

   b. Slope: $y' = g(3, 6) = 7 \cdot 3 - 2 \cdot 6 = 21 - 12 = 9$

   Since $g(x, y) = 7x - 2y$

9.4 Exercises

1. If the solution to $y' = 5x - 4y$ passes through the point $(4, 3)$, what is the slope of the solution at that point?

2. If the solution to $y' = x + 4y$ passes through the point $(3, 1)$, what is the slope of the solution at that point?

3. For the initial value problem $\begin{cases} y' = 4xy \\ y(1) = 3 \end{cases}$, state the initial point (x_0, y_0) and calculate the slope of the solution at this point.

4. For the initial value problem $\begin{cases} y' = x/y \\ y(6) = 2 \end{cases}$, state the initial point (x_0, y_0) and calculate the slope of the solution at this point.

For each initial value problem, calculate the Euler approximation for the solution on the interval $[0, 1]$ using $n = 4$ segments. Draw the graph of your approximation. (Carry out the calculations "by hand" with the aid of a calculator, rounding to two decimal places. Answers may differ slightly, depending on when you do the rounding.)

5. $y' = 3x - 2y$
   $y(0) = 2$

6. $y' = x + 2y$
   $y(0) = 1$

7. $y' = 4xy$
   $y(0) = 1$

8. $y' = 8x^2 - y$
   $y(0) = 2$

9. $y' + 2y = e^{4x}$
   $y(0) = 2$

10. $y' + e^y = 8x$
   $y(0) = 0$

For each initial value problem:

a. Use EULER or a similar program (see the previous page) to find the estimate for $y(2)$. Use the interval $[0, 2]$ with $n = 50$ segments.

b. Solve the differential equation and initial condition *exactly* by separating variables or using an integrating factor.

c. Evaluate the solution that you found in part (b) at $x = 2$. Compare this *actual* value of $y(2)$ with the *estimate* of $y(2)$ that you found in part (a).

11. $y' = \dfrac{x}{y}$
   $y(0) = 1$

12. $y' = -xy$
   $y(0) = 1$

13. $\dfrac{dy}{dx} = 0.2y$
   $y(0) = 1$

14. $\dfrac{dy}{dx} = -y$
   $y(0) = 1$

15. $y' + y = 2e^x$
   $y(0) = 5$

16. $y' + y = e^{-x}$
   $y(0) = 0$

For each initial value problem, use EULER or a similar program (see the previous page) to find the approximate solution at the stated x-value, using

50 segments. [*Hint:* Use an interval that begins at the initial x-value and ends at the stated x-value.]

17. $y' = xe^{-y}$
   $y(1) = 0.5$
   Approximate the
   solution at $x = 3$

18. $y' = e^{x/y}$
   $y(1) = 0.4$
   Approximate the
   solution at $x = 3$

19. $y' = \dfrac{x + 1}{y + 1}$
   $y(1) = \dfrac{1}{2}$
   Approximate the
   solution at $x = 2.5$

20. $y' = 1 + \dfrac{y}{x}$
   $y(?) = -1$
   Approximate the
   solution at $x = 3.5$

21. $\dfrac{dy}{dx} = \sqrt{x + y}$
   $y(3) = 1$
   Approximate the
   solution at $x = 3.8$

22. $\dfrac{dy}{dx} = (x - y)^2$
   $y(2) = 0$
   Approximate the
   solution at $x = 2.8$

Use EULER or a similar program (see the previous page) to find the approximate solution at the stated x-value, using the given numbers of segments.

23. $y' = e^{x/y}$
   $y(0) = 1$
   Approximate the solution at $x = 1$ using:
   a. $n = 10$
   b. $n = 100$
   c. $n = 500$

24. $y' = x^2 + y^2$
   $y(0) = 0$
   Approximate the solution at $x = 2$ using:
   a. $n = 10$
   b. $n = 100$
   c. $n = 500$

APPLIED EXERCISES

25. BUSINESS: Sales Sales of a popular new computer game are growing so that the number $y(t)$ (in thousands) sold in the first t months satisfies

$$y' = 2e^{0.05y} + 2$$
$$y(0) = 0$$

Use Euler's method with $n = 3$ to estimate $y(3)$, the number sold in the first 3 months. (Carry out the calculations "by hand" with

the aid of a calculator, rounding to two decimal places.)

26. BUSINESS: Value of a Company A biotechnology company is growing so that its value $y(t)$ (in millions of dollars) after t years satisfies

$$y' = \dfrac{1}{5}e^{\sqrt{y}} + 1$$
$$y(0) = 1$$

(Initially worth
$1 million)

(continues)

Use Euler's method with $n = 4$ to estimate $y(2)$, the company's value after 2 years. (Carry out the calculations "by hand" with the aid of a calculator, rounding to two decimal places.)

27. **GENERAL: Dam Sediment** Sediment is flowing into a lake behind a hydroelectric dam at the rate of 1000 tons per year, while 10% of the upper layer of accumulated sediment is flushed through the turbines each year. The amount $y(t)$ of sediment after t years satisfies

$$\frac{dy}{dt} = 1000 - 0.1y^{2/3}$$

$y(0) = 0$ (Initially no sediment)

Use EULER or a similar program with $n = 50$ to estimate $y(2)$, the amount of sediment after 2 years.

28. **GENERAL: Environmental Science** A crop parasite is spreading so that the affected area $y(t)$ (in square miles) in t years is growing according to

$$\frac{dy}{dt} = \sqrt{y^2 + 100}$$

$y(0) = 5$ (Initially affects 5 square miles)

Use EULER or a similar program with $n = 50$ to estimate the area affected by the parasite after 1 year.

29. **BIOMEDICAL: Genetics** Population genetics involves initial value problems of the form

$$q' = -0.01q^2(1 - q)$$

$$q(0) = 0.9$$

in which $q(t)$ is the frequency of a gene after t generations of evolution. Use EULER or a similar program with $n = 100$ to approximate $q(200)$, the gene frequency after 200 generations.

30. **BIOMEDICAL: Epidemics** An epidemic on a campus of 5000 students begins with 5 cases. An immunization program begins immediately, with vaccine shots given to 100 students per day. The total number of cases $y(t)$ after t days satisfies

$$y' = 0.0001y(5000 - y - 100t)$$

$$y(0) = 5$$

a. Use EULER or a similar program with $n = 50$ to estimate $y(4)$, the total number of cases after 4 days.

b. Find the answer to part (a) if 300 students (instead of 100) are vaccinated each day.

Chapter Summary with Hints and Suggestions

Reading the text and doing the exercises in this chapter have helped you to master the following skills, which are listed by section (in case you need to review them) and are keyed to particular Review Exercises. Answers for all Review Exercises are given at the back of the book, and full solutions can be found in the Student Solutions Manual.

9.1 Separation of Variables

▪ Find the general solution of a differential equation by separation of variables. *(Review Exercises 1–10.)*

▪ Find a particular solution of a differential equation with an initial condition. *(Review Exercises 11–14.)*

▪ Solve an applied problem involving a differential equation. *(Review Exercises 15–18.)*

▪ Use a program to graph the slope field of a differential equation, and sketch the solution through a given point. *(Review Exercises 19–20.)*

9.2 Further Applications of Differential Equations: Three Models of Growth

■ Choose an appropriate differential equation for an applied problem, and use it to solve the problem. (*Review Exercises 21–28.*)

9.3 First-Order Linear Differential Equations

■ Find the general solution of a first-order linear differential equation. (*Review Exercises 29–34.*)

$$y' + p(x) \cdot y = q(x) \qquad I(x) = e^{\int p(x)\, dx}$$

■ Find a particular solution of a first-order linear differential equation. (*Review Exercises 35–38.*)

■ Solve an applied problem involving a first-order linear differential equation. (*Review Exercises 39–42.*)

9.4 Approximate Solutions of Differential Equations: Euler's Method

■ Find an Euler approximation for the solution of a differential equation. Then find the *exact* solution and compare the results. (*Review Exercises 43–44.*)

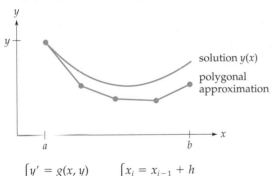

$$\begin{cases} y' = g(x, y) \\ y(a) = y_0 \end{cases} \qquad \begin{cases} x_i = x_{i-1} + h \\ y_i = y_{i-1} + g(x_{i-1}, y_{i-1}) \cdot h \end{cases}$$

$$\text{for } i = 1, 2, \ldots, n$$

■ Find an Euler approximation for the solution of a differential equation. (*Review Exercises 45–46.*)

■ Solve an applied problem by finding an Euler approximation for the solution of a differential equation. (*Review Exercises 47–48.*)

Hints and Suggestions

■ A differential equation is an equation involving derivatives (rates of change). A solution involving an arbitrary constant is called a *general solution,* while a solution with the arbitrary constant replaced by a number is called a *particular solution.* The constant is determined by an *initial condition,* specifying the value of the solution at a particular point.

■ Solving a differential equation by separation of variables involves moving the x's and y's to opposite sides of the equation and integrating both sides. Many useful differential equations can be solved by this technique, but many cannot. In fact, many differential equations cannot be solved by *any* method.

■ A graphing calculator with a slope field program can show a "picture" of a differential equation of the form $\dfrac{dy}{dx} = f(x, y)$, drawing little slanted dashes with the correct slopes at many points of the plane. A solution can then be drawn through a given point following the indicated slopes.

■ With practice, you can *guess* the integrating factor that makes a first-order linear differential equation integrable. If not, use the formula on page 712.

■ Euler's method is easiest when carried out on a graphing calculator or a computer.

■ **Practice for Test:** Review Exercises 1, 13, 15, 19, 23, 25, 29, 35, 41, 45, 47.

Review Exercises for Chapter 9

Practice test exercise numbers are in green.

9.1 Separation of Variables

Find the general solution of each differential equation.

1. $y^2 y' = x^2$ **2.** $y' = x^2 y$ **3.** $y' = \dfrac{x^3}{x^4 + 1}$

4. $y' = xe^{-x^2}$ **5.** $y' = y^2$ **6.** $y' = y^3$

7. $y' = 1 - y$ **8.** $y' = \dfrac{1}{y}$

9. $y' = xy - y$ **10.** $y' = x^2 + x^2 y$

Solve each differential equation and initial condition.

11. $y^2 y' = 3x^2$ **12.** $y' = \dfrac{y}{x^2}$

     $y(0) = 1$ $y(1) = 1$

13. $y' = \dfrac{y}{x^3}$ **14.** $y' = \sqrt[3]{y}$

     $y(1) = 1$ $y(1) = 0$

15. GENERAL: Accumulation of Wealth Suppose that you now have $10,000 and that you expect to save an additional $4000 during each year, and all of this is deposited in a bank paying 5% interest compounded continuously. Let $y(t)$ be your bank balance (in thousands of dollars) after t years.

   a. Write a differential equation and initial condition to model your bank balance.

   b. Solve your differential equation and initial condition.

   c. Use your solution to find your bank balance after 10 years.

16. ENVIRONMENTAL SCIENCE: Pollution A town discharges 4 tons of pollutant into a lake annually, and each year bacterial action removes 25% of the accumulated pollution.

   a. Write a differential equation and initial condition for the amount of pollution in the lake.

   b. Solve your differential equation to find a formula for the amount of pollution in the lake after t years.

17. GENERAL: Fever Thermometers How long should you keep a thermometer in your mouth to take your temperature? *Newton's Law of Cooling* says that the thermometer reading rises at a rate proportional to the difference between your actual temperature and the present reading. For a fever of 106 degrees Fahrenheit, the thermometer reading $y(t)$ after t minutes in your mouth satisfies $y' = 2.3(106 - y)$ with $y(0) = 70$ (initially at room temperature). (The constant 2.3 is typical for household thermometers.) Solve this differential equation and initial condition.

18. GENERAL: Fever Thermometers (*continuation*) Use your solution $y(t)$ in Exercise 17 to calculate $y(1)$, $y(2)$, and $y(3)$, the thermometer readings after 1, 2, and 3 minutes. Do you see why 3 minutes is the usually recommended time for keeping the thermometer in your mouth?

 The following exercises require the use of a slope field program.

For each differential equation and initial condition:

   a. Use SLOPEFLD or a similar program to graph the slope field for the differential equation on the window $[-5, 5]$ by $[-5, 5]$.

   b. Sketch the slope field on a piece of paper and draw a solution curve that follows the slopes and that passes through the point $(0, -2)$.

   c. Solve the differential equation and initial condition.

   d. Use SLOPEFLD or a similar program to graph the slope field and the solution that you found in part (c). How good was the sketch that you made in part (b) compared with the solution graphed in part (d)?

19. $\begin{cases} \dfrac{dx}{dy} = \dfrac{x^2}{y^2} \\ y(0) = -2 \end{cases}$ **20.** $\begin{cases} \dfrac{dy}{dx} = \dfrac{x}{y^2} \\ y(0) = -2 \end{cases}$

9.2 Further Applications of Differential Equations: Three Models of Growth

For each situation, write an appropriate differential equation (unlimited, limited, or logistic). Then find its solution and solve the problem.

21. GENERAL: Postage Stamps On average, the price of a first-class postage stamp grows continuously at the rate of about 7% per year. If in 2002 the price was 37 cents, estimate the price in the year 2010.

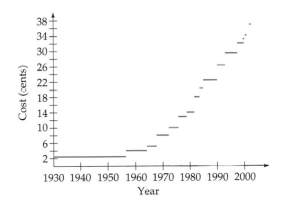

22. ECONOMICS: Computer Expenditure The amount spent by Americans on computers and related hardware is increasing continuously at the rate of 12% per year. If in 2000 the amount was $16.8 billion, estimate the amount in the year 2010.

23. BIOMEDICAL: Epidemics A virus spreads through a university community of 8000 people at a rate proportional to both the number already infected and the number not yet infected. If it begins with 10 cases and grows in a week to 150 cases, estimate the size of the epidemic after 2 weeks.

24. SOCIAL SCIENCE: Rumors A rumor spreads through a school of 500 students at a rate proportional to both the number who have heard and the number who have not heard the rumor. If the rumor began with 2 students and within a day had spread to 75, how many students will have heard the rumor within 2 days?

25. BUSINESS: Total Sales A manufacturer estimates that he can sell a maximum of 10,000 videocassette recorders in a city. His total sales grow at a rate proportional to how far they are below this upper limit. If after 7 months the total sales are 3000, find a formula for the total sales after t months. Then use your answer to estimate the total sales at the end of the first year.

26. GENERAL: Learning Suppose that the maximum rate at which a mail carrier can sort letters is 60 letters per minute, and that she learns at a rate proportional to her distance from this upper limit. If after 2 weeks on the route she can sort 25 letters per minute, how many weeks will it take her to sort 50 letters per minute?

27. BUSINESS: Advertising A new product is advertised extensively on television to a city of 500,000 people, and the number of people who have seen the ads increases at a rate proportional to the number who have not yet seen the ads. If within 2 weeks 200,000 have seen the ads, how long must the product be advertised to reach 400,000 people?

28. BUSINESS: Sales A company estimates the maximum market for fax (facsimile transmission) machines in a city to be 40,000. Sales are growing in proportion to both the number already sold and the size of the remaining market. If 1000 fax machines have been sold at time $t = 0$, and after 1 year 4000 have been sold, estimate how long it will take for 20,000 to be sold.

9.3 First-Order Linear Differential Equations

Solve each first-order linear differential equation.

29. $xy' - 5y = 4x^7$ **30.** $xy' + \frac{1}{2}y = 3x$

31. $y' + xy = x$ **32.** $y' - 2xy = -4x$

33. $xy' + y = xe^x$ **34.** $xy' + y - x \ln x$

[*Hint for Exercises 33 and 34:* Use integration by parts or an integral table.]

Solve each first-order linear differential equation and initial condition.

35. $y' - xy = 0$
$y(0) = 3$

36. $y' + 4xy = 0$
$y(0) = 2$

37. $xy' + 2y = 6x$
$y(1) = 0$

38. $xy' - y = 4$
$y(1) = 6$

39. **BUSINESS: Investment Fund** An investment fund grows in value by 5% each year, and also attracts new capital at the rate of $10e^{0.05t}$ thousand dollars per year, where t is the number of years since the fund was established. Therefore, its value $y(t)$ (in thousands of dollars) satisfies

$$y' = 0.05y + 10e^{0.05t}$$

$y(0) = 100$ \qquad (Initial value $100 million)

a. Solve this differential equation and initial condition.

b. Use your answer to part (a) to find the value of the fund after 5 years.

c. Graph the solution you found in part (a) on a graphing calculator, using an appropriate window. Find when the fund will reach a quarter of a million dollars. [*Hint:* Use INTERSECT with $y_2 = 250$.]

40. **GENERAL: Population Growth with Immigration** A country's population has a natural growth rate of 2%, and also accepts immigrants amounting to 1% of a neighboring population of $10e^{0.01t}$ million at time t years. Therefore, the population $p(t)$ (in millions) satisfies

$$p' = 0.02p + 0.1e^{0.01t}$$

$p(0) = 100$ \qquad (Initial population)

a. Solve this differential equation and initial condition.

b. Use your answer to part (a) to estimate the population at time $t = 10$ years.

c. Graph the solution you found in part (a) on a graphing calculator, using an approprite window. Find when the population will reach 150 million. [*Hint:* Use INTERSECT with $y_2 = 150$.]

41. **ENVIRONMENTAL SCIENCE: Air Quality** The air in a stuffy 10,000-cubic-foot lecture hall is only 10% oxygen. Each minute, 500 cubic feet of air are exchanged with the outside (containing 20% oxygen). Assume that the air in the room mixes thoroughly.

a. Write a differential equation and initial condition to describe the total amount $y(t)$ of oxygen in the room after t minutes.

b. Solve the differential equation and initial condition.

c. Use your solution to find the amount of oxygen after 15 minutes of ventilation.

d. Use your solution to find the long-run oxygen content of the room.

e. Graph the solution you found in part (b) on a graphing calculator, using an appropriate window. Find when the oxygen level will reach 19%. [*Hint:* Use INTERSECT with y_2 equal to the amount of oxygen for the 19% level.]

42. **ENVIRONMENTAL SCIENCE: Bird Population** The natural growth rate of a population of bald eagles in a nature sanctuary is 20% per year, and two more eagles are added to the population each year. Therefore, the population $p(t)$ t years from now satisfies

$$p' = 0.2p + 2$$

$p(0) = 5$ \qquad (Initial population)

a. Solve this differential equation and initial condition.

b. Use your solution to predict the eagle population after 2 years.

c. Graph the solution you found in part (b) on a graphing calculator, using an appropriate window. Find when the eagle population will reach 24.

9.4 Approximate Solutions of Differential Equations: Euler's Method

43–44. For each initial value problem:

a. Calculate the Euler approximation on the interval $[0, 1]$ using $n = 4$ segments. Carry out the calculations "by hand" with the aid

REVIEW EXERCISES FOR CHAPTER 9 **737**

of a calculator, rounding calculations to two decimal places. Draw the graph of your approximation.

b. Solve the differential equation and initial condition *exactly* by separating variables or using an integrating factor.

c. Evaluate at $x = 1$ the solution that you found in part (b). Compare this actual value of $y(1)$ (rounded to two decimal places) with the *estimate* for $y(1)$ that you found in part (a). [*Note:* Expect low accuracy using $n = 4$.]

43. $y' = 2 - 2y$

$y(0) = 2$

44. $y' = \dfrac{4x}{y}$

$y(0) = 1$

45–46. For each initial value problem, use EULER or a similar program (see page 730) to find an estimate for the solution at the stated *x*-value, using 50 segments.

45. $y' = \dfrac{1}{y^2 + 1}$, $y(0) = 1$

Approximate the solution at $x = 2$

46. $y' = ye^{-x}$, $y(1) = 5$

Approximate the solution at $x = 2.5$

47. BUSINESS: Advertising If a new product becomes known *both* by word of mouth *and* by advertising, then $p(t)$, the proportion of people who have heard of the new product after t weeks, satisfies a differential equation and initial condition of the form

$$p' = 0.1(1 - p) + 0.1p(1 - p)$$

$p(0) = 0.3$ \quad (Initially 30% know the product)

a. Show that the differential equation simplifies to $p' = 0.1(1 - p^2)$

b. Use EULER or a similar program (see page 730) with 90 segments to estimate $p(4)$, the proportion of people who have heard of the product after 4 weeks.

48. ENVIRONMENTAL SCIENCE: Insect Population A colony of flying insects occupying a spherical nest grows so that the number $y(t)$ of insects after t weeks satisfies

$$y' = 0.2y - 0.1\sqrt[3]{y}$$

$y(0) = 100$ \quad (Initially 100 insects)

Use EULER or a similar program (see page 730) with 80 segments to estimate the population after 2 weeks.

10 Sequences and Series

Application Preview

Saving Pennies

Suppose that you were to save one penny on January 1, two pennies on January 2, four pennies on January 3, and so on, doubling the number each day. How much would you save by the end of January?

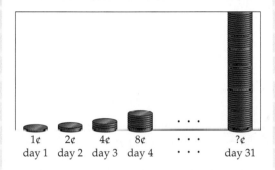

The surprising answer is more than 21 million dollars. This result (see Exercise 59), and many others, will follow from a formula for the sum of a "geometric series" developed in this section.

10.1   GEOMETRIC SERIES

Introduction

The decimal expansion of $\frac{1}{3}$ is the "infinitely long decimal" 0.333 ... , which represents an "infinitely long" sum:

$$0.3333... = \frac{3}{10} + \frac{3}{100} + \frac{3}{1000} + \frac{3}{10000} + \cdots$$

Such sums are called *infinite series* and are the subject of this chapter. This first section discusses *geometric* series, with applications from the multiplier effect in economics to the cumulative effect of long-term drug dosage. Later sections discuss Taylor series and Newton's method for approximating the solution to an equation.

Finite Geometric Series

A *finite series* is a sum of numbers in a particular order. The numbers are called the *terms* of the series. A finite *geometric* series is a series in which each successive term is obtained from the previous term by multiplying by a fixed number, denoted *r* (for *ratio*).

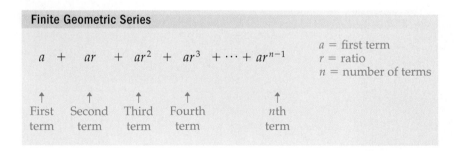

Observe that the *number* of terms is always one more than the final exponent of *r* (so, the *n*th term has exponent $n - 1$). Some examples of finite geometric series are given below.

$$2 + 4 + 8 + 16 + 32$$
First term is $a = 2$
Ratio is $r = 2$
Number of terms is $n = 5$

$$3 - 30 + 300 - 3000$$
$a = 3, \quad r = -10, \quad n = 4$

$$5 + \frac{5}{2} + \frac{5}{2^2} + \frac{5}{2^3} + \frac{5}{2^4} + \cdots + \frac{5}{2^{100}}$$
$a = 5, \quad r = \frac{1}{2}, \quad n = 101$
(*n* is one more than the final power of *r*)

A series is a sum, and so it may be evaluated by adding the terms. For the first two examples above, the sums are easily found to be 62 and

−2727 (as you may check). For the last example, adding up the 101 terms would be extermely tedious. Shortly we will develop a *formula* for summing such a series.

Not all series are geometric. The following series is *not* geometric, since successive terms are found by adding 1 to the previous denominator, *not* by multiplying by a fixed ratio.

$$\frac{1}{1} + \frac{1}{2} + \frac{1}{3} + \frac{1}{4} + \frac{1}{5} + \frac{1}{6}$$

Practice Problem 1

For each series below, if it is geometric, find the first term a, the ratio r, and the number of terms n. If the series is not geometric, say so.

a. $1 - \pi + \pi^2 - \pi^3 + \pi^4 - \pi^5$

b. $2 + \dfrac{2}{10} + \dfrac{2}{10^2} + \dfrac{2}{10^3} + \cdots + \dfrac{2}{10^8}$

c. $0.01 + 0.02 + 0.03 + 0.04 + 0.05$ ➤ Solutions on page 750

To find a formula for the *sum* of a finite geometric series, we denote the sum by S_n:

$$S_n = a + ar + ar^2 + \cdots + ar^{n-1}$$

From this we subtract r times S_n:

S_n	$= a + ar + ar^2 + \cdots + ar^{n-1}$		S_n (sum of first n terms)
rS_n	$= \quad\quad ar + ar^2 + \cdots + ar^{n-1} + ar^n$		S_n multiplied by r
$S_n - rS_n = a$		$- ar^n$	Subtracting (most terms cancel)
$S_n(1 - r) = a(1 - r^n)$			Factoring each side of $S_n - rS_n = a - ar^n$
$S_n = a\dfrac{1 - r^n}{1 - r}$			Dividing each side by $1 - r$ (for $r \neq 1$)

This formula is very useful for summing a finite geometric series.

Sum of a Finite Geometric Series

$$a + ar + ar^2 + \cdots + ar^{n-1} = a\frac{1 - r^n}{1 - r} \qquad (r \neq 1)$$

Memorization hint: The power n on the right is the first power *left out* of the series on the left.

EXAMPLE **1**

SUMMING CUMULATIVE FINES

In the summer of 1988, the city of Yonkers, New York, was fined for refusing to build low-income housing. The fine was $100 for the first day, $200 for the second day, $400 for the third day, and so on, with the fines doubling each day until the city council approved the housing. Find the total fine for the 14 days until the housing was approved.

Solution

$$\binom{\text{Total}}{\text{fine}} = \underbrace{100}_{\text{1st day}} + \underbrace{2 \cdot 100}_{\text{2nd day}} + \underbrace{4 \cdot 100}_{\text{3rd day}} + \cdots + \underbrace{2^{13} \cdot 100}_{\text{14th day}}$$

Geometric series (final power is 1 less than the final day)

The sum formula gives

$$\binom{\text{Total}}{\text{fine}} = 100 \frac{1 - 2^{14}}{1 - 2} = 100 \underbrace{\frac{1 - 16{,}384}{-1}}_{16{,}383} = 1{,}638{,}300$$

$a \dfrac{1 - r^n}{1 - r}$ with $a = 100$, $r = 2$, and $n = 14$

The total fine was $1,638,300.

Practice Problem **2** Sum the finite geometric series $3 + 3 \cdot 2 + 3 \cdot 2^2 + \cdots + 3 \cdot 2^5$.

➤ Solutions on page 750

Annuities

An *annuity* is a series of equal payments at regular intervals. The total value or *amount* of an annuity, including interest, may be found from the formula for the sum of a geometric series.

EXAMPLE **2**

FINDING THE VALUE OF AN ANNUITY

At the end of each month for 5 years you deposit $200 in a bank at 6% interest compounded monthly. Find the amount of this annuity.

Solution

At 6% interest compounded monthly, you receive $\frac{1}{2}$% per month, so payments are multiplied by $(1 + 0.005)$ or (1.005) for each month on deposit. In 5 years there will be $5 \cdot 12 = 60$ monthly deposits, beginning with the first \$200 (at the end of the first month, and so earning interest for 59 months) and ending with the final \$200 (at the end of the last month, and so earning *no* interest). We sum these amounts *in reverse order:*

$$\underbrace{200}_{\substack{\text{Final} \\ \text{payment}}} + 200(1.005) + 200(1.005)^2 + \cdots + \underbrace{200(1.005)^{59}}_{\substack{\text{First} \\ \text{payment}}}$$

Geometric series with $a = 200$, $r = 1.005$, and $n = 60$ (final power plus 1)

The formula gives

$$\begin{pmatrix} \text{Amount} \\ \text{of annuity} \end{pmatrix} = 200 \, \frac{1 - 1.005^{60}}{1 - 1.005} \approx \underbrace{13,954}_{\substack{\text{Using a} \\ \text{calculator}}}$$

$a \dfrac{1 - r^n}{1 - r}$ with the stated values of a, r, and n

The amount of the annuity is \$13,954.

Note that five years of monthly payments of \$200 adds up to only $60 \cdot 200 = \$12,000$. The remainder here is interest.

The amount of a *continuous* annuity can be found using differential equations (see page 691).

Infinite Series

An infinite series is an expression written as a sum of infinitely many terms.*

$$a_1 + a_2 + a_3 + a_4 + \cdots$$

The following are examples of infinite series:

$$1 + 2 + 3 + 4 + 5 + \cdots$$

$$1 + \frac{1}{2} + \frac{1}{4} + \frac{1}{8} + \cdots + \frac{1}{2^n} + \cdots \qquad \frac{1}{2^n} \text{ represents a "typical" term}$$

We cannot sum an infinite series simply by adding the terms, because adding infinitely many numbers would be impossible. Instead, we sum an infinite series by first finding the *partial* sum S_n of the first n

---

* We will sometimes begin with a term a_0, writing the series as $a_0 + a_1 + a_2 + \cdots$. In Section 10.3 we will consider infinite series whose terms are *functions*.

terms, and then taking the *limit* as n approaches infinity. The partial sums of the infinite series $a_1 + a_2 + a_3 + \cdots$ are defined as follows:

$S_1 = a_1$ First partial sum is the first term

$S_2 = a_1 + a_2$ Second partial sum is the sum of the first *two* terms

$S_3 = a_1 + a_2 + a_3$ Third partial sum is the sum of the first *three* terms

$\vdots$

$S_n = a_1 + a_2 + a_3 + \cdots + a_n$ nth partial sum is the sum of the first n terms

$\vdots$

If the *limit* of this sequence $S_1, S_2, S_3, \ldots, S_n, \ldots$ as n approaches infinity *exists* and equals a number S, then the series is said to *converge* to S, and S is the *sum of the series*.

Convergence of an Infinite Series

For an infinite series $a_1 + a_2 + a_3 + \cdots$, if the partial sums S_n approach a limit S,

$$\lim_{n \to \infty} S_n = S$$

then the series is *convergent*, and *converges* to S:

$$S = a_1 + a_2 + a_3 + \cdots$$

An infinite series that does *not* converge is said to *diverge*, or be a *divergent* series, and therefore has no sum. The procedure of first summing a *finite* number n of terms and then letting n approach infinity is another example of dealing with the infinite by *dropping back to the finite and taking the limit* (see page 469).

Infinite Geometric Series

An infinite series in which each successive term is a fixed multiple of the previous term is called an *infinite geometric series*.

Infinite Geometric Series

$$a + ar + ar^2 + ar^3 + \cdots$$

a = first term
r = ratio

The partial sums of an infinite geometric series are *finite* geometric series, for which we have a formula.

$$S_n = a\frac{1 - r^n}{1 - r}$$

Partial sum of an infinite geometric series (from page 741)

The sum of an *infinite* geometric series is then the *limit* of S_n as n approaches infinity. If the ratio r satisfies $|r| < 1$, then r^n will approach zero as $n \to \infty$.

Graphing Calculator Exploration

Verify this last statement by choosing an r satisfying $|r| < 1$ and raising it to a high power. Try a few different values of r and n.

```
(.5)^25
  2.980232239E-8
(-.9)^200
  7.055079116-10
```
←—means 0.0000000298 . . .
←—means 0.0000000007055 . . .

The sum S of the infinite series is then found by taking the limit of the partial sums.

Approaches zero for $|r| < 1$

$$S = \lim_{n \to \infty} S_n = \lim_{n \to \infty}\left(a\frac{1 - r^n}{1 - r}\right) = a\frac{1 - 0}{1 - r} = a\frac{1}{1 - r} = \frac{a}{1 - r}$$

$$\underbrace{\phantom{a\frac{1 - r^n}{1 - r}}}_{S_n}$$

The last expression gives a formula for the sum of a convergent infinite geometric series.

Sum of an Infinite Geometric Series

An infinite geometric series with $|r| < 1$ is convergent, and its sum is given by

$$a + ar + ar^2 + ar^3 + \cdots = \frac{a}{1 - r} \qquad \text{for} \quad |r| < 1$$

An infinite geometric series with $|r| \geq 1$ is divergent.

The divergence of infinite geometric series with $|r| \geq 1$ will be shown in Exercise 50.

EXAMPLE 3

SUMMING CONVERGENT SERIES

a. $\dfrac{1}{2} + \dfrac{1}{4} + \dfrac{1}{8} + \dfrac{1}{16} + \cdots = \dfrac{\dfrac{1}{2}}{1 - \dfrac{1}{2}} = \dfrac{\dfrac{1}{2}}{\dfrac{1}{2}} = 1$ $\qquad$ $\dfrac{a}{1-r}$ with first term $a = \frac{1}{2}$ and ratio $r = \frac{1}{2}$

b. $8 - \dfrac{8}{3} + \dfrac{8}{9} - \dfrac{8}{27} + \cdots = \dfrac{8}{1 - \left(-\dfrac{1}{3}\right)} = \dfrac{8}{\dfrac{4}{3}}$ $\qquad$ $\dfrac{a}{1-r}$ with $a = 8$ and $r = -\frac{1}{3}$

$= 8 \cdot \dfrac{3}{4} = 6$

c. $1 + \dfrac{3}{2} + \dfrac{9}{4} + \dfrac{27}{8} + \cdots$ $\quad$ *diverges.* $\qquad$ $r = \dfrac{3}{2}$ so $|r| > 1$

Practice Problem 3 $\quad$ Find the sum of the series $\quad 2 + 2\left(\dfrac{3}{5}\right) + 2\left(\dfrac{3}{5}\right)^2 + 2\left(\dfrac{3}{5}\right)^3 + \cdots$

➤ Solution on page 750

For medications taken over an extended period of time, the amount in the bloodstream at any time is the sum of small contributions from all past doses. This amount can be approximated by an infinite series.

EXAMPLE 4

FINDING LONG-TERM DRUG DOSAGE

A patient takes a daily dose of 150 milligrams (mg) of a drug, and each day 70% of the accumulated drug in the bloodstream is absorbed by the body. Find the long-term maximum and minimum amounts of the drug in the bloodstream.

Solution

One day after taking a dose, 70% of that dose will have been absorbed, so 30% of it will remain. After a second day, only 30% of that 30% will remain, with proportional reductions on subsequent days. The maximum amount in the bloodstream will occur just after a new dose, and it will be the most recent 150 mg, plus 30% of the previous 150 mg, plus $(0.30)^2$ of the dose previous to that, and so on, giving the infinite series

$$\begin{pmatrix} \text{Maximum} \\ \text{amount} \end{pmatrix} = 150 + 150(0.3) + 150(0.3)^2 + \cdots$$

$$= \frac{150}{1 - 0.3} = \frac{150}{0.7} \approx 214 \qquad \text{Using } \frac{a}{1 - r} \text{ with } a = 150 \text{ and } r = 0.3.$$

Therefore, the long-term amount in the bloodstream just after a dose will be 214 mg. Just before the next dose, 70% of this amount will have been absorbed, leaving 30%, or $0.30 \cdot 214 \approx 64$ mg.

The long-term amount in the bloodstream will vary between 214 mg and 64 mg.

These maximum and minimum amounts are important in calculating dosages that avoid toxicity and maintain efficacy. Notice that we used an *infinite* series even though the number of past doses is certainly finite. The *approximation of the finite by the infinite* is permissible because the contribution from earlier doses rapidly becomes negligible. It also has the advantage that the infinite formula is *simpler* than the finite formula.

Graphing Calculator Exploration

```
sum(seq(150*.3^X
,X,0,4,1))
          213.765
```

(Requires series operations)

Show that the sum of the first five terms of the series in the preceding example is already very close to the sum of the *infinite* series. [*Hint:* On some calculators, you sum a sequence of terms of the form $150 \cdot (0.3)^x$ with x going from 0 to 4 by increments of 1, as shown on the left.]

The long-term amount for medication given continuously (intravenously) may be found using differential equations (see page 691).

Multiplier Effect in Economics

Suppose that you pay someone $1000. The recipient may spend a part of it (say 80%, or $800) and save the rest. The $800 spent goes to other recipients, who in turn spend a part (say 80%) for purchases of their own and save the rest. If this continues, with each recipient spending 80% of what is received, the result will be an infinite series of decreasing expenditures throughout the economy. We will see that the sum of all of these expenditures is a *multiple* of the original expenditure (hence the term "multiplier effect"). The percentage spent (here 80%) is called the "marginal propensity to consume" or *MPC*.

EXAMPLE **5**

FINDING THE MULTIPLIER

If the marginal propensity to consume is 80%, find the total of all of the expenditures generated by an initial expenditure of $1000, and find the multiplier.

Solution

The sum of all of the expenditures, beginning with the original $1000, is

$$1000 + 1000(0.8) + 1000(0.8)^2 + \cdots = \frac{1000}{1 - 0.8}$$

$\dfrac{a}{1 - r}$ with

$a = 1000$ and

$r = 0.8$

$$= \frac{1000}{0.2} = 5000$$

The original expenditure of $1000 generated a total of $5000 of expenditures in the economy, so the multiplier is 5.

Notice that the $1000 can be replaced by *any* initial expenditure, showing that the multiplier is independent of the initial amount, depending only on the *MPC*. The multiplier is used by economists to assess the effects of tax increases and cuts.

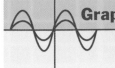

Graphing Calculator Exploration

(Requires series operations)
How many terms in the series in the preceding example are required for the sum to reach 4500? To reach 4900?

Summation Notation

A series may be written very compactly using the symbol Σ (the Greek capital letter "sigma" corresponding to our S). Σ is followed by a formula for the terms, using an "index variable" that takes the values between those indicated at the bottom and top of the Σ (similar to the notation for definite integrals). An infinity symbol (∞) at the top indicates an *infinite* series.

Terms written out　　　**Sigma notation**

$$2 + 4 + 8 + \cdots + 2^{10} = \sum_{i=1}^{10} 2^i$$

$\leftarrow$ Last value of index i is 10
$\leftarrow$ Formula for terms
$\leftarrow$ First value of index i is 1

$$\frac{1}{3} + \frac{1}{4} + \frac{1}{5} + \frac{1}{6} + \cdots = \sum_{k=3}^{\infty} \frac{1}{k}$$

$\leftarrow$ ∞ means an *infinite series*
$\leftarrow$ Formula for terms
$\leftarrow$ First value of index k is 3

Read: the sum of $1/k$ from k equals 3 to infinity

10.1 Section Summary

A series is *geometric* if multiplying any term by a fixed number gives the next term. The sum of a *finite* geometric series is given by the following formula:

First power omitted

$$\sum_{k=0}^{n-1} ar^k = a + ar + ar^2 + \cdots + ar^{n-1} = a\,\frac{1 - r^n}{1 - r} \qquad \text{for } r \neq 1$$

In sigma notation

Terms written out
(n = number of terms)

An *infinite* geometric series converges if the ratio r satisfies $|r| < 1$, and diverges if $|r| \geq 1$. The sum of a convergent infinite geometric series is given by the following formula:

$$\sum_{k=0}^{\infty} ar^k = a + ar + ar^2 + ar^3 + \cdots = \frac{a}{1 - r} \qquad \text{for } |r| < 1$$

In sigma notation

Terms written out

Notice that the "infinite" formula is simpler than the "finite" formula. These formulas for the sums of *geometric* series are particularly important since there are no such formulas for most (nongeometric) series.

➤ **Solutions to Practice Problems**

1. a. $a = 1$, $r = -\pi$, $n = 6$

b. $a = 2$, $r = \frac{1}{10}$, $n = 9$

c. Not geometric

2. $3 + 3 \cdot 2 + 3 \cdot 2^2 + \cdots + 3 \cdot 2^5 = 3\dfrac{1 - 2^6}{1 - 2}$

$$= 3\dfrac{1 - 64}{-1} = 3 \cdot 63 = 189$$

3. $2 + 2\left(\dfrac{3}{5}\right) + 2\left(\dfrac{3}{5}\right)^2 + 2\left(\dfrac{3}{5}\right)^3 + \cdots$

$$= \dfrac{2}{1 - \dfrac{3}{5}} = \dfrac{2}{\dfrac{2}{5}} = 2 \cdot \dfrac{5}{2} = 5$$

10.1 Exercises

Write out each finite series.

1. $\displaystyle\sum_{i=1}^{5} \dfrac{1}{i + 1}$

2. $\displaystyle\sum_{i=1}^{6} \dfrac{i}{i + 1}$

3. $\displaystyle\sum_{k=1}^{4} \left(-\dfrac{1}{3}\right)^k$

4. $\displaystyle\sum_{k=1}^{5} \dfrac{(-1)^k}{k^2}$

5. $\displaystyle\sum_{n=1}^{6} \dfrac{1 - (-1)^n}{2}$

6. $\displaystyle\sum_{n=1}^{4} (-n)^n$

Write each infinite series in sigma notation, beginning with $\sum_{i=1}^{\infty}$

7. $\dfrac{1}{3} + \dfrac{1}{9} + \dfrac{1}{27} + \dfrac{1}{81} + \cdots$

8. $\dfrac{1}{7} + \dfrac{2}{7} + \dfrac{3}{7} + \dfrac{4}{7} + \dfrac{5}{7} + \cdots$

9. $-2 + 4 - 8 + 16 - \cdots$

10. $-3 + 9 - 27 + 81 - \cdots$

11. $-1 + 2 - 3 + 4 - \cdots$

12. $-1 + 4 - 9 + 16 - \cdots$

Find the sum of each finite geometric series.

13. $1 + 2 + 2^2 + 2^3 + \cdots + 2^9$

14. $1 + 2 + 2^2 + 2^3 + \cdots + 2^{10}$

15. $3 + 3 \cdot 4 + 3 \cdot 4^2 + \cdots + 3 \cdot 4^5$

16. $2 + 2 \cdot 3 + 2 \cdot 3^2 + \cdots + 2 \cdot 3^9$

17. $3 - 3 \cdot 2 + 3 \cdot 2^2 - 3 \cdot 2^3 + \cdots + 3 \cdot 2^6$

18. $4 - 4 \cdot 3 + 4 \cdot 3^2 - 4 \cdot 3^3 + \cdots + 4 \cdot 3^6$

Determine whether each infinite geometric series converges or diverges. If it converges, find its sum.

19. $4 + \dfrac{4}{5} + \dfrac{4}{25} + \dfrac{4}{125} + \cdots$

20. $5 + \dfrac{5}{6} + \dfrac{5}{36} + \dfrac{5}{216} + \cdots$

21. $2 - \dfrac{2}{3} + \dfrac{2}{9} - \dfrac{2}{27} + \dfrac{2}{81} - \cdots$

22. $27 - \dfrac{27}{2} + \dfrac{27}{4} - \dfrac{27}{8} + \dfrac{27}{16} - \cdots$

23. $\dfrac{1}{3} + \dfrac{2}{3} + \dfrac{4}{3} + \dfrac{8}{3} + \dfrac{16}{3} + \cdots$

24. $\dfrac{1}{100} - \dfrac{3}{100} + \dfrac{9}{100} - \dfrac{27}{100} + \dfrac{81}{100} - \cdots$

25. $\dfrac{3}{2} + \dfrac{3^2}{2^3} + \dfrac{3^3}{2^5} + \dfrac{3^4}{2^7} + \dfrac{3^5}{2^9} + \cdots$

26. $\dfrac{1}{3} + \dfrac{1}{3^3} + \dfrac{1}{3^5} + \dfrac{1}{3^7} + \dfrac{1}{3^9} + \cdots$

27. $8 + 6 + \dfrac{9}{2} + \dfrac{27}{8} + \cdots$

28. $9 + 6 + 4 + \dfrac{8}{3} + \cdots$

29. $\displaystyle\sum_{i=1}^{\infty} \dfrac{100}{5^i}$

30. $\displaystyle\sum_{i=1}^{\infty} 30 \cdot 4^{-i}$

31. $\displaystyle\sum_{j=1}^{\infty} \frac{3^j}{7 \cdot 2^j}$

32. $\displaystyle\sum_{j=1}^{\infty} \frac{5 \cdot 2^j}{3^j}$

33. $\displaystyle\sum_{j=0}^{\infty} \left(-\frac{1}{2}\right)^j$

34. $\displaystyle\sum_{n=0}^{\infty} 3^{-n}$

35. $\displaystyle\sum_{k=1}^{\infty} (1.01)^k$

36. $\displaystyle\sum_{n=1}^{\infty} \pi^n$

37. $\displaystyle\sum_{k=0}^{\infty} (0.99)^k$

38. $\displaystyle\sum_{n=1}^{\infty} \pi^{-n}$

39. $\displaystyle\sum_{j=0}^{\infty} \frac{3^j}{4^{j+1}}$

40. $\displaystyle\sum_{i=0}^{\infty} \frac{(-1)^i}{1000}$

Find the value of each repeating decimal. [*Hint:* Write each as an infinite series. For example,

$$0.252525\ldots = \frac{25}{100} + \frac{25}{100^2} + \frac{25}{100^3} + \cdots$$

The bar indicates the repeating part.]

41. $0.363636\ldots = 0.\overline{36}$

42. $0.636363\ldots = 0.\overline{63}$

43. $2.\overline{54} \ (=2 + 0.\overline{54})$

44. $1.\overline{24} \ (=1 + 0.\overline{24})$

45. $0.\overline{027}$

46. $0.\overline{9}$

47. $0.\overline{0123456789}$

48. $0.\overline{987654320}$

(For Exercises 47 and 48, after finding the fraction, try dividing the numerator and denominator by 123456789.)

49. APPROXIMATING THE FINITE BY THE INFINITE For the patient in Example 4 (pages 746–747), the amount of the drug in the bloodstream just after the nth dose is

$$150 + 150(0.3) + 150(0.3)^2 + \cdots + 150(0.3)^{n-1}$$

Use a graphing calculator to sum this finite geometric series for the following values of n: 1, 2, 3, 4, 5, 6, and 7.

a. For which dose does the cumulative amount first rise above 200 milligrams?

b. For which dose does the cumulative amount first rise above 213 milligrams?

50. DIVERGENCE OF GEOMETRIC SERIES Show that the geometric series

$$a + ar + ar^2 + ar^3 + \cdots$$

diverges if $|r| \geq 1$, as follows (for $a \neq 0$):

a. If $r = 1$, the series becomes $a + a + a + \cdots$. Show that this series diverges by showing that the sum of the first n terms is $S_n = n \cdot a$, which does not approach a limit as $n \to \infty$ (since $a \neq 0$).

b. If $r = -1$, then the series becomes $a - a + a - \cdots$. Show that this series diverges by showing that the partial sums S_1, S_2, S_3, $S_4, \ldots$ take values a, 0, a, 0, $\ldots$, and so do not approach any limit.

c. If $|r| > 1$, use the fact that r^n does not approach a limit as $n \to \infty$ to show that the partial sum formula $a\dfrac{1 - r^n}{1 - r}$ does not approach a limit as $n \to \infty$.

51. Show that applying the formula $\dfrac{a}{1-r}$ to the geometric series $2 + 4 + 8 + 16 + \cdots$ gives a negative answer. Explain this seeming contradiction.

52. Derive the formula for the sum S of a geometric series as follows:

$$S = a + ar + ar^2 + ar^3 + \cdots$$
$$= a + r\underbrace{(a + ar + ar^2 + \cdots)}_{S} = a + rS$$

Now solve the equation $S = a + rS$ for S. (Note that this derivation *assumes* that the series converges.)

APPLIED EXERCISES

Finite Geometric Series

53. a. POLITICAL SCIENCE: Yonkers Fine Suppose that the fine in Example 1 (beginning at \$100 and doubling each day; see page 742) had continued for a total of 18 days. Find the total fine. *(continues)*

b. (Graphing calculator with series operations helpful) How many days would it have taken for the fine to reach a billion dollars?

54. GENERAL: Annuity A star baseball player expects his career to last 10 years, and for his retirement deposits \$4000 at the end of each *(continues)* month in a bank account earning 12% interest compounded monthly. Find the amount of this annuity at the end of 10 years.

55. GENERAL: Annuity For your retirement "nest egg," you deposit $300 at the end of each month into a bank account paying 6% interest compounded monthly. Find the amount of this annuity at the end of your 40-year working career.

56. BUSINESS: Present Value of an Annuity A retirement plan pays $1000 at retirement and every month thereafter for a total of 12 years. Find the sum of the present values of these payments (at an interest rate of 6% compounded monthly) by summing the series

$$1000 + \frac{1000}{1.005} + \frac{1000}{(1.005)^2} + \cdots + \frac{1000}{(1.005)^{143}}$$

57. GENERAL: Million Dollar Lottery Most state lottery jackpots are paid out over time (often 20 years), so the "real" cost to the state is the sum of the present values of the payments. Find the cost of a "million dollar" lottery by summing the present values of 240 monthly payments of $4167, beginning now, if the interest rate is 6% compounded monthly. [*Hint:* The present value of $4167 in k months is $4167/(1 + 0.005)^k$.]

58. BUSINESS: Sinking Fund A sinking fund is an annuity designed to reach a given value at a given time in the future (often to pay off a debt or to buy new equipment). A company's $100,000 printing press is expected to last 8 years. If equal payments are to be made at the end of each quarter into an account paying 8% compounded quarterly, find the size of the quarterly payments needed to yield $100,000 at the end of 8 years. [*Hint:* Solve $x + x(1.02) +$
$x(1.02)^2 + \cdots + x(1.02)^{31} = 100,000$.]

59. a. GENERAL: SAVING PENNIES You make a New Year's resolution to save pennies, promising to save on successive days 1¢, 2¢, 4¢, 8¢, and so on, doubling the number each day throughout January. How much will you have saved by the end of January? *(continues)*

b. (Graphing calculator with series operations helpful) How many days will it take for the savings to reach a million dollars?

60. GENERAL: Chessboards It is said that the game of chess was invented by the Grand Vizier Sissa Ben Dahir for the Indian King Shirham. When the king offered him any reward, Sissa asked for one grain of wheat to be placed on the first square of the chessboard, two on the second, four on the third, and so on, doubling each time for each of the 64 squares. Estimate the total number of grains of wheat required. (It is said that the king, upon learning that the total was many times the world production of wheat, executed Sissa for making a fool out of him.)

61. GENERAL: Cycling (Graphing calculator with series operations helpful) You plan to cycle across the United States, cycling 10 miles the first day, 10% further the second day (so 11 miles), increasing each day's distance by 10%. On which day will you reach the opposite coast 3000 miles away?

62. GENERAL: Payments (Graphing calculator with series operations helpful) A person gives you $1, waits 1 minute, gives you another dollar, waits 2 minutes, gives you another dollar, waits four minutes, etc., doubling the wait each time. How much money will you have at the end of the first year?

Infinite Geometric Series

63. a. BIOMEDICAL: Drug Dosage A patient has been taking a daily dose of 12 units of insulin for an extended period of time. If 80% of the total amount in the bloodstream is absorbed by the body during each day, find the long-term maximum and minimum amounts of insulin in the bloodstream.

b. (Graphing calculator with series operations helpful) How many daily doses of insulin does it take for the maximum amount in the blood to reach 14.9 units?

64. BIOMEDICAL: Drug Dosage A patient has been taking a daily dose of D milligrams of a drug for an extended period of time, and a proportion p $(0 < p < 1)$ of the total amount in the bloodstream is absorbed by the body during each day. Find the long-term maximum and minimum amounts in the bloodstream.

65. **ECONOMICS: Multiplier Effect** Find the total economic effect of a $3 billion income tax cut if the marginal propensity to consume (*MPC*) is 75%. Also find the multiplier.

66. **ECONOMICS: Multiplier Effect** If the marginal propensity to *consume* (*MPC*) is 70%, then the marginal propensity to *save* (*MPS*) is 30%. In general, $MPS = 1 - MPC$. Show that for a given *MPC*, the multiplier is $\frac{1}{MPS}$.

67–68. BUSINESS: Perpetuities A *perpetuity* is an annuity that continues forever. The *capital value* of a perpetuity is the present value of all (future) payments. (Perpetuities are also called permanent endowments, and the continuous case was treated on pages 465–466 using improper integrals.)

67. The capital value of a perpetuity that pays $300 now and every quarter from now on at an interest rate of 6% compounded quarterly is

$$300 + \frac{300}{1 + 0.015} + \frac{300}{(1 + 0.015)^2} + \cdots$$

Find this value.

68. The capital value of a perpetuity that pays $1000 annually, beginning now, at an interest rate of 8% compounded annually is

$$1000 + \frac{1000}{1 + 0.08} + \frac{1000}{(1 + 0.08)^2} + \frac{1000}{(1 + 0.08)^3} + \cdots$$

Find this value.

69. a. **GENERAL: Bouncing Ball** A ball dropped from a height of 6 feet bounces to two-thirds of its former height with each bounce. Find the total vertical distance that the ball travels. (*continues*)

 b. (Graphing calculator with series operations helpful) At which bounce will the total vertical distance traveled by the ball first exceed 29 feet?

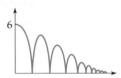

70. **GENERAL: Coupons** A 45¢ candy bar comes with a coupon worth 10% toward the next candy bar (which includes a coupon for 10% of the next, and so on). Therefore, the "real" value of a single 45¢ candy bar is

$$45 + 45(0.10) + 45(0.10)^2 + 45(0.10)^3 + \cdots$$

Find this value.

71–72. BIOMEDICAL: Long-Term Population A population (of cells or people) is such that each year a number a of individuals are added (call them "immigrants"), and a proportion p of the individuals who have been there die. Therefore, the proportion that survives is $(1 - p)$, so that just after an immigration the population will consist of a new immigrants plus $a(1 - p)$ from the previous year's immigration plus $a(1 - p)^2$ from the immigration before that, and so on. In the long run, the size of the population just after an immigration will then be the sum $a + a(1 - p) + a(1 - p)^2 + a(1 - p)^3 + \cdots$.

71. If the number of immigrants each year is 800 and the survival proportion is 0.95, find the long-run size of the population just after an immigration. Then find the long-run size just *before* an immigration.

72. Find the long-run size of the population just after an immigration for *any* number of immigrants a and *any* survival proportion $(1 - p)$. Then find the long-run size just *before* an immigration.

10.2 TAYLOR POLYNOMIALS

Introduction

Often in mathematics we approximate complicated objects by simpler objects, such as approximating a curve by its tangent line. In this section we will approximate functions such as e^x and $\sin x$ by *polynomials,* called *Taylor polynomials,* named after the English mathematician Brook Taylor (1685–1731), a student of Isaac Newton.

Factorial Notation

We begin by defining *factorial notation.* For any positive integer n, $n!$ (read: "n factorial") is defined as the product of all integers from n down to 1.

Factorial Notation	
$n! = n \cdot (n-1) \cdot \,\cdots\, \cdot 2 \cdot 1$	$n!$ is the product of all integers from n down to 1
$0! = 1$	Zero factorial is defined as 1

For example,

$$4! = 4 \cdot 3 \cdot 2 \cdot 1 = 24$$

A justification for defining $0!$ as 1 is given in Exercise 32.

First Taylor Polynomial at $x = 0$

We begin by finding the polynomial of degree 1 that best approximates a given function $f(x)$ near $x = 0$. A polynomial of degree 1 is a *linear* function $y = mx + b$, and it is clear from the following diagram that the best linear approximation to $f(x)$ at $x = 0$ will be the *tangent line* at $x = 0$. The tangent line has the same slope and y-intercept as the curve at $x = 0$, so that $m = f'(0)$ and $b = f(0)$. Substituting these into $y = mx + b$ gives $y = f'(0)x + f(0)$, or, writing it in reverse order, $y = f(0) + f'(0)x$.

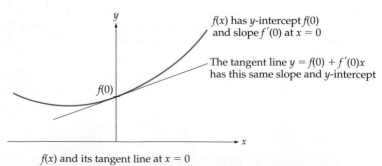

$f(x)$ and its tangent line at $x = 0$

The tangent line is called* the *first Taylor polynomial at* $x = 0$ for the function, and is denoted $p_1(x)$.

First Taylor Polynomial at $x = 0$

The first Taylor polynomial at $x = 0$ of $f(x)$ is

$$p_1(x) = f(0) + f'(0) \cdot x \qquad \text{Tangent line at } x = 0$$

EXAMPLE 1

FINDING A FIRST TAYLOR POLYNOMIAL

Find the first Taylor polynomial at $x = 0$ for $f(x) = e^x$.

Solution

The derivative of $f(x) = e^x$ is e^x, so $f(0) = f'(0) = e^0 = 1$. Therefore, the first Taylor polynomial is

$$p_1(x) = 1 + x \qquad \begin{array}{l} p_1(x) = f(0) + f'(0) \cdot x \\ \text{with } f(0) = 1 \text{ and } f'(0) = 1 \end{array}$$

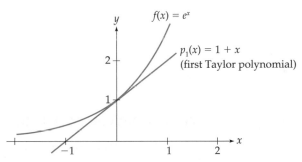

$f(x) = e^x$ and its first Taylor polynomial at $x = 0$

Higher Taylor Polynomials at $x = 0$

This idea of finding the best *linear* approximation at $x = 0$ by matching y-intercepts and slopes (derivatives) at $x = 0$ can be extended to finding the best *quadratic* approximation by also matching the "curl" or concavity (second derivative). In fact, we may find the best *polynomial* approximation by matching higher-order derivatives of the polynomial to those of the function. For example, to approximate $f(x)$ by a *third*-degree polynomial $p(x) = a_0 + a_1 x + a_2 x^2 + a_3 x^3$

---

* Taylor polynomials at $x = 0$ are sometimes called Maclaurin polynomials, although there is little historical justification for this terminology, and we will not use it.

at $x = 0$, we match derivatives up to the *third* order at $x = 0$, as is done in the right-hand column of the following table.

Polynomial $p(x)$ and Its Derivatives	$p(x)$ and Its Derivatives Evaluated at $x = 0$	$f(x)$ and Its Derivatives Evaluated at $x = 0$	Equating $p^{(n)}(0)$ and $f^{(n)}(0)$
$p(x) = a_0 + a_1 x + a_2 x^2 + a_3 x^3$	a_0	$f(0)$	$a_0 = f(0)$
$p'(x) = a_1 + 2a_2 x + 3a_3 x^2$	a_1	$f'(0)$	$a_1 = f'(0)$
$p''(x) = 2a_2 + 3 \cdot 2a_3 x$	$2 \cdot a_2$	$f''(0)$	$2! a_2 = f''(0)$
$p'''(x) = 3 \cdot 2a_3$	$3 \cdot 2a_3$	$f'''(0)$	$3! a_3 = f'''(0)$

The coefficients a_0, a_1, a_2, and a_3 of the polynomial are found by solving the equations in the extreme right column, giving

$$a_0 = f(0), \qquad a_1 = f'(0), \qquad a_2 = \frac{f''(0)}{2!}, \qquad a_3 = \frac{f'''(0)}{3!} \qquad a_n = \frac{f^{(n)}(0)}{n!}$$

Further coefficients follow the same pattern, and substituting them into the polynomial $p(x) = a_0 + a_1 x + a_2 x^2 + a_3 x^3 + \cdots + a_n x^n$ gives the Taylor polynomial of an n-times differentiable function.

Taylor Polynomials at $x = 0$

The nth Taylor polynomial at $x = 0$ of $f(x)$ is

$$p_n(x) = f(0) + \frac{f'(0)}{1!} x + \frac{f''(0)}{2!} x^2 + \frac{f'''(0)}{3!} x^3 + \cdots + \frac{f^{(n)}(0)}{n!} x^n$$

EXAMPLE 2

FINDING TAYLOR POLYNOMIALS

Find the first three Taylor polynomials at $x = 0$ for $f(x) = e^x$.

Solution

All derivatives of e^x are e^x, so $f(0) = f'(0) = f''(0) = f'''(0) = e^0 = 1$. Therefore, the first, second, and third Taylor polynomials for e^x at $x = 0$ are

$p_1(x) = 1 + x$ Found in Example 1

$p_2(x) = 1 + x + \dfrac{1}{2!} x^2$ $f(0) + \dfrac{f'(0)}{1!} x + \dfrac{f''(0)}{2!} x^2$
with $f(0) = f'(0) = f''(0) = 1$

$p_3(x) = 1 + x + \dfrac{1}{2!} x^2 + \dfrac{1}{3!} x^3$ $f(0) + \dfrac{f'(0)}{1!} x + \dfrac{f''(0)}{2!} x^2 + \dfrac{f'''(0)}{3!} x^3$
with $f(0) = f'(0) = f''(0) = f'''(0) = 1$

The following diagram shows e^x and these approximating Taylor polynomials.

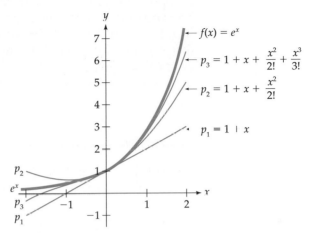

The original function $f(x) = e^x$ (heavier line) and
its first three Taylor polynomials at $x = 0$.
Which polynomial fits e^x most closely?

This graph shows two general properties of Taylor polynomials at $x = 0$:

1. Higher-order Taylor polynomials generally approximate functions more closely.

2. Each approximation is more accurate closer to $x = 0$ (and is *exact at* $x = 0$).

Practice Problem 1

Based on the pattern of the first three Taylor polynomials found in Example 2, find the fifth Taylor polynomial at $x = 0$ of e^x.

➤ Solution on page 765

Graphing Calculator Exploration

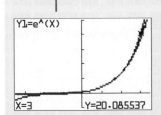

Graph the fifth Taylor polynomial found in Practice Problem 1 together with e^x on the window $[-3.5, 3.5]$ by $[-5, 25]$. Observe where the curves are close and where they diverge. Adjust the window to see how they diverge on a wider interval. Try adding another term, $\dfrac{x^6}{6!} = \dfrac{x^6}{720}$, to graph the *sixth* Taylor polynomial.

Error in Taylor Approximation at $x = 0$

An approximation is useful only if we can estimate how far "off" it is. The *error* or *remainder* in approximating $f(x)$ by $p_n(x)$ is defined as $|R_n(x)| = |f(x) - p_n(x)|$, the vertical distance between the curves at that x-value. If $f(x)$ is $n + 1$ times differentiable between 0 and x, the error can be estimated by the following formula, a proof of which may be found in a more theoretical calculus book.

Error in Taylor Approximation at $x = 0$

For the nth Taylor approximation $p_n(x)$ at $x = 0$ for $f(x)$, the error at x, $|R_n(x)| = |f(x) - p_n(x)|$, satisfies

$$|R_n(x)| \leq \frac{M}{(n + 1)!} |x|^{n+1}$$

Resembles the $n + 1$st term $\dfrac{f^{(n+1)}(x)}{(n + 1)!} x^{n+1}$

where M is any number such that $|f^{(n+1)}(t)| \leq M$ for all t between 0 and x.

EXAMPLE 3

APPROXIMATING BY A TAYLOR POLYNOMIAL

Approximate $e^{0.5}$ using the third Taylor polynomial for e^x, and estimate the error.

Solution

Using the third Taylor polynomial (found in Example 2),

$$p_3(0.5) = 1 + 0.5 + \frac{0.5^2}{2!} + \frac{0.5^3}{3!} \approx \underbrace{1.6458}_{\text{Estimate for } e^{0.5}}$$

$p_3(x) = 1 + x + \dfrac{x^2}{2!} + \dfrac{x^3}{3!}$ evaluated at $x - 0.5$

The error formula for the third Taylor polynomial requires a number M such that $|f^{(4)}(t)| \leq M$ for $0 \leq t \leq 0.5$. Since $|f^{(4)}(t)| = e^t$ is an increasing function, it is maximized at its right-hand endpoint $t = 0.5$, so

$$|f^{(4)}(t)| = e^t \leq e^{0.5} < e^1 < 3 \qquad \text{Since } e \approx 2.7 < 3$$

We therefore take $M = 3$ (although smaller estimates could be found). With this value of M, the error formula with $n = 3$, $M = 3$, and $x = 0.5$ becomes

$$|R_3(0.5)| \leq \frac{3}{4!} \, |0.5|^4 = \frac{3}{24} \, (0.0625) \approx 0.00781 \approx \underbrace{0.008}_{\text{Error estimate}}$$

Therefore,

$$e^{0.5} \approx 1.646 \quad \text{with an error less than } 0.008$$

We always round the error estimate *up* to the first nonzero digit (for example, an error of 0.0072 would be rounded up to 0.008) so as not to claim more accuracy than is justified. We then round the approximation to the same number of decimal places.

Why Bother to Calculate $e^{0.5}$?

You may wonder why we approximate $e^{0.5}$ when a calculator or computer will do so at the press of a few keys. How do you think computers actually evaluate such functions? They do so by using (modified) preprogrammed Taylor polynomials.* Therefore, this section explains what happens when you press the keys. Furthermore, at some time you may need to evaluate a new or more complicated function, and Taylor polynomials enable you to do this in terms of "easier" polynomials.

Not only can we find the accuracy of a particular approximation, but we can also determine the Taylor polynomial needed to *guarantee* a given accuracy.

EXAMPLE 4

APPROXIMATING WITHIN A GIVEN ACCURACY

Find the Taylor polynomial at $x = 0$ that approximates e^x with an error less than 0.005 on the interval $-1 \leq x \leq 1$.

Solution

We want the value of n that makes the error $\dfrac{M}{(n+1)!} \, |x|^{n+1}$ less than 0.005. This error formula with $M = 3$ (since $|e^1| < 3$, as in

---

* Most computers use "economized" Taylor series that make the error more uniform over an interval. Most calculators, on the other hand, use a system called CORDIC, for COordinate Rotation DIgital Computer.

Example 3) and $|x| = 1$ becomes $\dfrac{3}{(n+1)!}$. We evaluate this formula at $n = 1, 2, 3, \ldots$ until its value first falls below 0.005.

n	$\dfrac{3}{(n+1)!}$	
1	$\dfrac{3}{2!} = \dfrac{3}{2} = 1.5$	
2	$\dfrac{3}{3!} = \dfrac{3}{6} = 0.5$	We could have skipped some early values of n that clearly would not give small enough errors
3	$\dfrac{3}{4!} = \dfrac{3}{24} = 0.125$	
4	$\dfrac{3}{5!} = \dfrac{3}{120} = 0.025$	
5	$\dfrac{3}{6!} = \dfrac{3}{720} \approx 0.004$	< 0.005 for the first time

The first value of n for which the error is small enough (less than 0.005) is $n = 5$, so the fifth Taylor polynomial (found in Practice Problem 1) provides the desired accuracy:

$$e^x \approx 1 + x + \frac{x^2}{2!} + \frac{x^3}{3!} + \frac{x^4}{4!} + \frac{x^5}{5!}$$

For $-1 \le x \le 1$, with error less than 0.005

Graphing Calculator Exploration

(You should still have e^x and its fifth Taylor polynomial at $x = 0$ in your calculator.) Use TRACE to verify that e^x and the Taylor polynomial are indeed within 0.005 of each other between $x = -1$ and $x = 1$. How close are they actually? (Remember that the error formula generally *overstates* the error.)

EXAMPLE 5

FINDING AN APPROXIMATION IN AN APPLICATION

A company that produces seasonal goods produces sin 1.1 thousand units on a certain day. Approximate sin 1.1 by using the fifth Taylor polynomial at $x = 0$ for $\sin x$ and estimate the error.

Solution

The function $\sin x$ and its derivatives (from page 637) at $x = 0$ are

$$f(x) = \sin x \qquad\qquad f(0) = 0$$
$$f'(x) = \cos x \qquad\qquad f'(0) = 1$$
$$f''(x) = -\sin x \qquad\qquad f''(0) = 0$$
$$f'''(x) = -\cos x \qquad\qquad f'''(0) = -1$$
$$f^{(4)}(x) = \sin x \qquad\qquad f^{(4)}(0) - 0$$
$$f^{(5)}(x) = \cos x \qquad\qquad f^{(5)}(0) = 1$$

Therefore, the fifth Taylor polynomial is

$$p_5(x) = 0 + \frac{1}{1!}x + 0 - \frac{1}{3!}x^3 + 0 + \frac{1}{5!}x^5$$

$$f(0) + \frac{f'(0)}{1!}x + \cdots + \frac{f^{(5)}(0)}{5!}x^5$$
with the preceding values for
$f(0), f'(0), \ldots, f^{(5)}(0)$

$$= x - \frac{x^3}{3!} + \frac{x^5}{5!}$$

Simplifying (all even powers vanish)

Evaluating,

$$\sin 1.1 \approx 1.1 - \frac{(1.1)^3}{3!} + \frac{(1.1)^5}{5!} \approx 0.89159 \qquad \text{Evaluating at } x = 1.1$$

Therefore, $\sin 1.1 \approx 0.89159$. For the error formula, we need an M such that $|f^{(6)}(t)| \leq M$ for all t between 0 and 1.1. For the function $\sin x$, we have $|f^{(6)}(x)| = |-\sin x| \leq 1 = M$. With this M, the error formula becomes

$$|R_5(x)| \leq \frac{1}{6!}|1.1|^6 \approx 0.0024 \approx 0.003$$

$$\frac{M}{(n+1)!}|x|^{n+1} \text{ with}$$
$$M = 1, \quad n = 5, \quad x = 1.1$$
└── Rounded *up*

Therefore,

$$\sin 1.1 \approx 0.892 \quad \text{with an error less than } 0.003$$

These are production numbers in *thousands*, so the company should produce 892 units, with a possible error of 3 units. (To be safe, it might choose to produce 895 units.)

Practice Problem 2   As we saw in the preceding example, the Taylor polynomial for $\sin x$ at $x = 0$ consists only of *odd* powers of x. Therefore, the sixth Taylor polynomial will be the same as the fifth, so we can use the error formula for the sixth Taylor polynomial to estimate the accuracy of the approximation that we found. Use the error formula for $|R_6(x)|$ to estimate the accuracy of the approximation $\sin 1.1 \approx 0.89159$ from Example 5. ➤ Solution on page 765

The function $\sin x$ and its fifth Taylor polynomial are shown in the following graph. Notice that the approximation is quite accurate for x-values near 0.

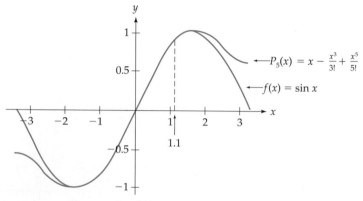

Sin x and its fifth Taylor polynomial at $x = 0$

Taylor Polynomials at $x = a$

We have been using Taylor polynomials centered *at* $x = 0$, meaning that they approximate $f(x)$ most accurately near $x = 0$. There are analogous formulas for Taylor polynomials centered at *any* number $x = a$ (provided that f is n times differentiable at a).

Taylor Polynomials at $x = a$

The nth Taylor polynomial at $x = a$ of $f(x)$ is

$$p_n(x) = f(a) + \frac{f'(a)}{1!}(x - a) + \frac{f''(a)}{2!}(x - a)^2 + \cdots + \frac{f^{(n)}(a)}{n!}(x - a)^n$$

For Taylor polynomials at $x = a$:

1. The polynomial is in powers of $(x - a)$ rather than powers of x.
2. The coefficients involve derivatives at $x = a$.

3. The polynomials approximate $f(x)$ more accurately nearer $x = a$ (and are *exact at* $x = a$).

4. If $a = 0$, we obtain the Taylor polynomials at $x = 0$.

EXAMPLE 6

FINDING A TAYLOR POLYNOMIAL AT $x = 4$

Find the second Taylor polynomial for $f(x) = \sqrt{x}$ at $x = 4$, and use it to approximate $\sqrt{4.2}$.

Solution

The function $f(x) = \sqrt{x}$ and its derivatives at $x = 4$ are

$$f(x) = x^{1/2} \qquad f(4) = 4^{1/2} = 2$$

$$f'(x) = \frac{1}{2} x^{-1/2} \qquad f'(4) = \frac{1}{2} 4^{-1/2} = \frac{1}{2} \cdot \frac{1}{2} = \frac{1}{4}$$

$$f''(x) = -\frac{1}{4} x^{-3/2} \qquad f''(4) = -\frac{1}{4} 4^{-3/2} = -\frac{1}{4} \cdot \frac{1}{8} = -\frac{1}{32}$$

The second Taylor polynomial at $x = 4$ is

$$p_2(x) = 2 + \frac{\frac{1}{4}}{1!}(x - 4) - \frac{\frac{1}{32}}{2!}(x - 4)^2$$

> $p_2(x) = f(a) + \dfrac{f'(a)}{1!}(x - a) +$
> $\dfrac{f''(a)}{2!}(x - a)^2$ with $a = 4$
>
> and the calculated values of
> $f(4), f'(4),$ and $f''(4)$

$$= 2 + \frac{x - 4}{4} - \frac{(x - 4)^2}{64} \qquad \text{Simplifying}$$

Evaluating,

$$p_2(4.2) = 2 + \frac{0.2}{4} - \frac{(0.2)^2}{64} = 2.049375 \qquad \text{Evaluating at } x = 4.2$$

Approximation for $\sqrt{4.2}$

$$\sqrt{4.2} \approx 2.0494 \qquad \text{Rounding to four decimal places}$$

Actually, $\sqrt{4.2} \approx 2.04939$, so the estimate above is correct in all four decimal places. The function $f(x) = \sqrt{x}$ and its second Taylor polynomial at $x = 4$ are graphed on the following page, showing that the approximation is very good near $x = 4$.

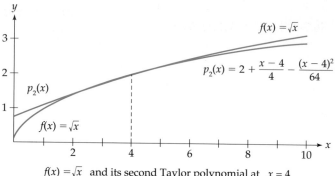

$$f(x) = \sqrt{x} \text{ and its second Taylor polynomial at } x = 4$$

Graphing Calculator Exploration

Some advanced graphing calculators can find Taylor polynomials. For example, the Texas Instruments *TI-89* graphing calculator finds the second Taylor polynomial at $x = 4$, for the function $f(x) = \sqrt{x}$ as follows.

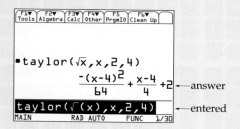

This answer agrees with that in the preceding example, except that it is written in reverse order.

If $f(x)$ is $n + 1$ times differentiable between a and x, the error in approximating $f(x)$ by $p_n(x)$ can be estimated as follows:

Error in Taylor Approximation at $x = a$

For the nth Taylor approximation $p_n(x)$ at $x = a$ for $f(x)$, the error at x, $|R_n(x)| = |f(x) - p_n(x)|$, satisfies

$$|R_n(x)| \le \frac{M}{(n + 1)!} |x - a|^{n+1}$$

where M is any number such that $|f^{(n+1)}(t)| \le M$ for all t between a and x.

10.2 Section Summary

Taylor polynomials approximate a curve near a single chosen point $x = a$. Any sufficiently differentiable function can be approximated by a polynomial, using the formulas

$$\sum_{k=0}^{n} \frac{f^{(k)}(0)}{k!} x^k$$

*n*th Taylor polynomial at $x = 0$
[first term simplifies to $f(0)$]

and

$$\sum_{k=0}^{n} \frac{f^{(k)}(a)}{k!} (x - a)^k$$

*n*th Taylor polynomial at $x = a$
[first term simplifies to $f(a)$]

Taylor polynomials can also be found from other Taylor polynomials by replacing the variable by a polynomial expression. The accuracy of the approximation can be found from the error formulas on pages 758 and 764.

➤ Solutions to Practice Problems

1. $1 + x + \dfrac{1}{2!} x^2 + \dfrac{1}{3!} x^3 + \dfrac{1}{4!} x^4 + \dfrac{1}{5!} x^5$, or $1 + x + \dfrac{x^2}{2!} + \dfrac{x^3}{3!} + \dfrac{x^4}{4!} + \dfrac{x^5}{5!}$

2. $|R_6(x)| \leq \dfrac{1}{7!} |1.1|^7 = \dfrac{1.9487}{5040} \approx 0.0004$

Therefore, $\sin 1.1 \approx 0.8916$ with an error of less than 0.0004.

10.2 Exercises

Find the third Taylor polynomial at $x = 0$ of each function.

1. $f(x) = e^{2x}$

2. $f(x) = \ln(1 - x)$

3. $f(x) = \sqrt{x + 1}$

4. $f(x) = \sqrt{2x + 1}$

5. $f(x) = 7 + x - 3x^2$

6. $f(x) = 2 - x + 3x^2 - x^3$

7. $f(x) = e^{x^2}$

8. $f(x) = \ln(\cos x)$

For each function in Exercises 9–12:

a. Find the fourth Taylor polynomial at $x = 0$.

b. Graph the original function and the Taylor polynomial together on the indicated window.

9. $f(x) = \ln(x + 1)$
(for (b), use the window $[-2, 2]$ by $[-2, 2]$)

10. $f(x) = e^{-2x}$
(for (b), use the window $[-2, 2]$ by $[-2, 8]$)

11. $f(x) = \cos x$
(for (b), use the window $[-\pi, \pi]$ by $[-2, 2]$)

12. $f(x) = \cos(x + \pi)$
(for (b), use the window $[-\pi, \pi]$ by $[-2, 2]$)

13. Find the fourth Taylor polynomial for e^{x^2} by taking the second Taylor polynomial for e^x (page 756) and replacing x by x^2.

14. Find the third Taylor polynomial for e^{-x} by taking the third Taylor polynomial for e^x (page 756) and replacing x by $-x$.

15. Find the fifth Taylor polynomial for $\sin 2x$ by taking the fifth Taylor polynomial for $\sin x$ (pages 760–761) and replacing x by $2x$.

16. Find the tenth Taylor polynomial for $\sin x^2$ by taking the fifth Taylor polynomial for $\sin x$ (pages 760–761) and replacing x by x^2.

17. a. For $f(x) = e^x$, find the fourth Taylor polynomial at $x = 0$.

   b. Use your answer in part (a) to approximate $e^{-1/2}$.

   c. Estimate the error by the error formula (page 758). [*Hint:* First calculate $f^{(5)}(x)$ and find an M such that $|f^{(5)}(x)| \leq M$ for $-\frac{1}{2} \leq x \leq 0$, using the fact that an increasing function is maximized at its right-hand endpoint.]

   d. Restate the approximation and error, rounded appropriately.

   e. Graph $f(x) = e^x$ and its Taylor polynomial on $[-3, 3]$ by $[0, 12]$.

18. a. For $f(x) = \ln(x + 1)$, find the fifth Taylor polynomial at $x = 0$.

   b. Use your answer in part (a) to approximate $\ln \frac{3}{2}$.

   c. Estimate the error by the error formula (page 758). [*Hint:* Calculate $f^{(6)}(x)$ and find an M such that $|f^{(6)}(x)| \leq M$ for $0 \leq x \leq \frac{1}{2}$, using the fact that a decreasing function is maximized at its left-hand endpoint.]

   d. Restate the approximation and error, rounded appropriately.

   e. Graph $f(x) = \ln(x + 1)$ and its Taylor polynomial on $[-2, 2]$ by $[-2, 2]$.

19. a. Find the Taylor polynomial at $x = 0$ that approximates $f(x) = \cos x$ on the interval $-1 \leq x \leq 1$ with error less than 0.0002.

   b. Graph $f(x) = \cos x$ and the Taylor polynomial on $[-2\pi, 2\pi]$ by $[-2, 2]$. Use TRACE to estimate the maximum difference between the two curves for $-1 \leq x \leq 1$.

20. For the function

$$f(x) = 1 + 2x + \frac{9}{2!}x^2 + \frac{5}{3!}x^3 - \frac{7}{4!}x^4$$

find $f'''(0)$ and $f^{(4)}(0)$. [*Hint:* No calculation is necessary.]

21. a. What linear function best approximates $\sqrt[3]{1 + x}$ near $x = 0$?

   b. Use the error formula to find an interval $[0, x]$ on which the error is less than 0.01.

   c. Graph $\sqrt[3]{1 + x}$ and the linear approximation on $[-2, 2]$ by $[-2, 2]$ and use TRACE to verify that the difference is less than 0.01 for the x-values found in part (b).

22. a. What quadratic function best approximates $\sqrt[3]{1 + x}$ near $x = 0$?

   b. Use the error formula to estimate the maximum error on $[-0.1, 0.1]$.

   c. Graph $\sqrt[3]{1 + x}$ and the quadratic approximation on $[-2, 2]$ by $[-2, 2]$ and use TRACE to verify that on $[-0.1, 0.1]$ the difference is less than the error estimate found in part (b).

Taylor Polynomials at $x = a$

23. a. For $f(x) = \sqrt{x}$, find the third Taylor polynomial at $x = 1$.

   b. Graph $\sqrt{x}$ and its Taylor polynomial on $[-1, 3]$ by $[-1, 2]$.

24. a. For $f(x) = 3x^2 - 4x + 5$, find the second Taylor polynomial at $x = 2$.

   b. Multiply out the Taylor polynomial found in part (a) and show that it is equal to the original polynomial.

25. a. For $f(x) = \sin x$, find the third Taylor polynomial at $x = \pi$.

   b. Graph $\sin x$ and its Taylor polynomial on $[0, 2\pi]$ by $[-2, 2]$.

26. a. For $f(x) = \cos x$, find the fourth Taylor polynomial at $x = \pi$.

   b. Graph $\cos x$ and its Taylor polynomial on $[0, 2\pi]$ by $[-2, 2]$.

27. a. For $f(x) = \sqrt[3]{x}$, find the second Taylor polynomial at $x = 1$.

   b. Use your answer from part (a) to approximate $\sqrt[3]{1.3}$.

   c. Estimate the error by the error formula (page 764). [*Hint:* Find an M such that $|f^{(3)}(x)| \leq M$ for $1 \leq x \leq 1.3$, using the fact that a decreasing function is maximized at its left-hand endpoint.]

   d. Restate the approximation and error, rounded appropriately.

28. a. For $f(x) = \ln x$, find the sixth Taylor polynomial at $x = 1$.

   b. Use your answer in part (a) to approximate $\ln \frac{3}{2}$.

   (continues)

c. Estimate the error by the error formula (page 764). [*Hint:* Find an M such that $|f^{(7)}(x)| \le M$ for $1 \le x \le 1.5$, using the fact that a decreasing function is maximized at its left-hand endpoint.]

d. Restate the approximation and error, rounded appropriately.

APPLIED EXERCISES

29. BUSINESS: Advertising Five days after the end of an advertising campaign, a company's daily sales (as a proportion of its peak sales) will be $e^{-0.25}$. Approximate $e^{-0.25}$ by using the second Taylor polynomial at $x = 0$ for e^x (rounding to two decimal places). Check your answer by using the $\boxed{e^x}$ key on your calculator.

30. GENERAL: Ecology A population of deer in a wildlife refuge varies with the season, and is predicted to be $300 + 50 \cos 0.52$ on the first day of summer. Approximate this number by using the second Taylor polynomial at $x = 0$ for $\cos x$. Then check your answer using the $\boxed{\cos x}$ key on your calculator (set for radians).

31. ECONOMICS: Supply and Demand The demand for a new toy is predicted to be $4 \cos x$,

and the supply is $x + 3$ (both in thousands of units), where x is the number of years ($0 \le x \le 2$). Find when supply will equal demand by solving the equation

$$4 \cos x = x + 3$$

as follows. Replace $\cos x$ by its second Taylor polynomial at $x = 0$ and solve the resulting quadratic equation for the (positive) value of x.

32. ZERO FACTORIAL The relationship $(n + 1)! = (n + 1)n!$, solved for $n!$, gives

$$n! = \frac{(n + 1)!}{n + 1}$$

Evaluate this at $n = 0$ to see the reason for defining $0! = 1$.

10.3 TAYLOR SERIES

Introduction

In this section we return to infinite series, but with the terms of the series now being *functions*. We begin by discussing *power series*, then specializing to *Taylor series*, which may be thought of as "infinitely long" Taylor polynomials. Taylor series lead to remarkably simple infinite series for functions such as e^x, $\ln x$, and $\sin x$, and enable us to find difficult integrals involving such functions.

Power Series

In this (and the next) section we will consider infinite series whose terms involve a *variable*, such as

$$1 + x + x^2 + x^3 + \cdots$$

Replacing x by $\frac{1}{2}$ gives

$$1 + \frac{1}{2} + \frac{1}{4} + \frac{1}{8} + \cdots \qquad \text{From } 1 + \frac{1}{2} + \left(\frac{1}{2}\right)^2 + \left(\frac{1}{2}\right)^3 + \cdots$$

which is a convergent geometric series with ratio $r = \frac{1}{2}$. If instead we replace x by 1, we obtain

$$1 + 1 + 1 + 1 + \cdots \qquad \text{From } 1 + 1 + (1)^2 + (1)^3 + \cdots$$

which clearly diverges since the sum becomes arbitrarily large. In general, a series involving a variable may converge for some values of the variable and diverge for others. It will therefore be important to determine the values of the variable for which the series converges. The most useful series is called a *power series,* which is simiar to the series $1 + x + x^2 + x^3 + \cdots$ but with the terms multiplied by coefficients. That is, a power series is of the form

$$a_0 + a_1 x + a_2 x^2 + \cdots + a_n x^n + \cdots$$

in which the coefficients $a_0, a_1, a_2, \ldots, a_n, \ldots$ are (real) numbers. Every power series in x converges for $x = 0$ (where its value is clearly a_0). It also converges at all points within some radius R around $x = 0$, with the understanding that R may be a positive number, or $R = 0$ (in which case the series converges only at $x = 0$), or $R = \infty$ (in which case the series converges for *all* values of x). Formally:

Power Series in *x* and Radius of Convergence

For a power series in x, $a_0 + a_1 x + a_2 x^2 + \cdots + a_n x^n + \cdots$, exactly one of the following statements is true:

1. There is a positive number R ("radius of convergence") such that the series converges for $|x| < R$ and diverges for $|x| > R$.
2. The series converges only at $x = 0$ (radius $R = 0$).
3. The series converges for *all* x (radius $R = \infty$).

These three cases can be shown graphically as follows.

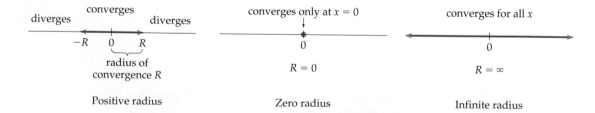

For a positive radius of convergence, the series may converge at *neither* endpoint, at just *one* endpoint, or at *both* endpoints. We will generally ignore these endpoints, finding only the radius of convergence by using the following *ratio test.*

> **Ratio Test**
>
> For an infinite series $c_0 + c_1 + c_2 + \cdots$ of nonzero terms such that the limit $r = \lim\limits_{n \to \infty} \left| \dfrac{c_{n+1}}{c_n} \right|$ exists or is infinite:
>
> 1. If $r < 1$, the series *converges*.
> 2. If $r > 1$ or if the limit is infinite, the series *diverges*.

The ratio test is related to the convergence of a geometric series (page 745) as follows. A geometric series converges if the ratio r of successive terms satisfies $|r| < 1$, while the ratio test says that even if a series is *not* geometric, as long as the *limiting* ratio of successive terms is less than 1 in absolute value, then the series converges. A proof of the ratio test is outlined in Exercises 48–49. Finding the radius of convergence of a power series by the ratio test involves four steps:

1. Find expressions for successive terms c_n and c_{n+1} (found from c_n by replacing each n by $n + 1$).

2. Write the quotient $\dfrac{c_{n+1}}{c_n}$ and simplify.

3. Calculate $r = \lim\limits_{n \to \infty} \left| \dfrac{c_{n+1}}{c_n} \right|$

4. Solve the relationship $r < 1$ for $|x|$ to find the radius of convergence R.

EXAMPLE 1

FINDING THE RADIUS OF CONVERGENCE

Find the radius of convergence of the power series

$$\frac{1}{2} x + \frac{2}{4} x^2 + \frac{3}{8} x^3 + \cdots + \frac{n}{2^n} x^n + \cdots$$

Solution

Successive terms are $c_n = \dfrac{n}{2^n} x^n$ and $c_{n+1} = \dfrac{n+1}{2^{n+1}} x^{n+1}$, and their ratio is

$$\frac{c_{n+1}}{c_n} = \frac{\dfrac{n+1}{2^{n+1}} x^{n+1}}{\dfrac{n}{2^n} x^n} = \frac{n+1}{2^{n+1}} \cdot \frac{x^{n+1}}{x^n} \cdot \frac{2^n}{n} = \frac{n+1}{n} \cdot \frac{2^n}{2^{n+1}} \cdot x = \left(1 + \frac{1}{n}\right) \cdot \frac{1}{2} \cdot x \longrightarrow \frac{x}{2}$$

Inverting $\longrightarrow x$; $1 + \dfrac{1}{n}$; $\dfrac{1}{2}$; Approaches 1 as $n \to \infty$

Therefore, $r = \lim\limits_{n \to \infty} \left| \dfrac{c_{n+1}}{c_n} \right| = \dfrac{|x|}{2}$. For convergence, this ratio must be less than 1, $\dfrac{|x|}{2} < 1$, which is equivalent to $|x| < 2$. Therefore, the radius of convergence is $R = 2$, meaning that the series converges for $|x| < 2$ and diverges for $|x| > 2$.

Taylor Series

We now consider a special type of power series called a *Taylor* series, which may be thought of as an "infinitely long" Taylor polynomial. We begin with a function f with derivatives of all orders at $x = 0$.

Taylor Series at $x = 0$

The Taylor series at $x = 0$ of f is

$$f(0) + \frac{f'(0)}{1!}x + \frac{f''(0)}{2!}x^2 + \frac{f'''(0)}{3!}x^3 + \cdots + \frac{f^{(n)}(0)}{n!}x^n + \cdots$$

Note that the coefficients are calculated just as in Taylor polynomials.

EXAMPLE 2

FINDING A TAYLOR SERIES

Find the Taylor series at $x = 0$ for e^x, and find its interval of convergence.

Solution

The Taylor series for e^x can be written down immediately from the Taylor polynomials for e^x in Example 2 on page 756 simply by letting the terms continue indefinitely:

$$1 + \frac{x}{1!} + \frac{x^2}{2!} + \frac{x^3}{3!} + \frac{x^4}{4!} + \cdots \qquad \text{Taylor series for } e^x$$

For the radius of convergence, we calculate the ratio of successive terms and use the ratio test:

$$\frac{\dfrac{x^{n+1}}{(n+1)!}}{\dfrac{x^n}{n!}} = \frac{x^{n+1}}{(n+1)!} \cdot \frac{n!}{x^n} = \frac{x^{n+1}}{x^n} \qquad \cdot \qquad \frac{n!}{(n+1)!}$$

$$c_n = \frac{x^n}{n!}$$

$$c_{n+1} = \frac{x^{n+1}}{(n+1)!}$$

Inverting ⟶ x $\dfrac{n\cdots 1}{(n+1)\cdot n \cdots 1} = \dfrac{1}{n+1}$

$$= x \cdot \frac{1}{n+1} \longrightarrow 0$$

Approaches 0 r
as $n \to \infty$

This ratio approaches $r = 0$ for *any* fixed value of x, so by the ratio test with $|r| < 1$, the series converges for *all* values of x (radius of convergence $R - \infty$).

We have shown that the series for e^x converges for every value of x, but we have not shown that it converges to e^x (it might instead converge to some other value). Proving convergence to the original function requires showing that the error $|R_n(x)|$ in the nth Taylor polynomial approaches zero as $n \to \infty$. Such proofs are difficult, and are relegated to the Exercises (see Exercises 46 and 47). For every Taylor series in this book, and for every Taylor series you are likely to encounter in applications, if it converges, it converges to its defining function. Based on this statement, we will write the series as *equal* to its defining function wherever the series converges. For example, the series for e^x *converges* for all x, and *equals* e^x:

> **Taylor Series at $x = 0$ for e^x**
>
> $$e^x = 1 + \frac{x}{1!} + \frac{x^2}{2!} + \frac{x^3}{3!} + \frac{x^4}{4!} + \cdots \qquad \text{for } -\infty < x < \infty$$

This series is called the "Taylor series expansion" of e^x. Evaluating at $x = 1$ gives a remarkable series for the constant e:

$$e = 1 + \frac{1}{1!} + \frac{1}{2!} + \frac{1}{3!} + \frac{1}{4!} + \cdots$$

e is the sum of the reciprocals of factorials

Graphing Calculator Exploration

```
sum(seq(1/N!,N,0
,7))
        2.718253968
```

(Sequence and series operations useful)

a. Sum the first seven terms of the series on the previous page. To how many decimal places does the result agree with $e \approx 2.71828$?

b. How many terms are required to get agreement to *five* decimal places?

EXAMPLE **3**

FINDING THE TAYLOR SERIES FOR sin x

Find the Taylor series at $x = 0$ for $\sin x$.

Solution

Differentiating $\sin x$ gives the following repeating pattern of positive and negative sines and cosines, which on the third line below are evaluated at $x = 0$.

f	f'	f''	f'''	$f^{(4)}$	$f^{(5)}$	$f^{(6)}$	$f^{(7)}$	$\cdots$
$\sin x$	$\cos x$	$-\sin x$	$-\cos x$	$\sin x$	$\cos x$	$-\sin x$	$-\cos x$	$\cdots$
0	1	0	-1	0	1	0	-1	$\cdots$

Substituting these into the Taylor series formula (page 770) gives

$$0 + \frac{1}{1!}x + \frac{0}{2!}x^2 + \frac{-1}{3!}x^3 + \frac{0}{4!}x^4 + \frac{1}{5!}x^5 + \frac{0}{6!}x^6 + \frac{-1}{7!}x^7 + \cdots$$

Omitting the zero terms and simplifying gives the following Taylor series for $\sin x$.

$$\sin x = x - \frac{x^3}{3!} + \frac{x^5}{5!} - \frac{x^7}{7!} + \cdots \qquad \text{Alternating signs, with only odd powers}$$

This series converges for *all x* (as is shown in Exercise 39).*

*This series gives another reason for expressing trigonometric functions in radians instead of degrees. The Taylor series for $\sin x$ for x *in degrees*, written $\sin x°$, is found by multiplying each x by $\frac{\pi}{180}$, obtaining

$$\sin x° = \frac{\pi}{180}x - \left(\frac{\pi}{180}\right)^3 \frac{x^3}{3!} + \left(\frac{\pi}{180}\right)^5 \frac{x^5}{5!} - \left(\frac{\pi}{180}\right)^7 \frac{x^7}{7!} + \cdots$$

This series lacks the simplicity of the radian version.

> **Taylor Series at** $x = 0$ **for** $\sin x$
>
> $$\sin x = x - \frac{x^3}{3!} + \frac{x^5}{5!} - \frac{x^7}{7!} + \cdots \qquad \text{for } -\infty < x < \infty$$

Operations on Taylor Series

In many ways, Taylor series behave like polynomials. For example, the Taylor series for $f(x)$ can be differentiated term by term to obtain the Taylor series for $f'(x)$ (but only for x *within* the interval of convergence, avoiding the endpoints). A similar statement holds for integration. A Taylor series can also be multiplied by a constant c or by a positive integer power x^n, or the variable may be replaced by an expression of the form cx^n. In each case, the resulting series will be the Taylor series for the function similarly modified, and the radius of convergence will remain the same (except in the case of replacing the variable, where a finite radius of convergence may change). Two series may be added or subtracted, giving the series for the sum or difference of the original functions, and the resulting series will have a radius of convergence equal to the smaller of the radii of the original series.

Any power series that converges to a function must be the *Taylor* series for that function. For example, according to the formula on page 745, the geometric series $1 + x + x^2 + x^3 + \cdots$ converges to $\frac{1}{1-x}$ for $|x| < 1$, and so must be the *Taylor* series for that function for $|x| < 1$. Proofs of all these statements may be found in more theoretical calculus books.

> **Taylor Series at** $x = 0$ **for** $\dfrac{1}{1-x}$
>
> $$\frac{1}{1-x} = 1 + x + x^2 + x^3 + \cdots \qquad \text{for } |x| < 1$$

EXAMPLE 4

FINDING ONE TAYLOR SERIES FROM ANOTHER

Find the Taylor series at $x = 0$ for $\cos x$.

Solution

The easiest way is to differentiate the series for $\sin x$ found in Example 3:

$$\sin x = x - \frac{1}{3!}x^3 + \frac{1}{5!}x^5 - \frac{1}{7!}x^7 + \cdots \qquad \text{for } -\infty < x < \infty$$

$$\cos x = 1 - \frac{1}{3!} 3x^2 + \frac{1}{5!} 5x^4 - \frac{1}{7!} 7x^6 + \cdots \qquad \text{Differentiating term by term by the Power Rule}$$

$$= 1 - \frac{1}{2!} x^2 + \frac{1}{4!} x^4 - \frac{1}{6!} x^6 + \cdots \qquad \text{Simplifying}$$

This gives the following Taylor series for $\cos x$.

$$\cos x = 1 - \frac{x^2}{2!} + \frac{x^4}{4!} - \frac{x^6}{6!} + \cdots \qquad \text{Alternating signs, with only even powers}$$

This series converges for *all* x (since term-by-term differentiation does not change the radius of convergence).

Taylor Series at $x = 0$ for $\cos x$

$$\cos x = 1 - \frac{x^2}{2!} + \frac{x^4}{4!} - \frac{x^6}{6!} + \cdots \qquad \text{for } -\infty < x < \infty$$

Practice Problem

Find the Taylor series at $x = 0$ for $\cos x^2$. [*Hint:* Use the series above with a substitution.]

➤ Solution on page 780

By integrating entire series, we can construct infinite series for difficult integrals that we could not otherwise evaluate.

Baby Booms and Echoes

An increase in births (a "baby boom") is usually followed in later generations by smaller fluctuations (called "echoes" of the boom) as the children grow up and in turn have more children. The resulting fluctuations in the number of births can be modeled by a function such as

$$f(t) = \frac{\sin t}{t}.$$

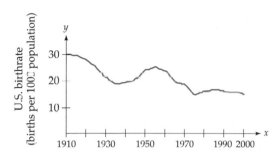

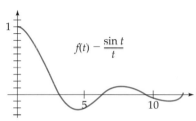

EXAMPLE **5**   **SUMMING EXCESS BIRTHS**

A country is predicted to have an excess of $\dfrac{\sin t}{t}$ million births per year (above the usual level) in year t. During the first x years, this means an additional $\displaystyle\int_0^x \frac{\sin t}{t}\, dt$ million births. Find the Taylor series for this integral. Then use the series to estimate the number of extra births during the first two years.

Solution

There is no simple antiderivative for the function $\dfrac{\sin t}{t}$, so we express it as a Taylor series (dividing the series for $\sin t$ by t), which we then integrate.

$$\frac{\sin t}{t} = \frac{1}{t}\left(t - \frac{1}{3!}t^3 + \frac{1}{5!}t^5 - \frac{1}{7!}t^7 + \cdots \right) \qquad \text{Using the series for } \sin t$$

$$= 1 - \frac{1}{3!}t^2 + \frac{1}{5!}t^4 - \frac{1}{7!}t^6 + \cdots \qquad \text{Simplifying}$$

Then

$$\int_0^x \frac{\sin t}{t}\, dt = \int_0^x \left(1 - \frac{1}{3!}t^2 + \frac{1}{5!}t^4 - \frac{1}{7!}t^6 + \cdots\right) dt$$

$$= \left(t - \frac{1}{3!}\frac{1}{3}t^3 + \frac{1}{5!}\frac{1}{5}t^5 - \frac{1}{7!}\frac{1}{7}t^7 + \cdots\right)\Bigg|_0^x \qquad \text{Integrating term by term by the Power Rule}$$

$$= x - \frac{x^3}{3\cdot 3!} + \frac{x^5}{5\cdot 5!} - \frac{x^7}{7\cdot 7!} + \cdots \qquad \text{Substituting } t = x \text{ (the evaluation at } t = 0 \text{ is zero)}$$

This infinite series gives the *exact* value of the integral $\displaystyle\int_0^x \frac{\sin t}{t}\, dt$. Evaluating at $x = 2$ and approximating by the first four terms gives

$$2 - \frac{2^3}{3\cdot 3!} + \frac{2^5}{5\cdot 5!} - \frac{2^7}{7\cdot 7!} \approx 1.605$$

Therefore, there will be about 1.605 million extra births during the first two years.

Although $\dfrac{\sin t}{t}$ is undefined at $t = 0$, its *limit* as $t \to 0$ is 1 (see pages 634–635), which is also the *value* of its Taylor series at $t = 0$.

Graphing Calculator Exploration

Try evaluating $\displaystyle\int_0^2 \frac{\sin t}{t}\, dt$ on your graphing calculator. You may get a "division by zero" error, showing that some other method (such as the Taylor series) is required. Even if you do get an answer, the Taylor series has the advantage of giving a formula for the integral (population gain) up to *any* year x.

Why Is Convergence So Important?

The radius of convergence of a Taylor series is important because only within that radius from the central point does the series *equal* the function. Beyond that radius the series will have no meaning, even though the *original function* may still be defined.

For example, the function $\dfrac{1}{1-x}$ is defined for all x-values except $x = 1$, but its Taylor series $1 + x + x^2 + x^3 + \cdots$ converges only on the interval $(-1, 1)$, as we saw on page 773. Outside of this interval the *function* is still defined (its value at $x = 2$ is $\frac{1}{1-2} = \frac{1}{-1} = -1$), but its Taylor series at $x = 2$ is $1 + 2 + 2^2 + 2^3 + \cdots$, which does not converge and is meaningless. That is, using a Taylor series outside of the radius of convergence leads to incorrect results.

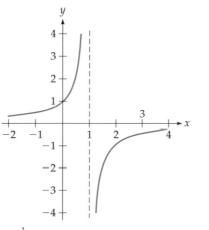

$\dfrac{1}{1-x}$ is defined for all x except $x = 1$.

$1 + x + x^2 + x^3 + \cdots$ converges only for $-1 < x < 1$.

Taylor Series at $x = a$

Just as with Taylor polynomials, we may define Taylor series in powers of $(x - a)$, shifting the "center point" from $x = 0$ to $x = a$. Let f be a function that has derivatives of all orders at $x = a$.

Taylor Series at $x = a$

The Taylor series at $x = a$ of f is

$$f(a) + \frac{f'(a)}{1!}(x - a) + \frac{f''(a)}{2!}(x - a)^2 + \cdots + \frac{f^{(n)}(a)}{n!}(x - a)^n + \cdots$$

Taylor *series* at $x = a$ are simply Taylor *polynomials* at $x = a$ with infinitely many terms. Taking $a = 0$ gives the Taylor series at $x = 0$, as defined earlier.

EXAMPLE 6

FINDING A TAYLOR SERIES AT $x = 1$

Find the Taylor series at $x = 1$ for $\ln x$.

Solution

Differentiating $f(x) = \ln x$ and evaluating at $x = 1$ gives

f	f'	f''	f'''	$f^{(4)}$	$\cdots$	
$\ln x$	x^{-1}	$-x^{-2}$	$2x^{-3}$	$-3!x^{-4}$	$\cdots$	$\ln x$ and its derivatives
0	1	-1	2	$-3!$	$\cdots$	Evaluated at $x = 1$

Therefore, the Taylor series at $x = 1$ for $\ln x$ is

$$(x - 1) - \frac{1}{2!}(x - 1)^2 + \frac{2}{3!}(x - 1)^3 - \frac{3!}{4!}(x - 1)^4 + \cdots$$

Taylor series with $a = 1$ and the derivatives found above

$$= (x - 1) - \frac{1}{2}(x - 1)^2 + \frac{1}{3}(x - 1)^3 - \frac{1}{4}(x - 1)^4 + \cdots \quad \text{Simplifying}$$

The radius of convergence is found from the ratio test (page 769), calculating the ratio of successive terms (and now including absolute value bars because of the alternating signs):

$$\left| \frac{\frac{1}{n + 1}(x - 1)^{n+1}}{\frac{1}{n}(x - 1)^n} \right| = \left| \underbrace{\frac{n}{n + 1}}_{\substack{\text{Approaches 1} \\ \text{as } n \to \infty}} \cdot \underbrace{\frac{(x - 1)^{n+1}}{(x - 1)^n}}_{(x - 1)} \right| \to \underbrace{| x - 1 |}_{r}$$

For convergence, this ratio must be less than 1: $|x - 1| < 1$, which is equivalent to $0 < x < 2$. Therefore, the Taylor series at $x = 1$ for $\ln x$ is

$$\ln x = (x - 1) - \frac{1}{2}(x - 1)^2 + \frac{1}{3}(x - 1)^3 - \frac{1}{4}(x - 1)^4 + \cdots$$

for $0 < x < 2$.

This series converges at the endpoint $x = 2$ but *not* at the endpoint $x = 0$, although we will not prove this.

> **Taylor Series at $x = 1$ for $\ln x$**
>
> $$\ln x = (x - 1) - \frac{(x-1)^2}{2} + \frac{(x-1)^3}{3} - \frac{(x-1)^4}{4} + \cdots \quad \text{for} \ \ 0 < x \leq 2$$

10.3 Section Summary

Every power series in x, $a_0 + a_1 x + a_2 x^2 + \cdots + a_n x^n + \cdots$, has a *radius of convergence R* (which may be zero, a positive number, or ∞), and the series converges for $|x| < R$ and diverges for $|x| > R$. We find R by using the ratio test, which may be thought of as testing whether the series is "asymptotically geometric."

A *Taylor series* is a type of power series whose coefficients are calculated from a given function f, using the same formulas as for Taylor polynomials:

$$f(x) = \sum_{n=0}^{\infty} \frac{f^{(n)}(0)}{n!} x^n \qquad\qquad \begin{array}{l}\text{Taylor series at } x = 0 \text{ for} \\ f(x) \text{ [the first term is } f(0)]\end{array}$$

$$f(x) = \sum_{n=0}^{\infty} \frac{f^{(n)}(a)}{n!} (x - a)^n \qquad\qquad \begin{array}{l}\text{Taylor series at } x = a \text{ for} \\ f(x) \text{ [the first term is } f(a)]\end{array}$$

A Taylor series can also be found by modifying an already existing Taylor series, as explained on pages 773–774. We derived the following Taylor series:

$$\frac{1}{1-x} = 1 + x + x^2 + x^3 + \cdots \qquad\qquad \text{for} \ \ -1 < x < 1$$

$$e^x = 1 + \frac{x}{1!} + \frac{x^2}{2!} + \frac{x^3}{3!} + \frac{x^4}{4!} + \cdots \qquad\qquad \text{for} \ \ -\infty < x < \infty$$

$$\sin x = x - \frac{x^3}{3!} + \frac{x^5}{5!} - \frac{x^7}{7!} + \cdots \qquad\qquad \text{for} \ \ -\infty < x < \infty$$

$$\cos x = 1 - \frac{x^2}{2!} + \frac{x^4}{4!} - \frac{x^6}{6!} + \cdots \qquad\qquad \text{for} \ \ -\infty < x < \infty$$

$$\ln x = (x - 1) - \frac{(x-1)^2}{2} + \frac{(x-1)^3}{3} - \frac{(x-1)^4}{4} + \cdots \qquad \text{for} \ \ 0 < x \leq 2$$

➤ **Solution to Practice Problem**

$$\cos x^2 = 1 - \frac{x^4}{2!} + \frac{x^8}{4!} - \frac{x^{12}}{6!} + \cdots \qquad \text{(Replacing } x \text{ by } x^2\text{)}$$

10.3 Exercises

Find the radius of convergence of each power series.

1. $1 - \dfrac{x}{2} + \dfrac{x^2}{2^2} - \dfrac{x^3}{2^3} + \cdots$

2. $1 + 2x + 2^2x^2 + 2^3x^3 + \cdots$

3. $\dfrac{x}{1 \cdot 2} + \dfrac{x^2}{2 \cdot 3} + \dfrac{x^3}{3 \cdot 4} + \dfrac{x^4}{4 \cdot 5} + \cdots$

4. $1 - \dfrac{x}{3} + \dfrac{x^2}{3^2} - \dfrac{x^3}{3^3} + \cdots + \dfrac{(-1)^n x^n}{3^n} + \cdots$

5. $\displaystyle\sum_{n=0}^{\infty} \dfrac{(-1)^n x^n}{n!}$ **6.** $\displaystyle\sum_{n=0}^{\infty} \dfrac{2^n x^n}{n!}$

7. $\displaystyle\sum_{n=0}^{\infty} \dfrac{n! x^n}{5^n}$ **8.** $\displaystyle\sum_{n=0}^{\infty} n! x^n$

9. $\displaystyle\sum_{n=0}^{\infty} \dfrac{x^{2n}}{n!}$ **10.** $\displaystyle\sum_{n=0}^{\infty} x^{n^2}$

Find the Taylor series at $x = 0$ for each function by calculating three or four derivatives and using the definition of Taylor series.

11. $\ln(1 + x)$ **12.** e^{2x}

13. $\cos 2x$ **14.** $\dfrac{1}{3 - x}$

15. $\sqrt{2x + 1}$ **16.** $2x^2 - 7x + 3$

Find the Taylor series at $x = 0$ for each function in *two* ways:

a. By calculating derivatives and using the definition of Taylor series
b. By modifying a known Taylor series

17. $e^{x/5}$ **18.** $\sin 2x$

19. Find the Taylor series at $x = \dfrac{\pi}{2}$ for $\sin x$.

20. Find the Taylor series at $x = 1$ for $\sqrt{x}$.

Find the Taylor series at $x = 0$ for each function by modifying one of the Taylor series from this section.

21. $\dfrac{x^2}{1 - x}$ **22.** $\dfrac{1}{1 + 5x}$

23. $\sin x^2$ **24.** e^{-x^2}

25. $\dfrac{e^x - 1}{x}$ **26.** $\sin x + \cos x$

(For Exercise 26, use two series from this section, writing the terms in order of increasing powers of x.)

27. $\dfrac{x - \sin x}{x^3}$ **28.** $\dfrac{1 - \cos x}{x^2}$

29. Find the Taylor series at $x = 0$ for $\ln(x + 1)$ by integrating both sides of

$$\dfrac{1}{1 + x} = 1 - x + x^2 - x^3 + \cdots \quad \text{for } |x| < 1$$

30. Find the Taylor series at $x = 0$ for $\dfrac{1}{(1 - x)^2}$ by differentiating both sides of

$$\dfrac{1}{1 - x} = 1 + x + x^2 + x^3 + \cdots \quad \text{for } |x| < 1$$

31. a. Find the Taylor series at $x = 0$ for $\dfrac{1}{1 + x^2}$ by modifying one of the series derived in this section.

b. Use the ratio test to find the radius of convergence of the series from part (a). Notice that although the original function $\dfrac{1}{1 + x^2}$ is defined for *all* values of x, the series is defined only for a much narrower set of x-values.

32. Use the Taylor series at $x = 0$ for e^x to find the series for $\dfrac{e^x - e^{-x}}{2}$. (This function is called the *hyperbolic sine*, written sinh x. Notice how similar its Taylor series is to that of sin x.)

33. Use the Taylor series at $x = 0$ for e^x to find the series for $\dfrac{e^x + e^{-x}}{2}$. (This function is called the *hyperbolic cosine*, written cosh x. Note how similar its Taylor series is to that of cos x. The graph of cosh x is called a "catenary." The St. Louis arch was built in this shape.)

34. SERIES CALCULATION OF sin 1

    a. Use a calculator set for *radians* to find sin 1 (rounded to five decimal places).

    b. Estimate sin 1 by using the first three terms of the Taylor series

$$\sin x = x - \frac{x^3}{3!} + \frac{x^5}{5!} - \cdots$$

evaluated at $x = 1$ (rounded to five decimal places). [*Note:* Computers use series similar to this to calculate values of trigonometric functions.]

35. SERIES CALCULATION OF $e^{0.5}$

    a. Use a calculator to find $e^{0.5}$ (rounded to five decimal places).

    b. Estimate $e^{0.5}$ by using the first four terms of the Taylor series

$$e^x = 1 + x + \frac{x^2}{2!} + \frac{x^3}{3!} + \frac{x^4}{4!} + \cdots$$

evaluated at $x = 0.5$ (rounded to five decimal places).

    c. (Requires sequence and series operations) Set your calculator to find the sum of the series in part (b) up to any number of terms. How many terms are required for

the sum (rounded to five decimal places) to agree with the value found in part (a)?

36. SERIES CALCULATION OF e^2 (Requires sequence and series operations)

    a. Use your calculator to find e^2 rounded to six decimal places.

    b. The Taylor series for e^x evaluated at $x = 2$ gives $e^2 = \displaystyle\sum_{n=0}^{\infty} \frac{2^n}{n!}$. Set your calculator to find the sum of this series up to any number of terms. How many terms are required for the sum (rounded to six decimal places) to agree with your answer to part (a)?

37. Differentiate the series

$$\cos x = 1 - \frac{x^2}{2!} + \frac{x^4}{4!} - \frac{x^6}{6!} + \cdots$$

and check that the resulting series is the negative of the series for sin x, showing (again) that that $\dfrac{d}{dx} \cos x = -\sin x$.

38. Differentiate the series

$$e^x = 1 + \frac{x}{1!} + \frac{x^2}{2!} + \frac{x^3}{3!} + \frac{x^4}{4!} + \cdots$$

showing (again) that e^x is its own derivative.

39. Use the ratio test to show that the Taylor series $\sin x = \displaystyle\sum_{n=0}^{\infty} \frac{(-1)^n x^{2n+1}}{(2n+1)!}$ converges for all x.

40. For a power series that converges to a function f,

$$f(x) = a_0 + a_1 x + a_2 x^2 + a_3 x^3 + \cdots$$

show that the coefficients must be $a_0 = f(0)$, $a_1 = f'(0)$, $a_2 = \dfrac{f''(0)}{2!}$, ... Since these are exactly the Taylor series coefficients, this shows that a power series that converges to a function *is* the Taylor series for that function. [*Hint:* For a_0, just evaluate the given equation at 0. For $a_1, a_2, \ldots$, differentiate one or more times and then evaluate at $x = 0$. We are assuming the fact that a power series can be differentiated term by term inside its radius of convergence.]

APPLIED EXERCISES

41. GENERAL: Bell Curve The "normal distribution" or "bell-shaped curve" is used to predict many things, including IQs, incomes, and manufacturing errors. The area under this curve from 0 up to any number x is given by

the integral $\dfrac{1}{\sqrt{2\pi}} \displaystyle\int_0^x e^{-t^2/2}\, dt$.

a. Find the Taylor series at 0 for $e^{-t^2/2}$. [*Hint:* Modify a known series.]

b. Integrate this series from 0 to x and multiply by 0.4 $\left(\text{which approximates } \dfrac{1}{\sqrt{2\pi}}\right)$,

obtaining a Taylor series for $\dfrac{1}{\sqrt{2\pi}} \displaystyle\int_0^x e^{-t^2/2}\, dt$.

c. Estimate $\dfrac{1}{\sqrt{2\pi}} \displaystyle\int_0^1 e^{-t^2/2}\, dt$ by using the first three terms of the series found in part (b) evaluated at $x = 1$. Your answer finds, for example, the proportion of people with IQs between 100 and 115 (see page 481).

42. SOCIAL SCIENCE: Baby Booms If the birthrate (thousands of births per year) rises by an amount $\dfrac{1 - \cos t}{t}$ (above the usual level) in year t, the number of excess births during the first x years will be $\displaystyle\int_0^x \dfrac{1 - \cos t}{t}\, dt$.

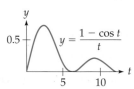

a. Find the Taylor series at 0 for $\dfrac{1 - \cos t}{t}$. [*Hint:* Modify a known series.]

b. Integrate this series from 0 to x, obtaining a Taylor series for the integral $\displaystyle\int_0^x \dfrac{1 - \cos t}{t}\, dt$

c. Estimate $\displaystyle\int_0^1 \dfrac{1 - \cos t}{t}\, dt$ by using the first

three terms of the series found in part (b) evaluated at $x = 1$.

43. BUSINESS: Sales Fluctuations As the pace of change in modern society quickens, popular fashions may fluctuate increasingly rapidly. Suppose that sales (above a certain minimum level) for a fashion item are $\cos t^2$ in year t, so that extra sales during the first x years are $\int_0^x \cos t^2\, dt$ (in thousands).

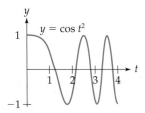

a. Find the Taylor series at 0 for $\cos t^2$. [*Hint:* Modify a known series.]

b. Integrate this series from 0 to x, obtaining a Taylor series for the integral $\int_0^x \cos t^2\, dt$.

c. Estimate $\int_0^1 \cos t^2\, dt$ by using the first three terms of the series found in part (b) evaluated at $x = 1$.

44. GENERAL: Temperature Under emergency conditions, the temperature in a nuclear containment vessel is expected to rise at the rate of $200e^{t^2}$ degrees per hour, so that the temperature change in the first x hours will be $200 \int_0^x e^{t^2}\, dt$ degrees. Estimate the temperature rise as follows:

a. Find the Taylor series at 0 for e^{t^2}. [*Hint:* Modify a known series.]

b. Integrate this series from 0 to x and multiply by 200, obtaining a Taylor series for $200 \int_0^x e^{t^2}\, dt$.

c. Estimate the temperature change in the first half hour by using the first three terms of the series found in part (b) evaluated at $x = \frac{1}{2}$.

45. nth TERM TEST For a convergent infinite series $S = c_1 + c_2 + c_3 + \cdots$, let S_n be the sum of the first n terms: $S_n = c_1 + c_2 + c_3 + \cdots + c_n$.

a. Show that $S_n - S_{n-1} = c_n$.

b. Take the limit of the equation in part (a) as $n \to \infty$ to show that $\lim\limits_{n \to \infty} c_n = 0$. This proves the nth term test for infinite series:

   In a convergent infinite series, the nth term must approach zero as $n \to \infty$.

   Or equivalently:

   If the nth term of a series does *not* approach zero as $n \to \infty$, then the series diverges.

46. Show that the Taylor series for e^x, $\displaystyle\sum_{n=0}^{\infty} \frac{x^n}{n!}$, converges to e^x by showing that the error for the nth Taylor approximation approaches zero as follows:

a. Use the ratio test to show that the series $\displaystyle\sum_{n=0}^{\infty} \frac{|x|^n}{n!}$ converges for all x.

b. Use the nth term test (Exercise 45) to conclude that $\dfrac{|x|^n}{n!} \to 0$ as $n \to \infty$.

c. Use part (b) to show that
$$|R_n(x)| \leq \frac{e^{|x|}}{(n+1)!}|x|^{n+1} \to 0$$
as $n \to \infty$ for any fixed value of x.

47. Show that the Taylor series for $\sin x$,
$$\sum_{n=0}^{\infty} \frac{(-1)^n x^{2n+1}}{(2n+1)!},$$ converges to $\sin x$ by showing that the error for the nth Taylor approximation approaches zero as follows:

a. Use the ratio test to show that the series $\displaystyle\sum_{n=0}^{\infty} \frac{|x|^n}{n!}$ converges for all x.

b. Use the nth term test (Exercise 45) to conclude that $\dfrac{|x|^{2n+1}}{(2n+1)!} \to 0$ as $n \to \infty$.
   [*Hint:* If the nth term approaches zero, then so does the $2n + 1$st.]

c. Use part (b) to show that
$$|R_{2n}(x)| \leq \frac{1}{(2n+1)!}|x|^{2n+1} \to 0$$
as $n \to \infty$ for any fixed value of x.

48. Prove part 1 of the Ratio Test (page 769) for positive terms as follows:

a. If $\lim\limits_{n \to \infty} \dfrac{c_{n+1}}{c_n} = r < 1$, and if s is a number between r and 1, so that $r < s < 1$, explain why there is an integer N such that
$$\frac{c_{n+1}}{c_n} \leq s \quad \text{for all} \quad n \geq N.$$

b. Show that this last inequality implies that
$$c_{N+1} \leq s \cdot c_N, \quad c_{N+2} \leq s^2 \cdot c_N,$$
$$c_{N+3} \leq s^3 \cdot c_N, \dots$$

c. Show that $\sum_{n=0}^{\infty} c_{N+n}$ converges by showing that
$$\sum_{n=0}^{\infty} c_{N+n} \leq \sum_{n=0}^{\infty} s^n \cdot c_N = c_N \sum_{n=0}^{\infty} s^n = c_N \frac{1}{1-s}$$

   It can be shown that if a series of positive terms remains below a (finite) number, then the series converges.

d. Lastly, show that $\sum_{n=0}^{\infty} c_n$ converges.
   [*Note:* It can be shown that if a series of positive terms converges, then the series obtained by changing the signs of some or all of the terms will also converge.]

49. Prove part 2 of the Ratio Test (page 769) as follows:

a. If $\lim\limits_{n \to \infty} \left|\dfrac{c_{n+1}}{c_n}\right| = r > 1$, explain why there is an integer N such that $\left|\dfrac{c_{n+1}}{c_n}\right| \geq 1$ for all $n \geq N$.

b. Show that this last inequality implies that
$$|c_{N+1}| \geq |c_N|, \quad |c_{N+2}| \geq |c_N|,$$
$$|c_{N+3}| \geq |c_N|, \dots$$

c. Then show that these inequalities imply that $\lim\limits_{n \to \infty} c_n \neq 0$, so $\sum_{n=0}^{\infty} c_n$ diverges by the nth term test (Exercise 45).

50. *Carry out the following steps to prove that if $a > 0$, then $\lim\limits_{x \to \infty} xe^{-ax} = 0$.

a. Show that $e^x > \dfrac{x^2}{2}$ for $x > 0$.

[*Hint:* Begin with the Taylor series $e^x = 1 + x + \dfrac{x^2}{2!} + \dfrac{x^3}{3!} + \cdots$ and drop all but the third term on the right.]

b. Replace x by ax to obtain $e^{ax} > \dfrac{a^2 x^2}{2}$.

c. Take the reciprocal of each side to obtain
$$e^{-ax} < \dfrac{2}{a^2 x^2}.$$

d. Multiply each side by x to obtain
$$xe^{-ax} < \dfrac{2}{a^2 x}.$$

e. Let $x \to \infty$ to obtain the claimed result.

51. *Prove that if $a > 0$ $\lim\limits_{x \to \infty} x^2 e^{-ax} = 0$.

[*Hint:* Use steps similar to those in Exercise 50, but in step (a) keep only the fourth term, and in step (d) multiply by x^2.]

* The results of Exercises 50 and 51 will be used to find the mean and variance of an exponential random variable on pages 836 and 844.

10.4 **NEWTON'S METHOD**

Introduction

Many times we have set a function equal to zero and solved for x. For example, setting the first or second derivative equal to zero is the first step in finding relative extreme points and inflection points. Newton's method is a procedure for *approximating* the solutions to such equations in case the *exact* solutions are difficult or impossible to find. The method was invented by Isaac Newton, one of the originators of calculus, and has many applications, such as finding the internal rate of return (interest rate) on a loan.

Zeros of a Function

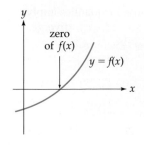

Recall from page 39 that an x-value at which a function $f(x)$ equals zero is called a *zero* of the function, or equivalently, a *root* of the equation $f(x) = 0$. A zero of a function is not simply the number 0, but is a value that, when substituted into the function, gives the *result* zero. Graphically, the zeros of a function are the x-values at which the graph meets the x-axis.

EXAMPLE 1

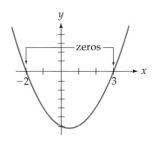

Zeros of $f(x) = x^2 - x - 6$ are the x-values where the curve crosses the x-axis.

FINDING ZEROS BY FACTORING

Find the zeros of $f(x) = x^2 - x - 6$.

Solution

$$f(x) = x^2 - x - 6$$
$$= (x - 3) \cdot (x + 2) = 0$$

Factoring and setting equal to zero

The zeros are

$$\begin{cases} x = 3 \\ x = -2 \end{cases}$$

From setting $(x - 3)$ and $(x + 2)$ equal to zero

These zeros are shown on the graph on the left.

Newton's Method

Only relatively simple functions can be factored. For more complicated functions, the zeros can only be *approximated,* and one of the most useful techniques is Newton's method. Newton's method begins with an initial approximation $x = x_0$, which may be merely a "guess" (more on this later). The function is then replaced by its first Taylor polynomial at x_0 (the tangent line at x_0), since it is the best linear approximation for $f(x)$ near x_0. That is, instead of solving $f(x) = 0$, we set the *Taylor polynomial* equal to zero:

$$f(x_0) + f'(x_0) \cdot (x - x_0) = 0$$

First Taylor polynomial at $x = x_0$ set equal to zero

$$f'(x_0) \cdot (x - x_0) = -f(x_0)$$

Subtracting $f(x_0)$ from each side

$$x - x_0 = -\frac{f(x_0)}{f'(x_0)}$$

Dividing each side by $f'(x_0)$ [assuming $f'(x_0) \neq 0)$]

$$x = x_0 - \frac{f(x_0)}{f'(x_0)}$$

Solving for x by adding x_0 to each side

This last formula gives the x-value (which we call x_1) where the first Taylor polynomial (the tangent line) crosses the x-axis. This x_1 should be a better approximation than the initial appoximation x_0 for the zero of the function $f(x)$, as shown in the diagram on the following page.

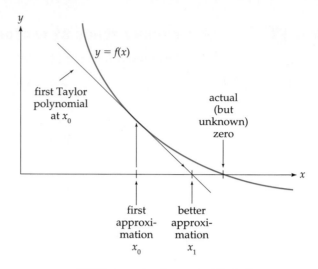

Initial approximation x_0, and the better approximation x_1, which is closer to the actual zero.

The same formula can then be applied to the "better" approximation x_1 to obtain an even better approximation x_2, and then again to x_2 for a still better x_3, and so on:

$$x_1 = x_0 - \frac{f(x_0)}{f'(x_0)}$$

$$x_2 = x_1 - \frac{f(x_1)}{f'(x_1)}$$ Same formula applied to successive approximations

$$x_3 = x_2 - \frac{f(x_2)}{f'(x_2)}$$

$$\vdots$$

Each use of the formula is called an "iteration" ("to iterate" means "to repeat"). If the results of the iterations $x_1, x_2, x_3, \ldots$ approach a limit, we say that the iterations "converge" to this limit. In summary:

Newton's Method

To approximate a solution to $f(x) = 0$, choose an initial approximation x_0, and calculate $x_1, x_2, x_3, \ldots$ using

$$x_{n+1} = x_n - \frac{f(x_n)}{f'(x_n)} \qquad \text{for } n = 0, 1, 2, \ldots$$

If the numbers $x_0, x_1, x_2, \ldots$ converge, they converge to a solution of $f(x) = 0$.

EXAMPLE 2

APPROXIMATING BY NEWTON'S METHOD

Approximate $\sqrt{2}$ by using three iterations of Newton's method.

Solution We approximate $\sqrt{2}$ by seeking a zero $f(x) = x^2 - 2$, which equals zero at $x = \sqrt{2}$. Since $\sqrt{2}$ lies between 1 and 2, we choose our initial approximation to be $x_0 = 1$ (we could just as well have chosen $x_0 = 2$). Calculating $f(x)$ and $f'(x)$:

$$f(x) = x^2 - 2 \qquad \text{The function whose zero we want}$$

$$f'(x) = 2x \qquad \text{Its derivative}$$

The iteration formula is

$$x_{n+1} = x_n - \frac{x_n^2 - 2}{2x_n} \qquad x_{n+1} = x_n - \frac{f(x_n)}{f'(x_n)} \text{ with the}$$

f and f' above

The approximations are

$x_0 = 1$ Initial approximation

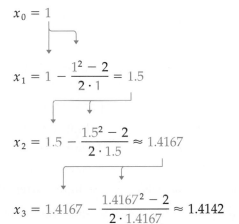

$x_1 = 1 - \dfrac{1^2 - 2}{2 \cdot 1} = 1.5$

$\qquad x_1 = x_0 - \dfrac{f(x_0)}{f'(x_0)}$

applied to $x_0 = 1$

$x_2 = 1.5 - \dfrac{1.5^2 - 2}{2 \cdot 1.5} \approx 1.4167$

$\qquad x_2 = x_1 - \dfrac{f(x_1)}{f'(x_1)}$

applied to $x_1 = 1.5$

$x_3 = 1.4167 - \dfrac{1.4167^2 - 2}{2 \cdot 1.4167} \approx 1.4142$

Iteration formula
applied to $x_2 = 1.4167$

The x_3 is our approximation for $\sqrt{2}$, the zero of $f(x) = x^2 - 2$:

$$\sqrt{2} \approx 1.4142$$

This estimate is correct to four decimal places (after only three iterations).

The approximations x_0, x_1, x_2, and x_3 to $\sqrt{2}$ are shown in the graph on the following page. The convergence is rapid enough that the second and third iterations require a magnified view.

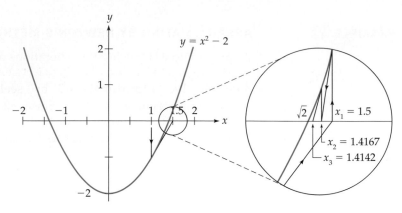

$f(x) = x^2 - 2$ and the iterations $x_0, x_1, x_2,$ and x_3 approaching $\sqrt{2}$.

Practice Problem 1   Carry out one more iteration "by hand" with the formula above, using $x_3 = 1.4142,$ keeping (if possible) nine decimal places of accuracy.

➤ **Solution on page 793**

Observe that the number of correct digits in successive approximations roughly *doubled* with each iteration:

$x_0 = 1$	initial "guess"
$x_1 = 1.5$	1 digit correct
$x_2 = 1.4167$	3 digits correct
$x_3 = 1.4142$	5 digits correct
$x_4 = 1.414213562$	10 digits correct (from Practice Problem 1)

How do you decide when to stop the iterations? Generally, you first choose a degree of accuracy, say 0.001, and stop when two successive approximations differ by less than your chosen accuracy.

How do you choose the function to use in the iterations? The number in Example 1, $\sqrt{2}$, satisfies $x^2 = 2$, or equivalently $x^2 - 2 = 0$, so we sought the zeros of the function $f(x) = x^2 - 2$. Other roots require different equations. For example, $\sqrt[3]{11}$ satisfies $x^3 = 11$, so we want to find a zero of $f(x) = x^3 - 11,$ beginning with, perhaps, $x_0 = 2$.

How do you choose the initial estimate x_0? There is no general rule—you just try to find a number close to the number that you are seeking. The better the initial estimate, the faster the given accuracy will be achieved. If a function has *several* zeros, the one that the iterations converge to will depend upon the initial choice x_0. In Example 2, the equation $x^2 - 2 = 0$ actually has *two* solutions, $\sqrt{2}$ and $-\sqrt{2}$, corresponding to the two places where the curve crosses the x axis (see the graph on the previous page). The initial choice of $x_0 = 1$ led to an approximation for $\sqrt{2}$, whereas choosing $x_0 = -1$ would have led to an approximation for $-\sqrt{2}$.

Practice Problem 2

To approximate $\sqrt[5]{30}$, what function $f(x)$ would you use, and what would be a good initial estimate x_0? ➤ Solution on page 793

Calculator and Computer Programs for Newton's Method

The program NEWTON for the *TI-83* graphing calculator carries out Newton's method for any number of iterations and is listed on the left in the following box.* A similar program in BASIC is listed to its right. In both, you choose the number of iterations in advance.

TI-83 Program NEWTON	*BASIC Program for Newton's Method*
Disp "USES FUNCTION Y$_1$" Input "NUM OF ITERNS? ", N Input "INIT GUESS? ", X For (I, 1, N) X $-$ Y$_1$/NDERIV (Y$_1$, X, X) $\rightarrow$ X Disp X End	The small print is explanation, not part of the program. x = fill in initial guess FOR i = 1 TO fill in number of iterations x = x $-$ $\left(\dfrac{\text{fill in}}{\text{function}}\right) / \left(\dfrac{\text{fill in}}{\text{derivative}}\right)$ PRINT x NEXT i END

The ZERO and SOLVE operations in a graphing calculator use Newton's method or one similar to it. The "initial guess" that these operations request is simply the initial x_0 of Newton's method.

EXAMPLE 3

APPROXIMATING WITHIN A GIVEN TOLERANCE

Approximate the solution to $e^x = 2 - 2x$, continuing until two successive iterations agree to nine decimal places.

---

* Although this program is short enough to be typed directly into a calculator, it may be obtained in the same way as the others, as described in the Preface.

On a Texas Instruments graphing calculator, Newton's method may also be carried out as follows. Enter the function in Y$_1$, then (if 0 is the initial guess) enter the two statements in the home screen 0 $\rightarrow$ X and X $-$ Y$_1$/NDERIV(Y$_1$, X, X) $\rightarrow$ X and then repeatedly press ENTER.

Solution We find the initial estimate x_0 from a quick sketch of the graphs of e^x and $2 - 2x$, as follows.

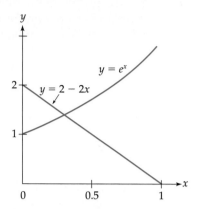

They intersect at an x-value closer to 0 than to 1, so we choose $x_0 = 0$ as our initial estimate (other choices are also possible). We write the equation $e^x = 2 - 2x$ with zero on the right:

$$e^x - 2 + 2x = 0$$
$$\underbrace{}_{f(x)}$$

Definition of $f(x)$

We could carry out the calculations "by hand," using the iteration formula

$$x_{n+1} = x_n - \frac{e^{x_n} - 2 + 2x_n}{e^{x_n} + 2} \qquad\qquad x_{n+1} = x_n - \frac{f(x_n)}{f'(x_n)}$$

beginning with $x_0 = 0$, but instead we use the graphing calculator program NEWTON (see the previous page). The results, using 5 iterations, are shown below.

```
INIT GUESS? 0
            .3333333148
            .3149922861
            .3149230588
            .3149230578  } agree
            .3149230578
            Done
```

The last two answers clearly agree to nine decimal places, giving the approximation

$$x = 0.3149230578 \qquad \text{Approximation to the solution of } e^x = 2 - 2x$$

Internal Rate of Return on a Loan

By federal law, any loan agreement must disclose the rate of interest. If the amount of the loan and the repayments are specified, finding the implied interest rate (called the *internal rate of return*) can involve an equation whose roots can only be approximated.*

EXAMPLE 4

FINDING THE INTERNAL RATE OF RETURN

A bank loan of $2500 is to be repaid in three payments of $1000 at the end of the first, second, and third years. Find the internal rate of return.

Solution
The internal rate of return is defined as the interest rate r for which the present values of the payments add up to the amount of the loan. To find the present value at the time of the loan, we divide the payment after n years by $(1 + r)^n$:

$$2500 = \frac{1000}{(1 + r)^1} + \frac{1000}{(1 + r)^2} + \frac{1000}{(1 + r)^3}$$

Amount   Present value   Present value   Present value
of loan   of first   of second   of third
        payment   payment   payment

Dividing by 1000 and replacing $1 + r$ by x:

$$2.5 = \frac{1}{x} + \frac{1}{x^2} + \frac{1}{x^3} \qquad\qquad x = 1 + r$$

$$2.5x^3 - x^2 - x - 1 = 0$$

$\underbrace{}$
$f(x)$

Multiplying by x^3 and moving all terms to the left

Definition of $f(x)$

Again, we could calculate the iterations "by hand" using the formula on page 786 with the $f(x)$ above and its derivative beginning with $x = 1$, but instead we use the graphing calculator program NEWTON.

---

* See H. Paley, P. Colwell, and R. Cannaday, "Internal Rates of Return," UMAP Module 640, COMAP, Arlington, Mass., 1984.

Graphing Calculator Exploration

```
NUM OF ITRNS? 4
INIT GUESS? 1
        1.111111049
        1.097250629
        1.097010329
        1.097010257
            Done
```

Use the graphing calculator program NEWTON with $y_1 = 2.5x^3 - x^2 - x - 1$, beginning with $x_0 = 1$ to obtain the results shown on the left.

The iterations give $x \approx 1.097$. To find r, we subtract 1 (since we defined $x = 1 + r$), giving $r \approx 0.097$.

The internal rate of return is 9.7%.   0.097 as a percent

The Iterations May Not Converge

Although Newton's method often converges rapidly, there are functions for which it converges very slowly, or not at all. For example, if the derivative is zero at any approximation, then the iteration formula will involve division by zero, and so the process will stop. (This can often be remedied by choosing a different starting point.) The following two diagrams illustrate other problems: the first shows what can happen if there is an inflection point between two successive approximations, and the second shows the difficulties caused by a nearby critical point.

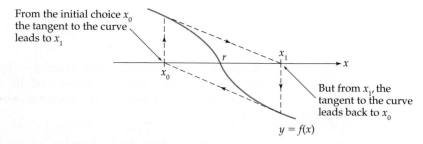

From the initial choice x_0 the tangent to the curve leads to x_1

But from x_1, the tangent to the curve leads back to x_0

$y = f(x)$

For this curve, the approximations will alternate between x_0 and x_1, never approaching the true root r in the center of the diagram.

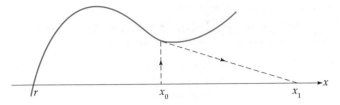

Starting at x_0, the tangent line to the curve
leads to x_1 far to the *right*, instead of to the
actual root r, which is to the left.

Although these diagrams show difficulties that can occur, Newton's
method is generally very successful in applications, usually doubling
the number of correct digits with each successive iteration.

10.4 Section Summary

Newton's method is a procedure for approximating the *zeros* of a function, or equivalently, the *roots* or *solutions* of $f(x) = 0$. An initial estimate x_0 is successively "improved" using the formula

$$x_{new} = x_{old} - \frac{f(x_{old})}{f'(x_{old})}$$

The convergence is often quite rapid, but is affected by such consider-
ations as the accuracy of the initial estimate, roundoff errors, and how
closely the function is approximated by its first Taylor polynomial (the
tangent line).

Besides finding internal rates of return, the method is useful for
solving many other equations, such as finding break-even points
(revenue = cost) or market equilibrium points (supply = demand), as
we will see in the exercises.

➤ **Solutions to Practice Problems**

1. $x_4 = 1.4142 - \dfrac{1.4142^2 - 2}{2 \cdot 1.4142} \approx 1.414213562$

2. Use $f(x) = x^5 - 30$ [or $f(x) = 30 - x^5$] and $x_0 = 2$ (although other
values are possible).

10.4 Exercises

Use Newton's method beginning with the given x_0 to find the first two approximations x_1 and x_2. Carry out the calculation "by hand" with the aid of a calculator, rounding to two decimal places.

1. $x^3 + x - 4 = 0$
$x_0 = 1$

2. $x^3 + 2x - 4 = 0$
$x_0 = 1$

3. $e^x - 3x = 0$
$x_0 = 0$

4. $x - \cos x = 0$
$x_0 = 0$

Use Newton's method to approximate each root, continuing until two successive iterations agree to three decimal places. Carry out the calculation "by hand" with the aid of a calculator, rounding to three decimal places.

5. $\sqrt{5}$

6. $\sqrt[3]{22}$

Use Newton's method to approximate the root of each equation, beginning with the given x_0 and continuing until two successive approximations agree to three decimal places. Carry out the calculation "by hand" with the aid of a calculator, rounding to three decimal places.

7. $x^3 + 3x - 8 = 0$
$x_0 = 2$

8. $x^3 - 3x + 3 = 0$
$x_0 = -2$

9. $e^x + 3x + 2 = 0$
$x_0 = -1$

10. $e^x + 4x - 4 = 0$
$x_0 = 0$

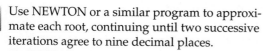 Use NEWTON or a similar program to approximate each root, continuing until two successive iterations agree to nine decimal places.

11. $\sqrt[3]{130}$

12. $\sqrt{130}$

Use NEWTON or a similar program to approximate the root of each equation in Exercises 13–20, beginning with the given x_0 and continuing until two successive approximations agree to nine decimal places.

13. $x^5 + 2x - 1 = 0$
$x_0 = 0$

14. $x^5 + 5x - 4 = 0$
$x_0 = 1$

15. $x + \ln x - 5 = 0$
$x_0 = 2$

16. $x^2 + \ln x - 3 = 0$
$x_0 = 2$

17. $e^{x^2} + x - 3 = 0$
$x_0 = 1$

18. $e^{2x} + 3x = 0$
$x_0 = 0$

19. $x + \sin x - 2 = 0$
$x_0 = 1$

20. $x^2 + \cos x - 4 = 0$
$x_0 = 1$

Use a graphing calculator to graph $f(x)$ and $g(x)$ together on a reasonable window and estimate the x-value where the curves meet. Then use Newton's method to approximate the solution of $f(x) = g(x)$, beginning with your estimate and continuing until two successive iterations agree to nine decimal places. (You may check your answer using INTERSECT, which uses a method similar to Newton's.)

21. $f(x) = x$, $g(x) = 2 \cos x$

22. $f(x) = \sin x$, $g(x) = 1 - x$

23. $f(x) = \ln x$, $g(x) = -x$

24. $f(x) = e^x$, $g(x) = 2 - x$

25. Use Newton's method (either "by hand" or using a graphing calculator) to solve $x^2 + 3 = 0$, beginning with $x_0 = 1$. Explain why it doesn't work.

26. Apply Newton's method (either "by hand" or using a graphing calculator) to the function $f(x) = x^{1/3}$ beginning with $x_0 = 1$. The graph below may help explain what goes wrong. [*Hint:* See the diagram on the page 792.] Does choosing another starting point help?

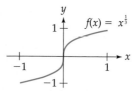

27. What happens when Newton's method is applied to a linear function $f(x) = mx + b$ with $m \neq 0$?

28. What happens in Newton's method if the initial estimate x_0 is actually a zero of the function?

29. Calculate the first 20 or so iterations for the zero of $f(x) = x^{10}$, beginning with $x_0 = 1$. Notice how slowly the values converge to the actual zero, $x = 0$. Can you see why from the following graph? How many iterations does it take to reach 0.001 (rounded)? Try using zero on your graphing calculator to find the root, starting at $x = 1$.

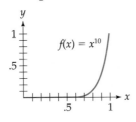

30. Show that finding $\sqrt{N}$ by applying Newton's method to the equation $x^2 - N = 0$ leads to the iteration formula

$$x_{n+1} = \frac{1}{2}\left(x_n + \frac{N}{x_n}\right).$$

APPLIED EXERCISES

31. ECONOMICS: Supply and Demand The supply and demand functions for a product are $S(p) = p$ and $D(p) = 10e^{-0.1p}$, where p is its price (in dollars). Find the "market equilibrium price" p at which $S(p) = D(p)$. Use Newton's method with two iterations, beginning with $x_0 = 4$. Carry out the calculation "by hand" with the aid of a calculator, rounding to two decimal places.

32. BUSINESS: Cost and Revenue The cost and revenue functions for a company are $C(x) = 50 + 2x$ and $R(x) = 2x^{3/2}$, where x is the quantity (in thousands). Find the "break-even" quantity x at which $C(x) = R(x)$. Use Newton's method with two iterations, beginning with $x_0 = 10$. Carry out the calculation "by hand" with the aid of a calculator, rounding to one decimal place.

33. ECONOMICS: Supply and Demand The supply and demand functions for a product are $S(p) = 2e^{0.5p}$ and $D(p) = 10 - p$, where p is the price of the product (in dollars) and $0 < p < 10$. Find the "market equilibrium price" p at which $S(p) = D(p)$. Use NEWTON or a similar program, beginning with some initial guess between 0 and 10, and continuing until the iterations agree to two decimal places.

34. BUSINESS: Cost and Revenue The cost and revenue functions for a company are $C(x) = 50 + 40\sqrt{x}$ and $R(x) = 3x$, where x is the quantity ($x \geq 100$). Find the "break-even" quantity x at which $C(x) = R(x)$. Use NEWTON or a similar program, beginning with an initial guess of at least 100, and continuing until the iterations agree when rounded to the nearest unit.

35. GENERAL: Internal Rate of Return A $15,000 loan is to be repaid in four installments of $5000 each at the end of the first, second, third, and fourth years. Use NEWTON or a similar program to find the internal rate of return (to the nearest tenth of a percent).

36. GENERAL: Internal Rate of Return An $8000 loan is to be repaid in installments of $1000, $2000, $3000, and $4000 at the end of the first, second, third, and fourth years, respectively. Use NEWTON or a similar program to find the internal rate of return (to the nearest tenth of a percent).

37. BIOMEDICAL: Epidemics An epidemic is subject to two effects: linear growth, as infected people move into the area, and exponential decline, as existing cases are cured. Find when the two effects will be in equilibrium by solving $5t = 30e^{-0.2t}$, where t is the number of weeks since the epidemic began. Use NEWTON or a similar program, beginning with some appropriate initial value and continuing until the iterations agree to two decimal places.

38. SOCIAL SCIENCE: Immigration Immigration into a country is occurring at the rate of

$2e^{0.1t}$ thousand people in year t, and the government sets an upper limit of $20 - t$ thousand people in year t (for $0 \le t \le 20$). Find when this limit will be reached by solving $2e^{0.1t} = 20 - t$. Use NEWTON or a similar program, beginning with some appropriate initial estimate and continuing until the iterations agree to two decimal places.

Chapter Summary with Hints and Suggestions

Reading the text and doing the exercises in this chapter have helped you to master the following skills, which are listed by section (in case you need to review them) and are keyed to particular Review Exercises. Answers for all Review Exercises are given at the back of the book, and full solutions can be found in the Student Solutions Manual.

10.1 Geometric Series

- Determine whether an infinite geometric series converges, and if so, find its sum. *(Review Exercises 1–8.)*

$$a + ar + ar^2 + ar^3 + \cdots = \frac{a}{1 - r} \quad \text{for} \ |r| < 1$$

- Find the value of a repeating decimal. *(Review Exercises 9–10.)*

- Find the value of an annuity. *(Review Exercises 11–12.)*

$$a + ar + ar^2 + \cdots + ar^{n-1} = a\frac{1 - r^n}{1 - r}$$

- Solve an applied problem using an infinite series. *(Review Exercises 13–14.)*

10.2 Taylor Polynomials

- Find a Taylor polynomial for a function. *(Review Exercises 15–18.)*

$$p_n(x) = f(0) + \frac{f'(0)}{1!}x + \frac{f''(0)}{2!}x^2 + \cdots + \frac{f^{(n)}(0)}{n!}x^n$$

- Find a Taylor polynomial for a function, use it to approximate a value, and estimate the error. *(Review Exercises 19–20.)*

$$|R_n(x)| \le \frac{M}{(n + 1)!}|x|^{n+1} \quad \text{for} \ |f^{(n+1)}| \le M$$

- Find the best linear approximation for a function. *(Review Exercise 21.)*

- Find a Taylor polynomial that approximates a function within a given accuracy. *(Review Exercise 22.)*

- Find a Taylor polynomial of $x = a$ for function and graph both. *(Review Exercises 23–24.)*

- Approximate a value using a Taylor polynomial and estimate the error. *(Review Exercise 25.)*

- Solve an applied problem using a Taylor polynomial and check using a calculator. *(Review Exercises 26–28.)*

10.3 Taylor Series

- Find the radius of convergence of a power series. *(Review Exercises 29–32.)*

$$r = \lim_{n \to \infty} \left| \frac{c_{n+1}}{c_n} \right|$$

- Use one Taylor series to find another. *(Review Exercises 33–36.)*

$$\frac{1}{1 - x} = 1 + x + x^2 + x^3 + \cdots \quad \text{on} \ (-1, 1)$$

$$e^x = 1 + \frac{x}{1!} + \frac{x^2}{2!} + \frac{x^3}{3!} + \frac{x^4}{4!} + \cdots \quad \text{on} \ (-\infty, \infty)$$

$$\sin x = x - \frac{x^3}{3!} + \frac{x^5}{5!} - \frac{x^7}{7!} + \cdots \quad \text{on } (-\infty, \infty)$$

$$\cos x = 1 - \frac{x^2}{2!} + \frac{x^4}{4!} - \frac{x^6}{6!} + \cdots \quad \text{on } (-\infty, \infty)$$

$$\ln x = (x-1) - \frac{(x-1)^2}{2} + \frac{(x-1)^3}{3} - \cdots$$
$$\text{on } (0, 2]$$

- Find a Taylor series in two ways: from the definition using derivatives and by modifying a known series. *(Review Exercises 37–38.)*

$$f(0) + \frac{f'(0)}{1!}x + \frac{f''(0)}{2!}x^2 + \frac{f'''(0)}{3!}x^3 + \cdots$$

- Find a Taylor series at $x = a$. Compare it to the original function. *(Review Exercise 39.)*

$$f(a) + \frac{f'(a)}{1!}(x-a) + \frac{f''(a)}{2!}(x-a)^2 + \cdots$$

- Verify two trigonometric identities using Taylor series. *(Review Exercise 40.)*

- Solve an applied problem by using a Taylor series. *(Review Exercise 41.)*

- Derive a formula for the radius of convergence of a power series. *(Review Exercise 42.)*

10.4 Newton's Method

- Use Newton's method to approximate a root within a given accuracy.
(Review Exercises 43–46.)

$$x_{\text{new}} = x_{\text{old}} - \frac{f(x_{\text{old}})}{f'(x_{\text{old}})}$$

- Use Newton's method to approximate a root within a given accuracy.
(Review Exercises 47–52.)

- Solve an applied problem by using Newton's method. *(Review Exercises 53–55.)*

Hints and Suggestions

- In a finite geometric series, the number of terms is always one more than the final power of x.

- In mathematics, the terms *series* and *sequence* mean different things. Terms in a *series* are separated by + or − signs, and those in a *sequence* by *commas* (as in a sequence of approximations from Newton's method).

- Compare the formulas for the sum of finite and infinite geometric series: the *infinite* formula is simpler than the *finite* formula.

- For an infinite geometric series, be sure to check convergence ($|r| < 1$) before using the formula, or you may get an incorrect answer.

- The *first* Taylor polynomial is the best *linear* approximation, the *second* Taylor polynomial is the best *quadratic* approximation (at a point), and so on. Taylor polynomials are found by matching values and derivatives at a point.

- A Taylor polynomial at $x = a$ is *exact at* $x = a$, and less accurate further from $x = a$ (as judged by the error formula). The accuracy generally improves as the degree increases.

- The error formula for Taylor polynomials is easy to remember, since it closely resembles the first term left off.

- Taylor series are most easily found by modifying existing Taylor series.

- In practice, Newton's method for approximating the root of an equation generally converges very quickly, typically doubling the number of digits of accuracy with each iteration.

- A graphing calculator helps by summing many terms of a series, but it cannot sum an *infinite* series (unless you give it a formula). It can also *graph* a function together with several Taylor polynomials to show how close they are. It is also useful for calculating iterations of Newton's method.

- **Practice for Test:** Review Exercises 1, 3, 7, 13, 19, 23, 29, 35, 41, 47, 53.

Review Exercises for Chapter 10

Practice test exercise numbers are in green.

10.1 Geometric Series

Determine whether each infinite geometric series converges or diverges. If it converges, find its sum.

1. $\frac{3}{10} + \frac{9}{100} + \frac{27}{1000} + \cdots$ 2. $\frac{2}{5} - \frac{4}{25} + \frac{8}{125} - \cdots$

3. $\frac{3}{100} - \frac{9}{100} + \frac{27}{100} - \cdots$ 4. $\frac{2}{3} - \frac{4}{9} + \frac{8}{27} - \cdots$

5. $10 + 8 + \frac{32}{5} + \frac{128}{25} + \cdots$ 6. $\sum_{i=0}^{\infty} \left(-\frac{3}{4}\right)^i$

7. $\sum_{k=0}^{\infty} 6(-0.2)^k$ 8. $\sum_{n=0}^{\infty} \frac{2^n}{3}$

Find the value of each repeating decimal.

9. $0.545454\ldots = 0.\overline{54}$

10. $0.727272\ldots = 0.\overline{72}$

11. **GENERAL: Annuity** A star basketball player preparing for his parents' retirement plans to deposit $6000 at the end of each month for 8 years into an account paying 6% compounded monthly. Find the amount of this annuity.

12. **BUSINESS: Annuities** Derive the formula

$$\text{Amount} = D\left[\frac{(1+i)^n - 1}{i}\right]$$

for the amount of an annuity consisting of n payments of D dollars each, with payments made at the end of each period, at compound interest rate i per period. (The formula in the brackets is denoted $S_{\overline{n}|i}$ (read "s angle n at i") and is calculated in tables.)

13. **BIOMEDICAL: Drug Dosage** A patient has been taking daily doses of 0.26 milligram (mg) of digoxin for an extended period of time, and 65% of the drug in the bloodstream is absorbed by the body during each day.

a. Find the maximum and minimum long-term amounts in the bloodstream.

b. How many doses does it take for the cumulative amount in the bloodstream to reach 0.39 mg? To reach 0.399 mg?

14. **ECONOMICS: Banking Reserves** When a bank receives a deposit, it keeps only a small part (say 10%) as its "reserves," lending out the rest. These loans result in other purchases, with the sums eventually deposited in other banks, which in turn keep 10% reserves and lend out the other 90%. This process continues, with an original deposit of $1000 generating deposits totaling

$$1000 + 1000(0.90) + 1000(0.90)^2 + \cdots$$

a. Sum this series and find the "multiplier" by which the original deposit is multiplied. This illustrates how the banking system multiplies money.

b. (Graphing calculator with series operations helpful) How many terms of the preceding series are required for the sum to surpass $9000? To surpass $9900?

10.2 Taylor Polynomials

For each function:

a. Find the third Taylor polynomial at $x = 0$.

b. Graph the original function and the Taylor polynomial together on the indicated window.

15. $f(x) = \sqrt{x+9}$
(for (b), use window $[-15, 15]$ by $[-6, 6]$)

16. $f(x) = e^{3x}$
(for (b), use window $[-1, 1]$ by $[-1, 8]$)

17. $f(x) = \ln(2x+1)$
(for (b), use window $[-1, 1]$ by $[-5, 5]$)

18. $f(x) = \sin^2 x$
(for (b), use window $[-3, 3]$ by $[-2, 2]$)

19. a. For $f(x) = \sqrt{x+1}$, find the third Taylor polynomial at $x = 0$.

b. Use your answer to part (a) to approximate $\sqrt{2}$.

(continues)

c. Estimate the error by the error formula. [*Hint:* Calculate $f^{(4)}(x)$ and find an M such that $|f^{(4)}(x)| \le M$ for $0 \le x \le 1$.]

d. Restate the approximation and error, rounded appropriately.

20. a. For $f(x) = \cos x$, find the fourth Taylor polynomial at $x = 0$.

b. Use your answer to part (a) to approximate $\cos 1$.

c. Estimate the error by the error formula. [*Hint:* Calculate $f^{(5)}(x)$ and find an M such that $|f^{(5)}(x)| \le M$ for $0 \le x \le 1$.]

d. Restate the approximation and error, rounded appropriately.

21. What linear function best approximates $\sqrt[n]{1 + x}$ near $x = 0$?

22. Find the Taylor polynomial at $x = 0$ that approximates e^x on the interval $-1 \le x \le 1$ with error less than 0.0001.

23. a. For $f(x) = \sin x$, find the fourth Taylor polynomial at $x = \pi/2$.

b. Graph $\sin x$ and its Taylor polynomial together on the window $[-2, 5]$ by $[-2, 2]$.

24. a. For $f(x) = x^{3/2}$, find the third Taylor polynomial at $x = 4$.

b. Graph $x^{3/2}$ and its Taylor polynomial together on the window $[-2, 15]$ by $[-5, 60]$.

25. a. For $f(x) = x^{2/3}$, find the second Taylor polynomial at $x = 1$.

b. Use your answer to part (a) to approximate $(1.2)^{2/3}$.

c. Calculate $f^{(3)}(x)$ and find an M such that $|f^{(3)}(x)| \le M$ for $1 \le x \le 1.2$.

d. Estimate the error by the error formula.

e. Restate the approximation and error, rounded appropriately.

26. BIOMEDICAL: Drug Dosage Two hours after a patient takes a 1-gram dose of penicillin, the amount of penicillin in the patient's bloodstream will be $e^{-0.22}$ gram. Approximate $e^{-0.22}$ by using the second Taylor polynomial for e^x (rounded to two decimal places). Check your answer by using $\boxed{e^x}$ on a calculator.

27. GENERAL: Temperature The mean temperature in New York City on December 25 is predicted to be $55 - 24 \cos 0.45$. Approximate this number by using the second Taylor polynomial at $x = 0$ for $\cos x$. Then check your answer using the $\boxed{\cos x}$ key on a calculator.

28. BUSINESS: Sales Just after introducing a new product, a company's weekly sales are predicted to be $200e^{0.1t^2}$ units.

a. Find the fourth Taylor polynomial at $t = 0$ for this function. [*Hint:* Begin with the second Taylor polynomial for e^x and use substitution, then multiply by 200.]

b. Use your Taylor polynomial to approximate the sales in week $t = 2$.

c. Use a calculator to evaluate the original function at $t = 2$ and compare answers.

10.3 Taylor Series

Find the radius of convergence of each power series.

29. $\displaystyle\sum_{n=0}^{\infty} \frac{nx^n}{3^n}$

30. $\dfrac{x}{\sqrt{1}} + \dfrac{x^2}{\sqrt{2}} + \dfrac{x^3}{\sqrt{3}} + \dfrac{x^4}{\sqrt{4}} + \cdots$

31. $\displaystyle\sum_{n=0}^{\infty} \frac{n!x^n}{6^n}$ **32.** $\displaystyle\sum_{n=1}^{\infty} \frac{(-1)^n x^n}{(n-1)!}$

For each function, find the Taylor series at $x = 0$ by modifying a known Taylor series.

33. $\dfrac{x^3}{1 - x^2}$ **34.** $x \cos 2x$

35. xe^{x^2} **36.** xe^{-x}

For each function, find the Taylor series at $x = 0$ in two ways:

a. By calculating derivatives and using the definition of Taylor series.

b. By modifying a known Taylor series

37. $\dfrac{1}{1 - 2x}$ **38.** $e^{-x/2}$

39. a. Find the Taylor series at $x = \pi/2$ for $\cos x$.

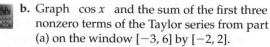 **b.** Graph $\cos x$ and the sum of the first three nonzero terms of the Taylor series from part (a) on the window $[-3, 6]$ by $[-2, 2]$.

40. Verify the negative angle identities (page 621) $\sin(-x) = -\sin x$ and $\cos(-x) = \cos x$ by using the Taylor series for the sine and cosine functions.

41. GENERAL: Commuter Traffic The number of commuter cars passing through a toll plaza per hour is $\dfrac{1}{1 + t^2}$ thousand, where t is the total number of hours after 9 A.M. Therefore, the total number of cars (in thousands) within x hours of 9 A.M. is giving by the integral $\displaystyle\int_0^x \dfrac{1}{1 + t^2}\, dt$.

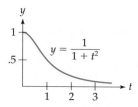

a. Find the Taylor series at $t = 0$ for $\dfrac{1}{1 + t^2}$ by modifying an existing series.

b. Integrate this series from 0 to x, thereby finding a Taylor series for $\displaystyle\int_0^x \dfrac{1}{1 + t^2}\, dt$.

c. Estimate the number of cars passing through the toll plaza during the first half hour after 9 A.M. by evaluating the first three terms of the series at $x = \tfrac{1}{2}$.

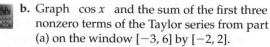 **d.** Compare your answer to that found by using FnInt or $\int f(x)\,dx$ on your graphing calculator to integrate the original function from 0 to 0.5.

42. Show that the power series $\sum_{n=0}^{\infty} a_n x^n$ has radius of convergence $R = \lim\limits_{n \to \infty} \left| \dfrac{a_n}{a_{n+1}} \right|$, provided that this limit exists. [*Hint:* Use the ratio test.]

10.4 Newton's Method

Use Newton's method to approximate each root, continuing until two successive iterations agree to three decimal places. Carry out the calculations "by hand" with the aid of a calculator, rounding to three decimal places.

43. $\sqrt{26}$ **44.** $\sqrt[3]{30}$

Use Newton's method to approximate the root of each equation, beginning with the given x_0 and continuing until two successive iterations agree to three decimal places. Carry out the calculations "by hand" with the aid of a calculator, rounding to three decimal places.

45. $x^3 + 3x - 3 = 0$ **46.** $e^x + 10x - 4 = 0$
      $x_0 = 1$ $x_0 = 0$

Use NEWTON or a similar program to approximate each root, continuing until two successive iterations agree to nine decimal places.

47. $\sqrt[3]{100}$ **48.** $\sqrt{10}$

Use NEWTON or a similar program to approximate the root of each equation, beginning with the given x_0 and continuing until two successive approximations agree to nine decimal places.

49. $e^x - 5x - 3 = 0$ **50.** $e^x + 4x - 100 = 0$
      $x_0 = 1$ $x_0 = 3$

51. $x^3 = 5 \ln x + 10$ **52.** $x^4 = 2 + \sin x$
      $x_0 = 2$ $x_0 = 0$

53. GENERAL: Internal Rate of Return For an investment of $20,000, you receive three payments of $9000 at the end of the first, second, and third years. Find the internal rate of return (to the nearest tenth of a percent).

54. ECONOMICS: Supply and Demand The supply and demand functions for a product are $S(p) = 20 + 3p$ and $D(p) = 50 - 8\sqrt{p}$ (for $0 \le p \le 25$), where p is the price (in dollars) of the product. Find the "market equilibrium price" p at which $S(p) = D(p)$. Use Newton's method, beginning with some reasonable initial guess and continuing until the iterations agree to two decimal places.

55. BUSINESS: Cost and Revenue The cost and revenue functions for a company are

$C(x) = 3x + 8000$ and $R(x) = 4x\sqrt{x}$, where x is the quantity. Find the "break-even" quantity, the x at which $C(x) = R(x)$. Use Newton's method, beginning with some reasonable initial value and continuing until the iterations agree to the nearest whole unit.

11 Probability

Application Preview

Coincidences

We often hear of "amazing" coincidences. For example, a woman named Evelyn Marie Adams won the New Jersey Lottery twice, in 1985 and in 1986, an event that was widely reported to have a probability of 1 in 17 trillion.

Actually, the figure 1 in 17 trillion is misleading; such events are not all that unlikely. In fact, given enough tries, the most outrageous things are virtually certain to happen. For example, if a coincidence is defined as an event with a one-in-a-million chance of happening to you today, then in the United States, with over 250 million people, we should expect more than 250 coincidences each day, and almost 100,000 in a year.

Returning to the supposed 1-in-17-trillion double lottery winning, that figure is the right answer to the wrong question: What is the probability that a *preselected* person who buys just *two* tickets for separate lotteries will win on both? The more relevant question is: What is the probability that *some* person, among the many millions who buy lottery tickets (most buying multiple tickets), will win twice in a lifetime? It has been calculated* that such a double winning is likely to occur

*See Persi Diaconis and Frederick Mosteller, "Methods for Studying Coincidences," *Journal of the American Statistical Association* 84(408):853–861, December 1989.

Coincidences
(continued)

once in seven years, with the likelihood approaching certainty for longer periods.

Some knowledge of probability is necessary for an understanding of the world, if only to cast doubt on the misleading statements that one often hears. In this first section we lay the foundations for the probability discussed in later sections. For further information on probability, see the two books listed below.*

11.1 DISCRETE PROBABILITY

Introduction

Probability measures the *likelihood* of an event: a 90% probability of rain means that rain is very likely, and should occur on about nine of ten similar days. Probability affects our lives in many ways: insurance rates are calculated from probabilities of disasters, election results are predicted from random samples, and airplanes are overbooked based on estimated likelihoods of "no-shows." Probability is a vast and fascinating subject, and in this chapter we will discuss only a few aspects of it that are related to calculus. We begin with *discrete* probability, in which probabilities are *added,* with later sections discussing *continuous* probability, in which probabilities are summed by *integration.*

Elementary Events and Their Probabilities

Probability begins with an experiment, real or imagined, whose outcome depends on chance. In the experiment of rolling one die (the singular of dice), the possible outcomes are the numbers of dots 1, 2, 3, 4, 5, and 6, which are called *elementary events.* An *event* is a set consisting of elementary events. For example, the event of *rolling at least a 5* on a die would be represented by

$$E_1 = \{5, 6\} \qquad \text{Rolling at least 5 means rolling 5 or 6}$$

Rolling an *even* number would be represented by

$$E_2 = \{2, 4, 6\} \qquad \text{Rolling evens means rolling 2, 4, or 6}$$

In general, the possible outcomes of an experiment, exactly one of which must occur, are called *elementary events,* and the collection of all

---

* For a readable introduction to probability, see Warren Weaver, *Lady Luck,* Dover Publications, 1982. For a more complete exposition, see William Feller, *An Introduction to Probability Theory and Its Applications,* vol. 1, 3rd ed., John Wiley, 1957.

of them is called the *sample space*. Some other experiments and their sample spaces are shown below.

Experiment	Sample Space
	Elementary events
Tossing a coin twice	{HH, HT, TH, TT}
Counting defects in a manufacturing process	{0, 1, 2, 3, ... }

"HT" means "heads then tails"

To each elementary event we assign a number that represents the *probability* of that event. In practice, we choose the number to be the long-run proportion of times that we expect the event to occur, but mathematically there are only two requirements: each probability must be between 0 and 1 (inclusive), and they must add to 1. Formally:

Elementary Events and Probabilities

Elementary events $E_1, E_2, ... , E_n$ (exactly one of which must occur) are assigned probabilities $P(E_i) = p_i$ (for $i = 1, 2, ... , n$) such that for each i,

$$0 \le p_i \le 1 \qquad \text{Probabilities between 0 and 1}$$

and

$$p_1 + p_2 + \cdots + p_n = 1 \qquad \text{Probabilities sum to 1}$$

If there are infinitely many elementary events $E_1, E_2, ... ,$ then the sum of the probabilities will be an *infinite series* that must sum to 1. The condition that the probabilities sum to 1 can be interpreted as meaning that *something* must happen.

EXAMPLE 1

ASSIGNING PROBABILITIES TO ELEMENTARY EVENTS

For the experiment of rolling a die, the six possible faces are equally likely, each expected to occur about one-sixth of the time, so we assign probability $\frac{1}{6}$ to each.

$$P(1) = \frac{1}{6}, \qquad P(2) = \frac{1}{6}, \qquad P(3) = \frac{1}{6},$$

P is read "the probability of"

$$P(4) = \frac{1}{6}, \qquad P(5) = \frac{1}{6}, \qquad P(6) = \frac{1}{6}$$

EXAMPLE **2**

ASSIGNING PROBABILITIES TO ELEMENTARY EVENTS

For the experiment of tossing a (fair) coin twice, the four elementary events HH, HT, TH, TT should each occur about one-quarter of the time, leading to the assignment

$$P(\text{HH}) = \frac{1}{4}, \qquad P(\text{HT}) = \frac{1}{4}, \qquad P(\text{TH}) = \frac{1}{4}, \qquad P(\text{TT}) = \frac{1}{4}$$

EXAMPLE **3**

ASSIGNING UNEQUAL PROBABILITIES

In America, newborn babies are slightly more often male than female, the ratio being about 52 to 48. Therefore, for the experiment of observing the sex of a newborn baby, with elementary events M and F, we would assign probabilities

$$P(\text{M}) = 0.52 \quad \text{and} \quad P(\text{F}) = 0.48 \qquad \text{Probabilities need not be equal}$$

An event consists of elementary events, and the *probability* of the event is found by adding the probabilities of the constituent elementary events.

> **Probability of an Event**
>
> The probability $P(E)$ of an event E is the sum of the probabilities of the elementary events in E.

EXAMPLE **4**

FINDING THE PROBABILITY OF AN EVENT

For one roll of a die, find the probability of rolling an even number.

Solution   The event "rolling an even number" is made up of the elementary events 2, 4, and 6, which have probability $\frac{1}{6}$ each (from Example 1). Therefore, the probability of rolling an even number is the *sum* of the probabilities of these elementary events:

$$P\left(\begin{array}{c}\text{Rolling an}\\\text{even number}\end{array}\right) = \underset{\uparrow}{P(2)} + \underset{\uparrow}{P(4)} + \underset{\uparrow}{P(6)} = \frac{1}{6} + \frac{1}{6} + \frac{1}{6} = \frac{3}{6} = \frac{1}{2}$$

Elementary events

EXAMPLE 5

FINDING THE PROBABILITY OF AN EVENT

For two tosses of a coin, find the probability of at least one head.

Solution The event "at least one head" consists of the elementary events HH, HT, and TH, with probability $\frac{1}{4}$ each (from Example 2). Adding these probabilities gives

$$P\begin{pmatrix} \text{At least} \\ \text{one head} \end{pmatrix} = P(\text{HH}) + P(\text{HT}) + P(\text{TH}) = \frac{1}{4} + \frac{1}{4} + \frac{1}{4} = \frac{3}{4}$$

Random Variables

An event may be described in terms of a *random variable,* a variable whose value depends on the particular outcome that occurs. For example, in the experiment of tossing a coin twice, if X stands for the *number of heads,* then the event "at least one head" is equivalent to $X \geq 1$. (Technically, a random variable is a *function* defined on the sample space, since for each elementary event it takes a numerical value.) Finding the *probability distribution* of a random variable means finding the probabilities that it takes each of its possible values.

Probability Distribution of a Random Variable

The probability distribution of a random variable X consists of the probabilities that X equals each of its possible values.

EXAMPLE 6

FINDING THE PROBABILITY DISTRIBUTION OF A RANDOM VARIABLE

For two tosses of a (fair) coin, find the probability distribution of $X = \begin{pmatrix} \text{Number} \\ \text{of heads} \end{pmatrix}$.

Solution The possible values of X are 0, 1, and 2 (the possible numbers of heads in two tosses). The probability that X takes any one of these values is calculated by finding the elementary events for which X takes that value and adding their probabilities.

$$P(X = 0) = \frac{1}{4}$$

$X = 0$ (no heads) only for TT, which occurs with probability $\frac{1}{4}$

$$P(X = 1) = \frac{1}{2}$$

$X = 1$ for HT and TH, which occur with probabilities $\frac{1}{4} + \frac{1}{4} = \frac{1}{2}$

$$P(X = 2) = \frac{1}{4}$$

$X = 2$ only for HH, which occurs with probability $\frac{1}{4}$

These three probabilities make up the probability distribution for X.

A probability distribution may be shown as a *bar graph*, with bars whose heights (and also areas) show the probability of each possible value. The probabilities adding to 1 means that the total area under the graph must be 1.

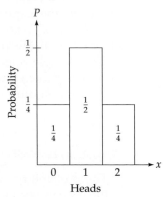

Probability distribution of $X = \begin{pmatrix} \text{Number} \\ \text{of heads} \end{pmatrix}$ in two tosses of a coin (from Example 6)

EXAMPLE 7

FINDING A PROBABILITY FOR A RANDOM VARIABLE

For two tosses of a coin and $X = \begin{pmatrix} \text{Number} \\ \text{of heads} \end{pmatrix}$, find $P(0 \le X \le 1)$.

Solution

$0 \le X \le 1$ means that $X = 0$ or $X = 1$, so we *add* these probabilities:

$$P(0 \le X \le 1) = P(X = 0) + P(X = 1) = \frac{1}{4} + \frac{1}{2} = \frac{3}{4}$$

The $\frac{1}{4}$ and $\frac{1}{2}$ are from Example 6

The probability $P(0 \le X \le 1)$ is given by the first two bars in the graph of the probability distribution.

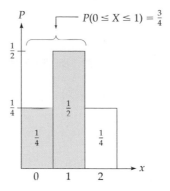

Calculating probabilities as areas under a graph will be very important in later sections.

Expected Value

Suppose that a game of chance pays $8 with probability $\frac{1}{4}$, and $4 with probability $\frac{3}{4}$. If you were to play this game many times, what would be your *long-term average winnings per play?* We multiply each prize by its probability (after all, winning $8 with probability $\frac{1}{4}$ means winning $8 about every fourth time, for an *average* of $\frac{1}{4} \cdot 8 = \$2$ each time).

$$\binom{\text{Average}}{\text{winnings}} = 8 \cdot \frac{1}{4} + 4 \cdot \frac{3}{4} = 2 + 3 = 5 \qquad \text{Each prize times its probability}$$

This answer, $5, is called the *expected value* of the game, meaning that *in the long run* your winnings should average about $5 per game. Therefore, $5 would be a *fair price* to charge for playing this game. For *any* random variable X, its *mean* or *expected value* is written $E(X)$ or μ ("mu," the Greek letter m), and is found by multiplying the possible values of X by their probabilities and adding.

Expected Value of a Random Variable

A random variable X taking values $x_1, x_2, \ldots, x_n$ with probabilities $p_1, p_2, \ldots, p_n$ has an *expected value* or *mean*

$$\mu = E(X) = \sum_{i=1}^{n} x_i \cdot p_i \qquad \qquad x_1 p_1 + \cdots + x_n p_n$$

The expected value is *weighted average,* with the values of X weighted by their probabilities. If a random variable has infinitely many possible values, then the sum will be an infinite series.*

EXAMPLE 8

FINDING THE EXPECTED VALUE OF A RANDOM VARIABLE

For two tosses of a coin and $X = \begin{pmatrix} \text{Number} \\ \text{of heads} \end{pmatrix}$, find $E(X)$.

Solution

Using the probabilities calculated in Example 6 on pages 807–808:

$$E(X) = 2 \cdot \frac{1}{4} \quad + \quad 1 \cdot \frac{1}{2} \quad + \quad 0 \cdot \frac{1}{4}$$

$$\underset{P(X=2)}{\uparrow} \qquad \underset{P(X=1)}{\uparrow} \qquad \underset{P(X=0)}{\uparrow}$$

Values times their probabilities, added

$$= \frac{1}{2} + \frac{1}{2} = 1$$

That is, in two tosses of a fair coin, the expected number of heads is $E(X) = 1$ (just as you might have thought).

For this random variable, the expected value 1 is the "middle number" of its possible values. In general, this will happen whenever the probability distribution is symmetric around a central value.

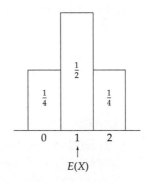

* In this case there is an added technicality that the series $\Sigma_{i=1}^{\infty} |x_i| \cdot p_i$ must converge.

EXAMPLE 9

FINDING THE EXPECTED VALUE OF REAL ESTATE

You are considering buying a piece of property whose value V depends on whether oil is found on it. You estimate that with probability 20% it will be worth $300,000 and with probability 80% it will be worth only $10,000. Find the expected value $E(V)$ of the property.

Solution

$$E(V) = 300{,}000 \cdot (0.2) + 10{,}000 \cdot (0.8)$$

Values times their probabilities, added

$$= 60{,}000 + 8000 = 68{,}000$$

$68,000

This expected value of $68,000 could then be compared with the selling price to decide whether it is a "good buy."

Variance

For a random variable X, the possible values may be clustered closely around the mean, or they may be distributed widely on either side of the mean. The *spread* or *dispersion* of a random variable around its mean is measured by the *variance*.

Variance of a Random Variable

A random variable X taking values $x_1, x_2, \ldots, x_n$ with probabilities $p_1, p_2, \ldots, p_n$, and having mean μ, has *variance*

$$\operatorname{Var}(X) = \sum_{i=1}^{n}(x_i - \mu)^2 p_i \qquad (x_1 - \mu)^2 p_1 + \cdots + (x_n - \mu)^2 p_n$$

The variance is the differences between the possible values and the mean, squared, multiplied by their probabilities, and added. It is sometimes called the *mean square deviation*. The squaring of the differences $x_i - \mu$ ensures that a positive difference does not cancel a negative one. Obviously, finding the variance requires first finding the mean μ. (An alternate formula for the variance is developed in Exercise 33.)

EXAMPLE 10

FINDING THE VARIANCE OF A RANDOM VARIABLE

For two tosses of a fair coin and $X = \begin{pmatrix} \text{Number} \\ \text{of heads} \end{pmatrix}$, find Var($X$).

Solution

In Example 6 we found the possible values of X and their probabilities, and in Example 8 we found the mean $\mu = 1$. From these,

$$\text{Var}(X) = (0-1)^2 \cdot \frac{1}{4} + (1-1)^2 \cdot \frac{1}{2} + (2-1)^2 \cdot \frac{1}{4}$$

Values minus $\mu = 1$, squared, and multiplied by their probabilities

$$= \frac{1}{4} + 0 + \frac{1}{4} = \frac{1}{2}$$

Simplified

A particular variance such as $\text{Var}(X) = \frac{1}{2}$ has little intuitive meaning by itself. Variance may be understood *comparatively:* a variance of 2 would indicate a greater spread away from the mean than a variance of $\frac{1}{2}$.

   If the units of the random variable are "dollars," then the variance has units "dollars squared," which is difficult to interpret. The *standard deviation of X*, written $\sigma(X)$, is the square root of the variance, and so has the same units as the random variable. [The symbol σ("sigma") is the Greek letter s.]

Standard Deviation of a Random Variable

The standard deviation of a random variable is the square root of the variance:

$$\sigma(X) = \sqrt{\text{Var}(X)}$$

The standard deviation is sometimes called the *root mean square deviation* of the random variable. Standard deviations will be used extensively in Section 11.4.

EXAMPLE 11

FINDING THE STANDARD DEVIATION OF A RANDOM VARIABLE

For two tosses of a fair coin and $X = \begin{pmatrix} \text{Number} \\ \text{of heads} \end{pmatrix}$, find $\sigma(X)$.

Solution

$$\sigma(X) = \sqrt{\text{Var}(X)} = \sqrt{0.5} \approx 0.707$$

Using $\text{Var}(X) = \frac{1}{2}$ from Example 10

Poisson Distribution

Until now we have defined random variables by imagining an experiment (rolling a die, tossing a coin) and finding the probability distribution. We now turn our attention to certain classic random variables that are useful in applications. The first of these is the *Poisson random variable*, named for the French mathematician Siméon Poisson (1781–1840). A Poisson random variable has an infinite number of possible values, 0, 1, 2, 3, ... , and its probability distribution depends on a constant or *parameter* $a > 0$, which is also its mean.

Poisson Random Variable

For a Poisson random variable X with mean $a > 0$,

$$P(X = k) = e^{-a}\frac{a^k}{k!} \qquad \text{for} \quad k = 0, 1, 2, \ldots$$

EXAMPLE 12

FINDING PROBABILITIES FOR A POISSON RANDOM VARIABLE

For the Poisson random variable X with mean $a = 3$, find $P(X = 0)$, $P(X = 1)$, and $P(X = 2)$.

Solution

$$P(X = 0) = e^{-3}\frac{3^0}{0!} = e^{-3}\frac{1}{1} \approx 0.05 \qquad e^{-a}\frac{a^k}{k!} \begin{array}{l} \text{with} \quad a = 3 \\ \text{and} \quad k = 0 \end{array}$$

$$P(X = 1) = e^{-3}\frac{3^1}{1!} = e^{-3}\frac{3}{1} \approx 0.15 \qquad e^{-a}\frac{a^k}{k!} \begin{array}{l} \text{with} \quad a = 3 \\ \text{and} \quad k = 1 \end{array}$$

$$P(X = 2) = e^{-3}\frac{3^2}{2!} = e^{-3}\frac{9}{2} \approx 0.22 \qquad e^{-a}\frac{a^k}{k!} \begin{array}{l} \text{with} \quad a = 3 \\ \text{and} \quad k = 2 \end{array}$$

The probability distribution of a Poisson random variable with mean $a = 3$ is graphed below.

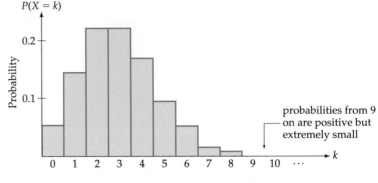

Probability distribution for the
Poisson random variable with mean $a = 3$

To show that the probabilities for a Poisson random variable sum to 1, we factor e^{-a} out of the sum and recognize the remaining sum as the Taylor series for e^a (from page 771):

$$\underbrace{\sum_{k=0}^{\infty} e^{-a} \frac{a^k}{k!}}_{\substack{\text{Sum} \\ \text{of the} \\ \text{probabilities}}} = \underbrace{e^{-a}}_{\substack{\text{Factored} \\ \text{out}}} \underbrace{\left(1 + \frac{a}{1} + \frac{a^2}{2!} + \frac{a^3}{3!} + \cdots\right)}_{\substack{\text{Taylor series for } e^a, \\ \text{which equals } e^a}} = \underbrace{e^{-a}(e^a)}_{\substack{\text{Replacing} \\ \text{the series} \\ \text{by } e^a}} = \underbrace{e^0 = 1}_{\substack{\text{Adding} \\ \text{exponents}}}$$

Therefore, the probabilities *do* sum to 1. Exercise 13 shows that a Poisson random variable X has mean $E(X) = a$. (The variance is also a, so the standard deviation is $\sqrt{a}$, although we will not use these facts.)

Poisson random variables are often used to model independent events that occur rarely, such as the number of defective items in a manufacturing process or people living to age 100.* The Poisson distribution is sometimes (although inaccurately) called the *law of rare events*.

EXAMPLE **13** **HURRICANE PREDICTION**

During the years 1900 through 2000, 168 hurricanes reached the United States, for an average of 1.66 per year during those 101 years.** Use the Poisson distribution with mean $a = 1.66$ to find the proba-

---

* Two events are *independent* if the occurrence of one does not make the occurrence of the other more or less likely. For further examples that fit the Poisson distribution, see F. Thorndike, "Applications of Poisson's Probability Summation," *Bell System Technical Journal* 5:604–624, 1926.
** *Source*: National Oceanic and Atmospheric Administration.

bility that three or fewer hurricanes will reach the United States during a particular year.

Solution

If $X = \begin{pmatrix} \text{Number of} \\ \text{hurricanes} \end{pmatrix}$, then we want to find $P(X \leq 3)$, where X is a Poisson random variable with $a = 1.66$.

$P(X \leq 3)$

$$- P(X = 0) + P(X = 1) + P(X = 2) + P(X = 3) \qquad \begin{array}{l} X \leq 3 \text{ means} \\ X = 0, 1, 2, \text{ or } 3 \end{array}$$

$$= e^{-1.66} \left(\frac{1.66^0}{0!} + \frac{1.66^1}{1!} + \frac{1.66^2}{2!} + \frac{1.66^3}{3!} \right) \qquad \begin{array}{l} e^{-a} \left(\dfrac{a^0}{0!} + \dfrac{a^1}{1!} + \dfrac{a^2}{2!} + \dfrac{a^3}{3!} \right) \\ \text{with } a = 1.66 \end{array}$$

$$\approx 0.913 \qquad\qquad\qquad\qquad\qquad\qquad\qquad \text{Using a calculator}$$

Therefore, with probability about 0.913, or about 91%, at most three hurricanes will reach the United States in any given year.

Graphing Calculator Exploration

A graphing calculator with a command like POISSONCDF (in the probability DISTRibution menu) finds the probability that a Poisson random variable takes values *less than or equal to* a given number. To find the probability in the preceding example, we would use the POISSONCDF command with the given mean (1.66) and the maximum number of occurrences (3), as shown below.

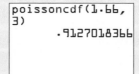

```
poissoncdf(1.66,
3)
          .9127018366
```

$P(X \leq 3) \approx 0.91$ (rounded), agreeing with the answer found in Example 13

What is the number such that with probability 99% at most that many hurricanes will reach the United States in a given year? [*Hint:* Experiment with POISSONCDF, replacing 3 by larger numbers.]

Calculations such as this are used to determine insurance premiums and land values in hurricane-prone areas. The Poisson distribution with $a = 1.66$ is graphed below. The entire area is 1, and the shaded portion shows the probability $P(X \leq 3)$.

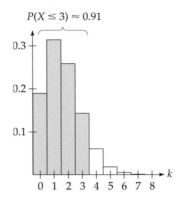

Practice Problem

In a shipment of 100 light bulbs, an average of 2 bulbs will be defective. Suppose that the number of defective bulbs in each shipment is modeled by a Poisson random variable.

a. What should be the value of the parameter a?

b. Find the probability that a given shipment has no defective bulbs.

➤ Solutions on page 817

11.1 Section Summary

For an experiment, the possible outcomes (exactly one of which must occur) are called *elementary events,* and collectively they make up the *sample space.* The elementary events are assigned nonnegative probabilities that sum to 1, usually based on long-run proportions. An *event* consists of one or more elementary events, and its probability is the sum of the probabilities of its constituent elementary events.

A *random variable* (represented by a capital letter, such as X, Y, or Z) takes values that depend on the elementary events, and its *probability distribution* consists of the probabilities that it takes these various values. The *mean* or *expected value* $E(X)$ gives an average value for X, and the *variance* $\mathrm{Var}(X)$ and *standard deviation* $\sigma(X)$ measure the *deviation* of the values away from the mean.

$$E(X) = \mu = \sum_{i=1}^{n} x_i p_i$$

For possible values $x_1, x_2, \ldots, x_n$ with probabilities $p_1, p_2, \ldots, p_n$

$$Var(X) = \sum_{i=1}^{n} (x_i - \mu)^2 p_i$$

$$\sigma(X) = \sqrt{Var(X)}$$

A *Poisson* random variable with mean $a > 0$ has probability distribution

$$P(X = k) = e^{-a} \frac{a^k}{k!} \qquad \text{for } k = 0, 1, 2, \ldots$$

Poisson random variables are used to model the frequencies of independent rare events, with the parameter a set equal to the average number of occurrences.

➤ **Solution to Practice Problem**

a. $a = 2$ **b.** $P(X = 0) = e^{-2} \dfrac{2^0}{0!} - e^{-2} \sim 0.135$

11.1 Exercises

1. For the event of rolling one die, find

   a. $P\begin{pmatrix} \text{Rolling at} \\ \text{least a 3} \end{pmatrix}$ **b.** $P\begin{pmatrix} \text{Rolling an} \\ \text{odd number} \end{pmatrix}$

2. For the event of tossing a coin twice, find

   a. $P\begin{pmatrix} \text{Tossing exactly} \\ \text{one head} \end{pmatrix}$ **b.** $P\begin{pmatrix} \text{Tossing at most} \\ \text{one head} \end{pmatrix}$

3–4. For the event of tossing a coin *three times*, find

   3. a. $P\begin{pmatrix} \text{Tossing exactly} \\ \text{one head} \end{pmatrix}$ **b.** $P\begin{pmatrix} \text{Tossing exactly} \\ \text{two heads} \end{pmatrix}$

   4. a. $P\begin{pmatrix} \text{Tossing exactly} \\ \text{three heads} \end{pmatrix}$ **b.** $P\begin{pmatrix} \text{Tossing more} \\ \text{heads than tails} \end{pmatrix}$

5–6. Find the expected value, variance, and standard deviation of each random variable. Notice how the variance and standard deviation are affected by the "spread" of the values.

   5. a. X takes values 9 and 11, each with probability $\frac{1}{2}$.

   b. Y takes values 0 and 20, each with probability $\frac{1}{2}$.

6. **a.** X takes values 19 and 21, each with probability $\frac{1}{2}$.

   b. Y takes values 0 and 40, each with probability $\frac{1}{2}$.

7–10. For each spinner, the arrow is spun and then stops, pointing in one of the numbered sectors, with probability proportional to the area of the sector. Let X stand for the number so chosen. Find

   a. The probability distribution of X.
   b. $E(X)$.
   c. $Var(X)$.

7.

8.

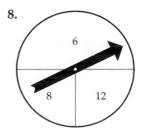

(See instructions on previous page.)

9.

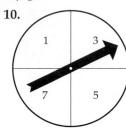

10.

11. For a spinner such as that described in Exercises 7–10, what is the probability that the arrow ends up pointing

a. in the top 30°? **b.** in the top 10°?

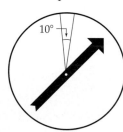

c. exactly upwards?

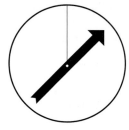

12. *(continuation)* If lines are drawn from the center of the spinner to 360 points evenly spaced around the circumference, what is the probability that the arrow ends up falling exactly on one of these lines?

13. Show that the mean of a Poisson random variable X with parameter a is $E(X) = a$.

14. A die is rolled, and X is defined as the number of dots on the face that lands upward. Find

  a. $E(X)$ (Your answer shows that the mean need not be a possible value.)
  b. $\text{Var}(X)$

15. For a Poisson random variable X with mean 2, find (to three decimal places)

  a. $P(X = 0)$ **b.** $P(X = 1)$
  c. $P(X = 2)$ **d.** $P(X = 3)$

16. For a Poisson random variable X with mean 1.5, find (to three decimal places)

  a. $P(X = 0)$ **b.** $P(X = 1)$
  c. $P(X = 2)$ **d.** $P(X = 3)$

17. For a Poisson random variable X with mean 2, find (to three decimal places)

  a. $P(X \le 1)$ **b.** $P(X \le 2)$
  c. $P(X \le 3)$ **d.** $P(X > 3)$

  [*Hint for (d):* Use your answer to (c).]

18. For a Poisson random variable X with mean 1.5, find (to three decimal places)

  a. $P(X \le 1)$ **b.** $P(X \le 2)$
  c. $P(X \le 3)$ **d.** $P(X > 3)$

  [*Hint for (d):* Use your answer to (c).]

APPLIED EXERCISES

19. GENERAL: Family Composition A young couple hopes to have four children.

  a. State the sample space for the sexes of the children listed in birth order, using B for boy and G for girl. (For example, GGBB means two girls then two boys.)

  b. If each possible outcome is equally likely, what is the probability of each?

  c. Find the probability distribution of the random variable $X = \left(\begin{array}{c}\text{Number}\\\text{of girls}\end{array}\right)$. Make a bar graph for the probability distribution.

(continues)

d. Find the probability $P(X \geq 1)$.

e. Find $E(X)$ and $\text{Var}(X)$.

20. BUSINESS: Expected Value of a Company
A venture capitalist buys a biotechnology company, estimating that with probability 0.2 it will be worth $2,000,000, with probability 0.7 it will be worth $300,000, and with probability 0.1 it will be worthless. Find the expected value of the company.

21. BUSINESS: Price of an Insurance Policy
An insurance company predicts that a policy it plans to offer will cost the company the following amounts in claims: $10,000 with probability $\frac{1}{10}$, $2000 with probability $\frac{1}{2}$, $1000 with probability $\frac{3}{10}$, and nothing at all with probability $\frac{1}{10}$. If it wants to sell the policy for $500 more than the expected cost of claims, what price should it set?

22. GENERAL: Lawsuits A lawyer in a court case estimates that with probability 0.25 she will win $100,000, with probability 0.4 she will win $20,000, and with probability 0.35 she will win nothing. Her client is offered a settlement of $45,000 to drop the suit. Based on expected value, should she advise her client to accept the settlement?

APPLIED EXERCISES USING POISSON RANDOM VARIABLES

23. BUSINESS: Quality Control An automobile assembly line produces 200 cars per day, and an average of 4 per day fail inspection. Use the Poisson distribution to find the probability that on a given day, 2 or fewer cars will fail inspection.

24. GENERAL: Misprints An 11-chapter book has 35 misprints, for an average of 3.2 per chapter. Use the Poisson distribution to find the probability that a given chapter has 2 or fewer misprints.

25. BIOMEDICAL: Bacilli In an examination of 1000 phagocytes (white blood cells), an average of 1.93 bacilli were found per phagocyte. Use the Poisson distribution to find the probability that a given phagocyte has 2 or fewer bacilli.

26. GENERAL: Spacial Poisson Distribution A paper company produces rolls of paper with an average of 0.2 flaw per square yard. Use the Poisson distribution to find the probability that a given square yard has no flaws.

27. GENERAL: Hurricanes An average of 1.66 hurricanes reach the United States each year. Use the Poisson distribution to find the probability that no hurricane reaches the United States in a given year.

28. GENERAL: Centenarians In a group of 100 Americans, an average of 1.76 will live to be 100 years old ("centenarians").* Use the Poisson distribution to find the probability that in a group of 100, at least one will be a centenarian. [Hint: Find the probability that *none* will be a centenarian, and subtract it from 1.]

For Exercises 29–32, a graphing calculator with series operations is useful but not necessary.

29. BUSINESS: Quality Control In a box of 100 screws, an average of 2 are defective. The manufacturer offers a refund if 5 or more are defective. Use the Poisson distribution to find the probability of a refund. [Hint: First find $P(X < 5)$.]

30. GENERAL: Parking A college parking lot has an average of six empty parking places at any time. If eight cars arrive at the same time, use the Poisson distribution to find the probability that they can all be accommodated. [Hint: First find the probability that the cars *cannot* be accommodated.]

31. BIOMEDICAL: Blood Type On the average, 5 people in 100 will have blood type AB. Use

* See United States Life Tables, 1998, U.S. Department of Health and Human Services.

the Poisson distribution to find the probability that a group of 100 will have 6 or fewer people with blood type AB.

32. **GENERAL: Airplane Accidents** During the years 1980 to 2000 there were, on average, 3.2 serious commercial airplane accidents per year. Use the Poisson distribution to find the probability that in any particular year the number of accidents will not exceed 5.

33. **ALTERNATIVE FORMULA FOR VARIANCE** For a random variable X with mean μ that takes values $x_1, x_2, \ldots, x_n$ with probabilities

$p_1, p_2, \ldots, p_n$, prove the following "alternative" formula for the variance

$$\text{Var}(X) = \sum_{i=1}^{n} x_i^2 \cdot p_i - \mu^2$$

by giving a justification for each numbered equality in the following derivation.

$$\text{Var}(X) \overset{1}{=} \sum_{i=1}^{n} (x_i - \mu)^2 p_i \overset{2}{=} \sum_{i=1}^{n} (x_i^2 - 2x_i\mu + \mu^2) p_i$$

$$\overset{3}{=} \sum_{i=1}^{n} x_i^2 p_i - \sum_{i=1}^{n} 2x_i\mu p_i + \sum_{i=1}^{n} \mu^2 p_i$$

$$\overset{4}{=} \sum_{i=1}^{n} x_i^2 p_i - 2\mu \sum_{i=1}^{n} x_i p_i + \mu^2 \sum_{i=1}^{n} p_i$$

$$\overset{5}{=} \sum_{i=1}^{n} x_i^2 p_i - 2\mu\mu + \mu^2 1 \overset{6}{=} \sum_{i=1}^{n} x_i^2 p_i - \mu^2$$

11.2 CONTINUOUS PROBABILITY

Discrete Random Variables

The random variables that we have encountered so far have taken values that are separated from each other, and can be written out in a list, such as 0, 1, 2, 3, … . Such random variables are called *discrete* random variables.* For example, the Poisson random variable (see page 813) is a discrete random variable. The probability distribution of a discrete random variable may be shown as a bar graph over the possible values, with total area 1.

Continuous Random Variables

Other random variables take values that make up a *continuous interval* of the real line. For example, the proportion of time that you study today can be any number in the interval [0, 1], and the duration of a telephone call can be any number in the interval (0, ∞). For a *continuous* random variable, the probability distribution is given by a nonnegative function over the interval of possible values, with total area 1. Such a function is called a *probability density function*.

---

* Do not confuse the word "discrete," which means *separated* or *distinct,* with the word "discreet," which means *prudent* or *showing self-restraint.*

Probability Density Function

A probability density function on an interval $[A, B]$ is a function f such that

1. $f(x) \geq 0$ for $A \leq x \leq B$

2. $\int_A^B f(x) \, dx = 1$

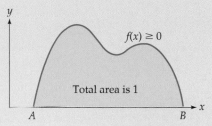

A probability density function on $[A, B]$

The interval may be infinite, in which case $A = -\infty$, or $B = \infty$, or both, with the integral above being an *improper* integral. For simplicity, we define $f(x)$ to be zero for all x-values outside of the interval $[A, B]$.

EXAMPLE **1**

CONSTRUCTING A PROBABILITY DENSITY FUNCTION

Find the value of the constant a that makes $f(x) = ax(1 - x)$ a probability density function on the interval $[0, 1]$.

Solution

The function is clearly nonnegative for $0 \leq x \leq 1$. To find the area under it, we integrate over the given interval:

$$a \int_0^1 x(1 - x) \, dx = a \int_0^1 (x - x^2) \, dx = a \left(\frac{1}{2} x^2 - \frac{1}{3} x^3 \right) \Big|_0^1$$

<u>Moved</u>
outside

<u>Multiplied</u>
out

<u>Integrating by</u>
the Power Rule

$$= a \left(\frac{1}{2} - \frac{1}{3} \quad - \quad 0 \right) \quad = \quad \frac{a}{6}$$

Must
equal 1

<u>Evaluating</u>
at 1

And
at 0

<u>Area under</u>
the curve

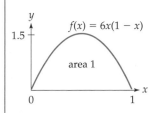

For this area $a/6$ to equal 1, we must have $a = 6$.   The probability density function with a replaced by 6 is then $f(x) = 6x(1 - x)$ on $[0, 1]$, which is shown in the graph on the left.

For *discrete* random variables, probabilities were given by *areas of bars*. For *continuous* random variables, probabilities are found by calculating *areas under curves*. More precisely, the probability that X takes values between two given numbers is found by integrating the probability density function between those two numbers.

Probabilities from Probability Density Functions

For a continuous random variable X with probability density function f,

$$P(a \le X \le b) = \int_a^b f(x)\, dx$$

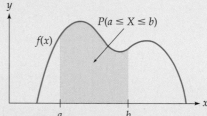

$P(a \le X \le b)$ is the area under the probability density function from a to b.

EXAMPLE 2

FINDING A PROBABILITY FOR A CONTINUOUS RANDOM VARIABLE

Let X be defined as the number of inches of rain in a town during the first week of April in each year. Suppose that a meteorologist has used past weather records to determine that the probability density function for X is $f(x) = 6x(1 - x)$ on [0, 1]. Find the probability that the rainfall is between 0 and 0.4 inch.

Solution

To find $P(0 \le X \le 0.4)$ we integrate the probability density function from 0 to 0.4:

$$P(0 \le X \le 0.4) = \int_0^{0.4} 6x(1 - x)\, dx \qquad \text{Integrating the probability density function}$$

$$= \int_0^{0.4} (6x - 6x^2)\, dx \qquad \text{Multiplying out}$$

$$= (3x^2 - 2x^3) \Big|_0^{0.4} \qquad \text{Using the Power Rule}$$

$$= (0.48 - 0.128) - 0 = 0.352 \quad \text{Evaluating}$$

The answer, $P(0 \le X \le 0.4) = 0.352$, means that with probability about 35% the rainfall will be between 0 and 0.4 inch.

The probability $P(0 \le X \le 0.4)$ found in the preceding example is the area under the probability density function $f(x) = 6x(1 - x)$ between $x = 0$ and $x = 0.4$, which is shown as the shaded area below.

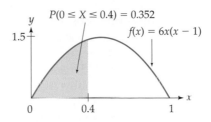

On pages 369–373 we defined definite integrals by summing areas of rectangles under curves. Therefore, finding continuous probabilitites as areas under curves is a natural extension of finding discrete probabilities as sums of areas of rectangles.

Practice Problem

a. Find the value of a that makes $f(x) = ax^2$ a probability density function on $[0, 1]$.

b. For X with this probability density function, find $P(0 \le X \le 0.5)$.

➤ Solutions on page 831

One-Point Probabilities Are Zero

For a continuous random variable X, the probability that X takes any particular single value is zero: $P(X = a) = 0$. This is because $P(X = a)$ is found by integrating the probability density function "from a to a," which gives zero:

$$P(X = a) = \underbrace{P(a \le X \le a)}_{\substack{\text{Equivalent} \\ \text{to } X = a}} = \int_a^a f(x)\, dx = 0$$

Same upper and lower limits

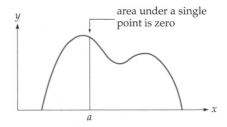

Therefore, adding one or both endpoints to the inequality in $P(a < X < b)$ does not change the probability:

$$P(a < X < b) = P(a \leq X \leq b) = \int_a^b f(x)\, dx \qquad \begin{array}{l} \text{Endpoints do not} \\ \text{change the probability} \end{array}$$

In continuous probability, what does zero probability mean? Zero probability does *not* mean that the event is impossible, only that it is "overwhelmingly unlikely." Think of choosing a number from the interval [0, 1] by some means such as throwing a dart at a line segment. The probability of choosing (hitting) exactly one single preselected number a is $P(X = a)$, which is zero, as we saw above. The probability being zero means only that predicting the exact point in advance, while theoretically possible, is overwhelmingly unlikely. (In *discrete* probability, however, zero probability *does* mean that the event is impossible. Intuitively, the difference is that in continuous probability a random variable has so many more possible values that the probability must be spread over them much more "thinly.")

Mean, Variance, and Standard Deviation

The mean and variance of a continuous random variable are defined by *integrals* that are analogous to the sums on pages 809 and 811.

Mean, Variance, and Standard Deviation

For a continuous random variable X with probability density function f on $[A, B]$,

$$\mu = E(X) = \int_A^B x\, f(x)\, dx \qquad \begin{array}{l} \text{Mean or expected value is the} \\ \text{integral of } x \text{ times } f(x) \end{array}$$

$$\text{Var}(X) = \int_A^B (x - \mu)^2 f(x)\, dx \qquad \begin{array}{l} \text{Variance is } x \text{ minus the mean,} \\ \text{squared, multiplied by } f(x), \\ \text{and integrated} \end{array}$$

$$\sigma(X) = \sqrt{\text{Var}(X)} \qquad \begin{array}{l} \text{Standard deviation is the} \\ \text{square root of the variance} \end{array}$$

If $A = -\infty$, or $B = \infty$, or both, then the integrals are *improper* integrals.* As before, the mean $\mu = E(X)$ gives an *average* value, and the variance $\text{Var}(X)$ and standard deviation $\sigma(X)$ measure the *spread* or *dispersion* of the values away from this average. Also as before, finding the variance involves first finding the mean.

---

* In this case there is an added technicality that the improper integral $\int |x|\, f(x)\, dx$ over the infinite interval must converge.

EXAMPLE **3**

FINDING THE EXPECTED VALUE OF A CONTINUOUS RANDOM VARIABLE

For the random variable in Example 2, $X = \begin{pmatrix} \text{Inches of} \\ \text{rainfall} \end{pmatrix}$ with probability density function $f(x) = 6x(1-x)$ on $[0, 1]$, find $E(X)$.

Solution

$$E(X) \;=\; \underbrace{\int_0^1 x \cdot 6x(1-x)\,dx}_{\substack{\text{Integrating } x \text{ times the} \\ \text{probability density function}}} \;=\; \underbrace{\int_0^1 (6x^2 - 6x^3)\,dx}_{\substack{\text{Multiplying} \\ \text{out}}}$$

$$=\; \underbrace{\left(2x^3 - \frac{6}{4}x^4\right)\Big|_0^1}_{\substack{\text{Integrating by} \\ \text{the Power Rule}}} \;=\; \underbrace{2 - \frac{3}{2}}_{\text{Evaluating}} \;=\; \underbrace{\frac{1}{2}}_{E(X)}$$

The expected value $\mu = E(X) = \frac{1}{2}$ means that the week's rainfall should average half an inch. It is reasonable that the mean is the "middle value" of the interval $[0, 1]$, since this probability density function is symmetric about this number.

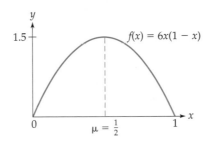

We could find the *variance* of the rainfall by evaluating the integral

$$\text{Var}(X) = \int_0^1 \left(x - \frac{1}{2}\right)^2 6x(1-x)\,dx \qquad \int_A^B (x - \mu)^2 f(x)\,dx \;\; \text{with} \;\; \mu = \frac{1}{2}$$

There is, however, an easier formula, which is proved on page 830. A discrete version of it was proved in Exercise 33 on page 820.

> **Alternative Formula for Variance**
>
> For a continuous random variable X with mean μ and probability density function f on $[A, B]$,
>
> $$\text{Var}(X) = \int_A^B x^2 f(x)\,dx - \mu^2 \qquad \text{Integrate } x^2 \text{ times } f(x), \text{ then subtract the square of the mean}$$

EXAMPLE 4

FINDING THE VARIANCE AND STANDARD DEVIATION

Find the variance and standard deviation of the random variable with probability density function $f(x) = 6x(1 - x)$ on $[0, 1]$.

Solution

In Example 3 we found that the mean of this random variable is $\mu = \frac{1}{2}$. To use the alternative formula for variance, we first find $\int_A^B x^2 f(x)\,dx$:

$$\int_0^1 x^2 \underbrace{6x(1 - x)}_{f(x)}\,dx = \int_0^1 (6x^3 - 6x^4)\,dx = \left(\frac{6}{4}x^4 - \frac{6}{5}x^5 \right)\Big|_0^1$$

$$= \frac{3}{2} - \frac{6}{5} = \frac{3}{10}$$

Then

$$\text{Var}(X) = \underbrace{\frac{3}{10}}_{\substack{\text{Preceding} \\ \text{integral}}} - \underbrace{\left(\frac{1}{2}\right)^2}_{\mu^2} = \frac{3}{10} - \frac{1}{4} = \frac{1}{20} = 0.05 \qquad \begin{array}{l} \int_A^B x^2 f(x)\,dx - \mu^2 \\ \text{with } \mu = \frac{1}{2} \end{array}$$

$$\sigma(X) = \sqrt{0.05} \approx 0.224 \qquad \text{Square root of the variance}$$

Therefore, $\text{Var}(X) = 0.05$ and $\sigma(X) \approx 0.224$

Graphing Calculator Exploration

Verify that the earlier formula (on page 824) for variance gives the same answer as in Example 4 by using a graphing calculator to evaluate $\int_0^1 (x - 0.5)^2 6x(1 - x)\,dx$.

```
fnInt((X-0.5)²6X
(1-X),X,0,1)
              .05
```

The variance and standard deviation will be used extensively in Section 11.4.

Cumulative Distribution Functions

Because we so often integrate a probability density function, it is useful to have a name for the integral up to a number x: it is called the *cumulative distribution function*.

Cumulative Distribution Function

For a continuous random variable X with probability density function f on $[A, B]$, the *cumulative distribution function* is

$$F(x) = P(X \le x) = \int_A^x f(t)\, dt$$

$F(x)$ is the probability that X is less than or equal to x, and is found by integrating the probability density function from A to x

---

We use uppercase letters (such as F, G, and H) for cumulative distribution functions, and lowercase letters (such as f, g, and h) for probability density functions. Graphically, the cumulative distribution function $F(x)$ is the *area* under the probability density function f up to x:

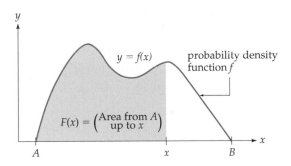

The curve $y = f(x)$ is the probability density function. The area under it up to x is the cumulative distribution function $F(x)$.

Since integrating f gives F, differentiating F gives f. (To be completely precise, this relationship holds at x-values where $f(x)$ is continuous.)

$$\frac{d}{dx} F(x) = f(x)$$

The derivative of the cumulative distribution function $F(x)$ is the probability density function $f(x)$.

Probabilities $P(a \le X \le b)$, which we found by integrating the probability density function, can also be found by *subtracting* values of the cumulative distribution function.

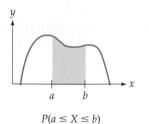

$P(a \le X \le b)$

equals

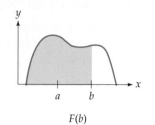

$F(b)$

minus

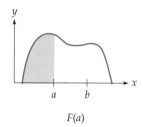

$F(a)$

In words:

Probabilities from the Cumulative Distribution Function

For a continuous random variable X with cumulative distribution function F,

$$P(a \le X \le b) = F(b) - F(a)$$ Either or both $\le$ may be replaced by $<$

EXAMPLE 5

FINDING AND USING A CUMULATIVE DISTRIBUTION FUNCTION

For a random variable W with probability density function $f(x) = \frac{1}{9}x^2$ on $[0, 3]$:

a. Find the cumulative distribution function $F(x)$.

b. Use $F(x)$ to find $P(1 \le W \le 2)$.

Solution

a. The cumulative distribution function is the integral of the probability density function up to x. We use t to avoid confusion with the upper limit x.

$$F(x) = \int_0^x \frac{1}{9} t^2 \, dt = \left(\frac{1}{9} \frac{1}{3} t^3 \right) \Big|_0^x = \frac{1}{27} x^3 \qquad F(x) = \int_A^x f(t) \, dt$$

Therefore, the cumulative distribution function is $F(x) = \frac{1}{27} x^3$ on $[0, 3]$.

b. To find $P(1 \leq W \leq 2)$, we evaluate $F(x) = \frac{1}{27}x^3$ at $x = 1$ and $x = 2$ and subtract:

$$P(1 \leq W \leq 2) = \underbrace{F(2) - F(1)}_{\substack{\text{Values of } F \\ \text{subtracted}}} = \underbrace{\frac{1}{27}2^3 - \frac{1}{27}1^3}_{\substack{\text{Evaluating } F \\ \text{at 2 and 1}}} = \frac{8}{27} - \frac{1}{27} = \frac{7}{27}$$

Therefore, $P(1 \leq W \leq 2) = \frac{7}{27}$.

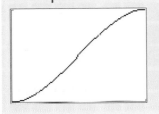

Graphing Calculator Exploration

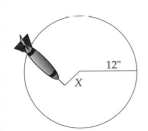

Graph the cumulative distribution function for the random variable with probability density function $f(x) = 6x(1 - x)$. [*Hint*: On some calculators, graph FnInt $(6x(1 - x), x, 0, x)$ on the window $[0, 1]$ by $[0, 1]$.]

Sometimes the cumulative distribution function can be found directly.

EXAMPLE 6

FINDING AND USING A CUMULATIVE DISTRIBUTION FUNCTION

A dart is thrown at a circular target of radius 12 inches, hitting a point at random. Let X be defined as

$$X = \left(\begin{array}{c} \text{Distance of dart} \\ \text{from center of circle} \end{array}\right)$$

Find: **a.** The cumulative distribution function $F(x)$

b. The probability density function $f(x)$

c. The mean $E(X)$

Solution

A point being chosen "at random" from a region means that the probability of the point's being in any subregion is proportional to the *area* of that subregion. (After all, the dart should have twice the probability of landing in an area twice as large.) The dart's landing within x inches of the center means that it lands inside a circle of radius x, and the probability of this, $P(X \le x)$, is the area of the inner circle (πx^2) divided by the area of the entire target ($\pi 12^2$):

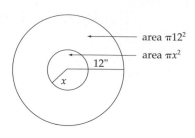

area $\pi 12^2$

area πx^2

12"

x

a. $P(X \le x) = \dfrac{\pi x^2}{\pi 12^2} = \dfrac{x^2}{144}$

Therefore, $F(x) = \dfrac{x^2}{144}$ for $0 \le x \le 12$.

b. $f(x) = \dfrac{d}{dx}\left(\dfrac{1}{144}x^2\right) = \dfrac{1}{72}x$ Differentiating $F(x)$ to find $f(x)$

Therefore, $f(x) = \dfrac{1}{72}x^2$ for $0 \le x \le 12$.

c. $E(X) = \displaystyle\int_0^{12} x \cdot \dfrac{1}{72} x\, dx = \dfrac{1}{72}\int_0^{12} x^2\, dx = \dfrac{1}{72}\cdot\dfrac{1}{3}x^3\Big|_0^{12} = \dfrac{12^3}{216} = 8$

The expected value $E(X) = 8$ means that if many darts are thrown, their average distance from the center will be about 8 inches.

Proof of the Alternative Formula for Variance

To prove the alternative formula for variance stated on page 826, we begin with the *definition* of variance given on page 824:

$$\text{Var}(X) = \int_A^B (x - \mu)^2 f(x)\, dx = \int_A^B (x^2 - 2x\mu + \mu^2)\, f(x)\, dx$$

Expanding $(x - \mu)^2$

$$= \int_A^B x^2 f(x)\, dx - 2\mu \underbrace{\int_A^B x f(x)\, dx}_{\substack{\mu \\ \text{(by definition} \\ \text{of mean)}}} + \mu^2 \underbrace{\int_A^B f(x)\, dx}_{\substack{1 \\ \text{(by property 2} \\ \text{on page 821)}}}$$

Separating into 3 integrals, and taking out constants

$$= \int_A^B x^2 f(x)\, dx \underbrace{- 2\mu^2 + \mu^2}_{-\mu^2} = \int_A^B x^2 f(x)\, dx - \mu^2$$

This last expression is the alternative formula for the variance

11.2 Section Summary

A random variable is *continuous* if its possible values form an *interval* [A, B], which may be infinite, and if its probabilities are found by integrating its *probability density function* (a nonnegative function whose integral over [A, B] is 1):

$$P(a \le X \le b) = \int_a^b f(x)\,dx \qquad \text{Either or both} \le \text{ may be replaced by } <$$

The *cumulative distribution function* for a random variable is

$$F(x) = P(X \le x) = \int_A^x f(t)\,dt \qquad A \text{ is the lowest possible value of the random variable}$$

Differentiating the cumulative distribution function $F(x)$ gives the probability density function $f(x)$. Probabilities $P(a < X \le b)$ may also be found by subtracting values of the cumulative distribution function:

$$P(a \le X \le b) = F(b) - F(a) \qquad \text{Either or both} \le \text{ may be replaced by } <$$

The mean gives the *average* value for the random variable, and the *variance* and *standard deviation* indicate how widely the values are spread around the mean:

$$\mu = E(X) = \int_A^B x f(x)\,dx$$

$$\text{Var}(X) = \int_A^B (x - \mu)^2\, f(x)\,dx = \int_A^B x^2 f(x)\,dx - \mu^2 \qquad \begin{array}{l}\text{Both formulas}\\\text{give the same}\\\text{answer}\end{array}$$

$$\sigma(X) = \sqrt{\text{Var}(X)}$$

The height of the probability density function at a point does not, by itself, give a probability, but it can be interpreted in a *relative* sense: if the probability density function is twice as high at one x-value as at another, the random variable is about twice as likely to be near the first x-value as near the other (where by "near" we mean "within a small fixed distance of").

➤ **Solutions to Practice Problem**

a. $a \displaystyle\int_0^1 x^2\, dx = a \frac{1}{3} x^3 \Big|_0^1 = \frac{a}{3}$

For the integral to equal 1, we must have $a = 3$.

b. $P(0 \le X \le 0.5) = \displaystyle\int_0^{0.5} 3x^2\, dx = x^3 \Big|_0^{0.5} = (0.5)^3 = 0.125$

11.2 Exercises

Find the value of the constant a that makes each function a probability density function on the stated interval.

1. $ax^2(1 - x)$ on $[0, 1]$ **2.** $a\sqrt{x}$ on $[0, 1]$

3. ax^2 on $[0, 3]$ **4.** $ax(2 - x)$ on $[0, 2]$

5. $\dfrac{a}{x}$ on $[1, e]$ **6.** $a \sin x$ on $[0, \pi]$

7. axe^x on $[0, 1]$ [*Hint:* Integrate by parts.]

8. $a \ln x$ on $[1, e]$ [*Hint:* Integrate by parts using $u = \ln x$.]

9. ae^{-x^2} on $[-1, 1]$ **10.** $\dfrac{a}{1 + x^2}$ on $[-1, 1]$

For each probability density function $f(x)$, find
a. $E(X)$ **b.** $\text{Var}(X)$ **c.** $\sigma(X)$

11. $f(x) = 3x^2$ on $[0, 1]$

12. $f(x) = \frac{1}{2}x$ on $[0, 2]$

13. $f(x) = 12x^2(1 - x)$ on $[0, 1]$

14. $f(x) = 20x^3(1 - x)$ on $[0, 1]$

15. $\dfrac{4}{\pi}\dfrac{1}{1 + x^2}$ on $[0, 1]$ **16.** $1.34e^{-x^2}$ on $[0, 1]$

17. If X has probability density function $f(x) = \frac{1}{12}x$ on $[1, 5]$, find
   a. The cumulative distribution function $F(x)$
   b. $P(3 \le X \le 5)$

18. If X has probability density function $f(x) = \frac{1}{10}$ on $[0, 10]$, find
   a. The cumulative distribution function $F(x)$
   b. $P(3 \le X \le 7)$

19. If X has probability density function $f(x) = \dfrac{2}{(1 + x)^2}$ on $[0, 1]$, find $P\left(X \le \dfrac{1}{3}\right)$.

20. If X has probability density function $f(x) = \frac{1}{6}x^{-1/2}$ on $[1, 16]$, find $P(4 \le X \le 9)$.

21. If X has cumulative distribution function $F(x) = \frac{1}{3}x - \frac{1}{3}$ on $[1, 4]$, find $P(2 \le X \le 3)$.

22. If X has cumulative distribution function $F(x) = \frac{1}{8}x^{3/2}$ on $[0, 4]$, find $P(1 \le X \le 4)$.

23. If X has cumulative distribution function $F(x) = x^2$ on $[0, 1]$, find
   a. $P(0.5 \le X \le 1)$.
   b. The probability density function $f(x)$

24. If X has cumulative distribution function $F(x) = \frac{1}{2}\sqrt{x} - \frac{1}{2}$ on $[1, 9]$, find
   a. $P(4 \le X \le 9)$
   b. The probability density function $f(x)$

25. For any cumulative distribution function $F(x)$ on the interval $[A, B]$, show that
   a. $F(A) = 0$
   b. $F(B) = 1$

26. For any cumulative distribution function $F(x)$, show that if $a \le b$, then $F(a) \le F(b)$.

27. If X has probability density function $f(x) = \dfrac{2}{\pi}\dfrac{1}{1 + x^2}$ on $[-1, 1]$, find $P\left(-\dfrac{1}{2} \le X \le \dfrac{1}{2}\right)$.

28. If X has probability density function $f(x) = 0.67e^{-x^2}$ on $[-1, 1]$, find $P\left(-\frac{1}{2} \le X \le \frac{1}{2}\right)$.

APPLIED EXERCISES

29. SOCIAL SCIENCE: Voter Turnout If the proportion of registered voters who actually vote in an election is a random variable X with probability density function $f(x) = 6x(1 - x)$, find $P(0.4 \le X \le 0.6)$.

30. BEHAVIORAL SCIENCE: Learning If the number of days required for a worker to learn a new skill is a random variable X with probability density function $f(x) = \frac{4}{27}x^2(3 - x)$ on $[0, 3]$, find the expected time $E(X)$.

31. BIOMEDICAL: Life Expectancy In a 4-year study, the number of years that a patient survives after an experimental medical procedure is a random variable X with probability density function $f(x) = \frac{3}{64}x^2$ on [0, 4]. Find

a. The expected survival time $E(X)$

b. $P(2 \leq X \leq 4)$

32. GENERAL: Milk Freshness If the shelf life of a carton of milk (in days) is a random variable X with probability density function $f(x) = \frac{1}{32}x$ on [0, 8], find

a. The expected shelf life $E(X)$

b. The time at which the probability of spoilage is only 25% (that is, find b such that $P(X < b) = 0.25$)

33. GENERAL: Darts A dart hits a point at random on a circular dart board of radius 9 inches. For

$$X = \left(\begin{array}{c} \text{Distance of dart} \\ \text{from center} \end{array} \right), \text{find}$$

a. The cumulative distribution function $F(x)$

b. The probability density function $f(x)$

c. $E(X)$

d. $P(3 \leq X \leq 6)$ using the *probability density function*

e. $P(3 \leq X \leq 6)$ using the *cumulative distribution function*

34. GENERAL: Product Failure A spherical ball bearing of radius 1 inch has a flaw located randomly somewhere within it. If the flaw is within half an inch of the center, the ball bearing will fracture when it is used. Find

a. The probability that the ball bearing will fracture. [*Hint:* Use the volumes of the "inner" and entire spheres.]

b. The expected distance from the flaw to the center of the ball bearing.

 35. BUSINESS: Profitability If the number of years until a new store makes a profit (or goes out of business) is a random variable X with probability density function $f(x) = 0.75x(2 - x)$ on [0, 2], find

a. The expected number of years $E(X)$

b. The variance and standard deviation

c. $P(X \geq 1.5)$

 36. BUSINESS: Newspaper Advertising If the proportion of a newspaper devoted to advertising is a random variable X with probability density function $f(x) = 20x^3(1 - x)$ on [0, 1], find

a. The expected proportion $E(X)$

b. The variance and standard deviation

c. The probability that more than 80% is devoted to advertising

 37. BIOMEDICAL: Weibull Aging Model The probability that the fruit fly *Drosophila melanogaster* lives longer than t days is given by the *Weibull survivorship function*

$$S(t) = e^{-(0.02t)^{3.3}}$$

a. Graph this function on the window $[-5, 105]$ by $[-0.5, 1.5]$.

b. What is the probability that a fly has a lifespan of more than 30 days?

11.3   UNIFORM AND EXPONENTIAL RANDOM VARIABLES

Introduction

The preceding section introduced the essentials of continuous probability: continuous random variables, probability density functions, cumulative distribution functions, mean, variance, and standard deviation. However, little was said about how to *use* random variables to model particular situations. Although a complete answer to this important question would require further courses in probability, statistics, and mathematical modeling, many situations can be modeled by just a few standard types of random variables. In this section we will study two of the most useful, *uniform* and *exponential* random variables, and in the next section we will study *normal* random variables.

Uniform Random Variables

Suppose that you want to choose a point at random from an interval $[0, B]$. If all points are to be "equally likely," the probability density function must be constant over the interval $[0, B]$, and if the area under it is to be 1, its uniform height must be $1/B$, as shown on the right.

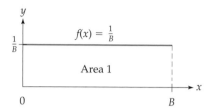

Uniform probability density function on $[0, B]$

The resulting probability density function, $f(x) = 1/B$ for $0 \le x \le B$, is called the *uniform probability density function on* $[0, B]$, and a random variable with this probability density function is called a *uniform random variable on* $[0, B]$. The mean of a uniform random variable on $[0, B]$ is easily calculated from the formula $E(X) = \int_A^B x f(x)\, dx$.

$$E(X) = \underbrace{\int_0^B x \frac{1}{B} dx}_{f(x)} = \underbrace{\frac{1}{B} \int_0^B x\, dx}_{\substack{\text{Moved} \\ \text{outside}}} = \underbrace{\left.\left(\frac{1}{B} \frac{1}{2} x^2\right)\right|_0^B}_{\substack{\text{Integrating by} \\ \text{the Power} \\ \text{Rule}}} = \underbrace{\frac{1}{B} \frac{1}{2} B^2 - 0}_{\substack{\text{Evaluating} \\ \text{at } B \text{ and at } 0}} = \underbrace{\frac{B}{2}}_{E(X)}$$

The mean $\mu = E(X) = B/2$ is the midpoint of the interval $[0, B]$, as should be expected. For the variance, we use the alternative formula from page 826, first integrating x^2 times the probability density function:

$$\int_0^B x^2 \frac{1}{B}\, dx = \frac{1}{B} \int_0^B x^2 dx = \left.\frac{1}{B} \frac{1}{3} x^3\right|_0^B = \frac{1}{B} \frac{1}{3} B^3 = \frac{B^2}{3}$$

Then

$$\text{Var}(X) = \underbrace{\frac{B^2}{3} - \left(\frac{B}{2}\right)^2}_{\substack{\text{Calculated } \mu \\ \text{above}}} = \frac{B^2}{3} - \frac{B^2}{4} = \underbrace{\frac{B^2}{12}}_{\text{Var}(X)} \qquad \text{Var}(X) = \int_A^B x^2 f(x)\, dx - \mu^2$$

Summarizing:

Uniform Random Variable

A uniform random variable X on the interval $[0, B]$ has probability density function

$$f(x) = \frac{1}{B} \quad \text{for } 0 \le x \le B$$

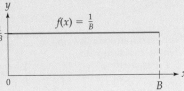

Its mean, variance, and standard deviation are

$$E(X) = \frac{B}{2} \qquad \text{Var}(X) = \frac{B^2}{12} \qquad \sigma(X) = \frac{B}{2\sqrt{3}}$$

EXAMPLE **1**

FINDING THE MEAN AND VARIANCE OF A UNIFORM RANDOM VARIABLE

A campus shuttle bus departs every 18 minutes. For a person arriving at a random time, find the mean and variance of the wait for the bus.

Solution   Arriving at a random time means that all waiting times between 0 and 18 minutes are equally likely, so $X = \begin{pmatrix} \text{Waiting} \\ \text{time} \end{pmatrix}$ is uniformly distributed on the interval $[0, 18]$. Therefore, the preceding formulas for the uniform distribution give

$$E(X) = \frac{18}{2} = 9 \qquad\qquad E(X) = \frac{B}{2} \text{ with } B = 18$$

$$\text{Var}(X) = \frac{18^2}{12} = 27 \qquad\qquad \text{Var}(X) = \frac{B^2}{12} \text{ with } B = 18$$

$$\sigma(X) = \sqrt{27} \approx 5.2 \qquad\qquad \text{Standard deviation}$$

The fact that the average wait is 9 minutes for buses that come every 18 minutes should be intuitively reasonable. The standard deviation of 5.2 minutes gives a rough indication of the expected deviation from this mean, although an exact interpretation of this number is difficult.

Exponential Random Variables

We next define *exponential* random variables, which are used to model waiting times that are potentially unlimited in duration, such as the time until an electronic component fails or the time between the ar-

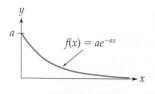

Exponential probability density function with parameter a.

rival of telephone calls. An exponential random variable takes values in the interval $[0, \infty)$ and has probability density function $f(x) = ae^{-ax}$ for $x \geq 0$ with parameter $a > 0$, as shown on the left.

To verify that $f(x)$ is a probability density function, we must show that the area under it is 1. The integral is improper, so we first integrate from 0 to B:

$$\int_0^B f(x)\, dx = a \int_0^B e^{-ax}\, dx = a\left(-\frac{1}{a}\right)e^{-ax}\ \Big|_0^B = -e^{-aB} + e^0 = 1 - e^{-aB}$$

Moved outside   Using formula 4 (inside back cover)   Evaluating   Reversing the order

Then we let $B \to \infty$:

$$\int_0^\infty f(x)\, dx = \lim_{B \to \infty}(1 - e^{-aB}) = 1 \qquad \text{Area is 1}$$

$\to 0$

To find the mean $E(X) = \int_0^\infty xf(x)\, dx$, we use integration by parts:

$$\int x\, ae^{-ax}dx = x(-e^{-ax}) - \int -e^{-ax}dx = -xe^{-ax} - \frac{1}{a}e^{-ax} \qquad \text{Indefinite integral}$$

u dv u v v du Using formula 4 (inside back cover)

$$\begin{bmatrix} u = x & dv = ae^{-ax}\, dx \\ du = dx & v = \int ae^{-ax}\, dx = -e^{-ax} \end{bmatrix}$$

Evaluating from 0 to B:

$$\left(-xe^{-ax} - \frac{1}{a}e^{-ax}\right)\Big|_0^B = -Be^{-aB} - \frac{1}{a}e^{-aB} + \frac{1}{a}$$

Finally, taking the limit as $B \to \infty$ (using $\lim_{B \to \infty} Be^{-aB} = 0$, from Exercise 50 on pages 783–784),

$$E(X) = \lim_{B \to \infty}\left(-Be^{-aB} - \frac{1}{a}e^{-aB} + \frac{1}{a}\right) = \frac{1}{a}$$

$\to 0$ as $B \to \infty$ $\to 0$ as $B \to \infty$ $E(X)$

A similar calculation (see Exercise 29) shows that $\text{Var}(X) = 1/a^2$, and therefore $\sigma(X) = 1/a$. These results are easy to remember: both the mean and standard deviation are $1/a$, the reciprocal of the parameter. Summarizing:

Exponential Random Variable

An exponential random variable X with parameter $a > 0$ takes values in $[0, \infty)$ and has probability density function

$$f(x) = ae^{-ax} \quad \text{for } x \geq 0$$

Its mean, variance, and standard deviation are

$$E(X) = \frac{1}{a}$$

$$\text{Var}(X) = \frac{1}{a^2}$$

$$\sigma(X) = \frac{1}{a}$$

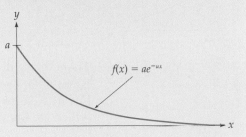

Practice Problem 1

Find the mean, variance, and standard deviation of the random variable with probability density function $f(x) = 2e^{-2x}$ for $x \geq 0$. [*Hint:* Solve by inspection.]

➤ Solution on page 842

Exponential random variables are used to model waiting times for events whose *nonoccurrence* up to any time t has no effect on the probability of the occurrence in any later time interval. (For example, your *not* winning the lottery last week has no effect on your chances this week, since each week's game is independent, so your time between winnings would be an exponential random variable.) In practice, the value of the parameter a is found by setting the theoretical mean $1/a$ equal to the observed average based on past experience.

EXAMPLE 2

FINDING PROBABILITIES OF COMPUTER FAILURE

A company's computer occasionally breaks down, and based on past records, the time X between failures averages 50 days. Model X by an exponential random variable and find

a. $P(X \leq 20)$ **b.** $P(X > 20)$ **c.** $P(30 \leq X \leq 60)$

Solution The value of the parameter a is found by setting the theoretical mean $1/a$ equal to the observed average, here 50:

$$\frac{1}{a} = 50 \quad \text{so} \quad a = \frac{1}{50} = 0.02 \qquad \text{Parameter } a = 0.02$$

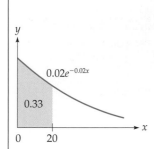

Therefore, the random variable X has probability density function

$$f(x) = 0.02e^{-0.02x} \quad \text{on } [0, \infty) \qquad f(x) = ae^{-ax} \text{ with } a = 0.02$$

For part (a), we integrate this function from 0 to 20 to find $P(X \le 20)$:

$$\underbrace{0.02}_{\substack{\text{Moved} \\ \text{outside}}} \int_0^{20} e^{-0.02x}\, dx = 0.02 \underbrace{\left(\frac{1}{-0.02}\right) e^{-0.02x}\Big|_0^{20}}_{\substack{\text{Using formula 4} \\ \text{(inside back cover)}}} = \underbrace{-e^{-0.4} + e^0}_{\text{Evaluating}} \underbrace{\approx 0.33}_{\substack{\text{Using a} \\ \text{calculator}}}$$

Therefore, with probability about $\frac{1}{3}$, the next failure will occur within 20 days. For part (b), $P(X > 20)$, we could integrate from 20 to ∞, but it is easier to find the probability of all *other* values, $P(X \le 20)$, which we found in part (a) to be 0.33, and subtract it from 1:

$$P(X > 20) = 1 - P(X \le 20) \approx 1 - 0.33 = 0.67$$

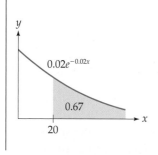

Therefore, with probability about $\frac{2}{3}$, the next failure will occur some time after 20 days. For part (c), $P(30 \le X \le 60)$, we could evaluate the integral $\int_{30}^{60} 0.02\, e^{-0.02x}\, dx$ "by hand." Instead, try it on a graphing calculator as follows.

Graphing Calculator Exploration

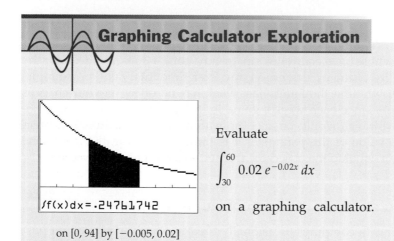

on [0, 94] by [−0.005, 0.02]

Evaluate

$$\int_{30}^{60} 0.02\, e^{-0.02x}\, dx$$

on a graphing calculator.

The value 0.248 means that with probability about $\frac{1}{4}$ the next failure of the computer will occur some time between 30 and 60 days (1 to 2 months) after the preceding failure. Such calculations are used to determine the need for a backup computer system.

EXAMPLE 3

FINDING AND USING A CUMULATIVE DISTRIBUTION FUNCTION

Find the cumulative distribution function for the probability density function in Example 2, and use it to find $P(30 \leq X \leq 60)$.

Solution For the cumulative distribution function $F(x) = P(X \leq x)$, we integrate the probability density function up to x:

$$F(x) = \int_0^x 0.02\, e^{-0.02t}\, dt = \left. -e^{-0.02t} \right|_0^x$$

$$= -e^{-0.02x} + 1 = 1 - e^{-0.02x}$$

$F(x) = 1 - e^{-0.02x}$

Therefore, the cumulative distribution function is $F(x) = 1 - e^{-0.02x}$ for $0 \leq x < \infty$.

We then find $P(30 \leq X \leq 60)$ by evaluating $F(x)$ at 60 and at 30 and subtracting:

$$P(30 \leq X \leq 60) = F(60) - F(30) = \underbrace{1 - e^{-0.02(60)}}_{F(60)} - \underbrace{(1 - e^{-0.02(30)})}_{F(30)}$$

$$= -e^{-1.2} + e^{-0.6} \approx 0.248 \qquad \text{Same answer as in Example 2}$$

Relationship Between Exponential and Poisson Random Variables

There is a useful relationship between the *Poisson* random variable discussed on pages 813–816 and the *exponential* random variables discussed here. If the time *intervals* between events are independent *exponential* random variables with parameter a, then the *total number* of events within t time units will be a *Poisson* random variable with parameter at. For instance, in the computer failure situation described in Examples 2 and 3, the times between failures are independent exponential random variables with parameter 0.02, so the total number of failures within t days will be Poisson with parameter (mean) $0.02t$. Such ideas are used in designing service queues, such as waiting lines for customers in a bank or docking facilities for ships in a port.

Median Value of a Random Variable

The mean or expected value of a random variable gives a "representative" or "middle" value for the random variable. There is another possible choice for a "representative" value—the *median*, which is a number m such that the random variable is equally likely to be above m as below m. More precisely:

Median of a Random Variable

The median of a random variable X is a number m such that x is equally likely to be on either side of m.

$$P(X \leq m) = P(X \geq m) = \frac{1}{2}$$

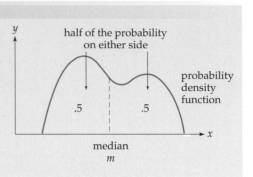

half of the probability on either side

probability density function

.5   .5

median
m

EXAMPLE 4

FINDING THE MEDIAN OF A CONTINUOUS RANDOM VARIABLE

Find the median time between failures for the computer in Example 2.

Solution   The median is most easily found by setting the cumulative distribution function $F(x)$ equal to $\frac{1}{2}$ and solving for x. We could do this "by hand," solving $1 - e^{-0.02x} = 0.5$ (using the cumulative distribution function found in Example 3), but instead we use a graphing calculator.

Graphing Calculator Exploration

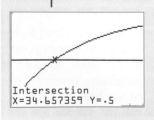

Intersection
X=34.657359 Y=.5

Define $y_1 = 1 - e^{-0.02x}$ and $y_2 = 0.5$, graph them on the window [0, 100] by [0, 1], and use INTERSECT to find where they meet.

The median waiting time is about 35 days, meaning that about half of the time the failure will occur *before* 35 days, and half of the time *after* 35 days.

Median Versus the Mean

In the computer failure example, the median $m \approx 35$ was smaller than the mean $\mu = 50$ (calculated from the formula $\mu = \frac{1}{a} = \frac{1}{0.02} = 50$). Intuitively, the mean is larger because such waiting times can occasionally be very long, and these very large values increase the mean but not the median. To clarify the difference between mean and median, think of the numbers 1, 2, 3, and 10. The mean (the average) is $\frac{1 + 2 + 3 + 10}{4} = \frac{16}{4} = 4$, but three of the four numbers are *below* this mean. For the median value, we could choose any number between 2 and 3 (such as 2.5), since two values are below and two are above this number. If we were to replace the 10 by an even larger number, this would increase the mean but not the median. In general, the mean is influenced by the values of the highest and lowest numbers, but the median is not.

Practice Problem 2

a. A news commentator recently read the statistic that the mean family income in the United States is about $40,000, and that 65% of families are below this level. He then said that this must be wrong, because only 50% of the families can be below the mean. Is he right?

b. In the 1995 major league baseball strike, it was revealed that the mean salary for major league baseball players was $1.2 million, while the median salary was only $500,000. What does this mean about the distribution of salaries among the players?

➤ Solutions on page 842

11.3 Section Summary

A *uniform* random variable takes values in a finite interval, with all values equally likely. An *exponential* random variable takes values in the infinite interval $[0, \infty)$, with decreasing probabilities for higher values.

Random Variable	Mean $E(X)$	Variance $Var(X)$	Probability Density Function	
Uniform on $[0, B]$	$\dfrac{B}{2}$	$\dfrac{B^2}{12}$	$f(x) = \dfrac{1}{B}$ for $0 \le x \le B$	
Exponential (parameter $a > 0$)	$\dfrac{1}{a}$	$\dfrac{1}{a^2}$	$f(x) = ae^{-ax}$ for $0 \le x < \infty$	

How do we choose the constants for these random variables in a given application? For the uniform distribution, B is chosen as the largest possible value of the random variable. For the exponential distribution, the parameter a is determined by setting the theoretical mean $1/a$ equal to the *observed average value* based on past experience.

The *median* of a random variable X is a number m such that the random variable is equally likely to be above it as below it: $P(X \geq m) = P(X \leq m)$. The median is most easily found by setting the cumulative distribution function equal to $\frac{1}{2}$ and solving.

➤ **Solutions to Practice Problems**

1. The probability density function is exponential with parameter $a = 2$, so

$$E(X) = \frac{1}{a} = \frac{1}{2}, \quad \text{Var}(X) = \frac{1}{a^2} = \frac{1}{4}, \quad \sigma(X) = \frac{1}{2}$$

2. **a.** He is wrong. It is the *median,* not the mean, that has 50% of the people below it and 50% above it.

   b. Baseball salaries are distributed very unequally, with a few players having extraordinarily high salaries that "skew" the mean, but most having much lower salaries, half of them under $500,000.

11.3 Exercises

EXERCISES ON UNIFORM RANDOM VARIABLES

1. If X is a uniform random variable on the interval $[0, 10]$, find

   a. The probability density function $f(x)$
   b. $E(X)$ **c.** $\text{Var}(X)$
   d. $\sigma(X)$ **e.** $P(8 \leq X \leq 10)$

2. If X is a uniform random variable on the interval $[0, 0.01]$, find

   a. The probability density function $f(x)$
   b. $E(X)$ **c.** $\text{Var}(X)$
   d. $\sigma(X)$ **e.** $P(X \geq 0.005)$

3. For a uniform random variable on the interval $[0, B]$, find

   a. The cumulative distribution function $F(x)$
   b. The median

APPLIED EXERCISES FOR UNIFORM RANDOM VARIABLES

4. GENERAL: Waiting Time Buses arrive at a station every 12 minutes, so if you arrive at a random time, your wait is uniformly distributed on the interval [0, 12]. Find

a. $E(X)$ **b.** $\text{Var}(X)$
c. $\sigma(X)$ **d.** $P(X \leq 3)$

5. ENVIRONMENTAL SCIENCE: Global Warming According to a 1990 report by the United Nations Intergovernmental Panel on Climate Change, global warming over the next century could melt polar ice and raise the sea level by 26 inches, flooding many coastal areas and submerging some islands. Suppose that the rise is modeled by a random variable X uniformly distributed on the interval [0, 30]. Find

a. $E(X)$ **b.** $\text{Var}(X)$
c. $\sigma(X)$ **d.** $P(X > 24)$

6. GENERAL: Waiting Time Your calculus teacher is often late to class, and the number of minutes you wait is uniformly distributed on the interval [0, 10] (since you leave after 10 minutes). Find

a. $E(X)$ **b.** $\text{Var}(X)$
c. $\sigma(X)$ **d.** $P(X \geq 9)$

7. Uniform Random Variable on [A, B] For any two numbers A and B with $A < B$, a *uniform random variable on the interval* [A, B] has probability density function

$$f(x) = \frac{1}{B - A} \quad \text{for} \quad A \leq x \leq B$$

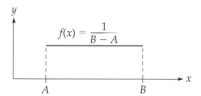

Find

a. The cumulative distribution function $F(x)$
b. The mean $E(X)$
c. The variance $\text{Var}(X)$

8. GENERAL: Roundoff Errors When rounding numbers to the nearest integer, the error X in the rounding is uniformly distributed on the interval [−0.5, 0.5]. Use Exercise 7 to find

a. $E(X)$ **b.** $\text{Var}(X)$
c. $\sigma(X)$ **d.** $P(-0.25 \leq X \leq 0.25)$

EXERCISES ON EXPONENTIAL RANDOM VARIABLES

For the exponential random variables with the following densities, find

a. $E(X)$ **b.** $\text{Var}(X)$ **c.** $\sigma(X)$
[*Hint:* Solve by inspection.]

9. $f(x) = 5e^{-5x}$ on [0, ∞)

10. $f(x) = e^{-x}$ on [0, ∞)

11. $f(x) = \frac{1}{2}e^{-x/2}$ on [0, ∞)

12. $f(x) = 0.001e^{-0.001x}$ on [0, ∞)

State the probability density function for an exponential random variable with

13. Mean 3   **14.** Mean 0.01   **15.** Variance $\frac{1}{4}$

16. Standard deviation $\frac{1}{4}$

17. For an exponential random variable with parameter a, find

a. The cumulative distribution function $F(x)$
b. The median

18. Find the probability that an exponential random variable is less than its mean.

APPLIED EXERCISES FOR EXPONENTIAL RANDOM VARIABLES

19. BUSINESS: Employee Safety Records indicate that the average time between accidents on a factory floor is 20 days. If the time between accidents is an exponential random variable, find the probability that the time between accidents is less than a month (30 days).

20. GENERAL: Emergency Calls The number of minutes between calls to an emergency 911 center is exponentially distributed with mean 8. There is one emergency operator who takes 2 minutes to deal with each call. What is the probability that the next call will get a busy signal?

21. BUSINESS: Quality Control The time until failure of a TV picture tube is an exponential random variable with mean 10 years. What is the probability that the tube will fail during the 2-year guarantee period?

22. POLITICAL SCIENCE: Supreme Court Vacancies The length of time between vacancies on the Supreme Court is exponentially distributed with mean 1.75 years. Find the probability that at least 4 years will elapse between vacancies.

23. POLITICAL SCIENCE: Russia Since the Russian Revolution, the length of time individual rulers have ruled Russia has been approximately exponentially distributed with mean 12 years. Find the probability that a ruler will last less than 10 years.

24. BUSINESS: Product Reliability The lifetime of a light bulb is exponentially distributed. What must the mean lifetime be if the manufacturer wants the bulb to last at least 750 hours with probability 0.90?

MEDIAN: Find the median of the random variable with the given probability density function.

25. $f(x) = 2x$ on $[0, 1]$

26. $f(x) = 3x^2$ on $[0, 1]$

27. $f(x) = \frac{1}{2}x$ on $[0, 2]$

28. $f(x) = \frac{1}{6}x^{-1/2}$ on $[0, 16]$

29. Find the variance of an exponential random variable with parameter a. [*Hint:* You will need to integrate by parts twice, and to use the results of Exercises 50 and 51 on pages 783–784.]

30. GENERAL: Telephone Calls You follow the telephone company's recommendation to let a number ring for ten rings (50 seconds) before hanging up. If the time it takes a government bureaucrat to answer his phone is exponentially distributed with mean 45 seconds, what is the probability that you will reach him?

31. GENERAL: Cash Machines The number of minutes between arrivals of customers at an automated teller machine is exponentially distributed with mean 7 minutes. If a customer's transactions take 3 minutes, what is the probability that the next arrival will have to wait?

32. BIOMEDICAL: ICU Admissions The number of hours between the arrivals of new patients in an intensive care unit is exponentially distributed with mean 8 hours. Find the median time between arrivals.

33. GENERAL: Airplane Maintenance The amount of flight time between failures of an airplane engine is exponentially distributed with mean 700 hours. If the engine is inspected every 100 hours of flight time, what is the probability that the engine will fail before it is inspected?

34. BUSINESS: Computer Failure The number of days between failures of a company's computer system is exponentially distributed with mean 10 days. What is the probability that the next failure will occur between 7 and 14 days after the last failure?

35. GENERAL: Light Bulbs The life of a light bulb is exponentially distributed with mean 1000 hours. What is the probability that the light bulb will burn out sometime between 800 and 1200 hours?

36–37. BIOMEDICAL: Survival Time For a person who has received treatment for a life-threatening disease, such as cancer, the number of years of life after the treatment (the *survival time*) can be modeled by an exponential random variable.

36. Suppose that the average survival time for a group of patients is 5 years. Find the probability that a randomly selected patient survives for no more than 7 years.

37. The *survival function* $S(x)$ gives the probability that a randomly selected patient will survive for at least x years after treatment has ended. Show that $S(x) = 1 - F(x)$, where $F(x)$ is the cumulative distribution function for the exponential random variable, and find a formula for $S(x)$ in terms of the exponential parameter a.

11.4 NORMAL RANDOM VARIABLES

Introduction

Studies show that heights of adults cluster around a mean of 65 inches, with larger and smaller heights occurring with lower probabilities. Similarly, IQs cluster around a mean of 100, with values farther away from the mean occurring with lower probabilities. Measurements that cluster around a mean μ, with values farther from the mean occurring with lower probabilities, are often modeled by a *normal* random variable. The probability density function of a normal random variable is often referred to as the *bell-shaped curve*, but more formally called the *Gaussian* probability density function, after the German mathematician Carl Friedrich Gauss (1777–1855). Although normal random variables are continuous, they are often used to model discrete quantities.

Normal Random Variables

The probability density function of a normal random variable depends on two parameters, denoted μ and $\sigma > 0$, which are, respectively, the mean and standard deviation of the random variable.

Normal Random Variable

A normal random variable with mean μ and standard deviation σ has probability density function

$$f(x) = \frac{1}{\sigma\sqrt{2\pi}} e^{-\frac{1}{2}\left(\frac{x-\mu}{\sigma}\right)^2} \qquad \text{for} \quad -\infty < x < \infty$$

A normal random variable is in *standard* form if it has mean $\mu = 0$ and standard deviation $\sigma = 1$. The standard normal random variable will be denoted by the letter Z. Any normal random variable X

can be put into standard form by subtracting the mean and dividing by the standard deviation:

$$Z = \frac{X - \mu}{\sigma}$$

The probability density function for a *standard* normal random variable Z takes the following simpler form:

Standard Normal Random Variable

A standard normal random variable Z has probability density function

$$f(x) = \frac{1}{\sqrt{2\pi}} e^{-\frac{1}{2}x^2} \qquad \text{for} \quad -\infty < x < \infty$$

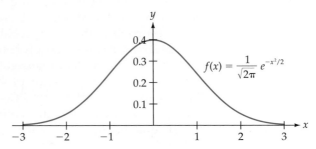

Normal probability density function with mean
$\mu = 0$ and standard deviation $\sigma = 1$

The area under the normal probability density function equals 1, although we will not prove this. The mean μ gives the location of the central peak of this probability density function. The following graph shows three normal probability density functions with different means.

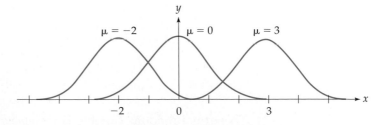

Three normal probability density
functions with different means μ

The standard deviation σ measures how the values *spread* away from the mean, so the peak of the probability density function will be higher and more pointed if σ is small, and lower and more rounded if σ is large.

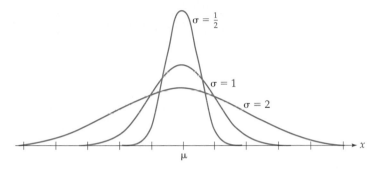

Three normal probability density
functions with different standard deviations σ

For those *not* using graphing calculators, turn now to the Appendix at the back of the book for normal probabilities calculated by *tables*.

Probabilities for Normal Random Variables

The probability that a normal random variable takes values between two numbers is the area under its probability density function between the two numbers.

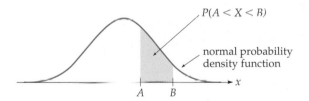

The integral of the normal probability density function cannot be expressed in terms of elementary functions, so we cannot find the area by integration. However, we can *approximate* the area by *numerical integration*, as in Section 6.4. The results can be listed in a table (as in the Appendix), or they can be found directly on a graphing calculator, using the probability DISTRibution menu and the NORMALCDF (or similar) command. The NORMALCDF command finds the probability that a normal random variable takes values between any two given numbers.

EXAMPLE **1**

FINDING A STANDARD NORMAL PROBABILITY

For a standard normal random variable Z, find $P(0 \leq Z \leq 1.24)$.

Solution

We use the NORMALCDF command with the given upper and lower limits 0 and 1.24.

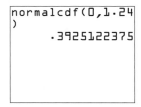

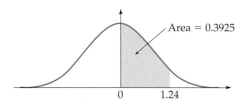

Area = 0.3925

Therefore, $P(0 \leq Z \leq 1.24) \approx 0.3925$.

Graphing Calculator Exploration

Use the NORMALCDF command to find $P(0 \leq Z \leq 0.52)$ for a standard normal random variable Z. (Answer: 0.1985)

On some calculators, you can show such results graphically using the SHADENORM command:

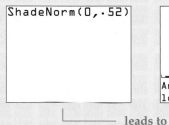

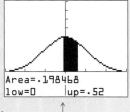

Window: $[-3, 3]$ by $[-0.2, 0.8]$ Use CLRDRAW afterward to clear the screen

leads to

The values of a normal random variable are said to be *normally distributed*.

EXAMPLE **2**   **FINDING THE PROBABILITY OF MEN'S HEIGHTS**

The heights of American men are approximately normally distributed, with mean $\mu = 68.1$ inches and standard deviation $\sigma = 2.7$ inches.* Find the proportion of men who are less than 6 feet tall.

Solution

If $X =$ height, we want $P(X < 72)$.   72 inches = 6 feet

For the lower limit, we enter a sufficiently small number to include essentially *all* of the area to the left of the upper limit (here, 0 will do). We use the NORMALCDF command, entering the given mean and standard deviation after the upper and lower limits.

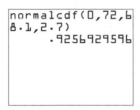

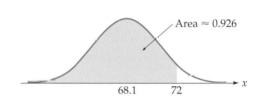

Therefore, approximately 93% of American men are under 6 feet tall.

Such calculations are used for designing automobiles and other consumer products.

Why did we enter the mean and variance in Example 2 but not in Example 1? Because the calculator will *assume* that $\mu = 0$ and $\sigma = 1$ (standard normal, which was the case in Example 1) unless told otherwise. How do we pick a lower limit to include essentially *all* of the area below the upper limit? As a "rule of thumb," choose a convenient number at least five standard deviations below the mean. In this example, five standard deviations means $5 \cdot 2.7 = 13.5$, and our lower limit of 0 was far more than this distance below the mean of 68.1. A similar rule of thumb applies when there is no upper limit: go at least five standard deviations *above* the mean.

* These and other data in this section are from Gavriel Salvendy, *Handbook of Human Factors*, John Wiley–Interscience, 1987.

Graphing Calculator Exploration

The weights of American women are approximately normally distributed, with mean $\mu = 134.7$ pounds and standard deviation $\sigma = 30.4$ pounds. Find the proportion of women who weigh more than 125 pounds. (Answer: 62.6%)

EXAMPLE 3

FINDING THE PROBABILITY OF ACCEPTABLE CHOLESTEROL LEVELS

Cholesterol is a fat-like substance that, in the bloodstream, can increase the risk of heart disease. Cholesterol levels below 200 (milligrams per deciliter of blood) are generally considered acceptable. A survey by the U.S. Department of Health found that cholesterol levels for one age group were approximately normally distributed, with mean $\mu = 222$ and standard deviation $\sigma = 28$. What proportion of the people in the study had acceptable cholesterol levels?

Solution

Letting $X = \left(\begin{matrix} \text{Cholesterol} \\ \text{level} \end{matrix} \right)$, we want $P(X < 200)$. We again use 0 for the lower limit. (Do you see why it satisfies the "at least five standard deviations below the mean" rule?) The NORMALCDF command with these numbers gives the following result.

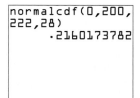

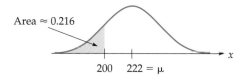

Therefore, only about 22% of the people in the study had acceptable cholesterol levels.

Inverse Probabilities for Normal Random Variables

We may *reverse* the process of finding probabilities, finding instead the value that results in a *given* probability. A graphing calculator command like INVNORM finds the number such that, with a given probability, a normal random variable is *below* that number.

EXAMPLE 4

FINDING THE HEIGHT FOR A GIVEN PROBABILITY

Women's heights are approximately normally distributed, with mean $\mu = 63.2$ inches and standard deviation $\sigma = 2.6$ inches. Find the height that marks the tallest 10%.

Solution

The height that marks the tallest 10% means that 90% are below that height. The INVNORM command with the given numbers provides the following answer.

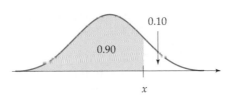
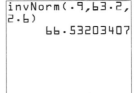

Therefore, a height of about 66.5 inches, or 5 feet $6\frac{1}{2}$ inches, marks the tallest 10% of women.

Measuring distance from the mean in *standard deviation units* leads to some useful probabilities for a normal random variable X.

With probability 68%, X is within 1 standard deviation of its mean.

With probability 95%, X is within 2 standard deviations of its mean.

With probability 99.7%, X is within 3 standard deviations of its mean.

These results are shown graphically as follows.

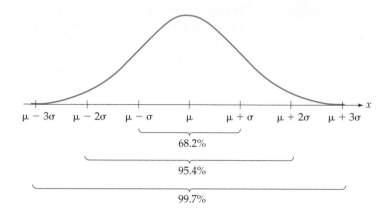

For example, from the fact that men's heights are normally distributed with $\mu = 68.1$ and $\sigma = 2.7$, we can conclude

About 68% of men have heights 68.1 ± 2.7 inches (that is, heights 65.4 inches to 70.8 inches).   Mean ± 1 standard deviation

About 95% of men have heights $68.1 \pm 2(2.7)$ inches (that is, heights 62.7 inches to 73.5 inches).   Mean ± 2 standard deviation

About 99.7% of men have heights $68.1 \pm 3(2.7)$ inches (that is, heights 60 inches to 76.2 inches).   Mean ± 3 standard deviation

11.4 Section Summary

Normal random variables are used to model values that cluster symmetrically around a mean, μ, with the probability decreasing as the distance from μ increases. The probability density function of a normal random variable with mean μ and standard deviation σ is

$$f(x) = \frac{1}{\sigma\sqrt{2\pi}}\, e^{-\frac{1}{2}\left(\frac{x-\mu}{\sigma}\right)^2} \qquad \text{for } -\infty < x < \infty$$

The mean μ gives the location of the central peak, and the standard deviation σ measures the *spread* of the values around this mean (larger σ meaning greater spread).

The probability density of a *standard* normal random variable ($\mu = 0$ and $\sigma = 1$) is

$$f(x) = \frac{1}{\sqrt{2\pi}} e^{-\frac{1}{2}x^2} \qquad \text{for } -\infty < x < \infty$$

Probabilities for a normal random variable can be found either in a *table* (by first *standardizing*, defining $Z = \dfrac{X - \mu}{\sigma}$ and then using the standard normal table, as explained in the Appendix) or by using a *graphing calculator* with an operation like normalCDF. Similarly, we can find the x-value such that with given probability the random variable is below that value either by using tables or a graphing calculator operation like INVNORM.

11.4 Exercises

Find each probability for a standard normal random variable Z.

1. $P(0 \le Z \le 1.95)$ **2.** $P(-1.23 \le Z \le 0)$

3. $P(-2.55 \le Z \le 0.48)$ **4.** $P(-1.5 \le Z \le 1.05)$

5. $P(1.4 \le Z \le 2.8)$ **6.** $P(1 \le Z \le 2)$

7. $P(Z < -3.1)$ **8.** $P(Z > 3)$

9. $P(Z \le 3.45)$ **10.** $P(Z \ge -3.5)$

11. If X is normal with mean 9 and standard deviation 2, find $P(10 < X < 11)$.

12. If X is normal with mean 3 and standard deviation $\frac{1}{2}$, find $P(2 \le X \le 3.5)$.

13. If X is normal with mean -5 and standard deviation 2, find $P(X \le -6)$.

14. If X is normal with mean 0.05 and standard deviation 0.01, find $P(X > 0.04)$.

15. Find the x-value at which the standard normal probability density function $f(x) = \dfrac{1}{\sqrt{2\pi}} e^{-x^2/2}$ is maximized. Also find the value of the function at this x-value.

16. Find the x-value at which the (nonstandard) normal probability density function $f(x) = \dfrac{1}{\sigma \sqrt{2\pi}} e^{-\frac{1}{2}\left(\frac{x-\mu}{\sigma}\right)^2}$ (for constants μ and σ) is maximized. Also find the maximum value of the function at this x-value.

17. Find the inflection points of the standard normal probability density function $f(x) = \dfrac{1}{\sqrt{2\pi}} e^{-x^2/2}$.

18. Find the x-coordinates of the inflection points of the (nonstandard) normal probability density function $f(x) = \dfrac{1}{\sigma \sqrt{2\pi}} e^{-\frac{1}{2}\left(\frac{x-\mu}{\sigma}\right)^2}$ (for constants μ and σ).

APPLIED EXERCISES

19. SOCIAL SCIENCE: Workload A motor vehicle office can process only 250 license applications per day. If the number of applications on a given day is normally distributed with mean 210 and standard deviation 30, what is the probability that the office can handle the day's applications? [*Hint:* Find $P(0 \leq X \leq 250)$.]

20. BIOMEDICAL: Hospital Stays A patient's medical insurance will pay for at most 20 days in the hospital. Past experience with similar cases shows that the length of hospital stay is normally distributed with mean 17 days and standard deviation 2 days. What is the probability that the insurance coverage will be enough? [*Hint*: Find $P(0 \leq X \leq 20)$.]

21. BUSINESS: Advertising The percentage of television viewers who watch a certain show is a normal random variable with mean 22% and standard deviation 3%. The producer guarantees to advertisers that the viewer percentage will be at least 20%, or else the advertisers will receive free air time. What is the probability that the producers will have to give free air time? [*Hint:* Find $P(X < 20)$.]

22. PSYCHOLOGY: Learning The time that a rat takes to "solve" a maze is normally distributed with mean 3 minutes and standard deviation 30 seconds. Find the probability that the rat will take less than 2 minutes.

23. BUSINESS: Orders The number of orders received by a company each day is normally distributed with mean 2000 and variance 160,000. If the orders exceed 2500, the company will have to hire extra help. Find the probability that extra help will be needed. [*Hint:* First find σ.]

24. BUSINESS: Airline Overbooking In order to avoid empty seats, airlines generally sell more tickets than there are seats. Suppose that the number of ticketed passengers who show up for a flight with 100 seats is a normal random variable with mean 97 and standard deviation 6. What is the probability that at least one person will have to be "bumped" (denied space on the plane)?

25. GENERAL: Weights Weights of men are approximately normally distributed with mean 163 pounds and standard deviation 28. If the minimum and maximum weights in order to be a city fireman are 128 and 254 pounds, respectively, what proportion of men qualify?

26. GENERAL: SAT Scores Test scores on the 2001 SAT I: Reasoning Test in Mathematics were approximately normally distributed with mean 514 and standard deviation 113. What score would put you in the top 20%?

27. SOCIAL SCIENCE: IQs Although intelligence cannot be measured on a single scale, IQ (Intelligence Quotient) scores are still used in some situations. IQ scores are normally distributed with mean 100 and standard deviation 15. Find the IQ score that places you in the top 2%.

28. BUSINESS: Product Reliability The number of years that an electrical generator will last is a normal random variable. Company A makes generators with a mean lifetime of 8 years and a standard deviation of 1 year. Company B makes generators with a mean lifetime of 7 years and a standard deviation of 2 years. Which company's generators have a higher probability of lasting at least 10 years?

29. BUSINESS: Quality Control A soft drink bottler fills 32-ounce bottles with a high-speed bottling machine. Assume that the amount of the soft drink put into the bottle is a normal random variable. The mean amount that the machine puts into the bottles is adjustable, but the standard deviation is 0.25 ounce. What should be the mean amount such that with probability 98% a bottle contains at least 32 ounces?

Chapter Summary with Hints and Suggestions

Reading the text and doing the exercises in this chapter have helped you to master the following skills, which are listed by section (in case you need to review them) and are keyed to particular Review Exercises. Answers for all Review Exercises are given at the back of the book, and full solutions can be found in the Student Solutions Manual.

11.1 Discrete Probability

- Find $E(X)$, $\text{Var}(X)$, and $\sigma(X)$ for a discrete random variable.   *(Review Exercises 1–2.)*

$$\mu = E(X) = \sum_{i=1}^{n} x_i p_i$$

$$\text{Var}(X) = \sum_{i=1}^{n} (x_i - \mu)^2 p_i = \sum_{i=1}^{n} x_i^2 p_i - \mu^2$$

$$\sigma(X) = \sqrt{\text{Var}(X)}$$

- Solve an applied problem using expected value.   *(Review Exercises 3–4.)*

- Solve an applied problem using a Poisson random variable.   *(Review Exercises 5–10.)*

$$P(X = k) = e^{-a}\frac{a^k}{k!} \quad k = 0, 1, 2, \ldots$$

$$E(X) = a$$

11.2 Continuous Probability

- Make a function into a probability density function.   *(Review Exercises 11–16.)*

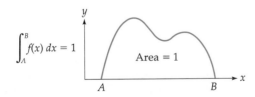

- Find a probability and the cumulative distribution function.   *(Review Exercise 17.)*

$$P(a \le X \le b) = \int_{a}^{b} f(x)\, dx$$

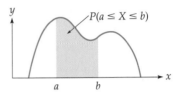

$$F(x) = \int_{A}^{x} f(t)\, dt$$

- Find a probability and the probability density function.   *(Review Exercise 18.)*

$$P(a \le X \le b) = F(b) - F(a) \qquad f(x) = \frac{d}{dx}F(x)$$

- Find $E(X)$, $\text{Var}(X)$, and $\sigma(X)$ for a continuous random variable.   *(Review Exercises 19–20.)*

$$\mu = E(X) = \int_{A}^{B} x\, f(x)\, dx$$

$$\text{Var}(X) = \int_{A}^{B} (x - \mu)^2 f(x)\, dx$$

$$= \int_{A}^{B} x^2 f(x)\, dx - \mu^2$$

$$\sigma(X) = \sqrt{\text{Var}(X)}$$

- Solve an applied problem involving a continuous random variable.   *(Review Exercises 21–22.)*

- Find $E(X)$, Var(X), and $\sigma(X)$ for a continuous random variable.
  (*Review Exercises 23–24.*)

- Solve an applied problem involving a continuous random variable.
  (*Review Exercises 25–26.*)

11.3 Uniform and Exponential Random Variables

- Find the probability density function, $E(X)$, Var(X), $\sigma(X)$, and a probability for a uniform random variable.   (*Review Exercises 27–30.*)

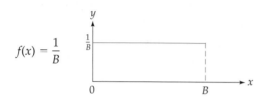

$f(x) = \dfrac{1}{B}$

$$E(X) = \frac{B}{2} \qquad \text{Var}(X) = \frac{B^2}{12} \qquad \sigma(X) = \frac{B}{2\sqrt{3}}$$

- Solve an applied problem using an exponential random variable.
  (*Review Exercises 31–32.*)

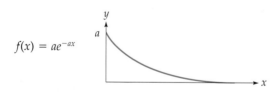

$f(x) = ae^{-ax}$

$$E(X) = \frac{1}{a} \qquad \text{Var}(X) = \frac{1}{a^2} \qquad \sigma(X) = \frac{1}{a}$$

- Find the median of continuous random variable.   (*Review Exercises 33–34.*)

$$P(X \le m) = P(X \ge m) = \frac{1}{2}$$

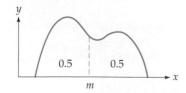

- Solve an applied problem involving an exponential random variable.
  (*Review Exercises 35–36.*)

11.4 Normal Random Variables

- Find a probability for a standard normal random variable.   (*Review Exercises 37–40.*)

$$f(x) = \frac{1}{\sqrt{2\pi}} e^{-x^2/2}$$

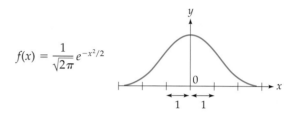

- Find a probability for a normal random variable with a given mean and variance.
  (*Review Exercises 41–42.*)

$$f(x) = \frac{1}{\sigma\sqrt{2\pi}} e^{-\frac{1}{2}\left(\frac{x-\mu}{\sigma}\right)^2}$$

$$Z = \frac{X - \mu}{\sigma}$$

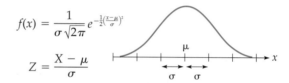

- Solve an applied problem using a normal random variable.
  (*Review Exercises 43–46.*)

Hints and Suggestions

- In discrete probability, probabilities are found by *adding*. In continuous probability, they are found by *integrating* a probability density function (or by subtracting values of a cumulative distribution function).

- The *expected value* or *mean* of a random variable gives a "representative" value, whereas the *variance* and *standard deviation* measure the "spread" or "dispersion" of the values away from the mean.

- Poisson random variables are often used to model the number of occurrences of "rare events," with the parameter a set equal to the *average number of occurrences*.

■ The alternative formula for the variance (page 826) is generally easier to use than the definition (page 824). The corresponding alternative variance formula for a *discrete* random variable is given in Exercise 33 on page 820.

■ The mean and the median are both used as "representative" numbers for a random variable. The mean is influenced by the size of the largest value, whereas the median is not. For example, if X takes values $\{1, 2, 99\}$ with equal probability, the median would be 2, whereas the mean would be 34 (showing the influence of the 99).

■ Uniform random variables are used to model waiting times for which all numbers in a *fixed interval* are *equally likely*. The parameter B is set equal to the largest possible value.

■ Exponential random variables are used to model waiting times that can be *arbitrarily long*. The parameter a is found by setting the mean $1/a$ equal to the observed average value based on past experience.

■ Normal random variables are used to model measurements that cluster symmetrically around the mean. About 68% of the probability is within one standard deviation of the mean, and about 95% is within two standard deviations of the mean.

■ A graphing calculator helps by finding medians and integrating probability density functions to find probabilities, especially for the normal distribution.

■ **Practice for Test:** Review Exercises 1, 5, 13, 17, 19, 21, 23, 25, 27, 29, 31, 33, 35, 41, 45.

Review Exercises for Chapter 11

Practice test exercise numbers are in green.

11.1 Discrete Probability

For each spinner, the arrow is spun and then stops, pointing in one of the numbered sectors. Let X stand for the number so chosen. If the probability of each number is proportional to the area of its sector, find **a.** $E(X)$ **b.** $\text{Var}(X)$ **c.** $\sigma(X)$

1.

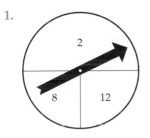

2.

(See instructions in column 1.)

3. **GENERAL: Land Value** A builder estimates a parcel of land to be worth $2,000,000 if he can get a zoning variance, and only $50,000 if he cannot. He believes that the probability of getting the variance is 40%. What is the expected value of the land?

4. **BUSINESS: Land Value** A land speculator must choose between two pieces of property, whose values depend on whether they contain oil. Property A will be worth $1,200,000 with probability 10% (that is, if oil is found), and only $2000 with probability 90%. Property B will be worth $1,000,000 with probability 15%, and only $1000 with probability 85%. On the basis of expected value, which property should the speculator buy?

5. **BUSINESS: Quality Control** Computer disks are manufactured in batches of 1000, with an average of two defective disks per batch. Use the Poisson distribution to find the probability that a batch has two or fewer defective disks.

6. **BUSINESS: Insurance** According to a 1990 insurance company study, the number of major natural "catastrophes" (such as earthquakes and floods) resulting in at least $5 million of claims has recently averaged 2.63 per month. Use the Poisson distribution to estimate the probability that the number X in a given month satisfies $2 \leq X \leq 4$.

7. **GENERAL: Mail Delivery** On an average day you receive 6 letters. Use a Poisson random variable to find the probability that on a given day you receive no mail.

8. **GENERAL: Cookies** Suppose that 100 cookies are to be made from dough containing 500 chocolate chips. Use a Poisson random variable to find the probability that a given cookie has no chips.

9. **GENERAL: Earthquakes** (Graphing calculator with series operations helpful) On the average, 18 "major" earthquakes (Richter 7 or higher) occur each year in the world. Use a Poisson random variable to estimate the probability that, in a given year, the number of earthquakes will be

a. 6 or fewer
b. 12 or fewer
c. 18 or fewer
d. 24 or fewer
e. more than 24

10. **BUSINESS: Quality Control** In a shipment of 500 television sets, an average of two sets will be defective. The dealer will return the shipment if more than four sets are defective. Use the Poisson distribution to find the probability of a return. [*Hint:* First find $P(X \leq 4)$ and then subtract it from 1.]

11.2 Continuous Probability

Find the value of a that makes each function a probability density function on the given interval.

11. $ax^2(2 - x)$ on $[0, 2]$ 12. $a\sqrt{9 - x}$ on $[0, 9]$

13. $\dfrac{a}{(1 + x)^2}$ on $[0, 1]$ 14. $a \sin x$ on $[0, \pi]$

15. $a\sqrt{1 - x^2}$ on $[-1, 1]$ 16. $ae^{-|x|}$ on $[-2, 2]$

17. For the probability density function $f(x) = \frac{1}{9}x^2$ on $[0, 3]$, find
   a. $P(1 \leq X \leq 2)$ using the probability density function
   b. The cumulative distribution function $F(x)$
   c. $P(1 \leq X \leq 2)$ using the cumulative distribution function

18. For the cumulative distribution function $F(x) = \frac{1}{4}\sqrt{x} - \frac{1}{4}$ on $[1, 25]$, find
   a. $P(4 \leq X \leq 9)$ using the cumulative distribution function
   b. The probability density function $f(x)$
   c. $P(4 \leq X \leq 9)$ using the probability density function

19. For the probability density function $f(x) = \frac{3}{2}(1 - x^2)$ on $[0, 1]$, find
   a. $E(X)$ b. $\text{Var}(X)$ c. $\sigma(X)$

20. For the probability density function $f(x) = \frac{2}{3}x^{-3/2}$ on $[1, 16]$, find
   a. $E(X)$ b. $\text{Var}(X)$ c. $\sigma(X)$

21. **GENERAL: Electrical Demand** The amount of electricity (in millions of kilowatt-hours) used by a city is a random variable X with probability density function $f(x) = \frac{1}{32}x$ on $[0, 8]$. Find
   a. The expected demand $E(X)$

b. The power level that, with probability 0.95, will be sufficient [that is, find b such that $P(X \leq b) = 0.95$]

22. **BEHAVIORAL SCIENCE: Learning** In a test of memory, the proportion X of items recalled is a random variable with probability density function $f(x) = 12x(1 - x)^2$ on $[0, 1]$. Find
    a. $P\left(X \leq \frac{1}{2}\right)$ **b.** $E(X)$

23. For the probability density function
    $$f(x) = \frac{2}{\pi}\sqrt{1 - x^2} \text{ on } [-1, 1], \text{ find}$$
    a. $E(X)$ **b.** $\text{Var}(X)$ **c.** $\sigma(X)$

24. For the probability density function
    $f(x) = 504x^3(1 - x)^5$ on $[0, 1]$, find
    a. $E(X)$ **b.** $\text{Var}(X)$ **c.** $\sigma(X)$

25. **BUSINESS: Manufacturing Flaws** A flaw is randomly located in a computer disk. The x-coordinate of the flaw has probability density function $f(x) = \dfrac{1}{2\pi}\sqrt{4 - x^2}$ on $[-2, 2]$. Find the probability that the x-coordinate of the flaw lies between $x = -1$ and $x = 1$.

26. **GENERAL: Time to Complete a Task** The number of minutes that a person takes to complete a task has probability density function $f(x) = 105x^2(1 - x)^4$ for x in $[0, 1]$. Find the probability that the time lies between $x = 0.25$ and $x = 0.5$ minute.

11.3 Uniform and Exponential Random Variables

27. If X is a uniform random variable on the interval $[0, 50]$, find
    a. The probability density function $f(x)$
    b. $E(X)$
    c. $\text{Var}(X)$
    d. $\sigma(X)$
    e. $P(15 \leq X \leq 45)$

28. **GENERAL: Rainfall** Rain is forecast, and you estimate that the actual number of inches X of rain is uniformly distributed on the interval $[0, 2]$. Find
    a. $E(X)$ **b.** $\text{Var}(X)$
    c. $\sigma(X)$ **d.** $P(X < 0.5)$

29. For the exponential random variable with probability density function $f(x) = 0.01e^{-0.01x}$ on $[0, \infty)$, find
    a. $E(X)$ **b.** $\text{Var}(X)$

30. State the probability density function for an exponential random variable with mean 4.

31. **GENERAL: Airplane Delays** The number of minutes that an airplane is late is an exponential random variable with mean 5 minutes. The airline must pay compensation if the flight is more than 15 minutes late. What is the probability that it has to pay compensation?

32. **BUSINESS: Quality Control** The "life" of a light bulb is exponentially distributed with mean 750 hours. Find the probability that the bulb will burn out within the first 500 hours.

Find the median of the random variable with the given probability density function.

33. $f(x) = \frac{3}{8}x^2$ on $[0, 2]$

34. $f(x) = 3e^{-3x}$ on $[0, \infty)$

35. **GENERAL: Fire Alarms** The time between calls to a fire department is an exponential random variable with mean 4 hours. If the fire company is out of the firehouse for 1 hour on each call, what is the probability that the company is out when the next call comes in?

36. **GENERAL: Ship Arrivals** The time between the arrival of ships in a port is an exponential random variable with mean 5 hours. If each arrival requires 1 hour of docking time with the harbor's tugboat, what is the probability that the next arriving ship will have to wait?

11.4 Normal Random Variables

Find the indicated probability for a standard normal random variable Z.

37. $P(0.95 \leq Z \leq 2.54)$

38. $P(-1.11 \leq Z \leq 1.44)$

39. $P(Z < 2.5)$ **40.** $P(Z > 1)$

41. If X is normal with mean 12 and standard deviation 2, find $P(10 < X \leq 15)$.

42. If X is normal with mean 1 and standard deviation 0.4, find $P(0 < X \leq 1.8)$.

43. BUSINESS: Refunds The "life" of an automobile tire is normally distributed with mean 50,000 miles and standard deviation 5000. If the company guarantees its tires for 40,000 miles, what is the probability that a tire will outlast its guarantee?

44. SOCIAL SCIENCE: Smoking In a large-scale study carried out in the 1960s among male smokers 35 to 45 years of age, the number of cigarettes smoked daily was approximately normal with mean 28 and standard deviation 10. What proportion had a two-pack-a-day habit (40 or more cigarettes)?

45. GENERAL: Heights Heights of women are approximately normally distributed, with mean 63.2 inches and standard deviation 2.6. If to be a police officer a woman must be at least 62 inches tall, what proportion of women qualify?

46. GENERAL: SAT Scores Test scores on the 2001 SAT I: Reasoning Test (Verbal) were approximately normally distributed with mean 506 and standard deviation 111. What score will put you in the top 30%?

Cumulative Review for Chapters 1–11

The following exercises review some of the basic techniques that you learned in Chapters 1–11. Answers to all of these Cumulative Review exercises are given in the answer section at the back of the book. A graphing calculator is suggested but not required.

1. **a.** Find an equation for the horizontal line through the point $(-2, 5)$.
   b. Find an equation for the vertical line through the point $(-2, 5)$.

2. Simplify: $\left(\frac{1}{4}\right)^{-3/2}$

3. Use the definition of the derivative,
$$f'(x) = \lim_{h \to 0} \frac{f(x + h) - f(x)}{h}, \quad \text{to find the derivative of } f(x) = 3x^2 - 7x + 1.$$

4. The temperature in an industrial refining tank after t hours is $T(t) = 12\sqrt{t^3 + 225}$ degrees Fahrenheit. Find $T'(4)$ and $T''(4)$ and interpret these values.

5. Find $\dfrac{d}{dx}\sqrt[3]{x^3 + 8}$.

6. Find the derivative of $f(x) = (2x + 3)^3(3x + 2)^4$.

7. Make sign diagrams for the first and second derivatives and then graph the function $f(x) = x^3 - 12x^2 - 60x + 400$, showing all relative extreme points and inflection points.

8. Make sign diagrams for the first and second derivatives and graph the function $f(x) = \sqrt[3]{x} + 1$ showing all relative extreme points and inflection points.

9. Make a sign diagram for the first derivative and graph the function $f(x) = \dfrac{1}{x^2 - 4}$ showing all asymptotes and relative extreme points.

10. A gardener wants to enclose two identical rectangular pens along an existing wall. If the side along the wall needs no fence, what is the largest total area that can be enclosed using only 300 feet of fence?

11. An open-top box is to be made from a square sheet of metal 12 inches on each side by cutting a square from each corner and folding up the sides. Find the volume of the largest box that can be made in this way.

12. For $xy - 3y = 5$, use implicit differentiation to find $\dfrac{dy}{dx}$ at the point $(4, 5)$.

13. A circular oil slick is expanding such that its area is growing at the rate of 3 square miles per day. Find how fast the radius is growing at the moment when the radius is 2 miles.

14. A sum of $5000 is deposited in a bank account earning 6% interest. Find the value after 4 years if the interest is compounded:
   a. Semiannually.   **b.** Continuously.

15. An investment grows by 9% compounded continuously. How long will it take to increase by 75%?

16. How much must you deposit now in a bank account paying 8% compounded quarterly to have $100,000 in 25 years?

17. Find
   a. $\dfrac{d}{dx}\left(\dfrac{e^{2x}}{x}\right)$ **b.** $\dfrac{d}{dx}\ln(x^3 + 1)$

18. Consumer demand for a product is $D(p) = 12{,}000e^{-0.04p}$ units, where p is the selling price in dollars. Find the price that maximizes consumer expenditure. [*Hint:* Expenditure is $E(p) = p \cdot D(p)$.]

19. The gross domestic product of a country after t years is $G(t) = 3 + 5e^{0.06t}$. Find the relative rate of change after 2 years.

20. Find

 a. $\int (12x^3 - 4x + 5)\, dx$

 b. $\int 6e^{-2x}\, dx$

21. Evaluate $\int_1^e \frac{2}{x}\, dx$.

22. A company's marginal cost function is $MC(x) = 3x^{-1/2}$ thousand dollars, where x is the number of units. Find the total cost of producing units 25 through 100.

23. Radioactive waste is being deposited into a new underground storage vault at the rate of $120e^{0.15t}$ tons per year, where t is the number of years that the depository has been open. Find a formula for the total amount of radioactive waste deposited during the first t years.

24. Find the area bounded by $y = x^2 + 4$ and $y = 6x + 4$.

25. The population of a city is predicted to be $P(t) = 60e^{0.04t}$ thousand people, where t is the number of years from now. Find the average population over the next 10 years (year 0 to year 10).

26. Find

 a. $\int \frac{x\, dx}{\sqrt{x^2 + 1}}$ b. $\int_0^1 xe^{-x^2}\, dx$.

27. Find by integration by parts: $\int x^2 \ln x\, dx$.

28. Use the integral table on the inside back cover to find $\int \frac{x}{\sqrt{x - 1}}\, dx$.

29. Evaluate $\int_0^\infty e^{-2x}\, dx$.

30. Use trapezoidal approximation with $n = 4$ trapezoids to approximate $\int_0^1 e^{\sqrt{x}}\, dx$.

 (If you are using a graphing calculator, compare your answer with that using FnInt.)

31. Use Simpson's Rule with $n = 4$ to approximate $\int_0^1 e^{\sqrt{x}}\, dx$.

 (If you are using a graphingcalculator, compare your answer with that using FnInt.)

32. Find the first partial derivatives of $f(x, y) = xe^y - y \ln(x^2 + 1)$.

33. Find the extreme values of $f(x, y) = 3x^2 + 2y^2 - 2xy - 8x - 4y + 15$.

34. Find the least squares line for the following points:

x	y
2	4
4	3
6	0
8	-1

35. Use Lagrange multipliers to find the maximum value of $f(x, y) = 100 - 2x^2 - 3y^2 + 2xy$ subject to the constraint $2x + y = 18$.

36. Find the total differential of $f(x, y) = 5x^2 - 2xy + 2y^3 - 12$.

37. Find the volume under the surface $f(x, y) = x^2 + y^2$ over the rectangle $R = \{ (x, y) \mid 0 \le x \le 3, 0 \le y \le 3 \}$.

38. a. Convert $135°$ into radians.

 b. Convert $\frac{5\pi}{6}$ into degrees.

39. Evaluate *without* using a calculator.

 a. $\sin \dfrac{3\pi}{4}$ b. $\cos \dfrac{3\pi}{4}$

 c. $\sin \dfrac{5\pi}{6}$ d. $\cos \dfrac{5\pi}{6}$.

40. Sketch the graph of $f(x) = 2 \cos 4x$.

41. Find $\dfrac{d}{dx}(\sin x \cos x)$.

42. Find $\displaystyle\int e^{\cos x} \sin x \, dx$

43. Find the area under $y = \cos x$ from $x = 0$ to $x = \pi/2$.

44. Evaluate *without* using a calculator.

   a. $\tan \dfrac{5\pi}{6}$ **b.** $\cot \dfrac{5\pi}{6}$

   c. $\sec \dfrac{3\pi}{4}$ **d.** $\csc \dfrac{3\pi}{4}$.

45. Find

   a. $\dfrac{d}{dx} \tan(x^2 + 1)$

   b. $\displaystyle\int \sec(2t + 1) \tan(2t + 1) \, dt$

46. **a.** Find the general solution to the differential equation $y' = x^3 y$.
   b. Then find the particular solution that satisfies $y(0) = 2$.

47. Solve the first-order linear differential
equation and intial condition $\begin{cases} y' + 6xy = 0 \\ y(0) = 4 \end{cases}$

48. Find the Euler approximation on the interval $[0, 1]$ using $n = 4$ segments for the following initial value problem, and draw a graph of your approximation. $\begin{cases} y' = \dfrac{12x}{y} \\ y(0) = 2 \end{cases}$

49. Determine whether the geometric series converges, and if so, find its sum:

$$\tfrac{4}{5} + \tfrac{16}{25} + \tfrac{64}{125} + \cdots$$

50. Find the third Taylor polynomial at $x = 0$ for $f(x) = e^{2x}$.

51. Find the Taylor series at $x = 0$ for $f(x) = \sin x^2$ by modifying a known Taylor series. What is its radius of convergence?

52. Use Newton's method to approximate $\sqrt{105}$, continuing until two successive iterations agree to five decimal places.

53. A department store receives umbrellas in batches of 120, and, on average, four are defective. Use a Poisson random variable to find the probability that three or fewer are defective.

54. If X is a uniform random variable on the interval $[0, 18]$, find:

   a. $E(X)$ **b.** $\text{Var}(X)$

55. Suppose that the waiting time for an elevator in a tall building is exponentially distributed with mean 1 minute. Find the probability that your wait will be $1\frac{1}{2}$ minutes or less.

56. If a random variable X is normally distributed with mean 12 and standard deviation 2, find $P(11 \le X \le 13)$.

Appendix

NORMAL PROBABILITIES FROM TABLES

Introduction

This Appendix is for readers of Section 11.4 who do not use graphing calculators. You should read Section 11.4 through "Normal Random Variables," which ends on page 847, then read this Appendix, and then return to page 851, just after Example 4.

Standard Normal Random Variables

A normal random variable with mean $\mu = 0$ and standard deviation $\sigma = 1$ is called a *standard* normal random variable. Its probability density function

$$f(x) = \frac{1}{\sqrt{2\pi}} e^{-\frac{1}{2}x^2} \qquad \text{for } -\infty < x < \infty$$

Probality density function for a standard normal random variable

was graphed on page 846. Throughout this section, the letter Z will always stand for a *standard normal random variable*. The probability that a standard normal random variable Z lies between two numbers is the area under the probability density function between the two numbers.

Probabilities for Standard Normal Random Variables

For a standard normal random variable Z,

$$P(a \le Z \le b) = \int_a^b \frac{1}{\sqrt{2\pi}} e^{-\frac{1}{2}x^2}\, dx$$

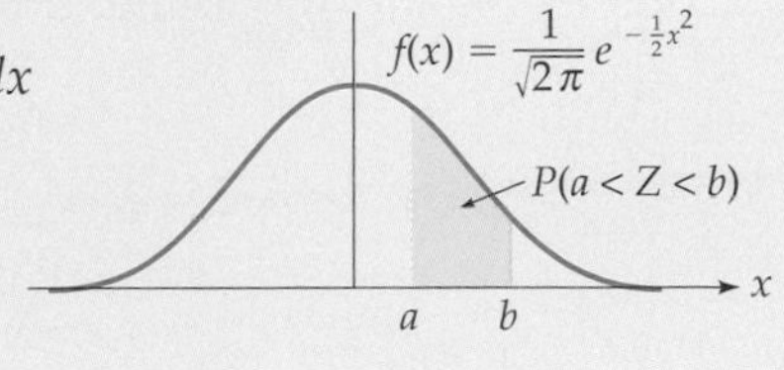

$$f(x) = \frac{1}{\sqrt{2\pi}} e^{-\frac{1}{2}x^2}$$

$P(a < Z < b)$

The integral of the normal probability density function cannot be expressed in terms of elementary functions, so we cannot find the area by integration. However, we can *approximate* the area by *numerical* integration as in Section 6.4, and then list the results in a table. One such table, on page A8 at the end of this Appendix, gives the areas under this curve from 0 to any positive number x. The first two examples show how to use the table.

EXAMPLE 1

FINDING A PROBABILITY FOR A STANDARD NORMAL RANDOM VARIABLE

Find $P(0 \leq Z \leq 1.24)$.

Solution

The table on page A8 gives the probability that a standard normal random variable Z is between 0 and any higher number. For the given number 1.24, we find the *row* headed 1.2 and the *column* headed .04 (the second decimal place), which intersect at the table value of .3925:

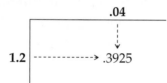

From the table at the end of the Appendix: the row headed 1.2 and the column headed .04 lead to the table value .3925

Therefore:

$$P(0 \leq Z \leq 1.24) = .3925$$

Also written 0.3925

row ⌟ ⌞ column ⌞ table

Practice Problem 1   Find $P(0 \leq Z \leq 0.52)$

➤ Solution on page A7

We will use the notation

$$A(x) = \left(\begin{array}{c} \text{Area under the standard} \\ \text{normal probability density} \\ \text{function from 0 to } x \end{array} \right)$$

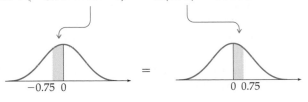

That is, $A(x)$ is the number found in the standard normal table in the row and column determined by x. For example, in Example 1 we found $A(1.24) = 0.3925$, and in Practice Problem 1 you found $A(0.52) = 0.1985$.

The probabilitiy that Z is between two numbers can be found by adding and subtracting values from the table, using the symmetry of the curve about $x = 0$ and the fact that the area under each half of the curve is 0.5 (since the total area is 1). The following example illustrates the technique, with the calculation shown graphically as well.

EXAMPLE 2

FINDING PROBABILITIES FROM A TABLE

Find: **a.** $P(-0.75 \le Z \le 0)$ **b.** $P(Z \le 1.44)$
   c. $P(-1.2 \le Z \le 0.68)$ **d.** $P(0.65 \le Z \le 1.45)$
   e. $P(Z > 0.95)$

Solution

a. $P(-0.75 \le Z \le 0)$ $= A(0.75) = 0.2734$

From the table on page A8

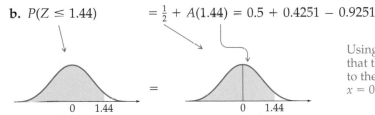

Using the symmetry of the curve about $x = 0$

b. $P(Z \le 1.44)$ $= \frac{1}{2} + A(1.44) = 0.5 + 0.4251 = 0.9251$

Using the fact that the area to the left of $x = 0$ is $\frac{1}{2}$

c. $P(-1.2 \le Z \le 0.68) = A(1.2) + A(0.68) = 0.3849 + 0.2518 = 0.6367$

Using symmetry

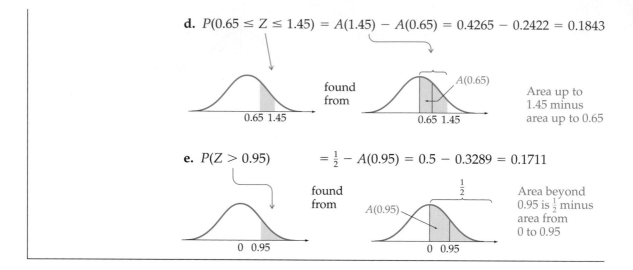

d. $P(0.65 \leq Z \leq 1.45) = A(1.45) - A(0.65) = 0.4265 - 0.2422 = 0.1843$

found from

A(0.65)

Area up to 1.45 minus area up to 0.65

0.65 1.45   0.65 1.45

e. $P(Z > 0.95)$ $= \frac{1}{2} - A(0.95) = 0.5 - 0.3289 = 0.1711$

found from

$A(0.95)$

$\frac{1}{2}$

Area beyond 0.95 is $\frac{1}{2}$ minus area from 0 to 0.95

0   0.95   0   0.95

Practice Problem 2 Find: **a.** $P(0.5 \leq Z \leq 1.25)$

b. $P(-2.1 \leq Z \leq 0.75)$

➤ Solutions on page A7

Probabilities for Nonstandard Random Variables

How do we find probabilities for a *non*standard normal random variable with mean μ and standard deviation σ? Instead of developing tables for all possible means and standard deviations, we use a process called *standardization* that changes a nonstandard random variable into one with mean 0 and variance 1, so we can use the standard normal table. Standardizing is achieved by defining a new random variable by subtracting the mean μ and dividing by the standard deviation σ.

Standardizing Normal Random Variables

A normal random variable X with mean μ and standard deviation σ is *standardized* by defining

$$Z = \frac{X - \mu}{\sigma}$$

Subtracting the mean and dividing by the standard deviation

Z is then a normal random variable with mean 0 and standard deviation 1.

The values of a normal random variable are said to be *normally distributed*.

EXAMPLE **3** **STANDARDIZING TO FIND A PROBABILITY**

The heights of American men are approximately normally distributed, with mean $\mu = 68.1$ inches and standard deviation $\sigma = 2.7$ inches.* Find the proportion of American men who are less than 6 feet tall.

Solution

If X = height, we want $P(X < 72)$ 72 inches − 6 feet
Standardizing:

$$X < 72 \text{ is equivalent to } \underbrace{\frac{X - \mu}{\sigma}}_{Z} < \underbrace{\frac{72 - 68.1}{2.7}}_{1.44}$$

Subtracting mean $\mu = 68.1$ and dividing by $\sigma = 2.7$

Therefore:

$$P(X < 72) = P(Z < 1.44) = 0.9251$$
$$\underbrace{\phantom{P(X < 72)}}_{\text{Normal}} \quad \underbrace{\phantom{P(Z < 1.44)}}_{\substack{\text{Standard} \\ \text{normal}}}$$

$P(Z < 1.44) = 0.9251$ from Example 2b

That is, approximately 93% of American men are less than 6 feet tall.

Such calculations are used for designing automobiles and other consumer products.

Practice Problem 3   The weights of American women are approximately normally distributed, with mean 134.7 pounds and standard deviation 30.4 pounds. Find the proportion of women who weigh more than 125 pounds.

➤ Solution on page A7

EXAMPLE **4** **FINDING THE PROBABILITY OF ACCEPTABLE CHOLESTEROL LEVELS**

Cholesterol is a fat-like substance that, in the bloodstream, can increase the risk of heart disease. Cholesterol levels below 200 (milligrams per deciliter of blood) are generally considered acceptable. A

* These and other data in this section are from Gavriel Salvendy, *Handbook of Human Factors*, John Wiley–Interscience, 1987.

survey by the U.S. Department of Health found that cholesterol levels for one age group were approximately normally distributed with mean $\mu = 222$ and standard deviation $\sigma = 28$. What proportion of people in the study had acceptable cholesterol levels?

Solution

Letting $X = \left(\begin{matrix} \text{Cholesterol} \\ \text{level} \end{matrix}\right)$, we want $P(X \leq 200)$. The inequality

$$X \leq 200$$

with X standardized becomes

$$\underbrace{\frac{X - \mu}{\sigma}}_{Z} \leq \underbrace{\frac{200 - 222}{28}}_{-0.79} \qquad \text{Using } \mu = 222 \text{ and } \sigma = 28$$

Therefore:

$$P(\underbrace{X \leq 200}_{\text{Normal}}) = P(\underbrace{Z \leq -0.79}_{\substack{\text{Standard} \\ \text{normal}}}) = \underbrace{\tfrac{1}{2} - A(0.79)}_{\substack{\text{Using the} \\ \text{graph below,} \\ \text{as in} \\ \text{Example 2}}} = 0.5000 - \underbrace{0.2852}_{\substack{\text{From the} \\ \text{standard} \\ \text{normal} \\ \text{table}}} = \underbrace{0.2148}_{\text{Answer}}$$

Therefore, only about 21% of the people in the study had acceptable cholesterol levels.

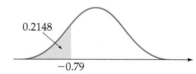

Inverse Probabilities for Normal Random Variables

The standard normal table can also be used "in reverse" to find the values that result in a given probability.

EXAMPLE 5

FINDING THE HEIGHT FOR A GIVEN PROBABILITY

Women's heights are approximately normally distributed, with mean $\mu = 63.2$ inches and standard deviation $\sigma = 2.6$ inches. Find the height that marks the tallest 10%.

Solution

We begin by finding the value of the *standard* normal random variable Z that marks the top $10\% = 0.10$ (or, equivalently, the number z with area 0.40 between 0 and z, as shown on the right).

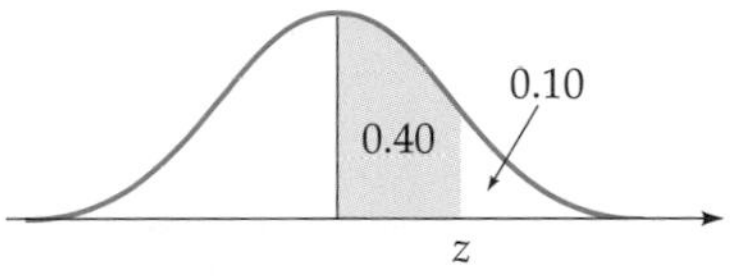

To find this z-value, we look for .40 *inside* the standard normal table. The value that comes closest to .40 is .3997, and the "outside" value that gives it is 1.28. Therefore:

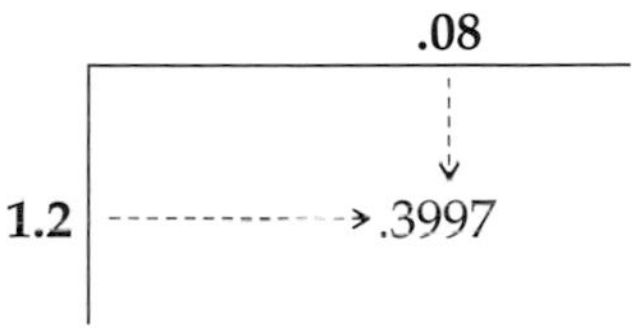

$$z = 1.28$$

z-value of the top 10% (using z to represent a value of the random variable Z)

$$\frac{x - 63.2}{2.6} = 1.28$$

Replacing z by $\dfrac{x - \mu}{\sigma}$ with the given μ and σ

$$x - 63.2 = 3.3$$

Multiplying by 2.6

$$x = 66.5$$

Adding 63.2

Therefore, a height of about 66.5 inches, or 5 feet $6\frac{1}{2}$ inches, marks the tallest 10% of women.

You should now return to Section 11.4, beginning just *after* Example 4 on page 851.

➤ **Solutions to Practice Problems**

1. $P(0 \leq Z \leq 0.52) = 0.1985$

2. a. $P(0.5 \leq Z \leq 1.25) = A(1.25) - A(0.5) = 0.3944 - 0.1915 = 0.2029$

   b. $P(-2.1 \leq Z \leq 0.75) = A(2.1) + A(0.75) = 0.4821 + 0.2734 = 0.7555$

3. $P(X > 125) = P\left(\dfrac{X - \mu}{\sigma} > \dfrac{125 - 134.7}{30.4}\right) = P(Z > -0.32) = \frac{1}{2} + A(0.32)$

$$= 0.5 + 0.1255 = 0.6255$$

The proportion is 62.6%.

Area Under the Standard
Normal Density from 0 to x

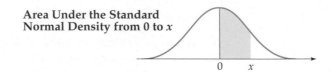

x	.00	.01	.02	.03	.04	.05	.06	.07	.08	.09
0.0	.0000	.0040	.0080	.0120	.0160	.0199	.0239	.0279	.0319	.0359
0.1	.0398	.0438	.0478	.0517	.0557	.0596	.0636	.0675	.0714	.0754
0.2	.0793	.0832	.0871	.0910	.0948	.0987	.1026	.1064	.1103	.1141
0.3	.1179	.1217	.1255	.1293	.1331	.1368	.1406	.1443	.1480	.1517
0.4	.1554	.1591	.1628	.1664	.1700	.1736	.1772	.1808	.1844	.1879
0.5	.1915	.1950	.1985	.2019	.2054	.2088	.2123	.2157	.2190	.2224
0.6	.2258	.2291	.2324	.2357	.2389	.2422	.2454	.2486	.2518	.2549
0.7	.2580	.2612	.2642	.2673	.2704	.2734	.2764	.2794	.2823	.2852
0.8	.2881	.2910	.2939	.2967	.2996	.3023	.3051	.3078	.3106	.3133
0.9	.3159	.3186	.3212	.3238	.3264	.3289	.3315	.3340	.3365	.3389
1.0	.3413	.3438	.3461	.3485	.3508	.3531	.3554	.3577	.3599	.3621
1.1	.3643	.3665	.3686	.3708	.3729	.3749	.3770	.3790	.3810	.3820
1.2	.3849	.3869	.3888	.3907	.3925	.3944	.3962	.3980	.3997	.4015
1.3	.4032	.4049	.4066	.4082	.4099	.4115	.4131	.4147	.4162	.4177
1.4	.4192	.4207	.4222	.4236	.4251	.4265	.4279	.4292	.4306	.4319
1.5	.4332	.4345	.4357	.4370	.4382	.4394	.4406	.4418	.4429	.4441
1.6	.4452	.4463	.4474	.4484	.4495	.4505	.4515	.4525	.4535	.4545
1.7	.4554	.4564	.4573	.4582	.4591	.4599	.4608	.4616	.4625	.4633
1.8	.4641	.4649	.4656	.4664	.4671	.4678	.4686	.4693	.4699	.4706
1.9	.4713	.4719	.4726	.4732	.4738	.4744	.4750	.4756	.4761	.4767
2.0	.4772	.4778	.4783	.4788	.4793	.4798	.4803	.4808	.4812	.4817
2.1	.4821	.4826	.4830	.4834	.4838	.4842	.4846	.4850	.4854	.4857
2.2	.4861	.4864	.4868	.4871	.4875	.4878	.4881	.4884	.4887	.4890
2.3	.4893	.4896	.4898	.4901	.4904	.4906	.4909	.4911	.4913	.4916
2.4	.4918	.4920	.4922	.4925	.4927	.4929	.4931	.4932	.4934	.4936
2.5	.4938	.4940	.4941	.4943	.4945	.4946	.4948	.4949	.4951	.4952
2.6	.4953	.4955	.4956	.4957	.4959	.4960	.4961	.4962	.4963	.4964
2.7	.4965	.4966	.4967	.4968	.4969	.4970	.4971	.4972	.4973	.4974
2.8	.4974	.4975	.4976	.4977	.4977	.4978	.4979	.4979	.4980	.4981
2.9	.4981	.4982	.4982	.4983	.4984	.4984	.4985	.4985	.4986	.4986
3.0	.4987	.4987	.4987	.4988	.4988	.4989	.4989	.4989	.4990	.4990
3.1	.4990	.4991	.4991	.4991	.4992	.4992	.4992	.4992	.4993	.4993
3.2	.4993	.4993	.4994	.4994	.4994	.4994	.4994	.4995	.4995	.4995
3.3	.4995	.4995	.4995	.4996	.4996	.4996	.4996	.4996	.4996	.4997
3.4	.4997	.4997	.4997	.4997	.4997	.4997	.4997	.4997	.4997	.4998
3.5	.4998	.4998	.4998	.4998	.4998	.4998	.4998	.4998	.4998	.4998

Answers to Selected Exercises

Exercises 1.1 page 14

1. $\{x \mid 0 \le x < 6\}$  **3.** $\{x \mid x \le 2\}$  **5. a.** Increase by 15 units

b. Decrease by 10 units **7.** $m = -2$ **9.** $m = \frac{1}{3}$ **11.** $m = 0$ **13.** Slope is undefined

15. $m = 3$, $(0, -4)$ **17.** $m = -\frac{1}{2}$, $(0, 0)$ **19.** $m = 0$, $(0, 4)$

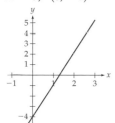 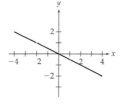 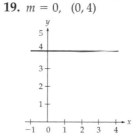

21. Slope and y-intercept do not exist. **23.** $m = \frac{2}{3}$, $(0, -4)$ **25.** $m = -1$, $(0, 0)$

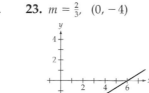 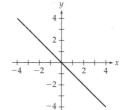

27. $m = 1$, $(0, 0)$ **29.** $m = \frac{1}{3}$, $\left(0, \frac{2}{3}\right)$ **31.** $m = \frac{2}{3}$, $(0, -1)$

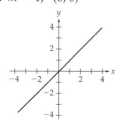 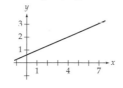 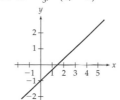

33.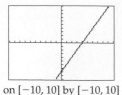

35.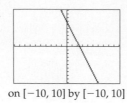

37.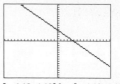

on $[-10, 10]$ by $[-10, 10]$ on $[-10, 10]$ by $[-10, 10]$ on $[-160, 160]$ by $[-160, 160]$

39. $y = -2.25x + 3$ **41.** $y = 5x + 3$ **43.** $y = -4$ **45.** $x = 1.5$ **47.** $y = -2x + 13$

49. $y = -1$ **51.** $y = -2x + 1$ **53.** $y = \frac{3}{2}x - 2$

55. $y = -x + 5$, $y = -x - 5$, $y = x + 5$, $y = x - 5$

57. Substituting $(0, b)$ into $y - y_1 = m(x - x_1)$ gives $y - b = m(x - 0)$, or $y = mx + b$.

59. $(-b/m, 0)$, $m \neq 0$ **61. a.** 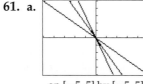   **b.**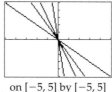

on $[-5, 5]$ by $[-5, 5]$ on $[-5, 5]$ by $[-5, 5]$

63. Low: $[0, 8)$; average: $[8, 20)$; high: $[20, 40)$; critical $[40, \infty)$ **65. a.** 3 minutes 38.28 seconds

b. In about 2033   **67. a.** $y = 4x + 2$ **b.** \$10 million   **c.** \$22 million   **69. a.** $y = \frac{9}{5}x + 32$

b. 68°   **71. a.** $V = 50{,}000 - 2200t$ **b.** \$39,000   **c.**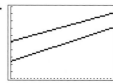

on $[0, 20]$ by $[0, 50{,}000]$

73. Smaller populations increase towards the carrying capacity   **75. a.**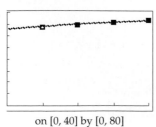

b. Men: 28.7 years; women: 27.2 years   **c.** Men: 30 years; women: 28.9 years

77. a. 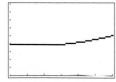   **b.** About 40% (from 39.8)   **c.** About 60% (from 60.4)

79. b. 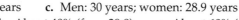 $y_1 = 0.203x + 65.92$ **c.** 79.1 years

on $[0, 40]$ by $[0, 80]$

Exercises 1.2 page 28

1. 64   **3.** $\frac{1}{16}$ **5.** 8   **7.** $\frac{8}{5}$ **9.** $\frac{1}{32}$ **11.** $\frac{8}{27}$ **13.** 1   **15.** $\frac{4}{9}$ **17.** 5   **19.** 125   **21.** 8

23. 4   **25.** -32 **27.** $\frac{125}{216}$ **29.** $\frac{9}{25}$ **31.** $\frac{1}{4}$ **33.** $\frac{1}{2}$ **35.** $\frac{1}{8}$ **37.** $\frac{1}{4}$ **39.** $-\frac{1}{2}$ **41.** $\frac{1}{4}$

43. $\frac{4}{5}$ **45.** $\frac{64}{125}$ **47.** -243 **49.** 2.14   **51.** 274.37   **53.** -128 **55.** 6.25   **57.** 0.5

59. 0.4   **61.** 0.977 (rounded)   **63.** 2.720 (rounded)   **65.** x^{10} **67.** z^{27} **69.** x^8 **71.** w^5

73. y^5/x **75.** $27y^4$ **77.** $u^2v^2w^2$ **79.** 25.6 feet   **81.** Costs will be multiplied by 2.3.

83.
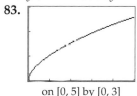
on [0, 5] by [0, 3]

Capacity can be multiplied by about 3.2.

85. 125 beats per minute   **87.**
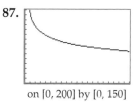
on [0, 200] by [0, 150]

Heart rate decreases more slowly as body weight increases.

89. About 42.6 thousand work-hours, or 42,600 work-hours, rounded to the nearest hundred hours.

91. a. About 32 times more ground motion   **b.** About 8 times more ground motion

93. About 312 mph

95.
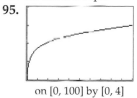
on [0, 100] by [0, 4]

$x \approx 18.2$.   Therefore, the land area must be increased by a factor of more than 18 to double the number of species.

97. b.
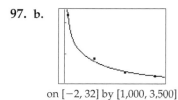
on [−2, 32] by [1,000, 3,500]

$y_1 = 3261x^{-0.267}$ **c.** 1147 work-hours

Exercises 1.3 page 45

1. Yes   **3.** No   **5.** No   **7.** No   **9.** Domain: $\{\,x \mid x \le 0 \text{ or } x \ge 1\,\}$; Range: $\{\,y \mid y \ge -1\,\}$

11. a. $f(10) = 3$ **b.** $\{\,x \mid x \ge 1\,\}$ **c.** $\{\,y \mid y \ge 0\,\}$ **13. a.** $h(-5) = -1$ **b.** $\{\,z \mid z \ne -4\,\}$ **c.** $\{\,y \mid y \ne 0\,\}$

15. a. $h(81) = 3$ **b.** $\{\,x \mid x \ge 0\,\}$ **c.** $\{\,y \mid y \ge 0\,\}$ **17. a.** $f(-8) = 4$ **b.** $\mathbb{R}$ **c.** $\{\,y \mid y \ge 0\,\}$

19. a. $f(0) = 2$ **b.** $\{\,x \mid -2 \le x \le 2\,\}$ **c.** $\{\,y \mid 0 \le y \le 2\,\}$

21. a. $f(-25) = 5$ **b.** $\{\,x \mid x \le 0\,\}$ **c.** $\{\,y \mid y \ge 0\,\}$

23. 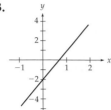   **25.** 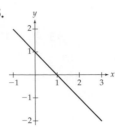   **27.** 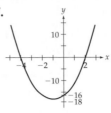   **29.**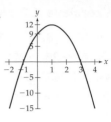

31. a. (20, 100) **b.** 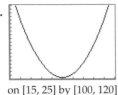   **33. a.** (−40, −200)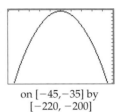

on [15, 25] by [100, 120]

on [−45, −35] by
[−220, −200]

35. $x = 7$, $x = -1$ **37.** $x = 3$, $x = -5$ **39.** $x = 4$, $x = 5$ **41.** $x = 0$, $x = 10$
43. $x = 5$, $x = -5$ **45.** $x = -3$ **47.** $x = 1$, $x = 2$ **49.** No solutions   **51.** No solutions
53. $x = -4$, $x = 5$ **55.** $x = 4$, $x = 5$ **57.** $x = -3$ **59.** No (real) solutions
61. $x = 1.14$, $x = -2.64$ **63. a.** The slopes are all 2, but the y intercepts differ   **b.** $y = 2x - 8$
65. $C(x) = 4x + 20$ **67.** $P(x) = 15x + 500$ **69. a.** 17.7 lbs/in.^{2} **b.** 15,765 lbs/in.^{2} **71.** 132 feet
73. a. 400 cells   **b.** 5200 cells   **75.** About 208 mph   **77.** 2.92 seconds
79. a. Break even at 40 units and at 200 units.
b. Profit maximized at 120 units. Max profit is $12,800 per week.
81. a. Break even at 20 units and at 80 units.
b. Profit maximized at 50 units. Max profit is $1800 per day.

83. $v = \dfrac{c}{w + a} - b$ **85. b.** 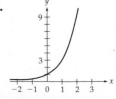   **c.** $y_1(5) = 4.575$,   so about 4.6 million units

on [0.5, 4.5] by [3.5, 4.5]

Exercises 1.4 page 64

1. Domain: $\{\, x \mid x > 0 \text{ or } x < -4 \,\}$; Range: $\{\, y \mid y > 0 \text{ or } y < -2 \,\}$ **3. a.** $f(-3) = 1$ **b.** $\{\, x \mid x \neq -4 \,\}$
c. $\{\, y \mid y \neq 0 \,\}$ **5. a.** $f(-1) = -\frac{1}{2}$ **b.** $\{\, x \mid x \neq 1 \,\}$ **c.** $\{\, y \mid y \leq 0 \text{ or } y \geq 4 \,\}$ **7. a.** $f(2) = 1$
b. $\{\, x \mid x \neq 0, x \neq -4 \,\}$ **c.** $\{\, y \mid y > 0 \text{ or } y \leq -3 \,\}$ **9. a.** $g(-\frac{1}{2}) = \frac{1}{2}$ **b.** $\mathbb{R}$ **c.** $\{\, y \mid y > 0 \,\}$
11. $x = 0$, $x = -3$, $x = 1$ **13.** $x = 0$, $x = 2$, $x = -2$ **15.** $x = 0$, $x = 3$ **17.** $x = 0$, $x = 5$
19. $x = 0$, $x = 3$ **21.** $x = -2$, $x = 0$, $x = 4$ **23.** $x = -1$, $x = 0$, $x = 3$ **25.** $x = 0$, $x = 3$
27. $x = -5$, $x = 0$ **29.** $x = 0$, $x = 1$ **31.** $x \approx -1.79$, $x = 0$, $x \approx 2.79$ **33.**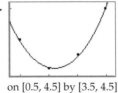

35. 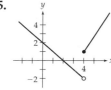 **37.** 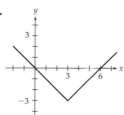 **39.** Polynomial **41.** Exponential

43. Polynomial **45.** Rational **47.** Piecewise linear **49.** Polynomial
51. None (not a polynomial because of the fractional exponent) **53. a.** y_4 **b.** y_1
c. 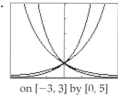 **d.** $(0, 1)$, because $a^0 = 1$ for any constant $a \neq 0$.

on $[-3, 3]$ by $[0, 5]$

55. a. 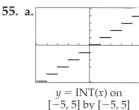 Note that each line segment in this graph includes its left endpoint but excludes its right endpoint, and so should be drawn like •———o .

$y = \text{INT}(x)$ on
$[-5, 5]$ by $[-5, 5]$

b. Domain: $\mathbb{R}$; Range: $\{..., -3, -2, -1, 0, 1, 2, 3,...\}$—that is, the set of all integers
57. a. $(7x - 1)^5$ **b.** $7x^5 - 1$ **59. a.** $\dfrac{1}{x^2 + 1}$ **b.** $\left(\dfrac{1}{x}\right)^2 + 1$

61. a. $(\sqrt{x} - 1)^3 - (\sqrt{x} - 1)^2$ **b.** $\sqrt{x^3 - x^2 - 1}$ **63. a.** $\dfrac{(x^2 - x)^3 - 1}{(x^2 - x)^3 + 1}$ **b.** $\left(\dfrac{x^3 - 1}{x^3 + 1}\right)^2 - \dfrac{x^3 - 1}{x^3 + 1}$

65. a. $f(g(x)) = acx + ad + b$ **b.** Yes **67.** $5x^2 + 10xh + 5h^2$ or $5(x^2 + 2xh + h^2)$
69. $2x^2 + 4xh + 2h^2 - 5x - 5h + 1$ **71.** $10x + 5h$ or $5(2x + h)$ **73.** $4x + 2h - 5$

75. $14x + 7h - 3$ **77.** $3x^2 + 3xh + h^2$ **79.** $\dfrac{-2}{(x + h)x}$

81. $\dfrac{-2x - h}{x^2(x + h)^2}$ or $\dfrac{-2x - h}{x^2(x^2 + 2xh + h^2)}$ or $\dfrac{-2x - h}{x^4 + 2x^3h + x^2h^2}$

83. a. 2.70481 **b.** 2.71815 **c.** 2.71828 **d.** Yes, 2.71828
85. Shifted left 3 units and up 6 units

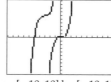

on $[-10, 10]$ by $[-10, 10]$

87. About 680 million people   **89. a.** $300   **b.** $500   **c.** $2000   **d.**

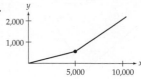

91. a $f(\frac{2}{3}) = 7$, $f(1\frac{1}{3}) = 14$, $f(4) = 29$, and $f(10) = 53$ **b.**

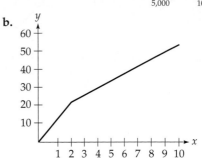

93. $R(v(t)) = 2(60 + 3t)^{0.3}$, $R(v(10)) \approx 7.714$ million dollars   **95. a.** About 1 million cells   **b.** No

97. a.   **b.** About $x = 27.9$ mpg

on [21.6, 40] by [0, 2000]

Chapter 1 Review Exercises page 70

1. $\{\, x \mid 2 < x \le 5 \,\}$ **2.** $\{\, x \mid -2 \le x < 0 \,\}$
3. $\{\, x \mid x \ge 100 \,\}$ **4.** $\{\, x \mid x \le 6 \,\}$
5. Hurricane: $[74, \infty)$; storm: $[55, 74)$; gale: $[38, 55)$; small craft warning: $[21, 38)$
6. a. $(0, \infty)$ **b.** $(-\infty, 0)$ **c.** $[0, \infty)$ **d.** $(-\infty, 0]$ **7.** $y = 2x - 5$ **8.** $y = -3x + 3$ **9.** $x = 2$
10. $y = 3$ **11.** $y = -2x + 1$ **12.** $y = 3x - 5$ **13.** $y = 2x - 1$ **14.** $y = -\frac{1}{2}x + 1$
15. a. $V = 25,000 - 3000t$ **b.** $13,000   **16. a.** $V = 78,000 - 5000t$ **b.** $38,000
17. b.   The regression line fits the data well.

on [0, 25] by [0, 30]

c. 13.7 million tons in the year 2010 [from $y_1(35)$]

18. b. 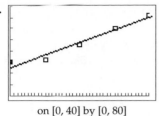 The regression line fits the data reasonably well.

on [0, 40] by [0, 80]

c. 81 to 1 in the year 2010 [from $y_1(50) = 81.4$]

19. 36   **20.** $\frac{3}{4}$ **21.** 8   **22.** 10   **23.** $\frac{1}{27}$ **24.** $\frac{1}{1000}$ **25.** $\frac{9}{4}$ **26.** $\frac{64}{27}$ **27.** 13.97   **28.** 112.32

29. a. $f(11) = 2$ **b.** $\{x \mid x \geq 7\}$ **c.** $\{y \mid y \geq 0\}$ **30. a.** $g(-1) = \frac{1}{2}$ **b.** $\{t \mid t \neq -3\}$ **c.** $\{y \mid y \neq 0\}$

31. a. $h(16) = \frac{1}{8}$ **b.** $\{w \mid w > 0\}$ **c.** $\{y \mid y > 0\}$ **32. a.** $w(8) = \frac{1}{16}$ **b.** $\{z \mid z \neq 0\}$ **c.** $\{y \mid y > 0\}$

33. Yes   **34.** No   **35.**    **36.**   **37.**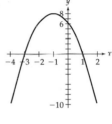

38. 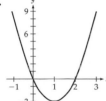   **39.** $x = 0, \ x = -3$ **40.** $x = 5, \ x = -1$ **41.** $x = -2, \ x = 1$

42. $x = 1, \ x = -1$ **43. a.** Vertex $(5, -50)$ **b.** 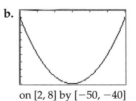   **44. a.** Vertex: $(-7, \ -64)$

on [2, 8] by [−50, −40]

b. 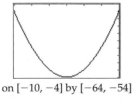   **45.** $C(x) = 45 + 0.12x$ **46.** $I(t) = 800t$ **47.** $T(x) = 70 - \dfrac{x}{300}$

on [−10, −4] by [−64, −54]

48. $C(t) = 27 + 0.58t$; In about 5 years from 2000, so in 2005.

49. a. Break even at 15 and 65 units.   **b.** Profit maximized at 40 units. Max profit is $1250 per week.

50. a. Break even at 150 and 450 units   **b.** Profit maximized at 300 units. Max profit is $67,500 per month.

51. a. $f(-1) = 1$ **b.** $\{x \mid x \neq 0 \text{ and } x \neq 2\}$ **c.** $\{y \mid y > 0 \text{ or } y \leq -3\}$

52. a. $f(-8) = \frac{1}{2}$ **b.** $\{x \mid x \neq 0 \text{ and } x \neq -4\}$ **c.** $\{y \mid y > 0 \text{ or } y \leq -4\}$

53. a. $g\left(\frac{3}{2}\right) = 27$ **b.** $\mathbb{R}$ **c.** $\{y \mid y > 0\}$ **54. a.** $g\left(\frac{5}{3}\right) = 32$ **b.** $\mathbb{R}$ **c.** $\{y \mid y > 0\}$

55. $x = 0, \ x = 1, \ x = -3$ **56.** $x = 0, \ x = 2, \ x = -4$ **57.** $x = 0, \ x = 5$ **58.** $x = 0, \ x = 2$

59.

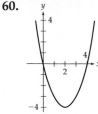

60.

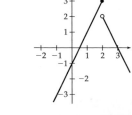

61.

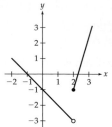

62.

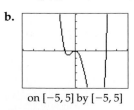

63. a. $f(g(x)) = \left(\dfrac{1}{x}\right)^2 + 1 = \dfrac{1}{x^2} + 1$ **b.** $g(f(x)) = \dfrac{1}{x^2 + 1}$

64. a. $f(g(x)) = \sqrt{5x - 4}$ **b.** $g(f(x)) = 5\sqrt{x} - 4$ **65. a.** $f(g(x)) = \dfrac{x^3 + 1}{x^3 - 1}$ **b.** $g(f(x)) = \left(\dfrac{x + 1}{x - 1}\right)^3$

66. a. $f(g(x)) = 2^{x^2}$ **b.** $g(f(x)) = (2^x)^2 = 2^{2x}$ **67.** $4x + 2h - 3$ **68.** $\dfrac{-5}{(x + h)x}$

69. $A(p(t)) = 2(18 + 2t)^{0.15}, \ A(p(4)) \approx \$3.26$ million   **70. a.** $x = -1, \ x = 0, \ x = 3$

b.

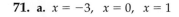

on $[-5, 5]$ by $[-5, 5]$

71. a. $x = -3, \ x = 0, \ x = 1$ **b.**

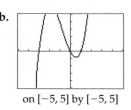

on $[-5, 5]$ by $[-5, 5]$

72. a. The points suggest a parabolic (quadratic) curve.   **b.**

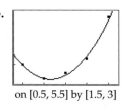

on $[0.5, 5.5]$ by $[1.5, 3]$

c. \$3.6 million, \$4.8 million

Exercises 2.1 page 91

1.

x	5x − 7
1.9	2.5
1.99	2.95
1.999	2.995

x	5x − 7
2.1	3.5
2.01	3.05
2.001	3.005

$\lim_{x \to 2} (5x - 7) = 3$

3.

x	$\dfrac{x^3 - 1}{x - 1}$
0.9	2.71
0.99	2.97
0.999	2.997

x	$\dfrac{x^3 - 1}{x - 1}$
1.1	3.31
1.01	3.03
1.001	3.003

$\lim_{x \to 1} \dfrac{x^3 - 1}{x - 1} = 3$

5. 7.389 (rounded)   **7.** −0.25   **9.** 1   **11.** 2   **13.** 8   **15.** 2   **17.** $\sqrt{2}$ **19.** 6   **21.** $5x^3$

23. 4   **25.** $\frac{1}{2}$ **27.** −9   **29.** $2x$ **31.** $4x^2$ **33. a.** 1   **b.** 3   **c.** Does not exist

35. a. −1   **b.** −1   **c.** −1   **37. a.** −1   **b.** 2   **c.** Does not exist

39. a. −2   **b.** −2   **c.** −2   **41. a.** 0   **b.** 0   **c.** 0

43. a. 1   **b.** 1   **c.** Does not exist   **45.** $\lim_{x \to 3^-} f(x) = \infty$, $\lim_{x \to 3^+} f(x) = -\infty$, $\lim_{x \to 3} f(x)$ does not exist

47. $\lim_{x \to 0^-} f(x) = \infty$, $\lim_{x \to 0^+} f(x) = \infty$, $\lim_{x \to 0} f(x) = \infty$ **49.** $\lim_{x \to -2^-} f(x) = -\infty$, $\lim_{x \to -2^+} f(x) = \infty$, $\lim_{x \to -2} f(x)$ does not exist

51. $\lim_{x \to 3^-} f(x) = \infty$, $\lim_{x \to 3^+} f(x) = \infty$, $\lim_{x \to 3} f(x) = \infty$ **53.** Continuous   **55.** Discontinuous, (3) is violated

57. Discontinuous, (1) is violated   **59.** Discontinuous, (2) is violated

61. a. 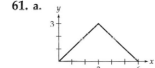   **b.** $\lim_{x \to 3^-} f(x) = 3$, $\lim_{x \to 3^+} f(x) = 3$ **c.** Yes

63. a. 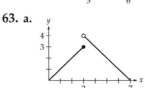   **b.** $\lim_{x \to 3^-} f(x) = 3$, $\lim_{x \to 3^+} f(x) = 4$ **c.** No, (2) is violated   **65.** Continuous

67. Discontinuous at $x = 1$ **69.** Discontinuous at $x = -1$, $x = 0$, and $x = 1$

71. Discontinuous at $x = 4$ **73.** Continuous   **75.** Continuous

77. The two functions are *not* equal to each other, since at $x = 1$ one is defined and the other is not (see pages 51–52).   **79.** 1.11 (dollars)   **81.** 100

Exercises 2.2 page 108

1. At P_1: positive slope   **3.** At P_1: positive slope   **5.** At P_1: slope is 3   **7.** Your graph should look roughly
At P_2: negative slope   At P_2: negative slope   At P_2: slope is $-\frac{1}{2}$ like the following:
At P_3: zero slope   At P_3: zero slope

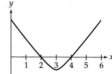

9. a. 5   **b.** 4   **c.** 3.5   **d.** 3.1   **e.** 3.01   **f.** 3
11. a. 13   **b.** 11   **c.** 10   **d.** 9.2   **e.** 9.02   **f.** 9
13. a. 5   **b.** 5   **c.** 5   **d.** 5   **e.** 5   **f.** 5
15. a. 0.2247   **b.** 0.2361   **c.** 0.2426   **d.** 0.2485   **e.** 0.2498   **f.** 0.25

17. 3   **19.** 5   **21.** 9   **23.** $\frac{1}{4}$ **25.** $f'(x) = 2x - 3$ **27.** $f'(x) = -2x$ **29.** $f'(x) = 9$

31. $f'(x) = \frac{1}{2}$ **33.** $f'(x) = 0$ **35.** $f'(x) = 2ax + b$ **37.** $f'(x) = 5x^4$ **39.** $f'(x) = \dfrac{-2}{x^2}$

41. $f'(x) = \dfrac{1}{2\sqrt{x}}$ **43.** $f'(x) = 3x^2 + 2x$ **45. a.** $y = x + 1$ **b.**

47. a.   **b.**

x=2
y=1X+1

X=1
y=⁻1X+4

on viewing window
$[-10, 10]$ by $[-10, 10]$

49. a. $f'(x) = 3$ **b.** The graph of $f(x) = 3x - 4$ is a straight line with slope 3.
51. a. $f'(x) = 0$ **b.** The graph of $f(x) = 5$ is a horizontal straight line with slope 0.
53. a. $f'(x) = m$ **b.** The graph of $f(x) = mx + b$ is a straight line with slope m.
55. a. $f'(x) = 2x - 8$ **b.** Decreasing at the rate of 4 degrees per minute (since $f'(2) = -4$)
     c Increasing at the rate of 2 degrees per minute (since $f'(5) = 2$)
57. a. $f'(x) = 4x - 1$
     b. When 5 words have been memorized, the memorization time is increasing at the rate of 19 seconds
     per word
59. a. $T'(x) = -2x + 5$ **b.** Increasing at the rate of 1 degree per day
     c. Decreasing at the rate of 1 degree per day   **d.** Deteriorating on day 2, improving on day 3

Exercises 2.3 page 124

1. $4x^3$ **3.** $500x^{499}$ **5.** $\frac{1}{2}x^{-1/2}$ **7.** $2x^3$ **9.** $2w^{-2/3}$ **11.** $-6x^{-3}$ **13.** $8x - 3$

15. $-\dfrac{1}{2}x^{-3/2} = -\dfrac{1}{2\sqrt{x^3}}$ **17.** $-2x^{-4/3} = -\dfrac{2}{\sqrt[3]{x^4}}$ **19.** $2\pi r$ **21.** $\frac{1}{2}x^2 + x + 1$ **23.** $\frac{1}{2}x^{-1/2} + x^{-2}$

25. $4x^{-1/3} + 4x^{-4/3}$ **27.** $-5x^{-3/2} - 15x^{2/3}$ **29.** $1 + 2x$ **31. a.** $f'(x) = 0$
     b. The graph of the constant function $f(x) = 2$ is a horizontal line and therefore has slope 0.
     c. Since $f(x)$ is a constant function, its rate of change is zero.

33. 80   **35.** 3   **37.** 27   **39.** 1   **41.** For $y_1 = 5$ and window $[-10, 10]$ by $[-10, 10]$, your calculator screen should look like the following:

43. a. $MP(x) = 0.03x^{1/2}$ **b.** $MP(10,000) = 3$. *Interpretation:* After 10,000 chips, the profit on each additional chip is about \$3.   **45.** 3.00007, which is close to \$3.

47. a. $P'(x) = -12,000 + 1200x + 300x^2$ **b.** Decreasing by about 10,500 per year
    c. Increasing by about 30,000 per year

49. Increasing by about 8000 people per additional day

51. Increasing by about 0.08 square centimeter per hour

53. Increasing by about 6 phrases per hour   **55. a.** $MU(x) = 50x^{-1/2}$ **b.** 50   **c.** 0.05

57. a. $f(12) \approx 40$. *Interpretation:* The probability of a high school graduate quitting smoking is about 40%.
    $f'(12) \approx 1.8$. *Interpretation:* The probability of quitting increases by about 1.8% for each additional year of education.
  b. $f(16) \approx 60$. *Interpretation:* The probability of a college graduate quitting smoking is about 60%.
    $f'(16) \approx 8.5$. *Interpretation:* The probability of quitting increases by about 8.5% for each additional year of education.

59. c. \$26,043   **e.** Tuition increasing at the rate of \$1191 per year

Exercises 2.4 page 140

1. $10x^9$ **3.** $9x^8 + 4x^3$ **5.** $5x^4 + 2x$ **7.** $15x^2 - 1$ **9.** $4x^3$ **11.** $9x^2 + 8x + 1$ **13.** 1

15. $36t + 8t^{1/3}$ (after simplification)   **17.** $7z^6 - 1$ (after simplification)   **19.** $6x^5$ **21.** $-\dfrac{3}{x^4}$ or $-3x^{-4}$

23. $\dfrac{x^4 - 3}{x^4}$ **25.** $-\dfrac{2}{(x-1)^2}$ **27.** $\dfrac{4t}{(t^2+1)^2}$ **29.** $\dfrac{2s^3 + 3s^2 + 1}{(s+1)^2}$ (after simplification)

31. $\dfrac{2x^5 + 4x^3}{(x^2+1)^2}$ (after simplification)   **33.** $y = 3x^{-1}, \dfrac{dy}{dx} = -3x^{-2}, \dfrac{dy}{dx} = -\dfrac{3}{x^2}$

35. $y = \dfrac{3}{8}x^4, \dfrac{dy}{dx} = \dfrac{3}{2}x^3, \dfrac{dy}{dx} = \dfrac{3x^3}{2}$ **37.** $\dfrac{d}{dx}(fgh) = f'(gh) + f(gh)' = f'gh + fg'h + fgh'$ **39.** $2f(x)f'(x)$

41. $3x^2 \dfrac{x^2 + 1}{x + 1} + (x^3 + 2)\dfrac{x^2 + 2x - 1}{(x + 1)^2}$ **43.** $\dfrac{3x^6 + 13x^4 + 18x^2 - 2x}{(x^2 + 2)^2}$ **45.** $\dfrac{x^{-1/2}}{(x^{1/2} + 1)^2} = \dfrac{1}{\sqrt{x}(\sqrt{x} + 1)^2}$

47. $MAR(x) = \dfrac{xR'(x) - R(x)}{x^2}$

49. a. $C'(x) = \dfrac{100}{(100 - x)^2}$ **b.** Increasing by 4¢ per additional percentage of purity
    c. Increasing by 25¢ per additional percentage of purity.

51. b. Rates of change are 4 and 25

53. $\dfrac{dy}{dx} = \dfrac{R}{\left(1 + \dfrac{R-1}{K}x\right)^2} > 0$; As the population increases, the number of offspring increases.

55. a. $AP(x) = \dfrac{12x - 1800}{x} = 12 - \dfrac{1800}{x} = 12 - 1800x^{-1}$

   b. $MAP(x) = \dfrac{1800}{x^2}$ or $1800x^{-2}$

   c. $MAP(300) = \dfrac{2}{100}$, so average profit is increasing by 2¢ per additional unit.

57. Increasing at the rate of 7 degrees per hour   **59. b.** 7   **c.** About 104.5 degrees

61. b.

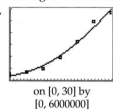

on [0, 30] by
[0, 6000000]

d.

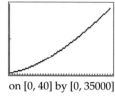

on [0, 40] by [0, 35000]

f.

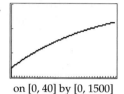

on [0, 40] by [0, 1500]

   f. (continued)
   $y_3(40) \approx 33{,}325$, so in 2010 the per capita national debt should be \$33,325.
   $y_4(40) \approx 1219$, so in 2010 the per capita national debt should be growing by \$1219 per year.

63. a. $y' = -2x^{-3} = -\dfrac{2}{x^3}$ (which is undefined at $x = 0$)

   b. Your calculator should give "Error" but may, incorrectly, give "0."

Exercises 2.5 page 155

1. a. $4x^3 - 6x^2 - 6x + 5$ **b.** $12x^2 - 12x - 6$ **c.** $24x - 12$ **d.** 24

3. a. $1 + x + \dfrac{1}{2}x^2 + \dfrac{1}{6}x^3 + \dfrac{1}{24}x^4$ **b.** $1 + x + \dfrac{1}{2}x^2 + \dfrac{1}{6}x^3$ **c.** $1 + x + \dfrac{1}{2}x^2$ **d.** $1 + x$

5. a. $\dfrac{5}{2}x^{3/2}$ **b.** $\dfrac{15}{4}x^{1/2}$ **c.** $\dfrac{15}{8}x^{-1/2}$ **d.** $-\dfrac{15}{16}x^{-3/2}$ **7. a.** $-\dfrac{2}{x^3}$ or $-2x^{-3}$ **b.** $-\dfrac{2}{27}$

9. a. $\dfrac{1}{x^3} = x^{-3}$ **b.** $\dfrac{1}{27}$ **11. a.** x^{-4} **b.** $\dfrac{1}{81}$ **13.** $12x^2 + 2$ **15.** $12x^{-7/3}$

17. $\dfrac{2x - 2}{(x^2 - 2x + 1)^2} = \dfrac{2}{(x - 1)^3}$ **19.** 2π **21.** 90   **23.** -720 **25.** 3

27. a. iii   **b.** i   **c.** ii   **29.** 0

31. $\dfrac{d^2}{dx^2}(fg) = \dfrac{d}{dx}(f'g + fg') = f''g + f'g' + f'g' + fg'' = f''g + 2f'g' + fg''$

33. a. 54 mph   **b.** -42 mph or 42 mph south   **c.** 24 mi/hr²   **35.** 310 ft/sec, 61 ft/sec²

37. a. 160 ft/sec   **b.** 32 ft/sec²   **39. a.** $-32t + 1280$ **b.** 40 seconds   **c.** 25,600 feet

41. $D'(8) = 24$: After 8 years the debt is growing by \$24 billion per year.
   $D''(8) = 1$: After 8 years the debt will be growing increasingly rapidly, with the rate of growth growing
        by about \$1 billion per year per year.

43. $L'(4) = \frac{1}{4}$: After 4 years the sea level will be rising by $\frac{1}{4}$ foot per year.
   $L''(4) = -\frac{3}{32}$: After 4 years the rate of growth will be slowing by about $\frac{3}{32}$ foot per year per year.

45. $P(3) \approx 4.87$, $P'(3) \approx -0.51$, $P''(3) \approx -0.39$
   Interpretation: In 3 years the profit will be about \$4.87 million, decreasing at the rate of \$0.51 million per
   year, and the decline of profit will be accelerating.

47. a. 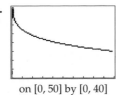 Approximately 22° and 18°

on [0, 50] by [0, 40]

b. Each successive 1-mph increase in wind speed lowers the windchill index, but less so as wind speed rises.

c. $y_2(15) \approx -0.4°$ and $y_2(30) \approx -0.2°$. *Interpretation:* At a wind speed of 15 mph, each additional mph decreases the windchill index by about 0.4°, whereas at a wind speed of 30 mph, each additional mph decreases the windchill index by only about 0.2°.

49. $20x^3 - 12x^2 + 6x - 2$ **51.** $\dfrac{2x^5 - 4x^3 - 6x}{(x^2 + 1)^4} = \dfrac{2x^3 - 6x}{(x^2 + 1)^3}$ **53.** $\dfrac{-32x - 16}{(4x^2 + 4x + 1)^2} = \dfrac{-16}{(2x + 1)^3}$

Exercises 2.6 page 168

Note: For Exercises 1 through 9, there are other possible correct answers.

1. $f(x) = \sqrt{x}, g(x) = x^2 - 3x + 1$ **3.** $f(x) = x^{-3}, g(x) = x^2 - x$ **5.** $f(x) = \dfrac{x + 1}{x - 1}, g(x) = x^3$

7. $f(x) = x^4, g(x) = \dfrac{x + 1}{x - 1}$ **9.** $f(x) = \sqrt{x} + 5, g(x) = x^2 - 9$ **11.** $6x(x^2 + 1)^2$

13. $4(3z^2 - 5z + 2)^3(6z - 5)$ **15.** $\frac{1}{2}(x^4 - 5x + 1)^{-1/2}(4x^3 - 5)$ **17.** $\frac{1}{3}(9z - 1)^{-2/3}(9) = 3(9z - 1)^{-2/3}$

19. $-8x(4 - x^2)^3$ **21.** $-12w^2(w^3 - 1)^{-5}$ **23.** $4x^3 - 4(1 - x)^3$

25. $-\frac{2}{3}(9x + 1)^{-5/3}(9) = -6(9x + 1)^{-5/3}$ **27.** $3[(x^2 + 1)^3 + x]^2[6x(x^2 + 1)^2 + 1]$

29. $6x(2x + 1)^5 + 30x^2(2x + 1)^4 = 6x(2x + 1)^4(7x + 1)$

31. $6(2x + 1)^2(2x - 1)^4 + 8(2x + 1)^3(2x - 1)^3 = 2(2x + 1)^2(2x - 1)^3(14x + 1)$

33. $-6\dfrac{(x + 1)^2}{(x - 1)^4}$ **35.** $2x(1 + x^2)^{1/2} + x^3(1 + x^2)^{-1/2}$ **37.** $\dfrac{1}{4}x^{-1/2}(1 + x^{1/2})^{-1/2}$

39. a. $4x(x^2 + 1)$ **b.** $4x^3 + 4x$ **41. a.** $-\dfrac{3}{(3x + 1)^2}$ **b.** $-3(3x + 1)^{-2}$

43. $\dfrac{d}{dx} L(g(x)) = L'(g(x)) g'(x) = \dfrac{1}{g(x)} g'(x) = \dfrac{g'(x)}{g(x)}$ **45.** $20(x^2 + 1)^9 + 360x^2(x^2 + 1)^8$

47. $MC(x) = 4x(4x^2 + 900)^{-1/2}$ $MC(20) = \frac{8}{5} = 1.60$ **49.** $x = 27$

51. $S'(25) \approx 2.1$. *Interpretation:* At income level \$25,000, status increases by about 2.1 units for each additional \$1000 of income.

53. $R'(50) = 32\frac{1}{3}$ **55.** 26 mg   **57.** $P'(2) = 0.24$; pollution is increasing by about 0.24 ppm per year.

59. b. 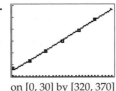   **e.** Slope ≈ 0.036 **f.** $\dfrac{1.8}{0.036} \approx 50$ years

on [0, 30] by [320, 370]

Exercises 2.7 page 176

1. $-2, 0$, and 2 **3.** -3 and 3

5. $\lim\limits_{h \to 0} \dfrac{f(x+h) - f(x)}{h}$ simplifies to $\lim\limits_{h \to 0} \dfrac{|2h|}{h}$, which gives $\begin{cases} 2 & \text{for } h > 0 \\ -2 & \text{for } h < 0' \end{cases}$ so the limit
(and therefore the derivative at $x = 0$) does not exist.

7. $\lim\limits_{h \to 0} \dfrac{f(x+h) - f(x)}{h}$ simplifies to $\lim\limits_{h \to 0} \dfrac{h^{2/5}}{h} = \lim\limits_{h \to 0} \dfrac{1}{h^{3/5}}$, which does not exist. Therefore, the derivative
at $x = 0$ does not exist.

9. If you got a numerical answer, it is wrong, since the function is undefined at $x = 0$, so the derivative
at $x = 0$ does not exist. (For an explanation, see the Graphing Calculator Exploration on page 173.)

11. a. The formula comes from substituting the given function and x-value into the definition of the
derivative and simplifying.

  b. 3.16, 31.6, 316 (rounded)   **c.** No, No   **d.**

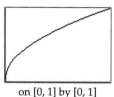

on $[0, 1]$ by $[0, 1]$

Chapter 2 Review Exercises page 179

1.

x	4x + 2	x	4x + 2
1.9	9.6	2.1	10.4
1.99	9.96	2.01	10.04
1.999	9.996	2.001	10.004

$\lim\limits_{x \to 2} (4x + 2) = 10$

2.

x	$\dfrac{\sqrt{x+1} - 1}{x}$	x	$\dfrac{\sqrt{x+1} - 1}{x}$
-0.1	0.513	0.1	0.488
-0.01	0.501	0.01	0.499
-0.001	0.500	0.001	0.500

$\lim\limits_{x \to 0} \dfrac{\sqrt{x+1} - 1}{x} = 0.5$

3. a. 3   **b.** -2 **c.** Does not exist   **4. a.** -1 **b.** -1 **c.** -1 **5.** 5   **6.** π **7.** 4
8. 2   **9.** $\frac{1}{2}$ **10.** -3 **11.** $2x^2$ **12.** $-x^2$ **13.** Continuous   **14.** Continuous
15. Discontinuous at $x = -1$ **16.** Continuous   **17.** Discontinuous at $x = 0$ and at $x = -1$
18. Discontinuous at $x = 3$ and at $x = -3$ **19.** Discontinuous at $x = 5$ **20.** Continuous
21. $4x + 3$ **22.** $6x + 2$ **23.** $-\dfrac{3}{x^2} = -3x^{-2}$ **24.** $\dfrac{2}{\sqrt{x}}$ **25.** $10x^{2/3} + 2x^{-3/2}$ **26.** $10x^{3/2} + 2x^{-4/3}$
27. -16 **28.** -9 **29.** 1   **30.** $\frac{1}{2}$
31. a. $C'(x) = 3 - 27x^{-3/2}$ **b.** 2. *Interpretation:* Costs are increasing by about $2 per additional unit.

32. $f'(10) = -2.3$ (thousand hours). *Interpretation:* After 10 planes, the construction time is decreasing by about 2300 hours for each additional plane built.

33. a. $\dfrac{dA}{dr} = 2\pi r$ **b.** As the radius increases, the area "grows by a circumference."

34. a. $V' = \frac{4}{3}\pi r^2 \cdot 3 = 4\pi r^2$ **b.** As radius increases, volume "grows by a surface area."

35. $40x^3 + 6$ (after simplification)   **36.** $15x^4 - 2x$ **37.** $2x(x^2 - 5) + (x^2 + 5)2x = 4x^3$ **38.** $4x^3$

39. $(4x^3 + 2x)(x^5 - x^3 + x) + (x^4 + x^2 + 1)(5x^4 - 3x^2 + 1) = 9x^8 + 5x^4 + 1$

40. $(5x^4 + 3x^2 + 1)(x^4 - x^2 + 1) + (x^5 + x^3 + x)(4x^3 - 2x) = 9x^8 + 5x^4 + 1$ **41.** $\dfrac{2}{(x + 1)^2}$ **42.** $\dfrac{-2}{(x - 1)^2}$

43. $\dfrac{-10x^4}{(x^5 - 1)^2}$ **44.** $\dfrac{12x^5}{(x^6 + 1)^2}$ **45. a.** $-\dfrac{1}{x^2}$ **b.** $-x^{-2}$ (after simplification)

  c. By simplifying to $f(x) = 2 + x^{-1}$,   the derivative is $-x^{-2}$.

46. $S'(6) = -10$,   so at a price of \$6 each, sales will decrease by about 10 for each dollar increase in price.

47. a. $AP(x) = \dfrac{6x - 200}{x}$ **b.** $MAP(x) = \dfrac{200}{x^2}$

  c. $MAP(10) = 2$, so average profit is increasing by about \$2 per additional unit.

48. a. $AC(x) = \dfrac{5x + 100}{x}$ **b.** $MAC(x) = \dfrac{-100}{x^2}$

  c. $MAC(20) = -0.25$, so average cost is decreasing by about 25¢ per additional unit.

49. $9x^{-1/2} + 2x^{-5/3}$ **50.** $4x^{-4/3} - 3x^{-1/2}$ **51.** $2x^{-4}$ **52.** $6x^{-5}$ **53.** -24 **54.** 60

55. 480   **56.** $\frac{3}{8}$ **57.** 15   **58.** 70

59. $P(10) = 200$, $P'(10) = 20$, $P''(10) = 9$. *Interpretation:* 10 years from now the population will be 200 thousand, growing at the rate of 20 thousand per year, and the growth will be accelerating.

60. Velocity 2500 ft/sec; acceleration 150 ft/sec^2 **61. a.** 347.25 feet

62. b.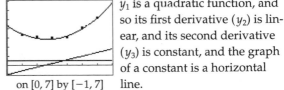

  c. $y_1(7) = 6.76$; at the end of year 7 the annual profit will be about \$6.76 million.

on [0.5, 6.5] by [2.5, 5.5]

  d. $y_2(7) \approx 1.77$. *Interpretation:* At the end of the seventh year, the annual profit will be growing by about \$1.77 million per year.

  e. $y_3(7) \approx 0.41$. *Interpretation:* Profit will be growing increasingly fast, with the rate of growth increasing by about \$0.41 million per year per year.   **f.** y_1 is a quadratic function, and so its first derivative (y_2) is linear, and its second derivative (y_3) is constant, and the graph of a constant is a horizontal line.

on [0, 7] by [−1, 7]

63. $3(4z^2 - 3z + 1)^2(8z - 3)$ **64.** $4(3z^2 - 5z - 1)^3(6z - 5)$ **65.** $-5(100 - x)^4$ **66.** $-4(1000 - x)^3$

67. $\frac{1}{2}(x^2 - x + 2)^{-1/2}(2x - 1)$ **68.** $\frac{1}{2}(x^2 - 5x - 1)^{-1/2}(2x - 5)$ **69.** $2(6z - 1)^{-2/3}$ **70.** $(3z + 1)^{-2/3}$

71. $-2(5x + 1)^{-7/5}$ **72.** $-6(10x + 1)^{-8/5}$ **73.** $2x(2x - 1)^4 + 8x^2(2x - 1)^3 = 2x(2x - 1)^3(6x - 1)$

74. $5(x^3 - 2)^4 + 60x^3(x^3 - 2)^3 = 5(x^3 - 2)^3(13x^3 - 2)$ **75.** $3x^2(x^3 + 1)^{1/3} + x^5(x^3 + 1)^{-2/3}$

76. $4x^3(x^2 + 1)^{1/2} + x^5(x^2 + 1)^{-1/2}$ **77.** $3[(2x^2 + 1)^4 + x^4]^2[16x(2x^2 + 1)^3 + 4x^3]$

78. $2[(3x^2 - 1)^3 + x^3][18x(3x^2 - 1)^2 + 3x^2]$ **79.** $\frac{1}{2}[(x^2 + 1)^4 - x^4]^{-1/2}[8x(x^2 + 1)^3 - 4x^3]$

80. $[(x^3 + 1)^2 + x^2]^{-1/2}[3x^2(x^3 + 1) + x]$ **81.** $12(3x + 1)^3(4x + 1)^3 + 12(3x + 1)^4(4x + 1)^2$

82. $6x(x^2 + 1)^2(x^2 - 1)^4 + 8x(x^2 + 1)^3(x^2 - 1)^3$ **83.** $\dfrac{-20}{x^2}\left(\dfrac{x + 5}{x}\right)^3 = \dfrac{-20(x + 5)^3}{x^5}$

84. $-20\left(\dfrac{x + 4}{x}\right)^4 \dfrac{1}{x^2} = -20\dfrac{(x + 4)^4}{x^6}$ **85.** $20(2w^2 - 4)^4 + 320w^2(2w^2 - 4)^3$

86. $24(3w^2 + 1)^3 + 432w^2(3w^2 + 1)^2$ **87.** $6z(z + 1)^3 + 18z^2(z + 1)^2 + 6z^3(z + 1)$

88. $12z^2(z + 1)^4 + 32z^3(z + 1)^3 + 12z^4(z + 1)^2$

89. a. $6x^2(x^3 - 1)$ **b.** $6x^5 - 6x^2$ **90. a.** $\dfrac{-3x^2}{(x^3 + 1)^2}$ **b.** $-(x^3 + 1)^{-2}(3x^2)$

91. $P'(5) = 3$. *Interpretation:* When producing 5 tons, profit increases by about 3 thousand dollars for each additional ton.

92. $V'(8) = 17.496$. *Interpretation:* Value increased by about $17.50 for each additional percentage of interest.

93. a. $P(5) - P(4) \approx 2.73$, $P(6) - P(5) \approx 3.23$, both of which are near 3   **b.** At about $x = 7.6$

94. $x \approx 16$ **95.** 0.08

96. $N'(96) = -250$. *Interpretation:* At age 96, the number of survivors is decreasing by about 250 people per year.

97. $x = -3, x = 1, x = 3$ **98.** $x = 2, x = -2$ **99.** $x = 0, x = 3.5$ **100.** $x = 0, x = 3$

101. $\lim\limits_{h\to 0}\dfrac{f(x + h) - f(x)}{h}$ simplifies to $\lim\limits_{h\to 0}\dfrac{|5h|}{h}$, which gives $\begin{cases} 5 & \text{for } h > 0 \\ -5 & \text{for } h < 0 \end{cases}$ and so the limit

(and therefore the derivative at $x = 0$) does not exist.

102. $\lim\limits_{h\to 0}\dfrac{f(x + h) - f(x)}{h}$ simplifies to $\lim\limits_{h\to 0}\dfrac{h^{3/5}}{h} = \lim\limits_{h\to 0}\dfrac{1}{h^{2/5}}$, which does not exist. Therefore, the derivative

at $x = 0$ does not exist.

Exercises 3.1 page 196

1. a. $(-\infty, -2)$ and $(0, \infty)$ **b.** $(-2, 0)$ **3.** All but 3 (where the function is undefined)   **5.** 4 and -4

7. $0, -4$, and 1   **9.** 3   **11.** No CNs   **13.** 1 and $\frac{1}{3}$

15.

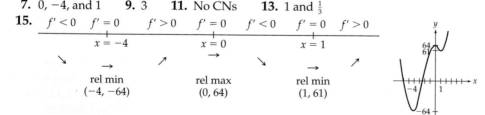

Open intervals of increase: $(-4, 0)$ and $(1, \infty)$; Open intervals of decrease: $(-\infty, -4)$ and $(0, 1)$

17.

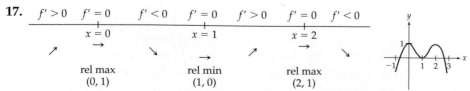

Open intervals of increase: $(-\infty, 0)$ and $(1, 2)$; Open intervals of decrease: $(0, 1)$ and $(2, \infty)$

19.

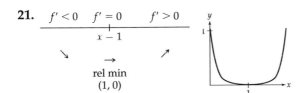

$f' < 0$	$f' = 0$	$f' > 0$	$f' = 0$	$f' > 0$
	$x = 0$		$x = 1$	
↘		↗	→	↗
	rel min		neither	
	$(0, 0)$		$(1, 1)$	

Open intervals of increase: $(0, 1)$ and $(1, \infty)$; Open interval of decrease: $(-\infty, 0)$

21.

$f' < 0$	$f' = 0$	$f' > 0$
	$x - 1$	
↘	→	↗
	rel min	
	$(1, 0)$	

Open interval of increase: $(1, \infty)$; Open interval of decrease: $(-\infty, 1)$

23.

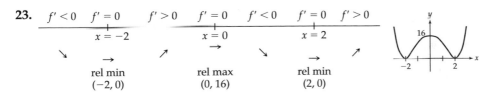

$f' < 0$	$f' = 0$	$f' > 0$	$f' = 0$	$f' < 0$	$f' = 0$	$f' > 0$
	$x = -2$		$x = 0$		$x = 2$	
↘	→	↗	→	↘	→	↗
	rel min		rel max		rel min	
	$(-2, 0)$		$(0, 16)$		$(2, 0)$	

Open intervals of increase: $(-2, 0)$ and $(2, \infty)$; Open intervals of decrease: $(-\infty, -2)$ and $(0, 2)$

25.

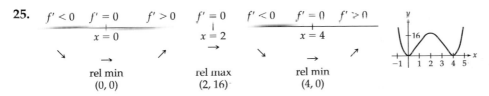

$f' < 0$	$f' = 0$	$f' > 0$	$f' = 0$	$f' < 0$	$f' = 0$	$f' > 0$
	$x = 0$		$x = 2$		$x = 4$	
↘	→	↗	→	↘	→	↗
	rel min		rel max		rel min	
	$(0, 0)$		$(2, 16)$		$(4, 0)$	

Open intervals of increase: $(0, 2)$ and $(4, \infty)$; Open intervals of decrease: $(-\infty, 0)$ and $(2, 4)$

27.

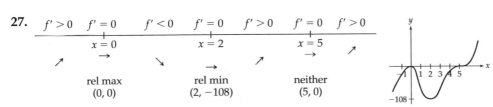

$f' > 0$	$f' = 0$	$f' < 0$	$f' = 0$	$f' > 0$	$f' = 0$	$f' > 0$
	$x = 0$		$x = 2$		$x = 5$	
↗	→	↘	→	↗	→	↗
	rel max		rel min		neither	
	$(0, 0)$		$(2, -108)$		$(5, 0)$	

Open intervals of increase: $(-\infty, 0)$, $(2, 5)$, and $(5, \infty)$; Open interval of decrease: $(0, 2)$

29.

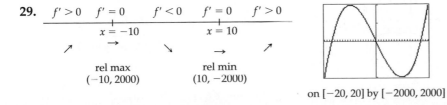
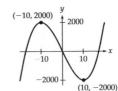

$f' > 0$	$f' = 0$	$f' < 0$	$f' = 0$	$f' > 0$
	$x = -10$		$x = 10$	
↗	→	↘	→	↗
	rel max		rel min	
	$(-10, 2000)$		$(10, -2000)$	

on $[-20, 20]$ by $[-2000, 2000]$

31.

$f' < 0$	$f' = 0$	$f' > 0$	$f' = 0$	$f' < 0$	$f' = 0$	$f' > 0$
	$x = -5$		$x = 0$		$x = 5$	

↘   →   ↗   →   ↘   →   ↗

rel min   rel max   rel min
$(-5, -650)$ $(0, -25)$ $(5, -650)$

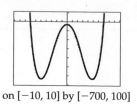

on $[-10, 10]$ by $[-700, 100]$

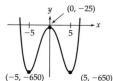

33.

$f' > 0$	$f' = 0$	$f' > 0$	$f' = 0$	$f' < 0$	$f' = 0$	$f' > 0$
	$x = 0$		$x = 1$		$x = 3$	

↗   →   ↗   →   ↘   →   ↗

neither   rel max   rel min
$(0, -23)$ $(1, -22)$ $(3, -50)$

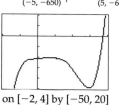

on $[-2, 4]$ by $[-50, 20]$

35.

$f' > 0$	$f' = 0$	$f' < 0$	$f' = 0$	$f' > 0$
	$x = -1$		$x = 1$	

↗   →   ↘   →   ↗

rel max   rel min
$(-1, 0.04)$ $(1, -0.04)$

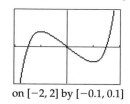

on $[-2, 2]$ by $[-0.1, 0.1]$

37.

$f' > 0$	$f' = 0$	$f' < 0$	$f' = 0$	$f' > 0$
	$x = \frac{1}{3}$		$x = 1$	

↗   →   ↘   →   ↗

rel max   rel min
$(\frac{1}{3}, 11.15)$ $(1, 11)$

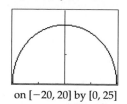

on $[-1, 3]$ by $[5, 15]$

39.

f' und	$f' > 0$	$f' = 0$	$f' < 0$	f' und
$x = -20$		$x = 0$		$x = 20$

↗   →   ↘

rel max
$(0, 20)$

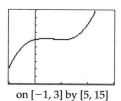

on $[-20, 20]$ by $[0, 25]$

41. Critical number: 1

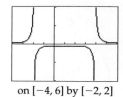

on $[-4, 6]$ by $[-2, 2]$

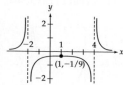

43. Critical number: 0

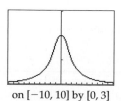

on $[-10, 10]$ by $[0, 3]$

45. Critical number: 0

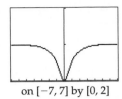

on $[-7, 7]$ by $[0, 2]$

47. Critical numbers: 0 and 6

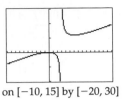

on $[-10, 15]$ by $[-20, 30]$

49. Critical number: 0

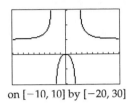

on $[-10, 10]$ by $[-20, 30]$

51. Critical numbers:
   $-1, 0,$ and 1

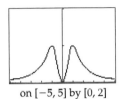

on $[-5, 5]$ by $[0, 2]$

53.

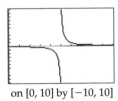

on $[0, 10]$ by $[-10, 10]$

55.

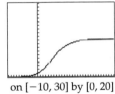

on $[-10, 30]$ by $[0, 20]$

57. $f'(x) = 2ax + b = 0$ at $x = \dfrac{-b}{2a}$

59.

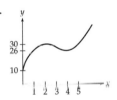

61. d.

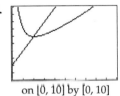

on $[0, 10]$ by $[0, 10]$

63. a.

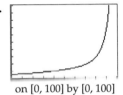

on $[0, 100]$ by $[0, 100]$

Exercises 3.2 page 211

1. Point 2   **3.** Points 3 and 5   **5.** Points 4 and 6

7. a.

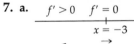

$f' > 0$	$f' = 0$	$f' < 0$	$f' = 0$	$f' > 0$
↗	→	↘	→	↗
	rel max		rel min	
	$(-3, 32)$		$(1, 0)$	

b.

$f'' < 0$	$f'' = 0$	$f'' > 0$
con dn		con up
	IP $(-1, 16)$	

c.

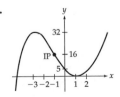

9. a.

$f' > 0$	$f' = 0$	$f' > 0$

$x = 1$

↗   →   ↗

neither
(1, 5)

b.

$f'' < 0$	$f'' = 0$	$f'' > 0$

$x = 1$

con dn   con up

IP (1, 5)

c.

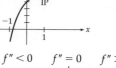

11. a.

$f' < 0$	$f' = 0$	$f' > 0$	$f' = 0$	$f' > 0$

$x = 0$ $x = 3$

↘   →   ↗   →   ↗

rel min   neither
(0, 2)   (3, 29)

b.

$f'' > 0$	$f'' = 0$	$f'' < 0$	$f'' = 0$	$f'' > 0$

$x = 1$ $x = 3$

con up   con dn   con up

IP (1, 13)   IP (3, 29)

c.

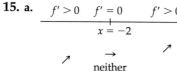

13. a.

$f' < 0$	$f' = 0$	$f' > 0$	$f' = 0$	$f' < 0$

$x = 0$ $x = 4$

↘   →   ↗   →   ↘

rel min   rel max
(0, 0)   (4, 256)

b.

$f'' > 0$	$f'' = 0$	$f'' > 0$	$f'' = 0$	$f'' < 0$

$x = 0$ $x = 3$

con up   con up   con dn

IP (3, 162)

c.

15. a.

$f' > 0$	$f' = 0$	$f' > 0$

$x = -2$

↗   →   ↗

neither
(−2, 0)

b.

$f'' < 0$	$f'' = 0$	$f'' > 0$

$x = -2$

con dn   con up

IP (−2, 0)

c.

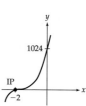

17. a.

$f' > 0$	$f' = 0$	$f' < 0$	$f' = 0$	$f' > 0$

$x = 1$ $x = 3$

↗   →   ↘   →   ↗

rel max   rel min
(1, 4)   (3, 0)

b.

$f'' < 0$	$f'' = 0$	$f'' > 0$

$x = 2$

con dn   con up

IP (2, 2)

c.

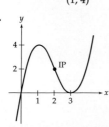

19. a.

$$
\begin{array}{ccc}
f' > 0 & f' \text{ und} & f' > 0 \\
\hline
& x = 0 &
\end{array}
$$

↗ ↗

neither
(0, 0)

b.

$$
\begin{array}{ccc}
f'' > 0 & f''\text{und} & f'' < 0 \\
\hline
& x = 0 &
\end{array}
$$

con up con dn

IP (0, 0)

c.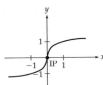

21. a.

$$
\begin{array}{ccc}
f' < 0 & f' \text{ und} & f' > 0 \\
\hline
& x = 0 &
\end{array}
$$

↘ ↗

rel min
(0, 2)

b.

$$
\begin{array}{ccc}
f'' < 0 & f''\text{und} & f'' < 0 \\
\hline
& x = 0 &
\end{array}
$$

con dn con dn

c.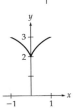

23. a.

$$
\begin{array}{cc}
f' \text{ und} & f' > 0 \\
\hline
x = 0 &
\end{array}
$$

↗

b.

$$
\begin{array}{cc}
f'' \text{ und} & f'' < 0 \\
\hline
x = 0 &
\end{array}
$$

con dn

c.

25. a.

$$
\begin{array}{ccc}
f' < 0 & f' \text{ und} & f' > 0 \\
\hline
& x = 1 &
\end{array}
$$

↘ ↗

rel min
(1, 0)

b.

$$
\begin{array}{ccc}
f'' < 0 & f''\text{und} & f'' < 0 \\
\hline
& x = 1 &
\end{array}
$$

con dn con dn

c.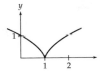

27.

$$
\begin{array}{ccccc}
f' > 0 & f' = 0 & f' < 0 & f' = 0 & f' > 0 \\
\hline
& x = 2 & & x = 10 &
\end{array}
$$

↗ → ↘ → ↗

rel max   rel min
(2, 76)   (10, −180)

$$
\begin{array}{ccc}
f'' < 0 & f'' = 0 & f'' > 0 \\
\hline
& x = 6 &
\end{array}
$$

con dn   con up
IP (6, −52)

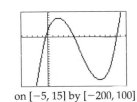

on [−5, 15] by [−200, 100]

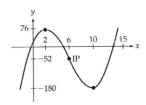

29.

$$
\begin{array}{ccccc}
f' < 0 & f' = 0 & f' < 0 & f' = 0 & f' > 0 \\
\hline
& x = 0 & & x = 12 &
\end{array}
$$

↘ → ↘ → ↗

neither   rel min
(0, 0)   (12, −6912)

$$
\begin{array}{ccccc}
f'' > 0 & f'' = 0 & f'' < 0 & f'' = 0 & f'' > 0 \\
\hline
& x = 0 & & x = 8 &
\end{array}
$$

con up   con dn   con up
IP (0, 0)   IP (8, −4096)

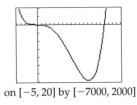

on [−5, 20] by [−7000, 2000]

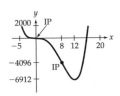

31.

$f' > 0$	$f' = 0$	$f' < 0$	$f' = 0$	$f' > 0$

$\qquad\qquad x = -2 \qquad\qquad\qquad x = 8$

$\nearrow \qquad \rightarrow \qquad \searrow \qquad \rightarrow \qquad \nearrow$

rel max $\qquad$ rel min
$(-2, 100)$ $\qquad$ $(8, -400)$

$f'' < 0$	$f'' = 0$	$f'' > 0$

$\qquad\qquad x = 3$

con dn $\qquad$ con up
$\qquad$ IP $(3, -150)$

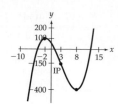

on $[-10, 15]$ by $[-500, 200]$

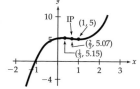

33.

$f' > 0$	$f' = 0$	$f' < 0$	$f' = 0$	$f' > 0$

$\qquad\qquad x = \frac{1}{3} \qquad\qquad\qquad x = 1$

$\nearrow \qquad \rightarrow \qquad \searrow \qquad \rightarrow \qquad \nearrow$

rel max $\qquad$ rel min
$(\frac{1}{3}, 5.15)$ $\qquad$ $(1, 5)$

$f'' < 0$	$f'' = 0$	$f'' > 0$

$\qquad\qquad x = \frac{2}{3}$

con dn $\qquad$ con up
$\qquad$ IP $(\frac{2}{3}, 5.07)$

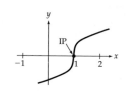

on $[-2, 3]$ by $[-5, 10]$

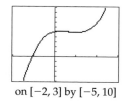

35.

$f' > 0$	f' und	$f' > 0$

$\qquad\qquad x = 1$

$\nearrow \qquad\qquad \nearrow$

$\qquad$ neither
$\qquad$ $(1, 0)$

$f'' > 0$	f'' und	$f'' < 0$

$\qquad\qquad x = 1$

con up $\qquad$ con dn
$\qquad$ IP $(1, 0)$

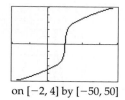

on $[-2, 4]$ by $[-50, 50]$

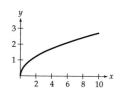

37.

f' und	$f' > 0$

$\qquad x = 0 \qquad \nearrow$

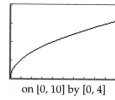

on $[0, 10]$ by $[0, 4]$

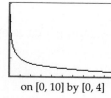

39.

f' und	$f' < 0$

$\qquad x = 0 \qquad \searrow$

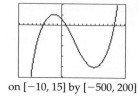

on $[0, 10]$ by $[0, 4]$

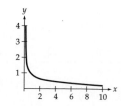

41.

$f' < 0$	f' und	$f' > 0$	$f' = 0$	$f' < 0$

$$x = 0 \qquad\qquad x = 1$$

↘   ↗   →   ↘

rel min   rel max
(0, 0)   (1, 3)

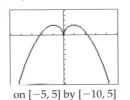

on $[-2, 10]$ by $[-10, 10]$

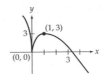

43.

$f' > 0$	f' und	$f' < 0$	$f' = 0$	$f' > 0$

$$x = 0 \qquad\qquad x = 1$$

↗   ↘   →   ↗

rel max   rel min
(0, 0)   (1, −2)

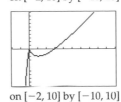

on $[-2, 10]$ by $[-10, 10]$

45.

$f' > 0$	$f' = 0$	$f' < 0$	f' und	$f' > 0$	$f' = 0$	$f' < 0$

$$x = -1 \qquad\qquad x = 0 \qquad\qquad x = 1$$

↗   →   ↘   ↗   →   ↘

rel max   rel min   rel max
(−1, 2)   (0, 0)   (1, ?)

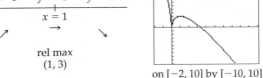

on $[-5, 5]$ by $[-10, 5]$

47. $f''(x) = 2a$, therefore: For $a > 0, f'' > 0$, so f is concave up.

For $a < 0, f'' < 0$, so f is concave down.

49. Where the curve is concave *up*, it lies *above* its tangent line, and where it is concave *down*, it lies *below* its tangent line, so *at* an inflection point it must cross its tangent line.   **51.** −0.77, 0, and 0.77

53. a.

$f' > 0$	$f' = 0$	$f' < 0$	$f' = 0$	$f' > 0$

$$x = 1 \qquad\qquad x = 5$$

↗   →   ↘   →   ↗

rel max   rel min
(1, 32)   (5, 0)

$f'' < 0$	$f'' = 0$	$f'' > 0$

$$x = 3$$

con dn   con up

IP (3, 16)

b.

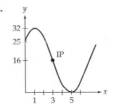

55. a.

$f' < 0$	$f' = 0$	$f' > 0$

$$x = 1$$

↘   →   ↗

rel min
(1, 109)

$f'' > 0$	$f'' = 0$	$f'' > 0$

$$x = 0$$

con up   con up

b.

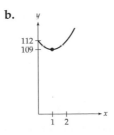

57.

$$\frac{f' \text{ und} \qquad f' > 0}{x = 0}$$

↗

rel min
(0, 0)

$$\frac{f'' \text{ und} \qquad f'' < 0}{x = 0}$$

con dn

59. a.

$$\frac{S' \text{ und} \qquad S' > 0}{i = 0}$$

↗

$$\frac{S'' \text{ und} \qquad S'' < 0}{i = 0}$$

con dn

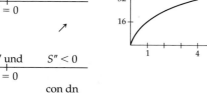

b. Concave down. Status increases more slowly at higher income levels.

61. b. $y = 1.96x^{0.66}$ **c.** Concave down   **d.** About 16   **e.** About 8.5

63. (50, 2.5). The curve is concave up (slope increasing) before $x = 50$ and concave down (slope decreasing) after $x = 50$. Therefore, the slope is maximized at $x = 50$.

Exercises 3.3 page 225

1. Max f is 12 (at $x = 1$), min f is -8 (at $x = -1$).   **3.** Max f is 16 (at $x = -2$), min f is -16 (at $x = 2$).

5. Max f is 9 (at $x = 1$), min f is 0 (at $x = 0$ and at $x = -2$).

7. Max f is 81 (at $x = 3$), min f is -16 (at $x = 2$).

9. Max f is 4 (at $x = 2$), min f is -50 (at $x = 5$).   **11.** Max f is 5 (at $x = 0$), min f is 0 (at $x = 5$).

13. Max f is 1 (at $x = 0$), min f is 0 (at $x = -1$ and at $x = 1$).

15. Max f is $\frac{1}{2}$ (at $x = 1$), min f is $-\frac{1}{2}$ (at $x = -1$).

17. a. The number is $\frac{1}{2}$.   **b.** The number is 3.

19. a. Both at endpoints

  b. One at a critical number (the maximum) and one at an endpoint (the minimum)

  c. Both at critical numbers   **d.** Yes; for example, [2, 10]

21. On the 20th day   **23.** 31 mph   **25.** 36 years   **27.** 12 miles from A toward B

29. Produce 40 per day, price = $400, max profit = $6500

31. 400 feet along the building and 200 feet perpendicular to the building

33. Each is 200 yards parallel to the river and 150 yards perpendicular to the river

35. 3 inches high with a base 12 inches by 12 inches; volume: 432 cubic inches

37. The numbers are 25 and 25.

39. $r = 2$ cm   **41.** $r = 110/\pi \approx 35$ yards, $x = 110$ yards   **43.** $x \approx 1.125$ inches, $y \approx 1.25$ megabytes

45. a. At time 10 hours; 1,500,000 bacteria (since $N(t)$ is in thousands)

  b. At time 5 hours; growing by 75,000 bacteria per hour (inflection point)

47. Remove a square of side $x \approx 0.96$ inch; volume ≈ 15 square inches

49. a. $p(x) = -4.5x + 54$ **b.** $R(x) = -4.5x^2 + 54x$ **c.** 6 bottles per hour, selling for $27 each

Exercises 3.4 page 236

1. Price: $14,400; sell 16 cars per day (from $x = 2$ price reductions)
3. Ticket price: $150; number sold: 450 (from $x = 5$ price reductions)
5. Rent the cars for $90, and expect to rent 54 cars (from $x = 2$ price increases)
7. 25 trees per acre (from $x = 5$ extra trees per acre)   **9.** Base: 2 feet by 2 feet; height: 1 foot
11. Base: 14 inches by 14 inches; height: 28 inches; volume: 5488 cubic inches
13. 50 feet along the driveway and 100 feet perpendicular to the driveway; cost: $800
15. 6.4%   **17.** 16 years

19. [*Hint:* If area is A (a constant) and one side is x, then show that the perimeter is $P - 2x + 2\dfrac{A}{x}$, which is
minimized at $x = \sqrt{A}$. Then show that this means the rectangle is a square.]

21. The page should be 8 inches wide and 12 inches tall.
23. $R' = 2cpx - 3cx^2 = xc(2p - 3x)$, which is zero when $x = \frac{2}{3}p$. The second-derivative test will show that
R is maximized.
25. e.

 f. Price: $325; quantity: 35 bicycles   **g.** Price: $350; quantity: 40 bicycles

Exercises 3.5 page 246

1. Lot size: 400 boxes; 10 orders during the year   **3.** Lot size: 500 bottles; 20 orders during the year
5. Lot size: 40 cars per order; 20 orders during the year
7. Produce 1000 games per run; 2 production runs during the year
9. Produce 40,000 tapes per run; 25 runs for the year
11. Population: 20,000; yield: 40,000 (from $p = 200$)
13. Population: 75,000; yield: 2250 (from $p = 75$)   **15.** Population: 625,000; yield: 625,000 (from $p = 625$)
17. Population: 3717; yield: 16,109 (from $x = 3.717$). (The yield exceeds the population because of reproduction later in the year.)

Exercises 3.6 page 258

1. $\dfrac{dy}{dx} = \dfrac{2x}{3y^2}$ **3.** $\dfrac{dy}{dx} = \dfrac{3x^2}{2y}$ **5.** $\dfrac{dy}{dx} = \dfrac{3x^2 + 2}{4y^3}$ **7.** $\dfrac{dy}{dx} = -\dfrac{x + 1}{y + 1}$ **9.** $\dfrac{dy}{dx} = -\dfrac{2y}{x}$ (after simplification)

11. $\dfrac{dy}{dx} = \dfrac{-y + 1}{x}$ **13.** $\dfrac{dy}{dx} = -\dfrac{y - 1}{2x}$ (after simplification)   **15.** $\dfrac{dy}{dx} = \dfrac{1}{3y^2 - 2y + 1}$

17. $\dfrac{dy}{dx} = -\dfrac{y^2}{x^2}$ (after simplification)   **19.** $\dfrac{dy}{dx} = \dfrac{3x^2}{2(y - 2)}$ **21.** $\dfrac{dy}{dx} = 2$ **23.** $\dfrac{dy}{dx} = -3$ **25.** $\dfrac{dy}{dx} = -1$

27. $\dfrac{dy}{dx} = -4$ **29.** $\dfrac{dp}{dx} = -\dfrac{2}{2p + 1}$ **31.** $\dfrac{dp}{dx} = \dfrac{1}{24p + 4}$ **33.** $\dfrac{dp}{dx} = -\dfrac{p}{3x}$ (after simplification)

35. $\dfrac{dp}{dx} = -\dfrac{p + 5}{x + 2}$

37. $\dfrac{dp}{dx} = -4$. *Interpretation:* The rate of change of price with respect to quantity is -4, so price decreases by about \$4 when quantity increases by 1.

39. a. $\dfrac{ds}{dr} = \dfrac{3r^2}{2s} = 8$ **b.** $\dfrac{dr}{ds} = \dfrac{2s}{3r^2} = \dfrac{1}{8}$ **c.** $\dfrac{ds}{dr} = 8$ means that the rate of change of sales with respect to research expenditures is 8, so that increasing research by \$1 million will increase sales by about \$8 million (at these levels of r and s).

$\dfrac{dr}{ds} = \dfrac{1}{8}$ means that the rate of change of research expenditures with respect to sales is $\dfrac{1}{8}$, so that increasing sales by \$1 million will increase research by about $\dfrac{1}{8}$ million dollars (at these levels of r and s).

41. $3x^2 \dfrac{dx}{dt} + 2y \dfrac{dy}{dt} = 0$ **43.** $2x \dfrac{dx}{dt} y + x^2 \dfrac{dy}{dt} = 0$ **45.** $6x \dfrac{dx}{dt} - 7 \dfrac{dx}{dt} y - 7x \dfrac{dy}{dt} = 0$

47. $2x \dfrac{dx}{dt} + \dfrac{dx}{dt} y + x \dfrac{dy}{dt} = 2y \dfrac{dy}{dt}$ **49.** Decreasing by $72\pi \approx 226$ in^3/hr

51. Increasing by $32\pi \approx 101$ cm^3/week **53.** Growing by \$16,000 per day

55. Increasing by 400 cases per year

57. Slowing by $\frac{1}{2}$ mm/sec per year **59.** Yes (65.8 mph)

Chapter 3 Review Exercises page 262

1.

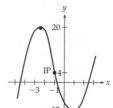

2.

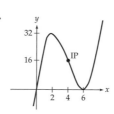

3.

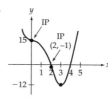

4.

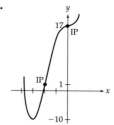

5.

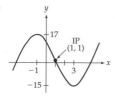

6.

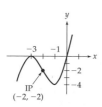

7.

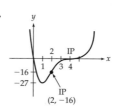

8.

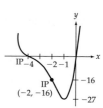

9.

10.

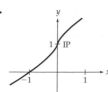

11.

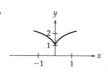

12.

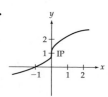

13.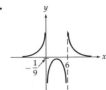

14.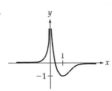

15.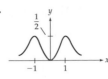

16.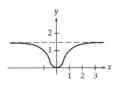

17.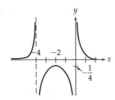

18.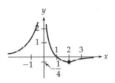

19. Max f is 220 (at $x = 5$), min f is -4 (at $x = 1$).   **20.** Max f is 130 (at $x = 5$), min f is -32 (at $x = 2$).

21. Max f is 64 (at $x = 0$), min f is -64 (at $x = 4$).

22. Max f is 6401 (at $x = 10$), min f is 1 (at $x = 0$ and $x = 2$).

23. Max h is 4 (at $x = 9$), min h is 0 (at $x = 1$).

24. Max f is 10 (at $x = 0$), min f is 0 (at $x = 10$ and $x = -10$).

25. Max g is 25 (at $w = 3$ and $w = -3$), min g is 0 (at $w = 2$ and $w = -2$).

26. Max g is 16 (at $x = 4$), min g is 0 (at $x = 0$ and $x = 8$).

27. Max f is $\frac{1}{2}$ (at $x = 1$), min f is $-\frac{1}{2}$ (at $x = -1$).

28. Max f is $\frac{1}{4}$ (at $x = 2$), min f is $-\frac{1}{4}$ (at $x = -2$).   **29.**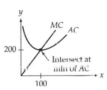

30. ① At the minimum point of $AC(x)$, the derivative of $AC(x)$ must be zero.

② Quotient Rule applied to $AC(x) = \dfrac{C(x)}{x}$

③ Factoring out $\dfrac{1}{x}$

④ Simplifying inside the square brackets

⑤ Recognizing $C'(x)$ as the marginal cost function and $\dfrac{C(x)}{x}$ as the average cost function.

⑥ $0 = \dfrac{1}{x}[MC(x) - AC(x)]$ means that the quantity in the square brackets must equal zero, and so the marginal cost must equal average cost at this x-value, where average cost is minimized.

31. $v = 2c$, which means that the tugboat should travel through the water at twice the speed of the current.

32. 3600 square feet   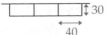   **33.** 1800 square feet

34. 15 cubits (gilded side) by 135 cubits   **35.** Base: 10 inches by 10 inches; height: 5 inches

36. Radius: 2 inches; height: 4 inches

37. Radius: 2 inches; height: 2 inches

38. $v = \sqrt[4]{\dfrac{aw^2}{3b}}$

39. Price: $2400 each; quantity: 9 per week   **40.** 5 weeks   **41.** $x = \frac{3}{4}$ mile
42. a. $t \approx 0.624 = 62.4\%$ **b.** $1.26   **43.** Radius $\approx$ 1.2 inches; height $\approx$ 4.8 inches
44. $x \approx 1.59$ inches; volume $\approx$ 33.07 cubic inches   **45.** $x \approx 1.13$ inches; volume $\approx$ 12.13 cubic inches
46. 600 per run; $1\frac{1}{2}$ runs per year (or 3 runs in 2 years)   **47.** Lot size: 50; 10 orders during the year
48. Population: 150,000 ($p = 150$); yield: 450,000
49. Population: $p = 900$ (thousand); yield: 900 (thousand)
50. $\dfrac{dy}{dx} = \dfrac{-6x - 4y}{4x + y}$ **51.** $\dfrac{dy}{dx} = \dfrac{-y^2}{2xy - 1}$ (after simplification)   **52.** $\dfrac{dy}{dx} = \dfrac{-2y^2 + 6xy}{4xy - 3x^2}$
53. $\dfrac{dy}{dx} = \dfrac{y^{1/2}}{x^{1/2}}$ (after simplification)   **54.** $\dfrac{dy}{dx} = -1$ **55.** $\dfrac{dy}{dx} = \dfrac{1}{7}$ **56.** $\dfrac{dy}{dx} = -\dfrac{1}{6}$ **57.** $\dfrac{dy}{dx} = 1$
58. 600 in^3/hr   **59.** Increasing by $4200 per day   **60.** Increasing by $45,000 per month
61. a. Decreasing by about 0.31 cm^3/min   **b.** Decreasing by about 0.05 cm^3/min

Cumulative Review for Chapters 1–3 page 266

1. $y = -\frac{1}{2}x + 1$ **2.** $\frac{5}{2}$ **3.** 20.085
4. a.

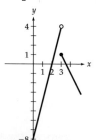

   b. 4   **5.** $f'(x) = 4x - 5$

   c. 1

   d. Does not exist

   e. Discontinuous at $x = 3$

6. $f'(x) = 12x^{1/2} + 6x^{-3}$ **7.** $f'(x) = 9x^8 + 10x^4 - 8x^3$ **8.** $f'(x) = \dfrac{11}{(3x - 2)^2}$

9. $P'(8) = 1200$, so in 8 years the population will be increasing by 1200 people per year.
   $P''(8) = -50$, so in 8 years the rate of growth will be slowing by 50 people per year per year.

10. $\dfrac{2x}{\sqrt{2x^2 - 5}}$ **11.** $12(3x + 1)^3(4x + 1)^3 + 12(3x + 1)^4(4x + 1)^2 = 12(3x + 1)^3(4x + 1)^2(7x + 2)$

12. $\dfrac{12(x - 2)^2}{(x + 2)^4}$

13.

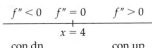

$f' > 0$	$f' = 0$		$f' < 0$	$f' = 0$	$f' > 0$

$x = -2$ $x = 10$

↗   →   ↘   →   ↗

rel max
$(-2, 464)$

rel min
$(10, -400)$

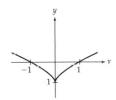

$f'' < 0$	$f'' = 0$	$f'' > 0$

$x = 4$

con dn   con up

IP (4, 32)

14.

$f' < 0$	f' und	$f' > 0$

$x = 0$

↘   ↗

rel min
$(0, -1)$

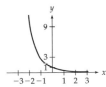

$f'' < 0$	f'' und	$f'' < 0$

$x = 0$

con dn   con dn

15. 15,000 square feet   **16.** Price: $170; quantity: 18 per day

17. $\dfrac{dy}{dx} = \dfrac{-x^2 - 3y^2}{6xy + 1}$ (after simplification). Evaluating at $(1, 2)$ gives $\dfrac{dy}{dx} = -1$.   **18.** $\dfrac{2}{\pi} \approx 0.64$ ft/min

Exercises 4.1 page 283

1. a. 7.389   **b.** 0.135   **c.** 1.649
3. a. e^3 **b.** e^2 **c.** e^5
5.

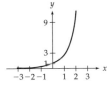

7.

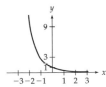

9. 5.697 (rounded)

11. a. e^x **b.** e^x **c.** e^x **d.** e^x **e.** e^x will exceed any power of x for large enough values of x.
13. a. $2144   **b.** $2204   **c.** $2226   **15. a.** $2196   **b.** $2096
17. The annual yield should be 9.69% (based on the nominal rate of 9.25%).   **19.** $8629   **21.** $72.65
23. 10% compounded quarterly (yielding 10.38%, better than 10.30%)   **25. a.** $2678   **b.** $12,093
27. 7.0 billion   **29. a.** 0.53 (the chances are better than 50–50)   **b.** 0.70 (quite likely)
31. a. 0.267 or 26.7%   **b.** 0.012 or 1.2%   **33. a.** 1.3 milligrams   **b.** 0.84 milligram   **35.** 208
37. a. About 153 degrees   **b.** About 123 degrees   **39.** 38   **41.** 6.5%   **43.** By about 25%
45. a.

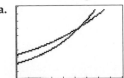

b. In about 2060 (from $x \approx 60.2$)

Exercises 4.2 page 301

1. a. 2   **b.** 4   **c.** -1 **d.** -2 **e.** $\frac{1}{2}$ **f.** $-\frac{1}{2}$
3. a. 10   **b.** $\frac{1}{2}$ **c.** $\frac{4}{3}$ **d.** 0   **e.** 1   **f.** -3
5. $\ln x$ **7.** $2\ln x$ or $\ln x^2$ **9.** $\ln x$ **11.** $3x$ **13.** $7x$
15. Domain: $\{\, x \mid x > 1 \text{ or } x < -1 \,\}$; Range: $\mathbb{R}$
17. a. 2.9 years (from 35 months)   **b.** 1.7 years (from 20.5 months)   **19. a.** 15.7 years   **b.** 3.2 years
21. 1.9 years   **23.** About 17.1 years   **25.** 77 days   **27.** 0.58 or 58%   **29.** About 4 weeks
31. About 31,400 years   **33.** 1.7 million years   **35.** About 138 days   **37. b.** About 39 hours
39. About 2.7 years

41. Solving $\;0.94 = (1 - \frac{1}{1000})^x\;$ gives $\;x = \dfrac{\ln 0.94}{\ln 0.999} \approx 61.84,\;$ so about 62 generations.

43. a. 35 years   **b.** 55.5 years
45. a. About 11.6 years (from 46.6 quarters)   **b.** About 6.8 years (from 27.2 quarters)
47. a. About 9 days   **b.** About 11 days
49. About 5300 years old. (For reference, this means that Iceman lived 2000 years before King Tutankhamen.)
51. About 6.5 years

Exercises 4.3 page 316

1. $2x\ln x + x$ **3.** $\dfrac{2}{x}$ **5.** $\dfrac{1}{2}x^{-1}$ **7.** $\dfrac{6x}{x^2 + 1}$ **9.** $\dfrac{1}{x}$ **11.** $\dfrac{xe^x - 2e^x}{x^3}$ or $\dfrac{e^x(x - 2)}{x^3}$
13. $(3x^2 + 2)e^{x^3+2x}$ **15.** $x^2 e^{x^3/3}$ **17.** $1 + e^{-x}$ **19.** 2   **21.** $e^{1+e^x}e^x$ or e^{1+x+e^x} **23.** ex^{e-1} **25.** 0
27. $\dfrac{4x^3}{x^4 + 1} - 2e^{x/2} - 1$ **29.** $2x\ln x + 2xe^{x^2}$ **31. a.** $\dfrac{1 - 5\ln x}{x^6}$ **b.** 1   **33. a.** $\dfrac{4x^3}{x^4 + 48}$ **b.** $\dfrac{1}{2}$
35. a. $\dfrac{e^x - 3}{e^x - 3x}$ **b.** -2 **37. a.** $5\ln x + 5$ **b.** $5\ln 2 + 5 \approx 8.466$
39. a. $\dfrac{xe^x - e^x}{x^2}$ **b.** $\dfrac{2e^3}{9} \approx 4.463$ **41.** $-4x^3 e^{-x^5/5} + x^8 e^{-x^5/5}$ or $x^3 e^{-x^5/5}(x^5 - 4)$
43. $f^{(n)}(x) = k^n e^{kx}$ **45.**

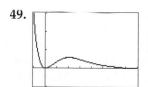

on $[-2, 2]$ by $[-1, 2]$

rel max: $(0, 1)$
IP: $\left(\frac{1}{2}, 0.61\right)$
     $\left(-\frac{1}{2}, 0.61\right)$

47.

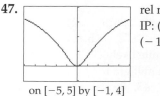

on $[-5, 5]$ by $[-1, 4]$

rel min: $(0, 0)$
IP: $(1, 0.69)$
     $(-1, 0.69)$

49.

on $[-1, 8]$ by $[-1, 3]$

rel min: $(0, 0)$
rel max: $(2, 0.54)$
IP: $(0.59, 0.19)$
     $(3.41, 0.38)$

51.

on $[-2, 2]$ by $[-2, 2]$

rel max: $(-0.37, 0.37)$
rel min: $(0.37, -0.37)$

53. $\dfrac{dy}{dx} = \dfrac{ye^x}{2y - e^x}$

55. a. Increasing by about \$50 per year   **b.** Increasing by about \$82.44 per year

57. Increasing by about 0.12 billion $(= 120$ million) people per year

59. a. Decreasing by 0.06 mg/hr   **b.** Decreasing by 0.054 mg/hr

61. a. Increasing by about 81.4 (thousand) sales per week
  b. Increasing by about 33 (thousand) sales per week

63. $p = \$100$ **65. a.** $R(x) = 400xe^{-0.20x}$ **b.** Quantity: $x = 5$ (thousand); price: $p = \$147.15$

67. When $N < K$, $\ln(K/N)$ is positive and the population is increasing, while when $N > K$, $\ln(K/N)$ is negative and the population is decreasing.

69. $r = a/b$

71. a. After 15 minutes the temperature of the beer is 57.5 degrees and increasing at the rate of 43.8 degrees per hour.
  b. After 1 hour the temperature of the beer is 69.1 degrees and increasing at the rate of 3.2 degrees per hour.

73. 2.3 seconds   **75.** $p = \$500$ **77.** At about 2.75 hours

79. a. $(\ln 10)10^x$ **b.** $(\ln 3)(2x)3^{x^2+1}$ **c.** $(\ln 2)3 \cdot 2^{3x}$ **d.** $(\ln 5)6x \cdot 5^{3x^2}$ **e.** $-(\ln 2)2^{4-x}$

81. a. $\dfrac{1}{(\ln 2)x}$ **b.** $\dfrac{2x}{(\ln 10)(x^2 - 1)}$ **c.** $\dfrac{4x^3 - 2}{(\ln 3)(x^4 - 2x)}$

Exercises 4.4 page 332

1. a. $2/t$ **b.** 2 and 0.2   **3. a.** 0.2   **b.** 0.2   **5. a.** $2t$ **b.** 20   **7. a.** $-2t$ **b.** -20

9. a. $\dfrac{1}{2(t - 1)}$ **b.** $1/10$ **11.** 0.0071 or 0.71%   **13. a.** 0.012 or 1.2%   **b.** Yes, in about 15.3 years

15. a. $E(p) = \dfrac{5p}{200 - 5p}$ **b.** Inelastic $\left(E = \frac{1}{3}\right)$ **17. a.** $E(p) = \dfrac{2p^2}{300 - p^2}$ **b.** Unitary elastic $(E = 1)$

19. a. $E(p) = 1$ **b.** Unitary elastic $(E = 1)$ **21. a.** $E(p) = \dfrac{3p}{2(175 - 3p)}$ **b.** Elastic $(E = 3)$

23. a. $E(p) = 2$ **b.** Elastic $(E = 2)$ **25. a.** $E(p) = 0.01p$ **b.** Elastic $(E = 2)$

27. Lower the price $(E = 8)$

29. No $(E = 1.25)$ **31.** Yes $\left(E = \frac{3}{8} = 0.375\right)$ **33.** Lower its price $(E = 1.2)$ **35.** $E = 0.112$

37. $E(p) = \dfrac{-pa(-c)e^{-cp}}{ae^{-cp}} = cp$ **39.** $E_s(p) = n$ **41. a.** $E \approx 0.35$ **b.** Raise the price   **c.** $p \approx \$20,400$

Chapter 4 Review Exercises page 336

1. a. $18,845.41 **b.** $18,964.81 **2.** The second bank $(1.015^4 \approx 1.0614 < e^{0.0598} \approx 1.0616)$
3. a. $V(t) = 800{,}000(0.8)^t$ **b.** $327,680
4. Drug B **5.** In about 2060 (from $x \approx 60.2$ years after 2000)
6. $4^7 = 16{,}384$ megabits, which is enough to hold sixty-four 16-volume encyclopedias on one chip.
7. a. 7.1 years (from 14.2 half-years) **b.** 4.2 years (from 8.3 half-years)
8. a. 9.9 years **b.** 5.8 years **9.** 50.7 million years **10.** 1.85 million years
11. 2.3 hours **12.** 13.7 years **13. a.** 12 years **b.** 11.9 years
14. a. 72 years **b.** 69.7 years **15.** $\dfrac{\ln k}{\ln(1 + r)}$ **16.** $\dfrac{\ln k}{r}$
17. a. In about 6.25 years (from $x \approx 25$ quarters) **b.** In about 6.2 years
18. a. About 11 days **b.** About 16 days
19. $\dfrac{1}{x}$ **20.** $\dfrac{4x}{x^2 - 1}$ **21.** $\dfrac{-1}{1 - x}$ or $\dfrac{1}{x - 1}$ **22.** $\dfrac{x}{x^2 + 1}$ **23.** $\dfrac{1}{3x}$ **24.** 1 **25.** $\dfrac{2}{x}$ **26.** $\ln x$
27. $-2xe^{-x^2}$ **28.** $-e^{1-x}$ **29.** $2x$ **30.** $2x(\ln x)e^{x^2 \ln x - x^2/2}$ **31.** $10x + 2 \ln x + 2$
32. $6x^2 + 3 \ln x + 3$ **33.** $6x^2 - 3e^{2x} - 6xe^{2x}$ **34.** $4 - 4xe^{2x} - 4x^2 e^{2x}$

35.
rel min: $(0, \ln 4) \approx (0, 1.4)$
IP: $(2, \ln 8) \approx (2, 2.1)$
$(-2, \ln 8) \approx (-2, 2.1)$

36.
rel max: $(0, 16)$
IP: $(2, 16e^{-1/2}) \approx (2, 9.7)$
$(-2, 16e^{-1/2}) \approx (-2, 9.7)$

37. a. Increasing by 136 thousand per week **b.** Increasing by 55 thousand per week
38. a. Decreasing by 0.12 mg per hour **b.** Decreasing by 0.08 mg per hour
39. Decreasing by 33.3% per second
40. a. Increasing by 3.5 degrees per hour **b.** Increasing by 2.1 degrees per hour
41. a. Increasing by 6667 per hour **b.** Increasing by 816 per hour **42.** 25 years
43. a. $R(x) = 200xe^{-0.25x}$ **b.** Quantity $x = 4$ (thousand); price $p = 73.58
44. a. $R(x) = 5x - x \ln x$ **b.** Quantity $x = e^4 \approx 54.60$; price = $1 **45.** Price = $50
46.
rel min: $(0, 0)$
rel max: $(4, 4.69)$
IP: $(2, 2.17)$
$(6, 3.21)$
on $[-2, 10]$ by $[-1, 10]$

47.
rel max: $(-0.72, 0.12)$
rel min: $(0.72, -0.12)$
IP: $(-0.43, 0.07)$
$(0.43, -0.07)$
on $[-2, 2]$ by $[-2, 2]$

48. $p \approx 769 **49.** $x \approx 130$ (in thousands) **50.** 0.0033 or 0.33% **51.** 0.0031 or 0.31%
52. Raise prices $(E = 0.8)$ **53.** Raise prices $(E = 0.7)$ **54.** $E = 0.44$ **55.** 0.104 or 10.4%
56. a. $E \approx 1.29$ **b.** Lower the price **c.** About $8700 (from $p \approx 8.7$)

Exercises 5.1 page 354

1. $\frac{1}{5}x^5 + C$ **3.** $\frac{3}{5}x^{5/3} + C$ **5.** $\frac{2}{3}u^{3/2} + C$ **7.** $-\frac{1}{3}w^{-3} + C$ **9.** $2\sqrt{z} + C$ **11.** $x^6 + C$
13. $2x^4 - x^3 + 2x + C$ **15.** $4x^{3/2} + \frac{3}{2}x^{2/3} + C$ **17.** $6x^{8/3} + 24x^{-2/3} + C$ **19.** $6t^{5/3} + 3t^{1/3} + C$
21. $\frac{1}{3}x^3 - x^2 + x + C$ **23.** $\frac{2}{3}w^{3/2} + 4w^{5/2} + C$ **25.** $2x^3 - 3x^2 + x + C$ **27.** $\frac{1}{3}x^3 + x^2 - 8x + C$
29. $\frac{1}{3}r^3 - r + C$ **31.** $\frac{1}{2}x^2 - x + C$ **33.** $\frac{1}{4}t^4 + t^3 + \frac{3}{2}t^2 + t + C$
35. a. $-\frac{1}{2}x^{-2} + C$ **b.** $\dfrac{x + C}{\frac{1}{4}x^4 + C_1}$ (where C_1 is another arbitrary constant) **37. b.**

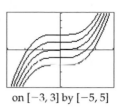

on $[-3, 3]$ by $[-5, 5]$

39. $C(x) = 8x^{5/2} - 9x^{5/3} + x + 4000$ **41.** $R(x) = 9x^{4/3} + 2x^{3/2}$
43. a. $D(t) = -0.03t^3 + 4t^2$ **b.** 370 feet

45. a. $6t^{1/2}$ **b.** 30 words **47. a.** $P(t) = 16t^{5/2}$ **b.** 512 tons **c.** No **49.** $\dfrac{1}{x}$

Exercises 5.2 page 366

1. $\frac{1}{3}e^{3x} + C$ **3.** $4e^{x/4} + C$ **5.** $20e^{0.05x} + C$ **7.** $-\frac{1}{2}e^{-2y} + C$ **9.** $-2e^{-0.5x} + C$ **11.** $9e^{2x/3} + C$
13. $-5\ln|x| + C$ **15.** $3\ln|x| + C$ **17.** $\frac{3}{2}\ln|v| + C$ **19.** $\frac{1}{3}e^{3x} - 3\ln|x| + C$
21. $6e^{0.5t} - 2\ln|t| + C$ **23.** $\frac{1}{3}x^3 + \frac{1}{2}x^2 + x + \ln|x| - x^{-1} + C$ **25.** $250e^{0.02t} - 200e^{0.01t} + C$
27. a. $360e^{0.05t} - 355$ **b.** About 624 cases
29. a. $50 \ln t$ (since $t > 1$, absolute value bars are not needed) **b.** No (about 170 sold)
31. a. $850e^{0.02t} - 850$ **b.** In about 2020 (20 years from 2000) **33. a.** $500e^{0.4x} - 500$ **b.** About $3195
35. a. $60e^{-0.2t} + 10$ **b.** In about 5 hours **37. a.** $-4000e^{-0.2t} + 4000$ **b.** About $3\frac{1}{2}$ years
39. a. $8000e^{0.05t} - 3000$ **b.** About $10,190 **41.** $\frac{1}{2}x^2 + 2x + \ln|x| + C$ **43.** $t + 2\ln|t| + 3t^{-1} + C$
45. $\frac{1}{3}x^3 - 3x^2 + 12x - 8\ln|x| + C$ **47. a.** $560e^{0.01t} - 560$ **b.** In about 2022 (22 years after 2000)

Exercises 5.3 page 383

1. 2.75 square units **3.** 4.15 square units **5.** 0.760 square unit
7. i. 2.8 square units **ii.** 3 square units
9. i. 4.411 square units (or 4.412, depending on rounding) **ii.** $\frac{14}{3} \approx 4.667$ square units
11. i. 0.719 square units **ii.** $\ln 2 \approx 0.693$ square units
13. i. for left rectangles: 2.9, 2.99, 2.999
    for midpoint rectangles: 3, 3, 3
    for right rectangles: 3.1, 3.01, 3.001
  ii. 3 square units
15. i. for left rectangles: 4.515, 4.652, 4.665
    for midpoint rectangles: 4.668, 4.667, 4.667
    for right rectangles: 4.815, 4.682, 4.668
  ii. $\frac{14}{3} \approx 4.667$ square units

17. i. for left rectangles: 0.719, 0.696, 0.693
for midpoint rectangles: 0.693, 0.693, 0.693
for right rectangles: 0.669, 0.691, 0.693
ii. ln 2 ≈ 0.693 square unit

19. 9 square units **21.** 8 square units

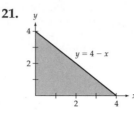

23. ln 2 square unit **25.** 160 square units **27.** 19 square units

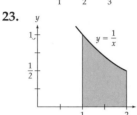

29. 2 square units **31.** 16 square units **33.** ln 5 square units **35.** ln 2 + $\frac{7}{3}$ square units
37. 4 square units (from $2e^{\ln 3} - 2$) **39.** $2e - 2$ square units **41.** 5 square units **43.** $\frac{15}{32}$ square unit
45. 4 + ln 2 square units **47.** $\frac{111}{100}$ **49.** 13 **51.** $\frac{3}{4}$ **53.** 1 (from ln e − ln 1) **55.** 78 + ln 3
57. $-3 \ln 2$ **59.** $4e^3 - 4$ **61.** $5e - 5e^{-1}$ **63.** 1 (from $e^{\ln 3} - e^{\ln 2}$) **65.** $\frac{7}{2} + \ln 2$ **67.** 1.107
69. 2.925 **71.** 92.744
73. a. 9 **b.** Completing the calculation: $= \left(\frac{1}{3}3^3 + C\right) - \left(\frac{1}{3}0^3 + C\right) = \frac{1}{3}3^3 + C - \frac{1}{3}0^3 - C = \frac{1}{3}\cdot 27 = 9$
c. The C always cancels because it is both added and subtracted in the evaluation step.
75. $\int_1^a \frac{1}{x}\,dx = \ln|x|\,\Big|_1^a = \ln a - \ln 1 = \ln a$ **77.** $3\frac{1}{2} + \frac{\pi}{4}$ **79.** 132 units **81.** 411 checks
83. $-300e^{-2} + 300 \approx \$259.40$ **85.** $225e^{0.8} - 225 \approx 276$, so about \$2.76
87. $24e^{0.1} - 24 \approx 2.5$ million tons **89.** $-30e^{-2} + 30 \approx 26$ words
91. $\frac{a}{-b+1}B^{-b+1} - \frac{a}{-b+1}A^{-b+1} = \frac{a}{1-b}(B^{1-b} - A^{1-b})$ **93.** $-4{,}000{,}000e^{-0.6} + 4{,}000{,}000 \approx \$1{,}804{,}753$
95. a. $\frac{1}{2}, \frac{1}{3}, \frac{1}{4}, \frac{1}{5}$ **b.** Area $= \frac{1}{n+1}$ **97.** 87 milligrams **99.** 586 cars **101.** 4023 people

Exercises 5.4 page 399

1. 3 **3.** 4 **5.** $\frac{1}{5}$ **7.** 5 **9.** $\frac{104}{3}$ or $34\frac{2}{3}$ **11.** 3 **13.** $e - 1$ **15.** ln 2 **17.** $\frac{1}{n+1}$
19. $a + b$ **21.** 0.845 **23.** 318 **25.** 70° **27.** About 25.6 tons **29.** \$3194.53
31. \$24.04 million **33.** 6 square units
35. $\frac{1}{2}e^4 - e^2 + \frac{1}{2}$ square units **37. a.** **b.** 9 square units

39. a.

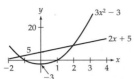

b. 18 square units   **41.** 4 square units   **43.** 32 square units

45. $\frac{1}{12}$ square unit   **47.** 1250 square units   **49.** 5.694 square units (rounded)   **51.** About 123 million
53. a. 10   **b.** $6000   **55.** About $529 billion   **57.** About $139 thousand

59. e.

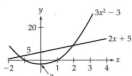

f. About 46,350 lives   **61.** $(2x + 5)e^{x^2+5x}$ **63.** $\dfrac{2x + 5}{x^2 + 5x}$

Exercises 5.5 page 412

1. $60,000   **3.** $10,000   **5.** $7500   **7.** $5285 (rounded)   **9.** $100   **11.** $160,000
13. a. $x = 500$ **b.** $50,000   **c.** $25,000   **15. a.** $x = 50$ **b.** $2500   **c.** $7500
17. a. $x \approx 119.48$ **b.** $10,065   **c.** $3446 (all rounded)   **19.** 0.52   **21.** 0.35   **23.** 0.2

25. $1 - \dfrac{2}{n + 1} = \dfrac{n - 1}{n + 1}$ **27.** 0.46   **29.** 0.28   **31.** $L(x) = x^{2.13}$, Gini index ≈ 0.36

33. $4(x^5 - 3x^3 + x - 1)^3(5x^4 - 9x^2 + 1)$ **35.** $\dfrac{4x^3}{x^4 + 1}$ **37.** $3x^2e^{x^3}$

Exercises 5.6 page 425

1. $\frac{1}{10}(x^2 + 1)^{10} + C$ **3.** $\frac{1}{20}(x^2 + 1)^{10} + C$ **5.** $\frac{1}{5}e^{x^5} + C$ **7.** $\frac{1}{6}\ln(x^6 + 1) + C$

9. $u = x^3 + 1, du = 3x^2\,dx$: the powers in the integrand and the du do not match.

11. $u = x^4, du = 4x^3\,dx$: the powers in the integrand and the du do not match.   **13.** $\frac{1}{24}(x^4 - 16)^6 + C$

15. $-\frac{1}{2}e^{-x^2} + C$ **17.** $\frac{1}{3}e^{3x} + C$ **19.** Cannot be found by our substitution formulas

21. $\frac{1}{5}\ln|1 + 5x| + C$ **23.** $\frac{1}{4}(x^2 + 1)^{10} + C$ **25.** $\frac{1}{5}(z^4 + 16)^{5/4} + C$

27. Cannot be found by our substitution formulas   **29.** $\frac{1}{24}(2y^2 + 4y)^6 + C$ **31.** $\frac{1}{2}e^{x^2+2x+5} + C$

33. $\frac{1}{12}\ln|3x^4 + 4x^3| + C$ **35.** $-\frac{1}{12}(3x^4 + 4x^3)^{-1} + C$ **37.** $-\frac{1}{2}\ln|1 - x^2| + C$ **39.** $\frac{1}{16}(2x - 3)^8 + C$

41. $\frac{1}{2}\ln(e^{2x} + 1) + C$ **43.** $\frac{1}{2}(\ln x)^2 + C$ **45.** $2e^{x^{1/2}} + C$ **47.** $\frac{1}{4}x^4 + \frac{1}{3}x^3 + C$

49. $\frac{1}{6}x^6 + \frac{2}{5}x^5 + \frac{1}{4}x^4 + C$ **51.** $\frac{1}{2}e^9 - \frac{1}{2}$ **53.** $\frac{1}{2}\ln 2$ **55.** $32\frac{2}{3}$ **57.** $-\ln 2$ **59.** $3e^2 - 3e$

61. a. u^nu' **b.** $\displaystyle\int u^nu'\,dx$ **63. a.** $\dfrac{u'}{u}$ **b.** $\displaystyle\int \dfrac{u'}{u}\,dx$ **65.** $\dfrac{1}{2}\ln(2x + 1) + 50$ **67.** $\dfrac{1}{2}$ million

⎣— agree —⎦   ⎣—agree—⎦

69. $\frac{1}{3}\ln 5 - \frac{1}{3}\ln 2 \approx 0.305$ million   **71.** $20\frac{1}{3}$ units   **73.** About 346   **75.** $\frac{1}{2}\ln 10 \approx 1.15$ tons

Chapter 5 Review Exercises page 429

1. $8x^3 - 4x^2 + x + C$ **2.** $3x^4 + 3x^2 - 3x + C$ **3.** $4x^{3/2} - 5x + C$ **4.** $6x^{4/3} - 2x + C$
5. $6x^{5/3} - 2x^2 + C$ **6.** $2x^{5/2} - 3x^2 + C$ **7.** $\frac{1}{3}x^3 - 16x + C$ **8.** $x^3 + x^2 + 4x + C$
9. $C(x) = 2x^{1/2} + 4x + 20{,}000$ **10. a.** $P(t) = 200t^{3/2} + 40{,}000$ **b.** 52,800 people
11. $2e^{x/2} + C$ **12.** $-\frac{1}{2}e^{-2x} + C$ **13.** $4\ln|x| + C$ **14.** $2\ln|x| + C$ **15.** $2e^{3x} - 6\ln|x| + C$
16. $\frac{1}{2}x^2 - \ln|x| + C$ **17.** $3x^3 + 2\ln|x| + 2e^{3x} + C$ **18.** $-x^{-1} + \ln|x| - e^{-x} + C$
19. a. $1000e^{0.05t} - 1000$ **b.** About 2044 (from $t \approx 44$)
20. a. $2000e^{0.1t} - 2000$ **b.** About 5.6 years   **21. a.** $450e^{0.02t} - 450$ **b.** About 2034 (from $t \approx 34$)
22. a. $200\ln x$ **b.** In about 20 months   **23.** 36   **24.** 90   **25.** 4   **26.** $\ln 5$
27. $1 - e^{-2}$ **28.** $2e - 2$ **29.** $20e^5 - 100e + 80$ **30.** $25e^{0.4} - 50e^{0.2} + 25$ **31.** 13 square units
32. 36 square units   **33.** $6e^6 - 6$ square units   **34.** $2e^2 - 2$ square units   **35.** $\ln 100$ square units
36. $\ln 1000$ square units
37.

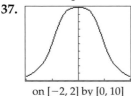

Area ≈ 21.4 square units   **38.**

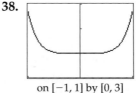

Area ≈ 2.54 square units

on $[-2, 2]$ by $[0, 10]$ on $[-1, 1]$ by $[0, 3]$

39. About 29.5 kilograms   **40.** 9 words   **41.** $1.5e - 1.5 \approx 2.6$ degrees   **42.** $1640   **43.** $653.39
44. About 143 pages   **45. a.** 2.28 square units   **b.** $\frac{8}{3} \approx 2.667$ square units
46. a. 4.884 square units   **b.** $\frac{16}{3} \approx 5.333$ square units
47. a. for left rectangles: 5.899, 7.110, 7.239   **b.** $e^2 - e^{-2} \approx 7.254$ square units
      for midpoint rectangles: 7.206, 7.253, 7.254
      for right rectangles: 8.801, 7.400, 7.268
48. a. for left rectangles: 1.506, 1.398, 1.387   **b.** $\ln 4 \approx 1.386$ square units
      for midpoint rectangles: 1.383, 1.386, 1.386
      for right rectangles: 1.281, 1.375, 1.385
49. $\frac{4}{3}$ square units   **50.** 108 square units   **51.** $\frac{1}{6}$ square unit   **52.** $\frac{3}{10}$ square unit
53. 2 square units   **54.** $\frac{1}{2}$ square unit   **55.** About 17.13 square units   **56.** About 0.496 square unit
57. $\frac{1}{3}\ln 4$ **58.** $\frac{28}{3}$ or $9\frac{1}{3}$ **59.** About 4.72   **60.** About 2.77   **61.** 9.2 billion   **62.** About $5800
63. About $629,000 (from 6.29 hundred thousand dollars)   **64.** About $49.95   **65.** 27 square meters
66. About $934 billion   **67.** About $372 million
68. b.

on $[0, 10]$ by $[0, 35]$

      c. About 93.61 million people   **69.** $480,000

70. $160,000   **71.** About $5623   **72.** About $611   **73.** About 0.56   **74.** About 0.43
75. About 0.23   **76.** About 0.68   **77.** $\frac{1}{4}(x^3 - 1)^{4/3} + C$ **78.** $\frac{1}{6}(x^4 - 1)^{3/2} + C$
79. Cannot be integrated by our substitution formulas
80. Cannot be integrated by our substitution formulas
81. $-\frac{1}{3}\ln|9 - 3x| + C$ **82.** $-\frac{1}{2}\ln|1 - 2x| + C$ **83.** $\frac{1}{3}(9 - 3x)^{-1} + C$ **84.** $\frac{1}{2}(1 - 2x)^{-1} + C$
85. $\frac{1}{2}(8 + x^3)^{2/3} + C$ **86.** $(9 + x^2)^{1/2} + C$ **87.** $-\frac{1}{2}(w^2 + 6w - 1)^{-1} + C$ **88.** $-\frac{1}{2}(t^2 - 4t + 1)^{-1} + C$

89. $\frac{2}{3}(1 + \sqrt{x})^3 + C$ **90.** $(1 + x^{1/3})^3 + C$ **91.** $\ln|e^x - 1| + C$ **92.** $\ln|\ln x| + C$ **93.** $\frac{61}{3}$ or $20\frac{1}{3}$
94. 2   **95.** 2   **96.** $\frac{5}{12}$ **97.** $\ln 7$ **98.** $-\ln 2$ **99.** $\frac{1}{4}e - \frac{1}{4}$ **100.** $\frac{1}{5}e - \frac{1}{5}$ **101.** 8 square units
102. 5 square units   **103.** $\frac{1}{4} - \frac{1}{4}e^{-4} = \frac{1}{4}(1 - e^{-4}) \approx 0.25$ **104.** $\frac{1}{4}\ln 13 \approx 0.64$
105. $C(x) = (2x + 9)^{1/2} + 97$ **106.** $\ln 28 \approx 3.33$ degrees

Exercises 6.1 page 445

1. $\frac{1}{2}e^{2x} + C$ **3.** $\frac{1}{2}x^2 + 2x + C$ **5.** $\frac{2}{3}x^{3/2} + C$ **7.** $\frac{1}{5}(x + 3)^5 + C$ **9.** $\frac{1}{2}xe^{2x} - \frac{1}{4}e^{2x} + C$
11. $\frac{1}{6}x^6 \ln x - \frac{1}{36}x^6 + C$ **13.** $(x + 2)e^x - e^x + C$ **15.** $\frac{2}{3}x^{3/2} \ln x - \frac{4}{9}x^{3/2} + C$
17. $\frac{1}{6}(x - 3)(x + 4)^6 - \frac{1}{42}(x + 4)^7 + C$ **19.** $-2te^{-0.5t} - 4e^{-0.5t} + C$ **21.** $-t^{-1} \ln t - t^{-1} + C$
23. $\frac{1}{10}s(2s + 1)^5 - \frac{1}{120}(2s + 1)^6 + C$ **25.** $-\frac{1}{2}xe^{-2x} - \frac{1}{4}e^{-2x} + C$ **27.** $2x(x + 1)^{1/2} - \frac{4}{3}(x + 1)^{3/2} + C$
29. $\frac{1}{a}xe^{ax} - \frac{1}{a^2}e^{ax} + C$ **31.** $\frac{1}{n + 1}x^{n+1} \ln ax - \frac{1}{(n + 1)^2}x^{n+1} + C$ **33.** $x \ln x - x + C$
35. $\frac{1}{2}x^2e^{x^2} - \frac{1}{2}e^{x^2} + C$ **37. a.** $\frac{1}{2}e^{x^2} + C$ (by substitution)   **b.** $\frac{1}{4}(\ln x)^4 + C$ (by substitution)
      c. $\frac{1}{3}x^3 \ln 2x - \frac{1}{9}x^3 + C$ (by parts)   **d.** $\ln(e^x + 4) + C$ (by substitution)
39. $e^2 + 1$ **41.** $9 \ln 3 - 3 + \frac{1}{9}$ **43.** $\frac{2^6}{30} = \frac{32}{15}$ **45.** $4 \ln 4 - 3$
47. a. $\frac{1}{6}x(x - 2)^6 - \frac{1}{42}(x - 2)^7 + C$ **b.** $\frac{1}{7}(x - 2)^7 + \frac{1}{3}(x - 2)^6 + C$
49. Using $u = x^n$ and $dv = e^x dx$, the result follows immediately.
51. $x^2e^x - 2xe^x + 2e^x + C$ **53. a.** The result follows immediately   **b.** [*Hint:* Think of the C.]
55. $R(x) = 4xe^{x/4} - 16e^{x/4} + 16$ **57.** $105.7 million   **59.** $-14e^{-2.5} + 4 \approx 2.85$ milligrams
61. $2 \ln 2 - 1 + \frac{1}{4} \approx 0.64$ square unit   **63.** $72 \ln 6 - 24 + \frac{1}{9} \approx 105$ thousand customers
65. $-x^2e^{-x} - 2xe^{-x} - 2e^{-x} + C$ **67.** $(x + 1)^2e^x - 2(x + 1)e^x + 2e^x + C$
69. $\frac{1}{3}x^3(\ln x)^2 - \frac{2}{9}x^3 \ln x + \frac{2}{27}x^3 + C$ **71.** $2e^2 - 2 \approx 12.78$
73. $-x^2e^{-x} - 2xe^{-x} - 2e^{-x} + C = -e^{-x}(x^2 + 2x + 2) + C$ **75.** $\frac{1}{2}x^3e^{2x} - \frac{3}{4}x^2e^{2x} + \frac{3}{4}xe^{2x} - \frac{3}{8}e^{2x} + C$
77. $\frac{1}{3}(x - 1)^3e^{3x} - \frac{1}{3}(x - 1)^2e^{3x} + \frac{2}{9}(x - 1)e^{3x} - \frac{2}{27}e^{3x} + C$

Exercises 6.2 page 457

1. Formula 12, $a = 5$, $b = -1$ **3.** Formula 14, $a = -1$, $b = 7$ **5.** Formula 9, $a = -1$, $b = 1$
7. $\frac{1}{6}\ln\left|\frac{3 + x}{3 - x}\right| + C$ **9.** $-\frac{1}{x} - 2\ln\left|\frac{x}{2x + 1}\right| + C$ **11.** $-x - \ln|1 - x| + C$ **13.** $\ln\left|\frac{2x + 1}{x + 1}\right| + C$
15. $\frac{x}{2}\sqrt{x^2 - 4} - 2\ln\left|x + \sqrt{x^2 - 4}\right| + C$ **17.** $-\ln\left|\frac{1 + \sqrt{1 - z^2}}{z}\right| + C$
19. $\frac{1}{2}x^3e^{2x} - \frac{3}{4}x^2e^{2x} + \frac{3}{4}xe^{2x} - \frac{3}{8}e^{2x} + C$ **21.** $-\frac{1}{100}x^{-100} \ln x - \frac{1}{10,000}x^{-100} + C$
23. $\frac{1}{3}\ln\left|\frac{x}{x + 3}\right| + C$ **25.** $\frac{1}{8}\ln\left|\frac{z^2 - 2}{z^2 + 2}\right| + C$ **27.** $\frac{x}{2}\sqrt{9x^2 + 16} + \frac{8}{3}\ln\left|3x + \sqrt{9x^2 + 16}\right| + C$
29. $-\frac{1}{2}\ln\left|\frac{2 + \sqrt{4 - e^{2t}}}{e^t}\right| + C$ **31.** $\frac{1}{2}\ln\left|\frac{e^t - 1}{e^t + 1}\right| + C$ **33.** $\frac{1}{4}\ln\left|x^4 + \sqrt{x^8 - 1}\right| + C$
35. $\frac{1}{3}\ln\left|\frac{\sqrt{x^3 + 1} - 1}{\sqrt{x^3 + 1} + 1}\right| + C$ **37.** $\frac{1}{2}\ln\left|\frac{e^t - 1}{e^t + 1}\right| + C$ **39.** $2xe^{x/2} - 4e^{x/2} + C$

41. $\dfrac{1}{4}\ln\left|\dfrac{e^{-x}+4}{e^{-x}}\right| + C = \dfrac{1}{4}\ln(1 + 4e^x) + C$ **43.** $\dfrac{15}{2} - 8\ln 8 + 8\ln 4 \approx 1.95$

45. $\dfrac{1}{2}\ln\dfrac{1}{2} - \dfrac{1}{2}\ln\dfrac{1}{3} \approx 0.203$ **47.** $-4 + 5\ln 3 \approx 1.49$ **49.** $\dfrac{1}{2}\ln|2x + 6| + C$

51. $\dfrac{x}{2} - \dfrac{3}{2}\ln|2x + 6| + C$ **53.** $-\dfrac{1}{3}(1 - x^2)^{3/2} + C$ **55.** $\sqrt{1 - x^2} - \ln\left|\dfrac{1 + \sqrt{1 - x^2}}{x}\right| + C$

57. $\dfrac{1}{2}\left(\ln|x + 1| - \dfrac{1}{3}\ln|3x + 1|\right) - \dfrac{1}{2}\ln\left|\dfrac{3x + 1}{x + 1}\right| + C$

59. $\ln\left|x + \sqrt{x^2 + 1}\right| - \ln\left|\dfrac{1 + \sqrt{x^2 + 1}}{x}\right| + C$ **61.** $x + 2\ln|x - 1| + C$

63. $-x^2e^{-x} - 2xe^{-x} - 2e^{-x} + 2$ million sales   **65.** 24 generations   **67.** $C(x) = \ln\left(x + \sqrt{x^2 + 1}\right) + 2000$

Exercises 6.3 page 470

1. 0   **3.** 1   **5.** Does not exist   **7.** Does not exist   **9.** $\frac{1}{2}$ **11.** $\frac{1}{8}$ **13.** Divergent   **15.** 100
17. 20   **19.** $\frac{1}{2}$ **21.** Divergent   **23.** $\frac{1}{3}$ **25.** $\frac{1}{3}$ **27.** Divergent   **29.** 1   **31.** Divergent

33. $\displaystyle\int_0^\infty e^{\sqrt{x}}\,dx$ diverges and $\displaystyle\int_0^\infty e^{-x^2}\,dx$ converges to 0.88623   **35.** \$200,000

37. a. \$10,000   **b.** \$9999.55   **39.** About \$3,963,000 (from 3963 thousand)

41. 1,000,000 barrels (from 1000 thousand)   **43.** 2 square units   **45.** $\dfrac{1}{a}$ square units   **47.** 0.61 or 61%

49. 0.30 or 30%   **51.** 20,000   **53.** \$40,992   **55.** D/r

Exercises 6.4 page 485

Some answers may vary depending on rounding.
1. a. 8.75   **b.** 8.667   **c.** 0.083   **d.** 1%   **3. a.** 0.697   **b.** 0.693   **c.** 0.004   **d.** 0.6%
5. 1.154   **7.** 0.743   **9.** 0.593   **11.** 8.6968   **13.** 2.925   **15.** 0.4772 or about 48%
17. \$17,300 (from 17.30 thousand)   **19.** 8.667   **21.** 0.693   **23.** 1.148   **25.** 0.747

27. 0.593   **29.** 8.69678496   **31.** 2.92530   **33. a.** $-\displaystyle\int_1^0 \dfrac{t}{1 + t^3}\,dt = \int_0^1 \dfrac{t}{1 + t^3}\,dt$ **b.** 0.374

35. 821 feet   **37.** The justification follows from carrying out the indicated steps.

Chapter 6 Review Exercises page 489

1. $\frac{1}{2}xe^{2x} - \frac{1}{4}e^{2x} + C$ **2.** $-xe^{-x} - e^{-x} + C$ **3.** $\frac{1}{9}x^9\ln x - \frac{1}{81}x^9 + C$ **4.** $\frac{4}{5}x^{5/4}\ln x - \frac{16}{25}x^{5/4} + C$
5. $\frac{1}{6}(x - 2)(x + 1)^6 - \frac{1}{42}(x + 1)^7 + C$ **6.** $\frac{1}{5}(x + 3)(x - 1)^5 - \frac{1}{30}(x - 1)^6 + C$ **7.** $2t^{1/2}\ln t - 4t^{1/2} + C$
8. $\frac{1}{4}x^4e^{x^4} - \frac{1}{4}e^{x^4} + C$ **9.** $x^2e^x - 2xe^x + 2e^x + C$ **10.** $x(\ln x)^2 - 2x\ln x + 2x + C$

11. $\dfrac{1}{n + 1}x(x + a)^{n+1} - \dfrac{1}{(n + 1)(n + 2)}(x + a)^{n+2} + C$

12. $-\dfrac{1}{n + 1}x(1 - x)^{n+1} - \dfrac{1}{(n + 1)(n + 2)}(1 - x)^{n+2} + C$ **13.** $4e^5 + 1$ **14.** $\frac{1}{4}e^2 + \frac{1}{4}$
15. $-\ln|1 - x| + C$ **16.** $-\frac{1}{2}e^{-x^2} + C$ **17.** $\frac{1}{4}x^4\ln 2x - \frac{1}{16}x^4 + C$ **18.** $(1 - x)^{-1} + C$

19. $\frac{1}{2}(\ln x)^2 + C$ **20.** $\frac{1}{2}\ln(e^{2x} + 1) + C$ **21.** $2e^{\sqrt{x}} + C$ **22.** $\frac{1}{8}(e^{2x} + 1)^4 + C$

23. $-15{,}000e^{-0.5} + 10{,}000 \approx 902$ million dollars **24.** 6.78 hundred gallons (from $25 - 10e^{0.6}$)

25. $\dfrac{1}{10}\ln\left|\dfrac{5 + x}{5 - x}\right| + C$ **26.** $\dfrac{1}{4}\ln\left|\dfrac{x - 2}{x + 2}\right| + C$ **27.** $2\ln|x - 2| - \ln|x - 1| + C$

28. $-\ln\left|\dfrac{x - 1}{x - 2}\right| + C$ or $\ln\left|\dfrac{x - 2}{x - 1}\right| + C$ **29.** $\ln\left|\dfrac{\sqrt{x + 1} - 1}{\sqrt{x + 1} + 1}\right| + C$ **30.** $\dfrac{2x - 4}{3}\sqrt{x + 1} + C$

31. $\ln\left|x + \sqrt{x^2 + 9}\right| + C$ **32.** $\ln\left|x + \sqrt{x^2 + 16}\right| + C$ **33.** $\dfrac{z^2 - 2}{3}\sqrt{z^2 + 1} + C$ (from formula 13)

34. $e^t - 2\ln(e^t + 2) + C$ **35.** $\frac{1}{2}x^2e^{2x} - \frac{1}{2}xe^{2x} + \frac{1}{4}e^{2x} + C$

36. $x(\ln x)^4 - 4x(\ln x)^3 + 12x(\ln x)^2 - 24x\ln x + 24x + C$ **37.** $\ln\left|\dfrac{2x + 1}{x + 1}\right| + 1000$

38. 1305 (from $750 + 800\ln 80 - 800\ln 40$) **39.** $\dfrac{1}{4}$ **40.** $\dfrac{1}{5}$ **41.** Divergent **42.** Divergent

43. $\frac{1}{2}$ **44.** $2e^{-2}$ **45.** Divergent **46.** Divergent **47.** 5 **48.** $10e^{-10}$ **49.** $\frac{1}{4}$ **50.** $\frac{1}{5}$

51. $\frac{1}{2}$ **52.** $\frac{1}{4}$ **53.** 1 **54.** 1 **55.** $\frac{1}{3}$ **56.** $\frac{1}{2}$ **57.** \$60,000 **58.** 0.35 or 35%

59. 240 thousand **60.** 7500 million tons **61.** $\displaystyle\int_1^{\infty}\dfrac{1}{x^3}\,dx$ converges to 0.5 **62.** $\displaystyle\int_1^{\infty}\dfrac{1}{\sqrt[3]{x}}\,dx$ diverges

63. 1.102 **64.** 1.09 **65.** 1.204 **66.** 0.852 **67.** 0.570 **68.** 1.313 **69.** 1.089 **70.** 1.075

71. 1.195 **72.** 0.856 **73.** 0.528 **74.** 1.348 **75.** 1.0894 **76.** 1.0747 **77.** 1.1951

78. 0.8556 **79.** 0.5285 **80.** 1.3357 **81.** 1.089429 **82.** 1.074669 **83.** 1.194958

84. 0.855624 **85.** 0.527887 **86.** 1.347855 **87. a.** $-\displaystyle\int_1^0\dfrac{1}{1 + t^2}\,dt = \int_0^1\dfrac{1}{1 + t^2}\,dt$ **b.** 0.783

88. a. $-\displaystyle\int_1^0\dfrac{1}{1 + t^4}\,dt = \int_0^1\dfrac{1}{1 + t^4}\,dt$ **b.** 0.862

Exercises 7.1 page 504

1. $\{(x, y)\,|\,x \neq 0, y \neq 0\}$ **3.** $\{(x, y)\,|\,x \neq y\}$ **5.** $\{(x, y)\,|\,x > 0, y \neq 0\}$
7. $\{(x, y, z)\,|\,x \neq 0, y \neq 0, z > 0\}$ **9.** 3 **11.** 4 **13.** -2 **15.** 1 **17.** $e^{-1} + e$ **19.** e^{-1}
21. 0 **23.** 0.0157 **25.** 45 minutes **27.** 472.7
29. $P(2L, 2K) = a(2L)^b(2K)^{1-b} = a\underset{2}{\underbrace{2^b L^b 2^{1-b}}}K^{1-b} = 2aL^bK^{1-b} = 2P(L, K)$ **31.** 1548 calls

33. $C(x, y) = 210x + 180y + 4000$ **35. a.** $V = xyz$ **b.** $M = xy + 2xz + 2yz$
37. a.

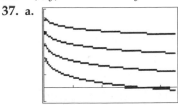

b. A given wind speed will lower the windchill further on a colder day than on a warmer day.

(continues)

c. For the lowest curve: $dy/dx \approx -0.63$, meaning that at 20 degrees and 10 mph of wind, windchill drops by about 0.63 degrees for each additional 1 mph of wind. For the highest curve: $dy/dx \approx -0.33$, meaning that at 50 degrees and 10 mph of wind, windchill drops by only about 0.33 degrees for each additional 1 mph of wind.

d. Yes—the effect of wind on the windchill index is greater on a colder day.

Exercises 7.2 page 519

1. a. $3x^2 + 6xy^2 - 1$ **b.** $6x^2y - 6y^2 + 1$ **3. a.** $6x^{-1/2}y^{1/3}$ **b.** $4x^{1/2}y^{-2/3}$

5. a. $5x^{-0.95}y^{0.02}$ **b.** $2x^{0.05}y^{-0.98}$ **7. a.** $-(x+y)^{-2}$ **b.** $-(x+y)^{-2}$ **9. a.** $\dfrac{3x^2}{x^3+y^3}$ **b.** $\dfrac{3y^2}{x^3+y^3}$

11. a. $6x^2e^{-5y}$ **b.** $-10x^3e^{-5y}$ **13. a.** ye^{xy} **b.** xe^{xy}

15. a. $\dfrac{x}{x^2+y^2}$ or $x(x^2+y^2)^{-1}$ **b.** $\dfrac{y}{x^2+y^2}$ or $y(x^2+y^2)^{-1}$ **17. a.** $3v(uv-1)^2$ **b.** $3u(uv-1)^2$

19. a. $ue^{(u^2-v^2)/2}$ **b.** $-ve^{(u^2-v^2)/2}$ **21.** $18, -10$ **23.** $0, 2e$ **25.** $1\frac{1}{2}$

27. a. $30x - 4y^3$ **b. and c.** $-12xy^2$ **d.** $-12x^2y + 36y^2$

29. a. $-2x^{-5/3}y^{2/3}$ **b. and c.** $2x^{-2/3}y^{-1/3} - 12y^2$ **d.** $-2x^{1/3}y^{-4/3} - 24xy$

31. a. ye^x **b. and c.** $e^x - \dfrac{1}{y}$ **d.** xy^{-2} **33.** All three are $36x^2y^2$.

35. a. y^2z^3 **b.** $2xyz^3$ **c.** $3xy^2z^2$

37. a. $8x(x^2+y^2+z^2)^3$ **b.** $8y(x^2+y^2+z^2)^3$ **c.** $8z(x^2+y^2+z^2)^3$

39. a. $2xe^{x^2+y^2+z^2}$ **b.** $2ye^{x^2+y^2+z^2}$ **c.** $2ze^{x^2+y^2+z^2}$ **41.** -14 **43.** $4e^6$

45. a. $P_x = 4x - 3y + 150$ **b.** \$50 (profit per additional tape deck)   **c.** $P_y = -3x + 6y + 75$

d. \$75 (profit per additional CD player)

47. a. 250 (the marginal productivity of labor is 250, so production increases by about 250 for each additional unit of labor)

b. 108 (the marginal productivity of capital is 108, so production increases by about 108 for each additional unit of capital)   **c.** Labor

49. $S_x = -0.1$ (sales fall by 0.1 for each dollar price increase)

$S_y = 0.4y$ (sales rise by $0.4y$ for each additional advertising dollar above the level y)

51. a. 0.52 (status increases by about 0.52 unit for each additional \$1000 of income)

b. 5.25 (status increases by 5.25 units for each additional year of education)

53. a. 97.2 (skid distance increases by about 97 feet for each additional ton)

b. 12.96 (skid distance increases by about 13 feet for each additional mph)

55. a. Rate at which butter sales change as butter prices rise

b. Negative: as prices rise, sales will fall.

c. Rate at which butter sales change as margarine prices rise

d. Positive: as margarine prices rise, people will switch to butter, so butter sales will rise.

Exercises 7.3 page 531

1. Rel min value: $f = 5$ at $x = 0, \ y = -1$ **3.** Rel min value: $f = -12$ at $x = -2, \ y = 2$

5. Rel max value: $f = 23$ at $x = 5, \ y = 2$ **7.** No rel extreme values [saddle point at $(2, -4)$]

9. No rel extreme values   **11.** Rel min value: $f = 1$ at $x = 0, \ y = 0$

13. Rel min value: $f = 0$ at $x = 0, \ y = 0$

15. Rel max value: $f = 3$ at $x = 1$, $y = -1$ [saddle point at $(-1, -1)$]

17. Rel max value: $f = 17$ at $x = -1$, $y = -2$ [saddle point at $(-1, 2)$]

19. No rel extreme values [saddle point at $(2, 6)$]

21. 10 units of product A, sell for $7000 each; 7 units of product B, sell for $13,000 each. Maximum profit: $22,000

23. a. $P = -0.2x^2 + 16x - 0.1y^2 + 12y - 20$

   b. 40 cars in America, sell for $12,000; 60 cars in Europe, sell for $10,000

25. 6 hours of practice and 1 hour of rest

27. a. $x = 1200$, $p = \$6$, $R = \$7200$ **b.** $x = 800$, $y = 800$, $p = \$4$, revenue $= \$3200$ for each

   c. Duopoly (1600 versus 1200)   **d.** Duopoly

29. Sell the sedans for $19,200, selling 12 per day, and sell the SUVs for $23,200, selling 7 per day.

31. a. $P = -0.2x^2 + 16x - 0.1y^2 + 12y - 0.1z^2 + 8z - 22$ **b.** 40 in America, 60 in Europe, 40 in Asia

33. Rel min value: $f = -1$ at $x = 1$, $y = 1$ [saddle point at $(0, 0)$]

35. Rel max value: $f = 32$ at $x = 4$, $y = 4$ [saddle point at $(0, 0)$]

37. Rel min value: $f = -162$ at $x = 3$, $y = 18$ and at $x = -3$, $y = -18$ [saddle point at $(0, 0)$]

Exercises 7.4 page 543

Note: Your answers may differ slightly depending on the stage at which you do the rounding.

1. $y = 3.5x - 1.67$ **3.** $y = -0.79x + 6.6$ **5.** $y = 2.4x + 6.9$ **7.** $y = -2.1x + 7.6$

9. $y = 2.2x + 5$; prediction: 16 million   **11.** $y = -8x + 125$; prediction: 85 arrests

13. $y = -0.009x + 0.434$; prediction: 0.380   **15.** $y = -2.5x + 38.1$; prediction: 20.6%

17. $y = -0.16x + 71.6$ **19.** $y = 1.09e^{0.63x}$ **21.** $y = 17.45e^{-0.47x}$ **23.** $y = 0.98e^{0.78x}$

25. $y = 16.95e^{-0.52x}$ **27.** $y = 459e^{0.215x}$; predictions: $1,085,000 (from 1085), 1,345,000 (from 1345)

29. $y = 30.33e^{0.198x}$; prediction: $148,000 per second (from 147.8 thousand)

31. c.

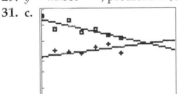

   d. In 2010 (from $x \approx 39.7$ years after 1970)

Exercises 7.5 page 559

1. Max $f = 36$ at $x = 6$, $y = 2$ **3.** Max $f = 144$ at $x = 6$, $y = 4$

5. Max $f = -28$ at $x = 3$, $y = 5$ **7.** Max $f = 6$ at $x = 2$, $y = -1$

9. Max $f = 2$ (from $\ln e^2$) at $x = e$, $y = e$ **11.** Min $f = 45$ at $x = 6$, $y = 3$

13. Min $f = -16$ at $x = -4$, $y = 4$ **15.** Min $f = 52$ at $x = 4$, $y = 6$

17. Min $f = \ln 125$ at $x = 10$, $y = 5$ **19.** Min $f = e^{20}$ at $x = 2$, $y = 4$

21. Max $f = 8$ at $x = 2$, $y = 2$ and at $x = -2$, $y = -2$; Min $f = -8$ at $x = 2$, $y = -2$ and at $x = -2$, $y = 2$

23. Max $f = 18$ at $x = 2$, $y = 8$; Min $f = -18$ at $x = -2$, $y = -8$

25. a. 1000 feet perpendicular to building, 3000 feet parallel to building

   b. $|\lambda| = 1000$; each additional foot of fence adds about 1000 square feet of area

27. $r \approx 3.7$ feet, $h \approx 3.7$ feet
29. End: 14 inches by 14 inches; length $= 28$ inches; volume $= 5488$ cubic inches
31. a. $L = 120$, $K = 20$ **b.** $|\lambda| \approx 1.9$; output increases by about 1.9 for each additional dollar
33. Base: 3 inches by 3 inches; height: 5 inches   **35.** Min $f = 24$ at $x = 4$, $y = 2$, $z = -2$
37. Max $f = 6$ at $x = 2$, $y = 2$, $z = 2$ **39.** Base: 50 feet by 50 feet; height: 100 feet

Exercises 7.6 page 571

1. $df = 2xy^3 \cdot dx + 3x^2y^2 \cdot dy$ **3.** $df = 3x^{-1/2}y^{1/3} \cdot dx + 2x^{1/2}y^{-2/3} \cdot dy$ **5.** $dg = \dfrac{1}{y} \cdot dx - \dfrac{x}{y^2} \cdot dy$

7. $dg = -(x - y)^{-2} \cdot dx + (x - y)^{-2} \cdot dy$ **9.** $dz = \dfrac{3x^2}{x^3 - y^2} \cdot dx - \dfrac{2y}{x^3 - y^2} \cdot dy$

11. $dz = e^{2y} \cdot dx + 2xe^{2y} \cdot dy$ **13.** $dw = (6x^2 + y) \cdot dx + (x + 2y) \cdot dy$

15. $df = 4xy^3z^4 \cdot dx + 6x^2y^2z^4 \cdot dy + 8x^2y^3z^3 \cdot dz$ **17.** $df = \dfrac{1}{x} dx + \dfrac{1}{y} dy + \dfrac{1}{z} dz$

19. $df = yze^{xyz} \cdot dx + xze^{xyz} \cdot dy + xye^{xyz} \cdot dz = e^{xyz}(yz \cdot dx + xz \cdot dy + xy \cdot dz)$ **21. a.** $\Delta f = 0.479$
   b. $df = 0.4$

23. a. $\Delta f \approx 0.112$ **b.** $df = 0.11$ **25. a.** $\Delta f = 0.1407$ **b.** $df = 0.14$
27. 125 square feet; 250 square feet   **29.** \$4300   **31.** About 113 feet   **33.** 2%
35. 0.5 liter per minute
37. a. f is being evaluated at two points along the curve; each of these points gives $f = c$, and $c - c = 0$.
   b. Subtracting and adding $F(x)$.
   c. Approximating the change $\Delta f = f(x + \Delta x, F + \Delta F) - f(x, F)$ by the total differential $df = f_x \Delta x + f_y \Delta F$.
   d. Subtracting $f_y \Delta F$ and dividing by f_y and Δx.
   e. Taking the limit as $\Delta x \to 0$ causes $\dfrac{\Delta F}{\Delta x}$ to approach $\dfrac{dF}{dx}$ and the approximation to become exact.

Exercises 7.7 page 585

1. $2x^9 - 2x$ **3.** $6y^4$ **5.** $2x^2$ **7.** 4   **9.** 2   **11.** $\frac{1}{2}$ **13.** 12
15. 14   **17.** $-9e^{-3} + 9e^3$ **19.** 0   **21.** -12 **23.** 0   **25.** 72   **27.** $\frac{1}{2}$
29. a. $\displaystyle\int_1^3\int_0^2 3xy^2 \, dx \, dy$ and $\displaystyle\int_0^2\int_1^3 3xy^2 \, dy \, dx$ **b.** Both equal 52
31. a. $\displaystyle\int_0^2\int_{-1}^1 ye^x \, dx \, dy$ and $\displaystyle\int_{-1}^1\int_0^2 ye^x \, dy \, dx$ **b.** Both equal $2e - 2e^{-1}$ **33.** 8 cubic units
35. $\frac{4}{3}$ cubic units   **37.** $\frac{1}{6}$ cubic unit   **39.** $\frac{1}{2}e^2 - e + \frac{1}{2}$ cubic units   **41.** 45 degrees $\left(\text{from } \frac{540}{12}\right)$
43. About 180,200 people   **45.** 900,000 cubic feet   **47.** 14   **49.** 10

Chapter 7 Review Exercises page 590

1. $\{ (x, y) \mid x \geq 0, y \neq 0 \}$ **2.** $\{ (x, y) \mid y > 0 \}$ **3.** $\{ (x, y) \mid x \neq 0, y > 0 \}$ **4.** $\{ (x, y) \mid x \neq 0, y > 0 \}$
5. a. $10x^4 - 6xy^3 - 3$ **b.** $-9x^2y^2 + 4y^3 + 2$ **c. and d.** $-18xy^2$
6. a. $12x^3 + 15x^2y^2 - 6$ **b.** $10x^3y - 6y^5 + 1$ **c. and d.** $30x^2y$

7. a. $12x^{-1/3}y^{1/3}$ **b.** $6x^{2/3}y^{-2/3}$ **c. and d.** $4x^{-1/3}y^{-2/3}$

8. a. $\dfrac{2x}{x^2 + y^3}$ **b.** $\dfrac{3y^2}{x^2 + y^3}$ **c. and d.** $\dfrac{-6xy^2}{(x^2 + y^3)^2}$

9. a. $3x^2e^{x^3-2y^3}$ **b.** $-6y^2e^{x^3-2y^3}$ **c. and d.** $-18x^2y^2e^{x^3-2y^3}$

10. a. $6xe^{-5y}$ **b.** $-15x^2e^{-5y}$ **c. and d.** $-30xe^{-5y}$

11. a. $-ye^{-x} - \ln y$ **b.** $e^{-x} - \dfrac{x}{y}$ **c. and d.** $-e^{-x} - \dfrac{1}{y}$

12. a. $2xe^y + yx^{-1}$ **b.** $x^2e^y + \ln x$ **c. and d.** $2xe^y + x^{-1}$ **13. a.** $\frac{1}{2}$ **b.** $\frac{1}{2}$

14. a. 0 **b.** $\frac{1}{2}$ **15. a.** 36 **b.** -24 **16. a.** 216 **b.** -216

17. a. 80: rate at which production increases for each additional unit of labor

b. 135: rate at which production increases for each additional unit of capital   **c.** Capital

18. $S_x = 222$: rate at which sales increase for each additional $1000 in TV ads.

$S_y = 528$: rate at which sales increase for each additional $1000 in print ads.

19. Min $f = -13$ at $x = -1$, $y = -4$ **20.** Min $f = -8$ at $x = 4$, $y = 1$

21. Max $f = 8$ at $x = 0$, $y = -1$ **22.** Max $f = 6$ at $x = 1$, $y = 0$

23. No rel extreme values $\left(\text{saddle point at } x = \frac{1}{2},\ y = -3\right)$

24. No rel extreme values $\left(\text{saddle point at } x = -\frac{1}{2},\ y = 1\right)$ **25.** Max $f = 1$ at $x = 0$, $y = 0$

26. Min $f - 1$ at $x = 0$, $y = 0$ **27.** Min $f = 0$ at $x = 0$, $y = 0$

28. Min $f = \ln 10$ at $x = 0$, $y = 0$

29. Max $f = 25$ at $x = -2$, $y = -3$ (saddle point at $x - 2$, $y = 3$)

30. Min $f = -20$ at $x = -2$, $y = 2$ (saddle point at $x = 2$, $y = 2$)

31. a. $C(x, y) = 3000x + 5000y + 6000$ **b.** $R(x, y) = 7000x - 20x^2 + 8000y - 30y^2$

c. $P(x, y) = -20x^2 + 4000x - 30y^2 + 3000y - 6000$

d. Make 100 18-foot boats, sell for $5000 each, and 50 22-foot boats, sell for $6500 each;

max profit: $269,000.

32. a. $P(x, y) = -0.2x^2 + 68x - 0.1y^2 + 52y - 100$

b. America: sell 170 for $46,000 each; Europe: sell 260 for $38,000 each (since prices are in thousands)

33. $y = 2.6x - 3.2$ **34.** $y = -1.8x + 8.4$ **35.** $y = 4.71x + 11.51$; prediction: 39.8 million

36. $y = -0.76x + 8.2$; prediction: 2.9% (from 2.88)   **37.** Max $f = 292$ at $x = 12$, $y = -24$

38. Max $f = 156$ at $x = 10$, $y = 8$ **39.** Min $f = 90$ at $x = 7$, $y = 4$

40. Min $f = -109$ at $x = -3$, $y = 3$ **41.** Min $f = e^{45}$ at $x = 3$, $y = 6$

42. Max $f = e^{-5}$ at $x = 2$, $y = 1$

43. Max $f = 120$ at $x = 2$, $y = -6$; Min $f = -120$ at $x = -2$, $y = 6$

44. Max $f = 64$ at $x = 4$, $y = 4$ and at $x = -4$, $y = -4$;

Min $f = -64$ at $x = 4$, $y = -4$ and at $x = -4$, $y = 4$

45. a. $40,000 for production, $20,000 for advertising

b. $|\lambda| \approx 159$: production increases by about 159 units for each additional dollar

46. a. $\frac{1}{2}$ ounce of the first and 7 ounces of the second

b. $|\lambda| = 9$: each additional dollar results in about 9 additional nutritional units

47. a. $L = 64$, $K = 8$ **b.** $40L^{-1/3}K^{1/3}$, $20L^{2/3}K^{-2/3}$

c. $\dfrac{40L^{-1/3}K^{1/3}}{20L^{2/3}K^{-2/3}} = \dfrac{25}{100}$ (and now simplify and substitute $L = 64$, $K = 8$)

48. Base: 12 inches by 12 inches; height: 4 inches   **49.** $df = (6x + 2y) \cdot dx + (2x + 2y) \cdot dy$

50. $df = (2x + y) \cdot dx + (x - 6y) \cdot dy$ **51.** $dg = \dfrac{1}{x}dx + \dfrac{1}{y}dy = \dfrac{dx}{x} + \dfrac{dy}{y}$

52. $dg = \dfrac{3x^2}{x^3 + y^3}\,dx + \dfrac{3y^2}{x^3 + y^3}\,dy$ **53.** $dz = e^{x-y}\,dx - e^{x-y}\,dy$ **54.** $dz = ye^{xy}\,dx + xe^{xy}dy$

55. Sales would decrease by about \$153,000 (from $dS = -153$); sales would decrease by about \$76,500.
56. 2% **57.** $8e^2 - 8e^{-2}$ **58.** 10 **59.** $\frac{4}{3}$ **60.** $\frac{32}{3}$ or $10\frac{2}{3}$ **61.** 40 cubic units

62. 24 cubic units **63.** $\frac{5}{6}$ cubic unit **64.** $\frac{8}{9}$ cubic unit **65.** $12{,}000\left(\text{from } \dfrac{192{,}000}{16}\right)$

66. \$640 hundred thousand, or \$64,000,000

Cumulative Review for Chapters 1–7 page 593

1.

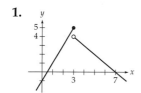

2. 4 **3.** $\dfrac{-1}{x^2}$ (but found using the *definition*) **4.** 4

5. $S'(12) = -2$: each \$1 price increase (above \$12) decreases sales by 2 per week
6. $3[x^2 + (2x + 1)^4]^2[2x + 8(2x + 1)^3]$
7.

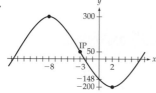

8.

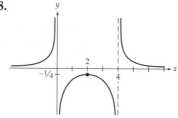

9. 40 feet parallel to wall, 20 feet perpendicular to wall
10. Base: 6 feet by 6 feet; height: 3 feet **11.** $\dfrac{2}{\pi} \approx 0.64$ foot per minute **12. a.** \$1268.24 **b.** \$1271.25

13. In about 14.4 years **14.** About 6.8 years
15.

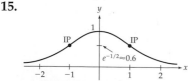

16. $4x^3 - 2x^2 + x + C$ **17.** $900e^{0.02t} - 900$ million gallons

18. $85\frac{1}{3}$ square units **19.** 16 **20. a.** $\frac{1}{3}\ln|x^3 + 1| + C$ **b.** $2e^{x^{1/2}} + C$

21. $\frac{1}{4}xe^{4x} - \frac{1}{16}e^{4x} + C$ **22.** $\sqrt{4 - x^2} - 2\ln\left|\dfrac{2 + \sqrt{4 - x^2}}{x}\right| + C$ **23.** $\dfrac{1}{2}$

24. 1.15148 [compared with actual (rounded) value of 1.14779]
25. 1.14778 [compared with actual (rounded) value of 1.14779]

26. a. $\{(x, y) \mid x \ge 0, y > 0, \ x \ne y\}$ **b.** 2   **27.** $f_x = \ln y + 2ye^{2x}, \ f_y = \dfrac{x}{y} + e^{2x}$

28. Min $f = 2$ at $x = 1, \ y = 4$; no relative max   **29.** $y = 1.75x - 4.75$

30. Min $f = 90$ at $x = 4, \ y = 7$ **31.** $df = (4x + y)\, dx + (x - 6y)\, dy$ **32.** 48 cubic units

Exercises 8.1 page 609

1. $90°, \dfrac{\pi}{2}$ **3.** $45°, \dfrac{\pi}{4}$ **5.** $540°, 3\pi$ **7.** $-45°, -\dfrac{\pi}{4}$ **9. a.** $\dfrac{\pi}{6}$ **b.** $\dfrac{5\pi}{4}$ **c.** $-\pi$

11. a. $\dfrac{\pi}{3}$ **b.** $\dfrac{7\pi}{4}$ **c.** $-\dfrac{2\pi}{3}$ **13. a.** 3π **b.** $\dfrac{3\pi}{4}$ **c.** $\dfrac{\pi}{180}$ **15. a.** $45°$ **b.** $120°$ **c.** $-150°$

17. a. $60°$ **b.** $225°$ **c.** $-210°$ **19. a.** $405°$ **b.** $-540°$ **c.** $\dfrac{900°}{\pi}$ **21.** $3\pi \approx 9.4$ inches

23. $\dfrac{3}{2}$ radians   **25.** 3600 miles   **27. i.** $(a + b)^2 = a^2 + 2ab + b^2$ **ii.** $c^2 + 4 \cdot \frac{1}{2}ab = c^2 + 2ab$

iii. $a^2 + 2ab + b^2 = c^2 + 2ab$ so $a^2 + b^2 = c^2$

Exercises 8.2 page 662

1. $\sin \theta = \frac{24}{25}$ **3.** $\sin \theta = \frac{5}{13}$ **5.** $\sin \theta = \frac{21}{29}$ **7.** $\sin \theta = -\frac{3}{5}$ $\left(\begin{array}{c}\text{after} \\ \text{simplification}\end{array}\right)$

$\cos \theta = \frac{7}{25}$ $\cos \theta = \frac{12}{13}$ $\cos \theta = -\frac{20}{29}$ $\cos \theta = -\frac{4}{5}$

9. $\sin \theta = -\frac{7}{25}$ $\left(\begin{array}{c}\text{after} \\ \text{simplification}\end{array}\right)$ **11. a.** $\dfrac{\sqrt{3}}{2}$ **b.** $\dfrac{1}{2}$ **c.** $\dfrac{\sqrt{2}}{2}$ **d.** $\dfrac{\sqrt{2}}{2}$

$\cos \theta = \frac{24}{25}$

13. a. $-\dfrac{\sqrt{2}}{2}$ **b.** $-\dfrac{\sqrt{2}}{2}$ **c.** $\dfrac{1}{2}$ **d.** $-\dfrac{\sqrt{3}}{2}$

15. a. $-\dfrac{\sqrt{3}}{2}$ **b.** $\dfrac{1}{2}$ **c.** $-\dfrac{\sqrt{2}}{2}$ **d.** $\dfrac{\sqrt{2}}{2}$ **17. a.** 0.31   **b.** 0.95   **c.** 0.84   **d.** 0.54

19. 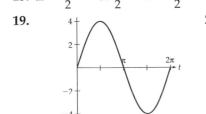   **21.** 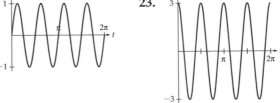   **23.**

25. 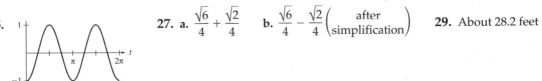   **27. a.** $\dfrac{\sqrt{6}}{4} + \dfrac{\sqrt{2}}{4}$ **b.** $\dfrac{\sqrt{6}}{4} - \dfrac{\sqrt{2}}{4}$ $\left(\begin{array}{c}\text{after} \\ \text{simplification}\end{array}\right)$ **29.** About 28.2 feet

31. About 299 feet   **33.** 1219 cones   **35. a.** 5.3 liters   **b.** 4.5 liters

37. a.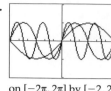

on $[-2\pi, 2\pi]$ by $[-2, 2]$

b.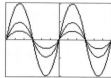

on $[-2\pi, 2\pi]$ by $[-2, 2]$

39. a.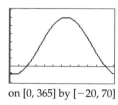

on $[0, 365]$ by $[-20, 70]$

b. $-12°$ **c.** $62°$ **41. c.** 395 ppm

43. The identities follow from the indicated arguments.

Exercises 8.3 page 637

1. $3t^2 \sin t + t^3 \cos t$ **3.** $\dfrac{-t \sin t - \cos t}{t^2}$ **5.** $3t^2 \cos(t^3 + 1)$ **7.** $5 \cos 5t$ **9.** $-6 \sin 3t$

11. $-\dfrac{\pi}{180} \sin\left(\dfrac{\pi}{180} t\right)$ **13. a.** $2t + \pi \cos \pi t$ **b.** π **15. a.** $-2 \sin \dfrac{t}{2}$ **b.** -1

17. a. $-\cos(\pi - t)$ **b.** 0 **19.** $-3(t + \pi)^2 \sin(t + \pi)^3$ **21.** $-3[\cos^2(t + \pi)][\sin(t + \pi)]$

23. $-2(\pi - t) \cos(\pi - t)^2$ **25.** $6t^2(\sin t^3)(\cos t^3)$ **27.** $-\frac{1}{2}(x + 1)^{-1/2} \sin(x + 1)^{1/2}$

29. $4 \cos x + 4 \cos 4x + 4(\sin^3 x)(\cos x)$ **31.** $e^x \cos e^x$ **33.** $(1 - \sin z) e^{z + \cos z}$ **35.** $4(1 - \cos z)^3 \sin z$

37. a. $\dfrac{\cos z}{\sin z}$ **b.** 0 **39.** $\left(1 + \dfrac{1}{z}\right) \cos(z + \ln z)$ **41.** $14 \sin x \cos x \left(\begin{array}{c}\text{after}\\\text{simplification}\end{array}\right)$

43. $-x^{-2} \cos x^2 - 2 \sin x^2$ **45.** $-e^{-x} \sin e^x + \cos e^x$ **47. a.** $\dfrac{-1}{\sin^2 x} \left(\begin{array}{c}\text{after}\\\text{simplification}\end{array}\right)$ **b.** $-\dfrac{4}{3}$

49. $4 \sin t \cos t \left(\begin{array}{c}\text{after}\\\text{simplification}\end{array}\right)$ **51. a.** $\dfrac{1 + \sin t}{\cos^2 t} \left(\begin{array}{c}\text{after}\\\text{simplification}\end{array}\right)$ **b.** 1 **53.** $-\cos(\cos x) \sin x$

55. $-\left(\cos \dfrac{x}{2} - 1\right) \sin \dfrac{x}{2}$ **57.** $x^2 \cos x$ **59.** $4t \sin(t^2 + 1) \cos(t^2 + 1)$

61. $2 \cos t - t \sin t \left(\begin{array}{c}\text{after}\\\text{simplification}\end{array}\right)$ **63.** $-(\sin x) e^{\sin x} + (\cos^2 x) e^{\sin x}$ **65.** $\cos z - z \sin z$

67. The formula follows from the indicated steps.   **69.** Use $f = t$ and $f' = 1$

71. The identity follows by differentiation, as indicated.

73. Increasing by about 45 sales per week   **75.** $\dfrac{\pi}{2}$ or $90°$

77. The result follows from the indicated steps.   **79. a.** $x \approx 0.7854$ (radian)   **b.** $45°$

81. a.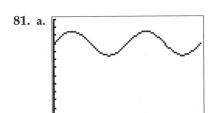

b. At $t = 0.2$ and $t = 1$ second, blood pressure is 105.

c. At $t = 0.6$ and $t = 1.4$ second, blood pressure is 75.

Exercises 8.4 page 648

1. $\sin t + \cos t + C$ **3.** $x - \cos x + C$ **5.** $\frac{1}{2}\sin(t^2 + 2t) + C$ **7.** $-\frac{1}{\pi}\cos \pi t + C$

9. $\frac{2}{\pi}\sin\frac{\pi t}{2} + C$ **11.** $\cos(\pi - t) + C$ **13.** $\frac{365}{2\pi}\sin\frac{2\pi(t + 20)}{365} + C$ **15.** $\frac{1}{3}\sin^3 t + C$

17. $\frac{1}{2}\cos^{-2}t + C$ **19.** $e^{1+\sin x} + C$ **21.** $-\ln|\cos t| + C$ **23.** $2(1 - \cos w)^{1/2} + C$

25. $(1 + \sin^2 y)^{1/2} + C$ **27.** $\frac{1}{3}x^3 - \ln|x| + \frac{1}{2}\sin 2x + C$ **29.** $\frac{1}{\pi} \approx 0.318$ **31.** $\frac{1}{2}\pi^2 \approx 4.935$

33. $\frac{1}{2} + \frac{2}{\pi} \approx 1.137$ **35.** -1 **37.** 2 square units   **39.** 2 square units

41. $\frac{d}{dt}\left(-\frac{1}{a}\cos at + C\right) = -\frac{1}{a}(-\sin at)\cdot a = \sin at$ and similarly for the other formula

43. \$204,000 (from $130 + \frac{234}{\pi}$, rounded to the nearest thousand)

45. a. About 101 tons $\left(\text{from } 90 + \frac{36}{\pi}\right)$ **b.** About 5.6 tons per month   **47.** 3600 birds

49. About \$6668   **51.** About 277 guests per week   **53. a.** About 32.3°   **b.** About 75.6°

Exercises 8.5 page 663

1. a. $\frac{5}{12}$ **b.** $\frac{12}{5}$ **c.** $\frac{13}{12}$ **d.** $\frac{13}{5}$ **3. a.** $\frac{8}{15}$ **b.** $\frac{15}{8}$ **c.** $-\frac{17}{15}$ **d.** $-\frac{17}{8}$

5. a. $\frac{1}{\sqrt{3}} = \frac{\sqrt{3}}{3}$ **b.** 2   **7. a.** Undefined   **b.** -1 **9. a.** 1   **b.** 1   **11. a.** $\frac{1}{\sqrt{3}} = \frac{\sqrt{3}}{3}$ **b.** -2

13. 1.376 (approx)   **15.** 3.864 (approx)   **17.** About 56 feet

19. Slope $= \frac{\Delta y}{\Delta x} = \frac{\text{opp}}{\text{adj}} = \tan\theta$ from diagram

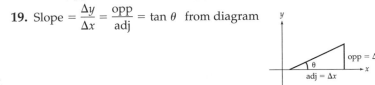

21. Yes, it will cross the goal post at a height of 14.2 feet.   **23.** 51.3° (for a height of 20.25 feet)

25.

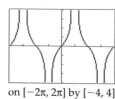

Period 2π; undefined at $\pm\dfrac{\pi}{2}, \pm\dfrac{3\pi}{2}, \ldots$

on $[-2\pi, 2\pi]$ by $[-4, 4]$

27. Result follows from the hint and the definitions of $\tan t$ and $\sec t$.   **29.** $\cot t - t \csc^2 t$

31. $3t^2 \sec^2(t^3 + 1)$ **33.** $\pi \sec(\pi x + 1) \tan(\pi x + 1)$ **35.** $-6z^2 \cot z^3 \csc^2 z^3$

37. Because of the identity $1 + \tan^2 t = \sec^2 t$ (differentiating each side)

39. Follows from the Generalized Power Rule (and the definitions of $\sec t$ and $\tan t$)

41. $S'(10) \approx -3.5$ (be sure your calculator is set for radians). Shadow shortens by about 3.5 feet for each additional degree of sun elevation (near $10°$ elevation).

43. $\dfrac{1}{3}\tan(t^3 + 1) + C$ **45.** $-\dfrac{1}{\pi}\csc \pi t + C$ **47.** $\ln|\cos(1 - t)| + C$ **49.** $\dfrac{1}{5}\ln|\csc x^5 - \cot x^5| + C$

51. $-\dfrac{1}{5}\cot^5 x + C$ **53.** $-\ln\dfrac{\sqrt{2}}{2} \approx 0.347$ **55.** $\ln(\sqrt{2} + 1) \approx 0.881$ square unit

57. a. $\dfrac{1}{2}\tan^2 t + C$ **b.** $\dfrac{1}{2}\sec^2 t + C$ **c.** The equality $\tan^2 t + 1 = \sec^2 t$ shows that $\dfrac{1}{2}\tan^2 t$ and $\dfrac{1}{2}\sec^2 t$ differ by only a constant, so the two answers are the same (because of the $+C$).

59. $-\ln|\cos t| + C$, $\ln|\sin t| + C$ **61.** $\ln|\csc t - \cot t| + C$

Chapter 8 Review Exercises page 669

1. a. $\dfrac{5\pi}{6}$ **b.** $20°$ **2. a.** $\dfrac{2\pi}{3}$ **b.** $36°$ **3.** $\dfrac{104\pi}{3} \approx 108.853$ feet   **4.** $39\pi \approx 122.46$ feet

5. a. $\dfrac{\sqrt{3}}{2}$ **b.** $-\dfrac{1}{2}$ **c.** $-\dfrac{\sqrt{2}}{2}$ **d.** $-\dfrac{\sqrt{2}}{2}$ **6. a.** $-\dfrac{1}{2}$ **b.** $-\dfrac{\sqrt{3}}{2}$ **c.** $\dfrac{\sqrt{2}}{2}$ **d.** $-\dfrac{\sqrt{2}}{2}$

7.

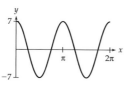

8.

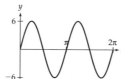

9. 24 feet   **10.** 20 feet

11. a.

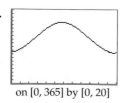

on $[0, 365]$ by $[0, 20]$

b. About 14.6 hours   **c.** About 8.7 hours

12. Daylight is maximized on day 172 (June 21) with 16 hours of daylight, and minimized on day 355 (December 21) with 8.5 hours of daylight.

13. $-8\sin(4t - 1)$ **14.** $-3\cos(2 - 3t)$ **15.** $x^3 \cos x$ **16.** $\dfrac{\cos t + \sin t + 1}{(1 + \cos t)^2}$ $\left(\begin{array}{c}\text{after} \\ \text{simplification}\end{array}\right)$

17. $6t \sin^2 t^2 \cdot \cos t^2$ **18.** $2x\, e^{x^2} \cos e^{x^2}$ **19.** $\dfrac{1}{2}(\sin t)^{-1/2} \cos t$ **20.** $\dfrac{1}{2}t^{-1/2} \cos t^{1/2}$

21. a. 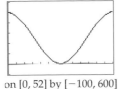 **b.** Increasing by about 29 sales per week

on $[0, 52]$ by $[-100, 600]$

22. Week 39, sales increasing by about 30 sales per week

23. a. 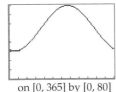 **b.** Decreasing by about 0.41 degree per day (from -0.411)

on $[0, 365]$ by $[0, 80]$

24. Day 111 (April 21), mean temperature rising by 0.41 degree per day   **25.** $\dfrac{\pi}{4}$ or $45°$

26. Follows from setting $R' = 0$ and solving   **27.** $\dfrac{5}{\pi} \sin \dfrac{\pi t}{5} + C$ **28.** $-\dfrac{1}{2}\cos(t^2 + 1) + C$

29. $\cos(\pi - t) + C$ **30.** $\cos e^{-x} + C$ **31.** $-e^{\cos x} + C$ **32.** $2(\sin t)^{1/2} + C$

33. $\frac{1}{3}\ln(1 + \sin 3t) + C$ **34.** $\dfrac{1}{4}x^4 + e^{-x} - \dfrac{1}{\pi}\sin \pi x + C$ **35.** $\sin(\ln y) + C$ **36.** $\cos z^{-1} + C$

37. $\dfrac{3}{4}$ **38.** $-\dfrac{1}{24}$ **39.** $\dfrac{2}{3\pi} \approx 0.212$ **40.** 200 **41.** $\frac{1}{3}$ square unit   **42.** $\sqrt{2} - 1 \approx 0.414$ square unit

43. a. 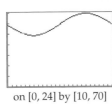 **b.** \$650,000 (from 650 thousand)

on $[0, 52]$ by $[-10, 50]$

44. a. **b.** About $53.6°$ **45. a.** $\sqrt{3}$ **b.** $\dfrac{1}{\sqrt{3}} = \dfrac{\sqrt{3}}{3}$ **c.** 2 **d.** $\dfrac{2}{\sqrt{3}} = \dfrac{2\sqrt{3}}{3}$

on $[0, 24]$ by $[10, 70]$

46. a. -1 **b.** -1 **c.** $\dfrac{-2}{\sqrt{2}} = -\sqrt{2}$ **d.** $\dfrac{2}{\sqrt{2}} = \sqrt{2}$ **47.** About 503 feet   **48.** About 220 feet

49. $\tan x^2 + 2x^2 \sec^2 x^2$ **50.** $2x \cot x - x^2 \csc^2 x$ **51.** $\csc(\pi - t)\cot(\pi - t)$

52. $\frac{1}{2}t^{-1/2}\sec t^{1/2}\tan t^{1/2}$ **53.** $(\pi \sec^2 \pi t)e^{\tan \pi t}$

54. The identity $1 + \cot^2 t = \csc^2 t$ shows that the functions differ by only a constant, so the derivatives must agree (the derivative of each is $-2 \cot t \csc^2 t$).

55. $\frac{2}{\pi}\tan\frac{\pi t}{2} + C$ **56.** $-\frac{1}{2}\ln|\cos x^2| + C$ **57.** $\frac{1}{3}\sec t^3 + C$ **58.** $\csc(\pi - t) + C$

59. $-\frac{2}{3}(\cot x)^{3/2} + C$ **60.** $2(\tan x)^{1/2} + C$ **61.** $\ln(\sqrt{2} + 1) \approx 0.881$

62. 1 **63.** $-\ln\frac{\sqrt{2}}{2} \approx 0.347$ square unit **64.** $-\ln(\sqrt{2} - 1) \approx 0.881$ square unit

65. 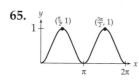 **66.** **67. a.** $y' = \frac{2\cos 2x}{\sin y}$ **b.** -2

68. a. $y' = -\frac{4\sin 4x}{\cos y}$ **b.** -4 **69.** $\frac{1}{80}$ radian per second **70.** 15 radians per hour

71. a. $2\cos 2x + y\sin x$ **b.** $-\cos x$ **72. a.** $-\frac{\sin(x - y)}{\cos(x - y)}$ **b.** $\frac{\sin(x - y)}{\cos(x - y)}$

73. a. $(\cos x)e^{\cos y}$ **b.** $-(\sin x)(\sin y)e^{\cos y}$ **74. a.** $3x^2y\cos x^3y$ **b.** $x^3\cos x^3y$
75. $-t\cos t + \sin t + C$ **76.** $t\sin t + \cos t + C$ **77.** $x\tan x + \ln|\cos x| + C$
78. $x\sec x - \ln|\sec x + \tan x| + C$ **79.** $\frac{1}{2}e^t(\sin t - \cos t)$ **80.** $\frac{1}{2}e^t(\sin t + \cos t)$
81. 6.209 **82.** 1.198 **83.** 6.208758036 **84.** 1.198

Exercises 9.1 page 690

1. Check that $(4e^{2x} - 3e^x) - 3(2e^{2x} - 3e^x) + 2(e^{2x} - 3e^x + 2) \overset{?}{=} 4$ **3.** Check that $kae^{ax} \overset{?}{=} a\left(ke^{ax} - \frac{b}{a}\right) + b$

5. $y = \sqrt[3]{6x^2 + c}$ **7.** Not separable **9.** $y = ce^{2x^3}$ Check that $c6x^2e^{2x^3} \overset{?}{=} 6x^2(ce^{2x^3})$

11. $y = cx$ (since $e^{\ln x} = x$) Check that $c \overset{?}{=} \frac{cx}{x}$ **13.** $y = \sqrt{4x^2 + c}$ and $y = -\sqrt{4x^2 + c}$

15. Not separable **17.** $y = 3x^3 + C$ **19.** $y = \frac{1}{2}\ln(x^2 + 1) + C$ **21.** $y = ce^{x^3/3}$

23. $y = \left(\frac{1 - n}{m + 1}x^{m+1} + c\right)^{1/(1-n)}$ **25.** $y = (x + c)^2$ **27.** Not separable **29.** $y = ce^{x^2/2} - 1$

31. $y = ce^{e^x} + 1$ **33.** $y = \frac{1}{c - ax}$ **35.** $y = ce^{ax} - \frac{b}{a}$

37. $y = \sqrt[3]{3x^2 + 8}$ Check that $(3x^2 + 8)^{\frac{2}{3}} \cdot \frac{1}{3}(3x^2 + 8)^{-\frac{2}{3}} \cdot 6x \overset{?}{=} 2x$ and $y(0) = \sqrt[3]{8} = 2$

39. $y = -e^{x^2/2}$ Check that $-xe^{x^2/2} \overset{?}{=} x(-e^{x^2/2})$ and $y(0) = -e^0 = -1$

41. $y = (1 - x^2)^{-1}$ Check that $-(1 - x^2)^{-2}(-2x) \overset{?}{=} 2x[(1 - x^2)^{-1}]^2$ and $y(0) = (1 - 0)^{-1} = 1$

43. $y = 3x$ (using $e^{\ln x} = x$) Check that $3 \overset{?}{=} \frac{3x}{x}$ and $y(1) = 3 \cdot 1 = 3$

45. $y = (x + 1)^2$ or $y = (x - 3)^2$ Check that $2(x + 1) \overset{?}{=} 2\sqrt{(x + 1)^2}$ and $y(1) = 2^2 = 4$ and that
$2(x - 3) \overset{?}{=} 2\sqrt{(x - 3)^2}$ and $y(1) = (-2)^2 = 4$

47. $y = \dfrac{1}{2 - e^x - x}$ Check that $\dfrac{e^x + 1}{(2 - e^x - x)^2} \stackrel{?}{=} \left(\dfrac{1}{2 - e^x - x}\right)^2 e^x + \left(\dfrac{1}{2 - e^x - x}\right)^2$ and $y(0) = \dfrac{1}{2 - 1} = 1$

49. $y = 2e^{ax^3/3}$ Check that $2ax^2 e^{ax^3/3} \stackrel{?}{=} ax^2 2e^{ax^3/3}$ and $y(0) = 2e^0 = 2$

51. $D(p) = cp^{-k}$ (for any constant c) **53.** $y = 20{,}000e^{0.05t} - 20{,}000$ **55. a.** $y = 28.6e^{-0.32t} + 70$
b. About 3.28 hours

57. $y = 150 - 150e^{-0.2t}$ **59. a.** $y' = 3 + 0.10y$ **b.** $y(0) = 6$ **c.** $y(t) = 36e^{0.1t} - 30$
d. $y(25) = 408.570$ thousand dollars, or \$408,570 (rounded)

61. a. $y' = 4y^{1/2}$ **b.** $y(0) = 10{,}000$ **c.** $y(t) = (2t + 100)^2$ **d.** 15,376

63. a. $y' = 8y^{3/4}$ **b.** $y(0) = 10{,}000$ **c.** $y(t) = (2t + 10)^4$ **d.** 234,256

65. a. $p(t) = Ce^{-Kt/R}$ for $t_0 \leq t \leq T$ **b.** $p(t) = p_0 e^{-K(t-t_0)/R}$ for $t_0 \leq t \leq T$
c. $p(t) = I_0 R - Ce^{-Kt/R}$ for $0 \leq t \leq t_0$ **d.** $p(t) = I_0 R - (I_0 R - p_0)e^{K(t_0-t)/R}$ for $0 \leq t \leq t_0$

67. c. $y = \sqrt[5]{10x^3 + 32}$ **69. c.** $y = \sqrt[4]{8x^2 + 16}$
d. 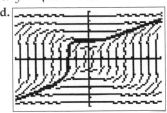 **d.**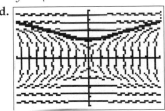

71. a. 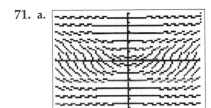 **b.**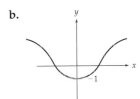

Exercises 9.2 page 706

1. $y' = cae^{at} = a(ce^{at}) = ay$ **3.** Unlimited **5.** Limited **7.** None **9.** Logistic **11.** Logistic
$y(0) = ce^0 = c$

13. $y = 1.5e^{6t}$ **15.** $y = 100e^{-t}$ **17.** $y = -e^{-0.45t}$ **19.** $y = 100(1 - e^{-2t})$ **21.** $y = 0.25(1 - e^{-0.05t})$

23. $y = 40(1 - e^{-2t})$ **25.** $y = 200(1 - e^{-0.01t})$ **27.** $y = \dfrac{100}{1 + 9e^{-500t}}$ **29.** $y = \dfrac{0.5}{1 + 4e^{-0.125t}}$

31. $y = \dfrac{10}{1 - \frac{1}{2}e^{-30t}} = \dfrac{20}{2 - e^{-30t}}$ **33.** $y = \dfrac{3}{1 + 2e^{-6t}}$

35. $y' = 0.08y$ **37.** $y' = a(100{,}000 - y)$ **39.** $y' = a(5000 - y)$
$y = 1500e^{0.08t}$ $y = 100{,}000(1 - e^{-0.021t})$ $y = 5000(1 - e^{-0.223t})$
About 22,276 About 7.2 weeks

41. $y' = ay(10{,}000 - y)$ **43.** $y' = ay(800 - y)$ **45.** $y' = ay(800 - y)$ **47.** $y = 5e^{-0.15t}$
$y = \dfrac{10{,}000}{1 + 99e^{-0.535t}}$ $y = \dfrac{800}{1 + 799e^{-0.558t}}$ $y = \dfrac{800}{1 + 7e^{-0.280t}}$ About 3.7
About 8612 About 675 people About 6.9 years

49. a. $y' = 0.1(200 - y)$ with $M = 200$ **b.** $y = 200(1 - e^{-0.1t})$
c. 30 years [from solving $200(1 - e^{-0.1t}) = 0.95 \cdot 200$]

51. $y = \dfrac{1}{c - at}$ **53.** $y = ce^{a \ln x} = cx^a$ **55.** The solution follows from the indicated steps.

57. a. About 16.6 feet per second **b.** About 0.6 foot per second **c.** About 0.006 foot per second
d. About $\dfrac{1}{0.006} \approx 167$ seconds, or about 2.8 minutes

Exercises 9.3 page 719

1. $y = 4 + Ce^{-2x}$ **3.** $y = -\frac{1}{4}e^{-2x} + Ce^{2x}$ **5.** $y = 3x^3 + Cx^{-5}$ **7.** $y = x^2 + Cx$ **9.** $y = 3 + Ce^{-x^3}$

11. $y = Ce^{x^2}$ **13.** $y = \dfrac{x^2 + C}{x + 1}$ **15.** $y = 3x^4 - 9x^3 + Cx^2$ **17.** $y = -x - 1 + Ce^x$

19. $y = 1 + Ce^{-\sin x}$ **21.** $y = 3e^x + 2e^{-3x}$ **23.** $y = 2x^5 - x^{-2}$ **25.** $y = x^2 \ln x + 3x^2$
27. $y = 2 - 2e^{-x^2}$ **29.** $y = Ce^x - 1$

31. a. $y = \dfrac{1}{x^2/2 + C}$ (by separation of variables) **b.** $y = \dfrac{1}{3}x^3 e^x + Ce^x$ (by using an integrating factor)

33. Using $e^{\int p(x)\,dx + C} = e^{\int p(x)\,dx}e^C$ multiplies the differential equation by a positive constant e^C, which, when divided out, leaves the same equation as when the C is omitted.

35. a. $y = 20te^{0.1t} + 5e^{0.1t}$ **b.** About \$55 million **37. a.** $y = 125t^3$ **b.** 421,875

39. a. $y = 3e^{t/2} - 1$ **b.** About 21 tons **c.** In about 2.3 weeks (2 weeks and 2 days)

41. a. $y' = 2500 - 0.05y, y(0) = 8,000,000$ **b.** $y = 50,000 + 7,950,000e^{-0.05t}$ **c.** About 40.1 hours

43. a. $y' = 10,000 - 0.05y, y(0) = 6,000,000$ **b.** $y = 200,000 + 5,800,000e^{-0.05t}$ **c.** About 35 minutes

45. a. $y' = 20 - 0.02y, y(0) = 10,000$ **b.** $y = 1000 + 9000e^{-0.02t}$ **c.** About 110 hours **d.** 1000 grams

47. a. $w(t) = 120 + 50e^{-0.005t}$ **b.** About 71 days **c.** 120 pounds

49. a. $y' = 30e^{-3t} - 0.2y$ **b.** $y = 10.7(e^{-0.2t} - e^{-3t})$ **c.** 7.1 milligrams **d.** $0.97 \approx 1$ hour (8.2 mg)

51. *Hint:* An intermediate step is $ye^{ct} = \dfrac{ab}{c - b}e^{(c-b)t} + K$

Exercises 9.4 page 730

1. 8 **3.** $(1, 3)$, slope $= 12$ **5.** **7.** **9.**

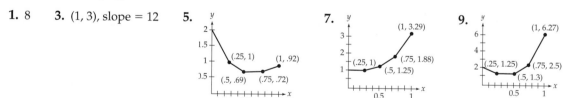

11. a. 2.226 **b.** $y = \sqrt{x^2 + 1}$ **c.** $y(2) = \sqrt{5} \approx 2.236$
13. a. 1.489 **b.** $y = e^{0.2x}$ **c.** $y(2) = e^{0.4} \approx 1.492$
15. a. 7.836 **b.** $y = e^x + 4e^{-x}$ **c.** $y(2) \approx 7.930$ **17.** 1.73 **19.** 2.24 **21.** 2.83
23. a. 2.3096 **b.** 2.3309 **c.** 2.3327 **25.** $y(3) \approx 13,490$ (from 13.49 thousand)
27. $y(2) \approx 1981$ tons **29.** $q(200) \approx 0.65$

Chapter 9 Review Exercises page 734

1. $y = \sqrt[3]{x^3 + c}$ **2.** $y = ce^{x^3/3}$ **3.** $y = \frac{1}{4}\ln(x^4 + 1) + C$ **4.** $y = -\frac{1}{2}e^{-x^2} + C$ **5.** $y = \frac{1}{c - x}$

6. $y = \frac{1}{\sqrt{c - 2x}}$ and $y = -\frac{1}{\sqrt{c - 2x}}$ **7.** $y = 1 + ce^{-x}$ or $y = 1 - e^{c-x}$

8. $y = \sqrt{2x + c}$ and $y = -\sqrt{2x + c}$ **9.** $y = ce^{\frac{1}{2}x^2 - x}$ **10.** $y = ce^{x^3/3} - 1$ **11.** $y = \sqrt[3]{3x^3 + 1}$

12. $y = e^{1-x^{-1}}$ (from $ee^{-x^{-1}}$)   **13.** $y = e^{\frac{1}{2} - \frac{1}{2}x^{-2}}$ (from $e^{\frac{1}{2}}e^{-\frac{1}{2}x^{-2}}$)   **14.** $y = \left(\frac{2}{3}x - \frac{2}{3}\right)^{3/2}$ or $y = -\left(\frac{2}{3}x - \frac{2}{3}\right)^{3/2}$

15. a. $y' = 4 + 0.05y$, $y(0) = 10$ **b.** $y = 90e^{0.05t} - 80$ **c.** 68.385 thousand or \$68,385

16. a. $y' = 4 - 0.25y$, $y(0) = 0$ **b.** $y = 16 - 16e^{-0.25t}$ **17.** $y = 106 - 36e^{2.3t}$

18. $y(1) \approx 102.4$ **19. c.** $y = \sqrt[3]{x^3 - 8}$ **20. c.** $y = \sqrt[3]{\frac{3}{2}x^2 - 8}$

$y(2) \approx 105.6$ **d.** 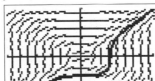   **d.**

$y(3) \approx 105.96$

21. $y' = 0.07y$ **22.** $y' = 0.12y$ **23.** $y' = ay(8000 - y)$ **24.** $y' = ay(500 - y)$

$y = 37e^{0.07t}$ $y = 16.8e^{0.12t}$ $y = \dfrac{8000}{1 + 799e^{-2.73t}}$ (t in weeks) $y = \dfrac{500}{1 + 249e^{-3.78t}}$

About 65¢   About \$55.8 billion   About 1819 cases   About 443

25. $y' = a(10,000 - y)$ **26.** $y' = a(60 - y)$ **27.** $y' = a(500,000 - y)$

$y = 10,000(1 - e^{-0.051t})$ $y = 60(1 - e^{-0.269t})$ $y = 500,000(1 - e^{-0.255t})$ (t in weeks)

About 4577   About 6.7 weeks   About 6.3 weeks

28. $y' = ay(40,000 - y)$ **29.** $y = 2x^7 + Cx^5$ **30.** $y = 2x + Cx^{-1/2}$ **31.** $y = 1 + Ce^{-x^2/2}$

$y = \dfrac{40,000}{1 + 39e^{-1.47t}}$

About $2\frac{1}{2}$ years

32. $y = 2 + Ce^{x^2}$ **33.** $y = e^x - x^{-1}e^x + Cx^{-1}$ **34.** $y = \frac{1}{2}x\ln x - \frac{1}{4}x + Cx^{-1}$ **35.** $y = 3e^{x^2/2}$

36. $y = 2e^{-2x^2}$ **37.** $y = 2x - 2x^{-2}$ **38.** $y = -4 + 10x$

39. a. $y = 10te^{0.05t} + 100e^{0.05t}$ **b.** \$192,604 [from $y(5) \approx 192.604$]   **c.** About 7.33 years

40. a. $p = -10e^{0.01t} + 110e^{0.02t}$ **b.** About 123 million people   **c.** About 19.4 years

41. a. $y = 100 - 0.05y$, $y(0) = 1000$ **b.** $y(t) = 2000 - 1000e^{-0.05t}$

c. About 1528 cubic feet (so the oxygen percentage has risen from 10% to 15.28%)

d. 2000 cubic feet   **e.** About 46 minutes

42. a. $p = 15e^{0.2t} - 10$ **b.** 12 eagles   **c.** About 4 years

43. a.   **b.** $y = 1 + e^{-2x}$

c. $y(1) = 1 + e^{-2} \approx 1.14$

44. a.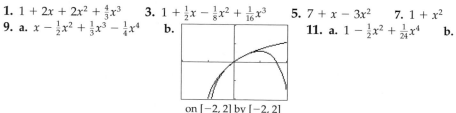

b. $y = \sqrt{4x^2 + 1}$

c. $y(1) = \sqrt{5} \approx 2.24$

45. $y(2) \approx 1.70$ **46.** $y(2.5) \approx 6.68$

47. a. $0.1(1 - p) + 0.1p(1 - p) = 0.1(1 - p)(1 + p) = 0.1(1 - p^2)$

    b. $p(4) \approx 0.61$, or 61%

48. $y(2) \approx 148$

Exercises 10.1 page 750

1. $\dfrac{1}{2} + \dfrac{1}{3} + \dfrac{1}{4} + \dfrac{1}{5} + \dfrac{1}{6}$ **3.** $-\dfrac{1}{3} + \dfrac{1}{9} - \dfrac{1}{27} + \dfrac{1}{81}$ **5.** $1 + 0 + 1 + 0 + 1 + 0$ **7.** $\displaystyle\sum_{i=1}^{\infty} \dfrac{1}{3^i}$ or $\displaystyle\sum_{i=1}^{\infty} \left(\dfrac{1}{3}\right)^i$

9. $\displaystyle\sum_{i=1}^{\infty} (-2)^i$ **11.** $\displaystyle\sum_{i=1}^{\infty} (-1)^i \cdot i$ **13.** $2^{10} - 1 = 1023$ **15.** $4^6 - 1 = 4095 \left(\text{from } 3 \cdot \dfrac{1 - 4^6}{1 - 4}\right)$

17. $1 + 2^7 = 129 \left(\text{from } 3 \cdot \dfrac{1 - (-2)^7}{1 - (-2)}\right)$ **19.** 5   **21.** $\frac{3}{2}$ **23.** Diverges   **25.** 6   **27.** 32   **29.** 25

31. Diverges   **33.** $\frac{2}{3}$ **35.** Diverges   **37.** 100   **39.** 1   **41.** $\frac{4}{11}$ **43.** $2\frac{6}{11}$ or $\frac{28}{11}$ **45.** $\frac{1}{37}$

47. $\frac{1}{81}$ **49. a.** Dose 3   **b.** Dose 5   **51.** $|r| > 1$, so the formula $\dfrac{a}{1 - r}$ does not apply.

53. a. \$26,214,300   **b.** 24 days   **55.** \$597,447   **57.** \$584,541

59. a. \$21,474,836.47   **b.** 27 days   **61.** On the 37th day   **63. a.** 15 units, 3 units   **b.** 4 doses

65. \$12 billion, multiplier = 4   **67.** \$20,300   **69. a.** 42 feet   **b.** At the 7th bounce

71. 16,000; 15,200

Exercises 10.2 page 765

1. $1 + 2x + 2x^2 + \frac{4}{3}x^3$ **3.** $1 + \frac{1}{2}x - \frac{1}{8}x^2 + \frac{1}{16}x^3$ **5.** $7 + x - 3x^2$ **7.** $1 + x^2$

9. a. $x - \frac{1}{2}x^2 + \frac{1}{3}x^3 - \frac{1}{4}x^4$ **b.** 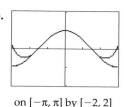   **11. a.** $1 - \frac{1}{2}x^2 + \frac{1}{24}x^4$ **b.**

on $[-2, 2]$ by $[-2, 2]$

on $[-\pi, \pi]$ by $[-2, 2]$

13. $1 + x^2 + \dfrac{1}{2}x^4$ **15.** $2x - \dfrac{2^3}{3!}x^3 + \dfrac{2^5}{5!}x^5$

17. a. $1 + x + \dfrac{1}{2!}x^2 + \dfrac{1}{3!}x^3 + \dfrac{1}{4!}x^4$ **b.** 0.60677

    c. $\left|R_4\left(-\dfrac{1}{2}\right)\right| \le \dfrac{1}{5!}\left|-\dfrac{1}{2}\right|^5 < 0.0003$ (using $M = 1$, from $f^{(5)}(x) = e^x$ evaluated at the right-hand endpoint 0)

d. $e^{-1/2} \approx 0.6068$ with error less than 0.0003 (actual value: $e^{-1/2} \approx 0.6065$)   **e.**

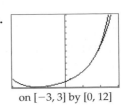

on $[-3, 3]$ by $[0, 12]$

19. a. $p_6(x) = 1 - \dfrac{1}{2!}x^2 + \dfrac{1}{4!}x^4 - \dfrac{1}{6!}x^6$ has error $\left|R_6(x)\right| \le \dfrac{1}{7!}\left|1\right|^8 < 0.0002$ on $[-1, 1]$.

b.   (Actual error for $-1 \le x \le 1$ is less than 0.00003.)

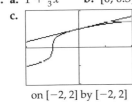

on $[-2\pi, 2\pi]$ by $[-2, 2]$

21. a. $1 + \frac{1}{3}x$ **b.** $[0, 0.3]$

c.

on $[-2, 2]$ by $[-2, 2]$

23. a. $1 + \frac{1}{2}(x - 1) - \frac{1}{8}(x - 1)^2 + \frac{1}{16}(x - 1)^3$

b.

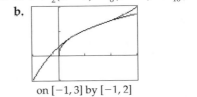

on $[-1, 3]$ by $[-1, 2]$

25. a. $-(x - \pi) + \frac{1}{6}(x - \pi)^3$ **b.**

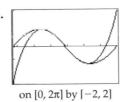

on $[0, 2\pi]$ by $[-2, 2]$

27. a. $1 + \frac{1}{3}(x - 1) - \frac{1}{9}(x - 1)^2$ **b.** 1.09

c. $\left|R_2(1.3)\right| \le \dfrac{10/27}{3!}\left|1.3 - 1\right|^3 < 0.002$ $\left(\text{using } M = \dfrac{10}{27}, \text{ from } f^{(3)}(x) = \dfrac{10}{27}x^{-8/3} \text{ evaluated at } x = 1\right)$

d. $\sqrt[3]{1.3} \approx 1.09$ with error less than 0.002 (actually, $\sqrt[3]{1.3} \approx 1.091$)   **29.** 0.78   **31.** $x = \frac{1}{2}$ year

Exercises 10.3 page 780

1. $R = 2$ **3.** $R = 1$ **5.** $R = \infty$ **7.** $R = 0$ **9.** $R = \infty$ **11.** $x - \dfrac{x^2}{2} + \dfrac{x^3}{3} - \dfrac{x^4}{4} + \cdots$

13. $1 - \dfrac{4x^2}{2!} + \dfrac{16x^4}{4!} - \cdots$ **15.** $1 + x - \dfrac{x^2}{2!} + \dfrac{3x^3}{3!} - \dfrac{5 \cdot 3x^4}{4!} + \cdots$ **17.** $1 + \dfrac{x}{5} + \dfrac{x^2}{5^2 \cdot 2!} + \dfrac{x^3}{5^3 \cdot 3!} + \cdots$

19. $1 - \dfrac{1}{2!}\left(x - \dfrac{\pi}{2}\right)^2 + \dfrac{1}{4!}\left(x - \dfrac{\pi}{2}\right)^4 - \cdots$ **21.** $x^2 + x^3 + x^4 + \cdots$ **23.** $x^2 - \dfrac{x^6}{3!} + \dfrac{x^{10}}{5!} - \dfrac{x^{14}}{7!} + \cdots$

25. $1 + \dfrac{x}{2!} + \dfrac{x^2}{3!} + \dfrac{x^3}{4!} + \cdots$ **27.** $\dfrac{1}{3!} - \dfrac{x^2}{5!} + \dfrac{x^4}{7!} - \cdots$ **29.** $x - \dfrac{1}{2}x^2 + \dfrac{1}{3}x^3 - \dfrac{1}{4}x^4 + \cdots$

31. a. $1 - x^2 + x^4 - x^6 + \cdots$ **b.** $R = 1$ **33.** $1 + \dfrac{x^2}{2!} + \dfrac{x^4}{4!} + \dfrac{x^6}{6!} + \cdots$

35. a. 1.64872   **b.** 1.64583   **c.** 7 terms (the last having exponent 6)

37. $\dfrac{d}{dx}\left(1 - \dfrac{x^2}{2!} + \dfrac{x^4}{4!} - \dfrac{x^6}{6!} + \cdots\right) = -2\dfrac{x}{2!} + 4\dfrac{x^3}{4!} - 6\dfrac{x^3}{6!} + \cdots = -x + \dfrac{x^3}{3!} - \dfrac{x^5}{5!} + \cdots$

$$= -\left(x - \dfrac{x^3}{3!} + \dfrac{x^5}{5!} - \cdots\right) = -\sin x$$

39. The absolute ratio of successive terms is $\left|\dfrac{x^2}{(2n+3)(2n+2)}\right| \to 0$ as $n \to \infty$.

41. a. $1 - \dfrac{1}{2}t^2 + \dfrac{1}{2^2 2!}t^4 - \dfrac{1}{2^3 3!}t^6 + \cdots$ **b.** $0.4x - \dfrac{0.4}{3 \cdot 2}x^3 + \dfrac{0.4}{5 \cdot 2^2 \cdot 2!}x^5 - \dfrac{0.4}{7 \cdot 2^3 \cdot 3!}x^7 + \cdots$

  c. Approximately 0.34, so about 34% have IQs in the specified range

43. a. $1 - \dfrac{t^4}{2!} + \dfrac{t^8}{4!} - \cdots$ **b.** $x - \dfrac{x^5}{5 \cdot 2!} + \dfrac{x^9}{9 \cdot 4!} - \cdots$

  c. Approximately 0.905 (thousand), so about 905 sales

45. a. $S_n - S_{n-1} = (c_1 + c_2 + c_3 + \cdots + c_{n-1} + c_n) - (c_1 + c_2 + c_3 + \cdots + c_{n-1}) = c_n$

  b. Taking limits: $\lim\limits_{n \to \infty}(S_n - S_{n-1}) = \lim\limits_{n \to \infty} c_n$. Since the series converges, the limit on the left-hand side is

  $S - S = 0$, so the limit on the right-hand side must also be zero: $\lim\limits_{n \to \infty} c_n = 0$.

47. The result follows from the indicated steps.

49. a. If $\left|\dfrac{c_{n+1}}{c_n}\right|$ approaches $r > 1$, then $\left|\dfrac{c_{n+1}}{c_n}\right|$ must eventually remain above 1.

  b. $\left|\dfrac{c_{n+1}}{c_n}\right| \geq 1$ for $n = N$ implies that $|c_{N+1}| \geq |c_N|$. Then, $\left|\dfrac{c_{n+1}}{c_n}\right| \geq 1$ for $n = N+1$ implies that $|c_{N+2}| \geq |c_{N+1}|$, which together with $|c_{N+1}| \geq |c_N|$ implies that $|c_{N+2}| \geq |c_N|$. Similar reasoning leads to the other inequalities.

  c. Terms beyond the Nth being (in absolute value) at least of size $|c_N| \neq 0$ means that the nth term cannot approach zero, so the series diverges.

51. The results of the steps are as follows: **a.** $e^x > \dfrac{x^3}{6}$ **b.** $e^{ax} > \dfrac{a^3 x^3}{6}$ **c.** $e^{-ax} < \dfrac{6}{a^3 x^3}$

  d. $x^2 e^{-ax} < \dfrac{6}{a^3 x}$ **e.** Letting $x \to \infty$ gives the claimed result.

Exercises 10.4 page 794

1. $x_1 = 1.5, x_2 = 1.39$ **3.** $x_1 = 0.5, x_2 = 0.61$ **5.** 2.236   **7.** 1.513   **9.** -0.814
11. 5.065797019   **13.** 0.486389036   **15.** 3.693441359   **17.** 0.869647810   **19.** 1.106060158
21. 1.029866529   **23.** 0.567143290
25. The iterations give $-1, 1, -1, 1, \ldots$. Reason: $x^2 + 3 = 0$ has no solutions (the left-hand side is always positive).

27. $x_1 = x_0 - \dfrac{mx_0 + b}{m} = x_0 - x_0 - \dfrac{b}{m} = -\dfrac{b}{m}$ which is the root of $mx + b = 0$, so the root is found exactly after just one iteration, regardless of the initial estimate.

29. 102 iterations   **31.** \$5.67   **33.** \$2.61   **35.** 12.6%   **37.** 3.18 weeks

Chapter 10 Review Exercises page 798

1. $\frac{3}{7}$ **2.** $\frac{2}{7}$ **3.** Diverges   **4.** $\frac{2}{5}$ **5.** 50   **6.** $\frac{4}{7}$ **7.** 5   **8.** Diverges   **9.** $\frac{6}{11}$ **10.** $\frac{8}{11}$

11. $736,971   **12.** Follows from summing $D + D(1 + i) + D(1 + i)^2 + \cdots + D(1 + i)^{n-1}$

13. a. 0.4 milligram and 0.14 milligram   **b.** Just after the 4th dose; just after the 6th dose

14. a. Sum = 10,000, multiplier = 10   **b.** 22 terms; 44 terms

15. a. $3 + \frac{1}{6}x - \frac{1}{216}x^2 - \frac{1}{3888}x^3$

b.
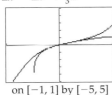
on $[-15, 15]$ by $[-6, 6]$

16. a. $1 + 3x + \frac{9}{2}x^2 + \frac{9}{2}x^3$

b.
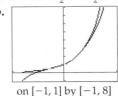
on $[-1, 1]$ by $[-1, 8]$

17. a. $2x - 2x^2 + \frac{8}{3}x^3$

b.
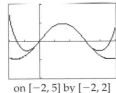
on $[-1, 1]$ by $[-5, 5]$

18. a. x^2

b.
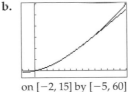
on $[-3, 3]$ by $[-2, 2]$

19. a. $1 + \frac{1}{2}x - \frac{1}{8}x^2 + \frac{1}{16}x^3$ **b.** 1.4375

c. $|R_3(1)| \le \dfrac{15/16}{4!}\, |1|^4 < 0.04$ (using $|f^{(4)}(x)| = \dfrac{15}{16}(x + 1)^{-7/2} \le \dfrac{15}{16} = M$ for $0 \le x \le 1$)

d. $\sqrt{2} \approx 1.44$ with error less than 0.04

20. a. $1 - \dfrac{1}{2}x^2 + \dfrac{1}{24}x^4$ **b.** 0.54167

c. $|R_4(1)| \le \dfrac{1}{5!}\, |1|^5 < 0.009$ (using $|f^{(5)}(x)| = |-\sin x| \le 1 = M$)

d. $\cos 1 \approx 0.542$ with error less than 0.009 (actual value: $\cos 1 \approx 0.540$)   **21.** $1 + \dfrac{1}{n}x$

22. $p_7(x) = 1 + x + \dfrac{1}{2!}x^2 + \dfrac{1}{3!}x^3 + \cdots + \dfrac{1}{7!}x^7$ has error $|R_7(x)| \le \dfrac{3}{8!}\, |1|^8 < 0.00008$ on $[-1, 1]$

23. a. $1 - \dfrac{1}{2!}\left(x - \dfrac{\pi}{2}\right)^2 + \dfrac{1}{4!}\left(x - \dfrac{\pi}{2}\right)^4$

b.
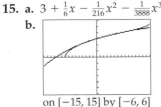
on $[-2, 5]$ by $[-2, 2]$

24. a. $8 + 3(x - 4) + \frac{3}{16}(x - 4)^2 - \frac{1}{128}(x - 4)^3$

b.
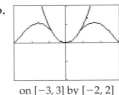
on $[-2, 15]$ by $[-5, 60]$

25. a. $1 + \dfrac{2}{3}(x-1) - \dfrac{1}{9}(x-1)^2$ **b.** 1.12889   **c.** $f^{(3)}(x) = \dfrac{8}{27}x^{-7/3}$, so $M = \dfrac{8}{27}$

d. $|R_2(1.2)| \le \dfrac{8/27}{3!} \, |0.2|^3 < 0.0004$

e. $(1.2)^{2/3} \approx 1.1289$ with error less than 0.0004 (actually, $(1.2)^{2/3} \approx 1.1292$)

26. 0.80 gram   **27.** 33 degrees   **28. a.** $200 + 20t^2 + t^4$ **b.** 296 units   **c.** $200e^{0.4} \approx 298$

29. $R = 3$ **30.** $R = 1$ **31.** $R = 0$ **32.** $R = \infty$ **33.** $x^3 + x^5 + x^7 + x^9 + \cdots$

34. $x - \dfrac{2^2 x^3}{2!} + \dfrac{2^4 x^5}{4!} - \cdots$ **35.** $x + x^3 + \dfrac{x^5}{2!} + \dfrac{x^7}{3!} + \cdots$ **36.** $x - x^2 + \dfrac{x^3}{2!} - \dfrac{x^4}{3!} + \cdots$

37. $1 + 2x + 2^2 x^2 + 2^3 x^3 + \cdots$ **38.** $1 - \dfrac{x}{2} + \dfrac{x^2}{2^2 2!} - \dfrac{x^3}{2^3 3!} + \cdots$

39. a. $-\left(x - \dfrac{\pi}{2}\right) + \dfrac{1}{3!}\left(x - \dfrac{\pi}{2}\right)^3 - \dfrac{1}{5!}\left(x - \dfrac{\pi}{2}\right)^5 + \cdots$ **b.**

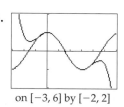

on $[-3, 6]$ by $[-2, 2]$

40. $\sin(-x) = -x - \dfrac{(-x)^3}{3!} + \dfrac{(-x)^5}{5!} - \dfrac{(-x)^7}{7!} + \cdots = -x + \dfrac{x^3}{3!} - \dfrac{x^5}{5!} + \dfrac{x^7}{7!} - \cdots = -\sin x,$ and similarly

for cosine

41. a. $1 - t^2 + t^4 - t^6 + \cdots$ **b.** $x - \frac{1}{3}x^3 + \frac{1}{5}x^5 - \frac{1}{7}x^7 + \cdots$

c. Approximately 0.465 (thousand), so about 465 cars   **d.** 0.464

42. The absolute ratio of terms is $|x|\left|\dfrac{a_{n+1}}{a_n}\right| \to |x| \lim\limits_{n\to\infty} \left|\dfrac{a_{n+1}}{a_n}\right|$ which will be less than 1 if and only if

$|x| < \lim\limits_{n\to\infty} \left|\dfrac{a_n}{a_{n+1}}\right| = R$

43. 5.099   **44.** 3.107   **45.** 0.818   **46.** 0.269   **47.** 4.641588834   **48.** 3.162277660
49. -0.475711445 **50.** 4.411048937   **51.** 2.435766883   **52.** -1.033503908 **53.** 16.6%
54. \$4.40   **55.** 165

Exercises 11.1 page 817

1. a. $\frac{2}{3}$ **b.** $\frac{1}{2}$ **3. a.** $\frac{3}{8}$ **b.** $\frac{3}{8}$

5. a. $E(X) = 10,\ \text{Var}(X) = 1\ \ \sigma(X) = 1$ **b.** $E(Y) = 10,\ \text{Var}(Y) = 100,\ \ \sigma(Y) = 10$

7. a. $P(X = 4) = \frac{1}{3},\ P(X = 7) = \frac{1}{3},\ P(X = 10) = \frac{1}{3}$ **b.** 7   **c.** 6

9. a. $P(X = 2) = \frac{1}{2},\ P(X = 4) = \frac{1}{4},\ P(X - 8) = \frac{1}{4}$ **b.** 4   **c.** 6

11. a. $\dfrac{1}{12}\left(\text{from } \dfrac{30°}{360°}\right)$ **b.** $\dfrac{1}{36}$ **c.** 0

13. $E(X) = e^{-a}\left(0\cdot 1 + 1\cdot\dfrac{a}{1} + 2\cdot\dfrac{a^2}{2!} + 3\cdot\dfrac{a^3}{3!} + \cdots\right) = e^{-a}\left(a + \dfrac{a^2}{1!} + \dfrac{a^3}{2!} + \cdots\right)$

$= e^{-a}a\left(1 + \dfrac{a}{1!} + \dfrac{a^2}{2!} + \cdots\right) = e^{-a}ae^a = a$

15. a. 0.135   **b.** 0.271   **c.** 0.271   **d.** 0.180   **17. a.** 0.406   **b.** 0.677   **c.** 0.857   **d.** 0.143
19. a. {BBBB, BBBG, BBGB, BBGG, BGBB, BGBG, BGGB, BGGG, GBBB, GBBG, GBGB, GBGG, GGBB,
   GGGB, GGGB, GGGG}

b. $\frac{1}{16}$

c. $P(X = 0) = \frac{1}{16}$
   $P(X = 1) = \frac{1}{4}$
   $P(X = 2) = \frac{3}{8}$
   $P(X = 3) = \frac{1}{4}$
   $P(X = 4) = \frac{1}{16}$

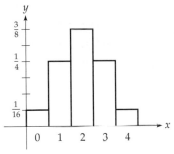

d. $\frac{15}{16}$ **e.** $E(X) = 2$, $\mathrm{Var}(X) = 1$

21. $2800   **23.** 0.238   **25.** 0.70   **27.** $e^{-1.66} \approx 0.19$, or about 19%.   **29.** 0.053, or less than 6%
31. 0.762, or about 76%
33. 1: Definition of variance   **2:** Expanding $(x_i - \mu)^2$ **3:** Separating into three sums
   4: Taking out constants   **5:** Using $\sum_{i=1}^{n} x_i p_i = \mu$ and $\sum_{i=1}^{n} p_i = 1$
   6: Simplifying and combining the last two terms

Exercises 11.2 page 832

1. $a = 12$ **3.** $a = \frac{1}{9}$ **5.** $a = 1$ **7.** $a = 1$ **9.** $a \approx 0.670$ **11. a.** $\frac{3}{4}$ **b.** $\frac{3}{80} = 0.0375$ **c.** 0.194
13. a. 0.6   **b.** $\frac{1}{25} = 0.04$ **c.** $\frac{1}{5} = 0.2$ **15. a.** 0.441   **b.** 0.079   **c.** 0.280
17. a. $\frac{1}{24}x^2 - \frac{1}{24}$ **b.** $\frac{2}{3}$ **19.** $\frac{1}{2}$ **21.** $\frac{1}{3}$ **23. a.** 0.75   **b.** $f(x) = 2x$ on $[0, 1]$
25. a. $F(A) = P(X \le A) = P(A \le X \le A) = 0$
   b. $F(B) = P(X \le B) - P(A \le X \le B) = 1$ (since A and B are the smallest and largest possible values)
27. 0.590   **29.** 0.296   **31. a.** 0   **b.** $\frac{7}{8} - 0.875$
33. a. $F(x) = \frac{1}{81}x^2$ on $[0, 9]$ **b.** $f(x) = \frac{2}{81}x$ on $[0, 9]$ **c.** 6   **d. and e.** $\frac{1}{3}$
35. a. 1   **b.** 0.2 and 0.447   **c.** 0.156, or about 16%   **37. a.** 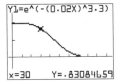   **b.** 0.83 (rounded)

Exercises 11.3 page 842

1. a. $f(x) = \dfrac{1}{10}$ on $[0, 10]$ **b.** 5   **c.** $\dfrac{25}{3}$ **d.** $\sqrt{25/3} \approx 2.89$ **e.** $\dfrac{1}{5} = 0.2$

3. a. $F(x) = \dfrac{x}{B}$ on $[0, B]$ **b.** $\dfrac{B}{2}$ **5. a.** 15   **b.** 75   **c.** 8.66   **d.** $\dfrac{1}{5} = 0.2$

7. a. $F(x) = \dfrac{x - A}{B - A}$ on $[A, B]$ **b.** $\dfrac{A + B}{2}$ **c.** $\dfrac{(B - A)^2}{12}$ (after simplification)

9. a. $\dfrac{1}{5}$ **b.** $\dfrac{1}{25}$ **c.** $\dfrac{1}{5}$ **11. a.** 2   **b.** 4   **c.** 2   **13.** $f(x) = \frac{1}{3}e^{x/3}$ on $[0, \infty)$

15. $f(x) = 2e^{-2x}$ on $[0, \infty)$ **17. a.** $F(x) = 1 - e^{-ax}$ on $[0, \infty)$ **b.** $\dfrac{\ln 2}{a} \approx \dfrac{0.693}{a}$ $\left(\text{from } \dfrac{\ln 0.5}{-a}\right)$

19. 0.777 **21.** 0.181 **23.** 0.565 **25.** 0.707 **27.** 1.41 **29.** $\dfrac{1}{a^2}$ **31.** 0.35 or 35%

33. 0.13 or 13% **35.** 0.15 or 15%

37. If $X = \left(\begin{smallmatrix}\text{Survival}\\\text{time}\end{smallmatrix}\right)$, then $S(x) = P(X \ge x) = 1 - P(X < x) = 1 - F(x) = 1 - (1 - e^{-ax}) = e^{-ax}$

Exercises 11.4 page 853

1. 0.4744 **3.** 0.679 **5.** 0.0782 **7.** 0.001 **9.** 0.9997 **11.** 0.15 **13.** 0.3085

15. $x = 0, f(0) = \dfrac{1}{\sqrt{2\pi}}$ **17.** $\left(1, \dfrac{1}{\sqrt{2\pi}} e^{-1/2}\right)$ and $\left(-1, \dfrac{1}{\sqrt{2\pi}} e^{-1/2}\right)$ $\left(\text{Note that the } y\text{-coordinates equal } \dfrac{1}{\sqrt{2\pi e}}\right)$

19. 0.909, or about 91% **21.** 0.2525, or about 25% **23.** 0.1056, or about 11%

25. 0.893, or about 89% **27.** 131 **29.** 32.51 oz.

Chapter 11 Review Exercises page 857

1. a. 6 **b.** 18 **c.** 4.24 **2. a.** 3 **b.** 5 **c.** 2.24 **3.** $830,000
4. Property B ($150,850 versus $121,800) **5.** 0.677 **6.** 0.612 **7.** 0.0025 or about one quarter of 1%
8. $e^{-5} \approx 0.0067$ or about seven tenths of 1 percent
9 a. 0.001 **b.** 0.09 **c.** 0.56 **d.** 0.93 **e.** 0.07 **10.** About 0.05 or 5% **11.** $a = \frac{3}{4}$
12. $a = \frac{1}{18}$ **13.** $a = 2$ **14.** $a = \frac{1}{2}$ **15.** $a \approx 0.637$ **16.** $a \approx 0.578$
17. a. $\frac{7}{27}$ **b.** $\frac{1}{27}x^3$ on $[0, 3]$ **c.** $\frac{7}{27}$ **18. a.** $\frac{1}{4}$ **b.** $\frac{1}{8}x^{-1/2}$ on $[1, 25]$ **c.** $\frac{1}{4}$
19. a. $\frac{3}{8} = 0.375$ **b.** $\frac{19}{320} \approx 0.0594$ **c.** 0.244
20. a. 4 **b.** 12 **c.** 3.46 **21. a.** 5.33 **b.** 7.8 (both in million kilowatt-hours)
22. a. $\frac{11}{16} = 0.6875$ **b.** $\frac{2}{5} = 0.4$ **23. a.** 0 **b.** 0.25 **c.** 0.5 **24. a.** 0.4 **b.** 0.0218 **c.** 0.1477
25. About 0.609 **26.** About 0.530 **27. a.** $\frac{1}{50}$ **b.** 25 **c.** $\frac{625}{3}$ $\left(\text{from } \frac{2500}{12}\right)$ **d.** 14.4 **e.** $\frac{3}{5} = 0.6$
28. a. 1. **b.** $\frac{1}{3}$ **c.** 0.577 **d.** $\frac{1}{4}$ **29. a.** 100 **b.** 10,000 **30.** $f(x) = 0.25e^{-0.25x}$ on $[0, \infty)$
31. 0.05 **32.** 0.49 **33.** 1.59 **34.** 0.231 **35.** About 0.22 or 22% **36.** About 0.18 or 18%
37. 0.1655 **38.** 0.7916 **39.** 0.9938 **40.** 0.1587 **41.** 0.7745 **42.** 0.971
43. 0.977 or about 98% **44.** 0.115 or about 12% **45.** 0.6772 or about 68% **46.** 564

Cumulative Review for Chapters 1–11 page 860

1. a. $y = 5$ **b.** $x = -2$ **2.** 8 **3.** $f'(x) = 6x - 7$ (but found from the *definition* of the derivative)
4. $T'(4) = 36$, so after 4 hours the temperature is rising at the rate of 36 degrees per hour. $T''(4) = \frac{9}{2} = 4.5$, so after 4 hours the temperature is rising increasingly fast, with the rate of increase rising by 4.5 degrees per hour per hour.
5. $f'(x) = x^2(x^3 + 8)^{-2/3}$
6. $f'(x) = 6(2x + 3)^2(3x + 2)^4 + 12(2x + 3)^3(3x + 2)^3 = 6(2x + 3)^2(3x + 2)^3(7x + 8)$

7.

$f' > 0$	$f' = 0$	$f' < 0$	$f' = 0$	$f' > 0$

$x = -2$ $x = 10$

↗   →   ↘   →   ↗

rel max   rel min
$(-2, 464)$ $(10, -400)$

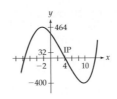

$f'' < 0$	$f'' = 0$	$f'' > 0$

$x = 4$

con dn   con up
    IP $(4, 32)$

8.

$f' > 0$	f' und	$f' > 0$

$x = 0$

↗   ↗

neither
$(0, 1)$

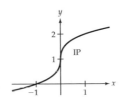

$f'' > 0$	f'' und	$f'' < 0$

$x = 0$

con up   con dn
    IP $(0, 1)$

9.

$f' > 0$	f' und	$f' > 0$	f' und	$f' < 0$	f' und	$f' < 0$

$x = -2$ $x = 0$ $x = 2$

↗   ↗   →   ↘   ↘

rel max
$(0, -1/4)$

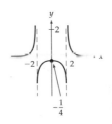

10. 7500 square feet   **11.** 128 cubic inches   **12.** $\dfrac{dy}{dx} = -5$ **13.** $\dfrac{3}{4\pi} \approx 0.24$ mile per day

14. a. \$6333.85   **b.** \$6356.25   **15.** About 6.22 years   **16.** \$13,803.30

17. a. $\dfrac{2xe^{2x} - e^{2x}}{x^2}$ **b.** $\dfrac{3x^2}{x^3 + 1}$ **18.** $p = \$25$ **19.** 0.039 or 3.9%

20. a. $3x^4 - 2x^2 + 5x + C$ **b.** $-3e^{-2x} + C$ **21.** 2   **22.** 30 thousand dollars or \$30,000

23. $800e^{0.15t} - 800$ **24.** 36 square units   **25.** 73.8 thousand people

26. a. $(x^2 + 1)^{1/2} + C$ **b.** $-\dfrac{1}{2}e^{-1} + \dfrac{1}{2} \approx 0.316$ **27.** $\dfrac{1}{3}x^3 \ln x - \dfrac{1}{9}x^3 + C$ **28.** $\dfrac{2x + 4}{3}\sqrt{x - 1} + C$

29. $\dfrac{1}{2}$ **30.** 1.98 (rounded) (compare with 2.000 from FnInt)

31. 1.99 (rounded) (compare with 2.000 from FnInt)   **32.** $\dfrac{\partial f}{\partial x} = e^y - y\dfrac{2x}{x^2 + 1}$; $\dfrac{\partial f}{\partial y} = xe^y - \ln(x^2 + 1)$

33. Min $f = 3$ at $x = 2$, $y = 2$. No max value.

34. $y = -0.9x + 6$ **35.** Max $f = 10$ at $x = 7$, $y = 4$ **36.** $df = (10x - 2y)\,dx + (-2x + 6y^2)\,dy$

37. 54 cubic units   **38. a.** $\dfrac{3\pi}{4}$ **b.** $150°$ **39. a.** $\dfrac{\sqrt{2}}{2}$ **b.** $-\dfrac{\sqrt{2}}{2}$ **c.** $\dfrac{1}{2}$ **d.** $-\dfrac{\sqrt{3}}{2}$

40.

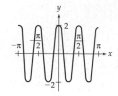

41. $\cos^2 x - \sin^2 x$ **42.** $-e^{\cos x} + C$ **43.** 1 square unit

44. a. $-\dfrac{\sqrt{3}}{3}$ **b.** $-\sqrt{3}$ **c.** $-\sqrt{2}$ **d.** $\sqrt{2}$ **45. a.** $2x\sec^2(x^2 + 1)$ **b.** $\frac{1}{2}\sec(2t + 1) + C$

46. a. $y = Ce^{x^4/4}$ **b.** $y = 2e^{x^4/4}$ **47.** $y = 4e^{-3x^2}$ **48.**   **49.** 4

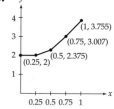

50. $1 + 2x + 2x^2 + \dfrac{4}{3}x^3$ **51.** $x^2 - \dfrac{1}{3!}x^6 + \dfrac{1}{5!}x^{10} - \dfrac{1}{7!}x^{14} + \cdots$ with $R = \infty$ **52.** 10.24695

53. 0.433   **54. a.** $E(X) = 9$ **c.** $\text{Var}(X) = 27$ **55.** 0.777   **56.** 0.383

Index

DEFINITION OF THE DERIVATIVE $\quad f'(x) = \lim\limits_{h \to 0} \dfrac{f(x+h) - f(x)}{h}$

DIFFERENTIATION FORMULAS

Power Rule: $\quad \dfrac{d}{dx} x^n = nx^{n-1}$

Constant Multiple Rule: $\quad \dfrac{d}{dx}[cf(x)] = cf'(x)$

Sum-Difference Rule: $\quad \dfrac{d}{dx}[f(x) \pm g(x)] = f'(x) \pm g'(x)$

Product Rule: $\quad \dfrac{d}{dx}[f(x)g(x)] = f'(x)g(x) + f(x)g'(x)$

Quotient Rule: $\quad \dfrac{d}{dx}\left[\dfrac{f(x)}{g(x)}\right] = \dfrac{g(x)f'(x) - g'(x)f(x)}{[g(x)]^2}$

Generalized Power Rule: $\quad \dfrac{d}{dx}[f(x)]^n = n[f(x)]^{n-1}f'(x)$

Chain Rule: $\quad \dfrac{d}{dx}[f(g(x))] = f'(g(x))g'(x) \qquad \dfrac{dy}{dx} = \dfrac{dy}{du}\dfrac{du}{dx} \qquad$ with $y = f(u)$ and $u = g(x)$

Logarithmic Formulas: $\quad \dfrac{d}{dx}\ln x = \dfrac{1}{x} \qquad\qquad \dfrac{d}{dx}\ln f(x) = \dfrac{f'(x)}{f(x)}$

Exponential Formulas: $\quad \dfrac{d}{dx}e^x = e^x \qquad\qquad \dfrac{d}{dx}e^{f(x)} = e^{f(x)}f'(x)$

Trigonometric Formulas: $\quad \dfrac{d}{dx}\sin x = \cos x \qquad\qquad \dfrac{d}{dx}\cos x = -\sin x$

$\qquad\qquad\qquad\qquad\qquad \dfrac{d}{dx}\tan x = \sec^2 x \qquad\qquad \dfrac{d}{dx}\cot x = -\csc^2 x$

$\qquad\qquad\qquad\qquad\qquad \dfrac{d}{dx}\sec x = \sec x \tan x \qquad \dfrac{d}{dx}\csc x = -\csc x \cot x$

AREA AND VOLUME FORMULAS

Rectangle

Area $= l \cdot w$

Perimeter $= 2l + 2w$

Circle

Area $= \pi r^2$

Circumference $= 2\pi r$

Rectangular solid

Volume $= l \cdot w \cdot h$

Cylinder

Volume $= \pi r^2 l$

Sphere

Volume $= \dfrac{4}{3}\pi r^3$

Surface area $= 4\pi r^2$

PROPERTIES OF NATURAL LOGARITHMS

1. $\ln 1 = 0$
2. $\ln e = 1$
3. $\ln e^x = x$
4. $e^{\ln x} = x$
5. $\ln(MN) = \ln M + \ln N$

6. $\ln\left(\dfrac{1}{N}\right) = -\ln N$

7. $\ln\left(\dfrac{M}{N}\right) = \ln M - \ln N$

8. $\ln(M^N) = N \ln M$